CHIRAL AUXILIARIES AND LIGANDS IN ASYMMETRIC SYNTHESIS

CHIRAL AUXILIARIES AND LIGANDS IN ASYMMETRIC SYNTHESIS

Jacqueline Seyden-Penne

With translation and scientific assistance from:
Dennis P. Curran

A WILEY-INTERSCIENCE PUBLICATION
JOHN WILEY & SONS, INC.
New York · Chichester · Brisbane · Toronto · Singapore

This book is a revised and updated English version of the 1994 French title *Synthèse et Catalyse Asymetriques,* by Jacqueline Seyden-Penne, published by InterEditions and CNRS Editions.

This text is printed on acid-free paper.

Library of Congress Cataloging in Publication Data:

Seyden-Penne, J.
[Synthèse et catalyse asymetriques. English]
Chiral auxiliaries and ligands in asymmetric synthesis
Jacqueline Seyden-Penne.
p. cm.
Includes bibliographical references (p. -) and index.
ISBN 0-471-11607-6 (cloth : acid-free)
1. Asymmetric synthesis. 2. Chirality. I. Title.
QD262.S48413 1995
547'.2—dc20 95-1600
CIP

10 9 8 7 6 5 4 3 2 1

CONTENTS

PREFACE

The discovery of chirality, during the last century, necessitated that chemists concern themselves with developing methods to obtain enantioenriched substances. The importance of chirality is underscored by the fact that nearly all natural products are chiral and that their physiological or pharmacological properties depend upon their recognition by chiral receptors. These chiral receptors will interact only with molecules of the proper absolute configuration.

For example, proteinogenetic α-aminoacids belong to the L-series. Aspartame (Figure P.1) is a potent sweetener that has the (S,S) configuration, while the (S,R) diastereoisomer has a bitter taste. The tragedy that occurred in the 1960s after racemic thalidomide (Figure P.2) was administered to pregnant women is a convincing example of the relationship of pharmacological activity to absolute chirality. The (R)-enantiomer of thalidomide does exhibit desirable analgetic properties; however, the (S)-enantiomer does not. Instead, this is teratogenetic and induces fetal malformations or deaths.

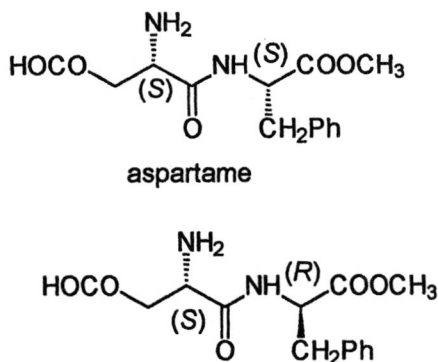

aspartame

Figure P.1

Following the thalidomide tragedy, the marketing regulations for synthetic drugs have become significantly more stringent. In order to commercialize a racemic mixture, the activity of each enantiomer of the racemate must be carefully evaluated and commercialization is only authorized if it can be shown that both enantiomers have similar potencies or that the non potent enantiomer is completely devoid of any side effects. The marketing of single enantiomers is becoming more and more popular.

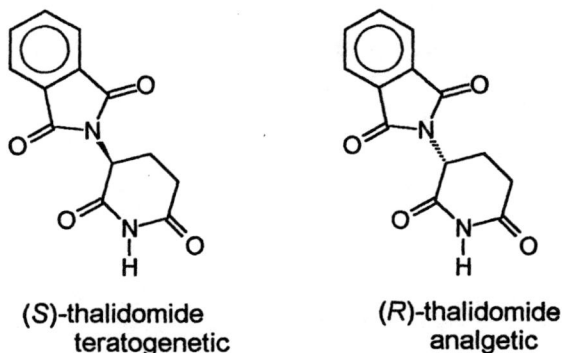

| (S)-thalidomide | (R)-thalidomide |
| teratogenetic | analgetic |

Figure P.2

The difference in biological activity between two enantiomers can be illustrated
the case of physostigmine. (-)-Physostigmine is a natural product which is an inhibitor
the cortex acetylcholinesterase. This natural enantiomer is 700 times more potent
vitro than the unnatural enantiomer, which was obtained by resolution of the synthet
racemate [Brossi 1990]. Recently, the development of inhibitors of HIV virus rever
transcriptase, as AIDS drugs, has become an active area of research. It has been four
[Vince and Brownell 1990] that (-)-carbovir triphosphate (Figure **P.3**) is a highly pote
inhibitor of HIV reverse transcriptase, but the antiviral activity of its enantiomer is n
gligible.

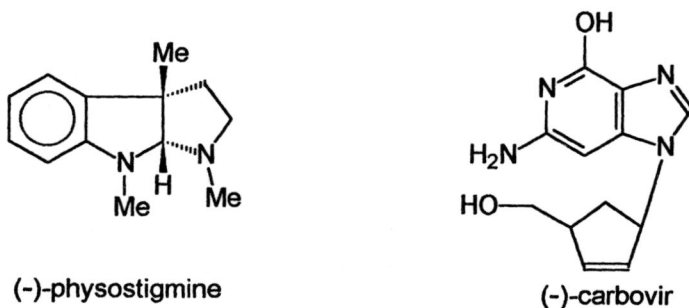

(-)-physostigmine   (-)-carbovir

Figure P.3

From these few examples, it is easy to understand why there has been amazi
growth in the development of new synthetic methods to obtain enantiopure or, mc
properly, highly enantioenriched compounds. The determination of "enantiopurity" c
pends on the precision of the analytical methods that are used, and this level of precisi
now reaches more than 99.5% [Rautenstrauch et al. 1993].

Enantiopure compounds are obtained by one of three strategies [Seebach 1990]:

(a) By resolution, either spontaneous (the way Pasteur resolved a tartaric acid salt) or with the aid of an enantiopure reagent. In a classical resolution, two diastereoisomers are formed, and their properties are sufficiently different so they can be separated by a conventional method such as fractional crystallization or chromatography. The desired enantiomer is then obtained from one of the purified diastereoisomers, while the other one is either recycled, used for another purpose or discarded. Classical methods are frequently applied on a large scale [Crosby 1991; Sheldon 1993].

Methods based on kinetic resolution are also frequently used. For example, enzymes and abzymes can recognize and transform a single enantiomer of a racemate, leaving the other one unchanged. [Chen and Sih 1988; Crosby 1991; Jones 1986; Klibanov 1990; Ohno and Otsuka 1989; Schultz 1989]. A chiral enantiopure reagent can also be used for a kinetic resolution provided that the rates of reaction of the reagent with each enantiomer of a racemate are significantly different. Horeau's method to determine the absolute configurations of alcohols by using α-phenylbutyric anhydride relies on this principle [Kagan and Fiaud 1988]. In asymmetric synthesis, kinetic resolution becomes advantageous when the slow reacting enantiomer rapidly racemizes under the reaction conditions. In such cases the yield of the kinetic resolution is not restricted to 50%. Example of such reactions are hydrogenation of β-ketoesters catalyzed by ruthenium complexes bearing chiral ligands (§ 6.2.1) [Nogradi 1986; Oppolzer 1987] or couplings of secondary Grignard reagents with halogen derivatives catalyzed by nickel or palladium complexes (§ 11.1).

(b) By the use of chiral substrates ("chirons") mainly of natural origin, which undergo highly stereoselective transformations leading to the desired enantiomeric target [Hanessian 1983].

(c) By conversion of a prochiral precursor into a chiral product (asymmetric synthesis). Biochemically, asymmetric synthesis can be performed by enzymes. Chemically, asymmetric synthesis can be performed using chiral auxiliaries, reagents or catalysts. These three methods are currently being actively developed. The requirements for a useful asymmetric synthesis include high (greater than 90%) regio-, diastereo- and enantioselectivities. Also important are the expense and accessibility of the reagents that are involved, the conditions of the reaction (solvent, temperature, pressure) and the ease of workup and purification. Because of these practical considerations, less selective methods are sometimes preferred over more selective ones for a large-scale asymmetric synthesis.

The design and study of the first chiral ligands for asymmetric hydrogenations catalyzed by transition metal complexes occurred in the early 1970s and the field of asymmetric catalysis is still expanding [Brunner and Zettlmeier 1993]. In the field of chiral reagents, important progress was made in the last 1980s when it was found that a number of types of reactions could be performed by using only a catalytic amount of the chiral rea gent alongside a stoichiometric amount or an excess of a less reactive achiral coreagen Examples of these types of reactions include reductions with boranes (§ 2.3), additioi of organozinc reagents catalyzed by chiral aminoalcohols (§ 2.5), epoxidation of allyl alcohols (§ 7.6) or asymmetric dihydroxylation of alkenes (§ 7.7). Another importai recent development has been the phenomenon of chirality amplification [Kagan ar Fiaud 1988], in which the use of a moderately enantioenriched reagent can lead to pro ducts of high enantiomeric purity (§ 2.5.1 and 6.10). From the standpoint of atom ec nomy [Trost 1991], these catalytic processes not only save money but also reduce chi mical waste.

The goal of this book is to summarize the present state of knowledge in the use of chiral auxiliaries, reagents and ligands in both stoichiometric and catalytic asymmetr syntheses. The book focuses on selective transformations that produce highly enantioei riched products from precursors that are readily available. Asymmetric transformatioi such as crystallization of an equilibrating mixture of diastereoisomers [Vedejs et a 1993] will not be described, nor will asymmetric synthesis performed in chiral crysta [Caswell et al. 1993; Inoue 1992]. Kinetic resolution [Kagan and Fiaud 1988]. will ge nerally only be considered when it is conducted on an equilibrating pair of enantiomers.

In a preliminary chapter, entitled Introduction, the underlying principles of physic organic chemistry, as applied to stereoselective reactions, are succintly recalled. Th three subsequent chapters describe the chiral auxiliaries, reagents, catalysts and ligani that are most commonly used in asymmetric synthesis. The remaining chapters are deve ted to the description and delineation of the scope of the main classes of asymmetric o ganic reactions. These include protonations and deprotonations; alkylations and relate reactions; additions to C=O, C=N and C=C double bonds; cycloadditions; rearrang ments; and transition metal-catalyzed reactions.

As often as possible, references will be given to important reviews; however, r cent references (post-1991) will mostly be to original papers. Two general treatises w also be referred to: Carey and Sundberg's *Advanced Organic Chemistry* [Carey ar Sundberg 1991] and *Comprehensive Organic Chemistry*, B. Trost and I. Fleming ed tors. Unfortunately, Eliel and Wilen's book *Stereochemistry of Organic Compoun* (Wiley 1994) was published as this book was nearing completion, so this important wo has not been cited in the text.

This book is the updated translation of the French version which was published the end of 1993. Once again, I am grateful to the members of the Orsay laboratory: Re bert Bloch, Yves Langlois, Tekla Strzalko, Marylise Calvié, Marie Louise Verrier, wl continuously helped me by providing the documents that I needed. I also found mu(

support at Aix-Marseille University, Centre de Saint-Jérôme where the library was open to me as often as I wanted. I thank all the library staff there, and I also thank Michel Chanon and Gilles Garosse for there help with electronic file transfer to and from Pittsburgh. I am especially grateful to my husband, Bob, who handled all the production aspects of the book from the typing of the English draft, to the drawing of the figures and to the production of the final copy. I also thank Suzanne Curran and Michele Russo for helping to correct the English draft. Finally, I am very happy that Dennis Curran agreed to participate in this joint venture. He not only did a great job of improving my English, but he also made helpful comments and suggestions on some aspects of the chemistry.

Jacqueline Seyden-Penne

Goult, November 1994

References

BROSSI, A., 1990, J. Med. Chem., **33**, 2311

BRUNNER, H., and ZETTLMEIER, W., 1993, Handbook of Enantioselective Catalysis VCH, Weinheim

CAREY, F. A., and SUNDBERG, R. J., 1991, Advanced Organic Chemistry 3rd edition, Plenum Press, New York

CASWELL, L., GARCIA-GARIBAY, M. A., SCHEFFER, J. R., and TROTTER, J., 1993, J. Chem. Ed., **70**, 785

CHEN, C. S, and SIH, C. J., 1988, Angew. Chem. Int. Ed. Engl., **28**, 711; 1989,Topics in Stereochemistry, ELIEL, E. L., and WILEN, S., Eds., **19**, 63

CROSBY, J., 1991, Tetrahedron, **47**, 4789

HANESSIAN, S.,1983, Total Synthesis of Natural Products: The Chiron Approach, Pergamon Press, New York

INOUE, Y., 1992, Chem. Rev., **92**, 741

JONES, J. B., 1986, Tetrahedron, **42**, 3351

KAGAN, H. B., and FIAUD, J. C., 1988, Topics in Stereochemistry, E. L. ELIEL and S. WILEN Ed., **18**, 249

NOGRADI, M., 1986, Stereoselective Synthesis, Verlag Chemie, New York

OHNO, M., and OTSUKA, M., 1989, Org. Reactions, **37**, 1

OPPOLZER, W., 1987,Tetrahedron, **43**, 1969 and quoted ref.

RAUTENSTRAUCH, V., LINDSTRÖM, M., BOURDIN, B., CURRIE, J., and OLIVEROS, E., 1993, Helv. Chim. Acta, **76**, 607

SCHULTZ, P. G., 1989, Angew. Chem. Int. Ed. Engl., **28**, 1283; and Acc. Chem. Res., **22**, 287

SEEBACH, D., 1990, Angew. Chem. Int. Ed. Engl., **29**, 1320 and quoted ref.

SHELDON, R. A., 1993, Chirotechnology, M. Dekker Ed., New York

TROST, B. M., 1991, Science, **254**, 1471

VEDEJS, E., FIELDS, S. C., and SCHRIMPF, M. R., 1993, J. Am. Chem. Soc., **115**, 11612

VINCE, R., and BROWNELL, J., 1990, Bioch. Bioph. Res. Comm., **168**, 912

ABBREVIATIONS

§	see section
Ac	acetyl
acac	acetylacetonyl
AIBN	azadiisobutyronitrile
Alk	alkyl
BOC	tertiobutoxycarbonyl
Bz	benzoyl
CIDNP	chemically induced dynamic nuclear polarization
cod	cyclooctadienyl
Cp	cyclopentadienyl
dba	dibenzylideneacetone
DBU	1,8-diazabicyclo-[5,4,0]-undec-7-ene
de	diastereoisomeric excess
DIP	diisocampheylborane
DMAC	*N,N*-dimethylacetamide
DMAP	4-dimethylaminopyridine
DME	dimethoxyethane
DMF	*N,N*-dimethylformamide
DMM	dimethoxymethane
DMPU	*N,N'*-dimethyl-*N,N'*-propyleneurea
DMSO	dimethylsulfoxyde
ee	enantiomeric excess
equiv	equivalent
Et	ethyl
EWG	electron-withdrawing group
HOMO	highest occupied molecular orbital
HMPA	hexamethylphosphorotriamide
i-Bu	isobutyl
i-Pr	isopropyl
LA	Lewis acid
LAH	lithium aluminohydride
LDA	lithium diisopropylamide
LHMDS	lithium hexamethyldisilylamide
LICA	lithium dicyclohexylamide
LUMO	lowest unoccupied molecular orbital
MCPBA	metachloroperbenzoic acid
Me	methyl
n-Bu	*n*-butyl
n-Pr	*n*-propyl
NBS	*N*-bromosuccinimide
NCS	*N*-chlorosuccinimide
NMO	*N*-methylmorpholine oxide

NMP	*N*-methylpyrrolidone
NMR	nuclear magnetic resonance
Np	naphthyl
Ph	phenyl
Py	pyridyl
s-Bu	secondary butyl
SOMO	singly occupied molecular orbital
TBDMS	tertiobutyldimethylsilyl
tert-Bu	tertiobutyl
Tf	trifluoromethanesulfonyl
THF	tetrahydrofurane
TMS	trimethylsilyl
Tol	paratolyl
Ts	paratoluenesulfonyl

INTRODUCTION

Principles of stereoselection have been uncovered mainly with racemates, so the examples in this chapter will include both non racemic and racemic substrates or reagents.

I.1 APPLICATION OF TRANSITION-STATE THEORY TO ASYMMETRIC INDUCTION

The interpretation of asymmetric induction relies on transition-state theory [1]. As a simple example, consider a reaction between a substrate and a reagent that leads to two diastereoisomeric primary reaction products in unequal amounts. Each reaction path will obey the same kinetic laws, and if the substrate Σ^* bears a chiral residue, two products, Σ^*-$\wp$ and Σ^*-$\Re$, will be generated. After processing of the reaction mixture and subsequent removal of the chiral auxiliary, an enantio-enriched compound Γ^* will be formed. The addition of an organometallic reagent to an α-ketoester of a nonracemic chiral alcohol (§ 6.5.2) shown in Figure I.1.a is an example of this type of reaction. The primary reaction products are the metal alcoholates, the processing is simple hydrolysis, and the removal of the chiral auxil-iary is performed by lithium aluminum hydride reduction. If the reagent bears a nonracemic chiral group, then the primary reaction products will incorporate this residue. Again, subsequent processing will generate the enantioenriched target compound. The reaction of a prochiral aldehyde with a non racemic chiral allylbo-rane shown in Figure I.1.b is a typical example of the reactions of a chiral reagent. The primary products of the reaction are diastereoisomeric boronates, which are transformed into an enantioenriched alcohol by exposure to H_2O_2 in the presence of base (§ 6.6.1). Similar considerations apply to catalyzed reactions. For in-stance, if the reaction is catalyzed by a chiral Lewis acid or Lewis base, one of the ground-state components will be a reagent-catalyst or substrate-catalyst complex and the primary reaction products will also include the catalyst. In the case of re-actions catalyzed by transition metal complexes, diastereoisomeric complexes are transient intermediates and a similar reasoning applies. Most of the examples dis-cussed in the following paragraphs will deal with substrates Σ^* bearing a remov-able chiral residue (substrate control). The reaction with a reagent $\Re$ will take place at a prostereogenic center of Σ^*, thus leading to diastereoisomeric primary products Σ^*-$\Re$ and Σ^*-$\wp$. However, the principles that emerge will apply to all types of asymmetric reactions.

1

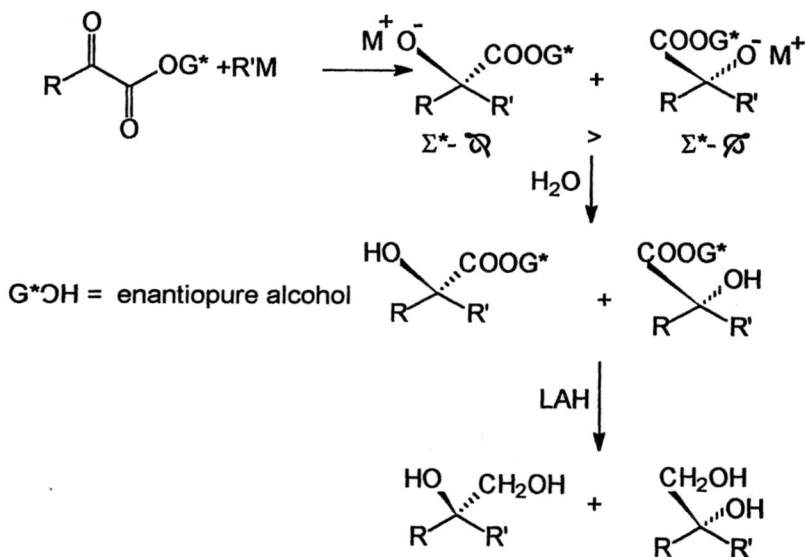

Figure I.1.a

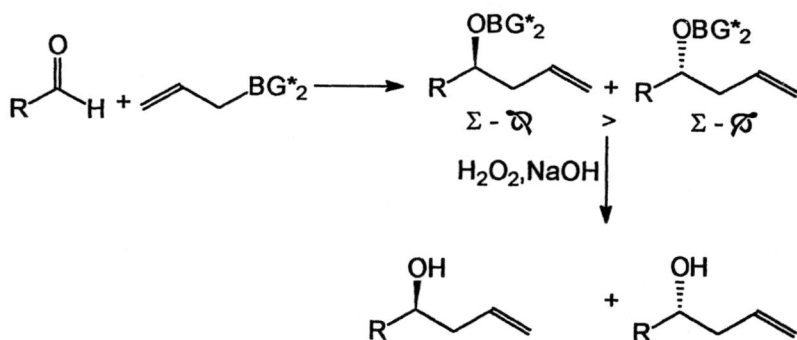

Figure I.1.b

I.1.1 One-Step Reactions

The reaction coordinate diagrams relating the formation of either diastereoisomer Σ^*-ℜ or Σ^*-℘ during the one-step reaction of a chiral substrate Σ^* with a reagent ℜ are given Figure **I.2.a** and **I.2.b**. In the first example, product Σ^*-ℜ is more stable than Σ^*-℘ (Figure **I.2.a**) while the reverse is true in the second example (Figure **I.2.b**). However, in both cases, the energy level of the transition state $[\Sigma^*$-ℜ$]^{\neq}$ is lower than that of transition state $[\Sigma^*$-℘$]^{\neq}$. Therefore, if the reaction is under kinetic control, the formation of Σ^*-ℜ will be favored in both cases. However, if the reaction is under thermodynamic control, Σ^*-ℜ will again predominate in the first case, but Σ^*-∅ will be the major product in the second case. Thus, in favorable situations, it is possible to obtain either diastereoisomer in reaction processes following the energy pathway given in Figure **I.2.b** by selecting the experimental conditions to favor kinetic or thermodynamic control.

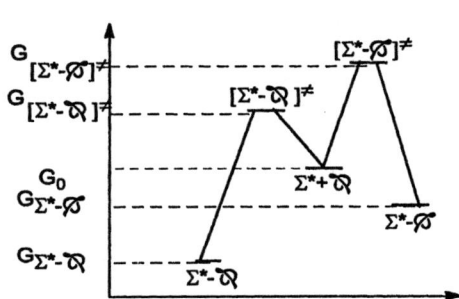

Figure I.2.a

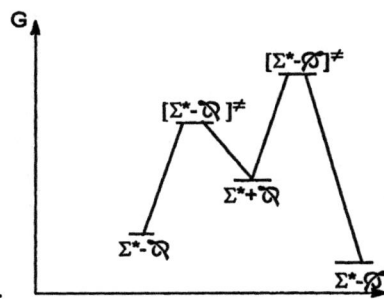

Figure I.2.b

Most reactions take place under kinetic control and the selectivity depends on the free-energy difference between the two transition states:

$$\Delta\Delta G^{\neq} = G^{\neq}_{\Sigma^*\text{-}ℜ} - G^{\neq}_{\Sigma^*\text{-}∅}$$

The magnitude of the difference in free energy depends on enthalpy and entropy terms:

$$\Delta\Delta G^{\neq} = \Delta\Delta H^{\neq} - T\Delta\Delta S^{\neq}$$

Frequently, the differences in entropy between two diastereomeric transition states are small, and the enthalpy term predominates. However, the entropy term cannot always be neglected, and sometimes it may even dominate according to the reaction conditions [2]. In some cases, the observed selectivity can be inverted by changing the temperature of the reaction, and the temperature at which $\Delta\Delta H^{\neq} = T\Delta\Delta S^{\neq}$ is called the isokinetic temperature (isoinversion). The occurrence of an isokinetic temperature in the normal operating range is not very frequent, but it is observed sometimes in multistep processes (see below). The ratio of the diastereoisomeric products formed in kinetically controlled reactions is given by the relationship

$$\frac{[\Sigma^*\text{-}ℜ]}{[\Sigma^*\text{-}∅]} = e^{-\Delta\Delta G^{\neq}/RT}$$

Table **I.1**

Ratio	de	$\Delta\Delta G^{\neq}$(kcal mol^{-1})
1.0	0.0	0.00
3.0	50.0	0.65
9.0	80.0	1.30
19.0	90.0	1.74
99.0	98.0	2.72
99.9	99.8	4.09

This ratio is temperature dependent, and the selectivity increases as the temperature decreases. The free-energy differences ($\Delta\Delta G^{\neq}$) are small compared to the total activation free energies ($\Delta G^{\neq}$), and Table **I.1** shows the product ratios that will form at + 25°C from competing transition states with $\Delta\Delta G^{\neq}$ between 0.0 and 4.09 kcal/mol [3].

I.1.2 Multistep Reactions [4]

When multistep reactions take place, stereoselection may or may not occur at the rate-determining step. Therefore, a careful analysis of the reaction mechanism is mandatory if the origin of stereoselection is to be understood.

Frequently, the formation of a single intermediate species I* is the rate-determining step in a transformation, while the subsequent rapid attack of either prochiral face of this species is the stereoselectivity determining step. The energy diagrams for this type of process are similar to the previously described diagrams, when interactions between intermediate I* and reagent $\mathfrak{R}$ are taken into account (Figure **I.3**). A more complex situation arises when two diastereoisomeric intermediates I*$_1$ and I*$_2$ are reversibly formed, and these respectively generate each primary reaction product Σ*-$\mathfrak{R}$ and Σ*-$\mathcal{S}$.

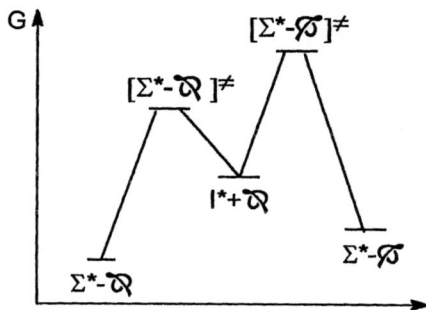

Figure **I.3**

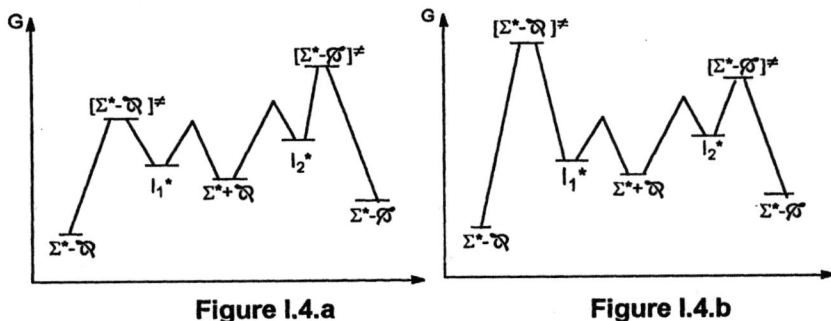

Figure I.4.a   Figure I.4.b

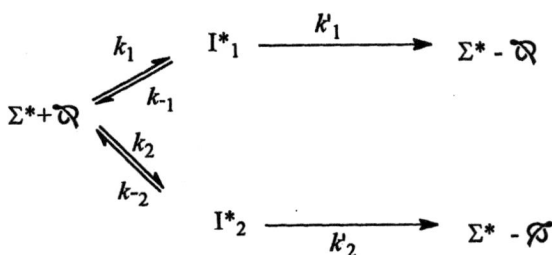

Figure I.4.c

Again, two possibilities can be envisioned, depending on whether or not the most stable intermediate I^*_1 is the precursor of the kinetically favored isomer (Figure **I.4.a** and **I.4.b**). The kinetic scheme related to these types of energy diagrams is shown in Figure **I.4.c**.

The Curtin-Hammett principle (see below) applies to this scheme when the rate constants for product formation k'_1 and k'_2 are small relative to rate constants for interconversion over the reactants and intermediates k_1, k_2, k_{-1} and k_{-2} (in other words, when I^*_1 and I^*_2 are in rapid equilibrium). A kinetic analysis of such a process shows that the selectivity depends only upon the free-energy difference between the transition states $[\Sigma^*\text{-}\mathbb{Q}]^{\neq}$ and $[\Sigma^*\text{-}\varnothing]^{\neq}$. The ratio of the diastereoisomeric primary products is given by the same relationship as before:

$$\frac{[\Sigma^*\text{-}\mathbb{Q}]}{[\Sigma^*\text{-}\varnothing]} = e^{-\Delta\Delta G^{\neq}/RT}$$

In some cases, the relative magnitudes of the rate constants do not fall within these bondary conditions. Scharf and coworkers [2] have analyzed reaction processes involving the fast transformation of one of the intermediates into the starting reagents ($k ->> k'$), while the reverse occurs for the other intermediate. Energy diagram **I.5.a** shows the situation when the more stable intermediate is also more reactive ($k'_1 > k'_2$), and diagram **I.5.c** shows the corresponding situation when the more stable intermediate is less reactive. Complex kinetic equations are needed to determine the ratio of the diastereoisomeric primary products, and such systems often give rise to isoinversion phenomena so that the selectivity can be temperature and/or pressure dependent. It has been shown that rhodium-catalyzed hydrogenations of carbon-carbon double bonds occur by this type of mechanism [2]. In precise temperature and pressure conditions (+25°C, 1 atm), Halpern has shown that the least stable intermediate complex is the most reactive one (§ 7.1).

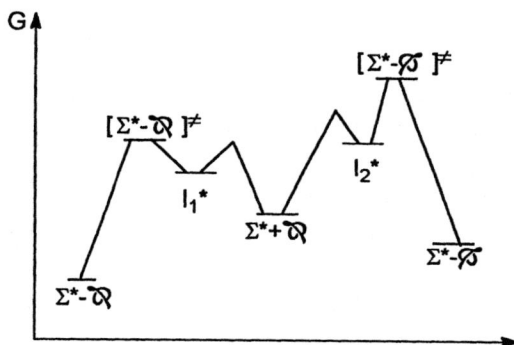

Figure I.5.a

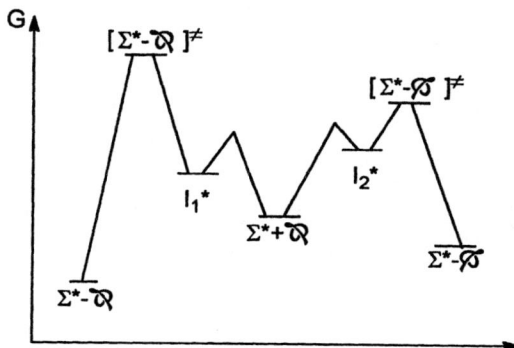

Figure I.5.b

I.2 STEREOELECTRONIC AND POLAR EFFECTS

As the reagent and the substrate approach each other, attractive and repulsive interactions occur. The discrimination between the various possible transition states depends upon the relative magnitudes of these interactions, which in turn depend upon the transition-state geometries. Unlike enzymatic reactions, which are usually controlled by attractive interactions due to site recognition, chemically induced discrimination is usually dependent upon differences in repulsive interactions.

I.2.1 Bonding Interactions

The interactions involved in the formation of one (or more) new bonds are attractive due to orbital mixing. The most important interactions are between the frontier orbitals [5], and in polar reactions, the interaction between the highest occupied molecular orbital (HOMO) of the electron donor and the lowest unoccupied molecular orbital (LUMO) of the acceptor dominates. For radical reactions, the singly occupied molecular orbital (SOMO) can interact with the HOMO and the LUMO of the substrate. According to frontier molecular orbital theory, the closer the energy level of the frontier orbitals, the lower the activation enthalpy of the reaction. The approach of the reagents takes place in an orientation that maximizes in-phase overlaps and minimizes out-of-phase overlaps. Secondary orbital interaction may also play a role. The directionality (obtuse angle of 109°) of the nucleophilic attack of double bonds advocated by Dunitz and Burgi [6] has been attributed to primary orbital effects: in-phase overlap of the HOMO of the nucleophile with one lobe of the π* LUMO of the double bond also minimizes out-of-phase overlap (Figure **I.6.a**). In the case of nucleophilic attack on a carbonyl group (Y = O), this approach also minimizes repulsive interactions of the nucleophile with the oxygen lone-pairs [7, 8, 9]. Similar considerations of optimized overlap suggest that electrophilic attack on double bonds should take place toward their center (acute angle) in order to favor the interaction of the two in-phase lobes of the π orbital (HOMO) with the LUMO of the electrophile [9] (Figure **I.6.a**). Radical reactions, however, do not follow such empirical rules, as calculations have indicated that both nucleophilic and electrophilic radicals approach the olefinic carbon atoms at angles between 104 and 109° [10] (Figure **I.6.a**).

These approach vectors have been supported by *ab initio* calculations of transition structures, performed with model reactions. (A transition structure is a saddle-point that separates two energy wells on a theoretical potential energy surface; it is related to the thermodynamically derived transition state [11]).

The presence of substituents on either the substrate or the reagent can affect the energy level of the frontier orbitals and the geometry of the reactive sites. For example, the presence of an electron-donating substituent in the α position of a double bond will raise the HOMO level due to a σ-π interaction when the substituent is *anti* to the π bond (Figure **I.6.b**). This will induce an increase in the nucleophilicity of the double bond [9, 12]. In contrast, the presence of an electron-withdrawing group in the same location will lower the HOMO level of the double bond due to a σ*-π interaction (Figure **I.6.b**). Similarly, the π* LUMO levels of double bonds will be raised by a σ-π* interaction and lowered by a σ*-π* interaction, so that the nucleophilicity of these double bonds can be either decreased or increased [7, 9] (Figure **I.6.c**).

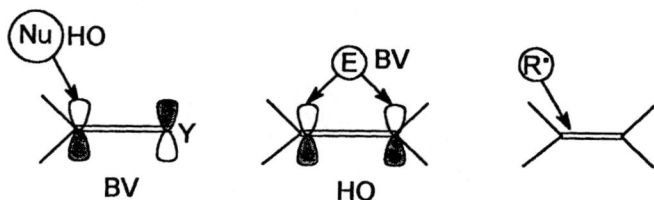

Figure I.6.a

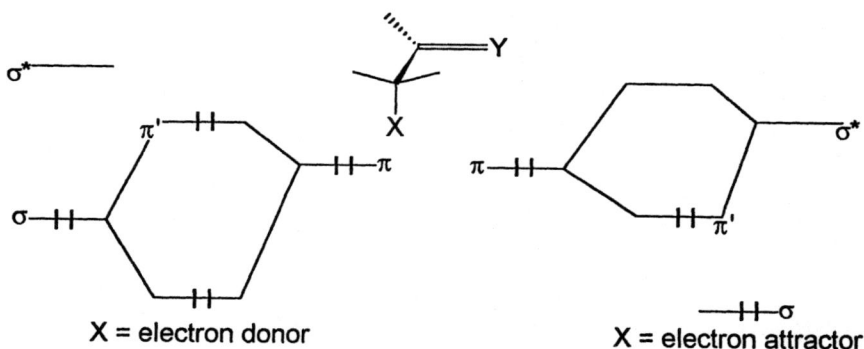

X = electron donor   X = electron attractor

Figure I.6.b

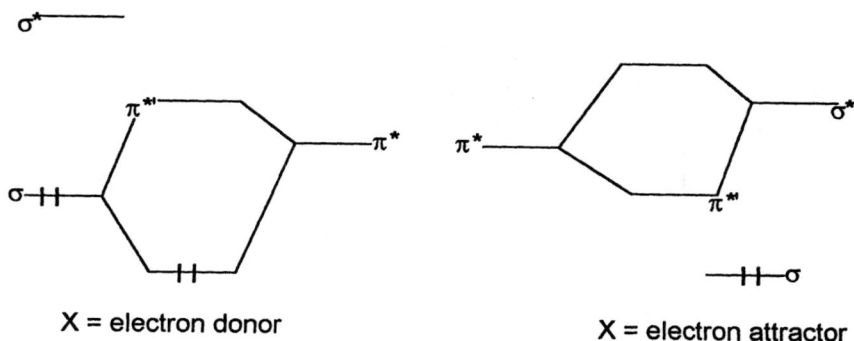

X = electron donor   X = electron attractor

Figure I.6.c

A related method to interpret the diastereofacial selectivities of the reactions of double bonds has been proposed by Dannenberg and coworkers [8, 13, 14,]. This method also relies on the π frontier orbitals of non symmetrical molecules, and proposes breaking the symmetry of the π or π^* orbitals due to polarization induced by the substituents. Application of frontier molecular orbital theory, taking into account only the substrate MOs, gives a qualitative trend of stereoselection in a number of nucleophilic (reductions of carbonyl compounds) and electrophilic reactions.

The geometric deformation of molecules can also induce orbital distortions. Seebach and coworkers [15] have shown that dioxanones **1** are distorted. In the ground state, carbon-4 is pyramidalized through the mean plane of the ring toward carbon-2. This distortion is enhanced in the transition state for a nucleophilic attack, so such reactions are stereoselective because of better overlap control (Figure **I.7**).

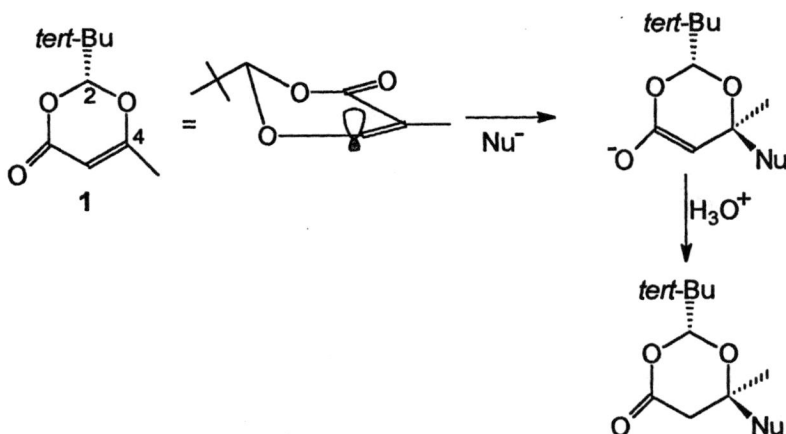

Figure I.7

The interactions between the substrate and the reagent increase or decrease in importance according to the position of the transition state along the reaction coordinate. As a rough approximation, the transition state can be described as very early, early, or late, according to the following guidelines:

(a) if the new bond formation is very weak (about 2.5 times longer than the final bond), the transition state is very early. The orbitals of the reagents will be very similar to those of ground state, and the conformation of the transition state will usually be very similar to that of the ground state (see below). Application of FMO or polarized FMO theory is often helpful [14]. The structure of the reagents can be easily distorted, and steric interactions are not very important. Such is the case in some radical reactions [16], although the conformation and the geometry of the radical at transition state might not be the same as at ground state [10, 17, 18] (see below).

(b) if the new bond is somewhat more formed (about 1.5 to 1.8 times its final length), the transition state is still early and the reagents maintain most of their ground-state orbital characteristics. Such a transition state is rigidified however, especially where the directionality of the attack is concerned [10]. The reactive conformations of the partners might differ from those of the ground state. These differences occur to improve in-phase orbital overlaps and to avoid repulsions of occupied orbitals (steric interactions). Such a situation prevails in the nucleophilic attack on C=O double bonds [19-24]. It has been shown that an electron-withdrawing substituent located on the carbon vicinal to the reacting site in *anti* position to the C=O axis lowers the LUMO level due to σ^*-π^* interaction. Therefore, this geometry of attack is favored (see below) [7, 24] (Figure **I.6.c**). A more refined analysis of the transition structure due to Anh and Eisenstein and Houk and coworkers indicates that there is a stabilization of the HOMO of the newly forming bond by interaction with the σ^* orbital of the *anti* bond located on the vicinal carbon [7, 9]. Another interpretation, due to Cieplak, takes into account the interaction of the vacant orbital of the incipient bond with σ orbitals of adjacent bonds [25]. Reactions occurring through such early transition states are often said to be under "steric approach control".

(c) if the new bond is nearly formed, the transition state resembles the primary reaction product, and it is considered to be late. Such is the case with some conjugate additions to α-enones. The transition state resembles the enolate products [26], but the reaction may still not be under thermodynamic control. Such a situation is termed "product development control".

The calculations of Dorigo and Morokuma [26] shown in Figure **I.8** exemplify the differences between early and late transition states. The C_1-C_2 bond length (1.267 Å) in the calculated transition structure for 1,2-addition of MeLi to acrolein is shorter than the C_1-C_2 bond length (1.277 Å) in the transition structure for 1,4-addition of MeCu, but both bonds are shorter than the corresponding C_1-C_2 bond (1.328 Å) of the enolate product.

I.2.2 Nonbonding Interactions

The most important types of nonbonding interactions are through space repulsions between filled orbitals (steric effects). These repulsions are dependent upon geometric factors, mainly distances and bond angles. This is why reagents approach each other through their "least hindered faces".

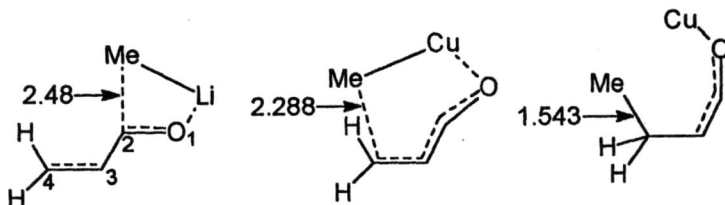

bond length Å	1,2 attack	1,4 attack	enolate
C_2O_1	1.267	1.277	1.328
C_2C_3		1.386	1.330
C_3C_4	1.318	1.397	1.512

Figure I.8

Estimation of such interactions in a transition state can be made by molecular mechanics calculations. This is similar to a conformational analysis [1], considering transition structures as supermolecules [11]. The four-term calculated energy E is therefore minimized:

$$E = E_r + E_\theta + E_\Phi + E_d$$

The first contribution E_r is related to the stretching or compression of bonds, and E_θ is the energy due to bond angle deformations, while E_Φ is the energy due to torsion, which is particularly significant. This torsion involves repulsions between the existing bonds of the reagent and the substrate (filled orbitals), as well as repulsion between existing and forming bonds. In most transition-state models, the approach of the reagents occurs with staggering of vicinal bonds in order to minimize gauche interactions [1, 3]. Moreover, the new bond formation(s) must occur with minimization of torsion interactions. Schleyer [27] underlined in 1967 the importance of torsional strain, and in 1968 Felkin emphasized the role of torsional interactions in understanding the stereoselectivity of a nucleophilic attack of carbonyl compounds [28]. The fourth contribution E_d is the dispersive force, which depends upon the attractive or repulsive interactions between uncharged atoms. To these contributions, one may add repulsions of lone pairs (filled orbitals repulsions) when heteroatoms are present [29], and electrostatic interactions between charged or dipolar species, which can be highly significant and which are only estimated by *ab initio* calculations [22, 24, 29-32].

Studies by Felkin and coworkers on $LiAlH_4$ reduction of cyclohexanones serve to underline the importance of torsional interactions [28]. These authors interpreted the predominant axial attack by minimization of torsional interactions relative to equatorial attack in the early transition state (case b) (Figure **I.9.a**). Calculations of Anh and Eisenstein and then Houk and coworkers corroborate such an interpretation [7, 25, 33]. Houk's group [21, 22, 24, 32, 34, 35], Fujita and coworkers [36], Shi and Boyd [37], and Coxon and Luibrand [38] have also emphasized the importance of orbital interactions, polar and electrostatic effects, and ring distortions on these reductions. These calculations have been extended to silicon hydride reductions [39] and to hydride reductions of norbornanones and related systems [25, 31, 34, 34a]. The presence of an axial fluorine on bicyclic ketone **2** enhances axial attack by hydrides, in agreement with the calculations which underline the influence of electrostatic and polar long-range effects (Figure **I.9.b**).

The aldol reaction (§ 6.8) and the Michael reaction (§ 7.13) are two of the most important carbon-carbon bond-forming reactions, and models of their transition states are shown in Figure **I.9.c** (X = O or CHEWG). These models take into account the minimization of torsion interactions; the existing and forming bonds are staggered [40-42].

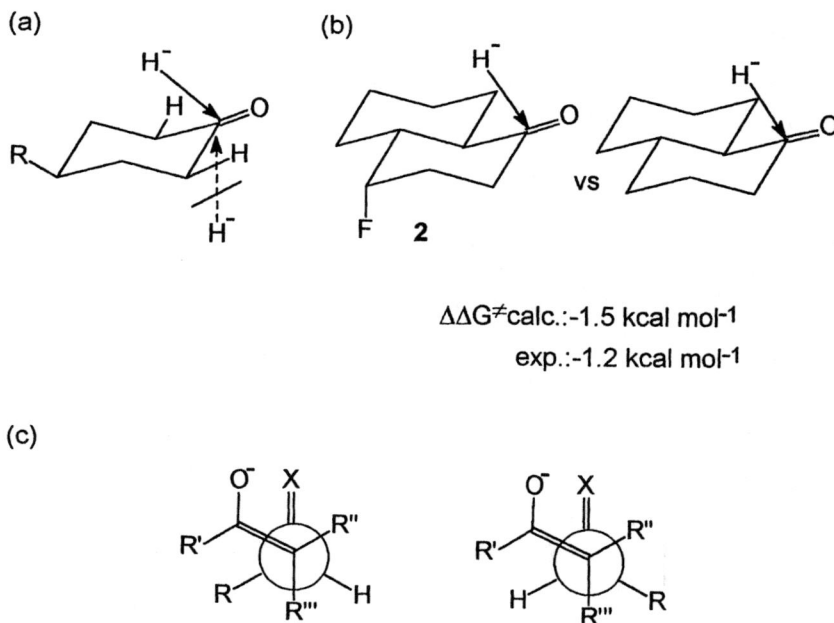

(a)   (b)

F **2**

$\Delta\Delta G^{\neq}$calc.:-1.5 kcal mol^{-1}

exp.:-1.2 kcal mol^{-1}

(c)

Figure I.9

When the substrate and the reagent bear unsaturation (double or triple bonds, aromatic rings), π-donor-acceptor (charge transfer) or π-stacking interactions can intervene. Such interactions are stabilizing, and they can direct the attack toward one face of the substrate, mimicking enzymatic processes. Guetté and coworkers [43] attempted to induce such interactions in asymmetric reductions of arylketones with chiral organomagnesium reagents bearing an aryl residue. Recently, Corey and coworkers [44, 45] designed chiral Lewis acids bearing a tryptophan residue **3** (§ 3.2.1 and 9.3). These Lewis acids form complexes with α,β-unsaturated aldehydes, and the π-stacking arrangement in these complexes leaves a single face of the unsaturated aldehyde open to the approach of the reagent (Figure **I.10.a**).

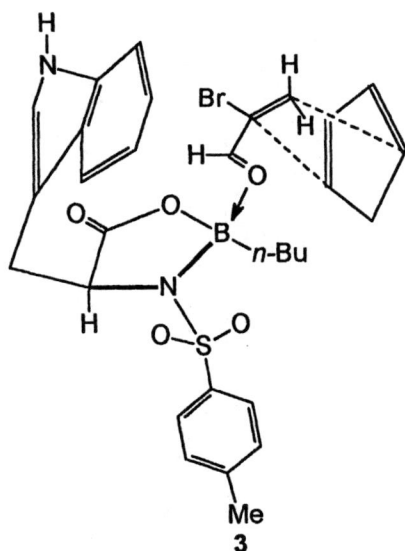

Figure I.10.a

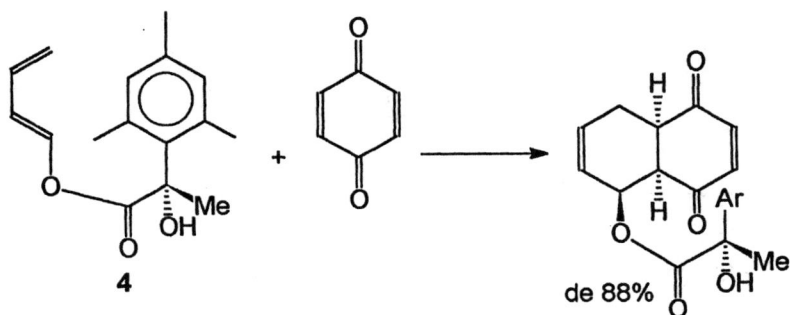

Figure I.10.b

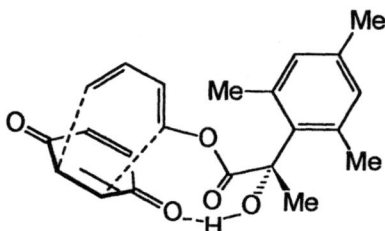

Figure I.10.b (continued)

However, the application of such a concept to the transition state of $[4\pi+2\pi]$ cycloadditions is controversial because the stabilizing energy involved may be too weak [46-49] (§ 9.3).

Inter- or intramolecular hydrogen bonding can also stabilize specific transition-state orientations. The Diels-Alder reaction [50] of diene **4** with benzoquinone is a good example [47]. Intermolecular hydrogen bonding induces a stereoselective cycloaddition, while no reaction takes place with the corresponding silyl ether. The occurrence of such a hydrogen bond promotes the intervention of a preferred conformer of the diene which induces a rigidification of the transition state (see below) (Figure **I.10.b**).

I.3 CONFORMATIONAL EFFECTS

When the substrate and the reagent have several low energy conformations, the ones that will be favored in the transition state minimize unfavorable interactions or maximize favorable ones. Examples involving π-π interactions or hydrogen bonding have already been presented. As a simple analysis, consider that one conformer $\wp_1$ is the precursor of the primary diastereoisomeric product Σ^*-$\wp$, and another conformer, $\wp_2$, is the precursor of diastereomer Σ^*-$\eth$. The kinetic equations corresponding to such a situation are given Figure **I.11.a**. Provided that the interconversion barrier between the two conformers is low relative to the activation free energies (k'_1 and $k'_2 \ll k_1, k_2, k_{-1}$ and k_{-2}), the Curtin-Hammett principle can be applied [1, 51]. The selectivity of the reaction only depends upon the free energy difference between the transition states.

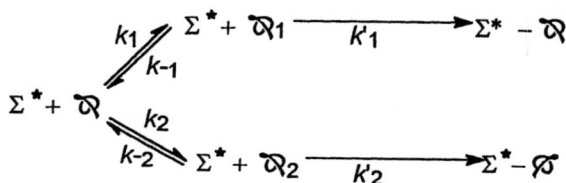

Figure I.11.a

$$k = \frac{k_1}{k_{-1}} \cdot \frac{k_{-2}}{k_2}$$

$$\frac{\Sigma^* - \mathcal{R}}{\Sigma^* - \mathcal{G}} = k\frac{k'_1}{k'_2} = e^{-\Delta\Delta G^{\neq}/RT}$$

Figure I.11.a (continued)

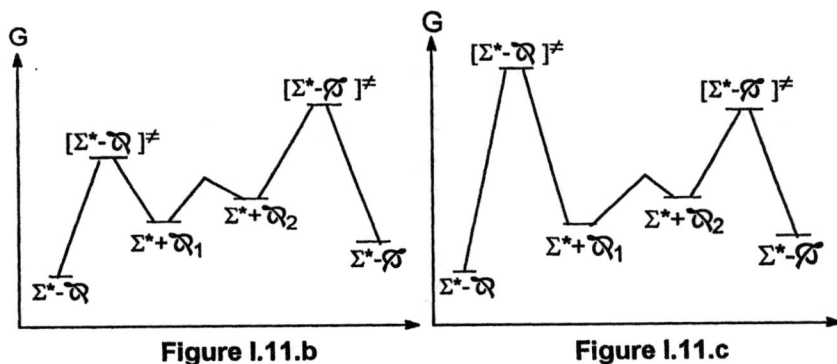

Figure I.11.b **Figure I.11.c**

Two possibilities can again be envisioned depending on whether the more stable conformer $\mathcal{R}_1$ is (Figure **I.11.b**) or is not (Figure **I.11.c**) the more reactive one. The difficulties in proposing the structure of the reactive conformer are often underestimated. Unless the interconversion barrier between the conformers is higher than the activation free energy, it is necessary to evaluate the conformational effects on transition states or transition structures. When barriers to interconversion are high, transition-state conformations usually resemble ground-state ones [51] .For example, Curran and coworkers [52] designed molecules **5** having axial chirality as substrates for radical reactions and cycloadditions. When R = *tert*-Bu, the barrier to rotation is high, the reaction is not under the conditions of Curtin-Hammett principle. The transition state presumably resembles the ground state and an interesting selectivity is observed.

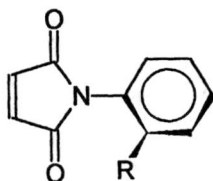

5

The favored ground-state conformations of many reagents and substrates have been determined experimentally or computationally. In addition to the previously described interactions, allylic (1,3) strain is frequently important [53, 54]. The favored conformation of 3-substituted allyl derivatives is that which staggers the C-R' and C-R" bonds relative to the C=C double bond. This arrangement minimizes gauche interactions (Figure I.12.a). When R = H, and R' = R" = Me, the energy difference between the two populated conformations C_1 and C_2 is 0.73 kcal mol^{-1}. However, when R = Me conformation C_2 energy lies 4 kcal mol^{-1} above C_1 and conformation C_3 is 3.4 kcal mole^{-1} above C_1 (Figure I.12.a). Amides can also suffer A(1,3) strain because the amide carbon-nitrogen bond has significant double-bond character. The exclusive formation of planar Z-amide enolates has been ascribed to destabilization of the E-isomers by A(1,3) strain (Figure I.12.b). When the E geometry is imposed by the structure of the molecule, distortions occur. Seebach and coworkers [55] have shown that amides 6 and their enolates have a pyramidal nitrogen in order to minimize A(1,3) strain. The ring is distorted and the *tert*-Bu group is in a pseudoaxial position (Figure I.12.c). Unsymmetrical α-branched enolate radicals also adopt preferred conformations in which A(1,3) strain is avoided, unless polar effects counterbalance this strain [56].

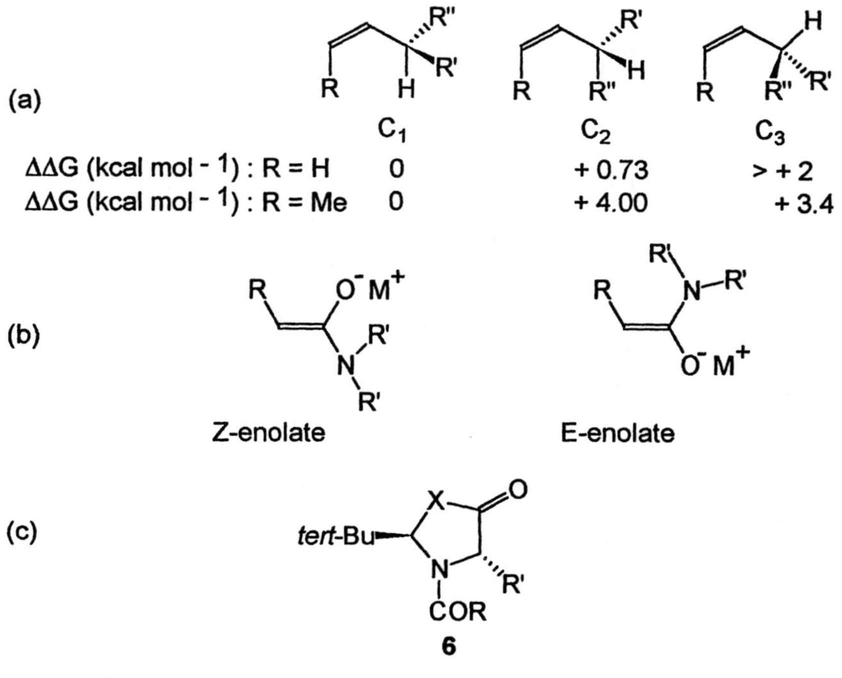

	C_1	C_2	C_3
$\Delta\Delta G$ (kcal mol^{-1}) : R = H	0	+ 0.73	> + 2
$\Delta\Delta G$ (kcal mol^{-1}) : R = Me	0	+ 4.00	+ 3.4

(b)

Z-enolate   E-enolate

(c)

6

Figure I.12

Steric, polar and A(1,3) strain effects can be of the same order of magnitude in ground states as in the transition states, so the formation of the new bond(s) does not necessitate that modifications to the relative energy difference between the two sets of conformers occur. For example, the pyramidalization of the nitrogen atom of the enolate of **6** is maintained throughout the transition state for alkylation (§ 5.3.2.3) so the stereoselectivity of such reactions is controlled by this conformational effect. Similarly, the ground-state conformation of π-conjugated radicals is maintained at transition state [16, 56]. Similar considerations hold for aldol reactions of some ketone enolates [57] and thermal [$4\pi+2\pi$] cycloadditions (§ 9.3). On the other hand, geometries of transition states can deviate significantly from those in the ground state, so a careful estimate of the various interactions in the transition state must be carried out. A few examples to illustrate this point will be discussed below.

Substrate control occurs in many reactions, and nucleophilic additions to racemic or nonracemic unsymmetrical α-branched aldehydes and ketones provide typical examples. Reaction on either *Re* or the *Si* face of a prochiral carbonyl group leads to two diastereoisomeric alcoholates **7** and **8**, as the primary products (Figure **I.13**). In 1952, Cram [1] proposed that the approach of the nucleophile takes place on the face bearing the smallest substituent (R_S) of the ketone through conformation C_1 (Figure **I.13**). Later, Felkin suggested that the nucleophilic attack occurs opposite to the larger (R_L) or to the more polar (R_P) bond of the carbonyl compound lying under conformation C_2 (Figure **I.13**). Felkin's approach minimized steric, torsion and polar interactions.

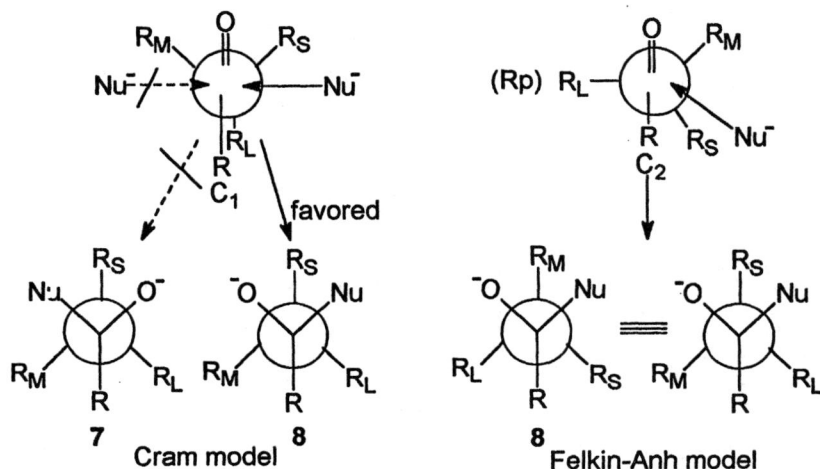

7   Cram model   8

8   Felkin-Anh model

Figure I.13

The Felkin model has been supported and refined by calculations of Anh and Eisenstein [7] as well as by Houk's group [23, 24]. The prevailing Felkin-Anh model takes into account a nonperpendicular Dunitz-Burgi approach of the nucleophile (see above) and it accounts for many stereochemical results.

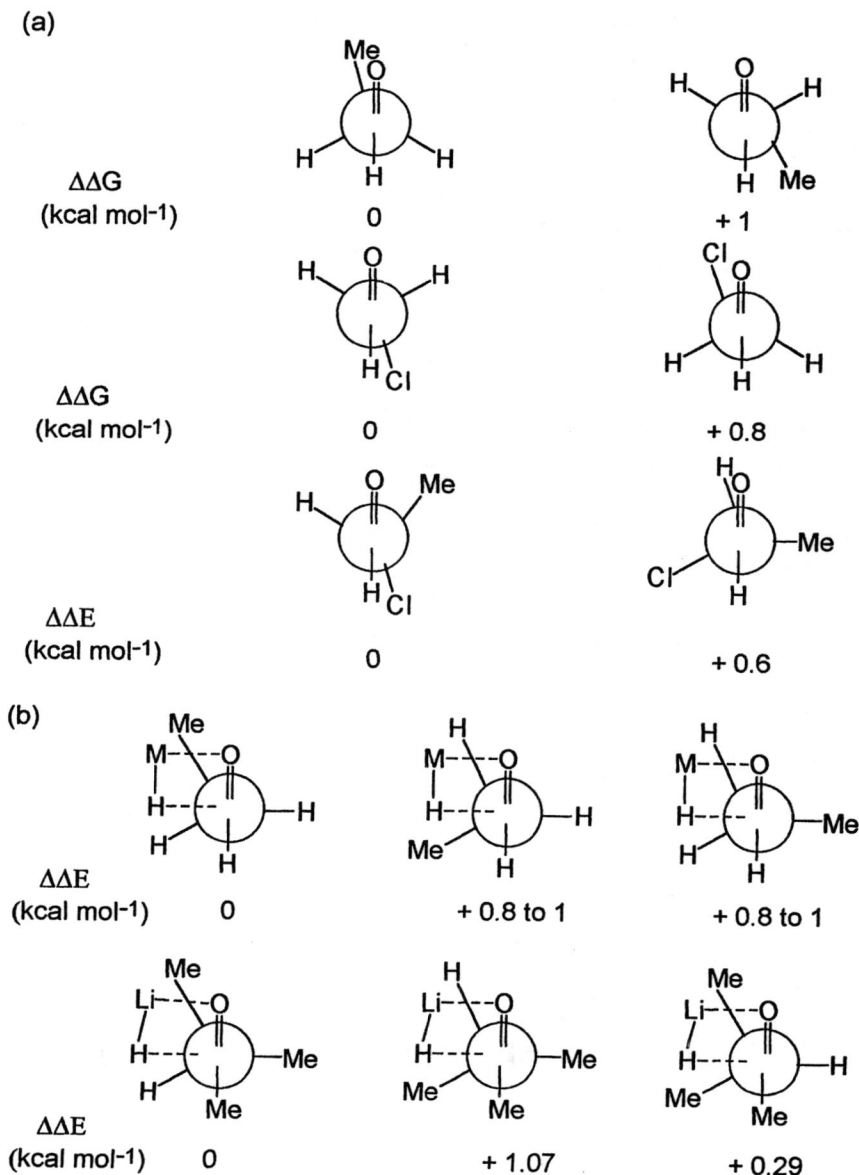

Figure I.14

The transition state is early, but has a fixed geometry where torsional strain is minimized. When a polar substituent (R$_P$) is present on the vicinal carbon, a stabilizing interaction between the HOMO of the forming bond and the low-lying σ* orbital of the substituent is important when the C-R$_P$ bond is perpendicular to the C=O bond axis [7, 23, 58, 59]. In the ground state, the favored conformation of α-bran-ched aldehydes or ketones has the C=O bond either eclipsed by a vicinal C-alkyl bond or antiperiplanar to a C-Cl bond [58, 60]. The most stable confor mers of propionaldehyde, chloracetaldehyde and 2-chloropropionaldehyde are shown in Figure I.14. Therefore, the Felkin-Anh model suggests that the carbonyl compound reacts through a transition-state conformation that is different from the favored ground-state conformation. Both Cieplak's proposals [25] (see above) and the interpretation of the antiperiplanar location of the C-R$_L$ bond in the Felkin-Anh model have been controversial. Some calculations of the reaction of propanal with LiH or NaH indicate that the most stable transition structure has the methyl residue gauche to the C=O bond (Figure I.14). However, the energy differences between this and other conformations are small [22, 24, 32, 58]. More recent calculations of Esterowicz and Houk [61] indicate that the most stable transition structure for the reaction MeCOCHMe$_2$ + LiH corresponds to the Felkin-Anh model (Figure I.14). The Felkin-Anh model is also used to interpret 1,2-asymmetric induction in S$_N$2' allylation of organometallic reagents [62]. Substrate control of 1,2-asymmetric induction in radical reactions has been examined by Curran, Giese and their coworkers [16, 17, 18, 56, 63, 64]. In the reactions of unsymmetrical α-branched π-conjugated radicals, there is an analogy between the ground-state conformation of the radical, as determined by EPR spectroscopy or calculation, and the transition-state conformation.

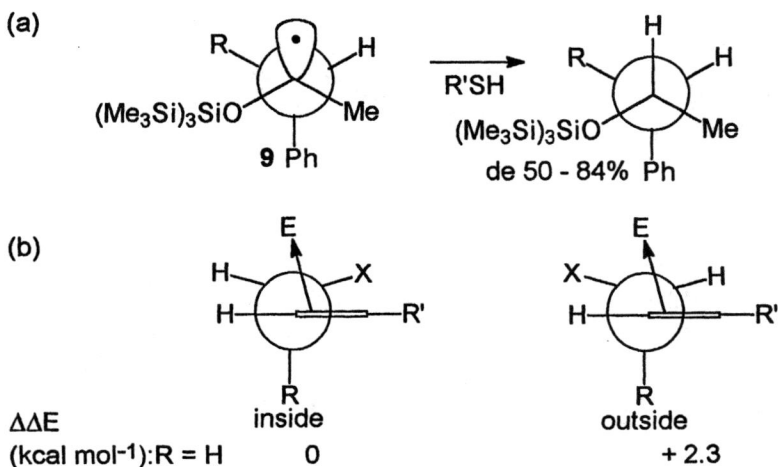

(a)

(b)

ΔΔE
(kcal mol^{-1}):R = H   0   + 2.3

Figure I.15

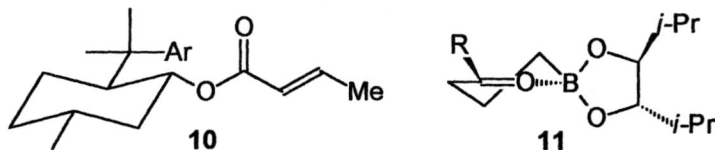

However, interaction of the radical with the trap must be taken into consideration, and radical models related to the Felkin-Anh model can provide valuable insights. The results generally support an early, rigidified transition state, which corresponds to Houk's calculations for addition of nucleophilic or electrophilic radicals to alkenes [10]. An example is given Figure **I.15.a**. The relative energies of the calculated transition structures are in agreement with the stereochemical results, provided that polar and A(1,3) strain effects are duely estimated [56]. In the case of alkyl-substituted radicals, which are easily pyramidalized, the transition state certainly lies earlier on the reaction coordinate and steric effects are less important [16]. The direction of pyramidalization of the radical is controlled by internal torsional strain and does not necessarily depend upon the ground-state conformation of the starting material or product.

Houk and coworkers [9] and Fujita and Ogura [12, 65] have interpreted the 1,2-asymmetric induction in electrophilic reactions of unsymmetrical α-branched double bonds bearing electron-withdrawing substituents in a similar fashion (Figure **I.15.b**). To minimize the destabilizing π-σ* interaction between the HOMO of the double bond and the low lying σ* orbital of the C–X single bond, the reactive conformations are those in which the C=C and the C–X bonds are gauche or skew. The calculations indicate that the favored transition structure is the upper case *inside* one (Figure **I.15.b**). In contrast, electron-donating substituents R will occupy the *anti* position because of the stabilizing π-σ interaction.

The influence of π-stacking on the conformation of substrates and transition states has been a matter of debate [44, 46-49] (see above). Nonetheless, the efficiency of 8-phenylsubstituted menthyl or *trans*-2-phenylsubstituted cyclohexyl esters of α,β-unsaturated acids as substrates in asymmetric induction has been ascribed to such effects [66]. Both NMR spectroscopy and calculations [67, 68] have shown that π-stacking can stabilize the *s-cis-syn* conformation of crotonates **10**, especially when Ar = 2-Np or 4-PhOC$_6$H$_4$. An X-ray determination shows that **10** (Ar = 2-Np) adopts the *s-cis* conformation in the solid state [67]. Interactions with remote substituents of the substrate can cause a preferred reactive conformation to intervene. For example, the reduction of γ-ketoboronates by borane is highly stereoselective and probably takes place via an internal complexed intermediate **11**, even though this could not be characterized in the ground state [69, 70].

The conformation of reagents may also be of prime importance, as illustrated by the reactions of prochiral aldehydes with α-substituted allyl boronates [53, 54]. The primary products of these reactions are homoallyl boronates, which are then hydrolyzed to provide diastereoisomeric homoallyl alcohols. Two six-membered cyclic transition states C_1 and C_2 involving two different conformers of the boronate are usually considered (see below) (Figure I.16). The *Re* or the *Si* face of the aldehyde is attacked according to which one minimizes the eclipsing 1,3-interactions between the C-R bond of the aldehyde and one of the boronate B-O bonds. If the boronate double bond is E (R_Z = H), the reaction is poorly stereoselective. The *syn* Z-12 and the *anti* E-13 alcohols are obtained in equal amounts, because the two transition states C_1 and C_2 have similar energies. However, the reaction of the Z-isomer (R_E = H) is highly stereoselective, favoring the *syn* alcohol E-14. Due to A(1,3) strain, transition state C_1 is strongly destabilized, as is the corresponding ground-state conformer of the Z-boronate (§ 6.6.1).

I.4 INFLUENCE OF THE REACTION CONDITIONS

The reaction conditions involve variables like temperature, pressure, solvent, the nature of associated metals or of counterions in the case of polar reactions, and the presence of additives such as salts, Lewis acids or bases. All these variables can modify the substrate-reagent interactions.

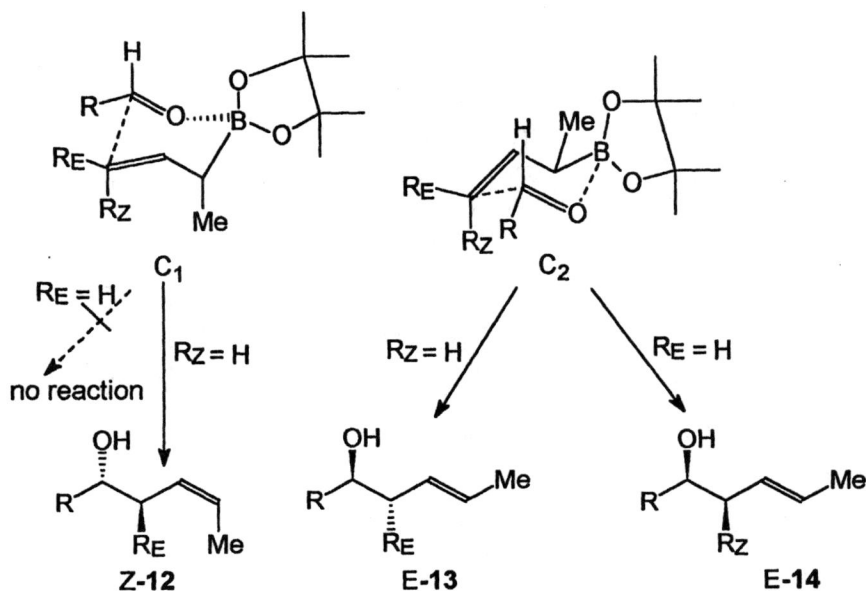

Figure I.16

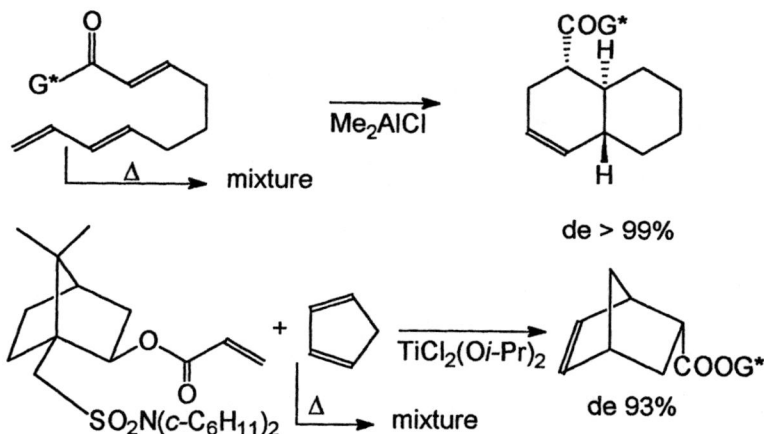

Figure I.17

Both the geometrical characteristics of the transition states (attack directionality or pyramidalization of the reacting sites) and the structure of the reactive
conformers can be altered by changing reaction conditions so that the relative energy level of the transition states, and therefore the stereoselection, can be
changed. A classic example of the effect of reaction conditions is the difference in
stereoselectivity of the Diels-Alder reaction of acrylates derivatives with dienes
when performed thermally or in the presence of Lewis acids [71-74] (Figure I.17)
(§ 9.3). In the thermal reaction, several conformers of the dienophile can intervene, while in the presence of Lewis acids, only the *s-trans* conformer of the ester
takes part to the reaction.

I.4.1 Nonconformational Effects

Lewis acid-Lewis base interactions are stabilizing. If the reagents themselves
or their associated cations behave as Lewis acids or if a Lewis acid is added to the
reaction mixture, the formation of a complex often induces a rate increase in the
reactions of saturated or α,β-unsaturated carbonyl compounds. This decrease in
transition-state free energy is due to the polarization of the double bond(s) and to
the lowering of their LUMO orbital(s) [5, 75]. Theoretical calculations of model
reactions [7, 21, 23] indicate that the transition state lies earlier on the reaction
coordinate of a Lewis acid complex and that the directionality of attack is modified
toward a less obtuse angle. Therefore, the various relative repulsive interactions
can be modified. The Lewis acid can also provide a sterically significant component to the stereochemical outcome of the reaction [76] (see below). Houk's calculations of transition structures of the reaction of formaldehyde with the enolate of
acetaldehyde are shown in Figure I.18. The geometry of the transition state in the
absence of a cation is different from that of the favored transition state when Li+ is

incorporated. The relative changes in characteristics of the formaldehyde fragment at transition structure show indeed that the transition structure lies earlier on the reaction coordinate when $\overset{+}{Li}$ is involved.

When Li^+ is associated in the transition state for addition, the forming bond is longer (2.368 compared to 2.02-2.06Å), the C=O bond is shorter (1.251 compared to 1.255-1.263), the attack angle less obtuse (107° compared to 113-118°) and there is a smaller degree of pyramidalization of the carbonyl group. Similar trends are seen in calculation of the transition structure of the reaction of LiH with MeCHO or $MeCOCHMe_2$ [61] (Figure I.18c). These calculations do not take into account the aggregation state of reagents [77, 78]; indeed, the calculations of Nakamura and Morokuma indicate that the open dimer of MeLi is a highly reactive species toward aldehydes [79]. Solvent and salt effects [77, 80, 81] can also modify transition states. As previously mentioned, inter- or intramolecular hydrogen bonding may also promote rate enhancements due to a decrease in transition-state free energy [50].

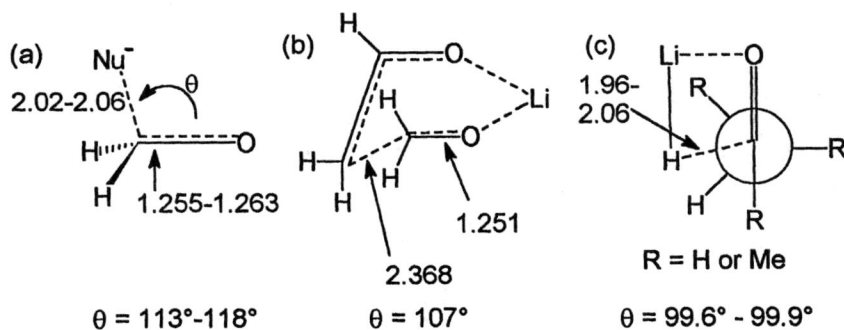

Figure I.18

I.4.2 Conformational Effects

The coordination of a substrate possessing one or several Lewis basic sites with a reagent having a Lewis acidic site or with an added Lewis acid may modify the favored conformation at ground state and at the transition state. Such a coordination can be mono- bi-, tri- or tetradentate according to the coordination number and geometry of the Lewis acid. Chelated structures may or may not be maintained at transition state The nature of the ligands other than the substrate around the Lewis acid site is important because these ligands can greatly influence the geometry of the complexes, especially when transition metals are involved. Many theoretical approaches are used to rationalize experimental results [75, 82, 83, 84].

I.4.2.1 Monodentate Interactions

It has been shown by X-ray crystallographic and NMR studies of aldehyde-Lewis acid complexes that a Lewis acid does not coordinate along the C=O bond axis [76, 85]. Moreover, according to the nature of the Lewis acid and the stoichiometry, several types of complexes can be generated [85, 86]. When considering the nucleophilic attack on complexed carbonyl compounds, the various interactions at transition state must be estimated. Indeed, the presence of the Lewis acid can provide new nonbonded interactions with the incoming reagent.

For instance, the reduction of unsymmetrical α-branched ketones by dialkylborane R'_2BH involves the coordination of the carbonyl group by the Lewis acidic boron atom. Hydride transfer takes place with minimization of repulsive interactions so that boronate 15 is the major diastereoisomeric product (Figure I.19). This diastereoisomer is the "anti-Cram" product, but the Houk model accounts for its generation. As in the Felkin-Anh model, nucleophilic attack occurs opposite to the C-R_L or C-R_P bond. But in order to minimize the repulsions with the borane bonds, the smallest group R_S is now located in a gauche (inside) relationship to the coordinated C=O bond [9, 87]. Therefore, nucleophilic tetracoordinate and electrophilic tricoordinate boron hydrides give opposite stereoselectivities in reductions of α-chiral carbonyl compounds. Midland's results for reductions of racemic ketone 16 are recorded Figure I.19. Reduction of 16 with Li s-Bu$_3$BH takes place according to the Felkin-Anh model and the tricoordinated borane reduction follows the Houk model.

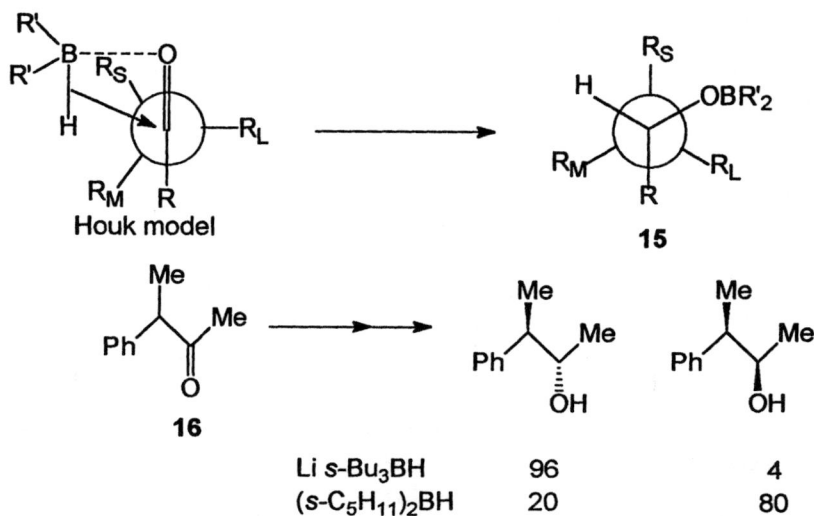

Houk model

15

16

Li s-Bu$_3$BH	96	4
(s-C$_5$H$_{11}$)$_2$BH	20	80

Figure I.19

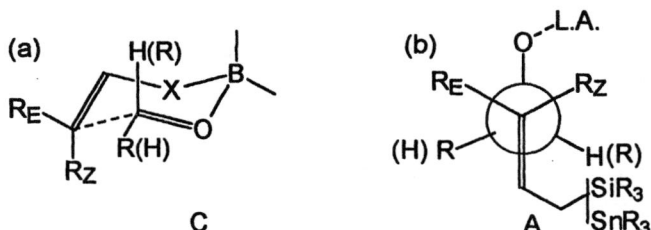

Figure I.20

The reactions of aldehydes with enolborinates and allylboranes also take place after coordination of the boron atom with the carbonyl group. An early cyclic transition model C ($X = O$ or CH_2) often accounts for the observed selectivity (Figure **I.20.a**) (§ 6.6.1 and 6.8).

On the other hand, the reactions of aldehydes with allylsilanes and allylstannanes require the addition of a separate Lewis acid. The stereoselectivity of these additions is best interpreted by acyclic transition model A (Figure **I.20.b**) (§ 6.6.1). The relative interactions involved between the *Re* and *Si* face attack of the aldehyde are estimated for both models, assuming that no major geometrical change takes place between ground state and transition state.

The predominant conformation of α,β-unsaturated carbonyl compounds (aldehydes, ketones, esters) can vary when Lewis acids are coordinated to the carbonyl group. It has been experimentally shown that the *s-trans* conformer of acrolein is more stable than the *s-cis* conformer, in the absence of any Lewis acid. The two conformers are of almost equal stability in the case of acrylic acid or methyl acrylate [83] (Figure **I.21**). The barriers between these conformers are low (4-9 kcal mol^{-1}). Interconversion is rapid, and the Curtin-Hammett principle (§ I.3) can be applied.

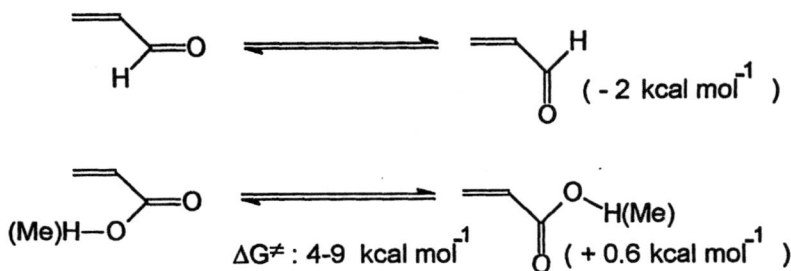

Figure I.21

In the presence of Lewis acids, 2-methacrolein, acrolein, E-2-heptenal, acrylic acid, methyl acrylate and ethyl cinnamate are more stable in the *s-trans* conformation [76, 83, 85, 88] by about 1.5-3.2 kcal mol^{-1}. The calculated barrier to conformational interconversion is higher (12 kcal mol^{-1}).

The *s-cis/s-trans* conformational problem has important repercussions on the facial discrimination of addition and cycloaddition reactions to conjugated double bonds. As previously stated (Figure I.17), thermal Diels-Alder reactions are poorly stereoselective, but Lewis-acid-catalyzed reactions are highly stereoselective [73, 74]. A theoretical study of the reaction between acrolein and butadiene was conducted by Birney and Houk [11, 89]. They found that the most stable conformations of both the noncatalyzed and the BH$_3$-catalyzed transition structures have the acrolein fragment in the *s-cis* geometry. Nevertheless, the energy differences with other transition structures are small (Figure I.22). These calculations indicate that the reactive conformer is not the most stable one in the ground state. They also indicate that, under BH$_3$ catalysis, the complexed transition structure is stabilized related to the noncomplexed one and that the formation of the two incipient bonds is more asynchronous (see the calculated bond lengths in figure I.22). Other calculations of Lewis-acid-promoted cycloadditions of acrylic esters [71, 90] indicate that the α,β-unsaturated ester fragment is in the *s-trans* orientation in both the ground state and the transition state. Similar calculations have been performed for conjugate additions [26] and other cycloadditions [11, 71].

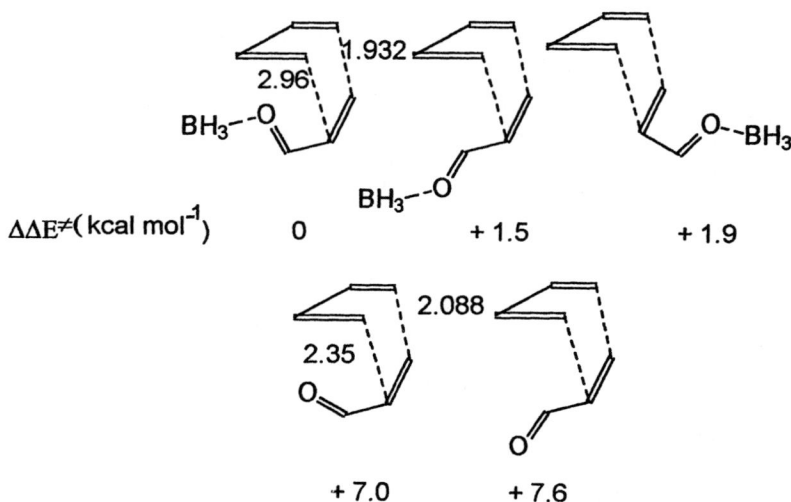

Figure I.22

Hydrogen-bonding solvents may also act as Lewis acids and modify the conformation of α,β-unsaturated esters. This effect has been proposed by d'Angelo and coworkers [91] to account for the stereoselectivity of the 1,4-addition of an amine to crotonate **10** (Ar = 2-Np). While this compound exhibits the *s-cis-syn* conformation in the crystal, the facial diastereodifferentiation observed is in agreement with the reaction through the *s-trans* conformer, possibly due to the protic nature of the solvent.

I.4.2.2 Polydentate Interactions

Substrates and reagents, whether chiral or prochiral, are often multifunctional so that chelates may be generated. Chelates are usually more stable than non-chelated species, and their ground-state structure can be maintained at transition state. In other words, chelates, although more stable, can also be more reactive. Most frequently, chelation involves a neutral or acidic Lewis acid site and several Lewis basic groups. Such basic groups can be located on substrates, on reagents, or as ligands on catalysts. The coordination number and the geometry of the Lewis acidic site are of prime importance. Among the most common Lewis acids, boron, lithium, aluminum and tin (II) have coordination number of 4. The coordination number of zinc and magnesium is 4 or 6, and that of titanium and tin (IV) is usually 6. The coordination number and geometry depends upon many factors [92] which will not be reviewed here. Five- or six-membered chelates are most frequently encountered.

I.4.2.2.1 *Chelation Between Substrates and Reagents*

The reactions of carbonyl compounds bearing in the α-, β- or γ-position another basic functional group with organometallic reagents such as organomagnesium or titanium species can take place through a bidentate chelate. The metal will be coordinated to both the carbonyl group and the other basic site. The cyclic Cram model accounts for the stereoselectivity of the reaction, leading to metal alcoholate **17** in place of **18**, which is formed under non chelation control (Felkin-Anh model) [93] (Figure **I.23**).

Eliel and coworkers [94] have shown that the occurrence of chelation-control provides a corresponding rate increase in the reactions of α-alkoxyketones with Me_2Mg (Figure **I.24**). To observe chelation control and the attendant rate enhancement, the solvent should be a poor Lewis base in order not to compete with the ether substituent. The ether R" substituent must not be too bulky; methyl or benzyl ethers are highly reactive and selective, while trimethylsilyl ethers are less useful. Triisopropylsilyl ethers react non stereoselectively and at the lowest rate (Figure **I.25**).

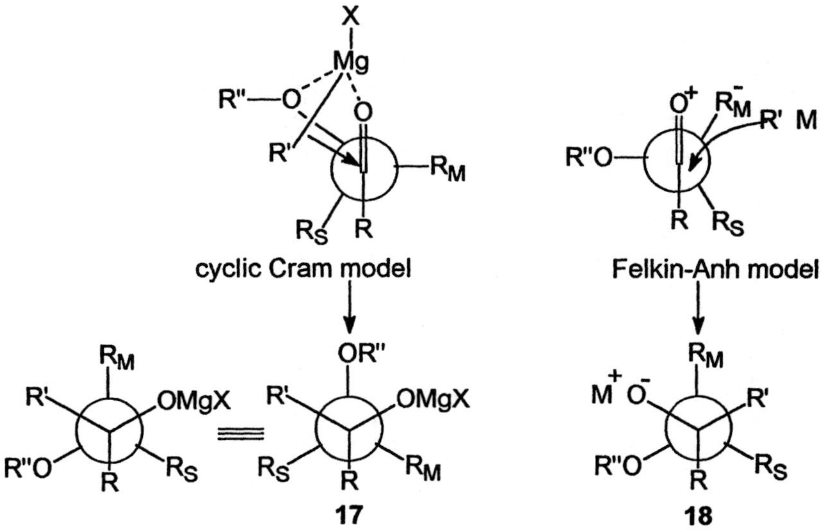
cyclic Cram model
Felkin-Anh model
17
18

Figure I. 23

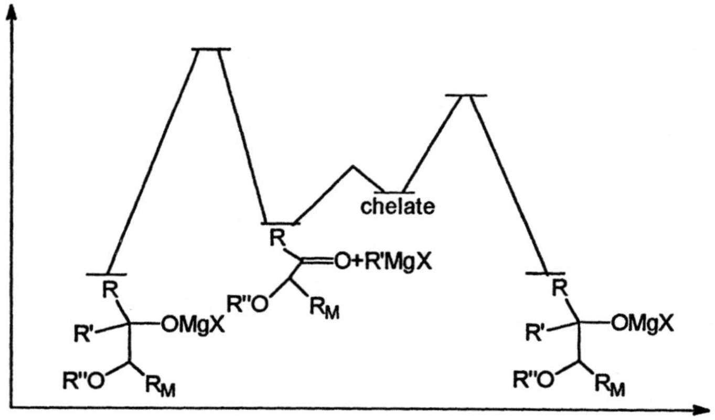
chelate

Figure I.24

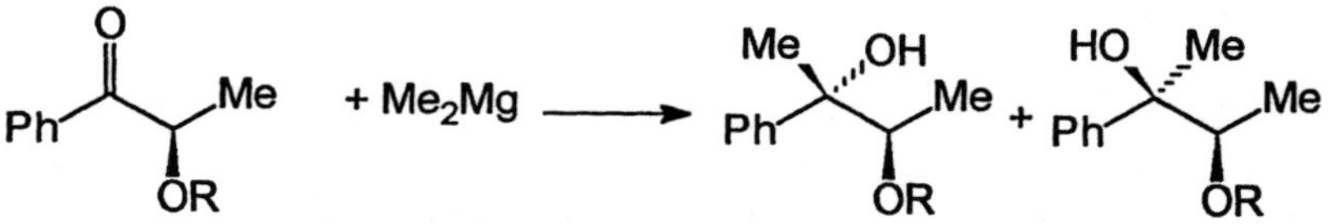

R =	rel. rate const.		
Me	2000	> 99	< 1
$SiMe_3$	200	99	1
$SiMe_2$*tert*-Bu	5	88	12
Si*i*-Pr_3	1	42	58

Figure I. 25

Reetz and coworkers have made similar observations in the reactions of methyltitanium reagents with α-alkoxyketones and aldehydes. They observed the formation of octahedral titanium complexes by ^{13}C NMR spectroscopy [93, 95]. These experimental results were supported by theoretical calculations [32, 93]. Aminoketones $R'_2N(CH_2)_nCOR$ behave in a similar fashion provided that the nitrogen basicity is high enough and that the R' substituent is not too bulky. Indeed, when R' = PhCH$_2$, chelation control does not operate [96]. Acetals, thioacetals, esters, amides, and sulfoxides may also behave as basic sites for bidentate chelation (§ 6.5.2, 6.5.3). The other substituents around the titanium atom of methyltitanium reagents may also promote or disfavor chelation control. Reetz and coworkers [93, 96-99] have shown that the reaction of unsymmetrical α- or β-alkoxyaldehydes is under chelation control when the reagent is the strongly Lewis acidic MeTiCl$_3$. In contrast, MeTi(Oi-Pr)$_3$ is a weaker Lewis acid and it does not promote chelation-control. However, this difference is not observed when such reactions are performed with related ketones.

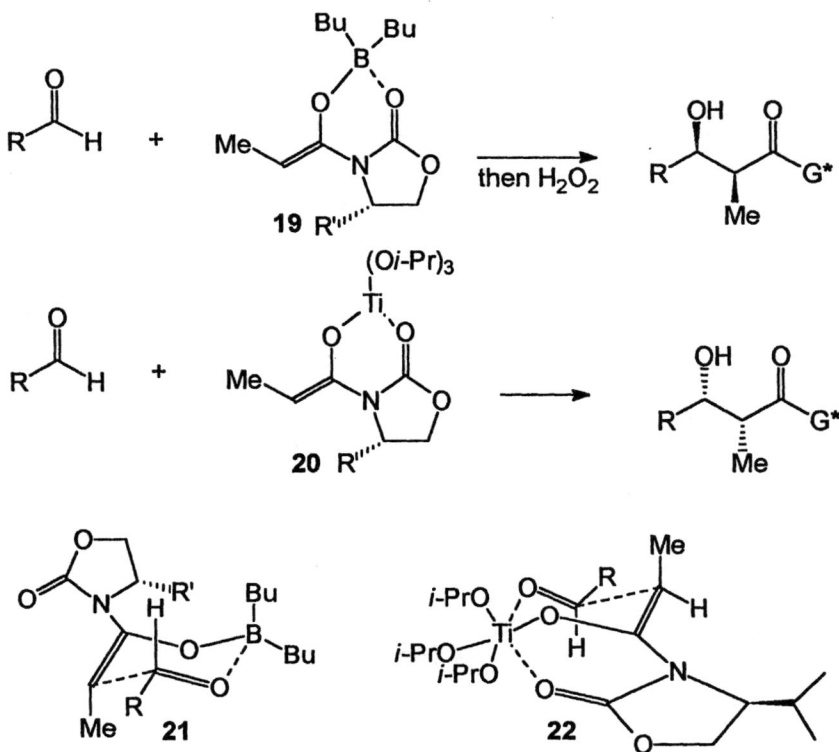

Figure I.26

Aldol and related reactions may also be chelation-controlled. Boron enolates of *N*-acyloxazolidinones **19** are chelated in the ground state. Their reactions with aldehydes will necessitate the coordination of the aldehyde with the boron atom at transition state, so that the initial bidentate chelate will be broken (Figure **I.26**). However, the titanium atom of related titanium enolates can accommodate hexa-coordination so that the initial titanium chelate **20** does not need to be disrupted. In each case, the aldol reaction leads to different *syn* stereoisomers (Figure **I.26**) via transition models **21** and **22**.

The introduction of several basic sites (such as methoxymethyl- $MeOCH_2O$ or methoxyethylethers $MeOCH_2CH_2O$) on the substrates induces a stronger pre-organization and rigidification around the metallic site of the organometallic rea-gent or the metal enolate. A higher stereoselectivity is promoted in this fashion. As shown in Figure **I.27**, the stereoselectivity of the alkylation of the lithium anion of chiral enamine **23** with $PhCH_2Br$ is solvent dependent. In toluene-THF, the reaction involves a tridentate lithium chelate **24**, which is attacked with a high selectivity (92%) on the face opposite to the *i*-Pr substituent. In the presence of HMPA, chelation is disrupted and alkylation takes place on the opposite face. Now the nucleophile reactive conformation is **25** due to dipolar interactions, and the selectivity is somewhat lower (80%). Choudury and Thornton [100] have also observed opposite diastereofacial selectivities in aldol reactions of α'-metho-xymethoxy-substituted lithium enolates of **26** ($R = MeOCH_2$), which promote chelate formation and benzoyloxy analogs **26** ($R = COPh$), which do not.

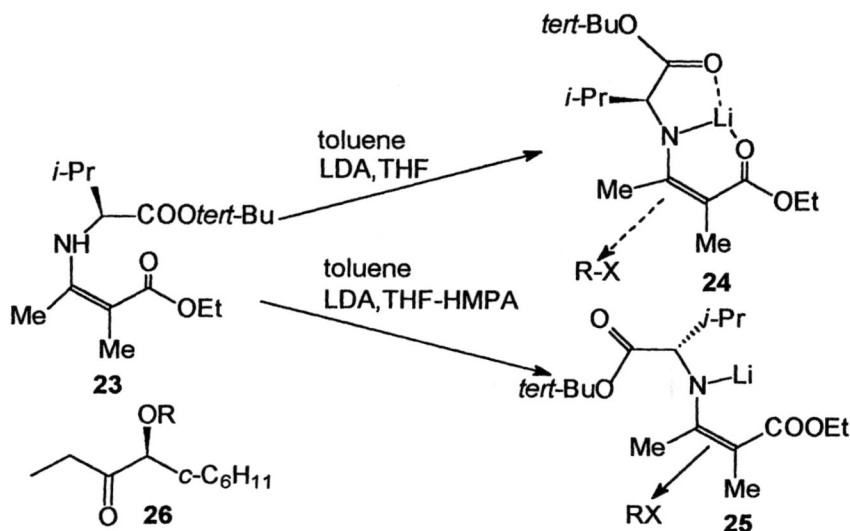

Figure I.27

A negative entropy component is associated with such rigidifications of ground states and transition states, so low temperatures favor these pathways over competing pathways.

I.4.2.2.2 Chelation with Added Lewis Acids

Precoordination of carbonyl groups bearing other basic sites can be performed prior to the addition of reagents. This is often necessary in reactions involving Lewis acid activation of carbonyl compounds toward nucleophilic attack by reagents such as enoxysilanes, allylsilanes and stannanes, Me_3SiCN [93]. In these cases, the chelate structure of the substrate will be maintained in the transition state and the least hindered face of the locked complex will be attacked. Thus a high facial discrimination often results (§ 6.6.2, 6.6.3, 6.8.2, 6.9), as illustrated by the example in Figure I.28 [99]. The stereoselectivity of the $TiCl_4$-catalyzed reaction of a chiral β-alkoxyaldehyde with an enoxysilane or an allylsilane is interpreted by attack of the least hindered face of chelate 27 by the nucleophile. This leads to a chelated trichlorotitanium alcoholate 28 as the primary product with an excellent diastereoselectivity.

A related method has been introduced by Posner and coworkers [101, 102] and Solladié and coworkers [103] to direct the facial discrimination of the reactions of chiral sulfoxides bearing another basic functional group. Addition of $ZnCl_2$ forms a chelate whose conformation is quite different from that of the free sulfoxide. The ensuing nucleophilic attack on the nonracemic α,β-unsaturated cyclic β-ketosulfoxides takes place on the face opposite to the tolyl group through either on conformer 29a or 29b, depending on whether $ZnCl_2$ is added to the reaction mixture. Similar models are proposed for hydride reductions of β-keto-sulfoxides. Conformer 30b or 30a is the reactive one (Figure I.29), depending on whether zinc is present. Chelation is indeed maintained in the transition state in both cases (§ 6.1.2, 7.10.2.3, 7.13.2.3).

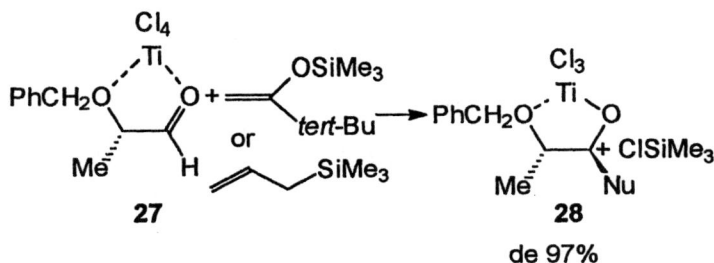

Figure I.28

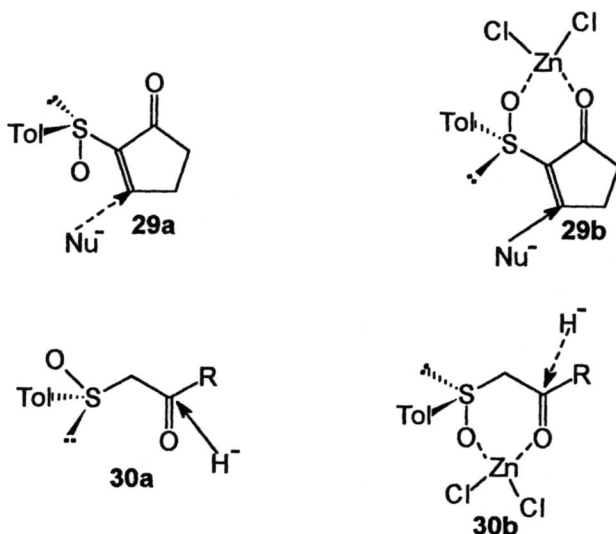

Figure I.29

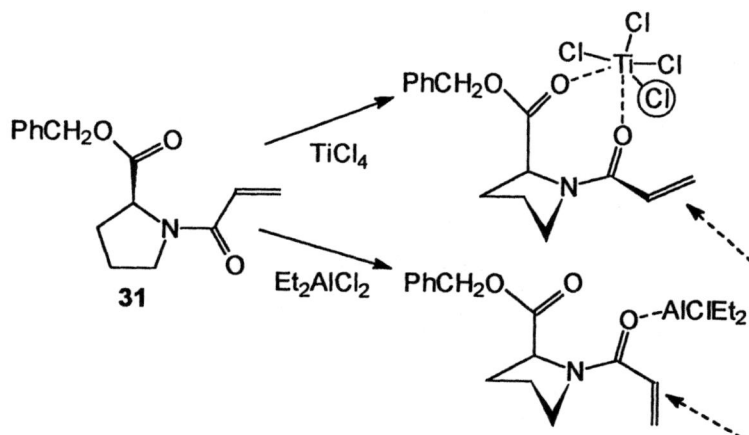

Figure I.30

As previously discussed for organotitanium reagents, the nature of the added Lewis acid influences the geometry of the reactive conformer. For example, $TiCl_4$ prefers hexacoordinated complexes, and it induces cycloadditions of benzyl ester of N-acryloylproline **31** through an *s-cis* chelate whose *Si* face is not shielded by a chlorine bound to titanium. The opposite facial selectivity is observed under $EtAlCl_2$ catalysis, probably because the *s-trans* monodentate complex is the reactive conformer [73, 104] (Figure **I.30**). The chelate structure may also be maintained in reagents. In aldol reactions performed with chelated boron enolates of *N*-acyloxazolidinones **19**, the introduction of another Lewis acid such as Bu_2BOTf or Et_2AlCl can both activate the aldehyde and prevent the disruption of the boron chelate. The stereoselectivity of the reaction no longer favors the *syn* aldol; instead, the *anti* aldol **32** is formed. Transition model **33** accounts for this result [105, 106] (Figure **I.31**) (§ 6.8.1.3).

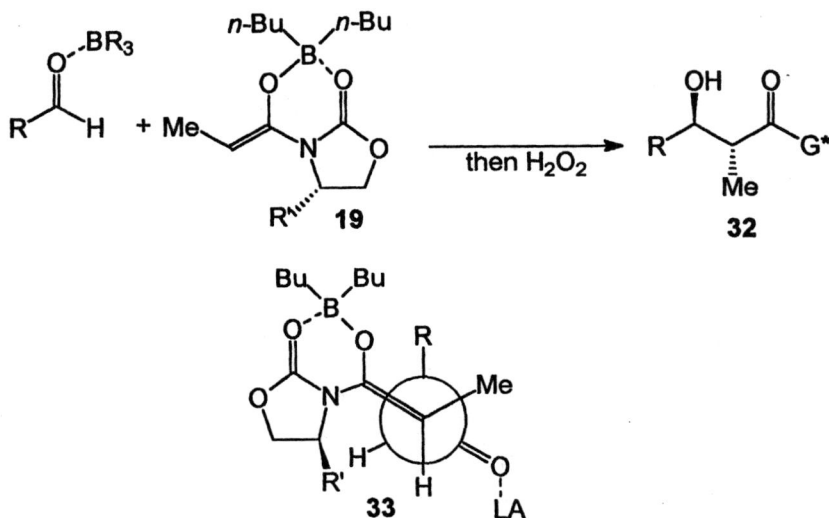

Figure I.31

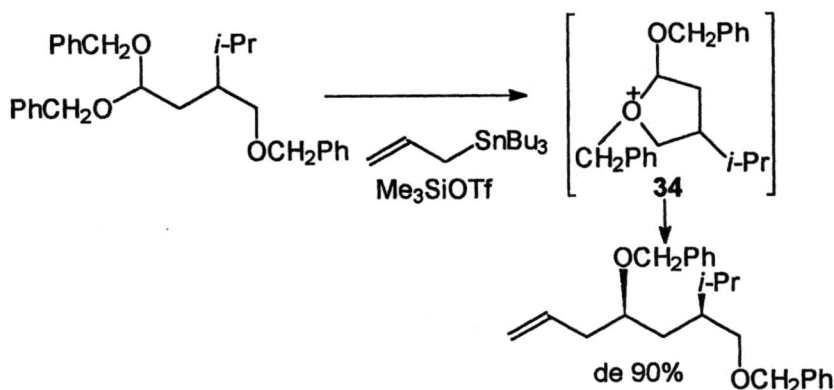

Figure I.32

Addition of external Lewis acids can also cause remote functional groups to form preorganized intermediates which induce a high stereochemical control. This strategy for stereocontrol has been proposed by Molander and Haar [107]. The reaction of allylstannanes with some β- or γ-alkoxyacetals under Me₃SiOTf catalysis likely involves the intermediate formation of a cyclic oxonium ion such as **34** which suffers preferential attack on one side controlled by its preferred conformation (Figure I.32). A related process is involved in SnCl₄-catalyzed reaction of similar acetals or γ-alkoxyaldehydes with Me₃SiCN, but the selectivities are not always so high.

I.5 INFLUENCE OF CHIRAL LIGANDS

The introduction of chiral ligands on a reagent or a catalyst induces asymmetry that will be transmitted to the corresponding diastereoisomeric transition states. In most of the cases, the ligands favor rigidified chelates, and there are only a few low-energy conformers to consider.

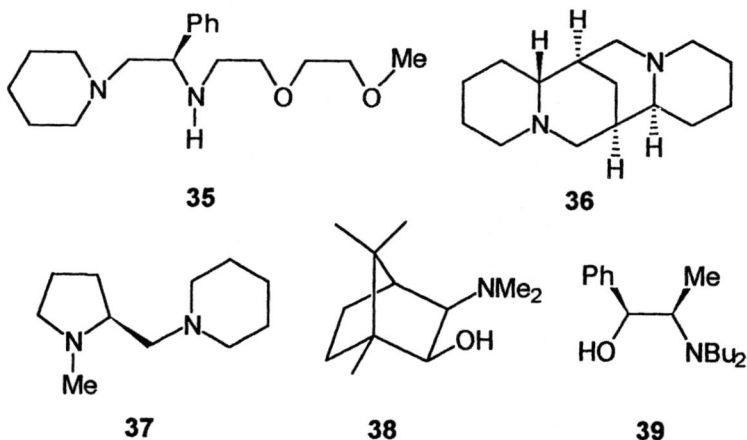

35   **36**

37   **38**   **39**

The reactions of organometallic reagents are quite sensitive to the influence of ligands bearing several basic sites. Enantiopure polyethers or aminoethers form unsymmetrical chelates with cations or metal atoms. For example, the aminoether **35** introduced by Koga and coworkers [108] and sparteine **36** [109] are potent ligands of lithium. Diamine **37** derived from (S)-proline is an efficient ligand for asymmetric reactions promoted by tin(II) (§ 2.6.1, 2.6.2, 6.9.1). The formation of aggregates can enhance asymmetry [77, 78]. Asymmetric condensation of organozinc reagents with aldehydes takes place in the presence of nonracemic aminoalcohols such as the bornane derivative **38** proposed by Noyori and coworkers [110] or the N,N-dibutylephedrine **39** recommended by Soai and coworkers [111].

Rigidified binuclear zinc chelates are formed, and their asymmetry promotes the facial diastereodifferentiation of the aldehyde carbonyl group in a highly enantiose-lective reaction (§ 2.5.1,6.5.1). Chiral aminoalcohols and diamines are the most efficient ligands to promote asymmetric conjugate additions of lithio- or magne-siocuprates to α,β-unsaturated carbonyl compounds [112] (§ 7.10.1).

Most ligands of transition metal complexes that are used in asymmetric ca-talysis are prone to chelate to the metallic center and generate di- or tricoordinated complexes. Nonracemic chiral diamines, diphosphines or aminophosphines are the most important classes of such ligands [113] (§ 3.3, 3.4). The other coordinative sites of the intermediate complexes involved in catalytic cycles are occupied by the reagent and the substrate. In many cases, the substrate is also bicoordinated to the metal. Therefore, the understanding of the conformations of the chiral ligands and of the substrates in such complexes will be of prime importance in determing the favored reaction process. Moreover, electronic effects can also be control ele-ments for the enantioselectivity of transition metal-catalyzed reactions. Such ef-fects operate in selectively stabilizing some reactive conformations of the com-plexes. For example, in asymmetric hydrogenation of double bonds under rhodium (I) catalysis, face diastereodifferentiation involves intermediate complex **40** in which a chiral diphosphine is coordinated to the rhodium atom in a bidentate fash-ion. The presence of electron-donating substituents on phosphorus P_2, located in the *trans* relationship to the bond to be hydrogenated, influences the reaction process in two ways [114] (Figure **I.33**).

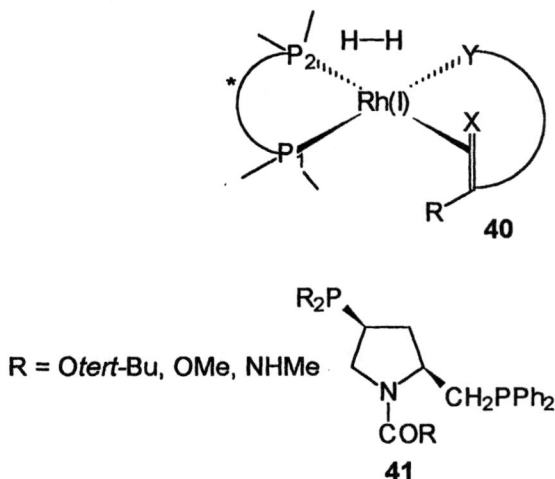

R = O*tert*-Bu, OMe, NHMe

Figure I.33

(a) it accelerates the oxidative addition of dihydrogen, which is the rate-determining step, due to an enhanced d-σ* interaction between the rhodium and H_2. Such an acceleration will decrease the amount of interconversion between the various conformers of the transition metal complex (§ 7.1.1).

(b) it strengthens the interaction between the rhodium atom and the electron-deficient substrate by d-π* back donation, so that facial discrimination will be increased.

These considerations have been used to design new rhodium ligands [114, 115]. For instance, Achiwa and coworkers replaced the phenyl R groups of ligand **41** with cyclohexyl groups. The usefulness of the corresponding complexes is quite broad (§ 7.1.1.1).

I.6 COOPERATIVITY, DOUBLE DIASTEREODIFFERENTIATION

Up to this point, the discussion has focused on the existence of just a single stereogenic center or asymmetry element on the substrate, the reagent or the catalyst. However, reacting systems may include more stereocenters or asymmetry elements located either on the same partner or on other components. The presence of the added elements of asymmetry can influence the relative energies of the various transition states and either augment or erode the diastereodifferentiation observed when a similar reaction is performed in the presence of a single asymmetry element. For the sake of simplicity, only the case of double diastereodifferentiation will be considered here. Two possibilities arise:

• a single partner, the substrate for example, bears the two stereocenters or elements of chirality.

• the reagent and substrate or the catalyst and substrate each bear one element of asymmetry and double asymmetric induction takes place. The concepts of double asymmetric induction were first introduced by Horeau, Kagan and Vigneron [115a] and further developed by Heathcock [116] and Masamune and coworkers [117].

When the substrate bears two centers of chirality, Tolbert and Ali [118] introduced the notion of cooperativity in asymmetric induction as a criterion of concertedness of reaction mechanism. In cooperative reactions, the introduction of the two identical stereocenters at independent sites of a symmetric molecule, will induce higher asymmetric induction than expected by simple additivity. To a first approximation, the free-energy difference between these transition states is doubled, which corresponds to an improved product ratio by the square of the effect. The main effect involved is conformational, as shown in model **42** (Figure **I.34.a**).

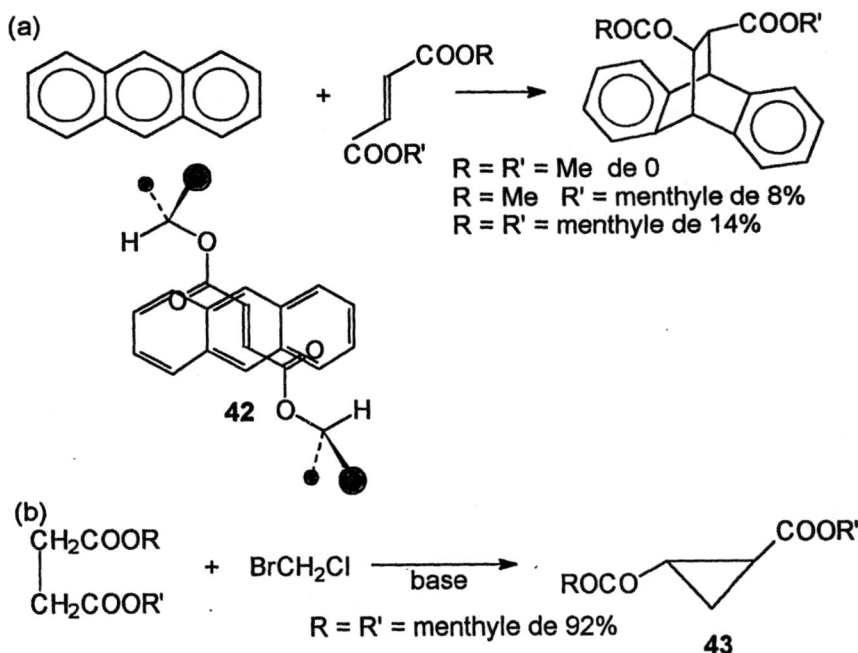

Figure I.34

Initially applied to thermal cycloaddition of fumarates of chiral alcohols with anthracene [118] (Figure **I.34.a**), cooperativity was extended to other reactions of symmetrical diesters. While reactions of mixed diesters bearing a single chiral alcohol residue are poorly stereoselective, those of symmetrical diesters can be quite useful. Indeed, the formation of cyclopropane diesters **43** from succinic acid diesters and $BrCH_2Cl$ is poorly diastereoselective when using a mixed methyl and menthyl ester. But it becomes highly diastereoselective when performed with symmetrical dimenthyl esters (Figure **I.34.b**).

When both the reagent (or catalyst) and the substrate bear a stereocenter, the diastereoselectivity can be increased relative to that observed when a single stereocenter is involved. In that case, the partners are said to be matched. If the diastereoselectivity is reduced, the partners are mismatched [117]. These trends are caused by a decrease (or an increase) of the free-energy difference between the two transition states due to new interactions promoted by the second stereocenter. An example of double diastereoselection is given in Figure **I.35**. The reaction of prochiral aldehyde PhCH$_2$OCH$_2$CH$_2$CHO with chiral boron enolate (S)-Z-**44** is stereoselective, favoring *syn* aldol **45** (de 92%). When the aldol reaction is run with chiral aldehyde **46** and the same boron enolate, the stereoselectivity is noticeably increased (de > 95%), showing that the chiral partners are matched for formation of the precursor of aldol **47**. The reaction of the same chiral aldehyde **46** with the (R)-enantiomer of Z-**44**, is less stereoselective (de 92%), and aldol **48** resulting from the attack on the other prochiral face of the carbonyl of **46** is the major product. Moreover, the reaction of **46** with achiral boron enolate **49** is poorly stereoselective (de 20%). Cyclic transition state models **50** and **51** account for these stereoselectivities (Figure **I.36**). In both models, the boron atom is coordinated to the carbonyl group of the aldehyde and the conformation of the enolate fragment is such that the polar Si-O-C and C-O-B bonds are antiperiplanar.

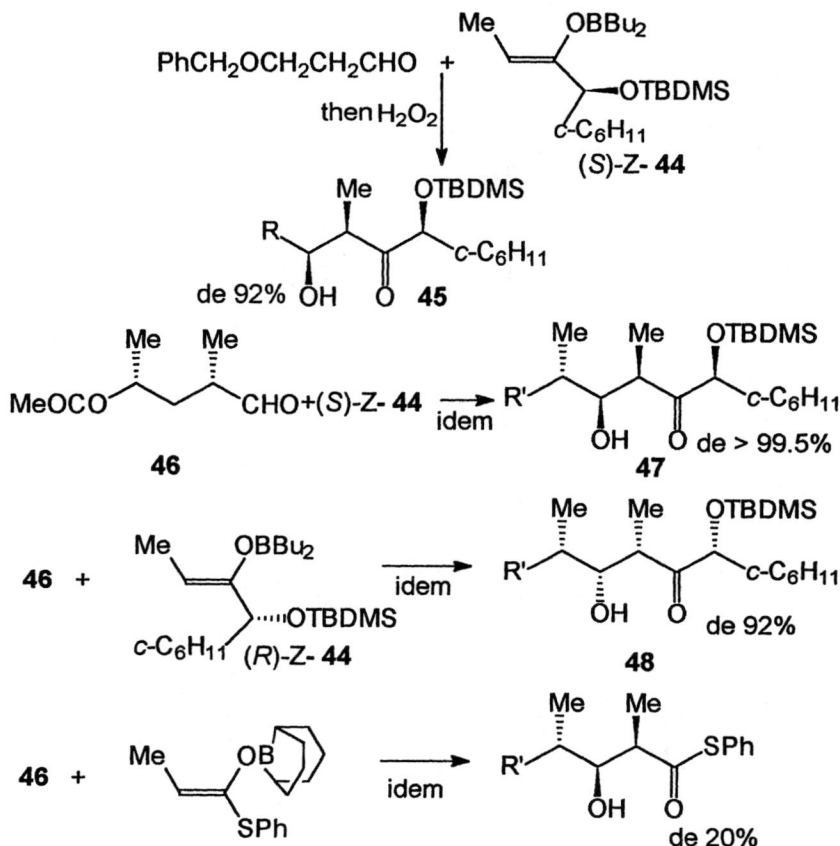

Figure I.35

The aldehyde fragment adopts a conformation in which gauche interactions as well as eclipsing interactions with the enolate C_2-Me and C_1-C_α bonds are minimized. In model **50**, which does not correspond to the Felkin-Anh model [119, 120], the *Re* face of the aldehyde is attacked, leading to **47**. The cyclohexyl substituent of the enolate lies in front and the aldehyde is introduced in the back of the enolate fragment. The new bond is formed *anti* to the bulkier substituent R. Model **51** accounts for the major formation of **48**. In this case, the *Si* face of the aldehyde being attacked is because of the location of the cyclohexyl substituent behind the plane of the enolate. However, the eclipsing interaction of the C_2-Me and C_α-Me bonds cannot be avoided, so a decrease in selectivity is observed. These examples show that the delicate balance between the various repulsive interactions can sometimes be difficult to estimate.

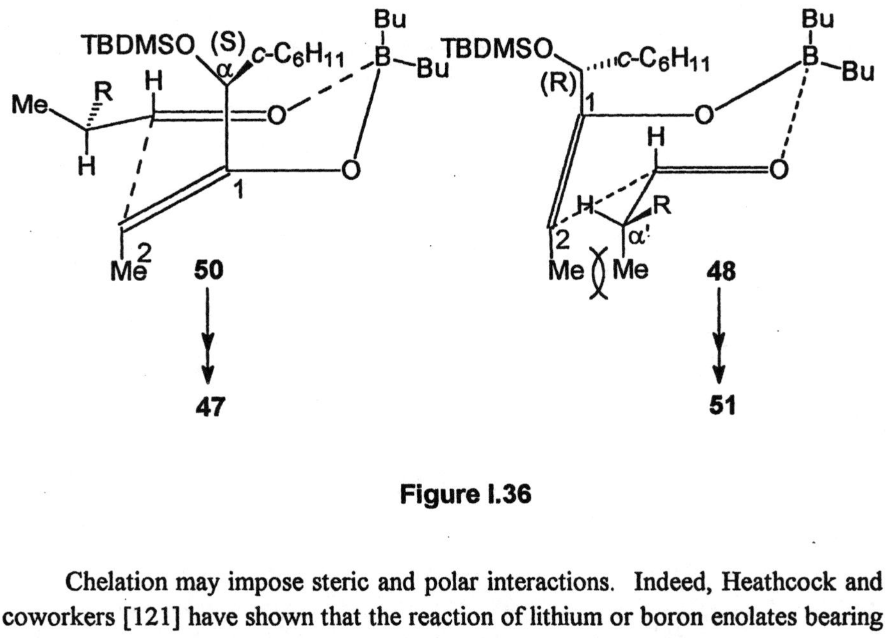

Figure I.36

Chelation may impose steric and polar interactions. Indeed, Heathcock and coworkers [121] have shown that the reaction of lithium or boron enolates bearing a chirality center Z-**52** (M = Li or BBu$_2$) with *i*-PrCHO yields two diastereoisomeric *syn* aldols **53** and **54**. The reaction of the lithium enolate occurs through a tridentate chelate **55** involving the aldehyde carbonyl and the OSiMe$_3$ groups. Therefore, the conformation of the enolate fragment is locked so that the *tert*-Bu group lies in the back. The aldehyde is then introduced on the front side of the enolate, whose *Si* face is attacked in order to minimize the eclipsing interactions (Figure I.37). In the reaction of the boron enolate, coordination to the boron atom involves only the carbonyl group. The enolate fragment adopts the conformation shown in model **56**, with the polar Si-O and O-B bonds antiperiplanar. The bulky *tert*-Bu substituent points toward the front, and the aldehyde is introduced on the back side through an *Re* face attack (Figure **I.37**).

These examples have been discussed in the framework of the Zimmermann-Traxler chair models, without considering the distortions of the chairs that can noticeably modify the relative free-energy differences. Moreover, in some aldol reactions, boat-shaped transition models better account for the observed stereoselectivities [122, 123]. *Ab initio* calculations have supported such boat transition models; however, the calculated energy differences are rather small [41, 124]. Molecular mechanics calculations have also been applied to these problems, by, among others, Bernardi, Gennari and Paterson [124, 125, 126], and some general trends were highlighted.

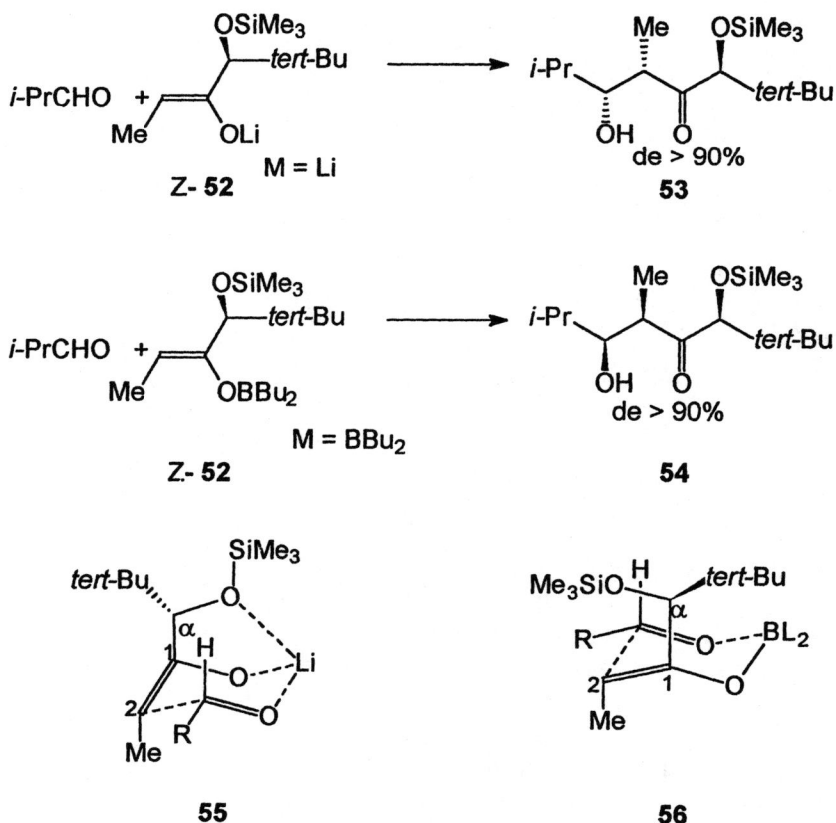

Figure I.37

Double diasterodifferentiation also occurs in kinetic resolution of racemates. Indeed, a chiral reagent will be matched with one of the enantiomers, which will react faster because the matched transition state is lower in energy than the mismatched one [127].

I.7 GENERAL CONSIDERATIONS

To conclude this succinct account, it is worthwhile to recall a few points related to the experimental procedures in asymmetric synthesis. Two determinations are of prime importance: the measurement of the enantiomeric (or diastereoisomeric) excess, and the assignment of the configuration of the major product. These determinations rely on many physicochemical methods such as NMR, high-performance chromatography, X-ray crystallography, circular dichroism. The significance of such determinations depends upon their precision, which is strongly linked to the method used, as recently underlined by Rautenstrauch and coworkers [128].

Enantiomeric mixtures are often transformed with a chiral reagent into diastereoisomers on which the determinations are carried out. The corresponding racemates must be reacted with the chiral reagent to verify that no kinetic resolution takes place. Moreover, the presence of impurities can cause large analytical errors. This is especially true in determinations of optical rotations, and this technique is not usually recommended for determinations of enantiomeric excesses. Racemizations can also occur during purifications or chromatographic analysis, so analytical methods require appropriate control experiments before application.

Chemical correlations toward a known enantiomer also sometimes take place with unwanted epimerizations so that scrutinity must be exerted in all the identification steps. Indeed, some data from the literature have given rise to controversial results.

Recovery and recycling of the chiral auxiliaries and ligands is another important concern in asymmetric synthesis, mainly when they are expensive. Therefore, the current *nec plus ultra* is the use of chiral inductors in catalytic amounts either as chiral reagents or as ligands of chiral catalysts, that is the practice of "atom economy" as coined by Trost [129]. Special emphasis will be given to the scope and limitations of this aspect of asymmetric synthesis.

CHAPTER 1

ENANTIOMERICALLY PURE CHIRAL AUXILIARIES

In asymmetric synthesis, the use of enantiomerically pure chiral auxiliaries involves the temporary introduction of a chiral group G* onto an achiral substrate R-Y. This modified substrate R-Y'-G* is subsequently transformed, ideally through a highly diastereoselective process, into a new product R-Z'-G*. After cleavage of the chiral auxiliary, the final product R-Z,* bearing a new stereocenter, is formed.

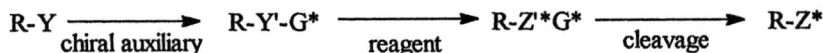

$$\text{R-Y} \xrightarrow[\text{chiral auxiliary}]{} \text{R-Y'-G*} \xrightarrow[\text{reagent}]{} \text{R-Z'*G*} \xrightarrow[\text{cleavage}]{} \text{R-Z*}$$

The requirements for methods based on chiral auxiliaries follow:

•The transformation R-Y'-G*$\longrightarrow$R-Z'*-G* must be highly stereoselective (de $\geq$ 90%), and the purification of R-Z'*-G* should be straightforward.

•The cleavage of R-Z'*-G* to R-Z* should occur under mild conditions, and no racemization at the newly introduced stereocenter should occur.

In general, the recycling is desirable, so the chiral auxiliary should be easy to recover without any loss of enantiomeric purity. However, if the auxiliary is inexpensive, it need not be recycled. In some cases, the chiral auxiliary is actually destroyed during the cleavage reaction to produce the final product. Such transformations are called immolative processes.

Each pure enantiomer of the chiral auxiliary should be readily available so that it is possible to prepare at will both enantiomers of the target molecules. Most chiral auxiliaries are therefore either natural product, or compounds derived from natural products by classical high yielding transformations.

This book will only consider methodologies in which the regeneration of the chiral auxiliary does not involve the cleavage of the carbon skeleton of the substrate. For example, the processes given in Figures **1.1** [130] and **1.2** [131] are not within the scope of this book.

In the present chapter, the most frequently used enantiomerically pure chiral auxiliaries will be described, along with some new and efficient auxiliaries. These auxiliaries will be classified according to their functional linkage to the substrate: aminoalcohols monoesters will be under the heading "alcohols" and their amides will be under the heading "amines", while "oxazolines", in which both functional groups are involved, are listed under "aminoalcohols". Occasionally, the syntheses of noncommercial or very expensive chiral auxiliaries will be summarized.

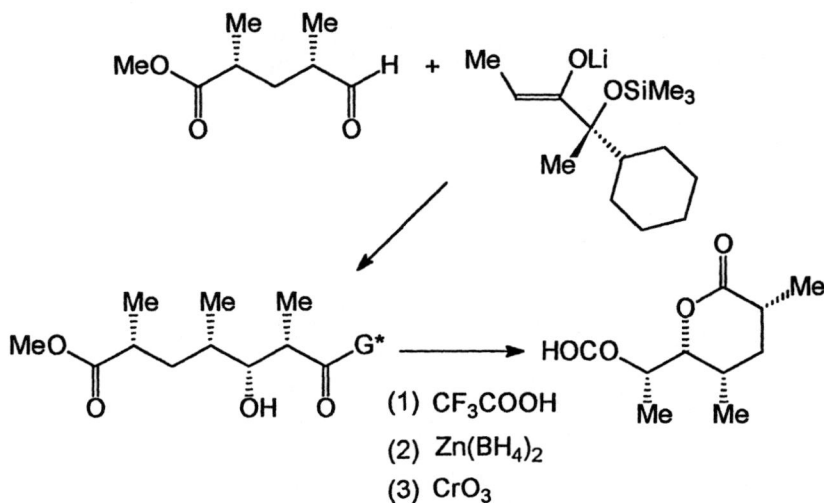

Figure 1.1

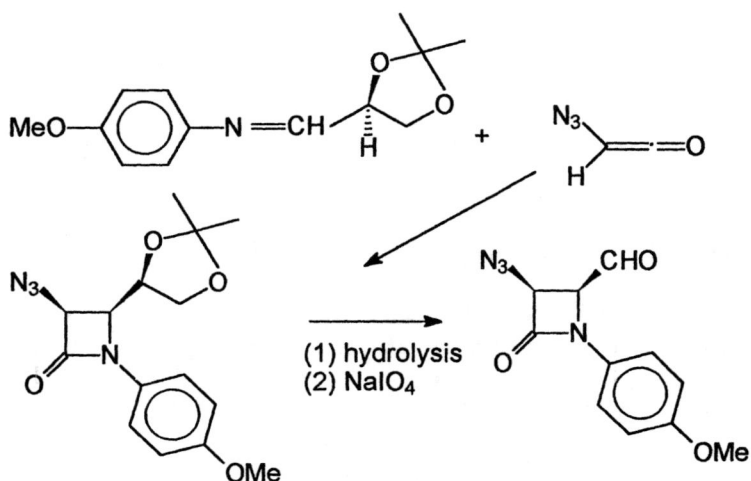

Figure 1.2

1.1 ALCOHOLS, DIOLS, DIPHENOLS AND DERIVATIVES

1.1.1 Alcohols and Derivatives

Most of the alcohols that are used as chiral auxiliaries are secondary. The few unfunctionalized enantiomerically pure acyclic alcohols that are used include: (R)- and (S)-1-phenylethanol **1.1** (Ar = Ph, R = Me), 1-phenyl-2-methylpropanol **1.1** (Ar = Ph, R = i-Pr), or (S)-1-naphthylethanol **1.1** (Ar = 1-Np, R = Me). Derivatives of mandelic acid **1.1** (Ar = Ph, R = COOCH$_2$C$_6$H$_4$-4-OMe) [132] are also useful. Among the cyclanols, (R)-2,2-diphenylcyclopentanol **1.3** has been proposed by d'Angelo and coworkers [133] to esterify enaminoacids for asymmetric catalytic hydrogenation. Menthols **1.4** (R = H) and 8-substituted menthols **1.4** (R = Ph, 4-tert-BuC$_6$H$_4$, 4-C$_6$H$_5$OC$_6$H$_4$) obtained from natural pulegone or its enantiomer have been used for a number of different kinds of applications [66, 133-140]. Also popular are derivatives of trans-(1S, 2R)- or (1R, 2S)-2-phenylcyclohexanols **1.5** (R = Ph) [66, 141, 142] or of the more recently elaborated trans-(1S, 2R)-2-(1-methyl-1-phenylethyl)-cyclohexanol **1.5** (R = CMe$_2$Ph) [140, 143]. These alcohols are related to menthol, and it is believed that conformers **1.4** (R ≠ H) and **1.5** are highly populated. The potential for good asymmetric induction exists with esters of these alcohols because the R group shields one face of the ester (see below). Isomenthol derivatives have also been used [2], but they usually exhibit less interesting selectivities. A very rigid bicyclic analog of phenmenthol **1.6**, prepared from natural podocarpic acid, has been recently introduced by Massy-Westropp and coworkers [144, 145]. Steroidal alcohols have also been used as auxiliaries in a few cases [144a, 146].

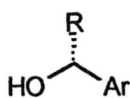

(R)-**1.1**

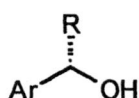

(S)-**1.1**

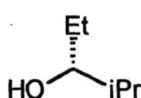

1.2

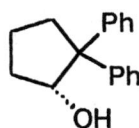

1.3

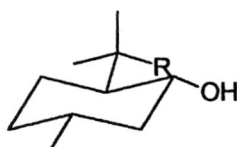

(1R,2S,5R)-**1.4**

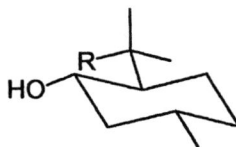

(1S,2R,5S)-**1.4**

(1S,2R)-**1.5**

(1R,2S)-**1.5**

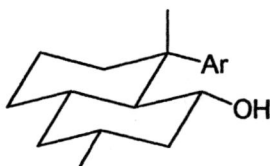

1.6

Alcohols bearing the bornane skeleton have often been used as chiral auxiliaries. Esters of natural products borneol **1.7** (R = H) and isoborneol **1.8** (R = H) usually give poor stereoselectivity and this has prompted the introduction of other substituents onto the rigid skeleton. For example, 2-(1-naphthyl)-bornanol **1.7** (R = 1-Np) [147] and *cis-exo* or *cis-endo*-3-alkylbornan-2-ols **1.8** and **1.9** (R = *tert*-BuCH$_2$CH$_2$ or Ph$_2$CH) [147] are useful auxiliaries. Either enantiomer of the chiral alcohol **1.10** is available from the appropriate enantiomer of 10-camphorsulfonic acid. The size and chelating ability of the sulphonamide substituent on the 10-position of these alcohols can induce highly stereoselective reactions [147].

Enantiopure β-alkoxyalcohols have also been used as chiral auxiliaries. Among the most useful auxiliaries in this class are the acyclic methylether **1.11** [148] or silylether **1.12** [149, 150], and the cyclic monobenzyl- or neopentylethers **1.13** (R = PhCH$_2$ or *tert*-BuCH$_2$) derived from *exo*-bornane-2,3-diols [147]. From inexpensive, commercially available ephedrine, (1R, 2S)-N-methylephedrine **1.14** is easily obtained, its (1S, 2R)-enantiomer is easily available too. The esters of these alcohols [151-154] have frequently been used. Among the cyclic alcohols, sulfonamide derivatives **1.8** and **1.9** (R = NHSO$_2$Ar or N(Ar)SO$_2$Ph) are especially useful [147, 155].

Other types of functionalized enantiopure alcohols also make good chiral auxiliaries. Esters of commercially available methyl or ethyl (S)-lactates **1.15** [156] and (R)-pantolactone **1.16** [157, 158] are popular. The (S)-hydroxysuccinimide derivative **1.17**, easily obtained from (S)-malic acid [158], provides access to the enantiomers of the products formed in the reactions of (R)-pantolactone **1.16**.

In most applications, the chiral alcohols are transformed into the corresponding monoesters, which in turn undergo stereoselective reactions. For example, metal enolates can be generated from propionates **1.18** (R = Me), benzyloxy- or silyloxyacetates **1.18** (R = PhCH$_2$O, R'$_3$SiO) and α-chloropropionates of chiral alcohols. These metal enolates can be alkylated [147, 159], or subjected to aldol reactions [147, 160] or Michael additions [161].

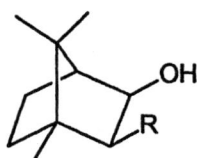

1.7

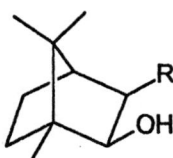

1.8

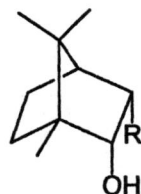

1.9

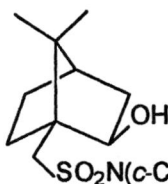

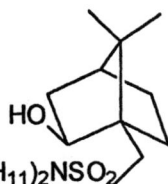

1.10

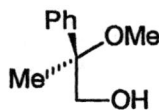

1.11

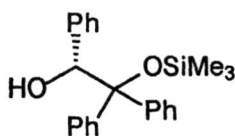

1.12

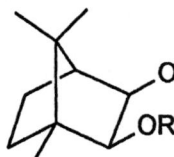

1.13

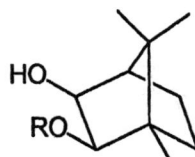

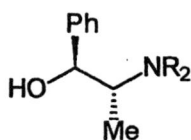

(1*S*,2*R*)-**1.14**

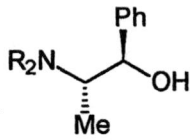

(1*R*,2*S*)-**1.14**

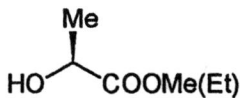

1.15

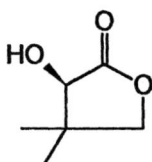

1.16

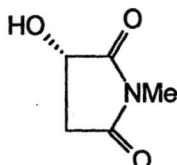

1.17

RCH$_2$COOG*

1.18

While the reactions of enolates derived from esters of **1.4** (R = H) are poorly stereoselective, reactions from esters of **1.4** (R = Ph) [160, 161] or **1.5** (R = Ph) provide major diastereoisomers with high selectivity. Ketene silylacetals **1.19** formed from esters of **1.10** [147] and **1.14** [151, 152, 153, 162] react with aldehydes and imines with high stereoselectivity. Ketene acetals of **1.6** also lead to interesting hetero-Diels-Alder reactions [140]. Alkylglycines [147, 154], halohydrins or acetoxyesters [147] are selectively obtained from **1.19**. Enolates of allyloxyacetates **1.18** (R = CH$_2$=CHCH$_2$O) derived from alcohols **1.4** (R = Ph) undergo [2,3]-sigmatropic rearrangements in a highly selective fashion [163]. The enolates of N-protected glycine esters **1.18** (R = Ph$_2$C=N, *tert*-BuOCONH) are alkylated with high selectivity when the auxiliary is **1.4** (R = H or Ph) or **1.5** (R = Ph) [131]. Deuteration and bromination reactions of these enolates are also very selective [164, 164a, 165].

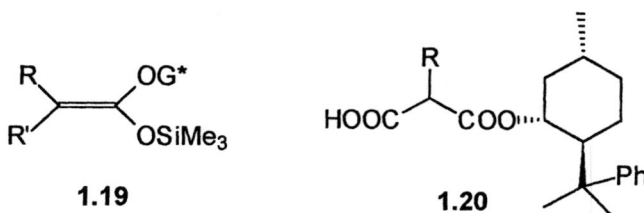

1.19   **1.20**

After these various transformations, the chiral alcohols are recovered for recycling by mild hydrolysis, transesterification or reduction with LiAlH$_4$ (Figure **1.3**).

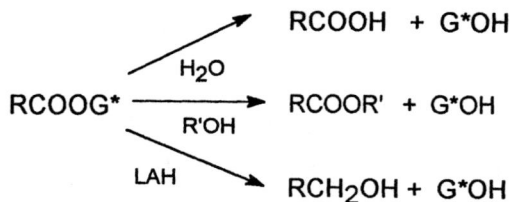

Figure 1.3

Esters of diacids are also useful in asymmetric synthesis. Alkylations of the dianions of 2-alkylmalonic acid half-esters **1.20** lead to products bearing a quaternary carbon with high diastereoselectivity. These products are precursors of chiral diols and aminoacids [166]. The corresponding half-esters of **1.4** (R = H) or **1.7**

(R = 1-Np) provide less selective reactions. The dialkylation of the enolate of (1*R*, 2*S*, 5*R*)-phenmenthyl dimalonate **1.21** (n = 1) by 1,4-dibromobut-2-ene is the key step of one of the first enantioselective syntheses [167] of estrone. Bornyl or menthyl esters are less efficient in this reaction. In contrast, dialkylation of the disuccinate (n = 2) dimenthyl ester **1.21** with $BrCH_2Cl$ under basic conditions gives interesting results due to a cooperative effect [168] (§ I. 6). α,β-Unsaturated esters of chiral alcohols are often used in thermal or Lewis-acid-catalyzed [4+2] cycloadditions [73, 144, 146, 147, 169, 170], dipolar cycloadditions [171, 172], or BF_3-catalyzed conjugate additions of organocopper reagents [147, 169, 173]. Esters of menthol **1.4** (R = H) usually give poor selectivities. However, good selectivities have been observed in a metal-catalyzed dipolar cycloaddition [171]. Cooperative effects in fumarate derivatives **1.22** [12, 73, 174, 175] can also lead to good selectivities. Very high diastereoselection is observed in [4+2] cycloadditions of α,β-unsaturated esters of **1.4** (R = Ph) [169, 170], **1.6** [144], (*R*)-pantolactone **1.16** and imide **1.17** [158, 176, 177, 178]. When using esters of **1.7** and **1.8**, only acrylates CH_2=CHCOOG* give cycloadducts in good yields [147]. Conjugate additions of amines to crotonates MeCH=CHCOOG* are more stereoselective if esters of **1.4** (R = 4-*tert*-BuC$_6$H$_4$ or 4-C$_6$H$_5$OC$_6$H$_4$) are used instead of those of **1.4** (R = Ph) [134].

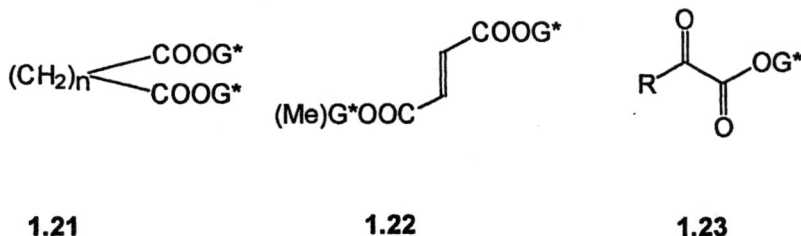

1.21   **1.22**   **1.23**

Esters bearing other functional groups have also been introduced. Chiral alcohol glyoxylates **1.23** (R = H) or phenylglyoxylates **1.23** (R = Ph) give high selectivities in asymmetric Friedel-Crafts [179] alkylations, ene-reactions [147], and photochemically induced [2+2] cycloadditions [2]. When R ≠ H, selective asymmetric reductions of the keto group of **1.23** occur [155, 180, 181, 182]. In all these cases, the best results are obtained with esters of **1.4** (R = Ph), **1.5** (R = Ph) [179, 183], and **1.6** [156]. Occasionally, esters of **1.7** and **1.8** [155] are useful. Very interesting selectivities are observed in reactions of **1.23** (R = H) with allyl- or vinylsilanes when the auxiliaries are **1.4** (R = Ph) and **1.5** (R = Ph) [147, 184]. Acyclic or cyclic carbamates of **1.4** (R = Ph) or **1.5** (R = CMe$_2$Ph) are used by Comins and coworkers for the synthesis of enantioenriched alkaloids [136, 185-

188]. Sulfinylcarbamates **1.24** are interesting reagents for [4+2] cycloadditions when attached to auxiliaries **1.4** (R = Ph) or **1.13** (R = *tert*-BuCH$_2$) [147, 189, 190]. Carbamates **1.25** suffer [2,3]-sigmatropic rearrangements leading to enantioenriched allylic alcohols [191]. For these reactions, the most interesting chiral auxiliary is **1.5** (R = Ph). The diazoester **1.26** of (*R*)-pantolactone **1.16** reacts with stilbene in the presence of a rhodium catalyst to give a precursor of 2-phenyl-1-aminocyclopropanecarboxylic acid, a biovegetal hormone [192, 193]. In these reactions, the chiral auxiliary is removed by hydrolysis, alcoholysis or reduction.

Ester-based chiral auxiliaries have also been used in other settings. β-Alkoxyesters **1.27** of (*R*)-1-phenylethanol **1.1** (R = Me, Ar = Ph) or (*S*)-1-naphthylethanol **1.1** (R = Me, Ar = 1-Np) are transformed into chiral synthons by reactions with a lithiated carbanion α to phosphorous followed by hydrogenolysis [194]. Ethers **1.28** of chiral alcohols **1.1** undergo selective alkylations or hydroxyalkylations [169]. The auxiliaries can be removed by hydrogenolysis. Enol or dienol ethers **1.29** and **1.30** suffer [2+2] [195, 196] or [4+2] cycloadditions [49, 197, 198, 199]. The best stereoselectivities are obtained when the chiral auxiliary is **1.1** (R = *i*-Pr, Ar = Ph), **1.4** (R = Ph), **1.5** (R = Ph), **1.10** or **1.13**. These auxiliaries are cleaved either by acid treatment [199] or by other means in subsequent steps. Acetylene ethers G*OC≡CR derived from **1.5** (R = Ph) [199a] can undergo stereoselective Pauson-Khand reactions [200, 201]. The auxiliaries are removed by treatment of the products with SmI$_2$ in THF-MeOH.

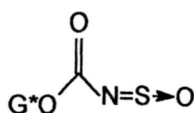

1.24

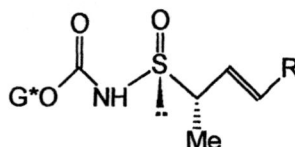

1.25

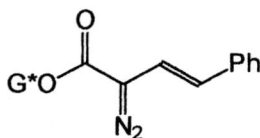

1.26

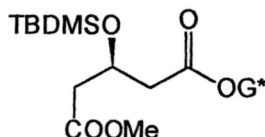

1.27

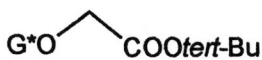

1.28

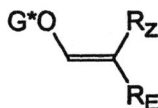

1.29

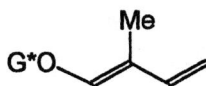

1.30

A cyclic compound **1.31** derived from menthol **1.4** (R = H) has seen interesting applications [73, 137, 202, 203] in [4+2] cycloaddition reactions and in 1,3-dipolar cycloaddition reactions, as well as in the synthesis of enantioenriched aminoalcohols or aminoacids. These reactions are highly stereoselective, and the chiral auxiliaries are easily recovered by methanolysis (Figure **1.4**).

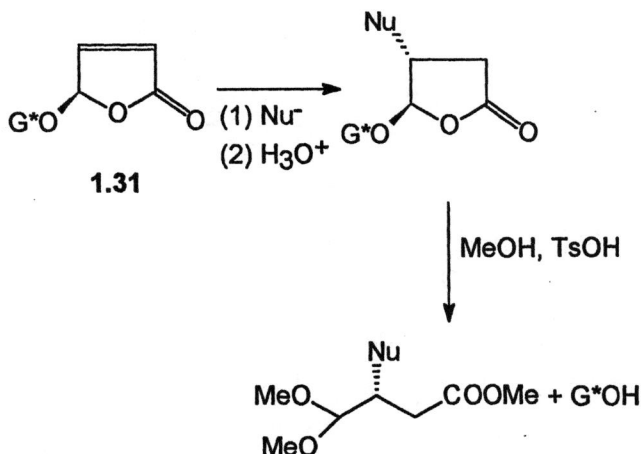

Figure 1.4

1.1.2 Diols, Diphenols and Related Compounds

A few secondary/tertiary diols have been used as chiral auxiliaries. The monoacetates of (*R*)- and (*S*)-**1.32** (R = Ph) [204] are interesting precursors for asymmetric aldol reactions [205-210]. (*S*)-1-Phenyl-3,3-bis(trifluoromethyl)pro-pan-1,3-diol **1.33** [211] induces highly diastereoselective alcoholysis of σ-symmetric dicarboxylic acid anhydrides.

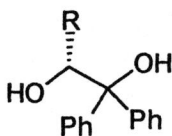

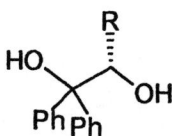

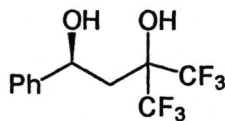

(R)-**1.32**   (S)-**1.32**   **1.33**

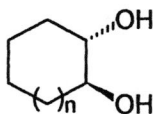

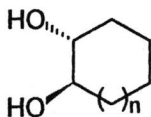

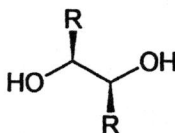

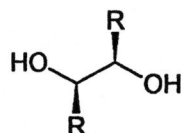

(S,S)-**1.34**   (R,R)-**1.34**   (S,S)-**1.35**   (R,R)-**1.35**

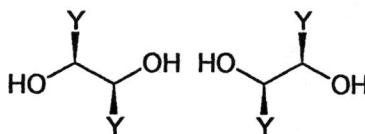

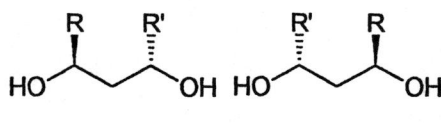

1.36   **1.37**

The most popular kinds of the diols for asymmetric synthesis are bissecondary diols that have a C_2 axis of symmetry [212]. The presence of the symmetry axis avoids the formation of diastereoisomeric esters or acetals [213]. (1R, 2R)-Cyclohexanediol **1.34** (n = 1) has been used as an auxiliary in asymmetric cyclopropanation [214] and (1S, 2S)-cycloheptanediol **1.34** (n = 2) in 1,4-addition of cuprates[157]. Dioxolane derivatives of **1.34** have been used for asymmetric β-ketoester alkylations [215] and cuprate 1,4-additions [216]. Linear 1,2-diols **1.35** (R = Me, i-Pr, c-C_6H_{11}, Ph) and functionalized 1,2-diols **1.36** (Y = COOalkyl, CONR'$_2$, CH$_2$OR') are readily available from optically active tartaric acids **1.36** (Y = COOH). Acetals derived from these diols are valuable reagents ın asymmetric synthesis [173, 213, 217], as the related 1,3-diols **1.37**. Acetals of 1,3-butanediol **1.37** (R = Me, R' = H) have also been used. When these acetals are formed from aldehydes under thermodynamic conditions, one 1,3-dioxane stereoisomer often predominates. In this favored isomer, the substituent from the aldehyde and the methyl group from **1.37** are both in equatorial orientations [213]. The chiral environment created by these cyclic acetals can promote a nucleophilic attack on one face of a nearby prochiral center. For example, reductions or organometallic additions to chiral acetals **1.38** (Y = COalkyl or COaryl) [173, 213] formed from α-ketoaldehydes often give interesting stereoselectivities, as do additions of ester enolates to the related imine analogs **1.38** (Y = C(alkyl)=NAr) [218]. α,β-Unsaturated derivatives **1.38** (Y = RCH=CH) are also valuable reagents in Simmons-Smith cyclopropanations and photoinduced [2+2]-cycloadditions [213, 219]. However, [4+2] cycloadditions or cuprate 1,4-additions with these substrates often give disappointing selectivities (exceptions include [4+2] cycloadditions of **1.39**). The reactions of organolithium or -magnesium reagents bearing chiral acetal residues with carbonyl compounds do not lead to interesting selectivities. The cyclic acetal products of these reactions are usually cleaved by mild acid hydrolysis or transacetalation [213]. The chiral diol can often be recovered and recycled.

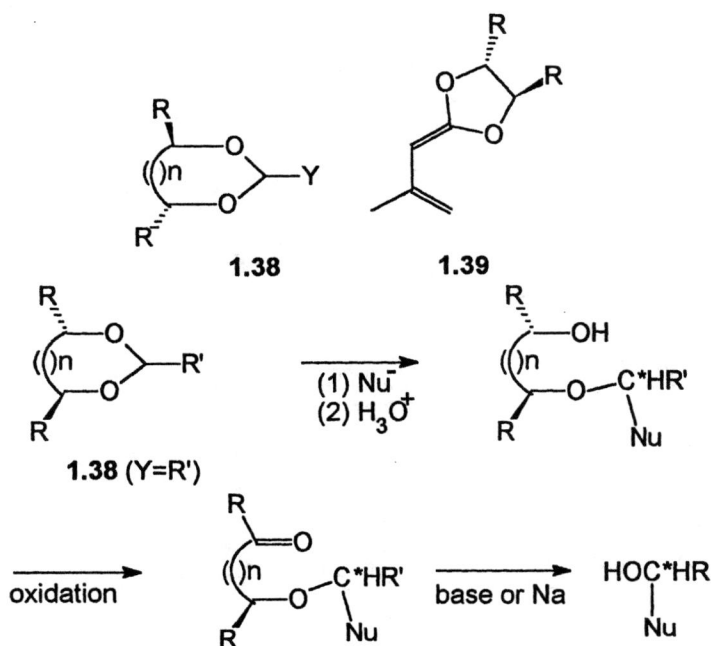

Figure 1.5

Many workers have cleaved chiral acetals with nucleophilic reagents such as metal hydrides, alkyl- or arylmetals, allyl- or alkynylsilanes or enoxysilanes in the presence of Lewis acids [213, 220, 221, 222]. Functionalized hydroxyethers are thus formed (Figure **1.5**), and these are converted into chiral alcohols by oxidation followed by α- or β-elimination. In such processes, the chiral diol auxiliary is not recycled, but due to the low cost of the diol, the method is popular nonetheless [213, 222, 223, 224].

Eliel and coworkers [94, 225, 226, 227] have introduced two 1,3-oxathianes **1.40** and **1.41** for use in asymmetric synthesis; **1.40** is readily prepared from natural pulegone and **1.41** from camphosulfonic acid. Reactions of the derived 2-lithio-1,3-oxathianes with aldehydes, followed by Swern oxidation [94], lead to 2-acyl derivatives such as **1.42**. These compounds can also be prepared by the reaction of the lithio derivatives with nitriles, followed by acid hydrolysis [94], or by the addition of lithium cuprates to acid chlorides [228]. Hydride reductions or organometallic additions to **1.42** are highly stereoselective. Chiral α-hydroxyaldehydes are obtained by treatment of the products [227] with NCS-AgNO$_3$, occasionally in the presence of 2,4,6-collidine [229]. The chiral auxiliary **1.43** is regenerated from the resulting sultim by treatment with LiAlH$_4$ (Figure **1.6**). Racemic 2-alkylcyclopentanones can be deracemized by acetalization with **1.43**, followed by thermodynamic equilibration and cleavage [229] of the acetal.

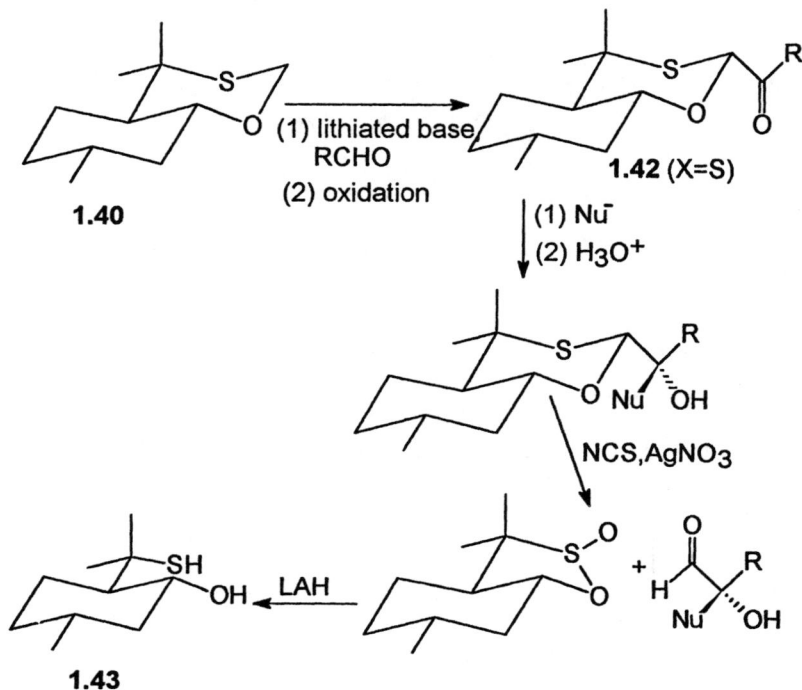

Figure 1.6

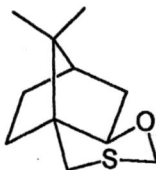

1.41

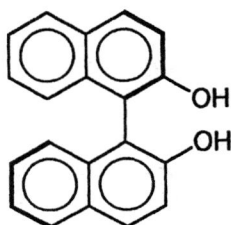

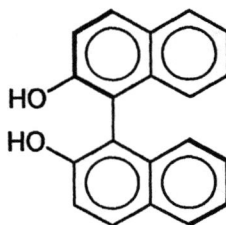

(*R*)-**1.44**   (*S*)-**1.44**

   (*R*)- and (*S*)-binaphthols **1.44** [230-235] have sometimes been used as chiral auxiliaries. Examples include reduction of γ-ketoester **1.45** [236], nucleophilic substitution of binaphthol ethers (G* = binaphthol) **1.46**, by organomagnesium reagents, organocuprate additions to binaphthol monocinnamates [237], and alkylations of arylacetic or crotonic esters [238, 239].

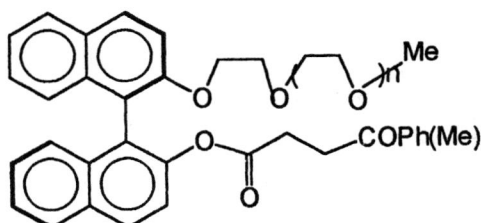

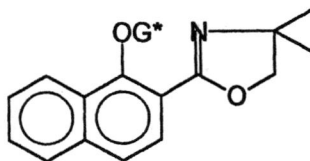

1.45   **1.46**

1.1.3 Carbohydrates and Derivatives

   Multifunctional carbohydrate derivatives have been used widely as chiral synthons [240], and they have more recently found application as chiral auxiliaries [241]. In this chapter, examples of carbohydrates in which an alcohol functional group is the point of attachment to the reagent will be presented. Alkylations of ester enolates of the D-allofuranose derivative **1.47** or of the diacetoneglucose derivative **1.48** provide mediocre selectivities [241, 242]. Organometallic additions to α-ketoester derivatives of quebrachitol bis-ketal **1.49** [243] however are stereoselective. Several α,β-unsaturated esters of **1.50**, **1.51**, **1.52**, **1.53** and **1.54** are interesting reagents in [4+2] cycloadditions [73, 241, 242, 244, 245, 246] or cuprate 1,4-additions [241, 247, 248]. [4+2] Cycloadditions of unsaturated enol ethers derived from diacetoneglucose **1.48** are also highly stereoselective [170, 249]. The cleavage of the carbohydrate group from the products of these reactions is usually accomplished by hydrolysis (LiOH/aq. THF, or enzyme hydrolysis) or by LiAlH$_4$ reduction.

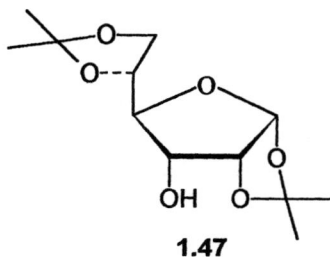

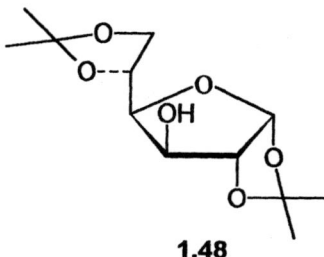

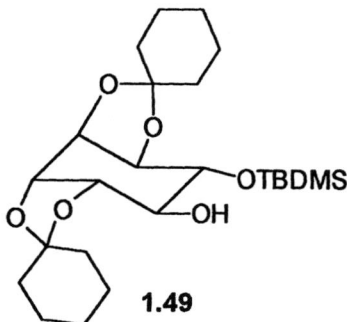

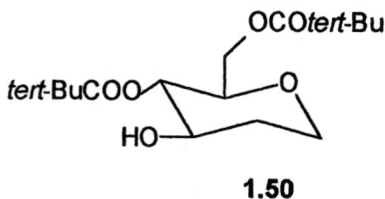

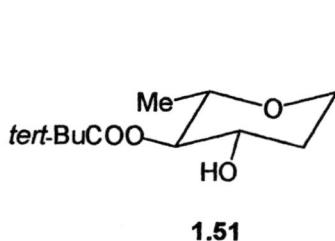

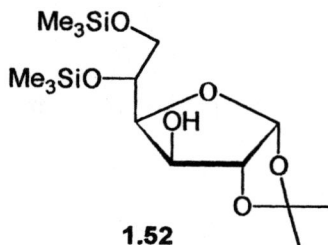

1.47

1.48

1.49

1.50

1.51

1.52

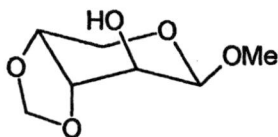
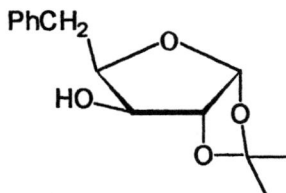

1.53   **1.54**

1.2 AMINES, DIAMINES, HYDRAZINES AND DERIVATIVES

1.2.1 Monoamines and Derivatives

1.2.1.1 *Primary Amines*

As early as 1968, Horeau used isobornylamine **1.55** to perform asymmetric alkylations of a cyclohexanone imine [250]. Among the nonfunctionalized primary amines that are useful, derivatives of 1-phenethylamine **1.56** (Ar = Ph) have met with numerous applications [162, 251-258]. These amines can even be grafted onto polymeric supports [253, 259]. β-Alkoxyphenethylamines **1.56** and **1.57** (R = Me) have also been used [169, 236, 252, 253, 260], and their derivatives give better selectivities than those of cyclic analogs [169, 253]. Among the aminoacid derivatives, those of valine *tert*-butyl ester **1.59** (R = *i*-Pr, R' = *tert*-Bu) [161, 167, 169, 252, 261-264], valinol **1.60** (R = *i*-Pr, R' = H) [253], and phenylglycinol **1.60** (R = Ph, R' = H) [265, 266] often lead to the most interesting results.

Chiral amines have been transformed into chiral imines RCH=NG*, which are usually in equilibrium with the tautomeric enamines. These enamines undergo asymmetric alkylations, and the best results are often obtained with ethers **1.58** or with valine derivatives **1.59** (R = *i*-Pr, R' = *tert*-Bu) [169, 173, 253] in the presence of bases. Enamines, lithioenamines and zinc enamines derived from imines are very potent Michael donors that often participate in highly stereoselective reactions [161, 162, 169, 173, 254, 257, 260, 262, 267]. Chiral imines can suffer very selective addition reactions of organomagnesium reagents [139, 253, 254] and allyl-metals [154, 258]. They also suffer stereoselective Ti-catalyzed silylcyanation [268], Strecker reaction [266], and [2+2] or [4+2] cycloadditions [131, 256, 263]. When the reaction produces an imine product, the chiral auxiliary is recovered after acidic hydrolysis. However, when an amine is obtained as the product, as is often the case from phenethylamine derivatives, the chiral residue is cleaved by hydrogenolysis. In such cases, the chiral amine is not, strictly speaking, a chiral auxiliary. But these processes will be discussed anyway because of their importance in asymmetric synthesis.

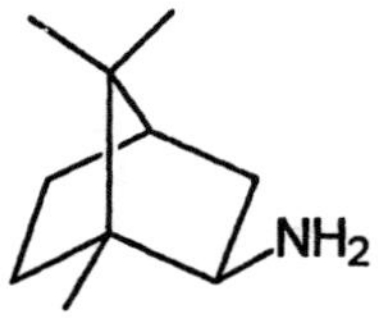

1.55 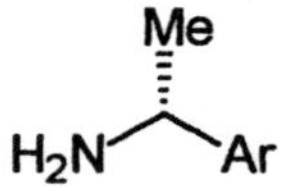   (*R*)-**1.56** 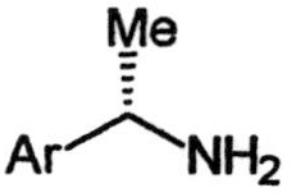   (*S*)-**1.56**

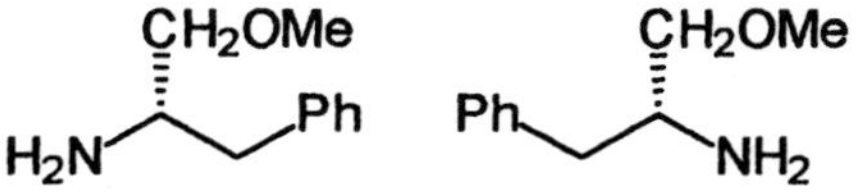

1.57 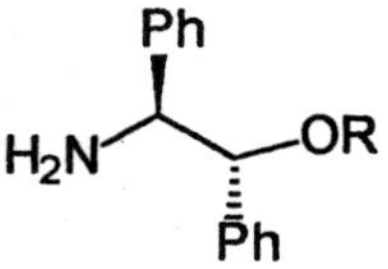   **1.58**

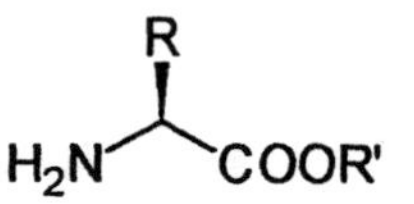

1.59 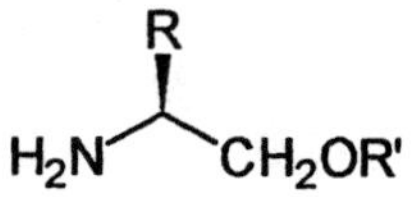   **1.60**

1.2.1.2 Secondary Amines

A number of acyclic secondary amines have been used in asymmetric synthesis. Ephedrine **1.61** (R = Me) has been transformed into α,β-unsaturated amides, which undergo highly stereoselective 1,4-addition reactions [161, 169, 173, 253, 262]. Indolylamines **1.62** generated from (*S*)-valine methyl ester **1.59** (R = *i*-Pr, R' = Me) have been used in highly stereoselective Pictet-Spengler reactions [269]. Chiral aminoethers **1.63** can be transformed into amides, whose enolates participate in selective α-alkylations [270, 271]. However, the most frequently used class of secondary amines are 2- and 2,5-substituted pyrrolidines **1.64** and **1.65** [252, 253, 261, 272]. Enantiopure monosubstituted pyrrolidines **1.64** are synthesized from (*S*)-proline **1.64** (R = COOH). Proline esters **1.64** (R = COOalkyl) and (*S*)-prolinols **1.64** (R = CH$_2$OH, CMe$_2$OH, C(Me)(Ph)OH, CPh$_2$OH) or their ethers **1.64** (R = CH$_2$OMe, CH$_2$OSiMe$_3$, CH$_2$OCH$_2$OMe) are the most popular auxiliaries [104, 161, 169, 261, 273].

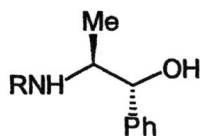

(1R,2S)-**1.61**

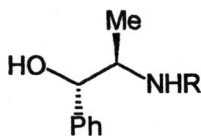

(1S,2R)-**1.61**

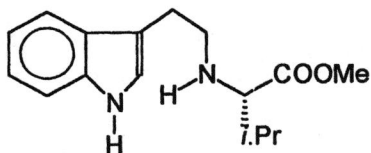

1.62

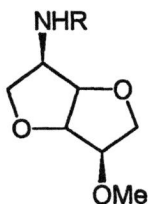

1.63

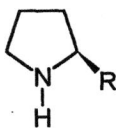

1.64

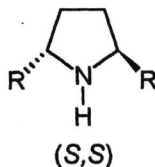

(S,S)

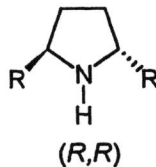

(R,R)

1.65

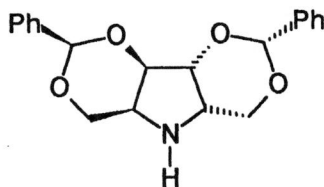

1.66

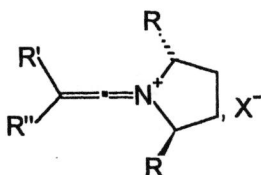

1.67

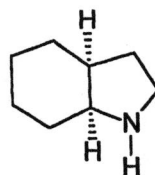

1.68

A number of 2,5-disubstituted pyrrolidines bearing a C_2 axis of symmetry have been used. The popular 2,5-dimethylpyrrolidine auxiliary **1.65** (R = Me) [274, 275] presents some limitations due to the difficulty of hydrolysis of its derived amides. However, radical additions to acrylamide derivatives of **1.65**, for example, are highly stereoselective [276, 277]. To solve the problem of cleavage, modified 2,5-disubstituted pyrrolidines such as **1.65** (R = CH_2OMe, CH_2OCH_2OMe, CH_2OSiR_3) [278] or **1.66** [279] have been introduced. Saturated or α,β-unsaturated amide derivatives of these pyrrolidines have been used in enolate alkylations [154, 159, 261, 280-283], aldol condensations [261, 284, 285] and Michael reactions [161], as well as in [4+2] cycloadditions [73, 104, 286-289], radical additions [276, 279] and organometallic additions [173, 261]. Starting from such amides, chiral ketenimminiums salts **1.67** can be prepared, and they undergo highly stereoselective [2+2] cycloaddition reactions [195, 272, 290, 291]. The chiral auxiliary is readily removed during the workup of these reactions. α-Ketoamides can be generated from **1.65** (R = CH_2OCH_2OMe, $CH_2OTBDMS$) and the reactions of their carbonyl groups with hydridic or organometallic reagents are highly stereoselective [217a, 292]. As in the previous cases, the auxiliary is cleaved by hydrolysis, although this process may sometimes be difficult [276, 293].

Reactions of enamine derivatives of auxiliary **1.64** [162, 252, 262] are often disappointing, except when R = CH_2OSiMe_3 [162]. However, the alkylation of enamines derived from **1.65** (R = Me) gives satisfactory results [253, 275], as do the Michael additions of enamines of bicyclic pyrrolidine **1.68** [294].

[4+2] Cycloadditions of acylnitropyrrolidine derivatives [272] of **1.64** or **1.65** give interesting results, as do reactions of alkoxyimminium salts [295] or of 2-pyrrolidinobutadienes [296]. The same pyrrolidine derivatives have been used in asymmetric Michael reactions of aminochromium carbene anions [297] and in nucleophilic additions to arenemanganese tricarbonyl complexes [298].

Kurth and coworkers [299] have proposed the use of (2R, 6R)-2,6-bis-benzyloxymethylpiperidine **1.69** as an auxiliary for alkylation of α,β-unsaturated amides.

Chiral aminonitriles such as **1.70** and **1.71** have been prepared either from the corresponding amino-1,3-diol [300] or from (1R, 2S)-ephedrine **1.61** (R = Me) [301]. Carbanion derivatives of these reagents undergo stereoselective alkylations, carbonyl additions [301] and Michael reactions [300]. Since the reaction products are aminonitriles, the chiral auxiliary is easily cleaved by mild hydrolysis.

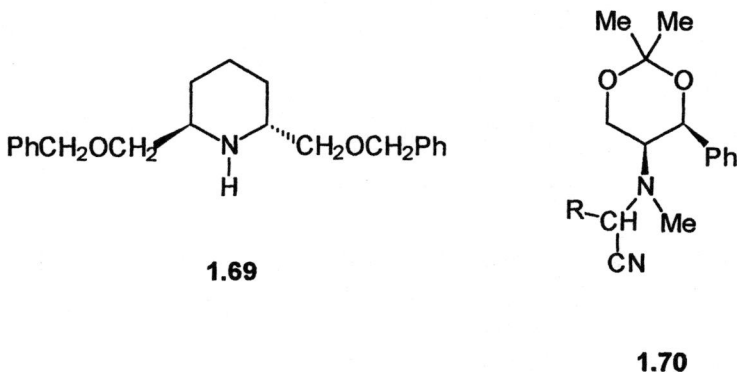

1.69

1.70

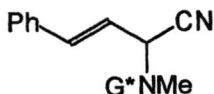

1.71

1.2.2 Diamines and Derivatives

Bis-secondary chiral diamines have been used to generate chiral aminals from functionalized aldehydes (Y=C(R')CHO). These aminal reagents are comparable to the chiral acetals previously described (§ 1.1.2). The cyclic aminals derived from (S)-proline derivative **1.64** (R = CH_2NHPh) [261] or C_2 symmetric diamine **1.72** (R = Ph or c-C_6H_{11}) [212, 302-305] bear an electrophilic double bond (Y=C = carbonyl, imine, C=C) that can react highly selectively with organometallic reagents [261, 302, 303, 306, 307] or can participate in diastereoselective 1,3-dipolar cycloaddition reactions [304]. The chiral auxiliaries from the products of these reactions are readily regenerated by mild acid hydrolysis or methanolysis. Starting from **1.72** (R' = H), chiral imidazolidines **1.73** are easily obtained, and their N-acyl derivatives have been used in stereoselective 1,3-dipolar cycloadditions [308, 309]. Starting from (R,R)- and (S,S)-N,N'-dialkyl-1,2-*trans*-cyclohexanediamines, Hanessian and coworkers [310-313] have prepared chiral phosphonamides **1.74**. The anions of **1.74** react with alkylating agents, 4-substituted cyclohexanones and α,β-unsaturated ketones and esters with high stereoselectivity.

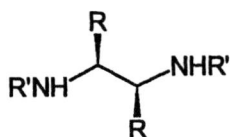

(S,S)-**1.72**

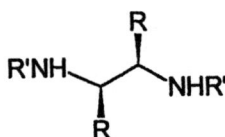

(R,R)-**1.72**

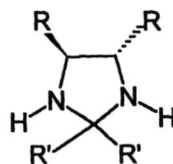

1.73

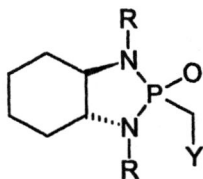

(R,R)-**1.74**

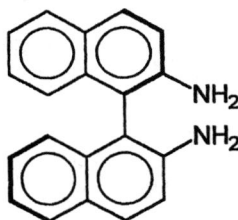

(R)-**1.75**

(R)- or (S)-Binaphthylamines 1.75 [231, 232, 233] have been used to perform asymmetric lactonization reactions of racemic symmetrical hydroxydiacids $HOCO(CH_2)_nCHOH(CH_2)_nCOOH$ [314]. The chiral auxiliary is cleaved after these lactonizations by acidic treatment.

1.2.3 Hydrazines and Derivatives

The use of hydrazines as chiral auxiliaries was initiated by Enders and coworkers [315]. They have developed the chemistry of hydrazones derived from epimeric 1-amino-2-methoxymethylpyrrolidines 1.76, Samp and Ramp [161, 169, 253, 261, 315, 316]. These compounds are commercially available, or they can easily be prepared from (S)-prolinol 1.64 (R = CH_2OH) or (R)-glutamic acid [261]. Hydrazones have some advantages over their related imine derivatives. First, they are formed in quantitative yield even from sterically hindered ketones. Second, their derived anions are often more reactive than the related aldehyde or ketone enolates.

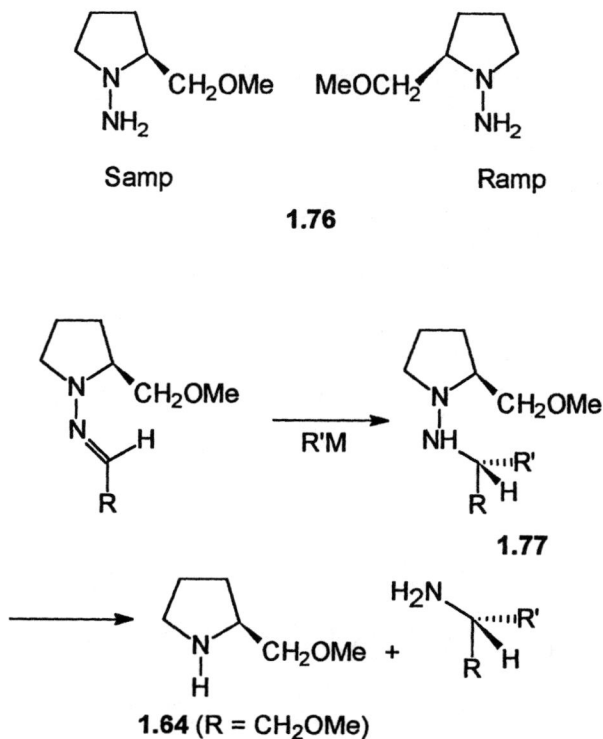

Samp   Ramp

1.76

1.77

1.64 (R = CH_2OMe)

Figure 1.7

Finally, they are easily cleaved in slightly acidic media, by ozonolysis, by sodium perborate treatment or, after quaternarization with MeI, by treatment with HCl under biphasic conditions.These cleavage methods typically provide precursors of the original chiral auxiliaries [315, 317]. The most important applications of hydrazone carbanions in asymmetric synthesis are alkylations [169, 253, 261, 315] and Michael additions [161, 173]. These reactions are highly diastereoselective, even though related aldol reactions often give disappointing results [315]. In a recent extension, chiral aldehyde hydrazones have been reacted with organolithium reagents, sometimes in the presence of CeCl$_3$ [316, 317a, 318] leading to hydrazines **1.77**. Treatment of these hydrazines with Raney nickel gives chiral primary amines and pyrrolidine **1.64** (R = CH$_2$OMe), from which the chiral hydrazine can be readily regenerated (Figure 1.7). Replacement of the CH$_2$OMe group of the Samp **1.76** hydrazone with CH$_2$OCH$_2$CH$_2$OMe or CMe$_2$OMe residues may improve the stereoselectivity of these processes [318, 319].

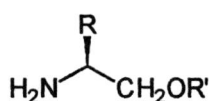

1.60

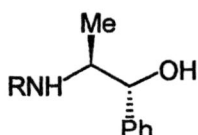

(1R,2S)-1.61

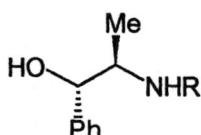

(1S,2R)-1.61

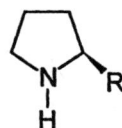

1.64

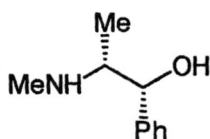

1.78

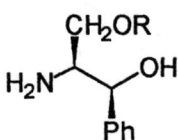

1.79

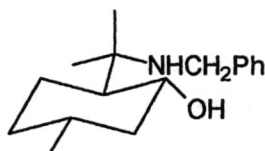

1.80

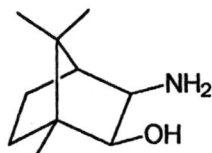

1.81

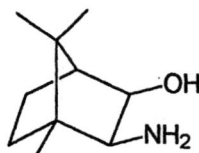

1.82

1.3 AMINOALCOHOLS AND DERIVATIVES

Reagents that encompass both the alcohol and the amine functional groups of a commercially available aminoalcohol are very useful. Ephedrine **1.61** (R = Me), pseudoephedrine **1.78**, norephedrine **1.61** (R = H) and their *N*-alkyl and *N*-acyl derivatives are commonly used, [253, 290, 320-324], as are aminoalcohols **1.79**. Also popular are aminoalcohols formed from the natural aminoacids (*S*)-alaninol **1.60** (R = Me, R' = H), (*S*)-phenylalaninol **1.60** (R = PhCH$_2$, R' = H), (*S*)-valinol **1.60** (R = *i*-Pr, R' = H), (*S*)-*tert*-leucinol **1.60**, (R = *tert*-Bu, R' = H) [261, 325, 326, 327], or (*S*)-prolinol **1.64** (R = CH$_2$OH) [328]. Aminoalcohol **1.80**, easily obtained from pulegone [329], and aminoalcohols **1.81** and **1.82**, easily prepared from camphor, have also been used [330, 331].

The reactions of these aminoalcohols with functionalized aldehydes (or aldehyde equivalents) give chiral 1,3-oxazolidines or 1,3-oxazines **1.83** (n = 0,1). When Y = R'''CO or R'''CH=CH, these compounds undergo highly selective reductions and organometallic additions [175, 253, 320, 321, 328, 329, 332]. 1,3-Oxazolidines **1.84** (EWG = COOCH$_2$Ph or Ts), prepared from orthoesters and *N*-substituted ephedrines, react with allylmetals or ketene acetals in the presence of Lewis acids with a good selectivity, though the chemical yields [322, 333, 333a] are sometimes poor. Oxazolidines (*S*)-**1.85**, (R = *i*-Pr, *tert*-Bu, Ph, R' = H, R" = Me), obtained from acetone and the corresponding aminoalcohols **1.60**, have been used as chiral auxiliaries by Porter and coworkers in radical reactions [334, 335]. Hegedus and coworkers have performed reactions of chiral chromium carbene complexes bearing (*R*)- and (*S*)-**1.85** (R = Ph, R' = H, R" = Me) [131, 336, 337, 338]. Other substituted oxazolidines **1.85** have been proposed as chiral auxiliaries for 1,3-dipolar cycloaddition reactions [308, 309]. The cleavage of such chiral auxiliaries is normally accomplished by acid hydrolysis, or in the case of **1.84**, by treatment by ethanedithiol/BF$_3$.

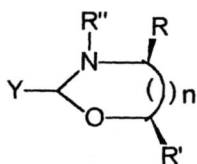 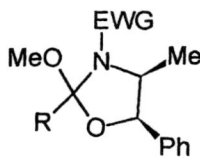 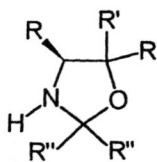 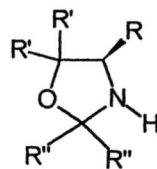

    1.83   **1.84**   (*S*)-**1.85**   (*R*)-**1.85**

The chemistry of chiral 1,3-oxazolines obtained by reaction of iminoesters with aminoalcohols has been studied extensively by Meyers and coworkers [159, 253, 324, 339]. The compounds derived from **1.79** (R = Me) have given the most interesting results, presumably because they bear a CH_2OMe substituent that is prone to chelate cations. Indeed, the lithium enolates of **1.86** (Y = $MeCH_2$ or $PhCH_2$) have been alkylated with good selectivity, although when Y = CH_2OMe, the selectivities are not so good. E-α,β-Unsaturated oxazolines **1.86** (Y = E-RCH=CH) can be easily prepared, and their reactions with organo-lithium reagents are highly stereoselective. After acid hydrolysis, the aminoalcohol is regenerated without racemization and chiral α,α- or β,β-dialkyl substituted diacids are obtained with a high enantiomeric excess. Starting from 2-aryl-oxazolines **1.86** (Y = Ar), chiral binapthols [35, 324, 327, 340] (or compounds resulting from C=C addition when Ar = Np or pyridyl [327, 340]) are obtained by reaction with organometallic reagents with an excellent asymmetric induction. Hydrolysis of the oxazolines provides acid products and regenerates the chiral auxiliary. Alternatively, sequential treatment of the oxazolines with MeOTf and $NaBH_4$ gives aldehydes. Mild hydrolysis followed by $LiAlH_4$ reduction gives primary alcohols [340, 341].

Oxazolines **1.87** (R' = H or Me) generated from amino alcohols **1.60** (R' = H) are very interesting reagents. Conjugate addition reactions of organolith-ium reagents to **1.87** (R = RCH=CH, R' = Me) are highly stereoselective. Treat-ment of the products of these reactions with MeOTf followed by $NaBH_4$ leads to β,β-dialkylaldehydes with a very high ee [173, 342]. Quaternization of **1.87** (R = Me, R' = H) by $CH_2=CHCH_2OTs$, followed by 3,3-sigmatropic rearrange-ment and hydrolysis, gives substituted acids with an excellent ee [261]. The chiral oxazoline can be attached to the N-atom of a tetrahydroquinoline or isoindoline by using reagent **1.88**, and such derivatives can be selectively alkylated in the presence of base. The chiral auxiliary is then cleaved by the action of hydrazine [326, 343]. Although alkylations of oxazolines bearing a camphor-like skeleton often lead to disappointing results [344], better selectivities are obtained in hydroxyalkylation reactions [345]. In the presence of Ac_2O, vinyloxazoline **1.89** (R = H) is a very potent dienophile and participates in asymmetric [4+2] cycloaddition reactions [331]. 1,3-Dipolar cycloaddition reactions of related N-oxides are also interesting [346]; however, the reactions of enoxysilanes with **1.89** give good selectivities only if R = Me [347].

The reactions of 1,2-aminoalcohols **1.60** (R = i-Pr, $tert$-Bu, Ph, R' = H) or **1.79** (R = H) with γ- or δ-ketoacids generate chiral bicyclic lactams **1.90** and **1.91**. The chemistry of this class of reagents has also been developed by Meyers and coworkers [261, 327, 348], and another access to these compounds has also been proposed [327].

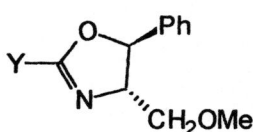

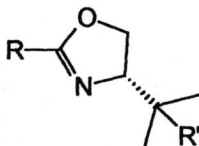

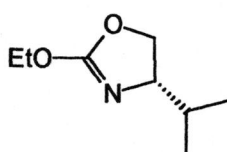

1.86   **1.87**   **1.88**

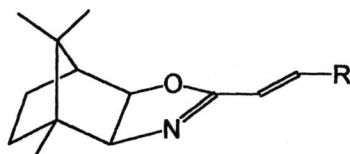

1.89

Alkylations of these heterocycles in basic media are stereoselective, but the acidic hydrolysis that is used to produce an α-monoalkylated acid can induce some epimerization. However, if a second alkylation is subsequently performed, the possibility for epimerization during hydrolysis is eliminated. This dialkylation methodology is useful for generation of γ- or δ-ketoacids bearing a stereogenic quaternary carbon atom in the α-position [327, 348]. The cleavage of the chiral auxiliary can also be accomplished by reduction. This produces chiral ketoaldehydes, which are precursors of cyclic enones. Unsaturated derivatives **1.92** are easily obtained by selenoxide elimination, and they participate in highly stereoselective [2+1], [2+2], [3+2] dipolar- and [4+2] cycloadditions [73, 327]. Applications of these compounds to the synthesis of a number of chiral pyrrolidines or pyrrolidinones and to an inhibitor of brain neurotransmission have been published [349, 350].

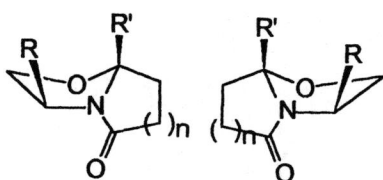

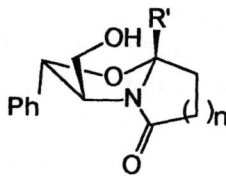

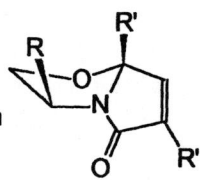

1.90   **1.91**   **1.92**

Oxepanes **1.93** derived from ephedrine **1.61** (R = Me) are interesting aminoalcohol derivatives that react very selectively with organomagnesium reagents [262]. After hydrolysis, chiral β,β-dialkylated acids are obtained. Compounds **1.93** also undergo stereoselective hetero-Diels-Alder reactions [351] and [3+2] cycloadditions [352]. Ketenamide acetal **1.94**, prepared from (*S*)-prolinol **1.64** (R = CH$_2$OH), gives excellent results in asymmetric aldol reactions [353]. Denmark and coworkers have proposed the use of 1,3-aminoalcohol **1.95** [354, 355] to generate 1,3,2-oxazaphosphorinanes **1.96**. The alkylation of reagents **1.96** in the presence of base, followed by hydrolysis, provides chiral alkylphosphonic acids. Reagents **1.96** can also participate in selective Wittig-Horner reactions [356].

Many groups have prepared chiral 1,4-oxazin-3-ones [154, 357] starting from (*S*)-phenylglycinol **1.60** (R = Ph, R' = H) or 1,2-diphenylglycinol **1.58** (R = H). These compounds undergo many different kinds of stereoselective transformations. However, because the subsequent treatment of the so-formed products is hydrogenolysis that cleaves the C–N bond of the original reagent, this interesting methodology is beyond the scope of this book.

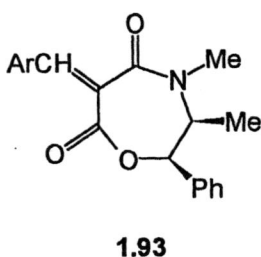

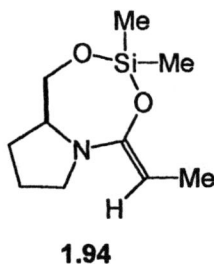

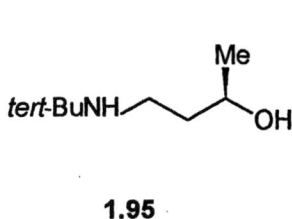

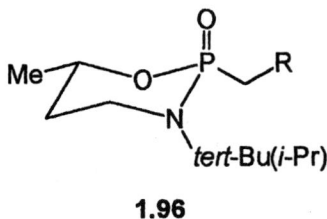

1.93   **1.94**

1.95   **1.96**

1.4 ALDEHYDES, KETONES AND DERIVATIVES

Most aldehyde-based chiral auxiliaries come from carbohydrates [248]. For example, aldehyde **1.97** is transformed into α-aminoester imines **1.98**, which undergo stereoselective alkylations in basic media. After acidic hydrolysis, the chiral auxiliary is regenerated, and chiral, nonepimerizable α,α-dialkylaminoesters are obtained [154, 358]. The most popular point of attachment of the acyl group to the carbohydrate is at the anomeric position. Useful reagents in this class include α- or β-glucopyranosides 1.99 and 1.100 (R = R '= H, Ac, PhCH$_2$, Me, *tert*-Bu), β-galactopyranosides 1.101 and arabinose derivative 1.102 [241, 358-365]. Less functionalized auxiliaries 1.103 and 1.104 have also been introduced by Charette and coworkers [366]. α-Glucopyranoside vinyl or dienylethers 1.99 (Y = OCH=CH$_2$ or OCH=CHCH=CH$_2$) give interesting results in cycloaddition reactions in aqueous media, although the related reactions of the β-isomers are less satisfactory [359, 362]. High diastereoselectivities are observed in cycloadditions conducted with nitrogen-substituted derivatives of 1.100 (Y = 1,2-dihydropyridyl or PhCH=N) [366a, 367] and with imine derivatives of 1.101 and 1.102. Useful results are also obtained in the synthesis of (R)- or (S)-aminoacids by the Strecker reaction or the isonitrile method, and in condensations with allylsilanes or -stannanes [241, 358, 361, 365]. Allylethers 1.100 (Y = OCH$_2$-CH=CH$_2$, R = PhCH$_2$) undergo highly stereoselective cyclopropanation reactions with Et$_2$Zn/CH$_2$I$_2$ provided that there is a free alcohol in the 3-position (R' = H) [248, 366, 368]. This auxiliary is less efficient in controlling the stereoselectivity of per-acid-promoted or transition-metal-catalyzed epoxidations [369]. In all these reactions, the sugar group is cleaved by acidic hydrolysis.

Charette and coworkers [366] found that the reactions of 1.103 and 1.104 (R' = CH$_2$CHO) with Bu$_3$SnCH$_2$CH=CH$_2$ in the presence of MgBr$_2$ are highly stereoselective and either stereoisomer can be obtained, depending upon whether or not the R-substituent (PhCH$_2$ or *i*-Pr$_3$Si) promotes chelation control. 1-Penten-4,5-diols are obtained after treatment of the products of these reactions with allyl alcohol under slightly acidic conditions. From nitrosochloride 1.105, chiral allylamines are obtained via ene-reactions [370] and chiral heterocycles are obtained via asymmetric [4+2] cycloaddition reactions on the N=O bond [241, 371, 372].

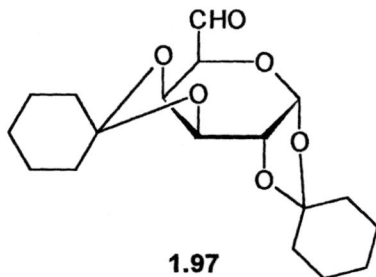

1.97

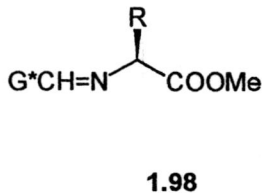

G*CH=N—CH(R)—COOMe

1.98

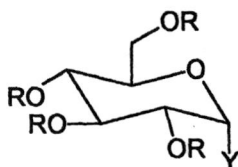

1.99

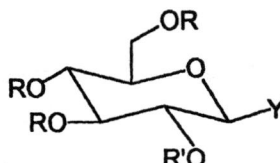

1.100

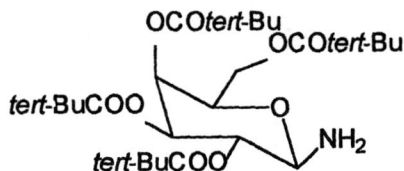

1.101

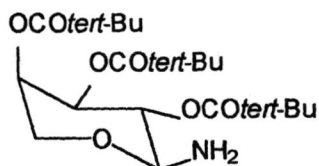

1.102

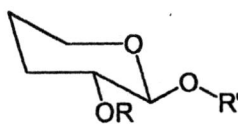

1.103

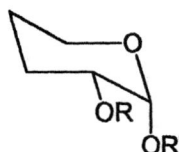

1.104

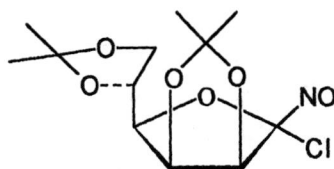

1.105

Naturally occurring ketones and their readily prepared derivatives have also been used as chiral auxiliaries. Menthone ketals have been the subject of some study [373, 374]. For example, enolates of compound **1.106**, obtained from 8-phenylmenthone, are alkylated with a good stereoselectivity [375]. Cleavage of the chiral auxiliary is accomplished by hydrolysis. Enolates of *tert*-butylglycine camphor imine **1.107** are alkylated [154, 159] or suffer Michael additions [376] with a high stereoselectivity, and the auxiliary is cleaved by treatment of the products with NH$_2$OH. However, the synthesis of **1.107** is not straightforward. *N*-Benzyl camphor imine enolates can be alkylated, but the stereoselectivities are only useful with benzyl halides as alkylating agents or in some sulfonamide alkylations [377]. Imines formed from α-amino acids and (1*S*, 2*S*, 5*S*)-2-hydroxypinan-3-one **1.108** or its enantiomer [154, 378, 379, 380] are also useful in asymmetric synthesis, and the chiral auxiliary is easily regenerated by acid hydrolysis. Belokon and coworkers have prepared imines from α-amino acids and (*S*)-2-[*N*-benzylprolyl]-aminobenzophenone **1.109** (R = Ph). Nickel complexes of these reagents undergo stereoselective alkylations [381, 382] and aldol and Michael reactions [383, 384]. The use of **1.109** (R = H or Me) as an auxiliary leads to less interesting selectivities [154, 261].

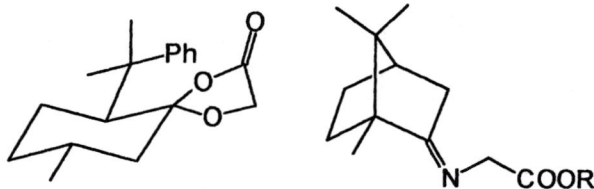

1.106   **1.107**

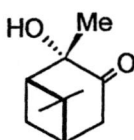

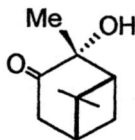

(1*S*,2*S*,5*S*)-**1.108**   (1*R*,2*R*,5*R*)-**1.108**

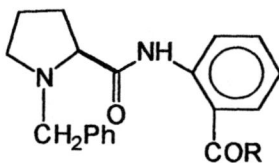

1.109

1.5 ACIDS, AMINOACIDS AND DERIVATIVES

Derivatives of enantiopure α- or β-hydroxyacids are useful reagents. The reactions of (*R*)- or (*S*)-3-hydroxybutanoic acid with aldehydes preferentially give *cis*-1,3-dioxan-4-ones **1.110**. The reactions of the Li enolates of these dioxanones with allyl- or ethynyltrimethylsilanes or with $CNSiMe_3$ in the presence of titanium salts leads to chiral secondary alcohols with a high enantiomeric excess. The inexpensive chiral auxiliary is not recovered in this process [386]. Similarly, from (R)- or (S)-mandelic acid **1.111**, cis-1,3-dioxolan-4-ones such as **1.112** are obtained. The reactions of these reagents with enoxysilanes in the presence of BF_3 provide chiral β-hydroxyketones. Applications of dioxanones **1.112** is limited because the stereoselectivity for formation of cis-dioxolan-4-ones are not as good as in the case of **1.110**, and because removal of the chiral controller requires an oxidation [385]. 1-O-Methylmandeloxydienes **1.113** are used in .[4+2] cycloadditions [73, 170, 387].

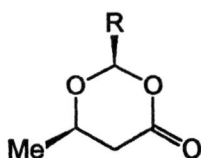

1.110

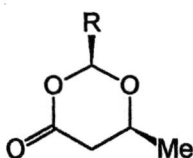

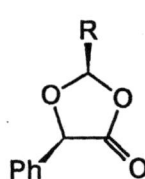

1.112

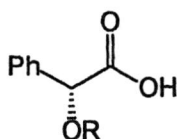

(R)-**1.111**

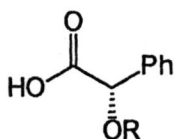

(S)-**1.111**

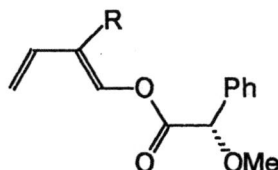

1.113

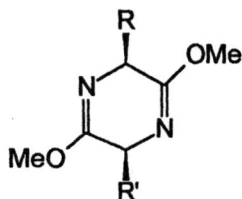

1.114 (R' = H or alkyl)

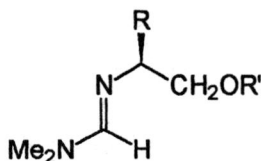

1.115

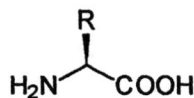

1.59 (R' = H)

Among the α-aminoacids, derivatives of proline **1.64** (R = COOH) have been used with satisfactory results in threonine synthesis or in intramolecular halolactonization reactions [261]. However, the most impressive application of these compounds as chiral auxiliaries is Schöllkopf's method, which uses cyclic lactim ethers **1.114** [154, 261]. Compounds **1.114** are formed by the condensation of two aminoacids, followed by reaction of the intermediate lactim with Me_3OBF_4. In these reagents, one of the aminoacids is the substrate, and the other is the chiral auxiliary. As chiral auxiliaries, the most useful amino acids are (*S*)-valine **1.59** (R = *i*-Pr, R' = H) or (*S*)-*tert*-leucine **1.59** (R = *tert*-Bu, R' = H). Treatment of **1.114** with one equivalent of base leads to a conjugated organometallic species that undergoes highly selective alkylations, condensations with carbonyl compounds or Michael additions [154, 159, 161, 167, 169, 173, 253, 261]. Further treatment of the products with acid leads to the methyl esters of two aminoacids, and this method provides chiral α-mono- or α,α-disubstituted aminoacids in ee's > 95%.

Chiral formamidines **1.115** have been developed by Meyers and coworkers [388-392]. These reagents are prepared from $HC(NMe_2)(OMe)_2$ and α-aminoethers **1.60** (R' = Me or *tert*-Bu) [393]. Once again, (*S*)-valinol and (*S*)-*tert*-leucinol derivatives **1.115** (R = *i*-Pr or *tert*-Bu) are the most effective chiral auxiliaries. The main applications of these reagents are enantioselective alkylations of tetrahydroquinolines, and the products of these alkylations are very useful in alkaloid synthesis [343, 391, 394]. The chiral auxiliary is regenerated by treating the products with hydrazine.

1.6 LACTAMS AND ANALOGS, SULTAMS

1.6.1 Oxazolidinones, Thiazolidinones, Thiazolidinthiones and Derivatives

Evans [167] initiated the use of chiral 1,3-oxazolidin-2-ones (*S*)-**1.116** (R = *i*-Pr, Ph, PhCH$_2$) and **1.117**, which are prepared by action of $COCl_2$ or diethylcarbonate [395] on (*S*)-valinol **1.60** (R = *i*-Pr, R' = H), (*S*)-phenylglycinol **1.60** (R = Ph, R' = H), (*S*)-phenylalaninol **1.60** (R = PhCH$_2$, R' = H), or (1*S*, 2*R*)-norephedrine **1.61** (R = H). Reactions of *N*-lithiated or *N*-silylated derivatives [396] of **1.116** and **1.117** with acid chlorides give *N*-acyloxazolidinones **1.118**. When starting from **1.117**, an excess of lithiated base must be avoided to prevent epimerization [397]. These acyloxazolidinones have been widely used in asymmetric synthesis. The lithium enolates of **1.118** (R = Me, SMe, Br, Ar) are alkylated, acylated, and hydroxyalkylated with a very high stereoselectivity [105, 106, 160, 167, 261, 398-403]. The metal (Li, Sn, Ti, Zn) or boron enolates of **1.118** (R = Me, Br, Cl, NCS) have found many very interesting applications in asymmetric aldol and Darzens reactions [123, 399, 403a-409], as α-aminoacid precursors by reaction with nitrogen nucleophiles [410, 411, 412], or in Michael additions [413]. α,β-Unsaturated N-acyloxazolidinones are used in aldol reactions [414], 1,4-addition reactions of trialkylaluminium reagents or thiols [415, 416], [4+2] cycloadditions [72, 73, 417, 418], and ene-reactions [419]. All these processes are highly stereoselective. The N-benzoyl derivative of (S)-1.116 (R = PhCH$_2$) has recently been used as a chiral acylating agent for alcohols [420]. Starting from acyl derivative 1.118 (R = Cl) of the oxazolidinone from (S)-1.116 (R = PhCH$_2$), chiral ketenes can be generated and their subsequent [2+2] cycloaddition reactions are highly stereoselective. These reactions have been used in β-lactams synthesis [131, 366a, 421]. The products of these reactions are usually cleaved to chiral acids by mild hydrolysis with H$_2$O$_2$/LiOH [167, 406]. Transesterification by LiOCH$_2$Ph or transamidation by MeONHMe in the presence of AlMe$_3$ [421a] are also valuable cleavage methods. The chiral auxiliaries are recovered in all cases and reaction conditions have been designed to minimize undesirable ring cleavage of the oxazolidinones [123, 397]. When β-lactams are the targets, cleavage of the auxiliary by Li/NH$_3$ or TMSI/HMDS treatment, followed by exposure to DABCO and careful acid hydrolysis, may be preferred [421]. A limitation of the use of these chiral auxiliaries is the poor selectivity observed in 1,3-dipolar cycloaddition reactions of α,β-unsaturated N-acyl derivatives [423]. More sterically hindered 1,3-oxazolidin-2-ones have been recently been devised. Some of these have been generated from cis-1-amino-2-hydroxyindane [424] or from 2-aminodeoxysugars 1.119 [241, 416]. Others, such as 1.120 and 1.121 (X = O), include the bornane skeleton [425-429]. N-Acyl derivatives of these oxazolidinones are useful for asymmetric alkylations and aldol reactions. Other very sterically congested oxazolidinones have also been prepared [430]. Camphor-derived chiral oxazinones 1.122 were designed so that the enolates of their N-acyl derivatives would participate in facially selective alkylations or aldol reactions [431, 432].

N-Acyl-1,3-thiazolidines-2-ones 1.123 (X = S, R = COOMe), obtained from cysteine methyl ether [261], have been introduced by Mukaiyama and coworkers for use in asymmetric aldol reactions [261, 433, 434, 435]. In reactions of related N-acyl-1,3-oxazolidines-2-thiones 1.123 (X = O, R = COOMe), each enantiomer can be obtained either from L- or D-serine [434] and the auxiliaries can easily be recovered by methanolysis. Similarly, N-acyl derivatives of 1.121 (X = S) have been used in asymmetric aldol reactions [429, 436], and N-acyl-1,3-thiazolidinethiones 1.123 (X = S, R = i-Pr) are useful in asymmetric acylation [437] and aldol and related reactions [437, 438]. Cleavage of the chiral auxiliary is accomplished by aminolysis with O-benzylhydroxylamine or by reduction with LiAlH$_4$.

A different methodology has been introduced by Seebach and coworkers to synthetize natural and nonnatural α-aminoacids [81, 154]. Strictly speaking, this method does not involve a chiral auxiliary.

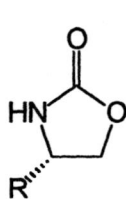

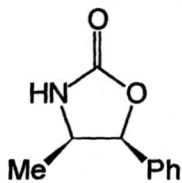

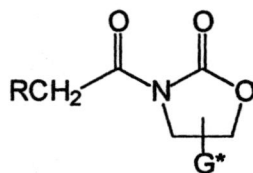

1.116   **1.117**   **1.118**

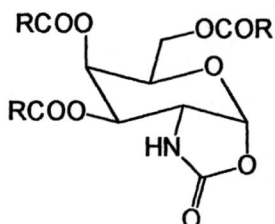

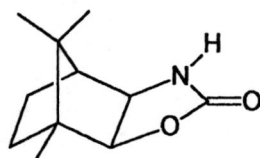

1.119 (R = *tert*-Bu)   **1.120**

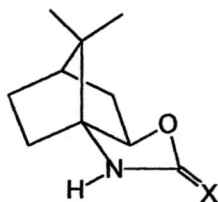

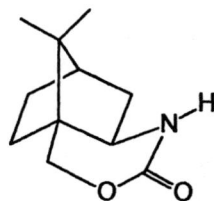

1.121 (X = O or S)   **1.122**

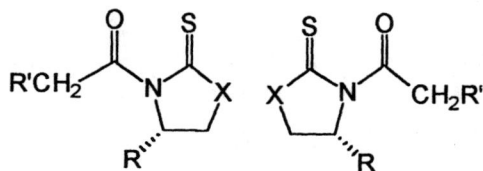

1.123

In this method, pivalaldehyde imines **1.124** are obtained starting from the *N*-methylamides of natural aminoacids. Treatment of these imines with HCl/MeOH followed by PhCOCl generates *anti* imidazolidinones **1.125** (X = NMe). In contrast, treatment with benzoic anhydride leads preferentially to the *syn* isomers **1.126** (X = NMe). The corresponding 1,3-oxazolidin-5-ones **1.125** (X = O) and **1.126** (X = O) are prepared from the appropriate α-aminoacid sodium salts. The glycine analogs (X = NMe or O, R = H) can be separated by chromatography on chiral support. Generation of the enolates from these heterocycles is accompanied by loss of the initial stereocenter of the α-aminoacid, and these enolates typically suffer highly diastereoselective alkylations and hydroxyalkylations [81, 154, 439, 440]. However, the cleavage of the chiral group requires hot aqueous HCl. The use of *N*-benzyloxy- or *tert*-butylcarbonyloxy analogs **1.127** (X = O or NMe) provides reaction products that can be cleaved by hydrogenolysis or mild hydrolysis [441, 442, 443].

1.6.2 Bicyclic Lactams

Boeckman and coworkers [444] have proposed the use of α,β-unsaturated acylimides **1.128** bearing a camphor like skeleton in [4+2] cycloaddition reactions.

The disappointing results obtained in 1,3-dipolar cycloadditions and in radical addition reactions to α,β-unsaturated esters linked to various auxiliaries (§ 1.6.1 and 1.6.4) prompted Curran and Rebek to design and prepare the new reagents **1.129** and **1.130** derived from Kemp's triacid [276, 445, 446]. These auxiliaries provide very high asymmetric inductions in a number of types of reactions, and can be cleaved either by H_2O_2/LiOH or Li *sec*-Bu$_3$BH.

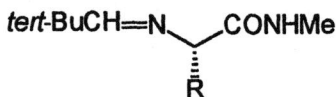

1.124

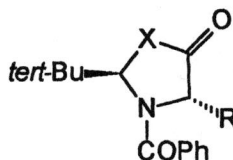

1.125

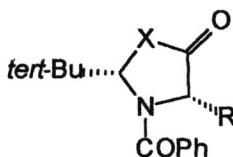

1.126

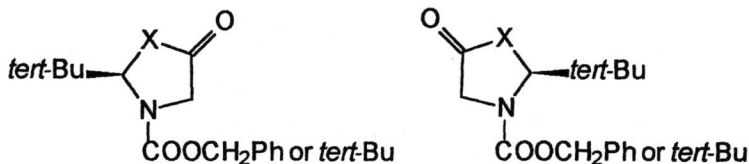

1.127

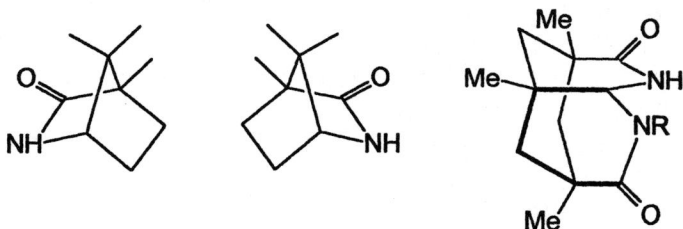

1.128   **1.129**

1.6.3 Imidazolidinones and Perhydropyrimidones

The enolates of *N*-acylimidazolidinones derived from **1.131** generated from ephedrine **1.61** (R = H) are useful in asymmetric alkylations [447, 448] and aldol reactions [449, 450] and cuprate additions to the α,β-unsaturated acyl analogs have recently been described [451]. These chiral auxiliaries are cleaved by MeONa/MeOH or LiEtBH₃. Recently, Davies and coworkers have suggested the use of symmetrical *N*,*N*-diacyl-1,3-imidazolidin-2-ones **1.132**, formed from diamines having a C₂ axis of symmetry [452], for asymmetric aldol reactions [449]. Juaristi [453] has used perhydropyrimidin-4-ones for related purposes.

1.6.4 Sultams

Oppolzer and coworkers [147, 454] have developed a class of reagents based on the enantiomeric bornane-2,10-sultam skeleton **1.133**. These chiral auxiliaries are easily prepared from the enantiomeric 10-camphosulfonic acids [455]. Saturated or α,β-unsaturated *N*-acylsultams **1.134**, occasionally prepared from *N*-silyl precursors [396], have been used very frequently. Asymmetric alkylations, aminations and aldol reactions of enolates or enoxysilane derivatives of **1.134** (R = R'CH₂) [147, 404, 407, 456-460] are highly selective. The α,β-unsaturated *N*-acylsultams **1.134** (R = R'R"C=CH) suffer highly stereoselective organocuprate 1,4-additions [147, 173], cyclopropanations [461], [4+2] and [3+2] cycloadditions [73, 276, 454, 462], OsO₄ promoted dihydroxylations [454, 463] and radical additions [276, 454]. When R' and R" are different from H, selective LiBEt$_3$H reduc-tions [147, 448, 464] or catalytic hydrogenations [465] give useful results. Hetero-Diels-Alder reactions can also be performed starting with appropriately substituted *N*-acylsultams [272, 466, 467]. The chiral auxiliary is cleaved to provide acids by treatment with aqueous LiOH, sometimes under phase transfer conditions [456], or to provide alcohols by reduction with LiAlH$_4$. The chiral auxiliary can always be recycled, though its high molecular weight and its cost may limit its applications.

More recently, Oppolzer has proposed the use of enantiomeric aryl sultams **1.135** (R = Me), each being easily obtained from saccharin [468]. In general, these chiral auxiliaries can be applied in the same types of reactions as the camphor sultam [469, 470]. However, the low selectivity observed in nitrile oxide cycloadditions led to the design and synthesis of a more bulky auxiliary **1.135** (R = *tert*-Bu), which is correspondingly more stereoselective [471]. Nevertheless, the cleavage of the chiral auxiliary necessitates the use of LiEt$_3$BH, and generates alcohols as the final products.

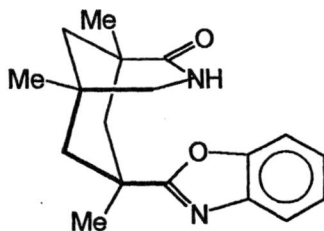

1.130

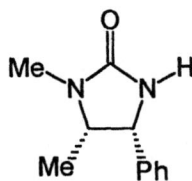 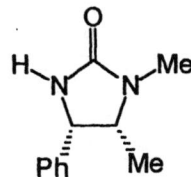

1.131

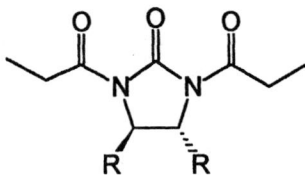

1.132

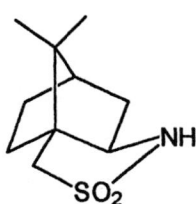 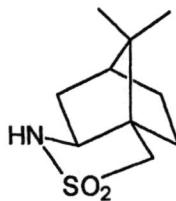

1.133

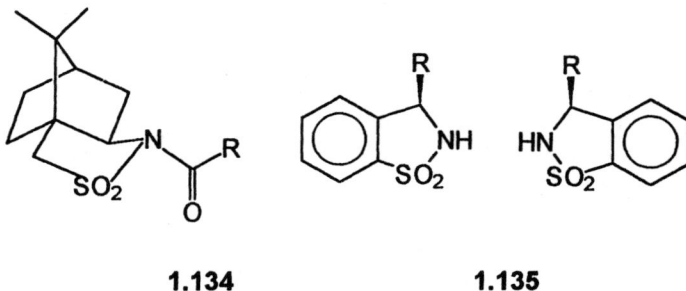

1.134   **1.135**

1.7 SULFOXIDES AND SULFOXIMINES

Chiral p-tolylsulfoxides **1.136** (Y = Tol) are easily prepared from menthyl (R)- or (S)-p-toluensulfinates **1.137**. The reactions of **1.137** with organometallic reagents or enolates occur with inversion of configuration and frequently give high yields. Many chiral sulfoxides **1.136** (Y = Tol) have been prepared, including those in which R = alkyl, ArCH$_2$, $tert$-BuOCOCH$_2$, Me$_2$NCOCH$_2$, PhSCH$_2$, R'CH=CHCH$_2$, R'C≡CCH$_2$ [472], RCOCH$_2$ or ArCOCH$_2$ [473, 474], Me$_2$NCSCH$_2$ [475], dihydrooxazolylmethyl [475a], $tert$-BuSO$_2$CH$_2$ [476], and 1,2-dehydropyrrolidylmethyl [477]. Replacement of the group Y = Tol by 2-pyridyl [478] or 1-Np residues [479] has been proposed. The electron-withdrawing ability of the sulfoxide group allows the easy generation of carbanions in the α-position. The alkylations of carbanion derivatives are highly stereoselective, as are hydroxyalkylations by symmetrical carbonyl compounds [472]. Also stereoselective are the reactions of carbanions formed from **1.136** (Y = Tol, R = Me) with imines [480], or from **1.136** (Y = Tol, R = $tert$-BuOCOCH$_2$, Me$_2$NCOCH$_2$, isoxazolyl CH$_2$) with aldehydes [209, 481]. However, disappointing results are obtained in the reactions of carbanions derived from **1.136** (Y = Tol, R = Me) with unsymmetrical carbonyl compounds [481]. Allylic carbanions formed from **1.136** (R = CH$_2$=CHCH$_2$) give variable selectivities, although some highly selective Michael additions have been described [161, 173, 477, 482]. Reductions of the carbonyl group of **1.136** (Y = Tol, R = RCOCH$_2$ or ArCOCH$_2$) are highly stereoselective [473, 474, 483], as are addition reactions with Me$_3$Al.

Chiral α,β-unsaturated sulfoxides **1.136** (Y = Tol, R = R'CH=CH) also have been used in asymmetric synthesis. These compounds are prepared either by treatment of **1.137** with vinylic organometallic reagents, or from saturated precursors by classical chemical transformations [102, 173, 476, 484-487]. Michael additions to these electrophiles are interesting only if R' = CF$_3$ [161]. Organometallic additions or [4+2] cycloadditions require the introduction of a second electron-withdrawing substituent [73, 102], and acyclic **1.138** and cyclic **1.139** gem-disubstituted sulfoxides have seen many interesting applications [101, 102,

169, 476, 484, 488-491]. α,β-Unsaturated β-functionalized chiral sulfoxides **1.140** have also been used in [4+2] cycloadditions, but only the Z-isomers react stereoselectively. The synthesis of reagents **1.140** requires tedious purifications, which limits their use [486, 492]. Kagan has recently proposed the activation of vinylsulfoxides in [4+2] cycloaddition reactions by quaternarization of the sulfur with Et_3OBF_4, and preliminary results are encouraging [493, 494]. p-Tolyl sulfoxide **1.136** (Y = Tol, R = 2'-i-PrOCO-1-Np) has been used in enantioselective synthesis of 1,1'-binaphthyls [495].

As an alternative to menthyl sulfinates, Alcudia and coworkers [496] have prepared chiral sulfoxides from the sulfinate **1.137** derived from diacetone glucose **1.48**. By varying the reaction conditions, either (R)- or (S)-**1.137** can be obtained so that the subsequent reactions with RMgX give, at will, either (S)- or (R)-**1.136**.

In most cases, the sulfoxides produced after the asymmetric transformation are reduced with sodium or aluminium amalgam, or Raney nickel. The sulfoxide may also be reduced by $LiAlH_4$ into a sulfide, which after quaternarization becomes a leaving group [474]. If the carbon skeleton of the substrate bears appropriate substituents, the sulfoxide can suffer [2,3] sigmatropic rearrangements [497, 498] or Pummerer rearrangements. In all these types of applications, the chiral auxiliary is never recovered.

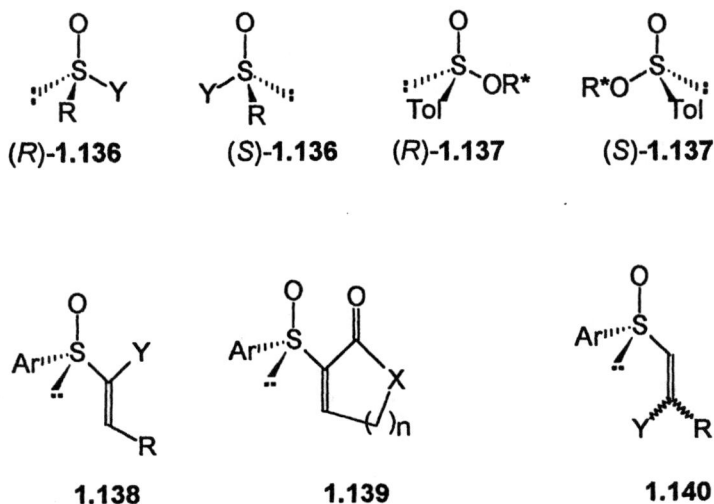

(R)-**1.136**   (S)-**1.136**   (R)-**1.137**   (S)-**1.137**

1.138   **1.139**   **1.140**

A new class of recoverable chiral sulfoxides **1.141** has been proposed by Wills and coworkers [499, 500, 501]. These reagents are obtained from a chiral sulfinamide **1.142** as shown in Figure **1.8**. After the asymmetric reaction, treatment of the products with aluminium amalgam generates a chiral thiol that can be recovered and retransformed into **1.142** by a two step-process. These reagents are useful in aldol reactions or imine reductions.

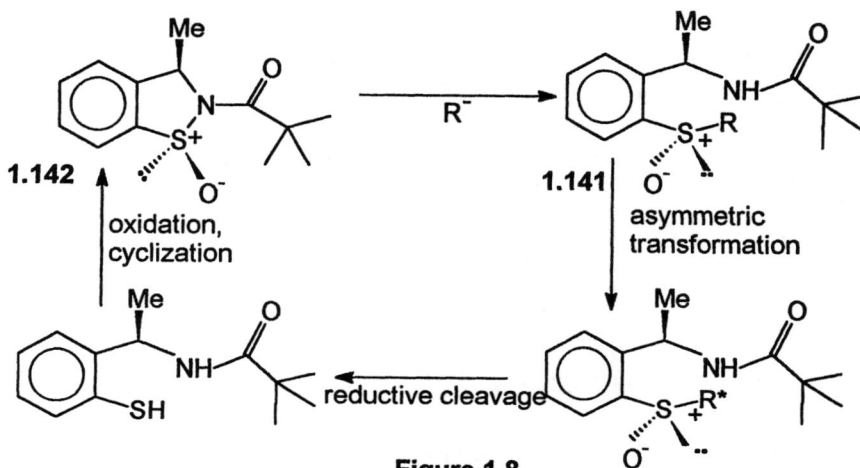

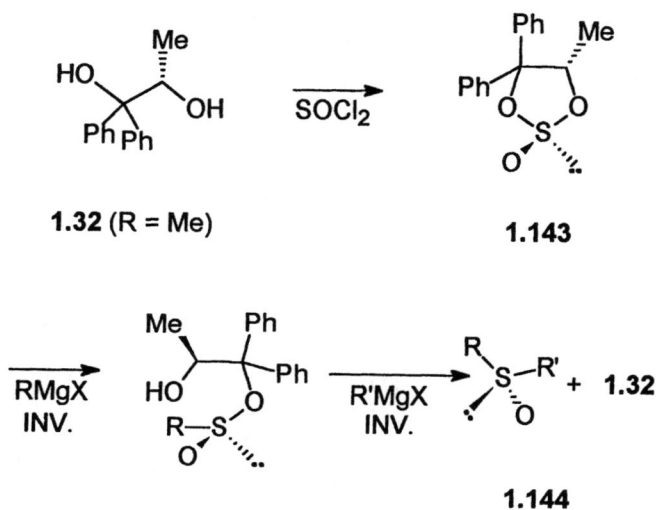

Figure 1.8

1.32 (R = Me)

1.143

1.144

Figure 1.9

1.145   **1.146**   **1.147**

(*R*)-**1.148**   (*S*)-**1.148**

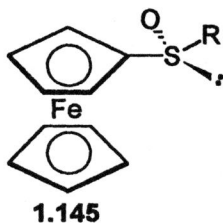

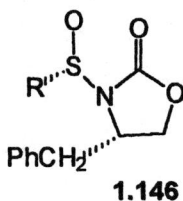

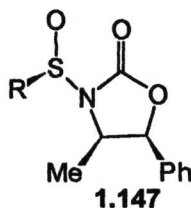

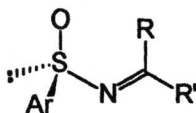

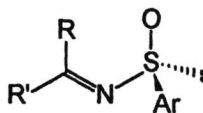

Quite recently, new methods for the synthesis of chiral sulfoxides have been discovered. Asymmetric oxidation of sulfides by peroxides in the presence of chiral titanium catalysts has been studied by Kagan and Modena [502-505] (§ 3.2.3), and the use of chiral oxaziridines (§ 2.8) as oxidizing agents has been introduced by Davis [506].

Kagan has also performed the sequential reaction of two organomagnesium reagents with a chiral cyclic sulfite **1.143** formed from (*S*)-1,2-diol **1.32** (R = Me) [503, 507] (Figure 1.9). By inverting the order of addition of RMgX and R'MgX, either enantiomeric sulfoxide **1.144** can be obtained starting from the same sulfoxide **1.143**. This route was applied to the synthesis of ferrocenyl derivatives **1.145** possessing planar chirality [508]. Evans and coworkers [509] have devised two chiral reagents that are able to transfer two sulfoxides group: *N*-sulfinyloxazolidinones **1.146** bearing auxiliary **1.116** (R = PhCH$_2$), and related reagent **1.147** bearing auxiliary **1.117**. These reagents react very rapidly with organomagnesium reagents with inversion of configuration to generate chiral dialkyl, arylalkyl, or other functionalized sulfoxides with high stereoselectivity. Davis, Hua and their coworkers have prepared chiral sulfinimines **1.148** by oxidation of related sulfenimines (§ 2.8) [510] with chiral oxaziridines, or by the reaction of the menthyl sulfinates (*R*)- or (*S*)-**1.136** with (Me$_3$Si)$_2$NLi followed by condensation with aromatic or α,β-unsaturated aldehydes [482, 511, 512]. These enantiomerically pure compounds are precursors of chiral amines by reduction, and of β-aminoesters by reaction with ester enolates [482, 510, 512].

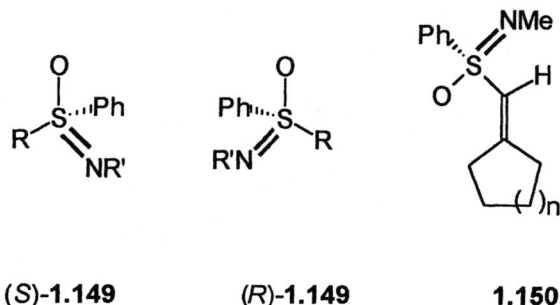

(S)-**1.149**   (R)-**1.149**   **1.150**

Asymmetric sulfoximine chemistry was introduced by Johnson and coworkers [513, 514, 514a], who resolved sulfoximines **1.149** (R = Me, R' = H). However, the applications of these sulfoximines in asymmetric synthesis have been limited [514]. Conjugate addition reactions of organocopper reagents to α,β-unsaturated sulfoximines are stereoselective [515]. In a new application, reaction of lithiated carbanions of **1.149** (R = R'CH$_2$) with substituted cyclic ketones, followed by dehydration, generates chiral unsaturated sulfoximines **1.150**. Their coupling with organometallic reagents takes place with retention of configuration [516, 517, 518]. These α,β-unsaturated sulfoximines **1.150** can be isomerized into allylic derivatives, which in turn suffer highly selective SN$_2$' reactions with organocuprates [519]. N-Nitrosulfoximines have also recently been used in asymmetric synthesis [520].

1.8 TRANSITION-METAL DERIVATIVES

The rapid development of organometallic chemistry [92] has led to applications of transition-metal reagents in asymmetric synthesis. In this book, two types of applications will be described:

• Asymmetric syntheses in which the transition-metal residue is part of a recycled chiral auxiliary.

• Dieneiron carbonyl and arenechromium carbonyl complexes in which the chirality is introduced by metal coordination and depends upon the structure of the complex. In these systems, the organometallic complex is destroyed in the cleavage reaction.

1.8.1 Cyclopentadienyliron Carbonyl Complexes

The uses of chiral (R)- and (S)-acetylcyclopentadienyl triphenylphosphinoiron carbonyl **1.151** (R = Me) have been developed by Davies and coworkers [521, 522, 523], and related achiral reagents were studied by Liebeskind [524].

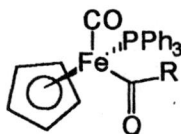 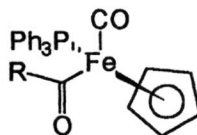

(R)-**1.151**   (S)-**1.151**

Deprotonation of complexes **1.151** (R = R'CH$_2$) generates enolates, which participate in highly stereoselective additions with carbonyl compounds, imines or epoxides [407, 408]. Acyl complexes **1.151** (R = R'CH=CH) are easily prepared, and they suffer very stereoselective conjugate addition reactions [173], cyclopropanations or [4+2] cycloadditions [73]. The cleavage of the chiral auxiliary is performed by addition of bromine or *N*-bromosuccinimide in the presence of water or an amine. These reactions generate acids or amides along with C$_5$H$_5$Fe(CO)(PPh$_3$)Br, from which the auxiliary can be recycled [522]. Brookhardt and coworkers [525] have replaced triphenylphosphine in **1.151** (R = Me) with triethyl- or trimethylphosphine, but disappointing stereoselectivities were observed in the reactions that were studied.

1.8.2 Irondiene Carbonyl Complexes

Unsymmetrically substituted irondiene carbonyl complexes are chiral, and some are easily accessible in an enantiomerically pure form by resolution. Complexes **1.152** (R = H, Me, *n*-Bu, Y = CHO) or **1.153** (Y = CHO), bearing an aldehyde functional group, can be resolved via chiral hydrazones [526], chiral aminals [527], or derivatives of ephedrine **1.61** [528]. The diastereoisomers thus formed are separated by chromatography, and the aldehydes are easily regenerated. The resolution of trimethylenemethane complexes **1.154** can be accomplished similarly [529]. Complexes bearing an ester functionality **1.152** (R = COOMe) are resolved through lactates, which are subsequently treated with KOH/MeOH [530]. From these enantiomerically pure complexes, classical reactions lead to other systems such as **1.152** (R = CH$_2$OH, Y = CHO) [531], **1.152** (R = *n*-Bu, Y = CH$_2$PPh$_3$, BF$_4$) [527, 532], **1.152** (R = Me, Y = COR' or CHOHR') [526], or **1.152** (R = Me, Y = H or alkyl) [533]. From these complexes, it is possible to generate carbocations which are stabilized by the metal, and these cations react with various nucleophiles [526, 528]. Complexes **1.152** (Y = CHO or COR) react with organometallic reagents or enolates, while complexes **1.152** (Y = R'CH=CH) participate in various cycloadditions [528]. However, all of these reactions are often poorly stereoselective so that chromatographic separations are needed [526, 528, 531, 534, 535]. In contrast, interesting results are obtained in the reactions of aldehyde **1.152** (R = MeOCO, Y = CHO) with allenylsilanes [536] and in reactions of keto derivatives **1.152** (R = Me, Y = COR') with hydrides [526, 537], organolithiums or organocuprates [526, 533]. Stereoselective osmylations of enynes are also described [538]. Decomplexation of the products is accomplished under mild oxidative conditions with $Ce(NO_3)_6(NH_4)_2$ in MeOH or with H_2O_2 in alkaline solution [528].

1.8.3 Arenechromium Tricarbonyl Complexes

o- or *m*-Substituted arenechromium tricarbonyl complexes **1.155** are chiral [539], and when Y = CO_2H or NH_2, the enantiomers can be resolved as chiral amine or acid salts. Aldehydes **1.155** (Y = CHO) and cyclic ketones **1.156** (Y = O) have been resolved through chiral hydrazones or imines [540]. Schmalz and coworkers [541] have recently proposed a new method to stereoselectively obtain cyclic complexes **1.156** (Y = H, OH or H, H) by enantioselective reduction of the uncomplexed ketone (§ 6.1.1) followed by diastereoselective complexation. Only *o*-substituted complexes **1.155** (R' = H, R = Me, OMe, Cl, $SiMe_3$) have given interesting results in asymmetric synthesis [539].

Cyclic ketones **1.156** (Y = O, n =1 or 2) react with hydrides [87, 539] or organomagnesium reagents [539] with very high selectivities. The derived enolates are alkylated with high selectivity [539], as are the related benzylic carbanions formed from **1.156** (Y = H, H) [541]. These types of alkylations have been applied in alkaloid synthesis [542, 543]. However, Michael reactions performed with the enolates give disappointing results [539]. The acid-catalyzed cyclization of diastereoisomerically pure complexes such as **1.157** bearing an alcohol group leads stereoselectively to precursors of alkaloids [542, 543]. In these compounds, the chromium carbonyl residue is removed by photolysis in slightly acidic media.

In acyclic systems, reductions of the ketone group of **1.155** (Y = COR) give poor selectivities. Reactions of *o*-substituted aldehydes **1.155** (Y = CHO) with organomagnesium reagents, perfluoroalkyllithiums or nitromethane [540] or chloracetophenone [540, 544] anions are very selective. Such is also the case for their reactions with functionalized isonitriles [540], silyl enolethers or thioketene acetals in the presence of Lewis acids [545, 546], or in Baylis-Hillmann reactions [547].

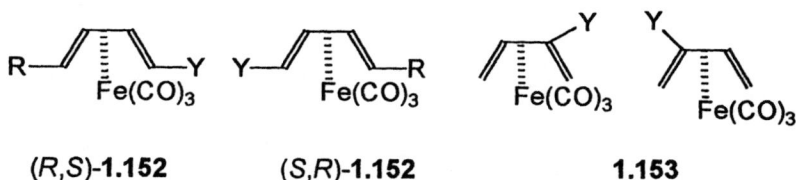

(R,S)-**1.152** (S,R)-**1.152**   **1.153**

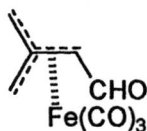

1.154

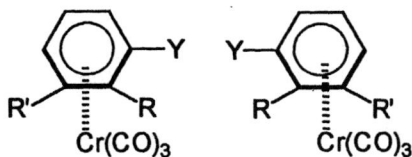

1.155

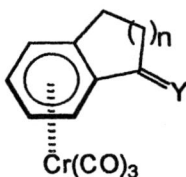

1.156

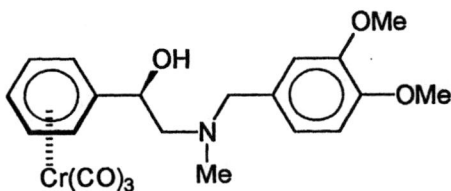

1.157

Alcohol derivatives **1.155** (Y = CHOHR) are useful as auxiliaries in α-keto-ester reductions [540] or [4+2] cycloadditions of acrylates [548]. β-Lactams are obtained from imines **1.155** (Y = CH=NAr) with a high enantiomeric excess after reaction with ester enolates and decomplexation [549]. Alkylations of benzylimine **1.155** (Y = Ph$_2$N=CH) give interesting results [539], and some 1,3-dipolar cycloaddition reactions with nitrones have been described [550].

CHAPTER 2

CHIRAL REAGENTS

Chiral reagents generally fall into one of two large classes: (1) reagents that are constructed only from covalent bonds, and (2) organometallic reagents. In the later class, a chiral ligand coordinates to the metal, and asymmetric induction can only occur if this ligand is not displaced from the coordination sphere during the reaction process. The chiral reagent or ligand is not transferred to the reacting substrate.

Chiral reagents may be used in stoichiometric amounts or, better, in catalytic amounts along with an achiral coreagent.

A catalytic cycle (Figure **2.1**) can operate provided that the following conditions are met:

• The chiral reagent R* or R–L* must react with the substrate Y–X more rapidly than the achiral coreagent R (step a), and

• The chiral reagent must be regenerated after the reaction process by interaction of the chiral primary product Y–Z'* with the achiral coreagent R.

These points will be further illustrated in three specific cases: (1) reductions of carbonyl compounds by chiral oxazaborolidines-borane complexes (§ 2.3), (2) reactions of carbonyl compounds with dialkylzinc reagents in the presence of chiral ligands (§ 2.5.4), and (3) dihydroxylation of olefins by OsO_4 in the presence of cinchona alkaloids (§ 2.9).

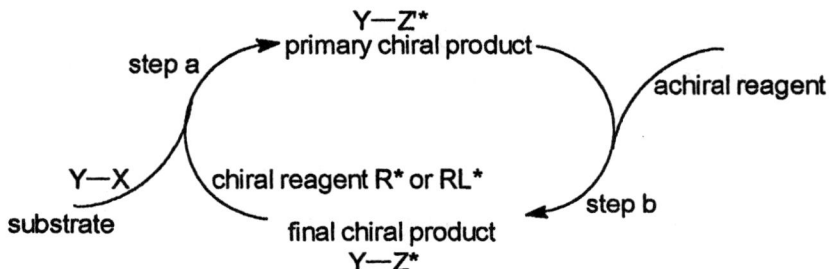

Figure 2.1

2.1. CHIRAL PROTON DONORS, CHIRAL BASES

Only reactions taking place under kinetic control will be considered. Asymmetric protonations resulting from second-order transformations will not be taken into account [127, 551].

Asymmetric protonations of prochiral ketenes, metal enolates or enamines are performed with chiral alcohols, amines or amine salts [552]. Recently, good enantiomeric excesses ($\geq$ 80%) have been obtained in ketene protonations with the following α-hydroxyesters: methyl (R)- or (S)-α-hydroxyisocaproate **2.1** (R = i-Bu, R' = Me) [553], ethyl (S)-lactate **2.1** (R = Me, R' = Et), i-propyl (R)-lactate **2.1** (R = Me, R' = i-Pr) or (R)-pantolactone **1.16** [554]. Excellent results were reported in asymmetric protonations of metal enolates by ephedrine **1.61** (R = Me) [555] or by triamine **1.64** (R = $CH_2NMeCH_2CH_2NMe_2$) in the presence of $BF_3 \cdot Et_2O$ [556]. L. Duhamel and P. Duhamel performed "deracemization" of ketones and aminoacids by enantioselective protonation of enolates or enamines by (R,R)-O,O-diacyltartaric acids **2.2** (R = OCOPh or OCO$tert$-Bu) [552].

Asymmetric deprotonation of carbonyl compounds has been accomplished with lithium amides generated from chiral amines [77, 552, 557, 558, 559]. The resulting lithium enolates can be trapped with trialkylsilylchlorides [557, 559, 560, 561]. Among the most efficient lithium amides that have been used for this purpose are those derived from linear amines having a C_2 axis of symmetry, (R,R) or (S,S)-**2.3** [557, 560]. The structures of several of these lithium amides have been determined by X-ray crystallography [562]. Lithium amides formed from bis-naphthylmethylamine [563], secondary-tertiary diamines **1.64** (R = $(CH_2)_4NCH_2$) and **2.4** (X = CH_2 or NMe, R = c-C_6H_{11} or $tert$-BuCH$_2$) [557, 559, 560, 561] are also recommended, as is s-BuLi in the presence of sparteine **2.5** [564]. Chiral lithium amides induce enantioselective rearrangements of epoxides to form chiral allyl alcohols [557]; the amide of diamine **1.64** (R = $(CH_2)_4NCH_2$) gives the best results. These reagents also promote Wittig rearrangements which, in some cases, are highly selective [565].

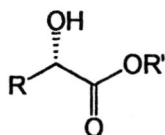   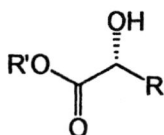   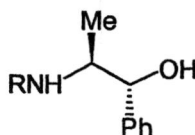

(*S*)-**2.1**   (*R*)-**2.1**   (1*R*,2*S*)-**1.61**

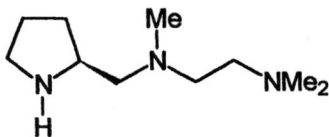 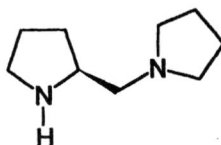

1.64 (R = CH$_2$NMeCH$_2$CH$_2$NMe$_2$)   **1.64** (R = CH$_2$N(CH$_2$)$_4$)

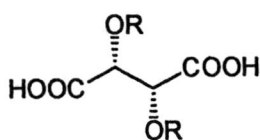 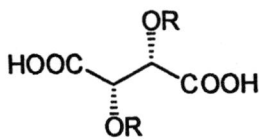

(*R,R*)-**2.2**   (*S,S*)-**2.2**

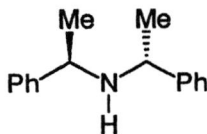 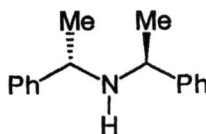

(*R,R*)-**2.3**   (*S,S*)-**2.3**

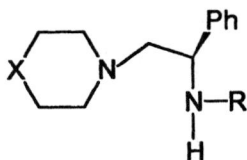 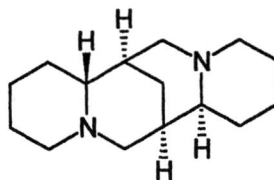

2.4   **2.5**

2.2 ALUMINUM AND BORON HYDRIDES

Aluminum and boron hydrides are extremely useful reducing agents [87, 566]. The aluminum or boron atom is tetracoordinated, so these reagents behave as nucleophiles.

As early as 1968, reduction of acetophenone by alkoxyaluminohydrides generated *in situ* by the action of *N*-methylephedrine **1.14** (R = Me) and a phenol or aniline on LiAlH$_4$ was shown to take place with a high enantiomeric excess [87, 169]. Similar reagents formed from aminoalcohol **2.6** or from secondary-tertiary diamines reduce several prochiral aralkylketones into secondary alcohols with very good enantiomeric excesses [87, 169, 567]. Unsymmetrical benzophenones are reduced into chiral benzhydrols by alkoxyaluminohydrides generated in the presence of aminoalcohols **2.7**. Alkoxyaluminohydrides **2.8** (R = Me, Et) are generated *in situ* by the reaction of (*R*)- or (*S*)-binaphtols **1.44,** alcohols, and LiAlH$_4$ in solution in the ratio 1:1:1. These reagents often have been used for asymmetric reductions of ketones [87, 231, 568-571]. In some cases, introduction of a crown ether onto the binaphtol skeleton [236, 572] improves the enantioselectivity. A chiral atropisomeric biphenol has also been used [573].

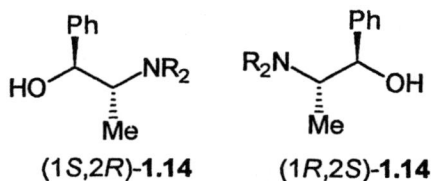

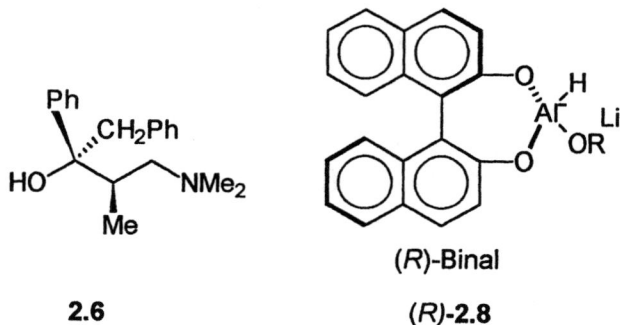

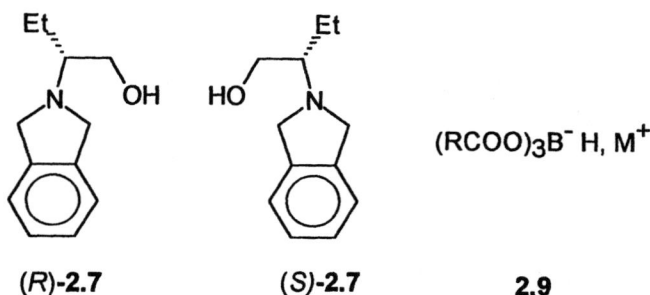

Ph

HO⟋⟍NR$_2$

Me

(1*S*,2*R*)-**1.14**

Ph

R$_2$N⟋⟍OH

Me

(1*R*,2*S*)-**1.14**

Ph   CH$_2$Ph

HO   NMe$_2$

Me

2.6

(*R*)-Binal

(*R*)-**2.8**

Et

N   OH

(*R*)-**2.7**

Et

HO   N

(*S*)-**2.7**

(RCOO)$_3$B$^-$ H, M$^+$

2.9

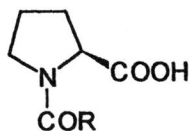

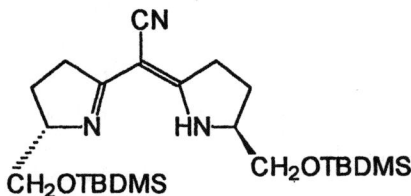

2.10   **2.11 (R = CH$_2$OTBDMS)**

Chiral substituents have also been introduced onto alkali borohydrides. Alkoxyborohydrides (RO)$_3$BHM or (RO)R'$_2$BHM derived from carbohydrates [169, 574], and acyloxyborohydrides **2.9** derived from tartaric acid **2.2** (R = H) [575] or from N-acylprolines **2.10** [169, 261] have been prepared. These reagents reduce α-functionalized ketones [575] or cyclic imines [169, 261]. Oxime ethers are transformed into chiral amines by reduction with NaBH$_4$/ZrCl$_4$ in the presence of chiral aminoalcohols [87, 575a]. Enantioselective reductions of prochiral C=C double bonds of α,β-unsaturated β,β-dialkylesters and -amides have been conducted with NaBH$_4$ in the presence of catalytic amounts of CoCl$_2$ and hemicorrin **2.11** [576, 577].

Reduction of ketones in homogeneous medium in the presence of cyclodextrins gives disappointing results. However, inclusion of pyridine-borane into crystalline β-cyclodextrins generates complexes that reduce prochiral ketones or activated C=C bonds with good enantiomeric excesses [578, 579].

Chiral ate-complexes prepared from chiral trialkylboranes and *tert*-BuLi [580, 581, 582] can reduce prochiral ketones. Among these reagents, borohydrides **2.12** (R = Et or PhCH$_2$OCH$_2$CH$_2$) prepared from α-pinene or nopol give interesting selectivities, although only at very low temperatures (−78 to −100°C).

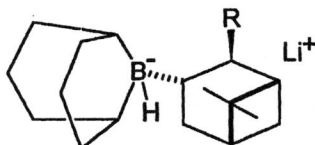

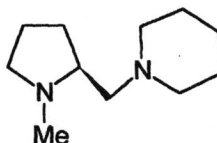

2.12   **2.13 (R = CH$_2$N(CH$_2$)$_5$)**

2.3 ALANES AND BORANES

The aluminum and boron atoms of these reagents are tricoordinated, so they exhibit electrophilic properties. Among the alanes, the most interesting reagent is a complex formed from i-Bu$_2$AlH, SnCl$_2$ and chiral diamine **2.13** (R = CH$_2$N(CH$_2$)$_5$) [87]. Chiral boranes have been widely developed, and they can be used in reductions and hydroborations [87, 169, 580, 583, 584]. Chiral allylboranes and enol boronates will be described later (§ 2.6 and 2.8).

Masamune and coworkers [212, 583] designed boranes **2.14** (R = H) bearing a C$_2$ axis of symmetry for asymmetric hydroboration. In the presence of catalytic amounts of **2.14** (R = MeSO$_3$), which behaves as a Lewis acid, boranes **2.14** (R = H) are very efficient reagents for asymmetric reductions of ketones [87]. A drawback to the use of boranes **2.14** (R = H) is their poor stability. In practice, they must be generated just prior to use from a precursor **2.14** (R = OMe).

By hydroboration of natural products such as α-pinene, H. C. Brown and coworkers have prepared mono- **2.15** (R = H) and diisopinocampheylboranes **2.16** (R = H). These reagents promote highly enantioselective hydroborations [580, 583]. The two α-pinene enantiomers are available, so both enantiomers of these reagents can be used. The intermediate di- or trialkylboranes formed in these hydroborations are treated with MeCHO. This forms a chiral boronate **2.17**, and the α-pinene is freed for recovery and recycling. From **2.17**, it is possible to obtain many functionalized compounds. Additionally, new chiral boranes **2.18** are available, and these are precursors of many chiral compounds bearing the R* group [169, 580, 583, 585-588] (Figure 2.2).

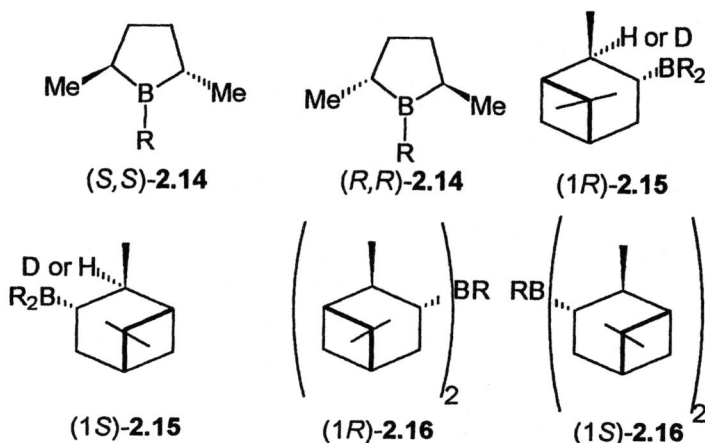

(S,S)-**2.14**   (R,R)-**2.14**   (1R)-**2.15**

(1S)-**2.15**   (1R)-**2.16**   (1S)-**2.16**

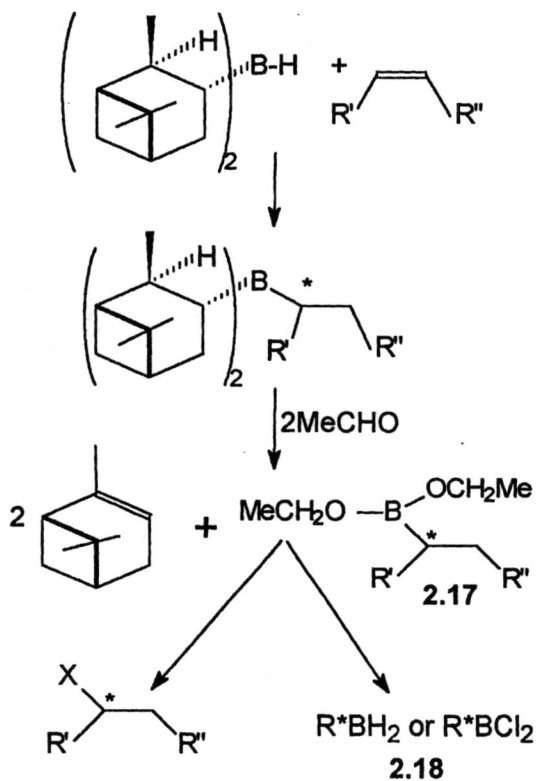

Figure 2.2

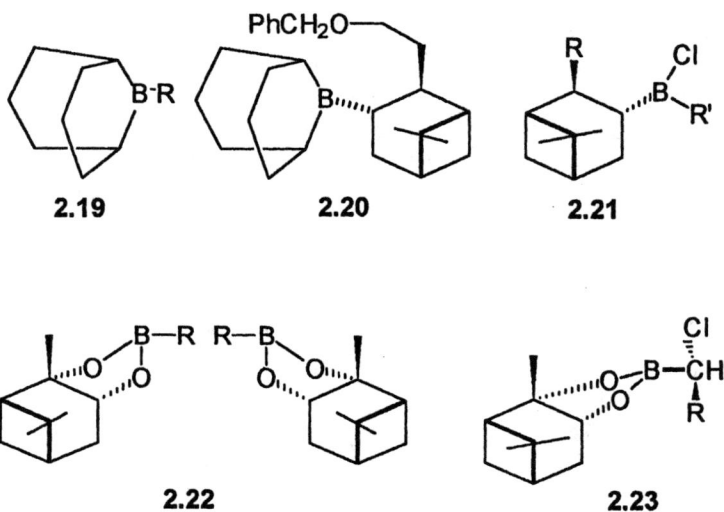

2.19   2.20   2.21

2.22   2.23

Another interesting reducing agent, Alpine-Borane [584] **2.15** (R_2 = 1,4-bi-cyclooctyl), is prepared from 9-BBN **2.19** (R = H) and α-pinene. This reagent is relatively unreactive, but it reacts with most ketones without solvent, occasionally under pressure. Two other reagents have been recommended as more reactive substitutes for Alpine-Borane: NB-enantrane **2.20**, or better yet, diisopinocampheyl chloroborane **2.16** (R = Cl) (DIP chloride). This last reagent is the most reactive, and it is easily prepared from (1*R*)- or (1*S*)-**2.16** (R = H) so that both enantiomers are available. Highly enantioselective reductions of saturated and unsaturated ketones to secondary alcohols can be performed with **2.16** (R = Cl) at –25°C [584, 589, 590, 591]. The mechanism of this reduction involves a hydride transfer to form a borate, from which the secondary alcohol is generated either by action of H_2O_2/NaOH or by displacement with ethanolamine. Deuteroboration of α-pinene gives deuteroborane **2.15** (D), which reduces aldehydes to deuterated primary alcohols with excellent enantioselectivities [584].

Bulky chloroboranes such as **2.21** (R = Me, Et, $PhCH_2OCH_2CH_2$, R' = *tert*-Bu) or analogs of **2.16** (R = Cl) can reduce even slightly unsymmetrical ketones [589, 592, 593], but these reactions are very slow. Chiral boronates **2.22** have been prepared by Matteson [594] by oxidation of α-pinene followed by trans-esterification. The reaction of **2.22** with $LiCHCl_2$ at low temperature forms chiral ate-complexes, which are transformed into α-chloroboronates **2.23**. These chloroboronates are precursors of numerous chiral compounds; however, their use is limited by the low temperatures required to generate them (–78 to –100°C) [595].

Chiral aminoboronates may be obtained from enantiopure aminoalcohols. In early work, Itsuno recommended the use of borane in the presence of aminoalcohol **2.24** for asymmetric reductions of ketones [87, 596]. Corey and coworkers later showed that this reducing reagent is an ate-complex of borane, and they introduced a new chiral 2,5-oxazaborolidine **2.25**. These authors broadly developed the chemistry of similar reagents [87, 597–601], and they showed that chiral 2,5-oxaza-borolidines could be used in catalytic amounts along with a slight excess of borane for enantioselective reductions of ketones. The catalytic cycle is shown in Figure **2.3** for the chiral oxazaborolidine **2.26**. The coordination of the ate-complex to the carbonyl compound takes place so as to minimize steric crowding. Hydride trans-fer from BH_3 leads to a chiral boronate **2.27**, which is displaced by excess BH_3 to provide a borate ester and regenerate **2.26**. For asymmetric induction to take place, the ate-complex must be more reactive than BH_3. In some cases, the less reactive catecholborane must be used instead of BH_3 as coreagent. The structure of the ate-complex has been determined by X-ray crystallography [603, 604] and a large-scale process for the preparation and use of **2.26** (R = Me, Ar = Ph) in enan-tioselective reductions of prochiral ketones has been published by a Merck group

[604]. A procedure for *in situ* formation and use of these reagents was also recently published [605]. Among all the oxazaborolidines that have been studied [35, 601], ate-complexes of bicyclic compounds **2.26** (R = Me, *n*-Bu or Ph, Ar = Ph or 2-Np) have shown the greatest efficiency. These reagents are easily obtained from (*R*)- or (*S*)-proline or from pyroglutamic acid, and they are used with BH_3 or catecholborane as the coreagents [87, 604, 606-612]. The reduction of many prochiral ketones or α-enones occurs with high enantiomeric excesses at room temperature or at 0°C [35, 598-601, 605-608, 613, 614]. The reduction of aldehydes by deuterocatecholborane occurs with satisfactory enantiomeric excesses at −120°C. After the reaction, the aminoalcohol precursor of the oxazaborolidine is easily recycled by precipitation of its hydrochloride salt [606, 607]. Among the other aminoalcohols that have been used as precursors of 1,3-oxazaborolidines are (*S*)-prolinol **1.64** (R = CH_2OH) [615], aminoalcohol **2.28** [616], indolinemethanols **2.29** [613], methionine-derived aminoalcohols [617], and (1*S*, 2*R*)-2-amino-1,2-diphenylethanol **1.58** (and its enantiomer) [614]. Chiral phosphine-boranes [601, 618, 619] and β-hydroxysulfoximine-borane complexes [620, 621] have also been shown to reduce ketones enantioselectively. These reagents have also been used to reduce *N*-aryl imines or oxime ethers into chiral amines [35, 601, 622], or to transform appropriate lactones into functionalized biaryls [623].

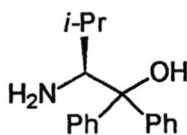

2.24

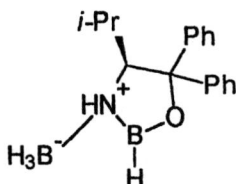

2.25

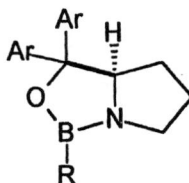

(*R*)-2.26

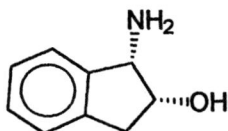

2.28

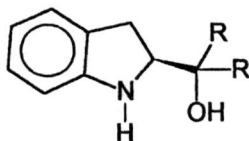

2.29

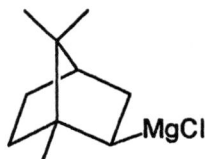

2.30

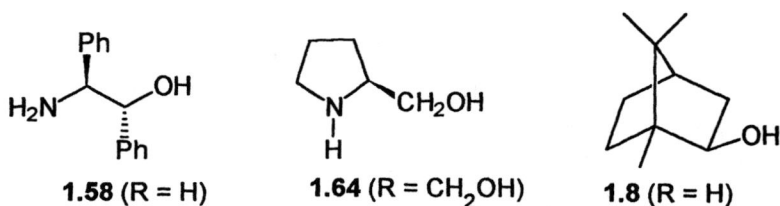

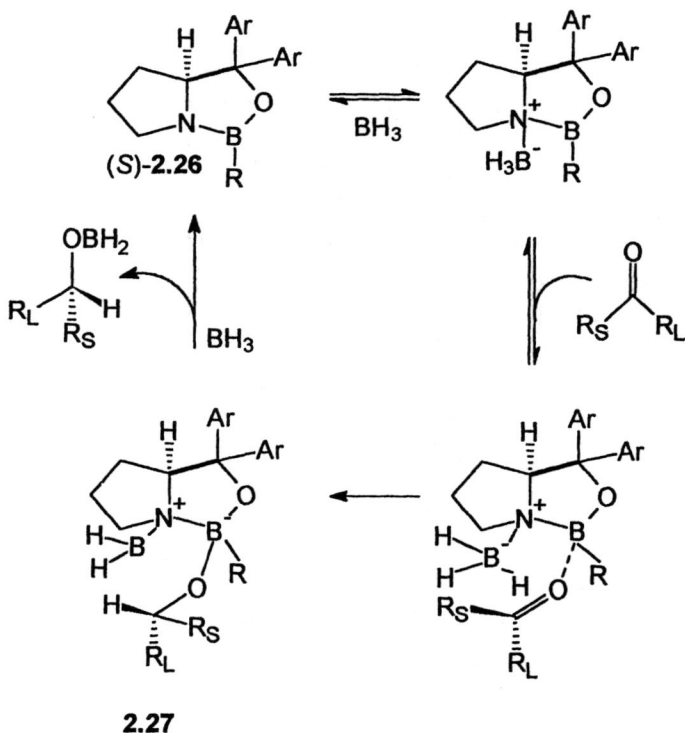

1.58 (R = H)   **1.64 (R = CH$_2$OH)**   **1.8 (R = H)**

Figure 2.3

2.4 OTHER HYDRIDE DONORS

Among the other hydride donors [591], organometallic reagents bearing a C–H bond in the β-position can be used for reduction of carbonyl compounds. In such reactions, direct addition of the organometallic substituents to the carbonyl group must be avoided. Meerwein-Verley-Pondorf reductions [624b] have also been studied, as have reductions based on models of NADH [625].

2.4.1 Organometals, Alcoholates, Amides

β-Branched organomagnesium, -zinc or -aluminum reagents, such as **2.30**, are able to transfer a hydride, and thereby reduce ketones or electrophilic double bonds. But these reactions usually take place with poor enantioselectivity [169, 551]. Similarly, Meerwein-Verley-Pondorf reductions of phenylalkylketones by the dichloroaluminum alcoholate of isoborneol **1.8** (R = H) give disappointing selectivities [551]. An improvement to this methodology has recently been described by Evans and coworkers, who introduced a new samarium catalyst (§ 3.2.5) [626]. The use of branched metal amides for similar purposes is also disappointing [169, 217a, 557].

2.4.2 Dihydropyridines [625, 627]

Enzyme-catalyzed oxido-reductions take place via the NADH/NAD⁺ couple, and biomimetic reduction of ketones by models of this system have been studied. The mechanism of an NADH reduction is summarized in Figure **2.4**. According to the specific enzyme, either the pro-(R) or pro-(S) hydride of the nicotamide skeleton is transferred. In the absence of an enzyme, some chiral 1,4-dihydropyridines will reduce activated ketones including α-ketoesters RCOCOOR' (R' = Me, Et), trifluoroacetophenone or 2-acetylpyridine with high enantiomeric excesses [625]. Several 1,4-dihydronicotinamides **2.31** (R = H or Me) have been used; the amide residues can either be chiral, derived from phenethylamine or (S)-prolinamide **1.64** (R = CONH₂), or achiral (R" = R'" = Me) [625, 628, 629]. The reagent **2.31** (R = R' = R" = R'" = Me) described by Vekemans [628, 629] is more reactive than the others, and this can even reduce unactivated ketones and imines [629]. Other 1,4-dihydronicotinamides such as **2.32**, **2.33** or **2.34**, also give useful results. Two other 1,4-dihydropyridines **2.35** (A = CH₂OH or S(O)Tol), which do not bear an amide group, also lead to good selectivities in ketone reductions. In all these nonenzymatic reactions, catalysis by a Lewis acid such as Mg(ClO₄)₂ is necessary. The use of Hantszch esters **2.35** (A = COOR') derived from carbohydrates gave disappointing results [630, 631].

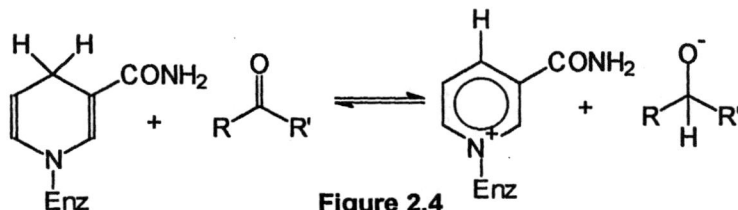

Figure 2.4

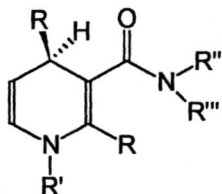

2.31

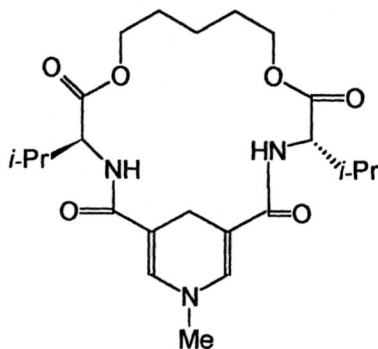

2.32

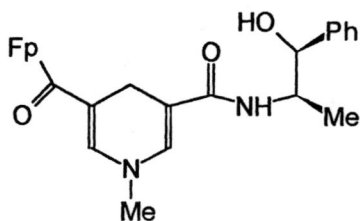

2.33

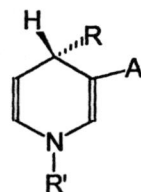

2.35

Fp = (η5-C5H5)Fe(CO)(PPh3)

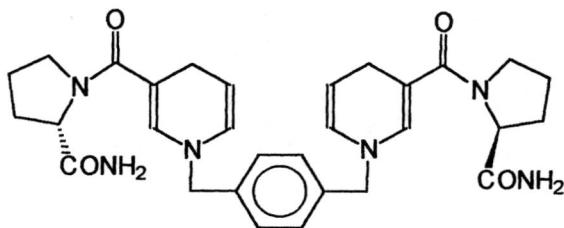

2.34

2.5 ORGANOMETALLIC REAGENTS

Organometallic reagents can be rendered chiral by coordination of a chiral ligand on the metal, or by covalent linkage of a protic chiral auxiliary (amine or alcohol) to the metal. In the second case, the auxiliary usually bears other basic sites that can also coordinate to the metal. In this chapter, nonfunctionalized organometallic reagents will be described; enolates are considered separately (§ 2.6).

2.5.1 Organolithium, Organomagnesium, Organozinc Reagents

Many chiral ligands have been introduced onto organolithium or organomagnesium reagents with the goal of inducing asymmetry in additions to saturated or α,β-unsaturated carbonyl compounds or imines [169, 253, 557, 559]. Unfortunately, most of these attempts have given disappointing results. However, Mukaiyama and Cram observed high enantioselectivities when adding alkyllithium or -magnesium reagents to benzaldehyde in the presence of aminoalcoholate **2.36** or diamine **2.37** at very low temperature [169, 559, 632]. Recently, Koga and coworkers used diamine **2.38** as a ligand for additions of organomagnesium reagents to aldehydes [633, 634]; the enantiomeric excesses observed are always lower than 75%. A mixed organometallic reagent generated from Et₂Mg and lithium binaphthoxide adds to aldehydes with a good enantioselectivity at −100°C [110]. The reactions of RLi with saturated or α,β-unsaturated imines in the presence of diether **2.39** [635] or aminoether **2.40** [636, 637, 638] give useful results. However, the use of sparteine **2.5** as a ligand for RLi in the asymmetric ring opening of oxabicyclic compounds leads to unsatisfactory selectivities [639].

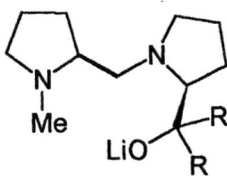

2.36

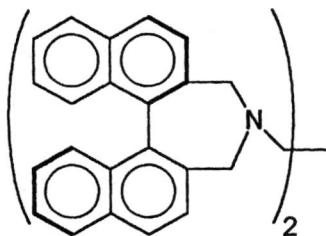

2.37

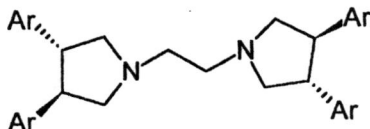

2.38

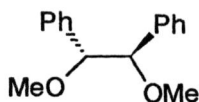

2.39

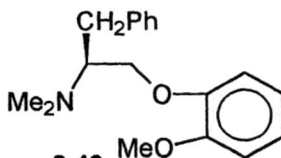

2.40

The use of organozinc reagents modified by aminoalcohols is a widely developed powerful method for asymmetric alkylation [110, 559, 571]. Dialkylzinc compounds (R₂Zn) do not react with aldehydes at room temperature; however, addition does occur in the presence of catalytic amounts of aminoalcohols [110]. Since this discovery by Oguni and Omi in 1984, many efficient systems have been developed by Noyori, Oguni, Soai and their coworkers. These methods allow the highly enantioselective addition of dialkylzincs to aldehydes [110, 640]. Noyori has shown that the chiral reagent is a binuclear zinc complex **2.41**. The system responds to the constraints of the catalytic cycle shown in Figure 2.5. The primary reaction product is a mixed binuclear zinc alcoholate **2.42**, which regenerates the complex **2.41** in the presence of excess R₂Zn. The driving force of the catalytic cycle is the irreversible formation of tetrameric zinc alcoholate **2.43**. Among the catalysts recommended, *N,N*-dialkylnorephedrines **1.14** (R = *n*-Bu), aminoalcohol **2.45**, (*S*)-prolinol derivative **2.13** (R = CPh₂OH) and 3-*exo*-dimethylamino-isoborneol **1.8** (R = NMe₂) and its enantiomer lead to highest enantioselectivities in reactions of dialkylzincs with aliphatic, aromatic or α,β-unsaturated aldehydes [110, 111, 640, 641]. These aminoalcohols can be supported on polymers so that their recycling is easier [110, 642, 643].

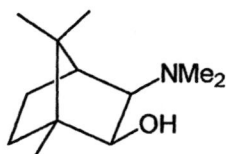

1.8 (R = NMe₂)

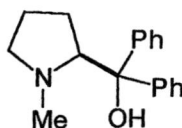

2.13 (R = CPh₂OH

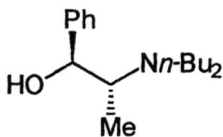

(1*S*,2*R*)-**1.14**

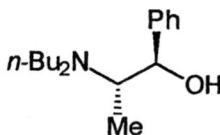

(1*R*,2*S*)-**1.14**

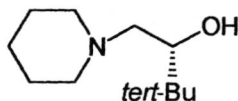

2.45

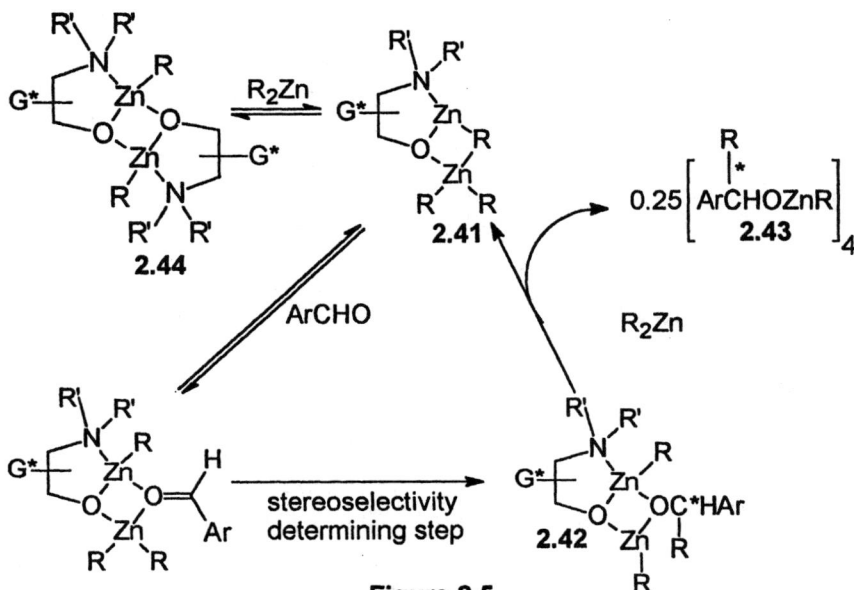

Figure 2.5

2,2-Bis-pyridine-derived aminoalcohols have recently been proposed as zinc ligands [644], as have sulfur derivatives of ephedra alkaloids [645], but the scope of their use is more restricted.

A remarkable feature of these processes is asymmetric amplification. Even when an aminoalcohol of low enantiopurity is used as a catalyst, the resulting secondary alcohol still often displays a high enantiomeric excess. This amplification has been interpreted [110] by differing relative stabilities of the two dimeric reagent precursors **2.44**. The two aminoalcohols present in these dimers can have the same or different absolute configurations (homo- or heterochiral dimers). Homochiral dimer **2.44** is less stable than its heterochiral analog, as shown by X-ray crystallography when the aminoalcohol is **1.8** (R = NMe$_2$). Therefore, the homochiral dimer is more easily dissociated by R$_2$Zn (Figure 2.5) leading to the reactive complex **2.41**. The minor enantiomeric ligand remains largely sequestered in the heterochiral dimer. The enantioselectivity of the reaction is thus considerably higher than in the case where both dimers would be equally dissociated.

Lithium aminoalcoholates have also been used as catalysts for additions of R$_2$Zn to aldehydes. In these reactions, the reagents are binuclear lithio-zinc complexes [646, 647]. Chiral diamines such as piperazine **2.46** or the derived lithium amides [110, 648] or else 2-[2-pyridyl]-pyrrolidine **1.64** (R = 2-Py) catalyze the reaction of dialkylzincs with aromatic aldehydes [649].

This method has been extended to the reaction of PhC≡CZnBr with aldehydes in the presence of stoichiometric amounts of the lithium alcoholate of **1.14** (R = Me). Divinylzincs (RCH=CH)$_2$Zn react under similar conditions, or also in the presence of other lithium alcoholates [650]. Under precise experimental conditions, PhMgBr adds to aliphatic or aromatic aldehydes in the presence of zinc salts and **1.14** (R = *n*-Bu). The enantiomeric excesses in these additions are higher than 75%. Diarylzincs react with aldehydes in the presence of aminoalcohols bearing a ferrocene skeleton **2.47** with a very high enantioselectivity [651, 652]. Schiff bases have also been used as catalysts in such reactions [367], as have some titanium complexes (see below) [559, 653, 654, 655].

Conjugate additions of Et$_2$Zn to α-enones catalyzed by aminoalcohol **1.14** (R$_2$ = (CH$_2$)$_5$) are highly enantioselective [655a]. Lower selectivities are observed in reactions of α-enones with RMgX catalyzed by chiral aminoalcohol-zinc complexes or with organozinc reagents in the presence of **1.14** (R = *n*-Bu) and Ni(acac)$_2$, or with sulfoximines [656, 657, 658]. When alcohol **1.8** is used as the catalyst in the presence of 2,2'-bipyridine, conjugate additions of R$_2$Zn to α-enones give more satisfactory results [659]. Diethyl tartrate is a good ligand for zinc in enantioselective Simmons-Smith cyclopropanation of allylic alcohols [660] with Et$_2$Zn/CH$_2$I$_2$.

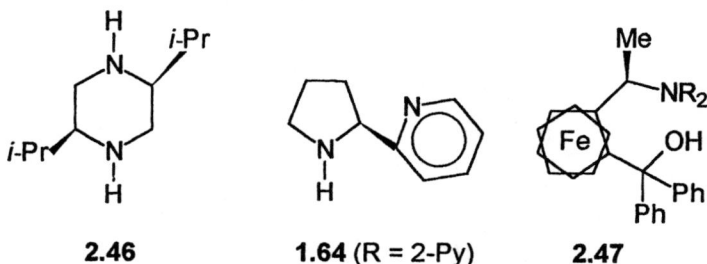

2.46   **1.64 (R = 2-Py)**   **2.47**

2.5.2 Organotitanium, Organocopper, Organolanthanide Reagents

The development of the chemistry of organotitanium reagents by Seebach [661, 662] and Reetz [97] encouraged the design of a chiral environment about this metal to induce enantioselective additions of aryl- or methylorganotitanium reagents to aldehydes. Chiral titanium reagents generated *in situ* from (*R*)- or (*S*)-binaphtol **1.44**, (*i*-PrO)$_3$TiCl and ArMgBr give highly enantioselective additions to ArCHO [110, 253, 663]. Reagent **2.48** includes an *N*-sulfonyl-norephedrine **1.61** skeleton, and its reactions with aromatic aldehydes yield good enantioselectivies; however, with aliphatic aldehydes, poor enantioselectivities are found [110, 664]. Enantioselective additions of allyltitanium reagents to aldehydes have been developed by Duthaler, Riediker and coworkers [358, 665, 666, 667]. The chiral reagents are generated by exchange between allylmagnesium or allyllithium reagents and cyclopentadienyldialkoxychlorotitanium **2.49**. The chiral alkoxy groups of **2.49** are derived either from diacetone glucose **1.48** or from diol **2.50**, easily prepared from (R,R)-tartaric acid **2.2** (R = H). The reactions of these two reagents with aldehydes lead to enantiomeric homoallylic alcohols with high enantioselectivities (ee 85 - 90%) [665, 666]. The reactions of other allyltitanium complexes bearing a single chiral alkoxy group or formed from other carbohydrates lead to lower selectivities [665, 667].

Related reactions also have been performed with crotyl analogs. Independent of the geometry of the starting crotyl chloride, the E-titanium species **2.51** is formed. In **2.51**, the titanium is bonded to the unsubstituted carbon [667] (Figure **2.6**). The reactions of complex **2.51** with linear or α-branched aldehydes are highly enantio- and (if possible) diastereoselective (ee > 94%, de > 95%). The chiral auxiliaries are recovered after the reaction by precipitation from hexane. Other titanium complexes **2.52** [260, 668] and **2.53** [654] have been used to promote the reaction of dialkylzincs [559] with aromatic [653, 654, 655, 669, 670], aliphatic or α,β-unsaturated aldehydes [653, 654, 655] in the presence of stoichiometric amounts of Ti(Oi-Pr)$_4$. Dialkylthiophosphoramidates **2.54** derived from norephedrine also catalyze the reaction of Et$_2$Zn with ArCHO in the presence of Ti(Oi-Pr)$_4$, and secondary alcohols are obtained with high enantiomeric excesses [671]. These reactions can be extended to organozinc reagents bearing functional groups by using the catalyst **2.52** [672]. The reagents in these reactions are probably binuclear titanium-zinc complexes. The enantioselectivities observed in such processes are much higher than those in reactions catalyzed by mixed titanium-lithium, -magnesium, or -lithiocuprates complexes [559, 673].

Homoallylic alcohols can be obtained with a high enantioselectivity by reacting aldehydes with a titanium complex generated from E-MeCH=CHCH$_2$OCO Ni-Pr$_2$ in the presence of sparteine **2.5** [109].

Lithium cuprates R$_2$CuLi, magnesium cuprates R$_2$CuMgX or high-order cuprates [624a] are key reagents in the arsenal of organic synthesis. Therefore different chiral ligands of copper have been examined to induce asymmetric additions of these reagents to α-enones [112, 173]. Among the reagents investigated to date, diaminoalcohol **2.55**, aminoether **1.64** (R = CH$_2$OMe), aminoalcohol **2.13** (R = CH$_2$OH) [559], and aminoalcohols bearing the bornane skeleton **2.56** (X = NMe or S) [674, 675] display the best efficiencies. Diamine **2.57** gives interesting results [676] as do the phosphorus-containing ligands **2.58**, recently proposed by Alexakis [677, 678] and **2.59**, proposed by Koga [679]. Chiral organocerium reagents formed *in situ* from (R)-binaphthol **1.44** and R$_3$Ce react at

−100°C with aromatic aldehydes to give secondary alcohols with good enantioselectivity [680]. Ytterbium analogs seem less useful.

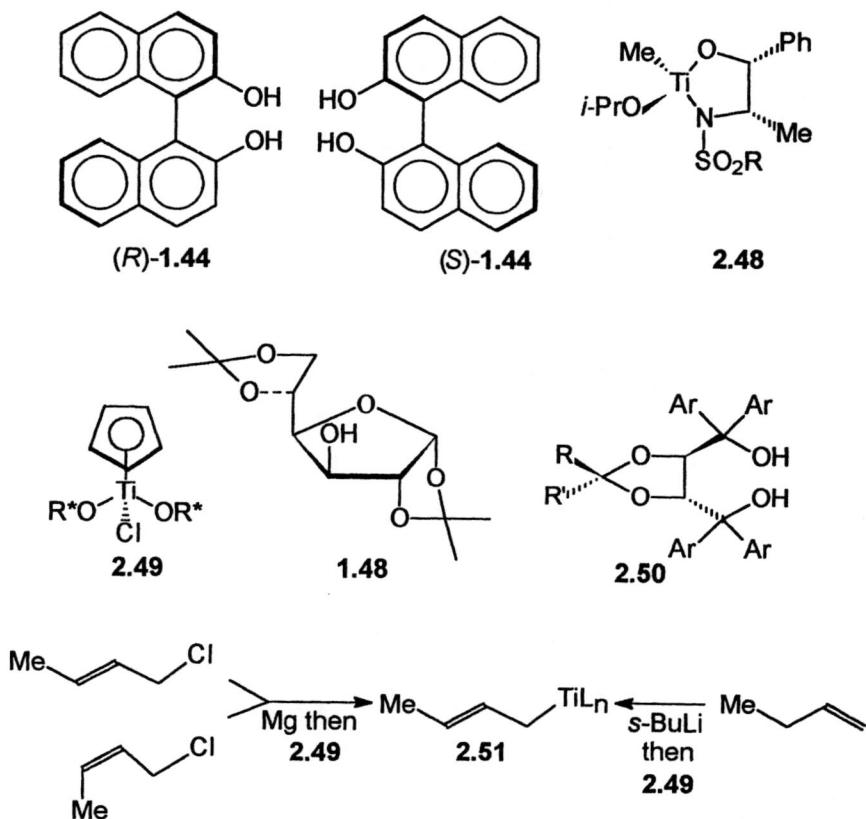

Figure 2.6

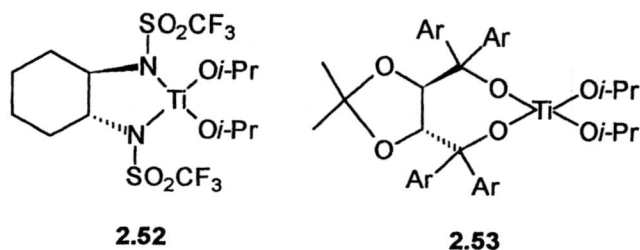

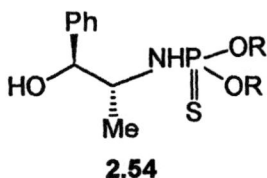

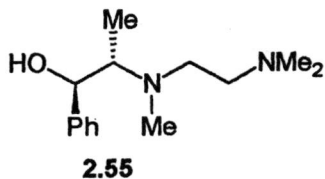

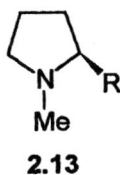

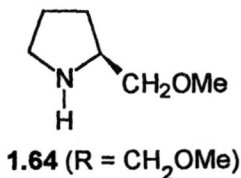

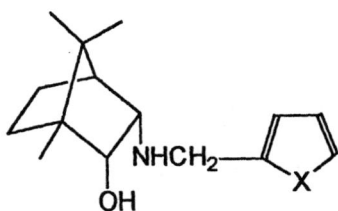

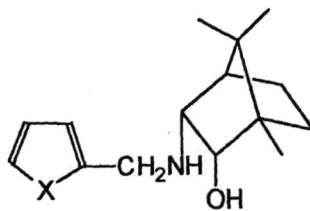

2.54

2.55

2.13

1.64 (R = CH₂OMe)

2.56

2.57

2.58

2.59

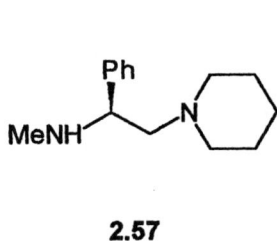

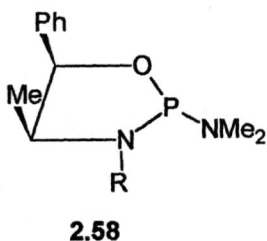

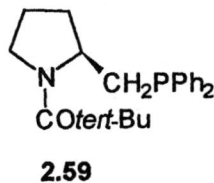

2.6 ENOLATES AND RELATED REAGENTS

2.6.1 Lithium and Zinc Enolates [209]

Lithium enolates of ketones and esters can be generated by the action of chiral lithium amides. If the base is used in stoichiometric amounts, the lithium cation of the enolate bears the chiral amine as a ligand. If the amide is used in excess, chiral mixed aggregates can be formed [77, 557, 558, 559]. These lithium enolates participate in alkylations, aldol reactions, and Michael reactions with moderate asymmetric induction [558]. However, Koga and coworkers have performed highly enantioselective aldol reactions of aromatic aldehydes with acetophenone lithium enolate in the presence of the lithium amide of **2.4** (R = c-C$_6$H$_{11}$, X = NMe). The same authors developed the asymmetric alkylation of aggregates of LiBr and lithium enolates of cyclic ketones in the presence of **2.4** (R = MeOCH$_2$CH$_2$OCH$_2$CH$_2$, X = CH$_2$) [108, 209, 559].

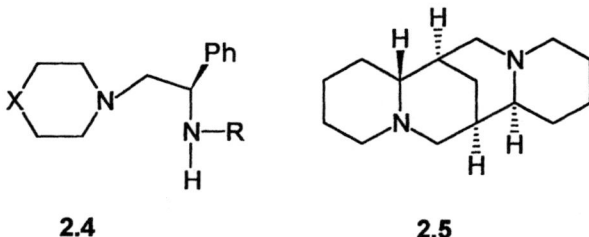

2.4   **2.5**

The Reformatsky reaction run in the presence of sparteine **2.5** is highly enantioselective, but its chemical yield is low [160, 253, 681].

2.6.2 Tin, Titanium and Other Transition-Metal Enolates [408]

Tin (II) enolates have been widely used by Mukaiyama and coworkers in aldol and Michael reactions. These enolates are generated by action of Sn(OTf)$_2$ on ketones, esters, N-acyl-1,3-oxazolidin-2-ones or -1,3-thiazolidin-2-thiones, thiolesters or dithioesters in the presence of chiral diamines **2.13** (R = CH$_2$NHAr or CH$_2$N(CH$_2$)$_5$) bearing the (S)-proline skeleton [253, 559, 682]. Diastereoisomeric and enantiomeric excesses observed in these reactions are often excellent.

Titanium enolates have frequently been used in aldol reactions [160, 253, 401, 403a, 404, 408, 426]. Chiral titanium complexes **2.49** have been employed by Duthaler, Riediker and coworkers [665, 666, 683] to generate Ti enolates bearing chiral ligands. Exchange of these complexes with lithium enolates of *tert*-Bu acetate or propionate, N-propionoyloxazolidinones or protected glycine esters is easily performed. Impressive enantioselectivities are obtained in aldol reactions when the two alkoxy groups of complexes **2.49** come from diacetoneglucose **1.48** [665, 666, 683].

2.6.3 Boron Enolates [407]

The powerful control of stereochemistry in aldol reactions that is available through boron enolates has been amply demonstrated [122, 160, 169, 253].

Hence, the control of absolute stereochemistry by the introduction of chirality in the vicinity of the boron atom is an attractive goal. E- or Z-Enol borinates **2.60** derived from α-pinene give highly selective condensations with aldehydes [684]. These borinates are generated by reacting chloroborane **2.16** (R = Cl) or triflate **2.16** (R = OTf) with methyl or ethyl ketones in the presence of tertiary amines [684, 685, 686]. Z-Enolborinates **2.60** can also be prepared by conjugate additions of chiral borane **2.16** (R = H) to α-enones [687, 688]. According to Reetz [689, 690] or Masamune [691], other chloroboranes **2.61** or boron triflates **2.14** (R = OTf) with a C_2 axis of symmetry [212] can be used to generate enol borinates **2.60** from ketones [689] or thiolesters. In the case of thiolesters, enolates Z-**2.60** (R' = SCEt$_3$) are formed [690, 691], and their reactions with aldehydes are particulary valuable. Unfortunately, enolborinates of esters or amides cannot be obtained by these methods [685]. Cleavage of the borate aldols formed in these reactions to β-hydroxycarbonyls is accomplished by alkaline H_2O_2, but the chiral auxiliary is not recovered.

Corey and coworkers have recently synthesized sulfonamidobromoboranes **2.62** (X = Br) bearing a C_2 axis of symmetry. From these reagents, boron enolates of diethylketone, phenylthiol acetate or thioacetate and *tert*-Bu or allyl propionates can be generated [601, 692-695], provided that the aryl group is properly substituted. By varying the ester substituents and the experimental conditions, both E- and Z-boron enolates **2.60** can be selectively formed. These reagents give highly stereoselective condensations with aldehydes [693, 694, 695] and imines [692], and they participate in asymmetric Ireland-Claisen rearrangements [696]. The chiral bis-sulfonamide precursor of reagents **2.62** can be recycled after workup. The use of boron azaenolates was recommended as early as 1981 by Meyers [697]. The selectivities of their aldol reactions were interesting, but the chemical yields were low.

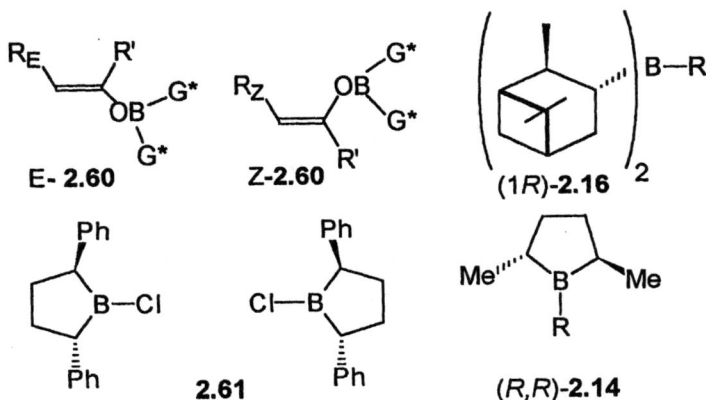

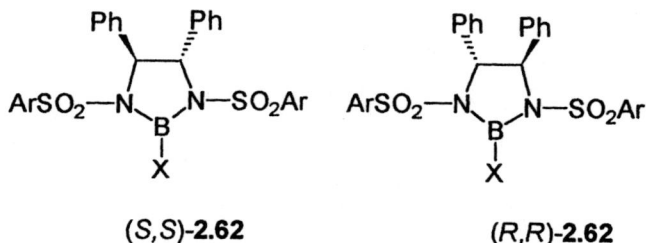

(S,S)-**2.62** (R,R)-**2.62**

2.7 ALLYL AND PROPARGYLBORANES, -SILANES AND -STANNANES

The reactions of many allylmetallic reagents are often poorly regioselective. In contrast, the reactions of allylboranes **2.63**, -silanes or -stannanes often lead to a single regioisomer. The reactions of these reagents with aldehydes lead to homoallylic alcohols **2.65** with double-bond migration (SE' process) (Figure 2.7). Lewis acid catalysis is necessary for the additions of silanes and stannanes [169, 253, 698], and similar reactions can be run with allenyl or propargyl analogs. In such systems, chirality may be introduced in two places:

• On the allylic carbon itself: during the reaction process there will be a chirality transfer between this site and the prochiral carbonyl carbon of the aldehyde.

• On the other substituents of the heteroatom (most common with boron derivatives).

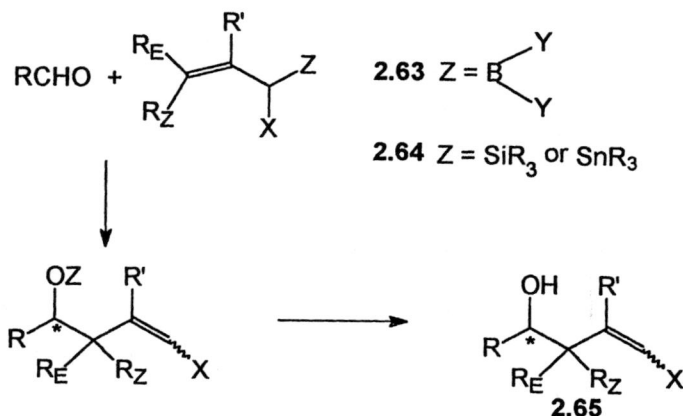

Figure 2.7

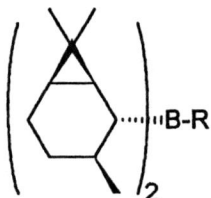 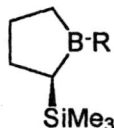 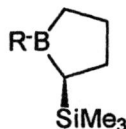

2.66   **2.67**

2.7.1 Allylboranes, -boronates and -boronamides [698]

The chemistry of chiral allylboranes has been mainly developed by Brown and coworkers [699-707]. The chiral boranes **2.16** (R = H) and **2.66** (R = H) can be obtained by hydroboration of α-pinene or 2-carene [707]. The corresponding allylboranes (R = CH_2=CHCH$_2$) are prepared *in situ* by sequential treatment with MeOH and CH_2=CHCH$_2$MgBr. Z- and E-Crotylboranes **2.16** and **2.66** (R = Z- or E-MeCH=CHCH$_2$) are similarly obtained by using an organometallic reagent formed from the corresponding but-2-ene and n-BuLi/*tert*-BuOK at −78°C [700, 701]. At higher temperatures, Z/E isomerization takes place. Other substituted allylboranes **2.16** (R = CH_2=C(Me)CH$_2$, Me$_2$C=CHCH$_2$, 3-cyclohexen-1-yl and higher homologs) have been generated in the same fashion [699, 705, 706]. The reactions of these allylboranes with aldehydes are highly stereoselective. After treatment of the initial adducts with alkaline H_2O_2, ethanolamine or 8-hydroxyquinoline [708], homoallylic alcohols **2.65** are obtained with an excellent enantio- and (if possible) diastereoselectivity [699, 702, 703, 705]. The chiral boranes can be recovered and recycled if the workup is performed with aminoalcohols [708]. Functionalized allylboranes **2.16** (R = Z-MeOCH=CHCH$_2$, E-*i*-Pr$_2$NMe$_2$SiCH=CHCH$_2$ or E-Ph$_2$NCH=CH-CH$_2$) have been prepared and used in asymmetric allylboration of aldehydes [709-712]. The availability of both enantiomers of **2.16** increases the usefulness of such reagents, although all these reactions must be performed at −78°C. Masamune and coworkers [713, 714] have proposed the use of allylboranes **2.14** and **2.67** (R = CH_2=CHCH$_2$, Z- or E- MeCH=CHCH$_2$, CH_2=C(Me)CH$_2$) for asymmetric allylborations. Homoallylic alcohols are obtained with high selectivities. The chemistry of chiral allylboronates has been simultaneously studied by the groups of Hoffmann [715, 716] and Roush [698]. The boronates **2.68** used by Roush and coworkers in asymmetric allylboration of aldehydes feature the tartaric acid skeleton, and isopropyl esters (R = *i*-Pr) are the most efficient reagents.

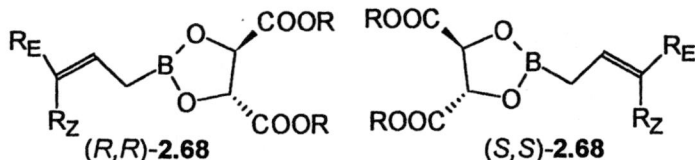

(R,R)-**2.68** (S,S)-**2.68**

These boronates are generated from allylmetals and isopropyl borate, followed by transesterification of the intermediate product with (R,R)- or (S,S)-diisopropyltartrate **2.69** (R = i-Pr). The reagent thus formed is used without any further purification [717, 718]. The Z- or E-crotylboronates **2.68** (R_Z or R_E = Me) do not interconvert even at room temperature, and this stability makes these reagents very practical. The reactions of boronates **2.68** with aliphatic aldehydes are highly stereoselective [717-721]. Aromatic or α,β-unsaturated aldehydes give less satisfactory results, but their temporary transformation into chromium complexes improves the stereoselectivity [722]. Functionalized boronates E-**2.68** (R_E = $Me_2C_6H_{11}OSi$, $PhMe_2Si$, R_Z = H) have also been prepared and used in allylboration of aldehydes [723, 724]. The workup of these reactions is a simple hydrolysis, and the inexpensive chiral reagents are not usually recovered.

A class of allylboronates **2.70** (X = O, R = Ph) [147, 715] and **2.70** (X = NSO_2Me, R = H) [725] bearing the bornane skeleton has been introduced by Hofmann and Reetz. However, because the access to these reagents is rather difficult, they have not yet seen many applications. Hofmann [715, 716] has examined the scope of reactions of α-substituted allylboronates **2.71** derived from chiral diols. The reagents with R = c-C_6H_{11} are the most interesting [715, 716, 726, 727, 728]. As indicated in Figure **2.8**, these boronates are prepared from dichloroboronates **2.72** either by reaction with CH_2=$CHMgCl/ZnCl_2$ or by sequential addition of $MeLi/ZnCl_2$ followed by a vinylmagnesium or -lithium reagent [716, 726, 728, 729, 730]. While the reaction of the unsubstituted analogs **2.71** (X = H) with aldehydes is poorly stereoselective [729], allylborations of aldehydes by **2.71** (X = Cl, Me) or related reagents **2.73** and **2.74** [715, 731, 732, 733] give very high diastereoselectivities. A limitation of the method is the low temperature ($-100°C$) required to generate reagent **2.72**.

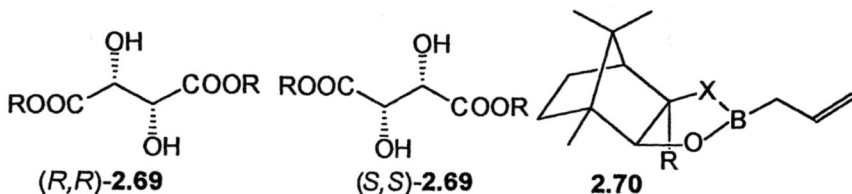

(R,R)-**2.69** (S,S)-**2.69**   **2.70**

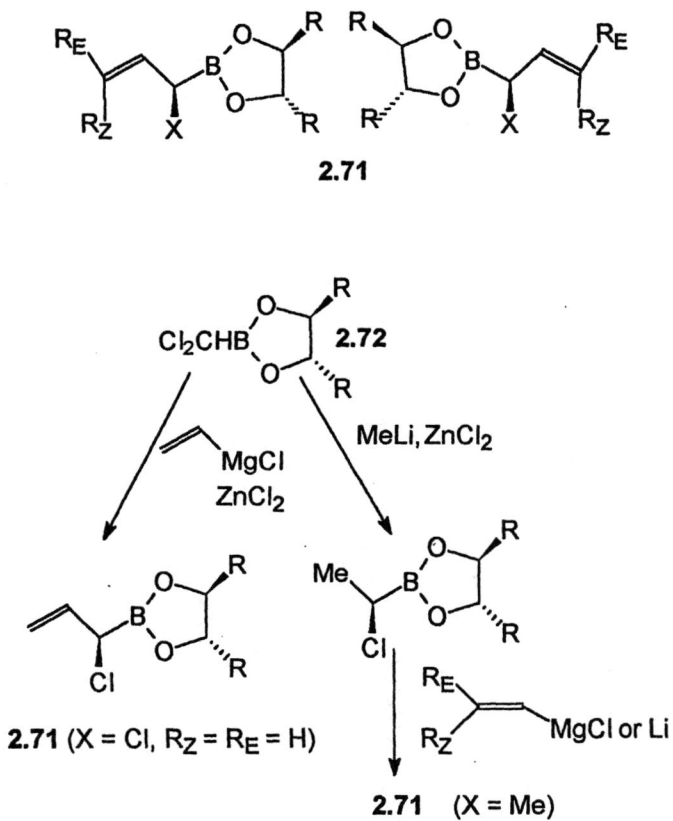

Figure 2.8

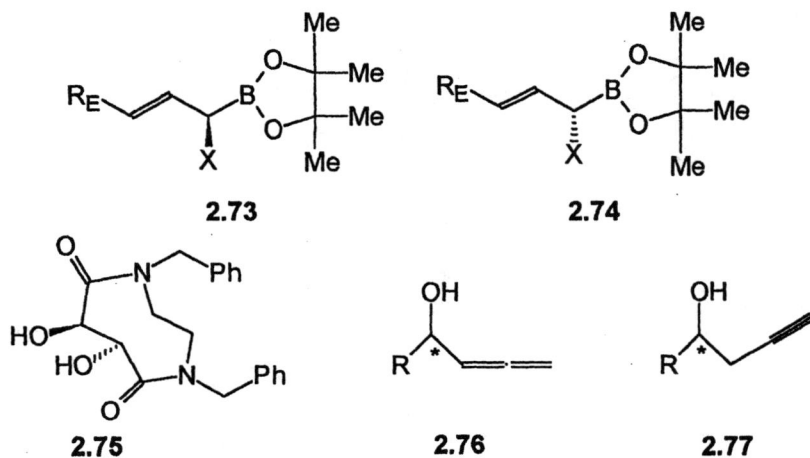

$$\underset{(S,S)\text{-}\mathbf{2.62}}{\text{ArSO}_2-N\underset{\overset{|}{X}}{\overset{\text{Ph}\quad\text{Ph}}{\underset{B}{\diagdown\diagup}}}N-\text{SO}_2\text{Ar}}\qquad\underset{(R,R)\text{-}\mathbf{2.62}}{\text{ArSO}_2-N\underset{\overset{|}{X}}{\overset{\text{Ph}\quad\text{Ph}}{\underset{B}{\diagdown\diagup}}}N-\text{SO}_2\text{Ar}}$$

(S,S)-**2.62**   (R,R)-**2.62**

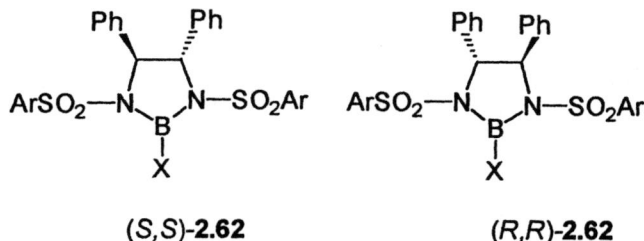

The disappointing results obtained by Roush and coworkers in allylboration of aromatic aldehydes led these authors to design and synthesize a chiral tartramide **2.75**, from which a new allylboronate was prepared. This reagent gives more stereoselective allylborations with aldehydes, but its reactions are extremely slow.

Corey and coworkers [734] prepared allylboronamides **2.62** (Ar = Tol, X = CH$_2$=CHCH$_2$, CH$_2$=C(Cl)CH$_2$, CH$_2$=C(Br)CH$_2$) from the bromoboranes **2.62** (Ar = Tol, X = Br) and the corresponding allyltin reagents CH$_2$=C(X)CH$_2$SnBu$_3$. Homoallylic alcohols are easily prepared from these reagents with a high enantiomeric excess. Adding to the attractiveness of this method, both enantiomers of **2.62** are available and the bromoborane precursor is recoverable. Reaction of bromoborane **2.62** (Ar = Tol, X = Br) with allenyltin CH$_2$=C=CHSnBu$_3$ leads to propargylboranes **2.62** (Ar = Tol, X = HC≡CCH$_2$). *In situ* reactions of these reagents with aldehydes give α-allenyl alcohols **2.76** with high enantiomeric excesses [423, 735]. Similar reactions can be performed starting from CH$_2$=C=C(Et)SnPh$_3$ or HC≡CCH$_2$SnBu$_3$ [735]; chiral homopropargylic alcohols **2.77** can be prepared in the latter case. Chiral homopropargylic alcohols **2.77** were earlier prepared by Yamamoto and coworkers [169] from aldehydes and allylboronates formed from CH$_2$=C=CHB(OH)$_2$ and dialkyltartrates **2.69**.

2.7.2. Allylsilanes and Allylstannanes

The reactions of allylsilanes and -stannanes with electrophiles in the presence of Lewis acids occur with clean allylic inversion [624d]. If the allylic carbon bearing the silicon or tin atom is chiral, then a good chirality transfer can be expected. For example, the reactions of chiral silane **2.78** (R$_E$ = Me, R$_Z$ = H) with aldehydes or of **2.78** (R$_Z$ or R$_E$ = Me) with *tert*-BuCl in the presence of TiCl$_4$ [54, 736] are highly selective. Less efficient chirality transfers are observed with other electrophiles [736]. Introduction of chirality on silicon gave disappointing results [737]; however, regio- and stereoselective alkylation of the lithium carbanion of **2.79** has been observed [738]. Enantiopure α-alkoxystannanes **2.80** are prepared by reductions of acylstannanes with (R)- or (S)-Binal **2.8** [569, 570], followed by etherification. The reactions of these alkoxystannanes with aldehydes in the presence of

BF$_3$•Et$_2$O are diastereo- and enantioselective if R' = Me and R" = PhCH$_2$OCH$_2$ [739]. When R' = H, good results are obtained only with aromatic aldehydes [740]. Other chiral allylstannanes **2.81** can be prepared by BF$_3$•Et$_2$O-mediated isomerization of **2.80** (R' = H, R" = MeOCH$_2$). The reactions of these allylstannanes with aldehydes can give valuable selectivities [569].

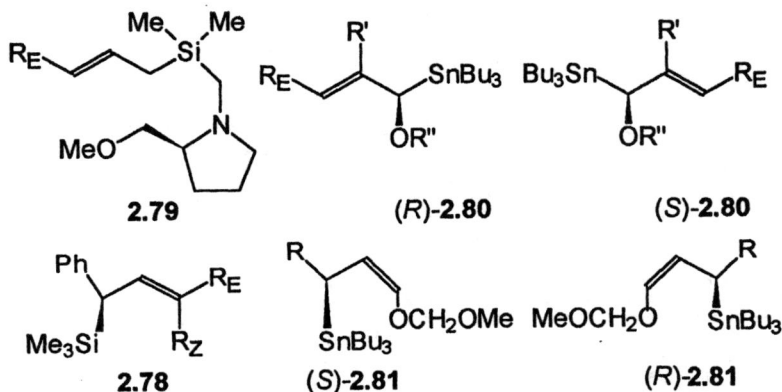

2.79   (R)-**2.80**   (S)-**2.80**

2.78   (S)-**2.81**   (R)-**2.81**

2.8 OXAZIRIDINES

Davis and coworkers have shown that *N*-sulfonyl oxaziridines are potent oxidants [741]. In order to induce asymmetry, these authors prepared rigid, bulky oxaziridines bearing the bornane skeleton. Oxaziridines **2.82** (X = H, Cl or Me) are very interesting reagents for the asymmetric oxidation of alkali metal enolates to α-hydroxyketones [147, 742-746]. In some cases, the presence of chlorine or methoxy substituents on reagents **2.82** is necessary to observe a high asymmetric induction [747, 748]. Both enantiomers are available, and these chiral reagents may be recycled.

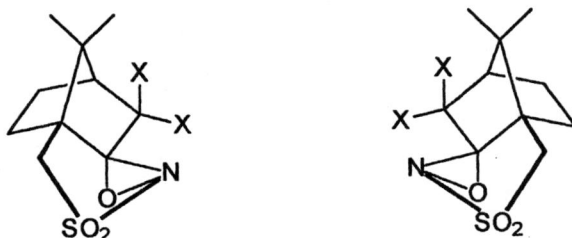

2.82

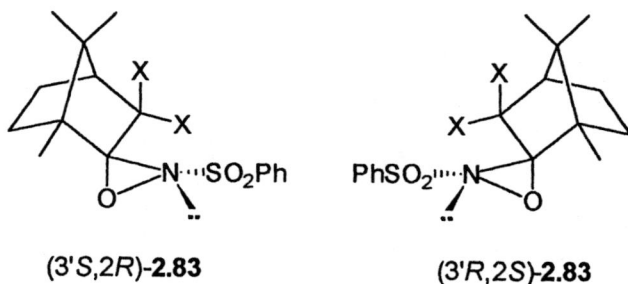

(3'S,2R)-**2.83**   (3'R,2S)-**2.83**

Related reagents **2.83**, obtained from camphor imine, are useful for asymmetric oxidation of sulfides and selenides to chiral sulfoxides and selenoxides [506, 749]. To obtain high enantioselectivities, the substituents on the sulfur or selenium must be of a sufficiently different size.

2.9 ASYMMETRIC OSMYLATION [750, 751, 752]

Osmium tetroxide (OsO$_4$) is the most popular reagent for effecting the *syn* dihydroxylation of olefins. The use of chiral ligands on osmium leads to highly enantioselective dihydroxylations. The use of stoichiometric amounts of OsO4 in the presence of chiral tertiary diamines such as **2.38**, **2.84** or **2.85** allows dihydroxylation of stilbene with high enantioselectivities [559]. Under the same conditions, the secondary diamine **2.86**, proposed as a ligand for osmium by Corey [559], also gives impressive results in dihydroxylation of numerous olefins. However, due to its cost and its toxicity, OsO$_4$ is better used in catalytic amounts along with co-oxidants such as amine oxides. Sharpless and coworkers [753-756] have extended this popular catalytic method to asymmetric dihydroxylation. The chiral ligands for osmium are derivatives of dihydroquinine **2.87** and dihydroquinidine **2.88,** which lead to enantiomeric products.

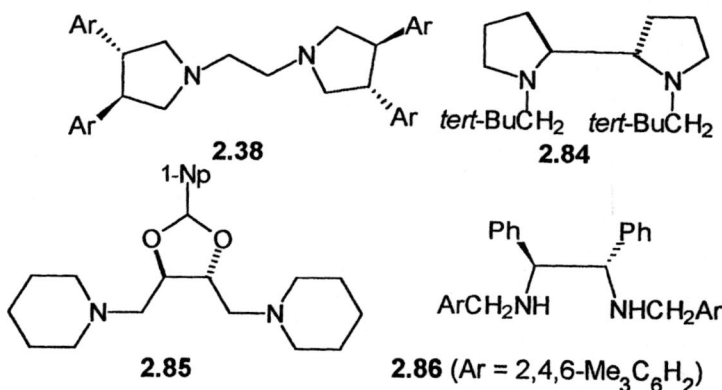

2.38

2.84

2.85

2.86 (Ar = 2,4,6-Me$_3$C$_6$H$_2$)

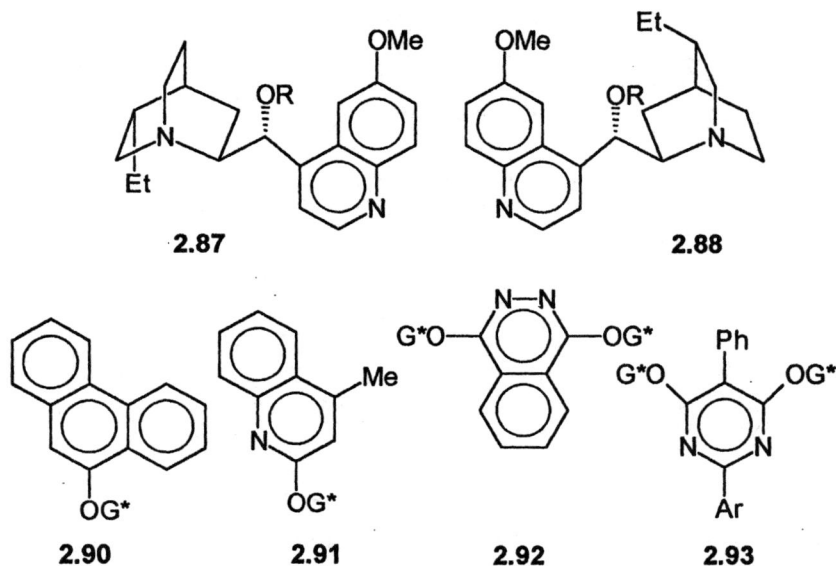

2.87   **2.88**

2.90   **2.91**   **2.92**   **2.93**

Typically, the alkene is reacted with 1-5% OsO_4 or $K_2OsO_2(OH)_4$ and 0.02 equivalents of ligand. Either N-methylmorpholine oxide [750] or better $K_3Fe(CN)_6$ is used in excess as a co-oxidant in the presence of K_2CO_3 in aqueous *tert*-BuOH [754-758]. The general catalytic scheme is shown in Figure **2.9**.

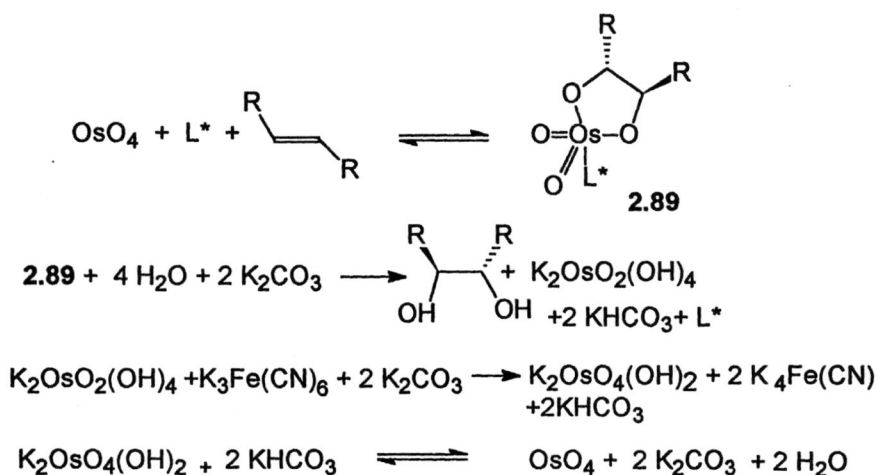

$$OsO_4 + L^* + \text{R}\diagdown\diagup\text{R} \rightleftharpoons \text{2.89}$$

$$\textbf{2.89} + 4\,H_2O + 2\,K_2CO_3 \longrightarrow \underset{OH\quad OH}{\text{R}\quad\text{R}} + K_2OsO_2(OH)_4 + 2\,KHCO_3 + L^*$$

$$K_2OsO_2(OH)_4 + K_3Fe(CN)_6 + 2\,K_2CO_3 \longrightarrow K_2OsO_4(OH)_2 + 2\,K_4Fe(CN) + 2KHCO_3$$

$$K_2OsO_4(OH)_2 + 2\,KHCO_3 \rightleftharpoons OsO_4 + 2\,K_2CO_3 + 2\,H_2O$$

Figure 2.9

An osmate ester **2.89**, likely formed by a stepwise [2+2]/rearrangement process [759] or by a concerted [3+2] cycloaddition process [760, 761], is the primary product of the reaction. This is hydrolyzed to a chiral diol and $K_2OsO_2(OH)_4$ (which is reoxidized by $K_3Fe(CN)_6$), with release of the chiral ligand for reuse.

Ligands for the asymmetric dihydroxylation have evolved quickly. The cinchona alkaloid ligands first recommended were esters **2.87** and **2.88** (R = MeCO, $4\text{-ClC}_6\text{H}_4\text{CO}$). Soon, the 9-hydroxyphenanthrene ethers **2.90** and 4-methyl-2-hydroxyquinolines **2.91** were found to be more efficient [755]. Currently, the best systems are the phthalazine-1,4-diol bis-ethers **2.92** derived either from dihydroquinine **2.87** (AD mix-α) or from dihydroquinidine **2.88** (AD mix-β) [33, 753, 757, 762]. The structures of **2.91** and **2.92** have been determined by X-ray crystallography. Pyrimidine diethers **2.93** recently have been proposed as advantageous ligands for some classes of alkenes [762]. Pyridazine 1,4-diol bis-ethers, proposed by Corey and coworkers, seem somewhat less efficient [757, 763], but new highly potent monocinchona alkaloid ligands were designed in 1994 by these authors [761]. The use of polymer grafted ligands has been proposed, but the reaction duration is longer [764, 765].

CHAPTER 3

CHIRAL CATALYSIS AND CATALYSTS BEARING CHIRAL LIGANDS

Chiral compounds can behave as catalysts in organic reactions. For example, chiral nitrogen bases, chiral crown ethers or Lewis acids bearing chiral residues catalyze diverse types of reactions. Asymmetry can also be induced in transition-. metal-catalyzed reactions if the metal bears chiral ligands. These possibilities will be described in sequence.

3.1 AMINOALCOHOLS, AMINOACIDS AND DERIVATIVES, CROWN ETHERS

A few base-catalyzed reactions may be performed in the presence of chiral amines in homogeneous media or in the presence of chiral phase-transfer agents in bi- or triphasic systems [253, 559, 766, 767]. Among the many catalysts used in these types of reactions, cinchona alkaloids **3.1** and **3.2** used in homogeneous phase or their *N*-benzylammonium salts used as phase-transfer catalysts give interesting results. For example, alkylations [559, 766, 768], Michael reactions [173, 253, 766], or [2+2] cycloadditions [769, 770] are catalyzed by these alkaloids, and the products are formed with high enantiomeric excesses. Thiol additions to electrophilic double bonds or epoxidations of α-enones with H_2O_2 lead to lower selectivities [767, 771, 772], as do aldol reactions [773]. Quaternary salts **3.3** derived from *N*-methylephedrine are recommended as phase-transfer catalysts in asymmetric Michael reactions (R = $ArCH_2$) [559, 602] or in reactions of carbonyl compounds with sulfur ylides (R = Me). Enantiomeric excesses in these reactions are 70 - 80% [559]. The presence of a free hydroxyl group on the catalyst is mandatory for asymmetric induction to occur [602, 766, 767, 774].

Impressive results have been obtained by Hajos, Wiechert and coworkers [261] in enantioselective Robinson annulations of triketones catalyzed by (*S*)-proline **1.64** (R = COOH). This type of asymmetric intramolecular aldol reaction is quite general under aminoacid catalysis [261, 775]. Asymmetric hydrocyanation of aldehydes is catalyzed by dipeptides, among which **3.4** is the most efficient. Asymmetric epoxidation of chalcone by alkaline H_2O_2 is catalyzed by polyaminoacids [578, 776], but this reaction is not very general [777].

Asymmetric Michael reactions can be catalyzed by KO*tert*-Bu in the presence of chiral crown ethers [253, 559, 766]. Crown ether **3.5** derived from binaphthol has given the best results.

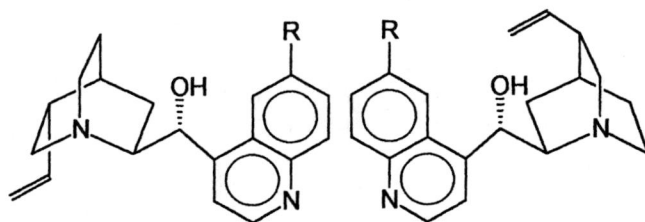

3.1   **3.2**

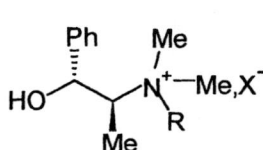

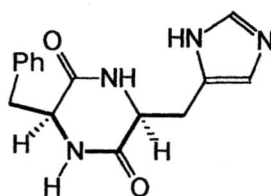

(1*R*,2*S*)- **3.3**   **3.4**

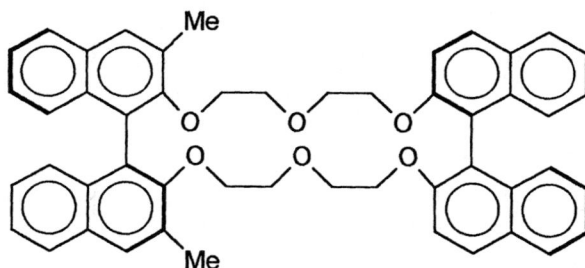

3.5

3.2 LEWIS ACIDS [777a]

Transformations such as Diels-Alder reactions, ene-reactions and condensations of enoxysilanes with carbonyl compounds often require Lewis acid catalysis. For this reason, many organometallic reagents bearing chiral residues have recently been used as Lewis acids in asymmetric synthesis. Many of these chiral Lewis acids feature structures similar to those described in Chapter 2.

3.2.1 Boron Derivatives

Chiral haloboranes such as diisopinocampheylchloroborane **2.16** (R = Cl) or cyclic borane **2.61** have been used as catalysts in Diels-Alder reactions or in hydrocyanation of aldehydes, but poor enantioselectivities were observed [778]. Catalyst **2.61** gives better results in reactions of aliphatic aldehydes with ketene acetals [689]. Enantiopure dichloroborane **3.6**, resolved by crystallization of its menthone complex, is useful as a catalyst for asymmetric Diels-Alder reactions of α,β-unsaturated esters [779, 780]; enantiomeric excesses are higher than 90%.

Chiral boronates are generated *in situ* by reaction of binaphthols **3.7** (R = H, Ph) [231] with BH$_3$ in the presence of acetic acid [778], with H$_2$BBr [781] or with B(OPh)$_3$ [782, 783]. Chiral borates are formed by reactions of substituted (*S*)-prolinol derivative **2.13** (R = CPh$_2$OH) and BBr$_3$ [784]. These boronates and borates are valuable catalysts in asymmetric Diels-Alder reactions [73, 231, 601, 780]. Tartaric acid derivatives, such as borate **3.8** and acyloxyboranes **3.9** recommended by Yamamoto and coworkers [73, 601, 778, 780, 785-791], are very efficient catalysts in asymmetric Diels-Alder reactions and in condensations of aldehydes with allylsilanes, enoxysilanes or ketene acetals. These catalysts are generated *in situ* from substituted monobenzoates of (*R,R*)- or (*S,S*)-tartaric acid and BH$_3$ (R = H) or an arylboric acid (R = Ar). The best asymmetric inductions are observed with catalysts **3.9**, R' = *i*-Pr. 1,3,2-Oxazaborolidines **3.10**, prepared from α-aminoacids [44, 601, 780, 792, 793], are efficient catalysts in asymmetric Diels-Alder reactions. The catalyst generated from *N*-tosyltrytophan **3.11** is more efficient than borolidines **3.10** (R = Et, *i*-Pr). The catalysts **3.10** prepared from **3.11**, **3.12** and **3.13** are also useful in asymmetric condensations of aldehydes with ketene acetals [794-797].

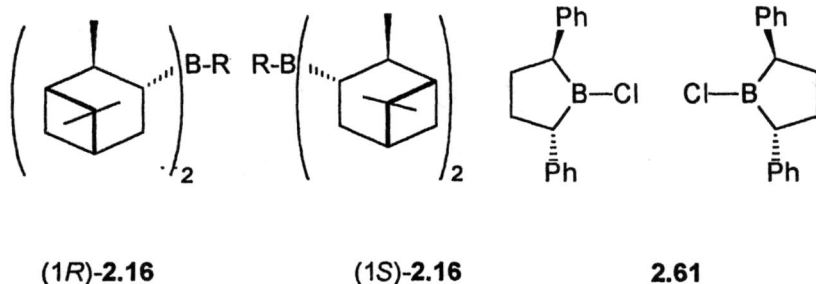

(1*R*)-**2.16**   (1*S*)-**2.16**   **2.61**

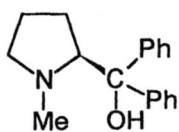   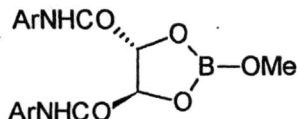

2.13 (R = CPh$_2$OH)   **3.8**

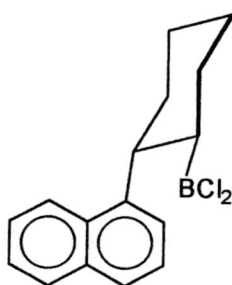

3.6

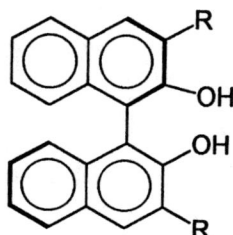

(*R*)-**3.7**

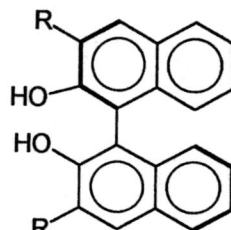

(*S*)-**3.7**

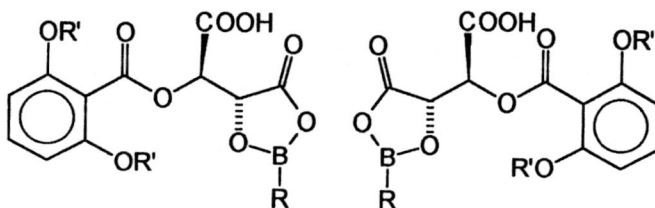

3.9

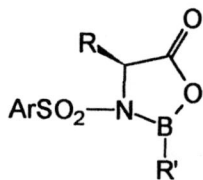

3.10

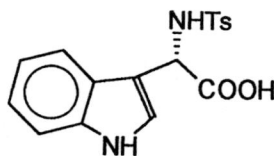

3.11

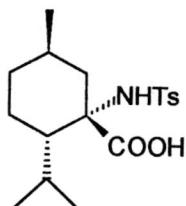

3.12

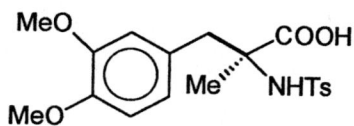

3.13

3.2.2 Aluminum and Zinc Derivatives

Aluminum trichloride, alkylaluminum chlorides and related zinc salts are efficient catalysts in Diels-Alder and ene-reactions [624c]. Therefore, several alkoxyaluminum mono- or dichlorides generated from chiral alcohols have been tested as catalysts in asymmetric Diels-Alder [253, 780, 798] or Friedel-Crafts [799] reactions. The catalysts generated *in situ* from menthol **1.4** (R = H), isoborneol **1.8** (R = H) or monoethers of diols **1.32** and AlCl$_3$ or EtAlCl$_2$ give disappointing results. Other catalysts have been generated from substituted binaphthols **3.7** (R = SiAr$_3$) [231, 249, 559, 778, 780] or more bulky systems [800], from diols **1.32** (R = Me) [801], **1.35** [559, 778], or from disulfamide **3.14** [693, 778] and DIBAH, Me$_3$Al, EtAlCl$_2$ or Et$_2$AlCl. These catalysts are very efficient in some asymmetric Diels-Alder [73, 693, 778, 780, 800, 801, 802] and ene-reactions [778], and in Claisen rerrangements [803]. Asymmetric cyanosilylation of aldehydes can be catalyzed by a Me$_3$Al/dipeptide complex, but the enantiomeric excesses observed are in the vicinity of 70% [804].

Catalysts formed from Me$_2$Zn and binaphthol **3.7** (R = H) have been used for asymmetric ene-reactions [778]. Enantioselective ring opening of *meso*-epoxides by *n*-BuSH is catalyzed by a potassium tartrate/ZnCl$_2$ complex [559, 778, 805]. Mukaiyama and coworkers have shown that reaction of Et$_2$Zn with chiral sulfamides **3.15** (R = PhCH$_2$, *i*-Pr) generates Lewis acids [806] that catalyze asymmetric reactions of aldehydes with ketene acetals.

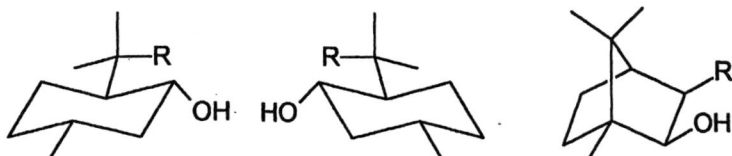

(1*R*,2*S*,5*R*)-**1.4**   (1*S*,2*R*,5*S*)-**1.4**   **1.8**

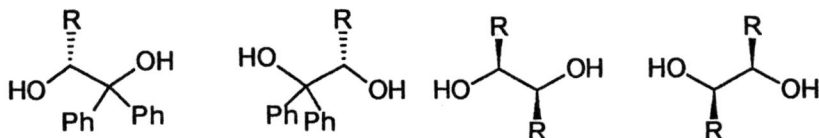

(*R*)-**1.32**   (*S*)-**1.32**   (*S*,*S*)-**1.35**   (*R*,*R*)-**1.35**

$$CF_3SO_2NH \overset{\overset{\displaystyle Ph}{\vdots}}{\underset{\underset{\displaystyle Ph}{\vdots}}{\diagup}} NHSO_2CF_3 \qquad CF_3SO_2NH \overset{\overset{\displaystyle Ph}{\vdots}}{\underset{\underset{\displaystyle Ph}{\vdots}}{\diagup}} NHSO_2CF_3$$

$$(R,R)\text{-}\textbf{3.14} \qquad\qquad\qquad (S,S)\text{-}\textbf{3.14}$$

$$\overset{\overset{\displaystyle NHSO_2CF_3}{\vdots}}{R \diagup \diagdown COOTBDMS}$$

3.15

3.2.3 Titanium and Zirconium Derivatives

Among the reactions catalyzed by titanium complexes, the asymmetric epoxidation of allylic alcohols developed by Sharpless and coworkers [752, 807-810] has found numerous synthetic applications. Epoxidation of allylic alcohols **3.16** by *tert*-BuOOH under anhydrous conditions takes place with an excellent enantioselectivity (ee > 95%) when promoted by titanium complexes generated *in situ* from Ti(O*i*-Pr)$_4$ and a slight excess of diethyl or diisopropyl (*R,R*)- or (*S,S*)-tartrates **2.69**. The chiral complex formed in this way can be used in stoichiometric or in catalytic amounts. For catalytic use, molecular sieves must be added. Because both (*R,R*)- and (*S,S*)-tartrates are available, it is possible to obtain either enantiomeric epoxide from a single allylic alcohol. Cumene hydroperoxide (PhCMe$_2$OOH) can also be used in place of *tert*-BuOOH. This method has been applied to industrial synthesis of enantiomeric glycidols [811, 812].

When racemic secondary allylic alcohols **3.17** are subjected to standard Sharpless epoxidation conditions, kinetic resolution takes place [127]. By choosing (*R,R*)- or (*S,S*)-tartrate, either enantiomer of the epoxyalcohol can be obtained with a maximum yield of 50%, alongside the unreacted allylic alcohol. The ratio of epoxidation rates of the enantiomeric allylic alcohols is usually high enough to obtain both the epoxyalcohol and the unreacted allylic alcohols in high enantiomeric excesses. In some cases, the use of dicyclohexyl- instead of diisopropyl tartrate improves the enantioselectivity. Homoallylic alcohols are also epoxidized, but the selectivities are significantly lower [808].

The structure of the catalyst has been studied by several spectroscopic methods [813], and it appears that a binuclear titanium complex **3.18** is the predominant species in solution. Tartramides **3.19** have been proposed as substitutes for dialkyl tartrates [808, 810], and their efficiency as catalysts is comparable to that of the diesters. However, with these diamides, the major enantiomer that is obtained depends on the ratio **3.19** (R = PhCH$_2$) to Ti(Oi-Pr)$_4$. If this ratio is 1/1, the predominate enantiomer is the same as that formed by using dialkyltartrates having the same configuration. But if this ratio is 2/1, the other enantiomer is formed.

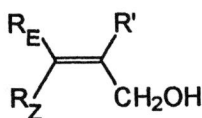

3.16

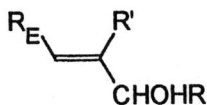

3.17

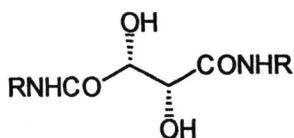

(*R,R*)-**3.19**

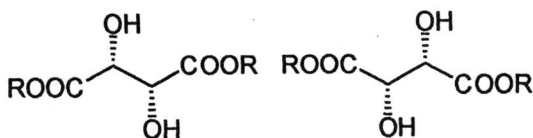

(*R,R*)-**2.69**   (*S,S*)-**2.69**

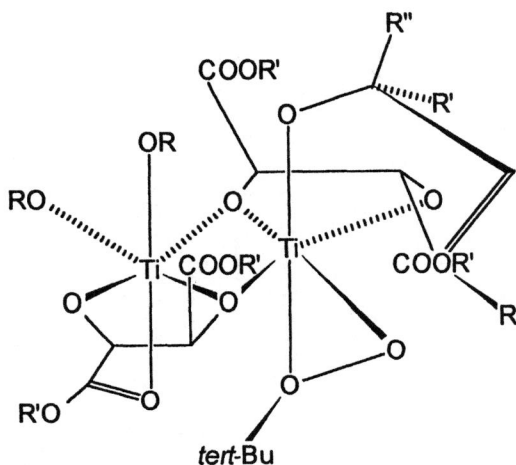

3.18

A modified Sharpless reagent has been developed by Kagan [503, 814], Modena [502, 814] and their coworkers. This new catalyst is formed by mixing water, Ti(Oi-Pr)$_4$, and diethyltartrate in a ratio of 1/1/2. The modified catalyst promotes enantioselective oxidation of arylalkylsulfides by $tert$-BuOOH, and chiral sulfoxides are produced with excellent enantiomeric excesses (> 90%). Lower selectivities are observed from dialkylsulfides. From (R,R)- or (S,S)-diethyl tartrate, either sulfoxide enantiomer can be obtained. The use of cumene hydroperoxide as the oxidant may improve the enantioselectivity. Uemura and coworkers obtained similar results by replacing the tartrates in these complexes with binaphthols [815].

Other titanium complexes derived from tartaric acid have been used as chiral catalysts. The complexes generated from diol 2.50 and TiCl$_2$(Oi-Pr)$_2$ are used as catalysts in asymmetric ene-reactions [778, 816], and in Diels-Alder [778, 780] or [2+2] cycloadditions of ketene thioacetals and unsaturated sulfides [778, 817]. The best enantiomeric excesses are observed with 2.50 (R = Me, R' = Ph) [778, 817] or 2.50 (R = R' = Et, Ar = 3,5-Me$_2$C$_6$H$_3$) [45]. These catalysts are also efficient in hydrophosphonylation [818] and in asymmetric hydrocyanation of aldehydes with Me$_3$SiCN [778]. These titanium complexes may be used in catalytic amounts provided that the reactions are run in the presence of molecular sieves [559, 816].

Mixing of (R)-binaphthol 3.7 (R = H) with TiCl$_4$, TiBr$_4$ or better yet TiCl$_2$ (Oi-Pr)$_2$ gives complexes that efficiently catalyze asymmetric ene-reactions [778, 819, 820] and some Diels-Alder reactions [821]. In a few cases, the selectivity is improved if the reaction is run in the presence of AgClO$_4$ [822] or if the titanium complex is formed from a substituted binaphthol 3.7 (R = Ph) [778, 823]. Other binaphthol/titanium complexes are recommended to catalyze reactions of aldehydes with thioketene silylacetals [824] or Me$_3$SiCN [778]. Silylcyanation is also catalyzed by titanium complexes formed from chiral Schiff bases 3.20 (R = i-Pr or $tert$-Bu) [268] or from chiral dipeptide 3.21 [825]. With 3.21, the best enantioselectivities are observed when R = i-Pr and R' = PhCH$_2$. This last complex is, however, inefficient in asymmetric epoxidation of allylic alcohols [826].

Titanium complexes that are similar to Duthaler's (§ 2.5.2) can be generated from TiCl$_4$, Ti(Oi-Pr)$_4$ and diacetoneglucose 1.48. These complexes catalyze asymmetric hetero-Diels-Alder reactions, and give high enantiomeric excesses [827]. Corey and coworkers [828] also prepared a chiral titanium catalyst derived from cis-N-sulfonyl-2-amino-1-indanol and used this to catalyze asymmetric Diels-Alder reactions. Buchwald and coworkers [829, 830] have proposed the use of titanocene-binaphthol catalysts for asymmetric hydrogenation of imines or trisubstituted olefins.

The use of zirconium complexes derived from tartramides **3.19** in asymmetric epoxidation of homoallylic alcohols does not result in any improvement over the related to titanium analogs [808]. A zirconium complex prepared from Zr(O*tert*-Bu)$_4$ and (*S*,*S*,*S*)-triisopropylamine **3.22** in the presence of water catalyzes the asymmetric ring opening of *meso*-epoxides by *i*-PrMe$_2$SiN$_3$ (ee ≥ 85%), while related titanium complexes are less efficient [805, 831].

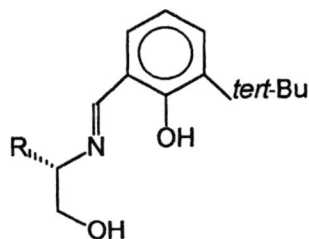

3.20

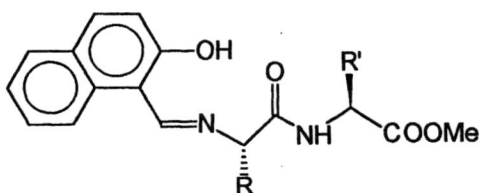

3.21

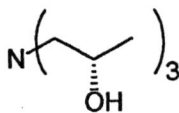

3.22

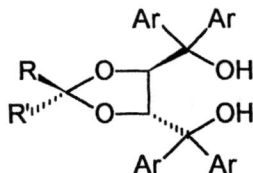

2.50

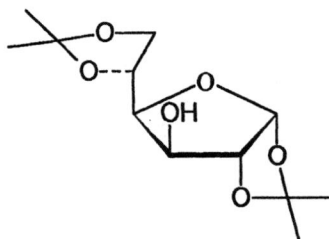

1.48

3.2.4 Tin Derivatives

Mukaiyama and coworkers have studied the reactions of aldehydes with enoxysilanes and ketene acetals in the presence of tin complexes [832]. These authors used tin (II) enolates, complexed to chiral amines, in asymmetric aldol reactions (§ 2.6.2). They extended this methodology to asymmetric catalysis of the reactions of aldehydes with ketene silylacetals **3.23** (R$_Z$ = H, Me, PhCH$_2$O). The most efficient catalysts are preformed *in situ* from Sn(OTf)$_2$, *n*-Bu$_2$Sn(OAc)$_2$ or better *n*-Bu$_3$SnF and chiral diamines **2.13** (R = CH$_2$N(CH$_2$)$_5$, CH$_2$NHAr) derived from (*S*)-proline [173, 833-839]. Sn(OTf)$_2$ may be replaced by a combination of Me$_3$SiOTf and tin oxide [840]. The best ligands **2.13** (R = NHAr) bear 1-Np or

1-[5,6,7,8-tetrahydronaphthyl] groups as Ar substituents [833, 835, 838]. Aminotetrahydrothiophene **3.24** can also be used as a ligand [841], and this complements **2.13** by providing the enantiomeric β-hydroxyester. Asymmetric allylations of aldehydes or epoxides by CH_2=$CHCH_2AlR_3$ in the presence of tin (II) halides and amines **2.13** gives interesting results [842], although propargylation is less selective [842].

Asymmetric hydrocyanation of aldehydes with Me_3SiCN is catalyzed by $Sn(OTf)_2$ in the presence of cinchonidine **3.1** (R = H).

3.2.5 Lanthanide Derivatives

Danishefsky and coworkers introduced the use of chiral europium complexes **3.25** [249], popular chiral shifts reagents in NMR spectroscopy [843], as Lewis acids in asymmetric [4+2] cycloadditions of silyloxydienes **3.26** with aldehydes [249, 778, 844]. Low enantiomeric excesses are observed if the diene group R is achiral; however, when this group is menthyl or 3-cholestanyl, the reaction is highly diastereoselective [249]. Other cycloadditions catalyzed by **3.25** give disappointing results [845], as does the reduction of methyl phenylglyoxylate by *N*-benzyldihydronicotamide [846].

Lanthanum catalysts were generated by Shibasaki and coworkers from lithium binaphthoxide and $LaCl_3$ in the presence of NaOH and water [847, 848, 849]. These reagents catalyze asymmetric nitroaldol reactions with a good enantioselectivity, but are less efficient in promoting asymmetric additions of dialkylphosphites to aldehydes [850].

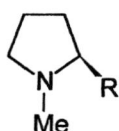

2.13

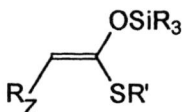

3.23

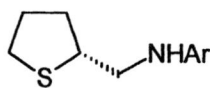

3.24

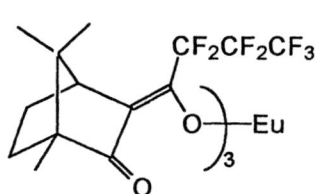

3.25

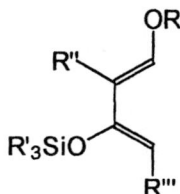

3.26

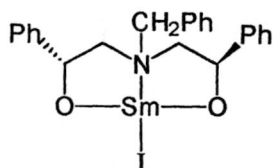

3.27

In 1993, Evans and coworkers [626] prepared a chiral samarium (III) complex **3.27** that catalyzes the asymmetric Meerwein-Verley-Pondorf reduction of aryl methylketones to secondary alcohols with a high enantiomeric excess (> 90%).

3.2.6 Iron and Copper Derivatives

Rigid ligands bearing a bis-oxazoline skeleton **3.28** (R = Ph, R' = H, R" = Me) readily derived from (*S*)-phenylglycinol **1.60** (R = Ph, R' = H) have been prepared by Corey and coworkers [780, 851] and used to generate iron complexes. These complexes are formed *in situ* from **3.28**, FeI$_3$ and iodine, and they catalyze some highly selective Diels-Alder cycloadditions with cyclopentene.

Evans and coworkers [852] recommended copper complexes generated from Cu(OTf)$_2$ and bisimines **3.29** derived from (1*R*, 2*R*)-cyclohexanediamine for similar purposes. The highest enantioselectivities are observed when R = 2,6-Cl$_2$C$_6$H$_3$, but the *endo/exo* selectivity depends on the structure of the acyclic dienophile.

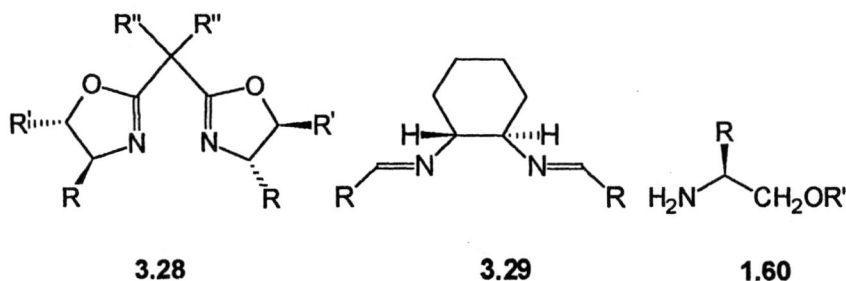

3.28   3.29   1.60

3.3 COLUMN VIII TRANSITION-METAL CATALYSTS: RHODIUM, RUTHENIUM, PALLADIUM, PLATINUM, IRIDIUM

Reactions catalyzed by complexes of these metals are highly interesting. Because these catalysts are often very efficient, only small amounts of chiral ligands are necessary for inducing asymmetry. Spurred by its industrial applications, this methodology has been broadly developed over the last fifteen years [811, 812, 853, 854]. The most impressive advances [113, 551, 752, 855] have been in the area of homogeneous catalysis. Recently, a novel strategy called "chiral poisoning" has been proposed by Faller and coworkers [856, 857]. In this strategy, a racemic ligand is used and an inexpensive chiral reagent is added to deactivate one of the enantiomers of the complex.

3.3.1 Heterogeneous Catalysis

In contrast to Raney nickel catalysts (§ 3.4.1), heterogeneous hydrogenation catalysts based on Pt, Rh or Pd do not induce asymmetry in the presence of tartaric acid [113, 578]. Platinum catalysts modified by cinchona alkaloids **3.1** and **3.2** cause asymmetric hydrogenation of the carbonyl group of α-ketoesters with a high enantiomeric excess (> 90%). From other types of ketones, the enantioselectivities are lower.

A few homogeneous catalysts have been grafted onto polymers, thus giving heterogeneous catalysts [113]. The uses of these catalysts will be presented in the following paragraph.

3.3.2 Homogeneous Catalysis: Phosphorus Ligands

3.3.2.1 *Diphosphines and Related Compounds*

The use of Wilkinson's catalyst, $Rh(PPh_3)_3Cl$, for hydrogenation of olefins under homogeneous conditions was first published in 1966 [92]. Very rapidly, the synthesis of chiral phosphines was initiated with the goal of replacing triphenylphosphine in this catalyst. The first significant results were obtained by Kagan and Knowles [113, 114, 169, 752, 853, 858, 859] who used diop **3.30** (Ar = Ph) and dipamp **3.31** as rhodium ligands in asymmetric hydrogenation of prochiral Z-enaminoacids **3.32** (R' = H). These authors prepared chiral aminoacids with high enantiomeric excesses (82 and 94%) at the beginning of the 1970s. These early results stimulated further work, and now numerous chiral diphosphines have been prepared [113, 114, 169, 752, 858, 860, 861, 862]. These diphosphines are used mainly as ligands for rhodium, ruthenium and palladium catalysts. They either feature a chiral carbon skeleton, such as diop **3.30**, or have phosphorus as center of chirality, such as dipamp **3.31**. Recently, Corey and coworkers [863] have proposed a general method to synthesize such ligands.

Diop **3.30** has been subjected to several modifications to improve its efficiency, and substituents have been introduced onto the aromatic ring to increase either the electron density on phosphorus [114, 752] or the steric hindrance [114, 864]. Water solubilization can be brought about by introducing appropriate substituents on the aromatic ring or on the ketal [752, 865, 866]. Boron analogs have also been prepared [867]. The main applications of diop and related diphosphines are as ligands in rhodium-catalyzed asymmetric hydrogenation of precursors of α-aminoacids or polypeptides [169, 752, 812, 853, 858, 859]. The diphosphines are also used in rhodium-catalyzed asymmetric hydrosilylation of ketones [566, 752, 868]. Dipamp **3.31** was patented by Monsanto for use in preparation of L-Dopa, a central nervous system drug, from Z-**3.32** (R_Z = 3,4-$(HO)_2C_6H_3$,

Z = Me). Several firms have patented other tetraaryldiphosphines or diphosphinites as rhodium ligands. Among these, **3.33**, **3.34** and **3.35** (X = NCH$_2$Ph) permit the enantioselective synthesis of α-aminoacids with enantiomeric excesses greater than 95% [169, 752, 853, 854]. Other chiral diphosphines are recommended as rhodium ligands for asymmetric hydrogenations. These include **3.35** (X = CH$_2$ or NCOO*tert*-Bu), **3.36** (R = Ar = Ph, R' = *tert*-BuO or NHPh), **3.37** (n = 0: chiraphos, n = 1: bdpp), **3.38** (R=Me, *c*-C$_6$H$_{11}$, PhCH$_2$) and **3.39** (norphos). Such ligands have found applications in the synthesis of α-aminoacids or dipeptides [113, 169, 752, 812, 853, 858, 869], α-aminophosphonic acids [752], and phosphonate derivatives [870]. Chiraphos **3.37** (n = 0) is a valuable ligand in rhodium complexes for asymmetric hydrosilylation [625, 868] or in palladium complexes for asymmetric allylation of malonates [752, 866, 871, 872]. However, high enantioselectivities are restricted to selected cases. The use of these tetraaryldiphosphines as iridium ligands in asymmetric reduction of ketones by hydrogen transfer from *i*-PrOH leads to modest selectivities [873]. Diphosphine **3.36** (R = Ar = Ph, R' = *tert*-BuO) is a good ligand for platinum complexes used in asymmetric hydroformylation of olefins ArCH = CH$_2$ [752, 853, 874, 875]. Grafting these ligands onto polymers can give interesting supported catalysts which are easily recovered [169, 752]. However, these tetraaryldiphosphines give disappointing results as transition metal ligands in enantioselective hydrogenations of imines and ketones and of olefins that do not bear a Z-NH aryl residue. Therefore, new families of chiral diphosphines and analogs have been designed.

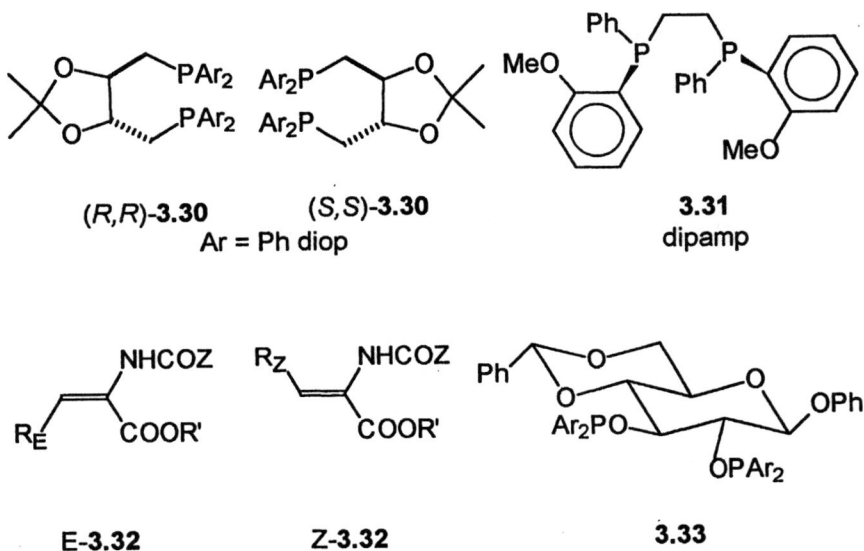

(*R,R*)-**3.30**   (*S,S*)-**3.30**   **3.31**
Ar = Ph diop   dipamp

E-**3.32**   Z-**3.32**   **3.33**

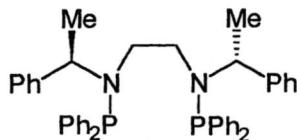

3.34

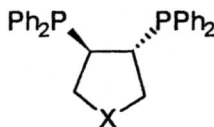

3.35

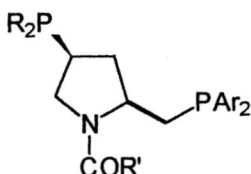

3.36

R = Ph:bppm   R = c-C6H11:bcpm

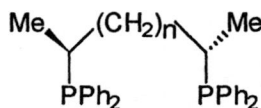

3.37

n = 0:chiraphos   n = 1:bdp

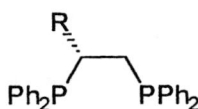

3.38

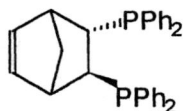

3.39 : norphos

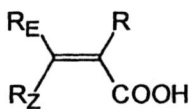

3.40

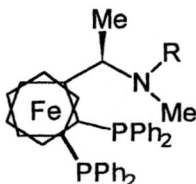

3.41

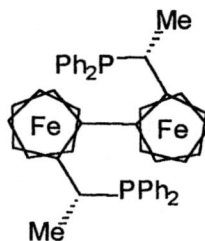

3.42

Replacement of aryl rings by cyclohexyl groups has shown encouraging results. Achiwa and coworkers [114] recommended diphosphines **3.36** (R = c-C$_6$H$_{11}$, Ar = Ph, R' = $tert$-BuO, MeO, MeNH) as ligands for rhodium in asymmetric hydrogenation of α,β-unsaturated acids **3.40** (R$_E$ = R$_Z$ = H). The appropriate ketolactone precursor can also be reduced to (R)-pantolactone **1.16**.

Introduction of electron-donating groups on **3.36** (R = 4-Me$_2$NC$_6$H$_4$, Ar = Ph, R' = *tert*-BuO) also leads to valuable ligands.

Catalytic hydrogenation of tetrasubstituted olefins is particularly difficult, but chiral aminoferrocenyldiphosphines **3.41** (R = R'$_2$NCH$_2$CH$_2$) are good ligands of rhodium complexes for such reductions [752]. Ferrocenyldiphosphine **3.41** (R = (HOCH$_2$)$_2$CH) is also an interesting ligand of palladium complexes in asymmetric allylation of acetylacetone [752, 871, 876]. Crown ethers have been introduced onto these ligands, and they increase reaction rates but do not improve the enantioselectivities [877]. Other R substituents give less interesting results [878]. Another ferrocenyldiphosphine **3.42** has been recommended as a rhodium ligand to catalyze asymmetric Michael reactions with satisfactory enantiomeric excesses [879].

Chiral atropisomeric biphosphines make up an important family of ligands for asymmetric catalysts. The most frequently used member, binap **3.43** (Ar = Ph), has been developed in the Ikariya, Noyori, Otsuka and Takaya groups [169, 752, 812, 859, 880-883]. Binap ligands are available in either (*R*)- or (*S*)-configurations, and they display a broader utility in asymmetric catalysis than the previously described diphosphines. As ligands of rhodium catalysts, they induce asymmetric hydrogenation of α-acylaminoacrylic acids into α-aminoacids, occasionally in water if the phenyl groups are sulfonated [261, 884]. Intramolecular hydrosilylation of silylethers of cinnamyl alcohol is similarly catalyzed with a high enantioselectivity [885]. Asymmetric hydroboration of styrenes (ArCH = CH$_2$) by catechol borane is catalyzed by binap-Rh complexes, and is regio- and enantioselective (ee > 95%) when performed at low temperature [886]. However, the selectivity is modest for β-substituted styrenes or phenylbutadiene [887], or when other boron reagents are used [888]. Binap-rhodium complexes also catalyze isomerization of allylamines **3.44** into chiral enamines with an excellent enantiomeric excess [752, 881, 882, 889, 890]. Industrial applications include syntheses of (+)-citronellol and (−)-menthol [811, 812, 853]. More efficient catalysts bear substituents such as 2-Me on the phenyl rings [882]. However, such catalysts give disappointing selectivities in hydroformylation of vinyl acetate [891].

Two ruthenium complexes, binap **3.43**-Ru(OCOR)$_2$(R = Me,CF$_3$) [892] and binap **3.43**-RuX$_2$ (X = Cl, Br, I) [893, 894], are quite useful. The acetate and trifluoroacetate complexes of **3.43** induce selective asymmetric hydrogenations of classes of prochiral olefins that are poorly selective with rhodium complexes. These classes include α,β- or β,γ-unsaturated acids and esters, allyl alcohols, β-acylaminoacrylates and enamide precursors of isoquinoline alkaloids [752, 853, 859, 881, 883, 895].

(R)-3.43   Ar = Ph : binap   (S)-3.43

3.44   3.45

Water-soluble catalysts have also been devised for such hydrogenations [896]. Hydrogenations of prochiral α,β-unsaturated acids can be also performed by $HCOOH/Et_3N$ with a good enantioselectivity [873, 897]. The halide complexes find their applications in asymmetric hydrogenations of ketones under H_2 pressure. For good results, the substrates must bear in β or γ position basic groups such as amines, alcohols, ethers, acids, amides or esters [752, 859, 881, 883, 894, 898, 899, 900]. These reactions are run in the presence of Et_3N and exhibit high enantioselectivities. Recently, binap **3.43**-iridium-aminophosphine systems have been shown to catalyze asymmetric hydrogenations of prochiral 1,2-benzocyclo-alkanones and of β-thiacycloalkanones [901].

Binap **3.43** is also a good ligand for palladium complexes. Although π-allyl palladium-binap complexes have met limited applications in asymmetric allylation of malonate like derivatives [872], binap-$PdCl_2$ complexes induce enantioselective Heck reactions [902] and 1,4-disilylations of α-enones [903].

Other atropisomeric phosphines **3.45** have been recommended as ligands in rhodium-catalyzed asymmetric hydrogenations of prochiral α,β-unsaturated acids or esters [892, 904] when R = c-C_6H_{11} or in rhodium-catalyzed asymmetric iso-merization of allylamines **3.44** [905] when R = Ph. These phosphines **3.45** (R = Ph or c-C_6H_{11}) are interesting ruthenium ligands in asymmetric hydrogenations of allylic alcohols, of enamides and of the keto group of α- or β-ketoesters [892, 906].

Very recently, Burk and coworkers proposed a new series of chiral diphosphines **3.46** (n = 2, 3) and **3.47** (R = Me, Et, i-Pr) bearing a C_2 axis of symmetry [907]. These electron-rich phosphines are excellent ligands for rhodium-catalyzed asymmetric hydrogenation of E- and Z-α-acylaminoacrylates **3.32** [907, 908], enol acetates [908], and α-substituted acrylates [909]. Furthermore, rhodium-catalyzed hydrogenations with these ligands are significantly faster than those with other diphosphines. However, lower selectivities are observed in ketone hydrogenations [909]. Ligand **3.47** (R = i-Pr) induces excellent enantiomeric excesses in rhodium-catalyzed intramolecular hydrosilylations of α-silyloxyketones [656].

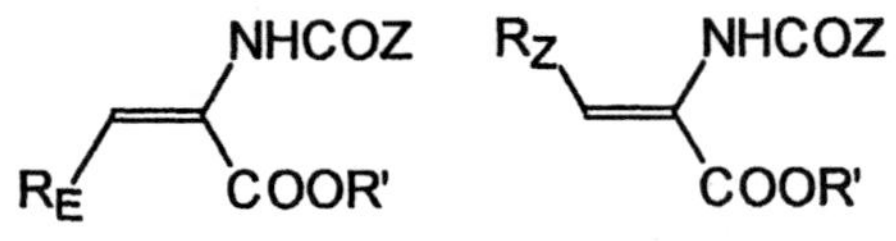

E-3.32   Z-3.32

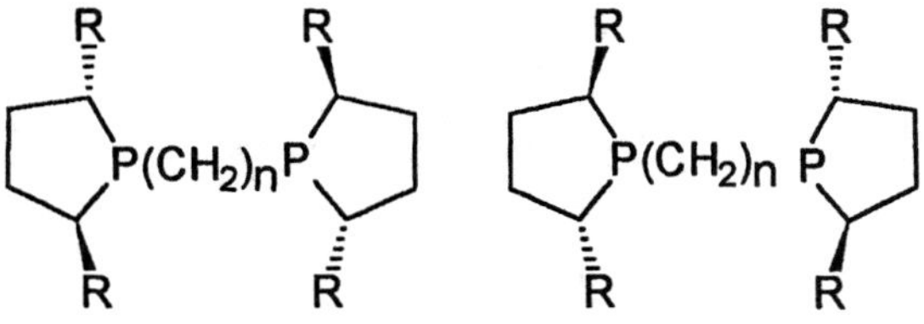

3.46 : duphos

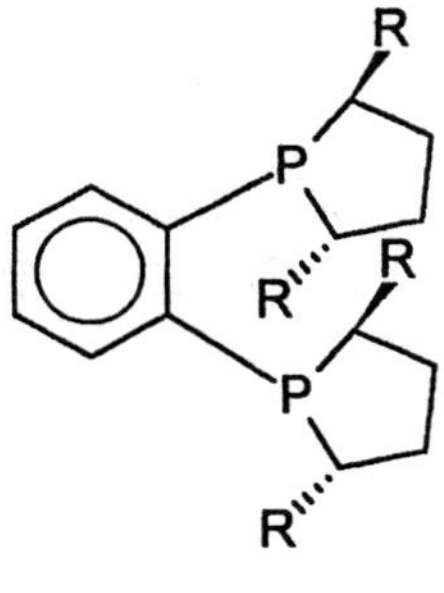

3.47

Asymmetric allylations of malonates and related nucleophiles catalyzed by palladium complexes bearing the previous diphosphines as ligands have yielded disappointing selectivities. To address this problem, Trost and coworkers have designed new diphosphines **3.48** prepared from 2-diphenylphosphinobenzoic acid [910, 911]. To decrease the degrees of freedom about the metal, these ligands have diester (X = O), or better yet, diamide (X = NH) groups, and they exhibit a C_2 axis of symmetry. The most efficient ligands are derived from diamines **1.72** (R = Ph, R' = H) or from *trans*-1,2-cyclohexanediamine. Amidophosphine-phosphinites prepared from chiral α-aminoalcohols **1.60** (R = *i*-Pr, R' = H) or **1.64** (R = CH$_2$OH) have also been used as rhodium ligands in enantioselective hydrogenations of precursors of aminoacids [861] or of (*R*)-pantolactone **1.16** [120a], and in hydroformylation of olefins [912] (although the regioselectivity of the latter process is not very high).

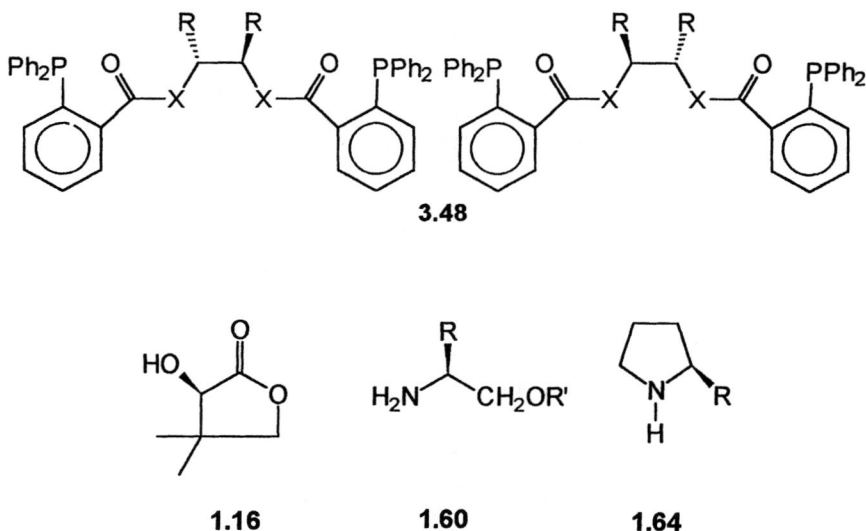

3.48

1.16   **1.60**   **1.64**

3.3.2.2. Monophosphines and Phosphorus Esters

Monophosphines are usually less efficient than diphosphines as ligands for transition-metal-catalyzed asymmetric reactions. However, in the case of palladium complexes, there are several exceptions to this generalization. Ferrocenylaminophosphines **3.49** (Y = NMe$_2$) or **3.50** [169, 752, 910, 913] are efficient ligands in enantioselective cross-coupling of organomagnesium or organozinc reagents with vinyl bromides. Monophosphines **3.51** (R = Me, PhCH$_2$, *i*-Pr) bearing a binaphthyl skeleton provide good selectivities in palladium-catalyzed hydrosilylation of alkenes [868, 914-917]. Novel triarylphosphines **3.52** (R = Me, *i*-Pr, *tert*-Bu, PhCH$_2$, Ph, R' = H, Ph) bearing a 1,3-oxazoline residue have been recently used by Helmchen, Pfaltz, Williams and their coworkers [871, 918-921] as palladium ligands for highly enantioselective allylations of malonates and analogs. Another aminophosphine, 1-(2-diphenylphosphino-1-naphthyl)isoquinoline **3.53**, is an interesting ligand of rhodium complexes that are used for asymmetric hydroboration of olefins [922].

Phosphates and phosphinates are also recommended as ligands. (*R*)- or (*S*)-Binaphthol phosphates **3.54** are used in palladium-catalyzed asymmetric hydrocarboxylation of olefins [923] or in rhodium-catalyzed cycloadditions of diazo compounds to olefins, albeit with modest selectivities in the latter case [924]. Seebach and coworkers [925] tested phosphinates and phosphites prepared from diol **2.50** (R = R' = Me, Ar = Ph) as ligands for rhodium and palladium in various enantioselective metal-catalyzed reactions [925]. Rhodium-catalyzed hydrosilylations of arylmethyl- or ethylketones by Ph$_2$SiH$_2$ were the only interesting reactions with these ligands.

Ferrocenylmonophosphines with a planar chirality have been prepared by Kagan and coworkers, and they await applications in asymmetric synthesis [926].

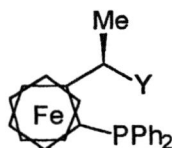

(*R*,*S*)-**3.49**

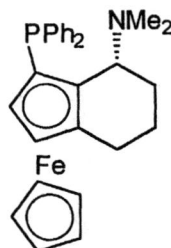

3.50

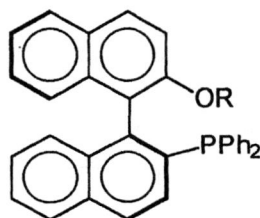

3.51
R = Me : (*S*)-mop

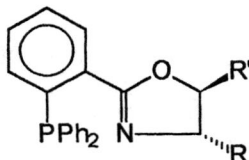

3.52 (X = PPh$_2$)

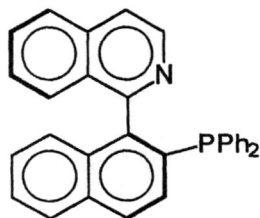

(*S*)-**3.53**

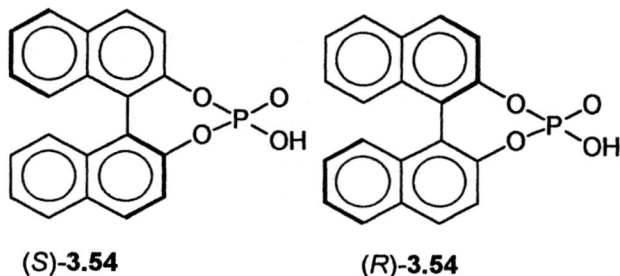

(S)-**3.54**   (R)-**3.54**

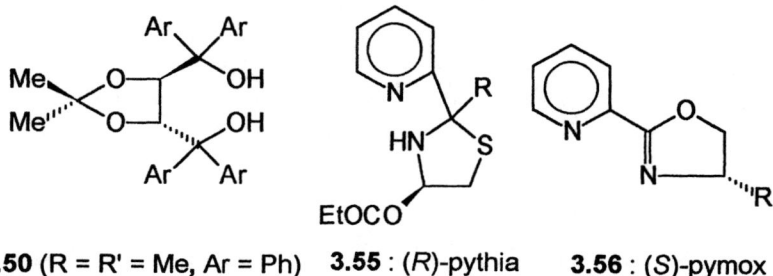

2.50 (R = R' = Me, Ar = Ph)   **3.55** : (R)-pythia   **3.56** : (S)-pymox

3.3.3 Homogeneous Catalysis: Nitrogen Ligands

Nitrogen ligands are very efficient in rhodium-catalyzed asymmetric hydrosilylation of ketones [566, 868, 927, 928, 929]. Some of the most efficient chiral ligands are pyridine derivatives **3.55** (R = H, Me) or better yet mono- or bis-1,3-oxazolines **3.56** (R = i-Pr, tert-Bu), **3.57** (R = i-Pr, PhCH$_2$), **3.58** (R = i-Pr, tert-Bu, PhCH$_2$, R' = H, Me$_2$N, MeO, Cl) and **3.59** [566, 868, 928-931]. However, asymmetric reduction of ketones by hydride transfer from isopropanol does not give interesting selectivities with these catalysts [873]. An interesting preliminary study has shown the potential of diamines **1.72** as ligands in rhodium-catalyzed hydride transfer to PhCOCOOMe [932].

Rhodium complexes generated from N-functionalized (S)-proline **3.60** [933, 934, 935] or from methyl 2-pyrrolidone-5-carboxylates **3.61** [936, 937, 938] catalyze the cyclopropanation of alkenes by diazoesters or -ketones. Diastereoisomeric mixtures of Z- and E-cyclopropylesters or -ketones are usually formed, but only the Z-esters exhibit an interesting enantioselectivity. However, if intramolecular cyclopropanation of allyl diazoacetates is performed with ligand **3.61**, a single isomer is formed with an excellent enantiomeric excess [936, 937]. The same catalyst also provides satisfactory results in the cyclopropanation of alkynes by menthyl diazoacetate [937, 939] or in the intramolecular insertion of diazoesters into C–H bonds [940].

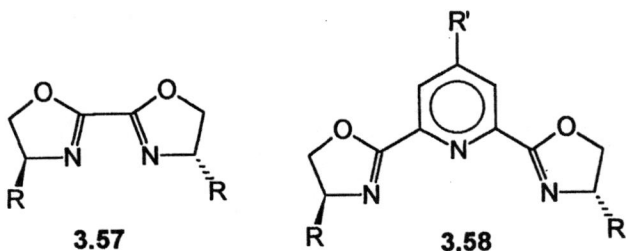

3.57

3.58

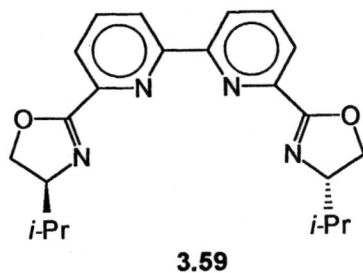

3.59

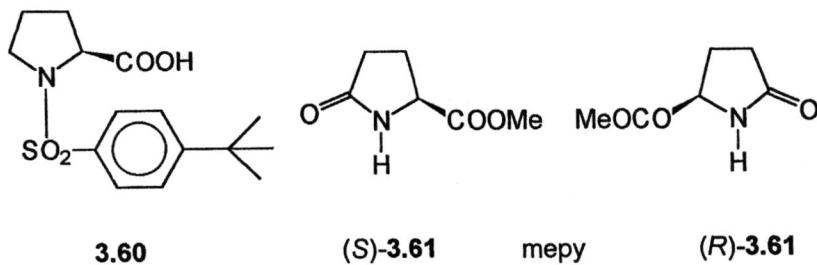

3.60

(S)-**3.61**

mepy

(R)-**3.61**

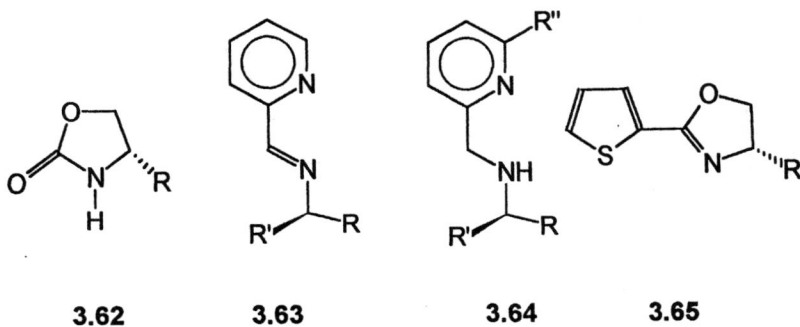

3.62

3.63

3.64

3.65

Enantiopure 4-alkyl-1,3-oxazolidin-2-ones **3.62** have also been reported as rhodium ligands in similar cyclopropanations [937, 938, 941], although they are less efficient than **3.61**.

Asymmetric reductions of arylalkylketones by hydride transfer from *i*-PrOH are catalyzed by iridium complexes, and the most efficient chiral ligands are bis-oxazolines **3.57** (R = *i*-Pr) and pyridine derivatives **3.63** and **3.64** [339, 873, 942].

In addition to 1,3-oxazolines bearing a phosphine functionality **3.52** (X = PPh$_2$), sulfur-substituted analogs like **3.52** (X = PhS) and 2-thienyl-1,3-oxazolines **3.65** [871, 918, 929, 943, 944, 945] have been examined as ligands in palladium-catalyzed asymmetric allylations, but oxazoline **3.52** (X = PPh$_2$) provides the highest efficiency. Sparteine **2.5**, and bis-oxazolines **3.28** (R = PhCH$_2$, R' = H, R" = Me), **3.57** (R = PhCH$_2$) have also given good enantioselectivities as ligands in these reactions [871, 929, 946]. However, the highest selectivities are obtained when the palladium ligands are azahemicorrin **3.66** (R = TBDMSOCH$_2$) or persubstituted 1,3-oxazoline **3.28** (R = TBDMSOCH$_2$, R' = Ph, R" = Me) [929, 946]. Koga and coworkers [947] proposed the use of chiral diamines **2.38** as palladium ligands in similar allylations, and high enantiomeric excesses were obtained from **2.38** (R = Me).

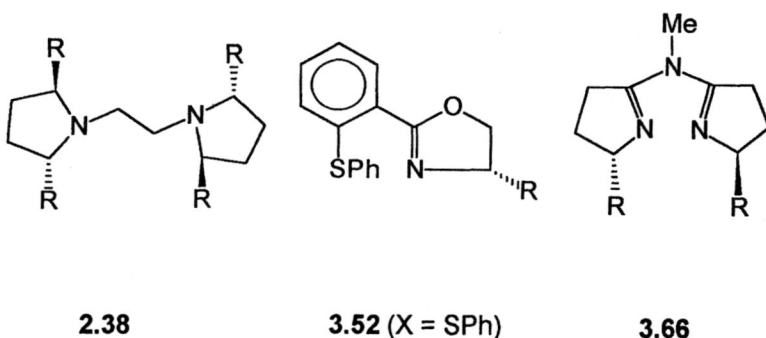

2.38   **3.52** (X = SPh)   **3.66**

3.4 NICKEL, COBALT, COPPER, MANGANESE AND GOLD CATALYSTS

3.4.1 Heterogeneous Catalysis

Raney nickel, modified by free tartaric acids **2.69** (R = H) or their salts, has frequently been used as a catalyst for asymmetric hydrogenations of carbonyl compounds [578, 948]. Several industrial applications have been described [578, 811, 812]. Neverless, hydrogenations of prochiral carbon-carbon double bonds in the presence of such catalysts gives disappointing results [578]. The use of tartrate-modified copper catalysts in cyclopropanation of styrenes by diazoketones takes place with a modest asymmetric induction [578, 936].

3.4.2 Homogeneous Catalysis: Phosphorus Ligands

Aminophosphines bearing a ferrocene skeleton **3.49** (Y = NMe$_2$, OMe) are interesting ligands not only in palladium-catalyzed processes, but also in nickel-catalyzed reactions such as asymmetric coupling of vinyl bromides and organomagnesium reagents [752, 910, 949]. However, better enantioselectivities have been observed in these couplings with the aminophosphines **3.67** (R = *i*-Pr, *tert*-Bu) easily prepared by Hayashi and coworkers from aminoacids [169, 752, 910, 923]. The use of other diphosphines such as **3.39** is less efficient [169, 752]. Similar nickel-catalyzed asymmetric coupling of allyl derivatives with organomagnesium reagents can be highly enantioselective when the ligands are diphosphines such as chiraphos **3.37** (n = 0) [169, 752, 858, 923].

Ferrocenyldiphosphines **3.41** (R = Me$_2$NCH$_2$CH$_2$) are used as ligands in gold-catalyzed asymmetric aldol reactions of α-isocyanoesters or -amides [408, 752, 858, 950]. Silver complexes can also be used with the modified phosphine **3.41** (R = (CH$_2$)$_5$ or (CH$_2$)$_6$) [951, 952].

3.4.3. Homogeneous Catalysis: Nitrogen Ligands

Asymmetric cyclopropanation is a very important reaction due to the applications of cyclopropanes as pesticides [811, 812, 936]. Following Nozaki's first attempts at asymmetric copper-catalyzed cyclopropanation of styrene by N$_2$CHCOOEt in the presence of the salicylaldimine ligand **3.68** (ee 10%), numerous analogs have been synthesized to improve the selectivity of this process [170, 936, 937, 953]. The most efficient ligands, designed by Aratani and coworkers, are salicylaldimines **3.69** (R = Me) in which the aromatic rings are substituted by bulky groups (R' = *tert*-Bu, R" = *n*-C$_8$H$_{17}$). This method is limited by the nature of ester substituents R on the diazoesters N$_2$CHCOOR. If R = Me or Et, mixtures of E- and Z-diastereoisomers are formed, but if R = *tert*-Bu or menthyl, E-esters are selectively formed with a good enantiomeric excess. Chiral bis-salicylidene ethylenediamines **3.70** and **3.29** are interesting manganese ligands in catalyzed asymmetric epoxidations of nonfunctionalized olefins by sodium hypochlorite or PhIO [954-960]. However, high enantioselectivities are observed only from Z-aryl substituted olefins, and the most efficient ligand is **3.29** (Ar = 2-OH, 3,5-(*tert*-Bu)$_2$C$_6$H$_2$). Manganese complexes bearing ligand **3.71** have recently been unsuccessfully tested as catalysts for PhI=NTs induced asymmetric aziridination [961]. Copper complexes bearing **3.29** (Ar = 2,6-Cl$_2$C$_6$H$_3$) as ligands are more efficient in such processes [962].

Hemicorrins **3.66** (R = Me$_3$SiOCMe$_2$) and **2.11** (R = CMe$_2$OH), designed by Pfaltz and coworkers [522, 929, 930, 942, 946, 953], are also interesting ligands in copper-catalyzed asymmetric cyclopropanations of styrene by diazoesters. Bis-oxazolines **3.28** (R = *tert*-Bu, R' = H, R" = Me), recommended by Evans, Masamune, and their coworkers [339, 929, 930, 963, 964], display a broader range of application with respect to the structure of the olefin partner, as does ligand **3.71** recently proposed by Masamune and coworkers [964]. However, in these reactions, the use of bulky alkyl diazoacetates(*tert*-Bu, dicyclohexylmethyl, 2,6-di-*tert*-Buphenyl) is required to selectively obtain one diastereoisomeric cyclopropane with a high enantioselectivity [963, 964]. Bis-oxazolines **3.28** (R = Ph, R' = H, R" = Me) are also efficient ligands in copper catalyzed asymmetric aziridination of cinnamates and related esters by PhI=NTs [965].

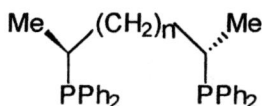

3.37

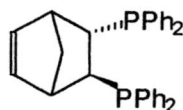

3.39

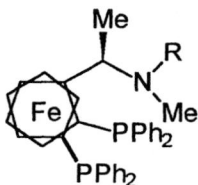

3.41

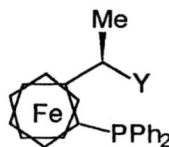

(*R,S*)-**3.49**

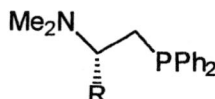

3.67

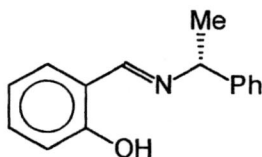

3.68

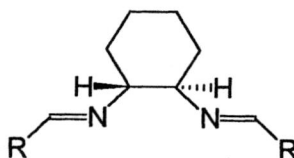

3.29

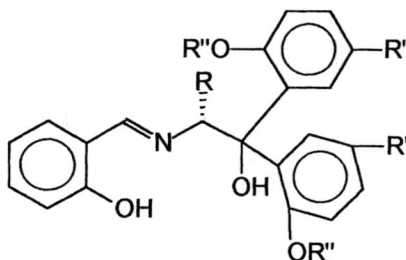

3.69

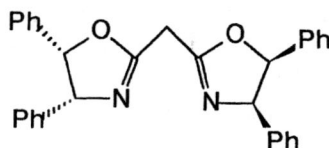

3.71

3.70

2.11 (R = CMe$_2$OH)   **3.28**

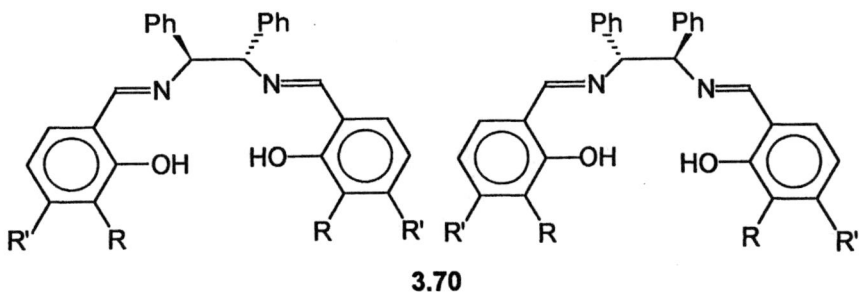

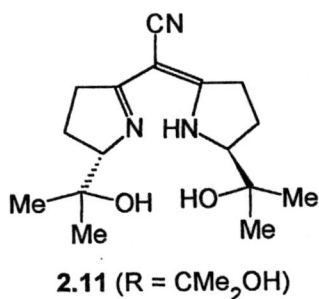

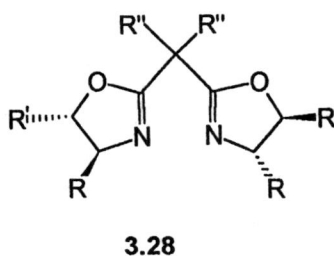

Chiral bipyridine-copper complexes have also been recommended as catalysts in cyclopropanations of styrene and E-substituted styrenes by diazoesters, but this process leads to mixtures of stereoisomers [966, 967]. The same limitations occur with bornanethione-derived chiral complexes [968].

Mercaptoaryl-1,3-oxazolines **3.52** (X = SH) have recently been recommended as ligands in enantioselective copper-catalyzed 1,4-addition of organomagnesium reagents to α-cyclenones [969], although interesting selectivities are only observed with 2-cycloheptenone.

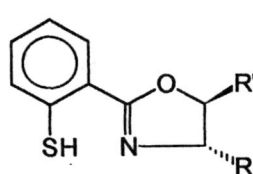

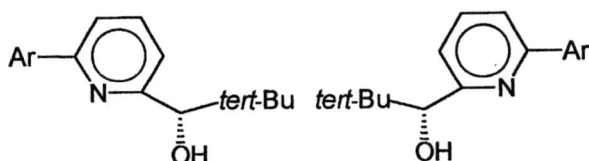

3.52 (X = SH)   **3.72**

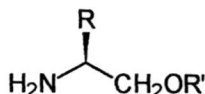

$$
\underset{\text{H}_2\text{N}}{} \overset{\text{R}}{\underset{}{\bigwedge}} \underset{\text{CH}_2\text{OR'}}{}
$$

1.60

Manganese and iron porphyrins substituted by chiral atropisomeric groups are models for cytochrome P-450 and several of such catalysts have been recommended by Groves, Kodadek, Mansuy and their coworkers in asymmetric epoxidations of olefins with hypochlorites or ArIO [957, 970]. Recently, asymmetric epoxidation of 2-nitrostyrene has been accomplished with 89% enantiomeric excess, but this result lacks generality. A preliminary report on the use of threitol-strapped manganese porphyrins in enantioselective epoxidations of unfunctionalized aryl substituted olefins appeared in 1993 [971]: enantiomeric excesses in the range of 80% were observed.

Substituted pyridylmethanols **3.72** have been recommended as ligands in nickel-catalyzed 1,4-additions of Et_2Zn to α,β-unsaturated carbonyl compounds [112, 972]. Asymmetric [2+2+2] cycloadditions of norbornene and acetylenes are catalyzed by a CoI_2/Zn/phosphine system and the most efficient ligands are amidophosphinephosphinites prepared from (*S*)-valinol **1.60** (R = *i*-Pr, R' = H) [973].

CHAPTER 4

ASYMMETRIC DEPROTONATIONS AND PROTONATIONS

4.1. ASYMMETRIC DEPROTONATIONS

Asymmetric deprotonation of a prochiral compound having a sufficiently acidic C-H bond can be performed by a lithium amide generated from an enantiopure secondary amine or by an organolithium reagent in the presence of a chiral tertiary amine [557, 559]. A chiral mixed aggregate is usually formed [77, 81, 974], and the reaction of this intermediate with electrophiles (including proton sources) can lead to a predominant enantiomer.

Deprotonation of prochiral cyclic ketones by chiral amides, followed by *in situ* Me$_3$SiCl trapping, can take place with an excellent selectivity. Koga and coworkers have recommended the use of lithium amides of **2.4** (X = CH$_2$ or NMe, R' = *i*-Pr, *tert*-BuCH$_2$, CF$_3$CH$_2$) in the presence of HMPA [557, 559, 561, 975]. The best results are obtained from 4-substituted cyclohexanones [975] or from 3-ketosteroids [976]. These deprotonations are performed at −100°C, and the chiral reagent is probably a five-membered monomeric chelate [561] (Figure **4.1**). Simkins and coworkers have advocated the use of lithium amide of **2.3** for enantioselective deprotonation of 4-*tert*-butylcyclohexanone and related ketones. The enolates are trapped by Me$_3$SiCl [557]. Substantial improvements are observed if this process is carried out in the presence of LiCl [977, 978, 979]. This method has been extended by Honda and Kimura to 3-phenylcyclobutanone, opening the way to nonracemic 3-phenylbutyrolactone [980]. Racemic prochiral 2-substituted cyclohexanones can suffer kinetic resolution when treated with these lithium amides [557]. Asymmetric deprotonation of six-membered thiane oxides has been carried out by Simkins and coworkers, although the selectivities are not very high [981]. Applications of these reagents to asymmetric aldol reactions will be considered in a subsequent chapter [§ 6.8.1.1].

Enantioselective deprotonations α to the oxygen or nitrogen atoms of carbamates **4.1** and **4.2** can be performed with *s*-BuLi in the presence of sparteine **2.5**. The *pro* (*S*)-hydrogen atom is selectively removed, and lithiated carbanions **4.3** are formed. These carbanions are stabilized by dipolar interactions [982] and coordination to sparteine [564, 983-987]. Such species are stable up to −30°C, and react with electrophiles at −78°C with retention of configuration [984-987].

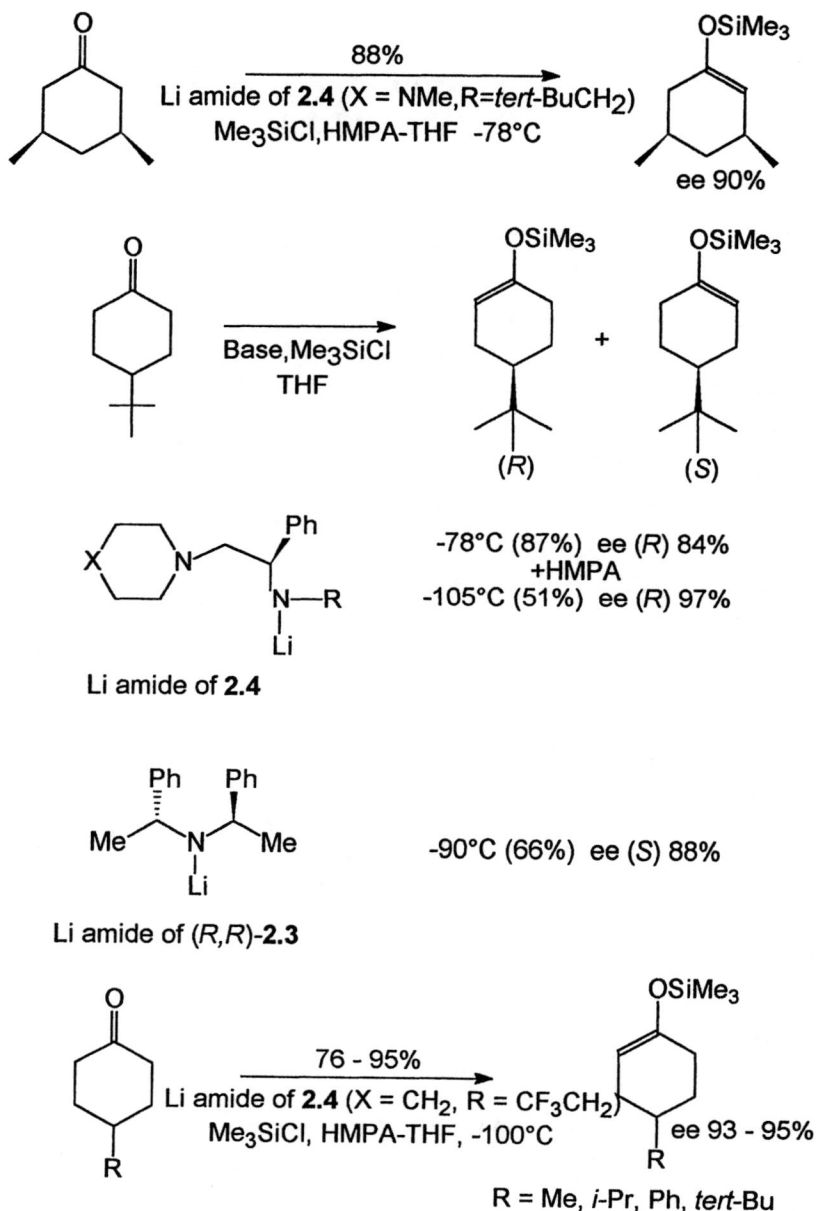

Figure 4.1

The carbamate residue designed by Hoppe and coworkers can be easily cleaved under mild conditions, providing a route to α-substituted alcohols with very high enantiomeric excesses (Figure 4.2). Bis-carbamates of 1,3- or 1,4-diols can be sequentially deprotonated [983, 988, 989] and deuterated. Because of the large kinetic H/D isotope effect [988], selective processes such as the kinetic resolution of racemic 2-phenylpropyl carbamates can be performed [990].

Asymmetric deprotonation of carbamates **4.2** under similar conditions, followed by trapping of the intermediate lithiated species by various electrophiles, also leads to 2-substituted pyrrolidines with very high enantiomeric excesses [564] (Figure **4.2**).

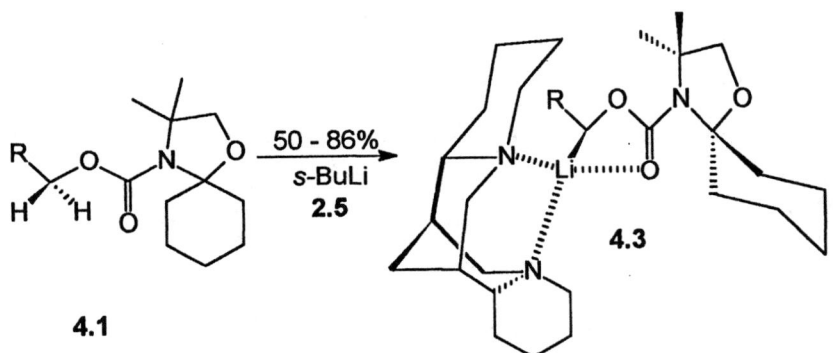

$R = Me, n\text{-}Pr, n\text{-}C_6H_{13}, i\text{-}Pr, MeO(CH_2)_4,$
$TBDMSO(CH_2)_4, (PhCH_2)_2N(CH_2)_2$

$EX = MeI, Me_3SiCl, Me_3SnCl, CO_2, AcOD$

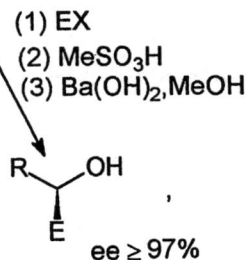

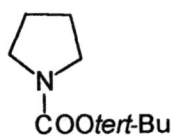

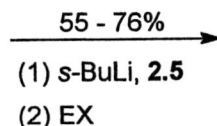

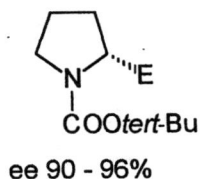

Figure 4.2

Rearrangements of epoxides to allyl alcohols (Figure **4.3**) can be induced by lithium amides. The mechanism of these reactions involves removal of an α-proton, *syn* to the epoxide oxygen. Reactions of epoxides **4.4** and **4.5** with chiral lithium amides lead to chiral allyl alcohols with high enantiomeric excesses [557, 559]. The most efficient chiral amide is derived from **1.64** (R = $CH_2N(CH_2)_5$) having the (*S*)-proline skeleton. Proton removal takes place through a chelated structure so that steric hindrance is minimized (Figure **4.3**). From other epoxides, the selectivity is lower. *N,O*-Dilithiated norephedrine **1.14** (R = H) also gives also a good selectivity (ee = 86%) in rearrangement of epoxide **4.5** conducted in THF [991]. Potassium alcoholates of *N,N*-dimethylephedrine **1.14** (R = Me) or -pseudoephedrine **1.78** (R = Me) promote the enantioselective dehydrobromination of prochiral β-bromoacid **4.6** [992] (Figure **4.3**).

4.2 ASYMMETRIC PROTONATIONS BY CHIRAL PROTON DONORS

Asymmetric protonations of prochiral enolates or enamines by enantiopure carboxylic acids typically occur with low enantioselectivity, but there are some exceptions. P. Duhamel, L. Duhamel and coworkers accomplished the " deracemization " of Schiff bases of α-aminoesters [552]. The best selectivities (ee 70%) are obtained when the substrates are deprotonated by Li (*R*)-*N*-ethylphenethylamide, and then reprotonated at –70°C by (*R,R*)-diacyltartaric acid **2.2** (R = *tert*-BuCO) [154] (Figure **4.4**). In another successful application, asymmetric protonation of the potassium enolate of racemic benzoin **4.7** by (*R,R*)-**2.2** (R = *tert*-BuCO) gives the (*S*)-enantiomer with a good enantiomeric excess [552] (Figure **4.4**).

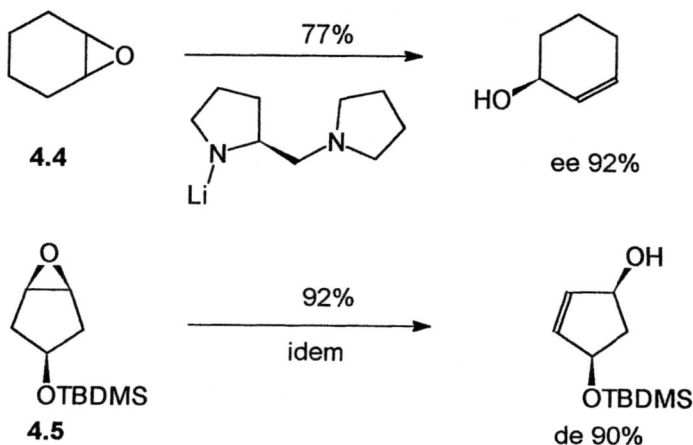

Figure 4.3

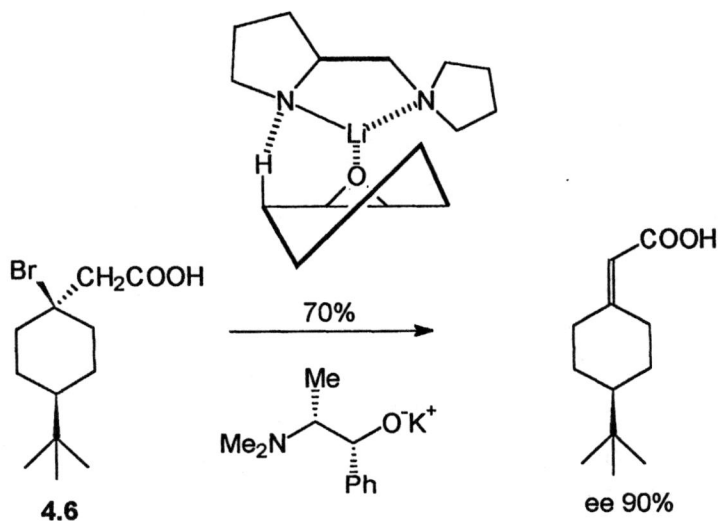

Figure 4.3 (continued)

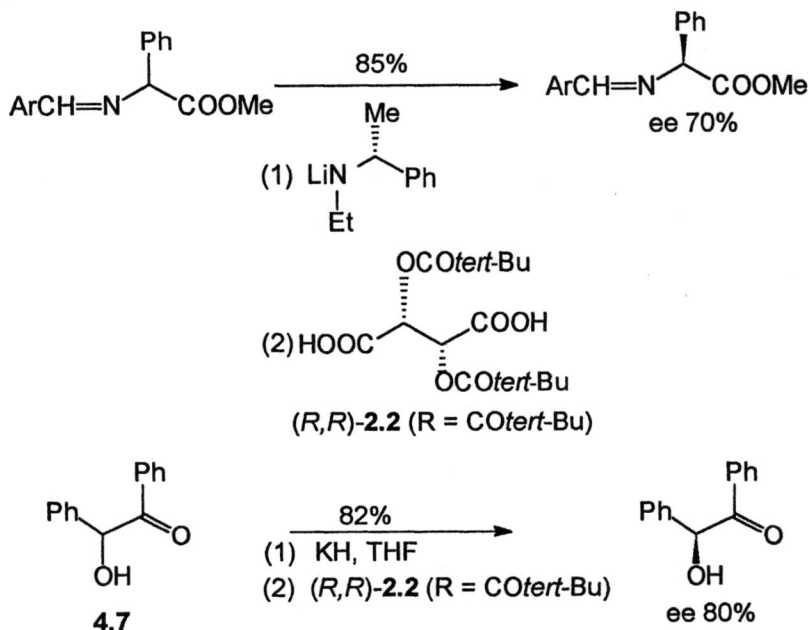

Figure 4.4

Enantioselective protonation of prochiral enolates **4.8** can be accomplished at −100°C by using enantiopure alcohols, the most efficient of which is (S)-**2.1**

(R = *i*-Pr, R' = Me) [584]. The highest selectivity is obtained when R = PhCH$_2$
(Figure **4.5**). Vedejs and Lee have performed asymmetric protonation of enolates
of α-arylacetamides **4.9** with chiral triamine **1.64** (R = CH$_2$NMeCH$_2$CH$_2$NMe$_2$).
These reactions are conducted at –78°C in the presence of BF$_3$•Et$_2$O. The use of
2 equivalents of *s*-BuLi and **1.64** first generates a mixed aggregate [77]. The
aggregate is likely the least hindered one, and the role of BF$_3$•Et$_2$O is to enhance
the acidity of the N–H bond (Figure **4.6**).

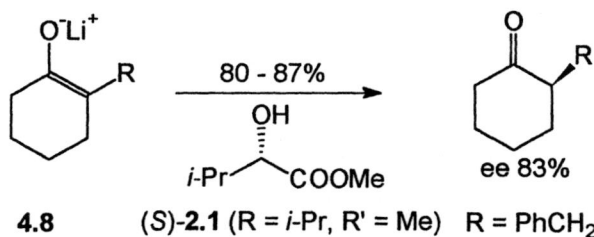

4.8   (S)-**2.1** (R = *i*-Pr, R' = Me) R = PhCH$_2$

Figure 4.5

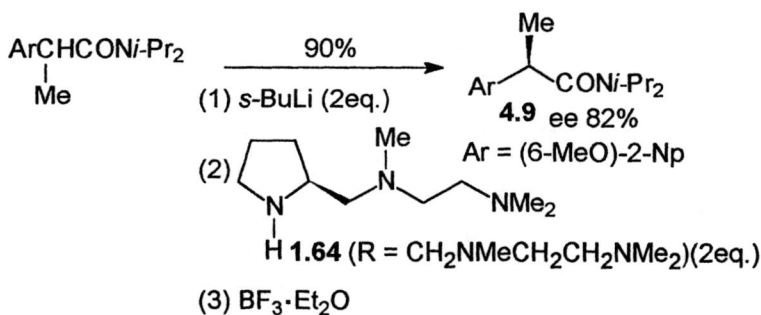

Figure 4.6

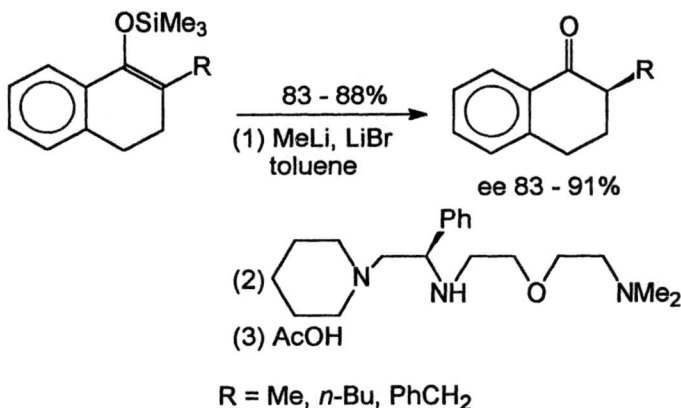

R = Me, n-Bu, PhCH₂

Figure 4.6 (continued)

Another enantiopure amine **2.4** (X = CH₂, R = CH₂CHPhNHCH₂CH₂OCH₂-NMe₂) has been introduced by Koga and Yasukata [993]. This amine effects the enantioselective protonation of achiral lithium enolates prepared from racemic 2-alkyl-1-tetralone trimethylsilylethers by the action of MeLi-LiBr in toluene. In the absence of LiBr, no asymmetric induction is observed (Figure **4.6**).

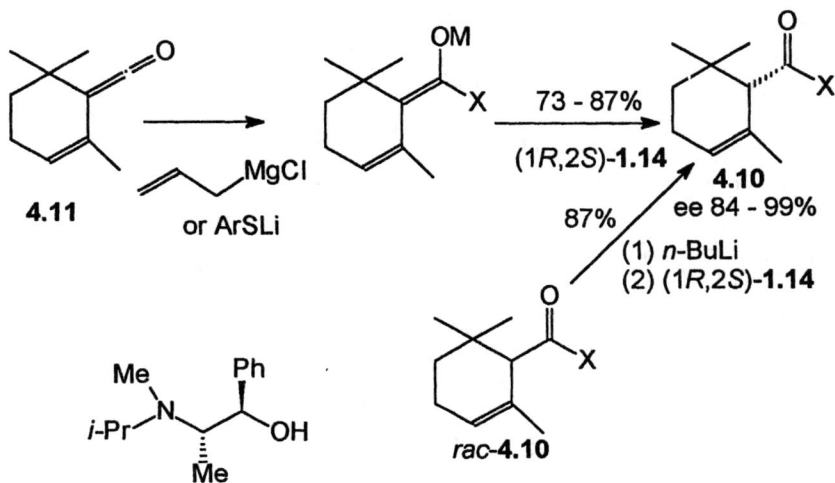

(1R,2S)-**1.14** (NR₂ = NMei-Pr) X = CH₂CH=CH₂, SAr

Figure 4.7

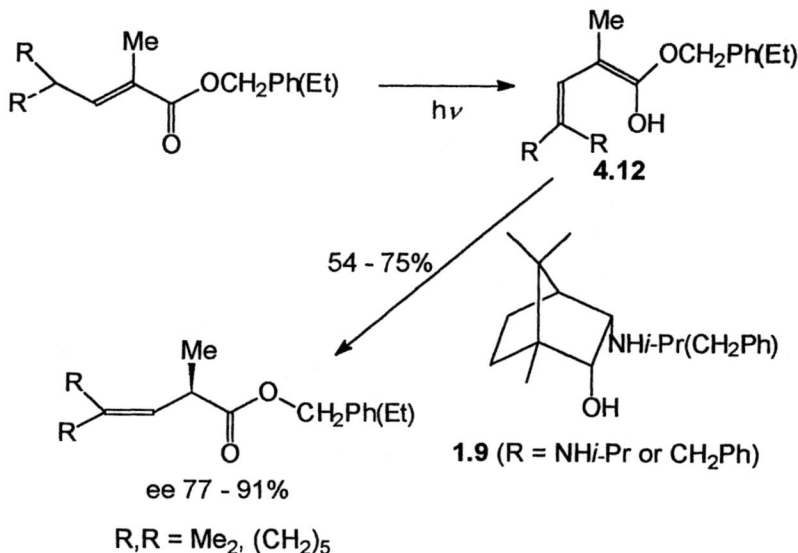

Figure 4.7 (continued)

Fehr and coworkers [555, 994] used (1R, 2S)- and (1S, 2R)-N-i-Pr ephedrines **1.14** (NR$_2$ = N-i-PrMe) as proton donors in the synthesis of (R)- or (S)-damascone, a fragance. Two methods were developed (Figure 4.7). In the first method, deprotonation of racemic arylthiolesters **4.10** (X = SPh, S2-Np) with n-BuLi at –100°C, followed by protonation with either enantiomer of **1.14** (NR$_2$ = Ni-PrMe) at –20°C, leads the corresponding thiolester with an excellent enantiomeric excess (99%). When using esters **4.10** (X = OPh), the enantiomeric excess is lower. In the second method, enolates are generated from the corresponding ketene **4.11** either by action of ArSLi or allylmagnesium chloride, and protonation by (1R, 2S)-**1.14** leads to the corresponding derivative (S)-**4.10** (X = SPh or CH$_2$CH=CH$_2$) with a very high selectivity. A similar method has been used by Takeuchi and coworkers [995]. They reacted arylketenes with allyl or benzyl halides/SmI$_2$, and protonated the enolates thus formed with a chiral diol.

Pete and coworkers [996, 997, 998] have generated prochiral dienols **4.12** by photoreactions of α,β-unsaturated esters and lactones bearing a proton in the γ-position. Asymmetric protonation of **4.12** is performed at –45°C with catalytic amounts of aminoalcohols such as **1.9** (R = NHi-Pr or NHCH$_2$Ph) (Figure 4.7). Good selectivities are observed only if the α,β-unsaturated ester bears a tertiary carbon in the γ-position.

While the reactions of ketenes with enantiopure alcohols usually give modest selectivities [769], the use of (S)-ethyl lactate (S)-2.1 (R = Me, R' = Et), (R)-isopropyl lactate (R)-2.1 (R = Me, R' = i-Pr) or (R)-pantolactone 1.16 as proton donors has allowed the highly enantioselective formation of 2-arylpropionic esters. A mild hydrolysis (AcOH/HCl or LiOH) leads to the corresponding acids, which are anti-inflammatory drugs [554, 923] (Figure 4.8). This method has been extended by Durst and Koh [861, 999] to the synthesis of enantioenriched α-halogenated esters, which are precursors of aminoacids (Figure 4.8).

4.3 PROTONATION OF SUBSTRATES BEARING CHIRAL RESIDUES

Pete and coworkers [1000, 1001] have irradiated α,β-ethylenic esters of enantiopure alcohols, and the intermediate dienols that form are protonated either with N,N-dimethylethanol at −35°C in hexane [1001] or with i-PrOH or tert-BuOH [1000]. The chiral auxiliaries are diacetoneglucose 1.48 or (S)-ethyl lactate 2.1 (R = Me, R' = Et), or better yet the corresponding acid (Figure 4.9). The cleavage of the chiral auxiliary is accomplished by treatment with PhCH$_2$OH/Ti(Oi-Pr)$_4$ or by mild hydrolysis. After the subsequent reaction with diazomethane, nonracemic benzyl or methyl α-alkylated-β,γ-unsaturated esters are obtained with high enantiomeric excesses (Figure 4.9).

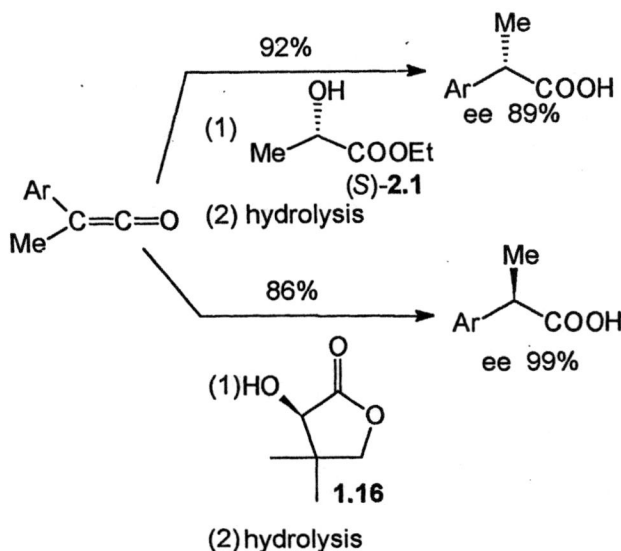

Figure 4.8

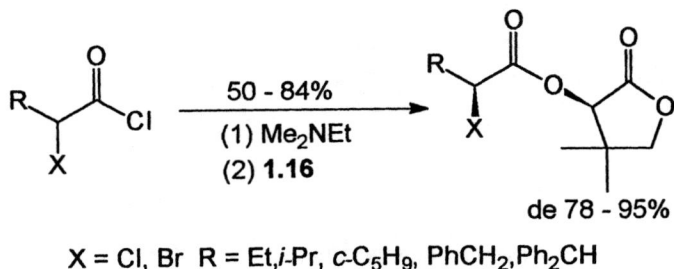

Figure 4.8 (continued)

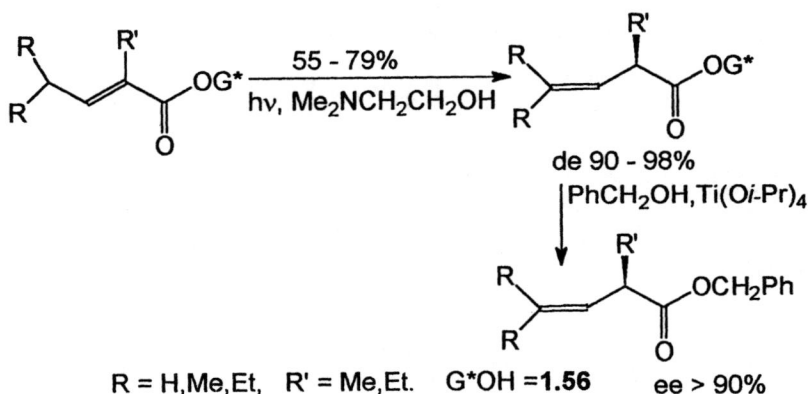

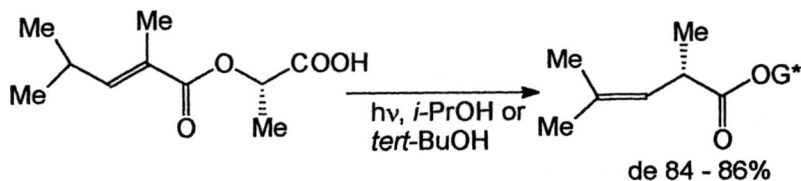

Figure 4.9

The protonation of lithiated anionic species formed from carbonyl compounds bearing a chiral group can be diastereoselective and can lead, after cleavage of the chiral group, to enantioenriched compounds. However, disappointing results are often obtained. Enantioselective protonation by ethanol of the anion of an imine of 2-methylcyclohexanone grafted on a chiral polymer has been observed [250, 552]. Dipole-stabilized organolithium reagents generated from chiral oxazolines **4.13** or formamidines **4.14** are deuterated with a good selectivity, however, the origin of the selectivity is difficult to interpret [391, 400, 1002, 1003]. Chiral dipole-stabilized α-lithiated 1,3-imidazolidinones are generated from the corresponding stannanes **4.15** (X = Bu$_3$Sn) bearing skeleton **1.131** [1004] by treatment with *n*-BuLi at −78°C. These intermediates suffer highly stereoselective deuteration by DCl/D$_2$O with retention of configuration leading to **4.15** (X = D) (Figure **4.10**).

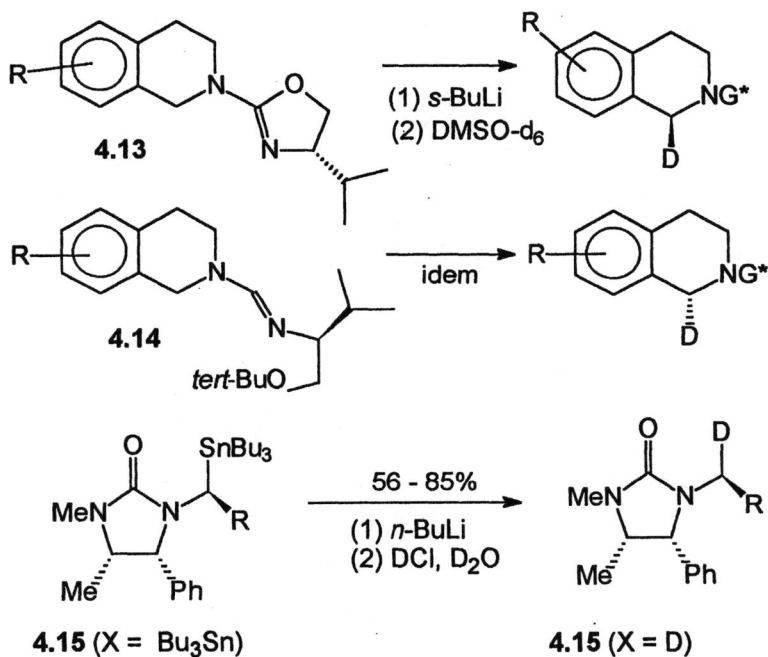

Figure 4.10

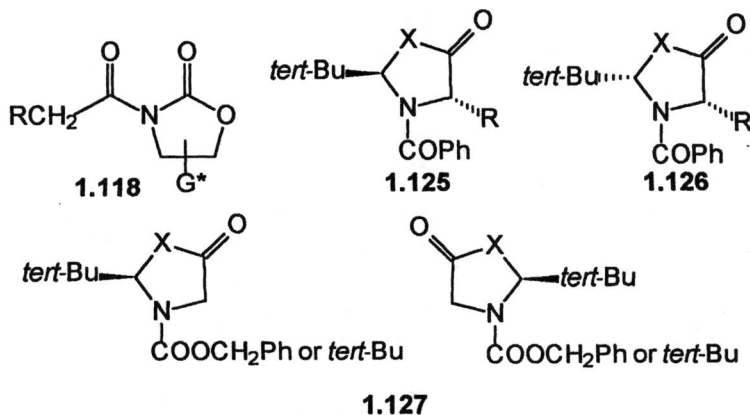

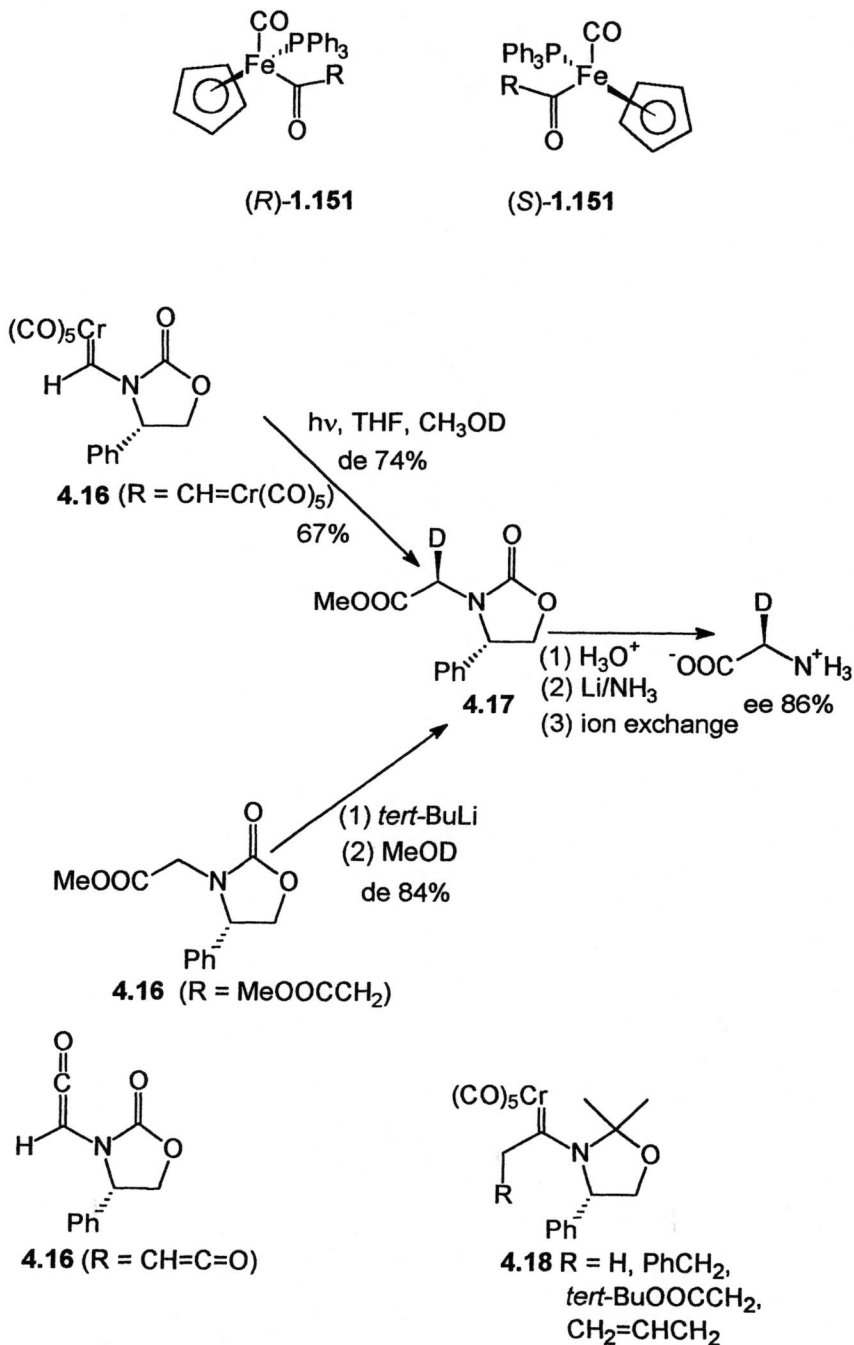

(R)-**1.151**　　　　　(S)-**1.151**

Figure 4.11

Stereoselective protonation of chiral boron enolates has been attempted by Hunig and coworkers [1005], but disappointing selectivities have been observed in most cases. Protonations or deuterations of alkali enolates of *N*-acyloxazolidinones **1.118** derived from **1.116** and **1.117**, of *N*-acylderivatives **1.125**, **1.126** and **1.127** [81], or of ironcarbonyl acylenolates of **1.151** (R=R'CH$_2$) [522] are highly stereoselective. Hegedus and coworkers [861, 1006] have observed that deuteriation of ketene **4.16** (R = CH=C=O) bearing a chiral 1,3-oxazolidinone group with MeOD is not stereoselective. The same reaction on the chromium-ketene complex generated by irradiation of **4.16** (R = CH=Cr(CO)$_5$ leads selectively to **4.17**. This diastereoisomer **4.17** is also obtained by protonation of the lithium enolate of **4.16** (R = MeOCOCH$_2$). Hydrolysis of the methyl ester, followed by reductive cleavage of the chiral auxiliary by Li/NH$_3$, gives (*R*)-(2-D)-glycine with a good enantiomeric excess (Figure 4.11). Other nonracemic α-amino-acids are available from similar chromium-ketene complexes of oxazolidines **4.18** bearing the skeleton of **1.85** (R' = H, R" = Me) [861].

CHAPTER 5

ALKYLATIONS AND RELATED REACTIONS

In this chapter, the reactions of carbon nucleophiles (carbanions, enolates, enamines, etc.) with electrophiles that are not prochiral will be described. These electrophiles include alkyl halides and related reagents, epoxides and analogs, symmetrical carbonyl compounds and CO_2. Also included are reagents inducing halogenation, amination and hydroxylation. In most cases, single diastereodifferentiation takes place, and chirality is introduced on a nucleophile bearing a chiral auxiliary.

The transition states of these reactions usually lie early on the reaction coordinate and the overlap of the nucleophile HOMO, which has some π character, with the electrophile LUMO is as large as possible. Repulsive interactions (steric interactions) are the most important factors in controlling stereochemistry, although the A (1,3) strain [54, 167] or hydrogen bonding sometimes come into play. Rigidification induced by chelation about the metal associated to the anionic nucleophile often increases the selectivity of the reactions. This is usually due to the high population of a single reactive conformer of the nucleophile. [167].

5.1 ALKYLATIONS OF NONCONJUGATED DIPOLE-STABILIZED CARBANIONS

Hoppe, Beak and their coworkers generated chiral lithiated carbanionic species by deprotonating carbamates **4.1** and **4.2** with *s*-BuLi in the presence of sparteine **2.5**. These dipole-stabilized species can be trapped by various electrophiles (MeI, Me$_3$SiCl, Me$_3$SnCl, CO_2), as described previously in § 4.1. Pearson and coworkers [1004] alkylated the dipole-stabilized lithiated carbanions formed by tin-lithium exchange from **4.15** and 1,3-oxazolidinone analogs **5.1**. Reactions of these α-lithio amides with Bu$_3$SnCl, ClCO$_2$Et and cyclohexanone are highly stereoselective (Figure 5.1). However, the removal of the chiral auxiliary from the product has not been carried out. Some isomerization of the intermediate carbanionic species occurs in the reactions of the α-isomers of **4.15**.

5.2 ALKYLATIONS OF BENZYL AND ALLYL CARBANIONS

Benzyl and allyl carbanions are usually planar, and several strategies for asymmetric alkylation have been proposed, including:

•Introduction of a chiral ligand on the metal, for example, sparteine **2.5**, or

•Introduction of a chiral group on the substrate in the vicinity of the carbanionic center, or

•In the case of benzylic anions, introduction of a bulky organometallic substituent, for example, the chromium tricarbonyl group, on one of the faces of the aromatic ring.

5.2.1 Use of a Chiral Ligand

Beak and Du [1007] performed the asymmetric alkylation of racemic dipole-stabilized dilithium reagents generated by benzylic metalation of **5.2** (X = H or n-Bu$_3$Sn) with s-BuLi (Figure 5.2). Sparteine **2.5** must be added to anionic species prior to the electrophile. These reactions are run in THF/*tert*-butyl methyl ether at −78°C. No asymmetric induction is observed when this experimental procedure is used for the alkylation of **4.2**.

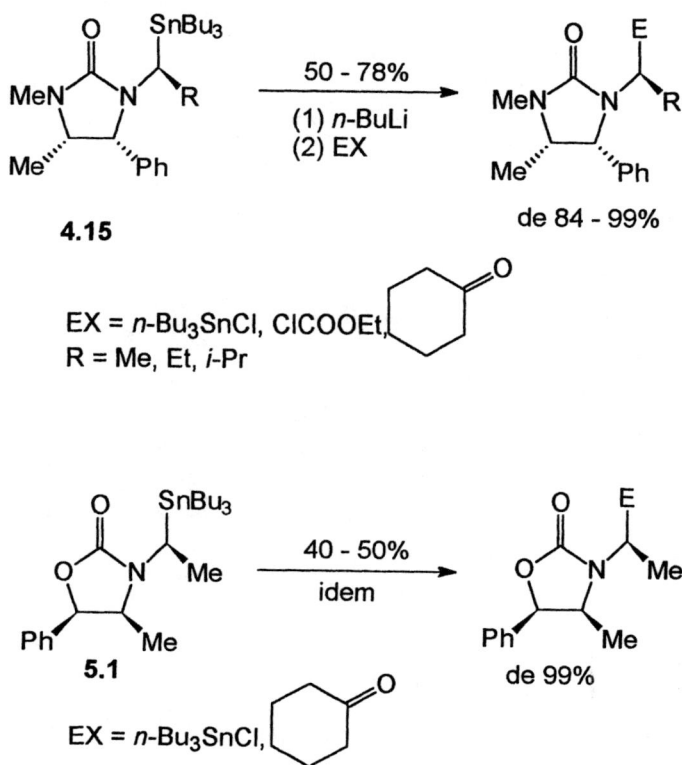

Figure 5.1

X = H, *n*-Bu$_3$Sn
EX = Me$_3$SiCl, MeI, *n*-BuI, PhCH$_2$Br, Ph$_2$CO

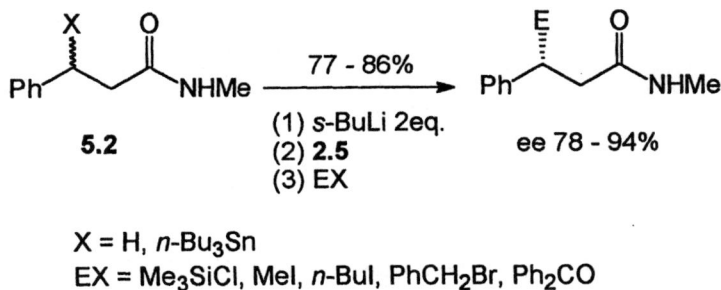

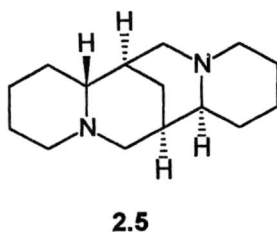

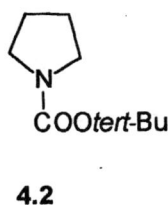

Figure 5.1 (continued)

5.2.2 Aminooxazolines and Formamidines [1008]

Several heterocyclic rings promote the formation of lithium carbanions α- to their heteroatoms. These carbanions can be stabilized by dipolar interactions [982], and pyramidalization of the anionic carbon center may sometimes occur [400].

Gawley and coworkers have studied the alkylations of lithiated derivatives of aminooxazolines **5.3** and **4.13** [326, 400], while Meyers and coworkers have developed alkylations of related derivatives of formamidines **5.4** and **4.14** [391, 392, 394, 1002, 1009]. The formamidines must bear an alkoxy group (preferably *tert*-BuO) on the carbon vicinal to the chiral substituent in order to observe a high asymmetric induction. The regioselectivity of alkylations of allylic derivatives is often problematic, so the chemical yields of alkylated products formed from **5.3** and **5.4** are often poor. However, interesting results are obtained in alkylation of the bicyclic system **5.4** (R,R = (CH$_2$)$_4$), which is a precursor of an analgetic drug [400].

Isoquinolines are precursors of many alkaloids, and asymmetric alkylations of these heterocycles have been successfully carried out with the same two auxiliaries

(Figure **5.2**). Both auxiliaries are derived from (S)-valine **1.59** (R = i-Pr, R' = H),
but they lead to opposite enantiomers after cleavage with N_2H_4.

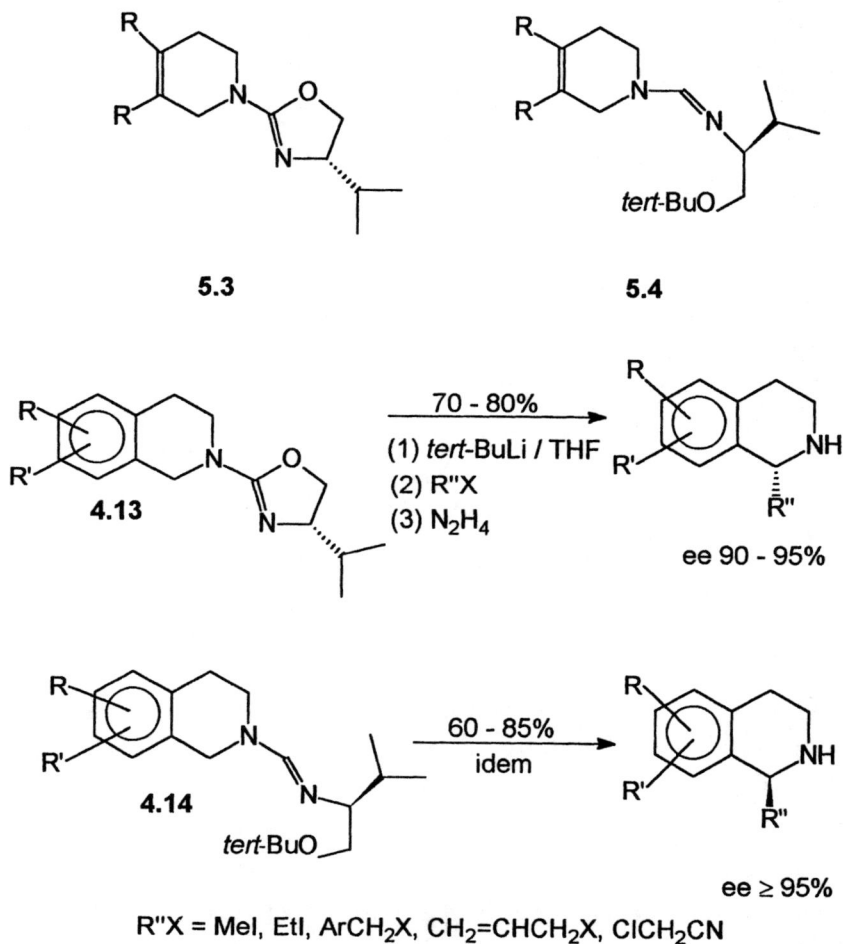

5.3   **5.4**

70 - 80%
(1) *tert*-BuLi / THF
(2) R"X
(3) N_2H_4

4.13

ee 90 - 95%

60 - 85%
idem

4.14

ee ≥ 95%

R"X = MeI, EtI, ArCH$_2$X, CH$_2$=CHCH$_2$X, ClCH$_2$CN

Figure 5.2

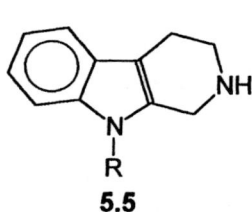

5.5

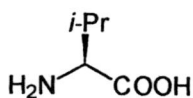

1.59 (R = i-Pr, R' = H)

This method has been extended to alkylation of β-carbolines **5.5** on the way to alkaloids of the yohimbine [1002] and corynantheine [1009] families. When carried out at the low temperatures (−78°C), these alkylations are highly stereoselective, especially when the aromatic ring is substituted by electron-donating groups. Rigidification of the lithiated species due to chelation is believed to play an important role. Gawley has suggested that the C–Li bond has some covalent character because the alkylation of the potassium anion species is not stereoselective [326]. The alkyl group is usually introduced on the same side of the molecule as the *i*-Pr group in the conformation shown in Figure **5.2**. However, when a second alkylation is subsequently performed, the opposite direction of approach is observed and the second alkyl group is introduced on the other face (Figure **5.3**). These observations have been interpreted by proposing different conformations of the lithium chelate according to its substitution pattern [1010].

Alkylation of the lithium anion of camphor benzylimine with benzyl halides takes place with high stereoselectivity [1011].

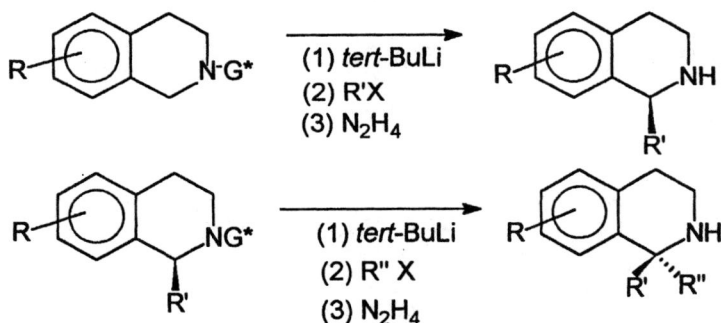

Figure 5.3

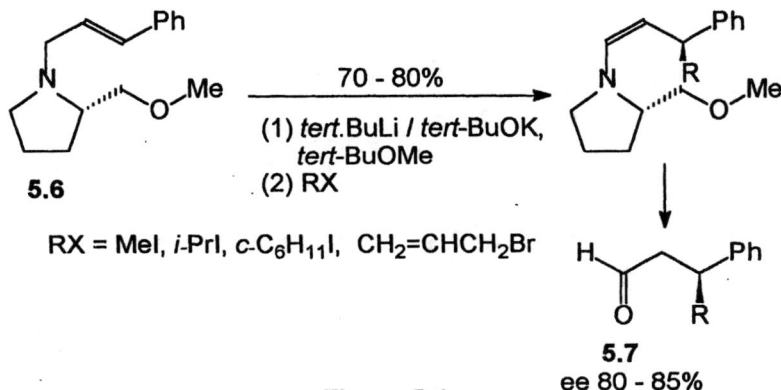

RX = MeI, *i*-PrI, *c*-C$_6$H$_{11}$I, CH$_2$=CHCH$_2$Br

5.7
ee 80 - 85%

Figure 5.4

Figure 5.4 (continued)

5.2.3 Amines, Ethers, Silanes and Phosphonamides

Asymmetric alkylations of anions of chiral allylamines **5.6** have been performed at −78°C by Ahlbrecht and Enders [1012]. These reactions form 2-enamines with high regio- and stereoselectivities. These enamines are hydrolyzed to generate 3-phenylalkanals **5.7** with a good enantioselectivity (Figure 5.4). The same types of nonracemic aldehydes **5.7** were already obtained with a similar enantioselectivity by Mukaiyama and coworkers [148] via selective alkylation of potassium anions of chiral allyl ethers **5.8** (Figure 5.4).

Alkylations of benzyl or allyl anions α- to a silicon atom bearing a chiral substituent such as **2.79** have been carried out by Chan and coworkers in diethylether or toluene at −78°C [738, 1013, 1014]. Subsequent oxidative cleavage of the C–Si bond gives access to benzyl **5.9** (R = Ph) or cinnamyl alcohols **5.9** (R = PhCH=CH) with an excellent enantiomeric excess. In the case of cinnamyl derivatives, the regioselectivity is excellent (Figure 5.5). This high selectivity has been attributed to the formation of a chelated anionic species involving the OMe group of the auxiliary. By using ethylene oxide or *gem*-substituted symmetrical epoxides as alkylating agents, (S)-1,3-diols have been prepared with an excellent enantiomeric excess (Figure 5.5). Monosubstituted epoxides lead to regioisomers. This general approach has also been extended to propargyl analogs [563].

Denmark and coworkers [354] have examined the alkylation of α-lithiated benzyl carbanions generated from diastereoisomeric chiral phosphonamides **1.96** and **5.10**. The stereoselectivity of this process depends on the configuration at phosphorus, and it is alway higher when starting from **1.96**. It is believed that planar benzylic anions are alkylated on their least sterically hindered face. Strong support for this rationale derives from the X-ray crystal structure of **A**. After hydrolysis and treatment of the products with diazomethane, (R)- or (S)-dimethyl phosphonates can be obtained (Figure 5.6). Hanessian and coworkers [311, 312] performed similar alkylations of phosphonamides **1.74** (Y = Me, Cl) derived from (R,R)- or (S,S)-*trans*-cyclohexanediamines bearing a C_2 axis of symmetry. The reaction of the intermediate anionic species on its less hindered face, followed by hydrolysis of the product, leads to (*R*)- or (*S*)-substituted phosphonic acids according to the absolute configuration of **1.74** (Figure 5.6).

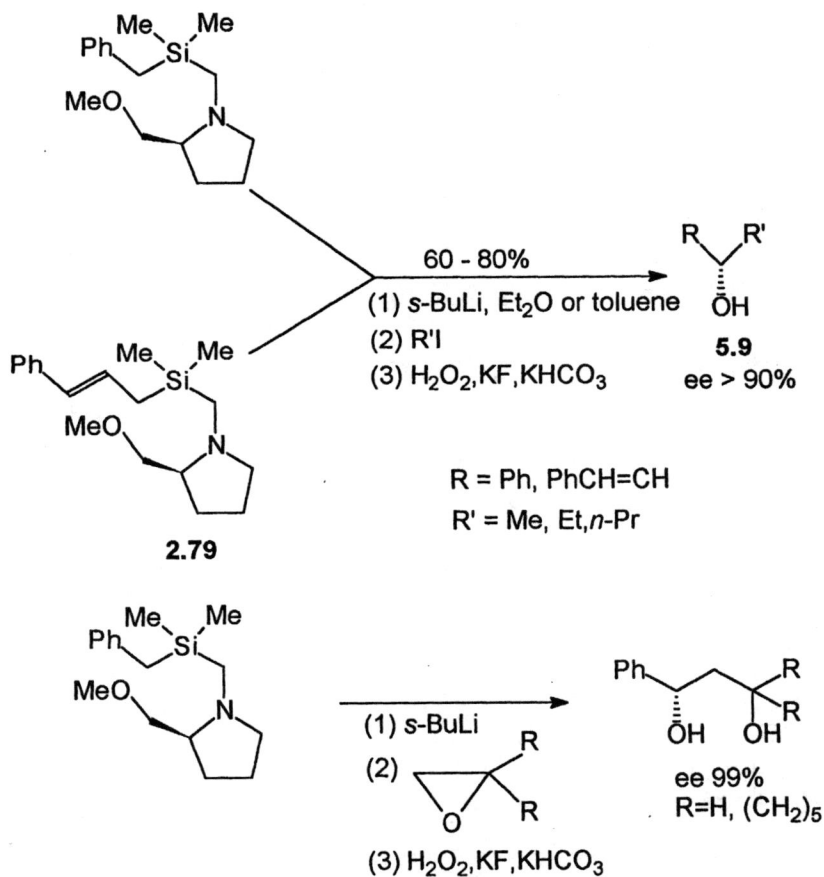

Figure 5.5

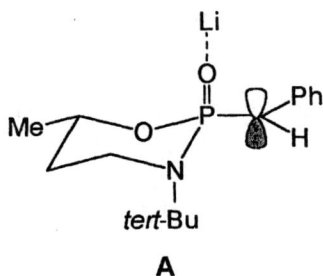

A

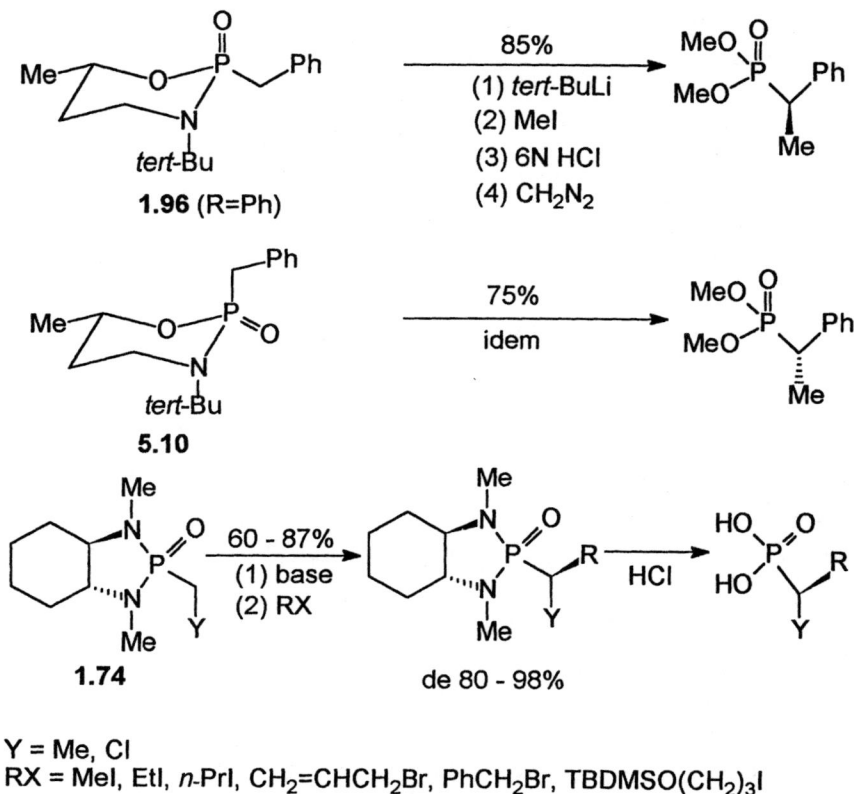

Figure 5.6

5.2.4 Arenechromium Tricarbonyl Complexes [540, 542]

A chromium tricarbonyl group shields one face of the aromatic ring to which it is attached, and it also increases the acidity of any benzylic protons. The reactions of benzyl anions of arene chromium tricarbonyls are usually highly stereoselective [543]. To perform enantioselective alkylations, it is mandatory to have enantiopure chromium tricarbonyl complexes. Complexes generated from enantiopure alkaloids such as **5.11** are deprotonated by *n*-BuLi and selectively alkylated on the face opposite the $Cr(CO)_3$ group (Figure 5.7). For substrate **5.11**, a temporary silylation of the aryl group is necessary to prevent ring alkylation. Benzylimines **5.12** prepared from resolved aldehyde complexes (§ 1.8.3) can be deprotonated and alkylated. After decomplexation, α-alkylbenzylamines are obtained with a high enantiomeric excess [540] (Figure 5.7). Davies and coworkers [1015] have methylated complex **1.155** (R = MeO, R' = H, Y = $MeOCH_2$) with a high selectivity.

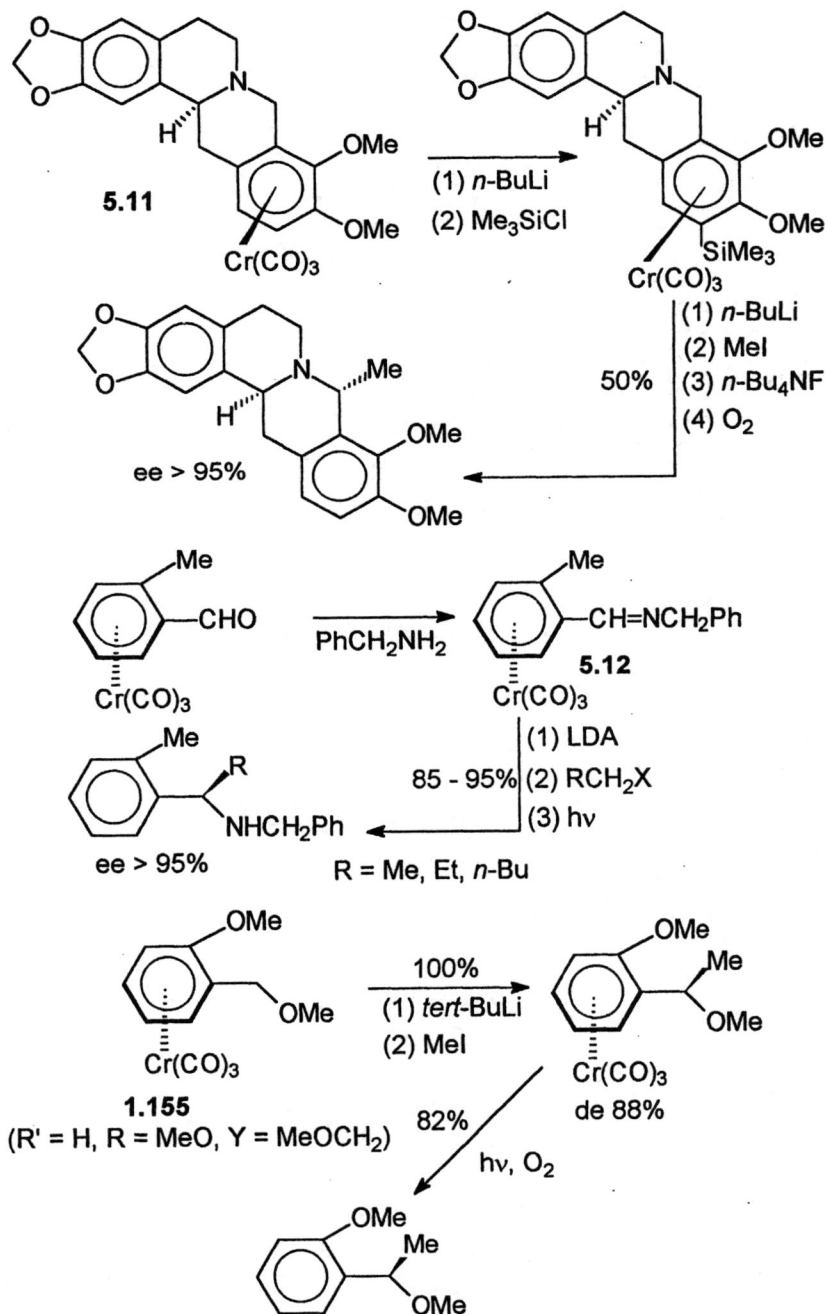

Figure 5.7

Deprotonation and methylation take place on the face of the molecule opposite to the $Cr(CO)_3$ group via a planar, nonchelated anionic species. After decomplexation, the corresponding nonracemic ether is obtained.

5.3 ALKYLATIONS OF METAL ENOLATES

Alkali enolates are among the most popular reagents in organic synthesis [624e] and their structures have been thoroughly examined [77, 81]. The reactivity of enolates is strongly dependent upon their interaction with the associated cation (solvation, aggregation state, etc.). The E- or Z-geometry is also important, and this is often related to the conditions of enolate generation (base, solvent) [1016]. Asymmetric induction in the alkylation of enolates may be effected by introduction of a chiral group on the enolate or on the alkylating agent, or by use of chiral additives or catalysts bearing chiral ligands.

5.3.1 Alkylations of Aldehydes, Ketones and Derivatives

Alkylations of lithium enolates of ketones in the presence of chiral bases has been widely studied [77, 559, 1008], but disappointing results were often obtained. However, Koga and coworkers performed asymmetric alkylations of cyclohexanone and tetralone lithium enolates in toluene at low temperatures [108, 1017]. The enolates are generated from the Li amide of chiral diamine 2.4 (X = CH_2, R = $MeOCH_2CH_2OCH_2CH_2$). The presence of LiBr is essential to observe a high enantioselectivity (Figure 5.8), and the involvment of mixed aggregates is implied.

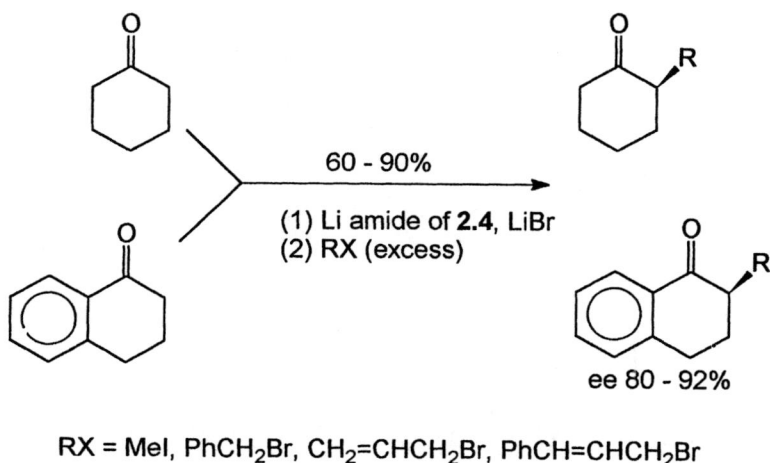

60 - 90%

(1) Li amide of 2.4, LiBr
(2) RX (excess)

ee 80 - 92%

RX = MeI, $PhCH_2Br$, CH_2=$CHCH_2Br$, $PhCH$=$CHCH_2Br$

Figure 5.8

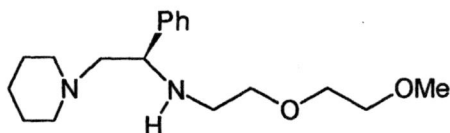

2.4 (X = CH$_2$, R = MeOCH$_2$CH$_2$OCH$_2$CH$_2$)

In most cases, asymmetric alkylations are performed on carbonyl derivatives such as imines or hydrazones. To rigidify the anionic species by chelation to the alkali cation, basic functional groups are incorporated in the imine or hydrazone skeleton [167, 169, 252, 253, 261, 262, 315, 316a, 317]. C$_2$-Symmetry can also be used as a conformational control element, as illustrated by the asymmetric alkylation of a chiral enamine of cyclohexanone described by Whitesell and coworkers [275] (Figure 5.9). Among the imines of cyclic ketones prone to form N-lithium chelates, those derived from 1-methoxymethylphenethylamines (R)- and (S)-**1.57**, from 1-phenyl-2-methoxyphenethylamine **1.58**, and from (S)-valine or (S)-*tert*-leucine *tert*-butyl esters **1.59** (R = *i*-Pr or *tert*-Bu, R' = *tert*-Bu) have given the most interesting results (Figure **5.9**). Alkylation is thought to take place by reaction of a five-membered chelate intermediate from the least hindered side, as shown on **5.13**. A similar chelate involving the ester group may be envisioned in the case of imines of **1.59**.

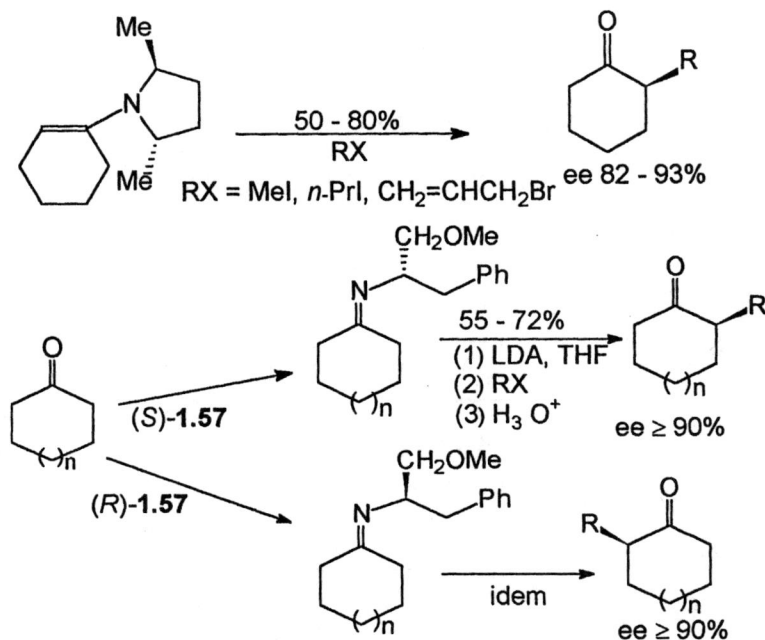

Figure 5.9

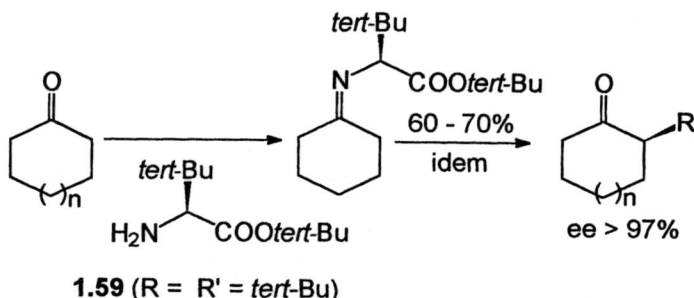

1.59 (R = R' = *tert*-Bu)

n = 0, 1, 2; RX = MeI, Me$_2$SO$_4$, *n*-PrI, CH$_2$=CHCH$_2$Br, PhCH$_2$Br

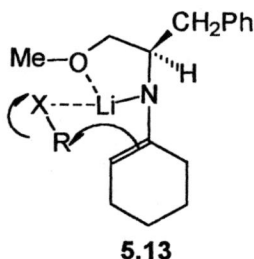

5.13

Figure 5.9 (continued)

This methodology can be applied to imines of 2-substituted cyclohexanones, which are selectively enolized from the most substituted side. Chiral 2,2-dialkyl cyclohexanones are obtained with an ee > 95%. With large ring ketones (n > 2) or prochiral linear ketones, the facile equilibration of Z- and E-lithium enamides is a problem. This isomerization results in less stereoselective alkylations, except in a few special cases [169, 252, 253]. However, the presence of a functional group such as an ester in the β-position leads to the formation of a single enamine **5.14**. A single chelated lithium enamide is formed on deprotonation of **5.14** with LDA in toluene. Alkylation in the presence of either THF or HMPA leads, after hydrolysis, to ethyl 2-alkyl-2-methylacetoacetate with a good enantiomeric excess [253, 1018, 1019] (Figure **5.10**). Similar results are obtained starting from 2-carboethoxycyclohexanone [1018]. Dilithium anionic species generated from chiral enaminones **5.15** also suffer stereoselective alkylation by alkyl halides at −100°C [1020]. After hydrolysis, substituted 1,3-diketones can be obtained with a high enantiomeric excess. Intramolecular chelation allowed Denmark and Ares [1021] to perform the asymmetric alkylation of chiral imine **5.16** via a complex of the anion with LDA in the presence of DMPU (Figure **5.11**).

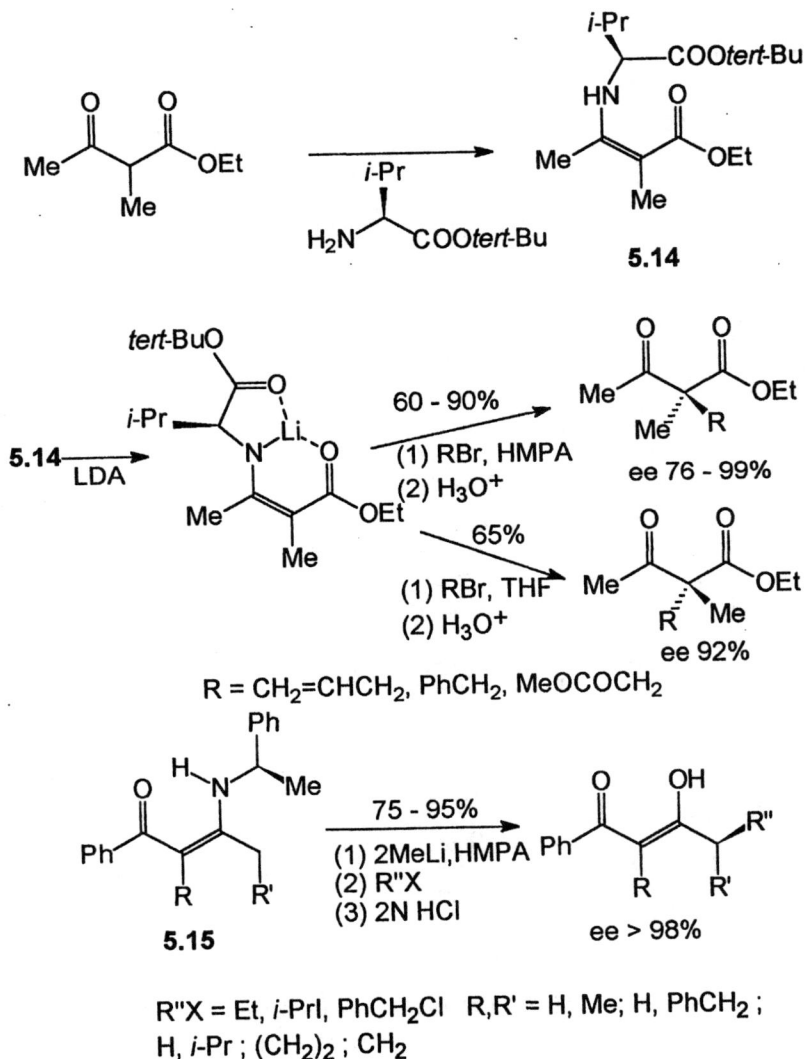

R"X = Et, *i*-PrI, PhCH₂Cl   R,R' = H, Me; H, PhCH₂ ;
H, *i*-Pr ; (CH₂)₂ ; CH₂

R"X = Et, *i*-PrI, PhCH₂Cl   R,R' = H, Me; H, PhCH₂ ; H, *i*-Pr ; (CH₂)₂ ; CH₂

Figure 5.10

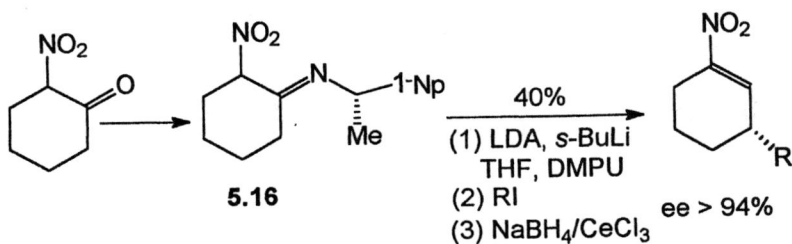

Figure 5.11

   Enders's Ramp and Samp (*R*)- and (*S*)-**1.76** hydrazones are among the most useful reagents for α-alkylations of aldehydes and ketones [169, 261, 315, 317, 1022]. Lithiated anions are very easily generated from aldehyde hydrazones **5.17** or from symmetrical ketone hydrazones **5.18**. The only limitation is the lack of regioselectivity in enolization of unsymmetrical hydrazones of RCH_2COCH_2R'. Alkylations by alkyl halides or epoxides are performed at –90°C in THF, and are highly stereoselective. After ozonolysis, α-branched aldehydes or ketones are obtained with an excellent enantiomeric excess (Figure **5.12**). The chelated anionic species has the E-C=C, Z-C=N geometry **5.19**, and the alkyl group is introduced on the least hindered face. In the presence of HMPA, the anionic species exhibits the Z-C=C, E-C=N geometry, but the alkylation stereoselectivity is lower. Alkylation of hydrazones of arylalkylketones or β-ketoesters is not very stereoselective. The Samp/Ramp reagents are also used for alkylation of cyclic ketones and enones, and α-ketoesters of bulky alcohols or phenols [1023, 1024]. With cyclic enones, alkylation takes place in the α'-position (Figure **5.13**).

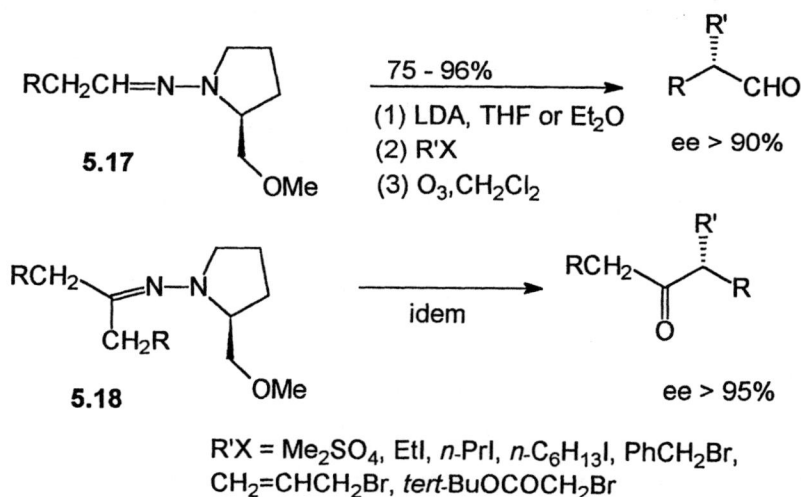

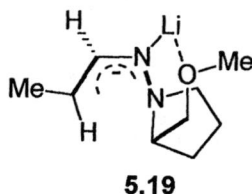

R'X = Me_2SO_4, EtI, *n*-PrI, *n*-$C_6H_{13}I$, $PhCH_2Br$,
CH_2=$CHCH_2Br$, *tert*-$BuOCOCH_2Br$

Figure 5.12

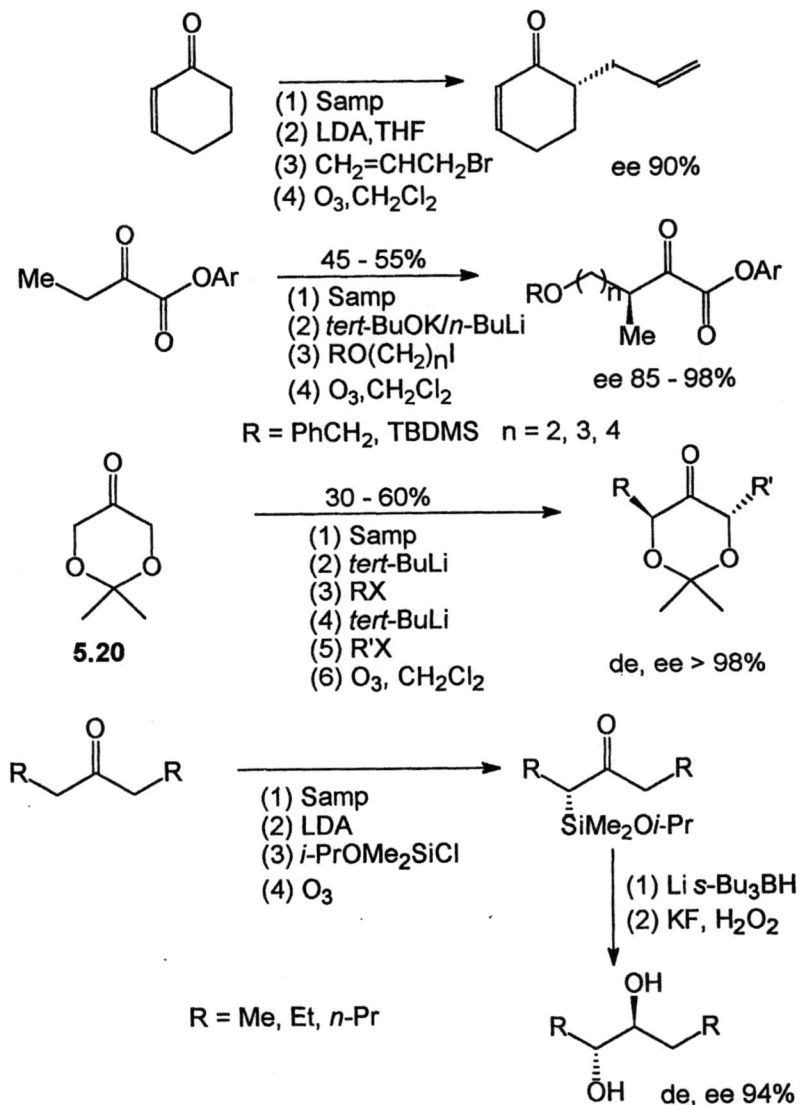

Figure 5.13

Starting from 2-cyanocycloalkanones, 2-alkyl-2-cyanocycloalkanones are obtained in high enantiomeric excess [1025]. 1,3-Dioxane-5-one **5.20** can be sequentially α,α'-di-alkylated with two alkyl halides, giving rise to nonracemic 1,3-diol derivatives [1026] from which deoxysugars polyols [1027] or an HIV-I protease inhibitor [1028] can be prepared with high selectivity. When trialkyl-chlorosilanes or i-PrOMe$_2$SiCl are used as silylating agents [1029, 1030], α-silicon substituted aldehydes or ketones are formed. These products can be transformed into nonracemic allylsilanes [1031] or 1,2-diols [1030] in a highly selective fashion (Figure **5.13**). Under trialkylsilyl triflate catalysis, tetrahydrofuran suffers ring opening by lithiated Samp or Ramp hydrazones at –78°C, giving rise to α-silyloxybutylated aldehydes and ketones with a high enantioselectivity (> 95%) [1032].

5.3.2 Alkylations of Carboxylic Acids and Derivatives [159]

Alkylations of esters, amides and related functionalities are usually carried out on molecules that bear a chiral group. This group can be fixed on the acyl substituent RCOG*, on another part of the molecule G*CH$_2$COY, or simultaneously on both sites. These three possibilities will be examined sequentially.

For the sake of simplicity, the nomenclature of ester enolates will be independent of the associated cation; O^-M^+ always takes priority over OR.

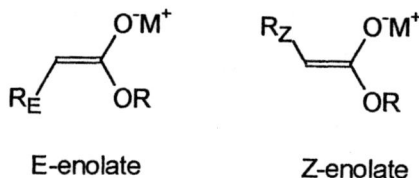

E-enolate   Z-enolate

5.3.2.1 *Derivatives Bearing a Chiral Acyl Group: Esters, Amides, Oxazolines, Oxazolidinones and Sultams*

Asymmetric alkylation of chiral esters RCH$_2$COOG* is usually poorly stereoselective, unless the alcohol residue is highly bulky. Useful ester derivatives bearing the bornane skeleton include **5.21**, **5.22** (R = PhMeNCOO, ArSO$_2$NPh) and **5.23** [147]. Deprotonation of **5.21** and **5.22** (R' = Me) with LICA in THF generates E-enolates, which are alkylated by primary halides on the least hindered face (Figure **5.14**). Z-Enolates are formed in the presence of HMPA, but their alkylation is less stereoselective. Similar results are obtained from **5.23** and analogs [1033]. When the alkylations are carried out with benzyloxyacetates PhCH$_2$OCH$_2$COOG*, chelated Z-enolates are generated and the opposite diastereoisomers are obtained after alkylation (Figure **5.14**). Reduction of the products by LiAlH$_4$ leads to chiral α-alkyl primary alcohols with moderate to high enantiomeric excesses [147]. Alkylation of binaphthol monoesters **5.24** is highly stereoselective [238, 239, 923] and occurs via reaction of a chelated dilithium E-enolate **5.25** on its less hindered face. After hydrolysis, α-alkylarylacetic acids are obtained with an excellent enantiomeric excess (Figure **5.15**).

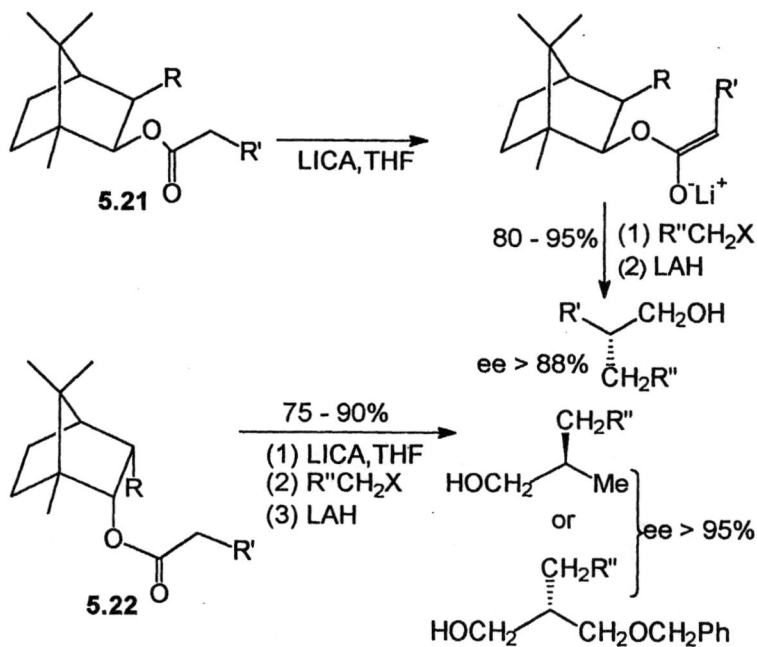

R' = Me or PhCH₂O (see text)

R' = Me or PhCH2O (see text)
R"CH₂X = EtI, MeI, n-C₁₄H₂₉I, n-C₈H₁₇I, PhCH₂Br

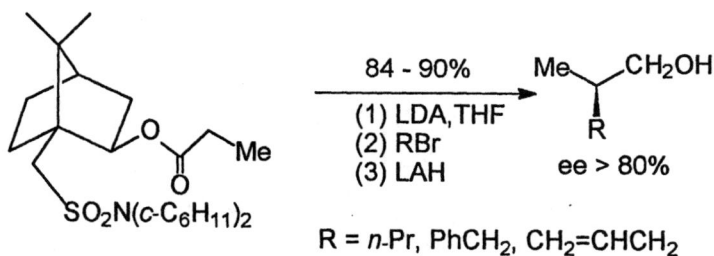

Figure 5.14

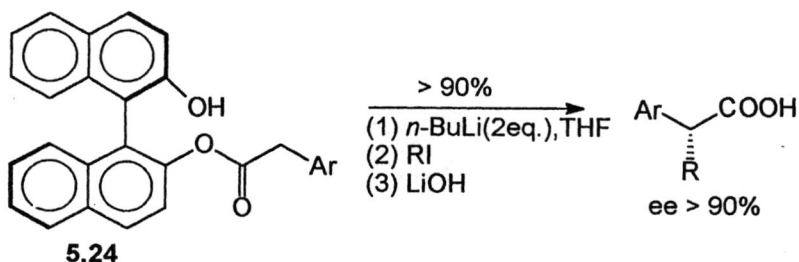

Figure 5.15

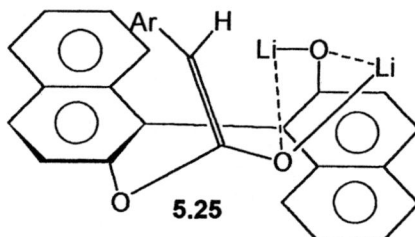

5.25

Figure 5.15 (continued

Phenmenthol monomalonates **1.20** are transformed by LDA into dilithiated anions, which also suffer diastereoselective alkylations on their least hindered face [166, 1008, 1034]. After LiAlH$_4$ reduction, enantioenriched 2,2-dialkyl-1,3-propanediols are obtained. Alternatively, Curtius reaction followed by hydrolysis gives rise to nonracemic α-substituted aminoacids (Figure **5.16**). The most interesting selectivities are observed when R and R' are PhCH$_2$ and Me. The highly stereoselective alkylation of the menthyl monoester of (1R, 2R)-cyclopentanedicarboxylic acid by methallyl bromide is the first step of the total synthesis of (−)-ptaquilosin [1035].

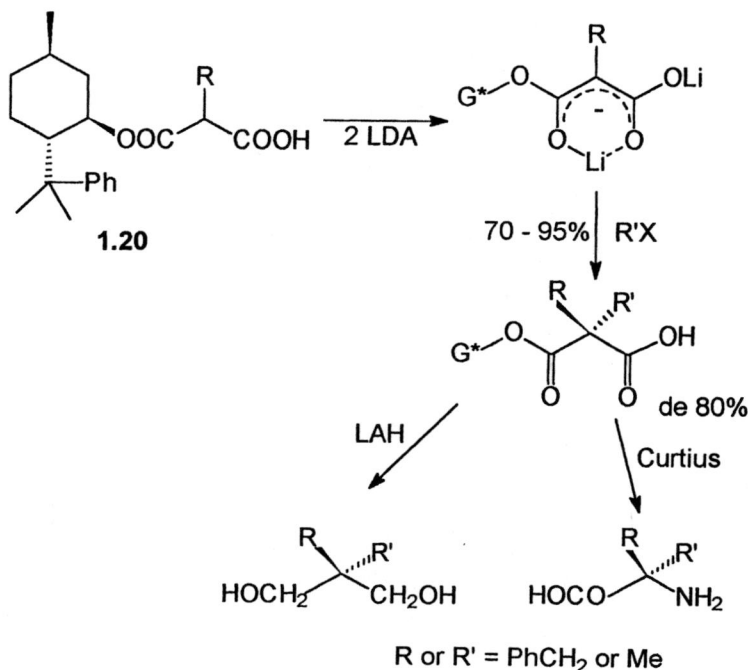

Figure 5.16

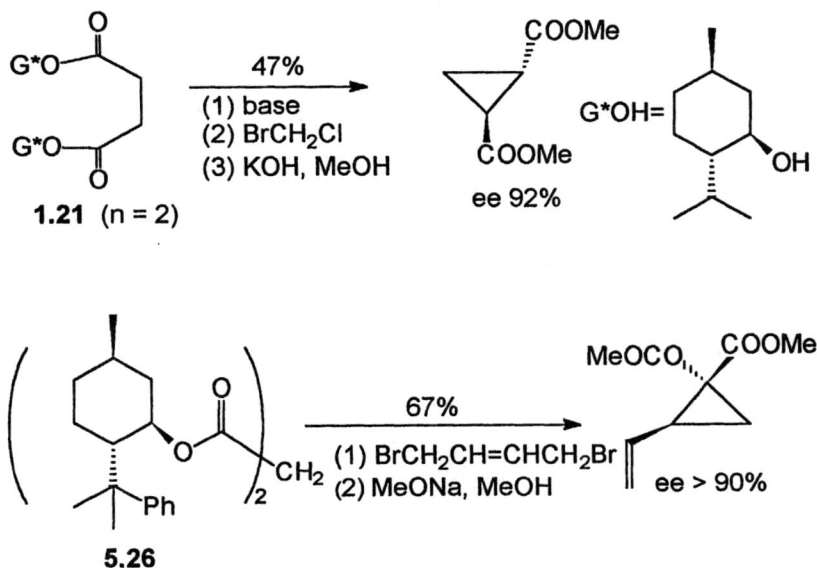

Figure 5.17

The cooperative effect (§ I.6) of two esters can improve selectivity. Alkylation of dimenthyl succinate **1.21** (n = 2) by BrCH$_2$Cl in the presence of base leads, after hydrolysis, to nonracemic *trans*-1,2-cyclopropanedicarboxylic acid with an excellent enantiomeric excess [12] (Figure **5.17**). However, on the way to estrone, Quinkert and coworkers had to use (1*S*,2*R*,5*S*)-phenmenthol dimalonate **5.26** to synthetize methyl 2-vinylcyclopropane-1,1-dicarboxylate with a good enantiomeric excess (Figure **5.17**).

Amide enolates exhibit the Z-configuration due to A(1,3) strain, which disfavors the E-isomers. The phenylacetamide derivative of chiral auxiliary **1.63** (R = *c*-C$_6$H$_{11}$) is methylated in THF/HMPA at −100°C with a good selectivity [270]. However, cyclic amides having a C$_2$ axis of symmetry and substituted by a CH$_2$OCH$_2$OMe group (to promote hydrolysis under mild conditions) have been used more often. Katsuki and coworkers [154, 280, 282, 283] used pyrrolidine amides **1.65** (R = CH$_2$OCH$_2$OMe) and Kurth used piperidine analogs **1.69** [299]. Lithium chelation is not involved in these reactions, and alkylations take place at −78°C on the least hindered side of enolate **5.27** (Figure **5.18**). This method has been applied to sequential dialkylation of cyanoamide **5.28** [281, 1008] (Figure **5.19**). In all these reactions, either enantiomer can be formed because both the (*R*,*R*) and (*S*,*S*) chiral auxiliaries are available.

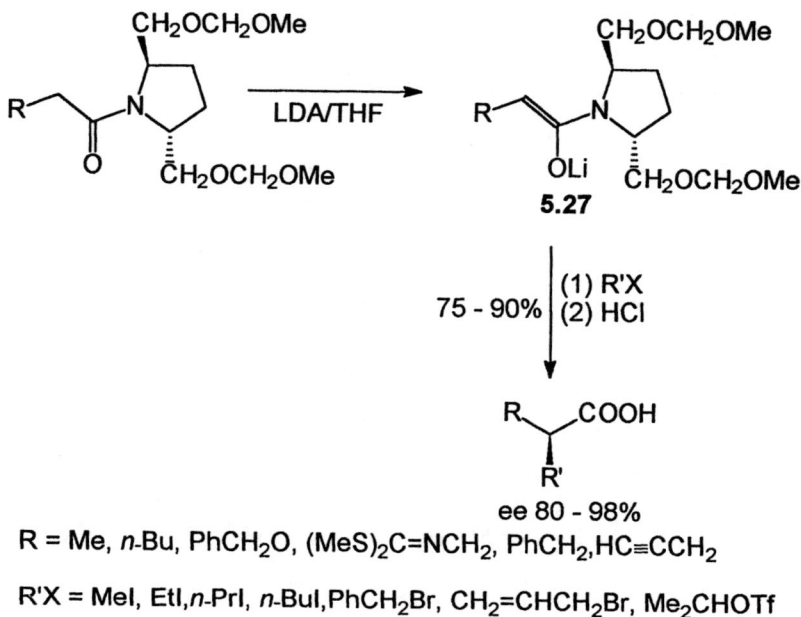

Figure 5.18

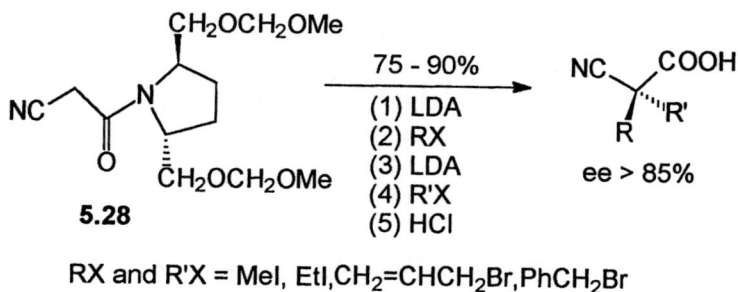

Figure 5.19

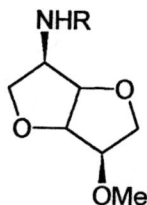

1.63

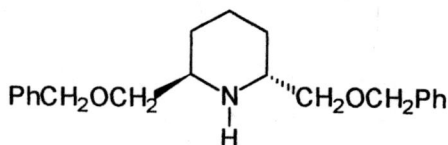

1.69

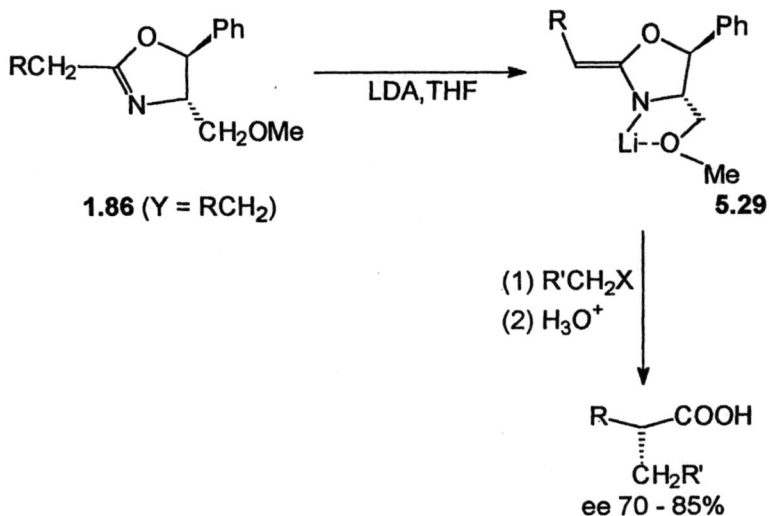

1.86 (Y = RCH$_2$)   **5.29**

(1) R'CH$_2$X
(2) H$_3$O$^+$

R⎯⎯COOH
     CH$_2$R'
ee 70 - 85%

R = Me, Et, n-Bu, PhCH$_2$
R'CH$_2$X = MeI, EtI, n-PrI, n-BuI, PhCH$_2$Cl, Me$_3$SiO(CH$_2$)$_n$I, CH$_2$=CHCH$_2$I

Figure 5.20

Meyers and coworkers [169, 253, 324] have proposed the use of oxazolines **1.86** (Y = RCH$_2$). The deprotonation of these oxazolines by LDA at –90°C gives a chelated lithium azaenolate **5.29**, whose upper face is shielded by the phenyl substituent. Alkylation takes place at –78°C on the open face and, after acidic hydrolysis, the corresponding chiral acids are obtained with good enantiomeric excesses if R = Me, Et, n-Bu or PhCH$_2$. Lower ee's are observed if R = OMe or Cl (Figure **5.20**).

Evans's oxazolidinones **1.116** and **1.117** are a class of chiral auxiliaries that has been widely applied [160, 167, 261, 411]. Deprotonation of N-acyl-1,3-oxa-zolidin-2-ones **5.30** and **5.31** smoothly gives chelated Z-enolates, which then suffer alkylation between –78 and –30°C on their least hindered face [167, 1036]. After hydrolysis, the corresponding enantiomeric acids are obtained according to the auxiliary that was used (Figure **5.21**). Due to the low reactivity of lithium enolates, sodium analogs are preferred in some cases [411, 862, 1036]. This methodology has been applied to the synthesis of chiral α-arylpropionic acid anti-inflammatory drugs [1037, 1038], natural products [1039, 1040], and α-substituted optically active β-lactams en route to nonracemic α,α-disubstituted aminoacids [136, 1041].

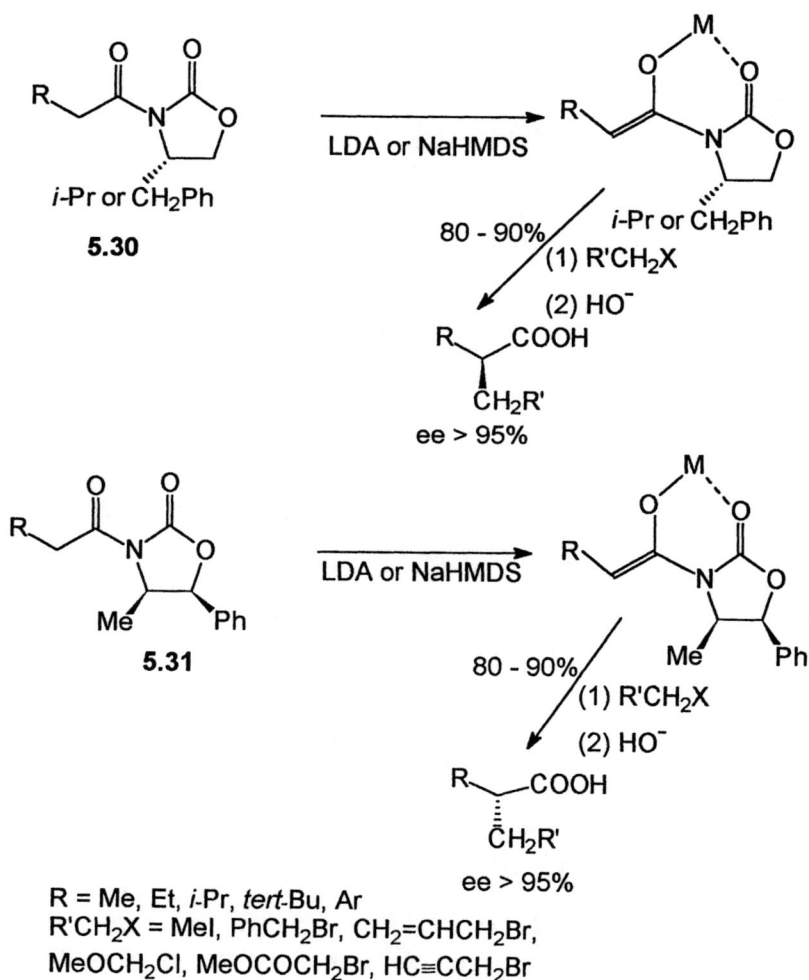

R = Me, Et, *i*-Pr, *tert*-Bu, Ar
R'CH₂X = MeI, PhCH₂Br, CH₂=CHCH₂Br,
MeOCH₂Cl, MeOCOCH₂Br, HC≡CCH₂Br

Figure 5.21

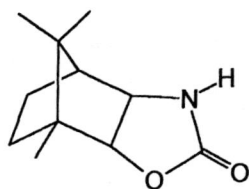

1.120

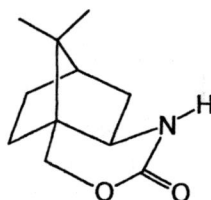

1.122

Titanium enolates of **5.30** have been alkylated by $ClCH_2OCH_2Ph$ and $ClCH_2NHCOPh$ at 0°C with high yield and stereoselectivity (> 90%) [1042]. Similar asymmetric alkylations can be carried out on N-acyloxazolidinones bearing the bornane skeleton **1.120** [425] or **1.122** [432], or on N-acyl-1,3-imidazolidin-2-ones derived from **1.131** [447, 448, 450, 1043]. Extended enolates formed from **5.30** (R = CH=dithianylidene) are regio- and stereoselectively alkylated by allyl or benzyl bromides or primary and secondary triflates. Cleavage of the chiral auxiliary by $LiAlH_4$ followed by TsOH treatment leads to β-alkyl-γ-butyrolactones with an excellent enantiomeric excess [1044]. The use of auxiliary **5.31** in this process leads to the opposite enantiomers (Figure **5.22**).

The N-acylsultams **1.134** (R = R'CH$_2$) introduced by Oppolzer and coworkers [147, 456, 458, 861, 1045, 1046] are deprotonated by NaHMDS or n-BuLi, and the corresponding enolates are alkylated on their less hindered side. After hydrolysis, nonracemic α-branched acids (R = Me, Et, PhCH$_2$, CH$_2$=CHCH$_2$)are obtained. α-Aminoacids are formed in high enantiomeric excesses starting from **1.134** (R = (PhS)$_2$C=N or Ph$_2$C=N) (Figure **5.22**).

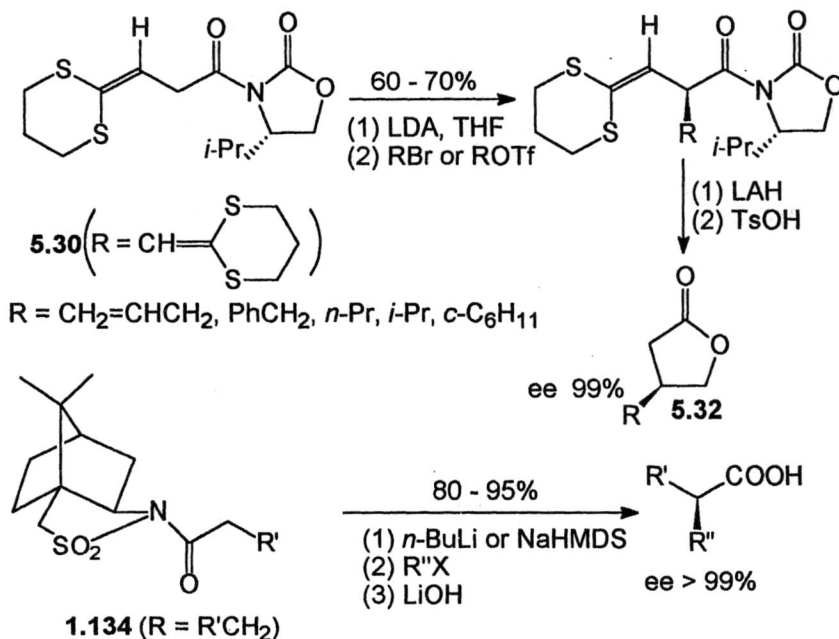

$$5.30\left(R = CH=\left<\begin{smallmatrix}S\\S\end{smallmatrix}\right>\right)$$

R = CH$_2$=CHCH$_2$, PhCH$_2$, n-Pr, i-Pr, c-C$_6$H$_{11}$

R' = (PhS)$_2$C=N, Ph$_2$C=N, PhCH$_2$, CH$_2$=CHCH$_2$, Me, Et
R"X = MeI, n-BuI, i-PrI, n.C$_5$H$_{11}$I, PhCH$_2$I, CH$_2$=CHCH$_2$I, $tert$-BuOCOCH$_2$Br

Figure 5.22

5.3.2.2 Derivatives Bearing Chiral Residues on Other Functional Groups

Asymmetry in enolate alkylations can be induced by the presence of chiral imines formed from aldehydes, ketones or aminoesters. Chiral acetals or oxazolidines of keto- or aldehydo-esters can also be efficient reagents.

Among imines derived from glycine esters [154, 261, 358, 378, 380, 1047], those of camphor **1.107** (R = *tert*-Bu or menthyl) have been recommended [154, 1048]. These imines are, however, difficult to hydrolyze. Imines of **1.108** are more easily cleaved [154, 253, 378, 380, 861, 1047]. Bulky ester substituents in **5.33** (R = *i*-Bu, *tert*-Bu) promote highly selective alkylations provided that the size of R' and R" are different enough (Figure **5.23**). Arylation of these enolates with fluorobenzenechromium tricarbonyl followed by decomplexation leads to nonracemic phenylsubstituted glycines [1047]. Alkylation of imines **1.98** (R = Me, *i*-Pr, *i*-Bu) formed from **1.97** by benzyl or allyl bromides also leads to α,α-dialkyl-aminoesters with a good enantiomeric excess [358]. Asymmetric alkylation of glycine may also be carried out via the nickel complex of the imine formed from functionalized benzophenone **1.109** [381, 382, 861, 1049, 1050]. These reactions are performed under phase transfer catalysis, or in DMF in the presence of solid KOH or NaOH. A predominant stereoisomer is formed under thermodynamic control. (*S*)-Aminoacids are obtained after treatment of the products with methanolic HCl (Figure **5.23**). This method has been extended to alkylation of glycine or alanine by PhCH$_2$Br or by fluorinated benzyl halides [1049, 1050].

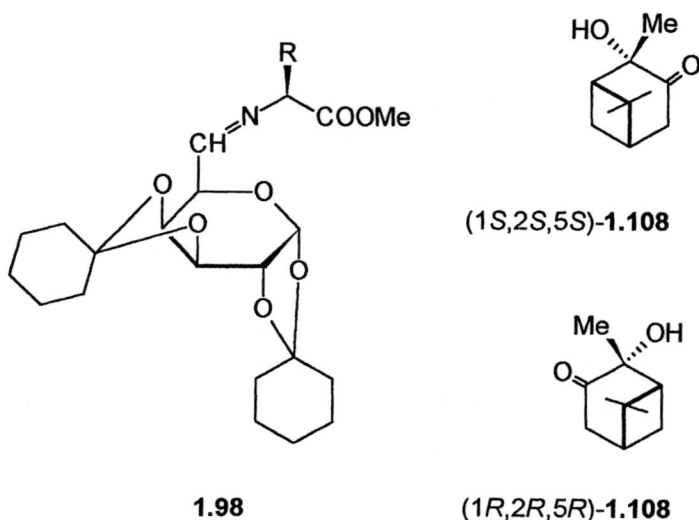

(1*S*,2*S*,5*S*)-**1.108**

(1*R*,2*R*,5*R*)-**1.108**

1.98

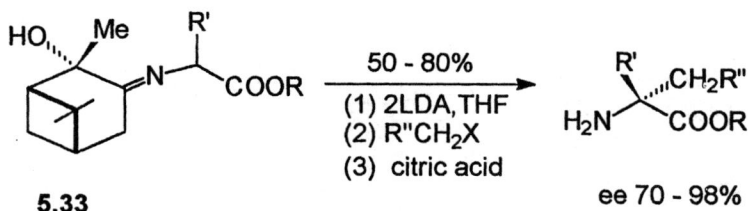

5.33

R' = H, Me, n-Pr, PhCH$_2$

R"CH$_2$X = MeI, n-PrI, I(CH$_2$)$_4$I, PhCH$_2$Br, CH$_2$=CHCH$_2$Br

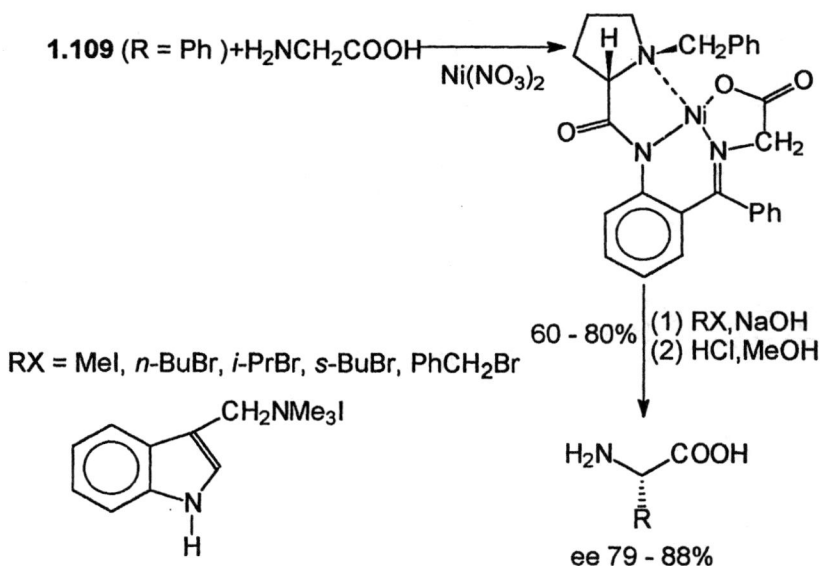

RX = MeI, n-BuBr, i-PrBr, s-BuBr, PhCH$_2$Br

Figure 5.23

A chiral group introduced by transformation of the carbonyl group of alde-
hydo- or ketoesters can induce stereoselective alkylations α- to the ester. Agami
and Couty [290] have performed the enantioselective methylation of **5.34** via chiral
oxazoline **5.35** at −78°C. The best result is observed with the menthyl ester, due to
double stereodifferentiation (Figure 5.24). Sakai and coworkers [215] prepared
the (R,R)-cycloheptanediol dioxolanes of β-ketoesters **5.36** and **5.37**. The alkylations of these dioxolanes are highly stereoselective and lead to enol ethers after
aqueous workup (Figure 5.24). Allylation of 2-carboethoxycyclohexanone takes
place with a high enantioselectivity (88%) if the keto group is transformed into a
chiral dioxolane. When the reaction is run on the corresponding chiral 1-phen-

ethylenamine or on the menthyl ester of the achiral dioxolane, the selectivity is lower [1051].

Aminonitrile **1.71** derived from ephedrine **1.61** can be deprotonated with two equivalents of LDA in the presence of LiI and HMPA. The resulting intermediate is regio- and stereoselectively alkylated at –78°C. After acidic hydrolysis, nonracemic β-alkylphenylpropionic acids are obtained with a good selectivity [301] (Figure **5.24**).

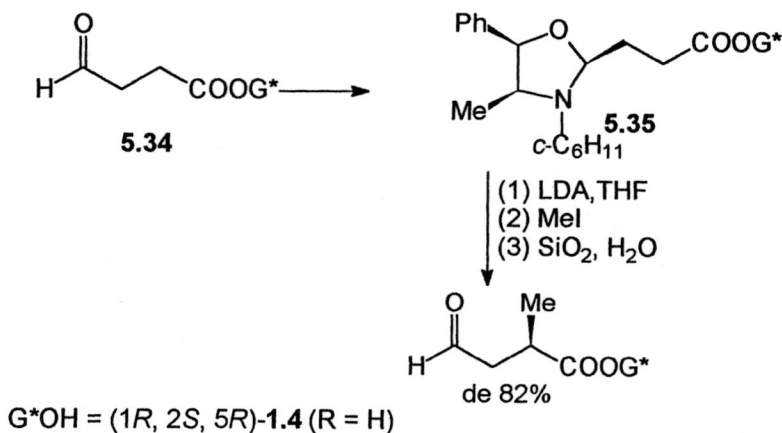

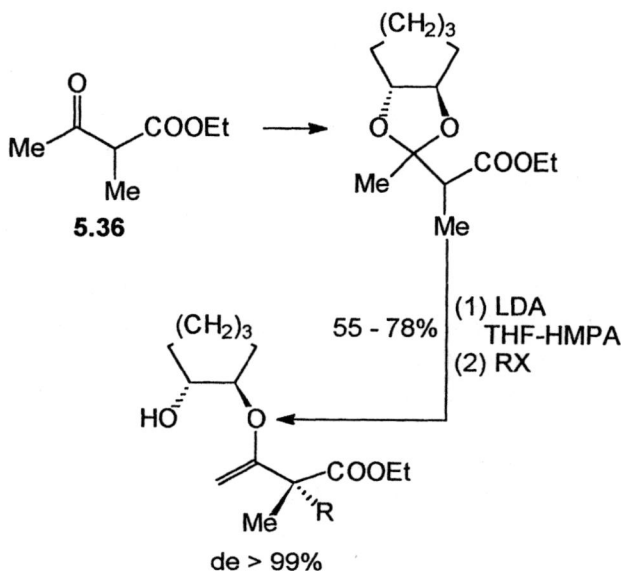

Figure 5.24

5.37

54 - 74%
idem

de > 99%

1.71

70 - 84%
(1) 2LDA
LiI, HMPA
(2) RX
(3) oxalic acid

ee 76 - 99%

RX = EtI, *n*-BuI, CH₂=CHCH₂Br, PhCH₂Br

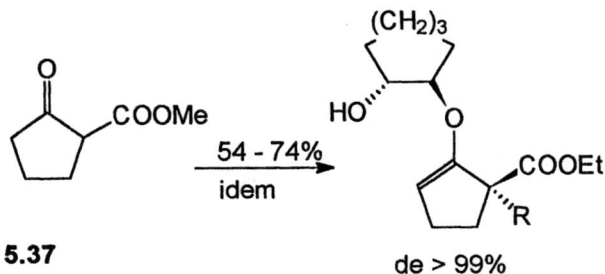

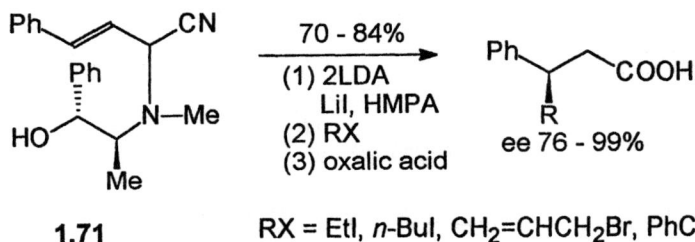

Figure 5.24 (continued)

5.3.2.3 *Derivatives Bearing the Chiral Residues on Two Functional Groups*

The alkylation of α-aminoacids has been performed by Schöllkopf and coworkers [154, 261, 861] via cyclic lactim ethers **1.114** formed from (*S*)-valine (R = *i*-Pr) or (*S*)-*tert*-leucine (R = *tert*-Bu). In this method, one aminoacid acts as a chiral auxiliary for alkylation of another. Deprotonation of **1.114** (R' = H or Me) by *n*-BuLi in THF takes place on the least hindered carbon leading to an anionic species **5.38** whose upper face is hindered by the R group. Alkylation is carried out at −78°C and takes place on the carbon bearing R' substituent, from the lower face (Figure 5.25). Acid hydrolysis leads to α-alkyl substituted aminoacids with a high enantiomeric excess. However, when R' = H, some unwanted epimerization can occur.

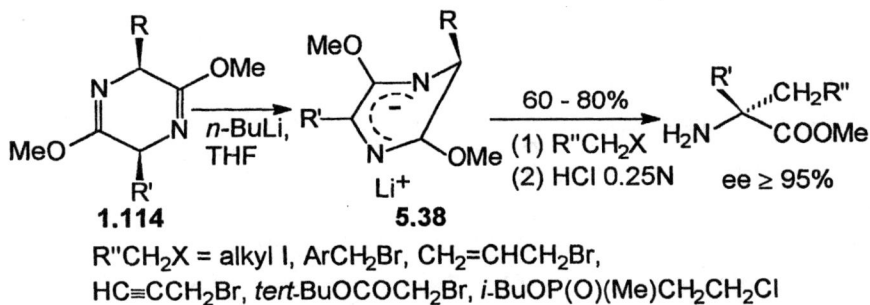

1.114   **5.38**

n-BuLi,
THF

60 - 80%
(1) R"CH₂X
(2) HCl 0.25N

ee ≥ 95%

R"CH₂X = alkyl I, ArCH₂Br, CH₂=CHCH₂Br,
HC≡CCH₂Br, *tert*-BuOCOCH₂Br, *i*-BuOP(O)(Me)CH₂CH₂Cl

Figure 5.25

Alkylations of anionic species generated from oxazolidinones and imidazolidinones 1.125 (X = O or NMe) and 1.126 (X = NMe) are recommended by
Seebach and coworkers [81, 154, 439, 440, 861, 1052] because they are highly
stereoselective. The most interesting results are obtained from 1,3-imidazolid-2-
ones 1.127 (n = 1) bearing easily cleavable COO*tert*-Bu or COOCH$_2$Ph groups on
nitrogen [440]. Alkylations occur from the least hindered side of enolates 5.39,
whose nitrogens are pyramidalized to some extent (Figure 5.26). Because protonation of these enolates takes place on the same face as alkylation, it is possible to
prepare the enantiomer of a given aminoacid by following the alkylation reaction
with a subsequent deprotonation-reprotonation (or deuteration) step (Figure 5.26).
Nonracemic β-aminoacids can be obtained from the six-membered analogs 1.127
(R = Ph, n = 2) [453, 1053].

Meyers and coworkers have transformed β- or γ-ketoacids into bicyclic lactams 1.90 and 1.91, and the derived lithium enolates are stereoselectively alkylated.
However, when the hydrolysis of the monoalkylated compound is carried out,
some epimerization occurs, so the method is most valuable if a subsequent enoliza-
tion-alkylation is performed. After acidic treatment, α,α-dialkyl-β- or γ-ketoacids
or esters are obtained with a high enantioselectivity (Figure 5.27). Either enantiomer of a given ketoacid may be prepared according to the alkylation sequence.
The alkyl halide used in the second step is introduced on the *endo* face of the
monoalkylated enolate 5.40 [327]. However, the stereoselectivity of the second
alkylation depends upon the nature of the alkyl halide. Ethylene oxide can also be
used as an electrophile [327]. This methodology has been applied to the synthesis
of nonracemic 5,5-disubstituted cyclopenten-2-ones [1054] and hydrinden-2-ones
[1055]. In a few cases, the use of 1.91 (n = 2) has given interesting results [327]
(Figure 5.27).

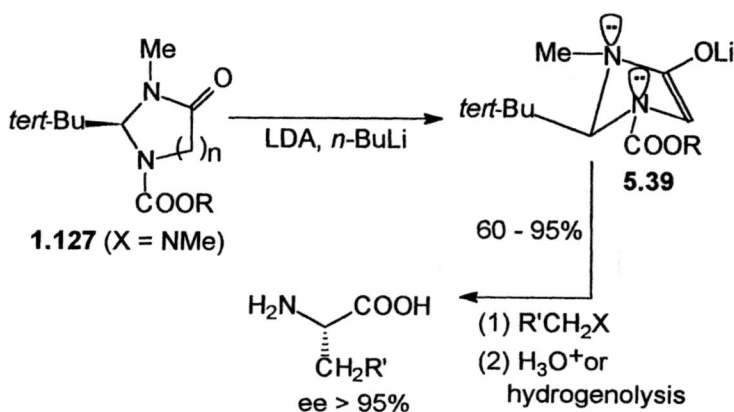

Figure 5.26

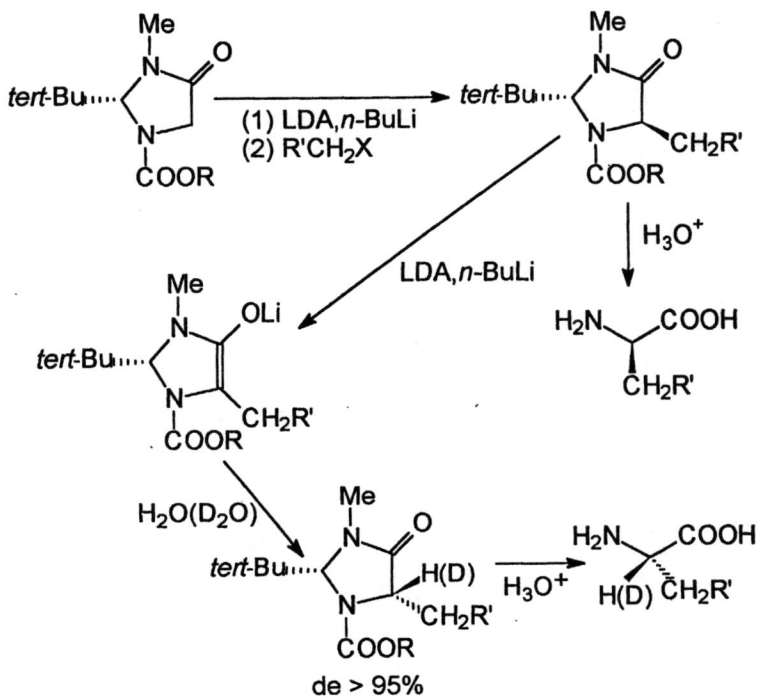

Figure 5.26 (continued)

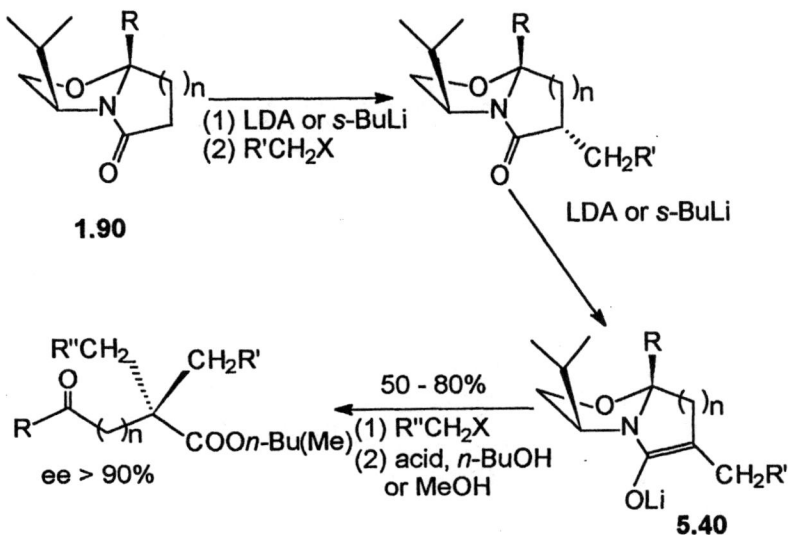

Figure 5.27

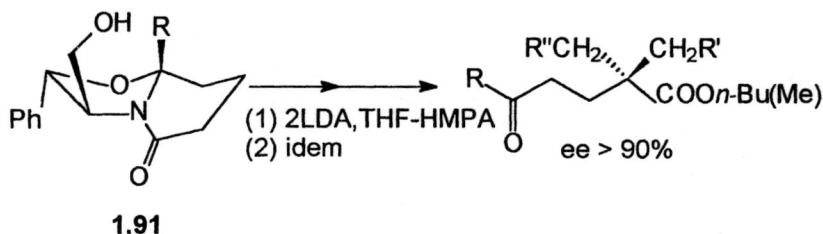

1.91

n = 1 or 2 R = Me, Ph
R'CH$_2$X and R"CH$_2$X = MeI, EtI, n-PrI, Me$_3$SiO(CH$_2$)$_2$CH$_2$I,
CH$_2$=CHCH$_2$Br, ArCH$_2$Br

Figure 5.27 (continued)

From 3-hydroxycarboxylic acids and (–)-menthone, 1,3-dioxan-4-ones can be prepared. Alkylation of this class of reagents is stereoselective and produces nonracemic 2-alkyl-3-hydroxycarboxylic acids after hydrolysis [374].

5.3.2.4 *Chiral Alkylating Reagents*

Enantioselective methylation of imine enolates of aminoesters **5.41** by diacetoneglucose **1.48** methyl sulfate has been carried out by Duhamel and coworkers [154, 358]. After hydrolysis, nonracemic α-alkyl alanines are obtained with enantiomeric excesses up to 76% (Figure **5.28**).

Stork and Schoofs [1056] have performed an intramolecular allylic substitution involving a chirality transfer from malonate derivative **5.42**. However, the resulting cyclic diester is obtained with a poor enantiomeric excess (Figure **5.28**). Seyden-Penne and coworkers conducted a regio- and stereoselective S$_N$2' allylic alkylation of *tert*-butyl phenylacetate enolate with bromoester **5.43**. After LiAlH$_4$ treatment, an *anti* diol is obtained with a high enantiomeric excess [68] (Figure **5.28**).

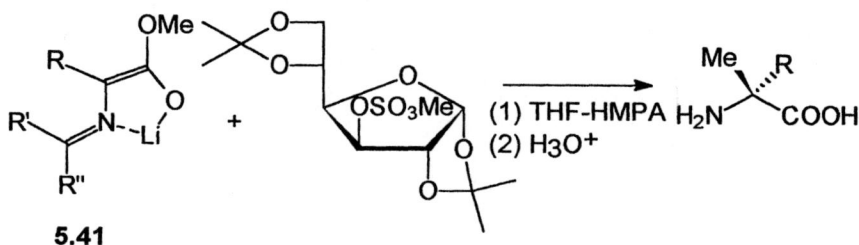

5.41

Figure 5.28

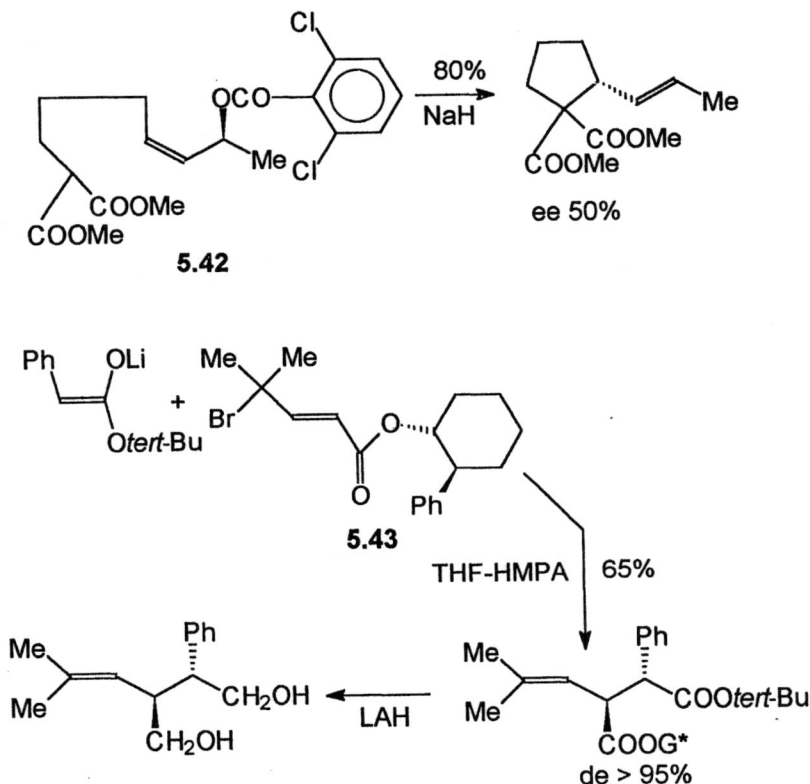

Figure 5.28 (continued)

5.3.3 Alkylation of Acyl Cyclopentadieneiron Carbonyl Complexes
[522]

Lithium enolates **5.44** of acyl cyclopentadieneiron carbonyl complexes **1.151** (R=R'CH$_2$) are generated by deprotonation with n-BuLi at $-78°C$. A spectroscopic study has shown that one phenyl ring of the Ph$_3$P group is positioned underneath the bottom side of **5.44**, so that alkylation takes place from the topside (Figure 5.29). When R' is an alkyl group, it is not possible to perform sequential dialkylations. The use of either (R)- or (S)-complexes allows the formation of either enantiomeric acid derivative (Nu = $tert$-BuO, PhCH$_2$ONH, etc.) with a high enantiomeric excess. If the alkylating agent is an epoxide, ring closure takes place to give a lactone when the chiral auxiliary is removed [523] (Figure **5.30**). Complexes **1.151** (R' = $tert$-BuOCOCH$_2$) also suffer diastereoselective alkylations which take place α- to the ester group, not α- to the ironacyl group [522, 1057]. This methodology has been applied to natural product synthesis [1057].

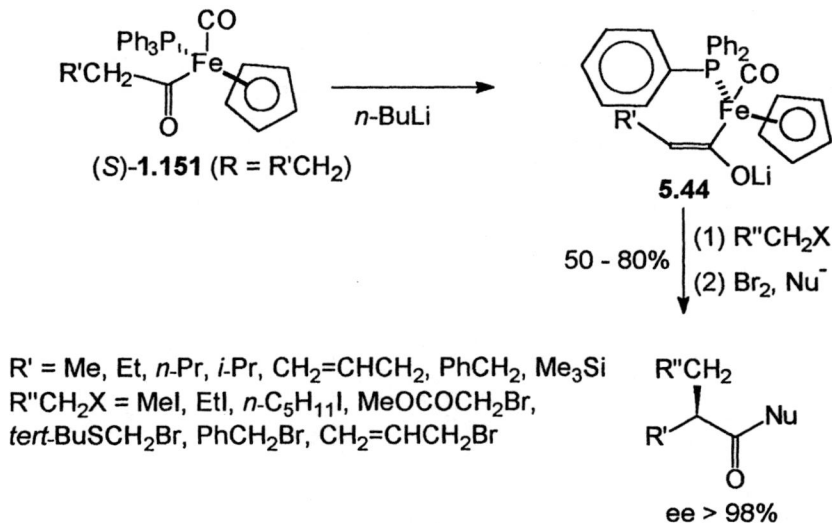

Figure 5.29

R' = Me, Et, n-Pr, i-Pr, CH$_2$=CHCH$_2$, PhCH$_2$, Me$_3$Si
R"CH$_2$X = MeI, EtI, n-C$_5$H$_{11}$I, MeOCOCH$_2$Br,
tert-BuSCH$_2$Br, PhCH$_2$Br, CH$_2$=CHCH$_2$Br

ee > 98%

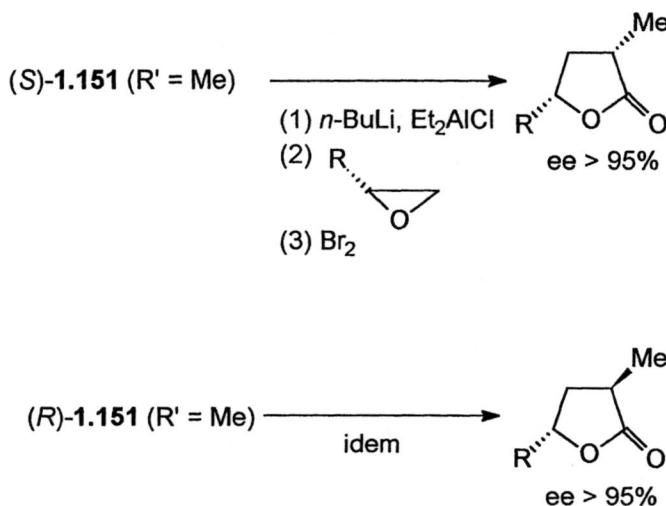

Figure 5.30

5.4 CATALYZED ALKYLATIONS

Asymmetric alkylations can be performed in the presence of catalytic amounts of chiral phase transfer catalysts or through the intermediacy of π-allyl-palladium complexes bearing chiral ligands. Both methods normally require relatively acidic carbon acids (pKa ≤ 17) such as β-diketones, β-keto- or cyanoesters or malonic acid derivatives.

5.4.1 Alkylations Catalyzed by Quaternary Salts of Cinchona Alkaloids [559, 766, 1058]

The first interesting results in this area are those of Dolling and coworkers, who performed the enantioselective methylation of indanone **5.45** under PTC conditions. The most efficient catalyst is cinchoninium bromide **5.46** (R = H, R' = CF$_3$). This methodology has been developed by industry [811, 812, 853]. Alkylation takes place through a hydrogen-bonded enolate/catalyst ion pair **5.47**, from the least hindered side [774] (Figure **5.31**). Alkylations of oxindoles with ClCH$_2$CN have been carried out under similar conditions [768]. The benzophenone imine of *tert*-butyl glycinate **5.48** has been alkylated with allyl or benzyl bromides in the presence of **5.46** (R = H, MeO, R' = H) as a phase-transfer catalyst. The enantioselectivity of this reaction is low (65%), but the racemate crystallizes and leaves behind in the filtrate the precursor of a chiral aminoacid with an excellent enantiomeric excess (98%) [154, 811, 861]. The other enantiomer can be obtained by using isomeric ammonium bromides **5.49** as phase-transfer catalysts (Figure **5.32**). The asymmetric synthesis of chiral α,α-dialkylaminoacids by a related method gives less satisfactory results [1059].

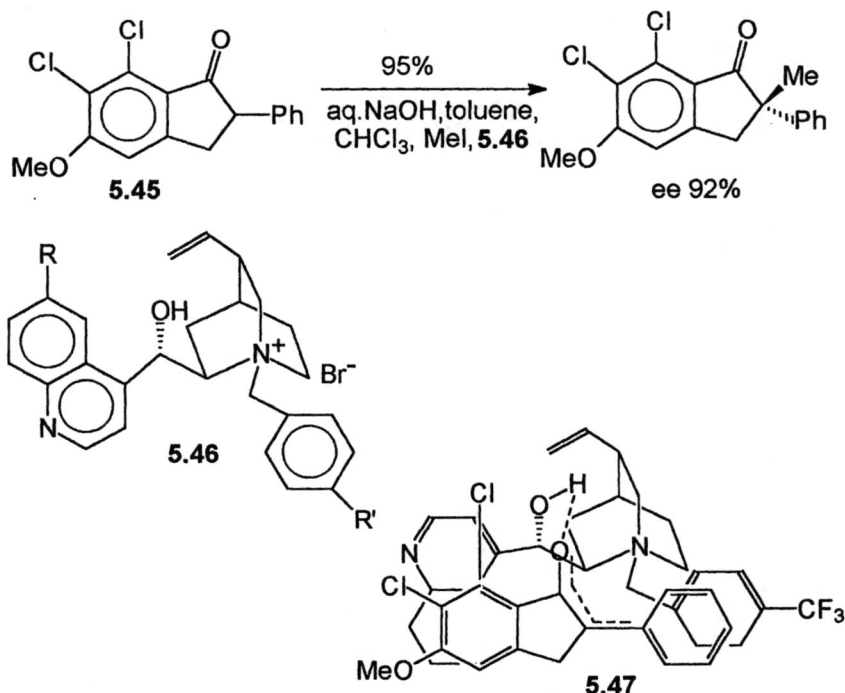

Figure 5.31

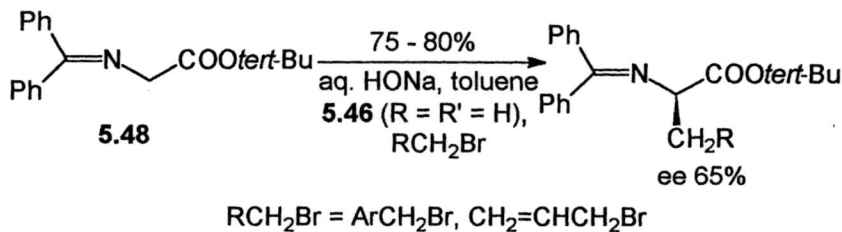

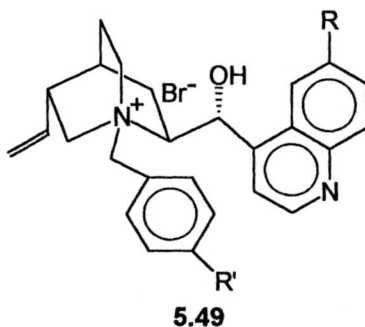

$RCH_2Br = ArCH_2Br, CH_2=CHCH_2Br$

Figure 5.32

5.4.2 Allylations of Enolates Catalyzed by Palladium Complexes

Enolates generated from malonates and related compounds undergo allylation under catalysis by Pd(0)-phosphine complexes. This chemistry has been widely developed by Tsuji, Trost and their coworkers [92, 253, 876]. To induce asymmetry, chiral phosphines have been used as palladium ligands [752, 858, 871, 876, 1008, 1060, 1061]. With a few exceptions, the use of chiral diphosphines gives disappointing selectivities. Bosnich and coworkers alkylated sodium ethyl malonates with allyl acetates 5.50 with a very good enantiomeric excess by using chiraphos 3.37 (n = 0) as the palladium ligand [752, 866, 871, 872, 1062] (Figure 5.33). Binap 3.43 has been recommended as the palladium ligand for similar reactions with 5.50 (R = H) and interesting enantiomeric excesses are observed when the malonate substituent R' is NHCOMe or NHCOCF$_3$[872]. Hayashi's ferrocenylphosphine 3.41 (R = (HOCH$_2$)$_2$CH) is a very efficient ligand when reactions are run at −60°C [1063]. The mechanism of a Pd(0)-catalyzed allylic alkylation involves the intermediate formation of a π-allylpalladium complex. This is attacked by a soft nucleophile on the face opposite the metal, so the interaction of the incoming nucleophile with the chiral ligand is therefore somewhat limited [752, 1061]. In the case of phosphine 3.41, Hayashi and Ito suggested the intervention of a secondary interaction between the nucleophile and an OH group, promoting some rigidification of the system [1063] (Figure 5.33).

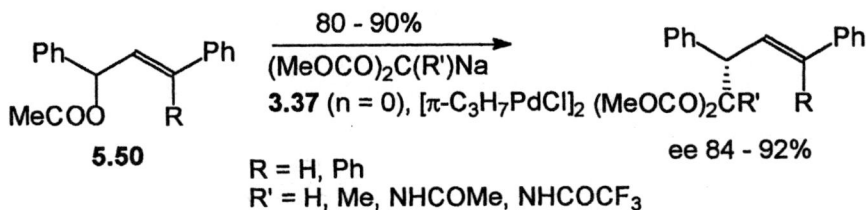

5.50

R = H, Ph
R' = H, Me, NHCOMe, NHCOCF₃

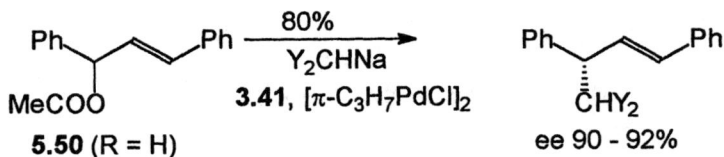

5.50 (R = H)

CHY₂ = CH(COOMe)₂, CH(COMe)₂

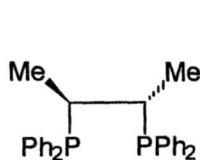

3.37 n = 0

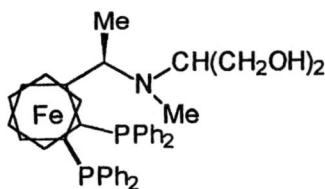

3.41 (R = (HOCH₂)₂CH)

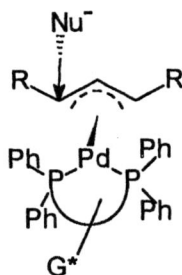

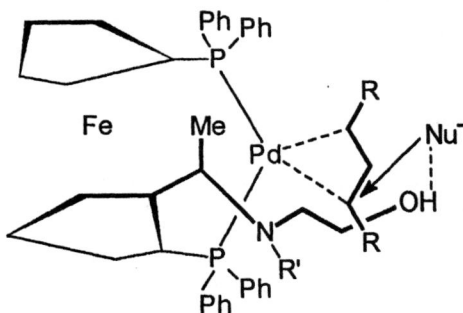

Figure 5.33

Introduction of a crown-ether group onto one of the nitrogen substituents of **3.41** increases the rate of allylation of potassium enolates of β-diketones by **5.50** (R = H), but this does not improve the moderate enantioselectivity (70% ee) [877, 1063].

Triarylmonophosphines bearing an oxazoline ring **3.52** (X = PPh₂, R' = H) have recently been introduced by Helmchen, Pfaltz, Williams and their coworkers [919, 920, 921, 1064]. These are very efficient palladium ligands for asymmetric allylation of methyl malonate and acetylacetone by **5.50** (R = H) at room temperature (Figure **5.34**). The best selectivity is observed when the reaction is run in the presence of *N,O*-bis-(trimethylsilyl)acetamide (BSA) and KOAc with **3.52** (X = PPh$_2$, R = *i*-Pr, R' = H) as the palladium ligand. The approach of the nucleophile from the least hindered side of the favored π-allylpalladium complex accounts for the observed selectivity [1064] (Figure **5.34**). Other monooxazolines such as **3.52** (X = MeS) [944] or **3.65** [943, 945] also have been recommended as palladium ligands. Higher amounts of these ligands are necessary to induce excellent selectivities than when **3.52** (X = PPh$_2$) is used. Bis-oxazolines **3.28** (R = PhCH$_2$, R' = H, R" = Me) and **3.57** also give excellent results, as does triamine **3.66** (R = TBMSOCH$_2$) [839, 871, 929, 946]. The use of sparteine **2.5** or bis-pyrrolidines **2.38** [304] as palladium ligands leads to lower enantioselectivities (75 - 91% ee). When using **3.52** (X = PPh$_2$), other allylic acetates E-RCH=CHCH(OAc)R can be employed as alkylating agents, and a high enantioselectivity (96%) is observed when R = *i*-Pr [920]. The limitation of this method is the poor regioselectivity of the reaction of unsymmetrical allylic derivatives RCH=CHCH(OAc)R' (R ≠ R') [1065].

Trost and coworkers [911, 1066] have examined the asymmetric allylation of sodium malonate with achiral cyclic allylic systems that lack a regioselectivity problem. Enantioselection occurs during the formation of the π-allyl complex.

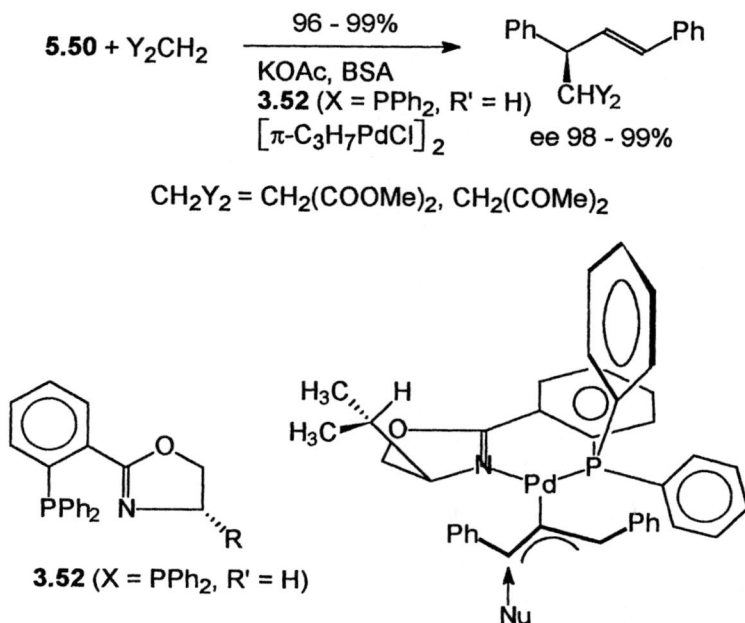

CH$_2$Y$_2$ = CH$_2$(COOMe)$_2$, CH$_2$(COMe)$_2$

3.52 (X = PPh$_2$, R' = H)

Figure 5.34

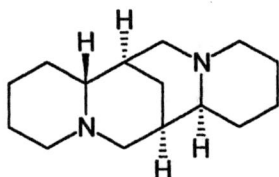

2.5

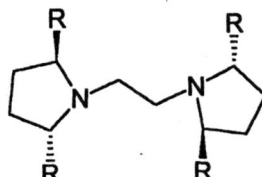

2.38

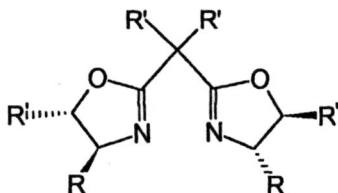

3.28

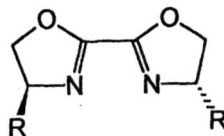

3.57

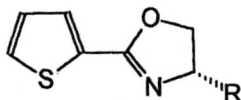

3.65

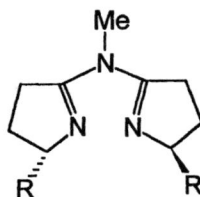

3.66

Intramolecular allylation with bis-carbamate **5.51** (R = NHTs) is carried out in the presence of a palladium(0) complex whose ligands are chiral diphosphines **3.48** (Figure **5.35**). The best enantioselectivities are observed with amides **3.48** (X = NH, R = Ph or R,R = $(CH_2)_4$), and the results have been interpreted in terms of a rigid, bidentate palladium complex. Both enantiomers of **3.48** are available, so either product can be prepared. The enantioselectivity is probably due to steric interactions between the propeller-like chiral ligand and the substrate that induce the selective rupture of the pro-*R* or pro-*S* carbon-oxygen bond (Figure **5.35**). Intermolecular asymmetric allylations of sodium or better yet tetracyclohexylammonium dimethyl malonate by **5.51** (R = Ph), 2-cyclopentenyl acetate and 2-cyclohexenyl carbamate have been performed at 0°C [1067, 1068]. In the presence of ligand (*S,S*)-**3.48** (X = NH, R = Ph), the expected monomalonates are obtained with a high enantiomeric excess (Figure **5.35**). Similar selectivities are observed with amines and with anions of 2-methylcyclohexane-1,3-dione and $PhSO_2CH_2NO_2$ [97, 911].

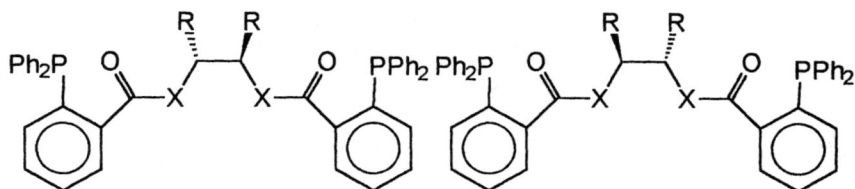

3.48

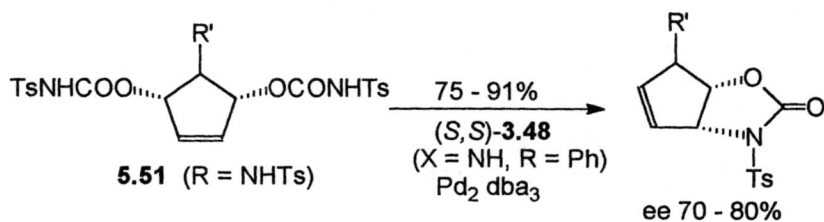

5.51 (R = NHTs)

75 - 91%

(S,S)-**3.48**
(X = NH, R = Ph)
Pd$_2$ dba$_3$

ee 70 - 80%

R' = H, PhCH$_2$OCH$_2$

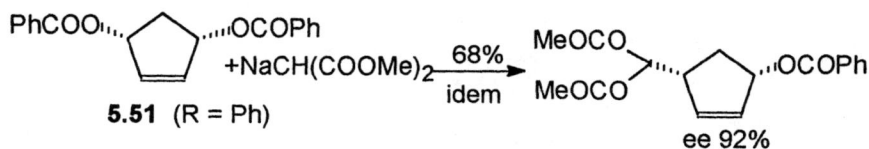

5.51 (R = Ph)

+NaCH(COOMe)$_2$

68%
idem

ee 92%

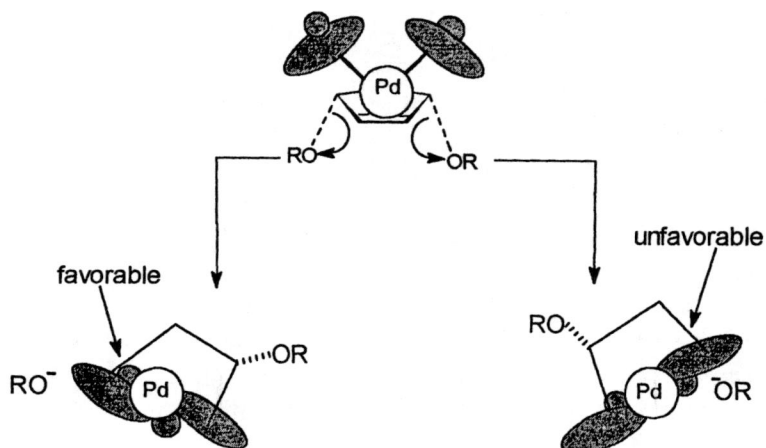

favorable

unfavorable

Figure 5.35

5.5 ALKYLATIONS OF CARBANIONS α TO SULFUR

Alkylations of lithiated carbanions α to sulfoxides or sulfoximines are usually poorly stereoselective unless another functionality such as an hydroxyl [472] or a nitro group [520] is introduced. For example, the bis-lithiated anions of non-racemic β-hydroxysulfoxides 5.52 and 5.53 are alkylated by primary iodides with a good stereoselectivity. The incoming alkyl group is introduced *anti* to the existing hydroxy group, regardless of the configuration of the sulfoxide (Figure 5.36). Further reactions allow the cleavage of the chiral group and lead to enantioen-riched compounds [472].

Davis and coworkers [377] have performed the asymmetric alkylation of sulfamides 5.54 having the bornane skeleton at −78°C. If alkyl sulfamides are used instead of benzyl sulfamides, the selectivity is very low (Figure 5.36).

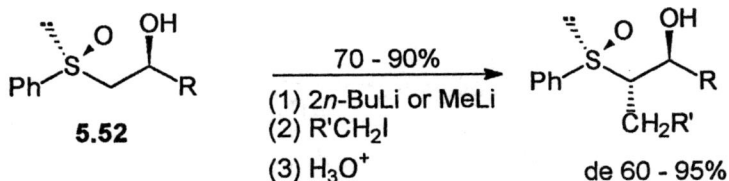

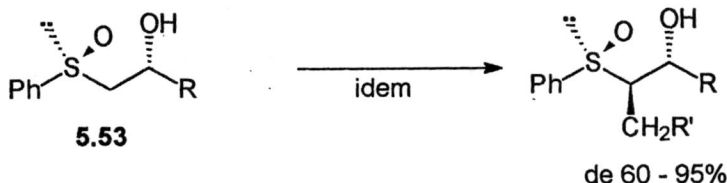

$$R = Me, PhCH_2, n\text{-}C_7H_{15}, n\text{-}C_9H_{19}$$
$$R'CH_2I = MeI, n\text{-}C_8H_{17}I, n\text{-}C_{10}H_{21}I$$

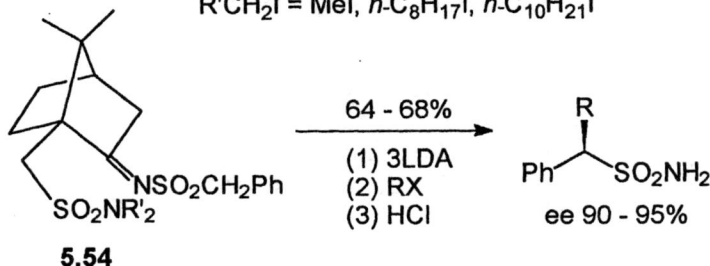

$$R' = i\text{-}Pr, c\text{-}C_6H_{11} \quad RX = PhCH_2Br, Me_2CHI$$

Figure 5.36

5.6 REACTIONS OF ENOLATES WITH OTHER ELECTRO-PHILES: HALOGENATIONS, AMINATIONS, ACYLA-TIONS AND OXIDATIONS

5.6.1 Enolates Bearing a Chiral Residue

Fluorination of lithiated *N*-acyloxazolidinones **5.55** at −78°C with *N*-fluoro-benzene-disulfonimide has been carried out by Davis and Han [1069]. After treatment of the products with $LiBH_4$, α-fluoroalcohols are obtained with a high enantiomeric excess (Figure **5.37**). Trifluoromethylation of lithium enolates of *N*-acyloxazolidinones **5.30** takes place with iodotrifluoromethane in the presence of Et_3B via a radical mechanism [1070]. $LiBH_4$ reduction leads to the corresponding alcohols with an interesting selectivity (Figure **5.37**). Evans and coworkers have studied the halogenation of the enolates of **5.30**. Bromination with NBS is highly stereoselective at −78°C provided that boron enolates are used instead of lithium enolates [411, 861, 1071].

Reaction of the bromide products with tetramethylguanidinium azide, followed by cleavage of the chiral auxiliary with LiOH, gives precursors of α-aminoacids with a very high enantiomeric excess, except when R = Ph (Figure **5.37**). Direct azide transfer to the potassium enolate of **5.30** can be carried out by treatment with trisyl azide **5.56**, followed by Me_4NOAc. A similar cleavage of the auxiliary by LiOH gives the other enantiomer of the α-azidoacid [411, 861, 1072] (Figure **5.37**). This method has been extended by Denmark and coworkers to the synthesis of nonracemic α-aminophosphonic acids [1073]. In all these cases, the electrophile is introduced on the least hindered side of the enolate.

Asymmetric iodination of 4-pentenamides derived from (2*R*,5*R*)-bis-(methoxymethyl)-pyrrolidine **1.65** (R = CH_2OMe) by I_2 and *s*-collidine has been carried out at room temperature by Taguchi and coworkers [1074] with a high selectivity.

Another approach to nonracemic α-aminoacids [154, 861, 1075] has been simultaneously proposed by Evans and coworkers [410] and Vederas and Trimble [412]. Lithium enolates of **5.30** are reacted with *tert*-Bu diazodicarboxylate. After treatment of the products with $PhCH_2OLi$, hydrazinocarboxylates **5.57** are obtained with a high selectivity (Figure **5.38**). Sequential treatment of **5.57** with CF_3COOH and Raney nickel gives nonracemic α-aminoesters with a high enantiomeric excess (Figure **5.38**). A chelated transition state model that minimizes steric repulsions accounts for the observed selectivity (Figure **5.38**).

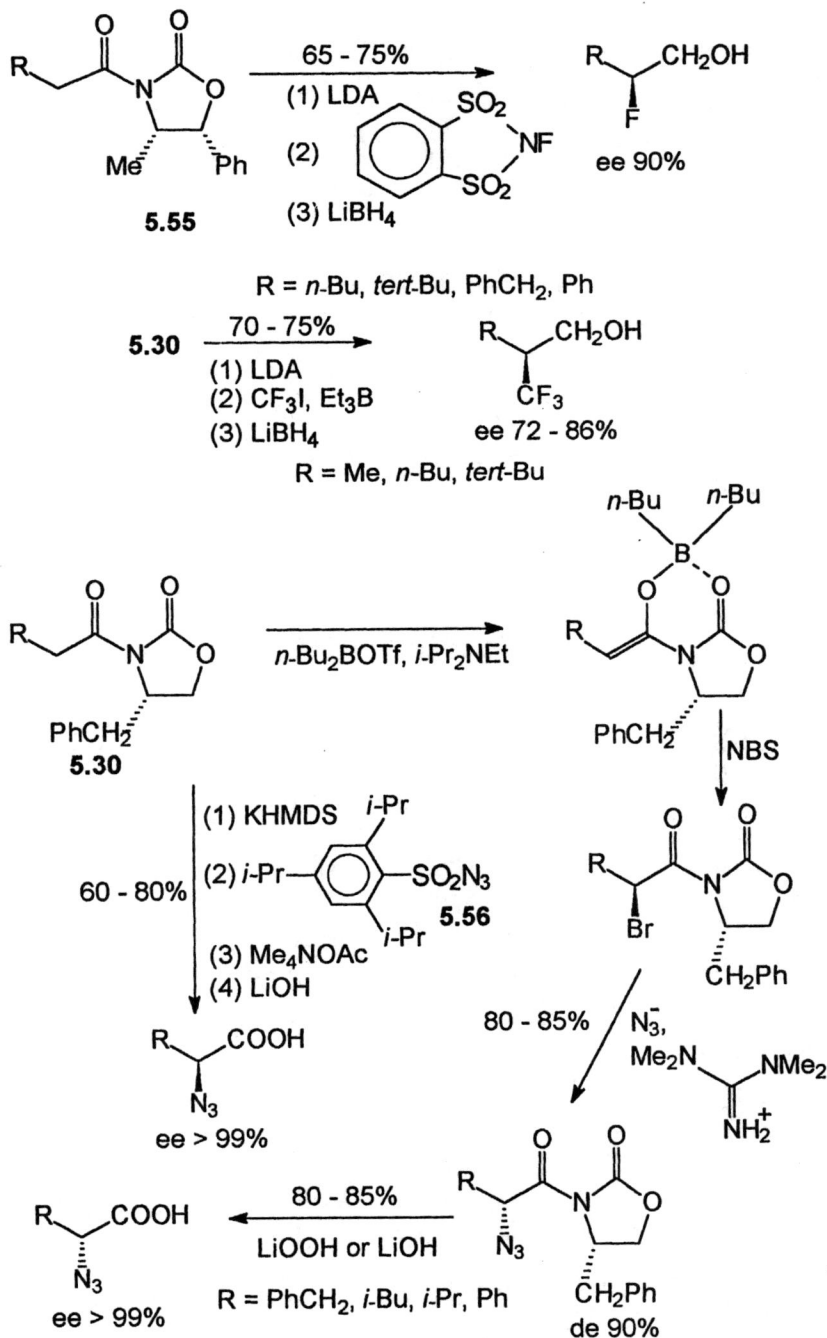

Figure 5.37

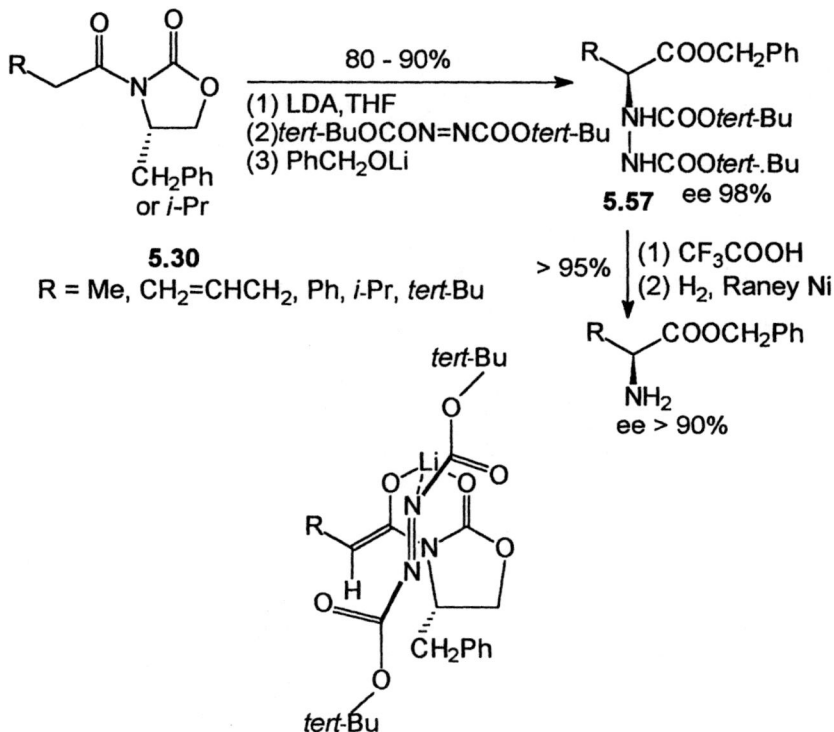

Figure 5.38

Oppolzer and Tamura [460, 861, 1076] have recommended 1-chloro-1-nitrosocyclohexane as an electrophile for preparation of nonracemic α-aminoacids. α-Aminoacids are formed with an excellent enantiomeric excess from sodium enolates of N-acylsultams **1.134** (R = R'CH$_2$) by a sequence of nitrosylation, hydrolysis and reduction of the intermediate hydroxylamine, and cleavage of the auxiliary with LiOH (Figure **5.39**). The stereoselectivity of this process is interpreted by attack of the electrophile on the face opposite to the nitrogen lone pair of the Z-chelated enolate **5.58** (Figure **5.39**).

Acylations of lithium enolates of N-acyloxazolidinones **5.30** and **5.31** by acetyl or benzoyl chloride at –78°C are highly diastereoselective (de > 90%) [167]. Such is also the case for reactions of Samp or Ramp **1.76** hydrazone enolates with ClCOOMe [1077] or MeNCS [1078]. Sodium enolates of N-acyloxazolidinones **5.30** and **5.31** can be oxidized on the least hindered face at –78°C by an oxaziridine [741]. After cleavage of the chiral auxiliary by magnesium methoxide, α-hydroxyesters are obtained with an excellent enantiomeric excess [742] (Figure **5.39**). Similar results are obtained from Samp and Ramp **1.76** hydrazone enolates [742] (Figure **5.39**). Oppolzer and coworkers [147] carried out the stereoselective α-hydroxylation of potassium enolates of esters **5.21** with MoO$_5$ in the presence of pyridine and HMPA. After basic hydrolysis, nonracemic α-hydroxyacids were obtained with an excellent enantiomeric excess (Figure **5.39**). These reactions take place on the least hindered face of the Z-enolate.

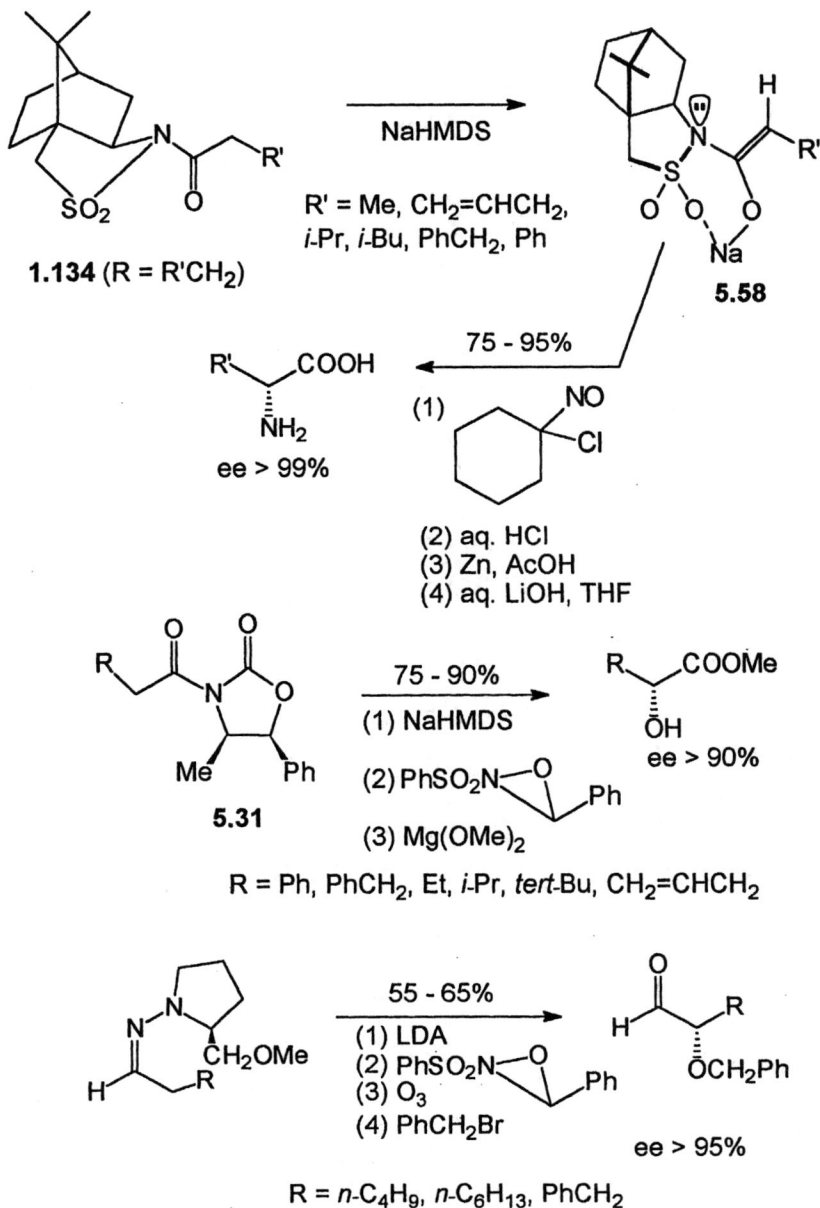

R' = Me, CH$_2$=CHCH$_2$, i-Pr, i-Bu, PhCH$_2$, Ph

1.134 (R = R'CH$_2$)

5.58

75 - 95%

(1)

(2) aq. HCl
(3) Zn, AcOH
(4) aq. LiOH, THF

ee > 99%

75 - 90%

(1) NaHMDS
(2) PhSO$_2$N—O—Ph
(3) Mg(OMe)$_2$

5.31

ee > 90%

R = Ph, PhCH$_2$, Et, i-Pr, tert-Bu, CH$_2$=CHCH$_2$

55 - 65%

(1) LDA
(2) PhSO$_2$N—O—Ph
(3) O$_3$
(4) PhCH$_2$Br

ee > 95%

R = n-C$_4$H$_9$, n-C$_6$H$_{13}$, PhCH$_2$

Figure 5.39

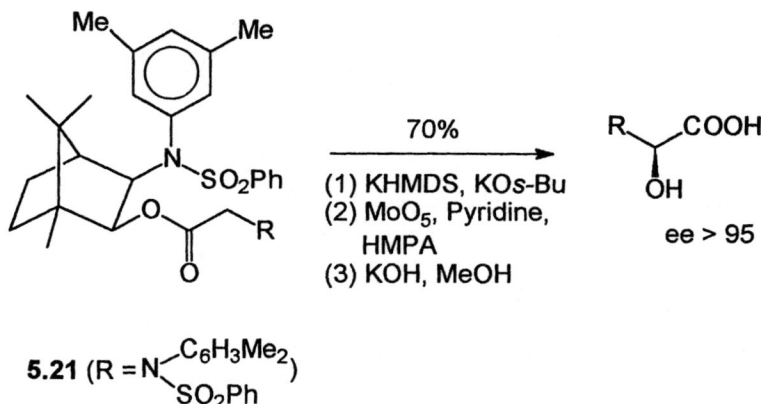

Figure 5.39 (continued)

5.6.2 Reactions with Chiral Electrophiles

Chiral camphorsulfonyloxaziridines **2.82** have been prepared by Davis and coworkers [741, 742, 744]. These authors also prepared nonracemic N-fluo-rodichlorocamphor sultams from precursors of **2.82**, and tested them as asymmet-ric fluorination reagents for enolates [1079]. Unfortunately, the enantiomeric excesses in these processes are not very high. Asymmetric oxidation of prochiral alkali enolates of ketones by oxaziridines **2.82** takes place at −78°C, and Z-enolates lead to higher enantioselectivities than E-isomers [742, 744, 747, 748] (Figure **5.40**). The best combination of enolate counterion (lithium, sodium or potassium) and reagent **2.82** (X = Cl, OMe or H) vary from case to case. The method has been applied to substituted tetralone **5.59** and to other daunomycinone precursors [743, 1080] (Figure 5.40). α-Hydroxy-β-aminoesters have also been prepared with a high selectivity by using these reagents to trap enolates obtained by conju-gate addition of lithium amides to α,β-unsaturated esters (§ 7.8) [1081, 1082]. However, asymmetric oxidations of ester and amide enolates do not give useful enantioselectivities, except in the case of dihydrooxazoles [1083].

Oppolzer and coworkers [1084] designed a bornane-based reagent **5.60** which reacts with zinc Z-enolates of ethylketones to generate α-ketohydro-xylamines with a high selectivity. Acidic hydrolysis, followed by reduction with NaBH$_4$ and then Zn/HCl, leads to *anti* 1,2-aminoalcohols with a high enantiomeric excess (Figure 5.40). The N=O group of the reagent is coordinated to the zinc atom of the enolate and the approach of **5.60** minimizes steric hindrance well for Z-enolates. There are additional repulsions with the bornane skeleton for E-eno-lates (Figure 5.40).

Asymmetric acylation of prochiral ester enolates is performed by using *N*-acetyl- or *N*-propionoylthiazolidine thiones **1.123** (X = S, R = *i*-Pr, R' = H or Me), but α,α-disubstituted β-ketoesters are obtained with a high enantiomeric excess only from substituted arylacetic esters [1085] (Figure **5.40**). Nonracemic *N*-acylbenzimidazoles have been proposed for asymmetric acylation of propionamides, but the enantiomeric excesses are low [1086].

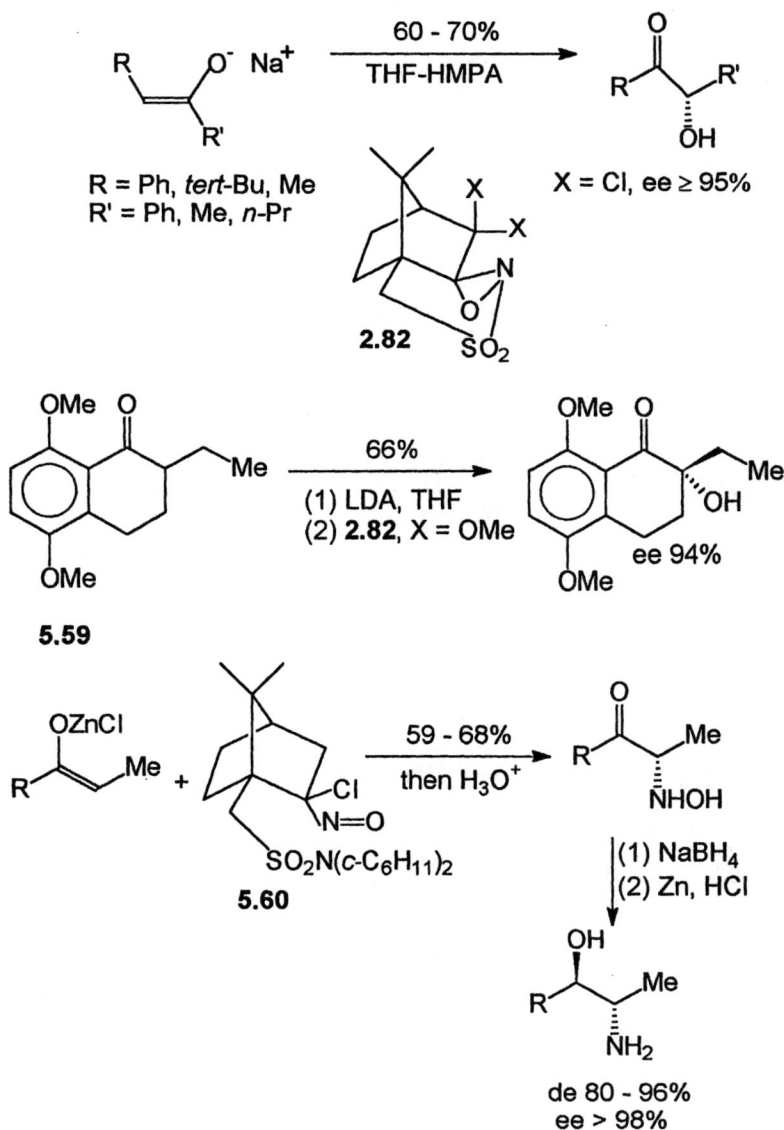

Figure 5.40

Z-enolate   E-enolate

Ar = Ph, 4-MeOC$_6$H$_4$

R' = H, Et

1.123 (X = S, R = *i*-Pr)

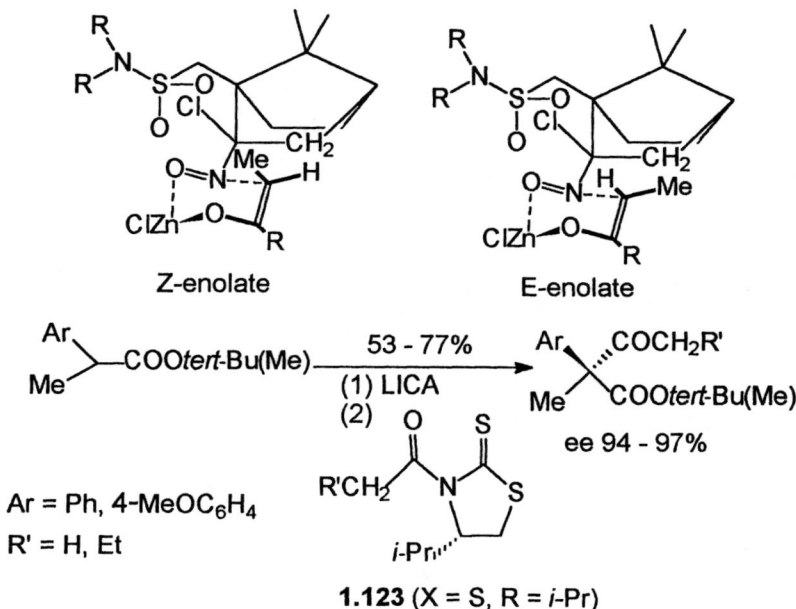

Figure 5.40 (continued)

5.7 REACTIONS OF KETENES ACETALS WITH VARIOUS ELECTROPHILES

E-Ketene silylacetals are prepared by kinetic deprotonation of esters with LDA in THF, followed by O-silylation with trialkylsilyl chlorides [161, 162, 1016]. Oppolzer and coworkers generated E-ketene acetals **5.61** *in situ* from esters of **1.10**. Addition of NBS or NCS to these compounds takes place on the least hindered side, leading to a halonium ion. Loss of the silyl group then leads to α-haloesters **5.62** with high stereoselectivity. The reduction of **5.62** leads to halohydrins with a very good enantiomeric excess (Figure **5.41**). Haloesters **5.62** are also converted to α-aminoacids with inversion of configuration by nucleophilic substitution with NaN$_3$. The reaction of E-ketene silylacetals **5.61** with *tert*-butyl diazodicarboxylate can be carried out in the presence of TiCl$_4$/Ti(O*i*-Pr)$_4$ at −78°C. Just as with metal enolates, this provides α-aminoacids of opposite configuration to those obtained by the halogenation/NaN$_3$ substitution procedure (yield 60 - 90%, ee > 95%). Similar results were obtained by Gennari and coworkers [147, 154] with ephedrine-derived ketene silylacetals.

The reaction of ketene acetals **5.61** with Pb(OAc)$_4$ in the presence of NEt$_3$•HF [147] leads to α-acetoxyesters **5.63** with an excellent stereoselectivity. Hydrolysis of **5.63** leads to nonracemic α-hydroxyacids with a high enantiomeric excess (Figure **5.42**).

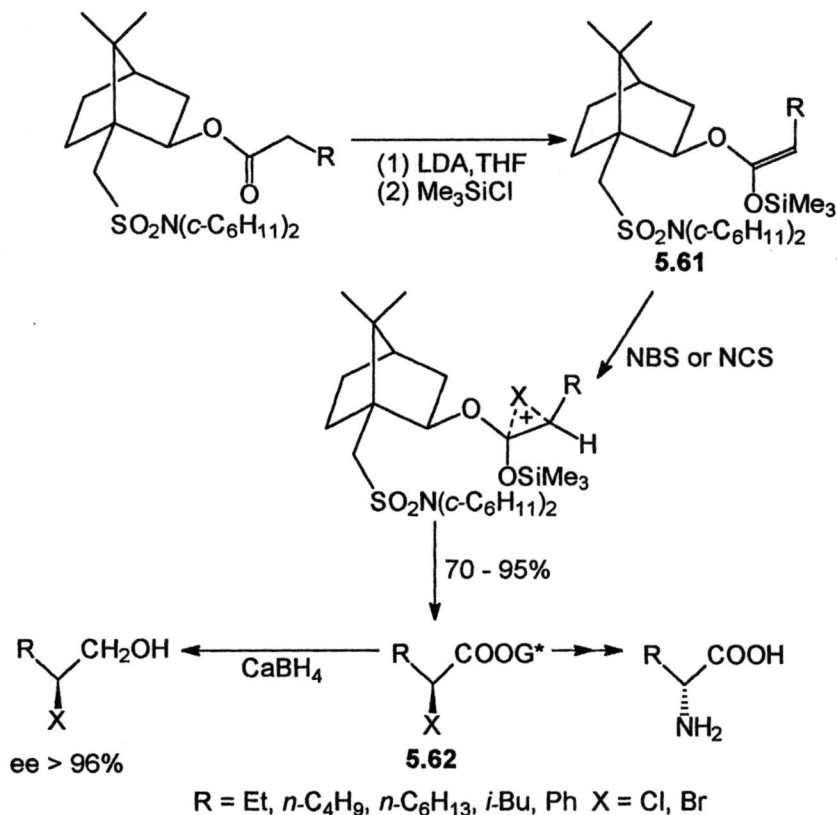

Figure 5.41

Evans's oxazolidinone *O*-silylaminals **5.64** react highly selectively with 1,3-dithienium fluoroborate or PhSCl, provided that the silicon substituents are bulky enough [1017a] (Figure **5.42**).

5.8 MISCELLANEOUS REACTIONS

5.8.1 Friedel-Crafts and Related Reactions

Lewis-acid-catalyzed reactions of phenol with alkyl glyoxylates are regioselective and 2-hydroxymandelic esters are formed. Starting from glyoxylates of chiral alcohols **1.23**, Bigi and coworkers obtained esters **5.65** with a high diastereoselectivity [179, 799]. Depending on whether the chiral auxiliary is menthol **1.4** (R = H) or phenmenthol **1.4** (R = Ph), the presence of either a chiral or an achiral titanium Lewis acid R"TiCl$_3$ is required (Figure **5.43**).

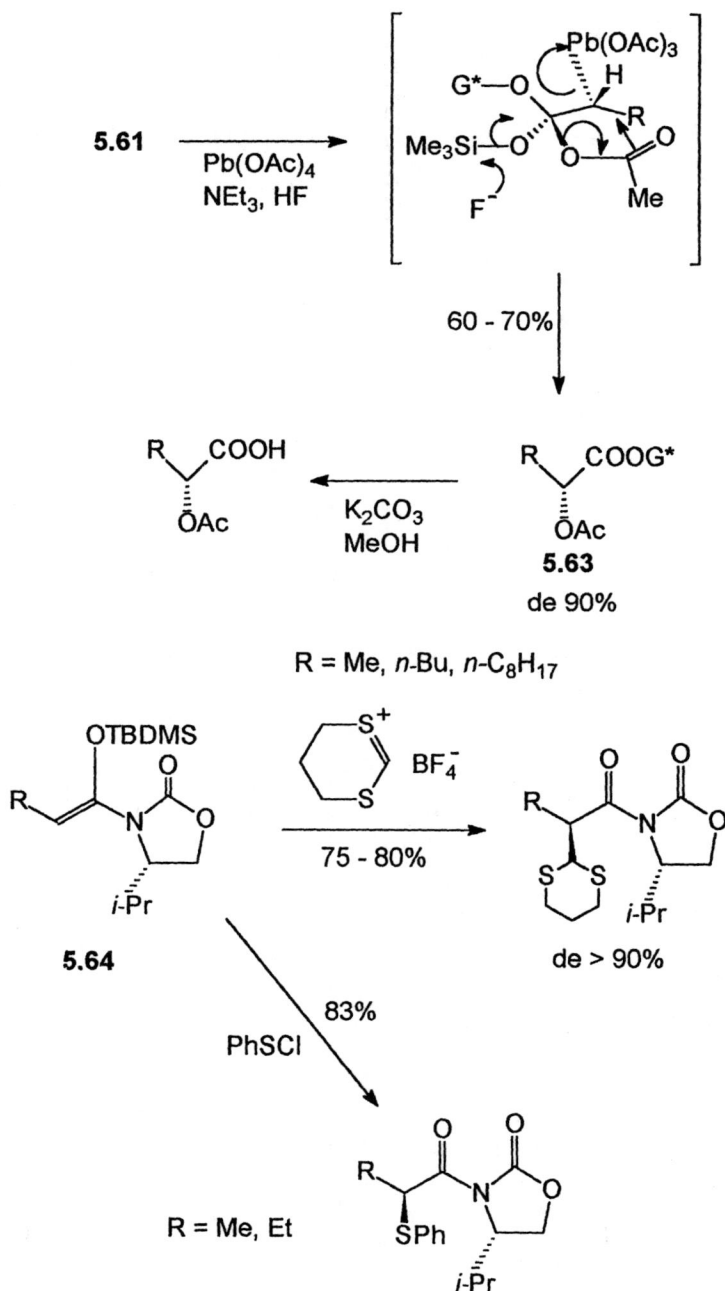

Figure 5.42

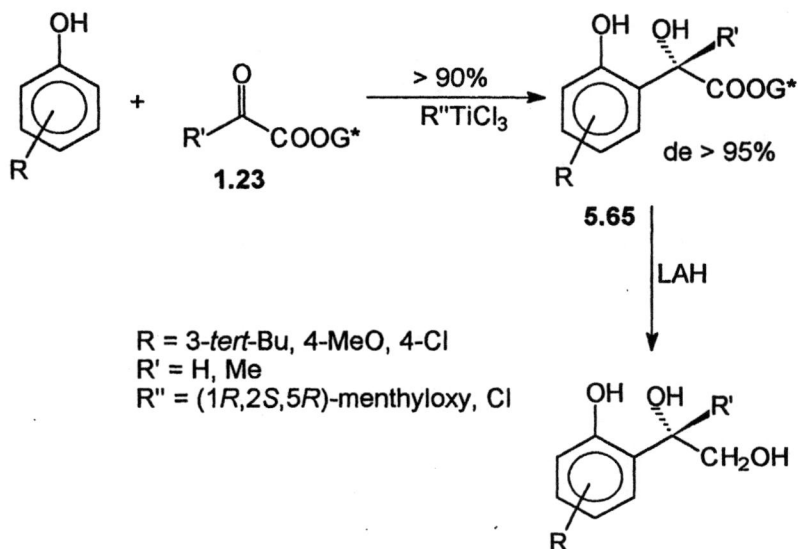

R = 3-*tert*-Bu, 4-MeO, 4-Cl
R' = H, Me
R" = (1*R*,2*S*,5*R*)-menthyloxy, Cl

Figure 5.43

Transition-metal-stabilized carbocations can be generated from functionalized butadieneiron carbonyl or arenechromium tricarbonyl complexes [92]. Reactions of such carbocations formed from chiral complexes have been studied, but low selectivities are usually observed [526, 528, 535]. However, chromium tricarbonyl complexes derived from ephedrine **5.66** suffer cyclization in acidic medium. After decomplexation, *cis*-tetrahydroquinolines are formed with a high diastereo- and enantioselectivity [540, 542] (Figure **5.44**).

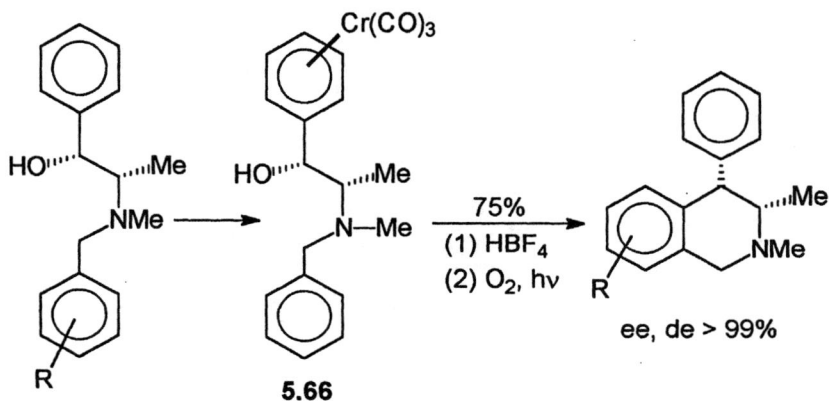

Figure 5.44

Arenemanganese tricarbonyl complexes bearing an (R,R)-2,5-dimethyl-pyrrolidine substituent on the phenyl ring react with PhMgBr with a good stereoselectivity (90%), but the results are not so useful with other nucleophiles [298].

5.8.2 Oxidative Coupling of Enolates

When Li enolates of N-acyloxazolidines **5.67** derived from (S)-**1.85** are treated with CuCl$_2$ at −78°C, dimer **5.68** is formed with high selectivity. In contrast, coupling promoted by iodine is less selective [1087] (Figure **5.45**).

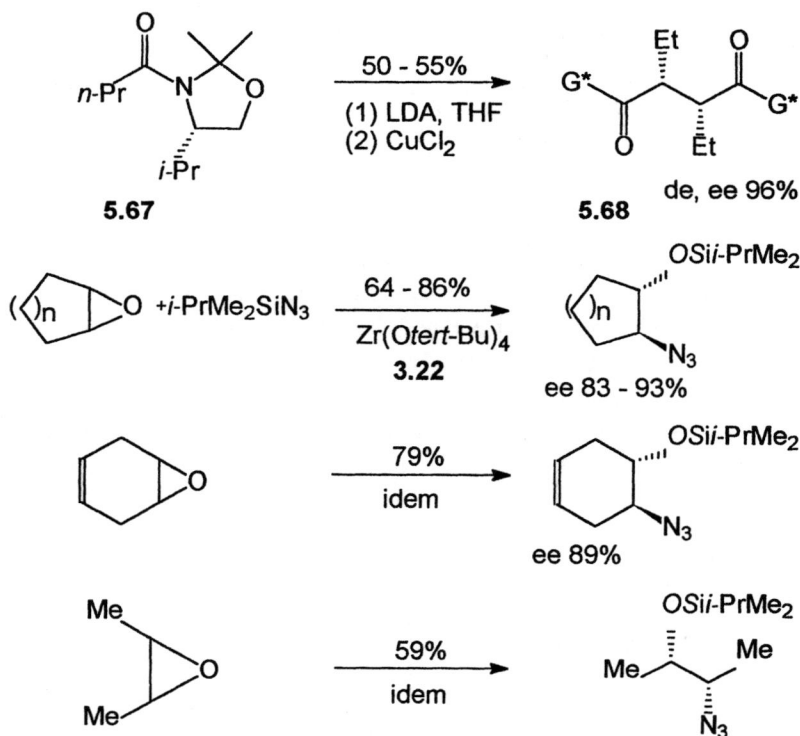

Figure 5.45

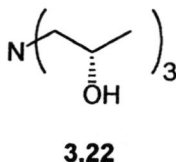

3.22

5.8.3 Ring Opening of *Meso* Epoxides [805]

Asymmetric openings of *meso* epoxides are catalyzed by chiral Lewis acids. Cyclohexene oxide reacts with *n*-butylthiol in the presence of zinc tartrate to give a β-hydroxysulfide with a good enantiomeric excess. However, ring openings with Me_3SiN_3 catalyzed by copper or titanium tartrates give disappointing results. Nugent [831] has proposed the use of $Zr(Otert\text{-}Bu)_4$ in the presence of (S,S,S)-tri-isopropanolamine **3.22** as a Lewis acid for epoxide openings with silyl azides at 0°C. Several *meso* epoxides were highly selectively transformed into 1,2-azido-silyloxy derivatives (Figure **5.45**). Attempts at kinetic resolution of oxiranes under chiral Lewis acid catalysis gave disappointing enantioselectivities [1088].

CHAPTER 6

ADDITIONS TO C=O AND C=N DOUBLE BONDS

Nucleophilic addition to C=O and C=N groups often occurs by an electrophilic assistance mechanism. Electrophilic assistance takes place when the n electrons on a double-bonded oxygen or nitrogen atom coordinate with a Lewis acid, thereby activating the double bond toward nucleophilic attack. The Lewis acid can be the cation associated with the nucleophile or it can be another compound added in stoichiometric or catalytic quantities. Electrophilic assistance decreases the frontier molecular orbital gap between the nucleophile and the electrophile by lowering the LUMO of the electrophile. Nucleophilic attack takes place along the familiar Bürgi-Dunitz trajectory (109°). This minimizes the unfavorable LUMO-HOMO overlap and the repulsive interaction between the nucleophile and the oxygen or nitrogen lone pairs (§ I.2.1). As with alkylations, when the reacting system can be rigidified by chelation, the corresponding process is favored. Charge transfer interactions, π stacking, and other factors (see Chapter I) can also play a rigidifying role.

6.1 REDUCTIONS BY HYDRIDES AND BORANES
[87, 566, 584, 589, 601, 1089]

When reductions are performed by tetracoordinated aluminum- or boron hydrides or by tricoordinated alanes or boranes, it is usually thought that asymmetric induction takes place during the nucleophilic hydride transfer step itself. With the Lewis acidic tricoordinated reagents, the asymmetric induction can also take place prior to the hydride transfer during the interaction of the reagent with the carbonyl or imine lone pair. Moreover, when the substrate bears a substituent prone to chelate formation with either a cation associated to the reducing agent or with an added metal salt, hydride transfer can take place on a rigid complex with both tricoordinated or tetracoordinated reagents.

6.1.1 Reductions of Nonfunctionalized and α-Unsaturated Aldehydes and Ketones

Asymmetric reductions of saturated and unsaturated carbonyl compounds are performed either with alkoxyaluminum hydrides or alkoxyboron hydrides bearing chiral substituents, with chiral boranes, or with achiral boranes in the presence of stoichiometric or catalytic amounts of oxazaborolidines [602, 604, 611] (see Chapter 2). The most efficient chiral alkoxyaluminum hydrides [87, 566] are generated *in situ* from solutions of LiAlH$_4$ by addition of (1R,2S)-N-methylephedrine

1.14 (R = Me) and 3,5-dimethylphenol (or ethylamine) in a 1:1:2 molar ratio, by addition of chirald **2.6** (or its enantiomer) in a 1:2 ratio, or by addition of (R)- or (S)-binaphthol **1.44** and EtOH in a 1:1:1 ratio. In this last case, the (R)- or (S)-Binal reagent **2.8** is formed. Diamine **1.64** (R = CH₂NHAr) can also be added to the LiAlH₄ solution, as well as 2-isoindolinylbutan-1-ol [1089a].

According to the additive, the reagent is rigid mono- or dihydrides **2.8**, **6.1** or **6.2** (Figure 6.1). Most of these reagents reduce arylalkylketones or α-enones with useful enantiomeric excesses; however, to obtain useful selectivities, the alkyl group should not be too bulky (Figure 6.2). The reagent generated from chirald **2.6** is useful for enantioselective reduction of α-ynones [567]. Compared to the other reagents, the (R)- and (S)-Binal reducing agents are more broadly applicable [1089] (Figure 6.3).

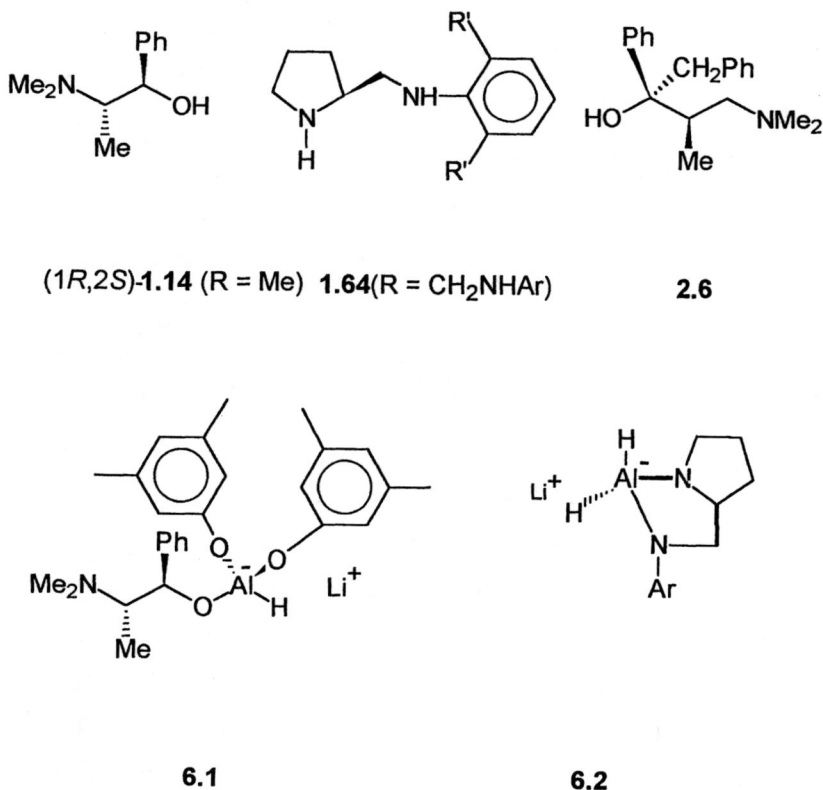

(1R,2S)-**1.14** (R = Me)   **1.64**(R = CH₂NHAr)   **2.6**

        6.1   **6.2**

Figure 6.1

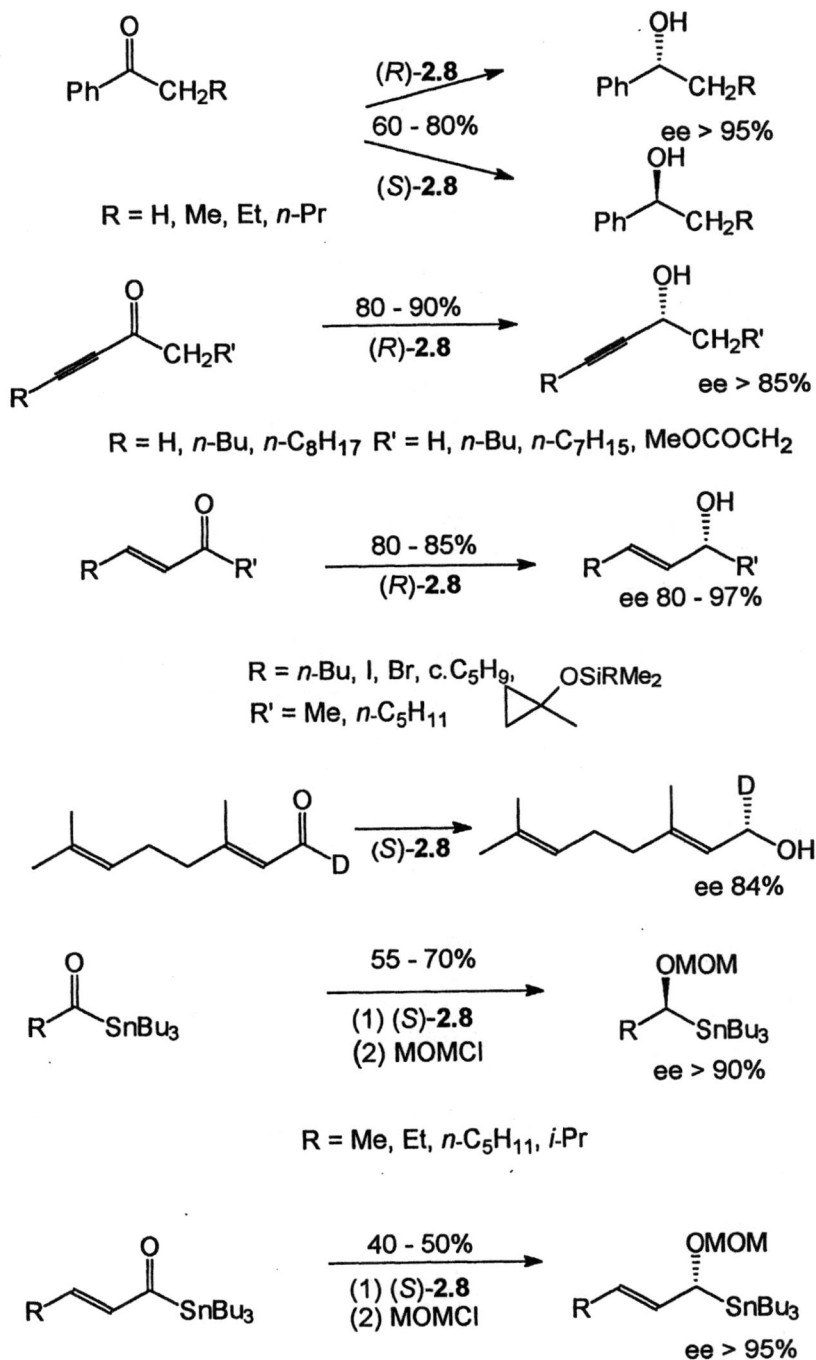

Figure 6.3

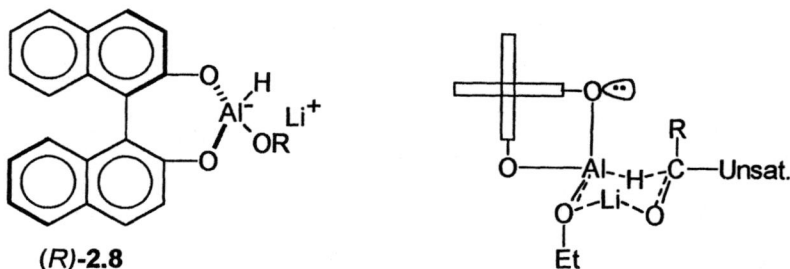

(R)-2.8

Figure 6.3 (continued)

Arylalkylketones, α-enones and α-ynones are reduced at −78°C by (R)-**2.8** to (R)-secondary alcohols and by (S)-**2.8** to (S)-enantiomers [566, 1090]. These reagents also reduce α-deuteroaldehydes at −100°C to nonracemic α-deuterated primary alcohols, and they reduce saturated or unsaturated acylstannanes with an excellent enantioselectivity [569, 570] (Figure **6.3**). The observed selectivity has been interpreted by repulsive interactions between the unsaturated group and binaphthyl oxygen lone pair [571]. If the R group of the ketone is bulky, repulsive interactions with this group decrease the selectivity (Figure **6.3**). Reductions of trifluoromethylarylketones occur with poor enantiomeric excesses; however, trifluoromethyl-9-anthrone leads to the corresponding secondary alcohol with a high selectivity (yield 89%, ee 91%) [568]. An application of the reduction of α-enones by Binal is a key step in Corey's prostaglandin synthesis [566]. Due to matched double diastereodifferentiation (§ I.6), the reduction of **6.3** with (S)-Binal is highly stereoselective, leading to (15R)-alcohol **6.4** (Figure **6.4**). The reduction of **6.3** with (R)-Binal is mismatched, and gives a mixture of isomers.

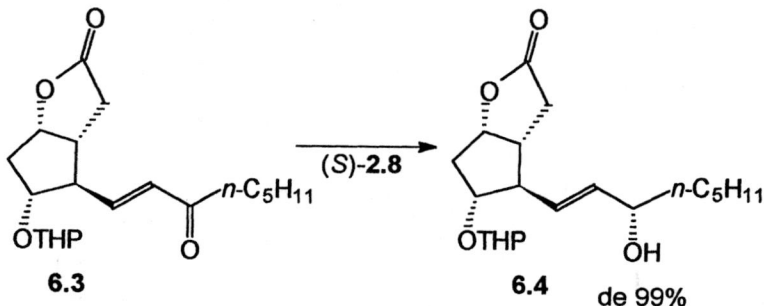

Figure 6.4

To overcome the disappointing results that are usually obtained in reductions of dialkylketones, Yamamoto and coworkers introduced crown-ether residues on the reagents and their preliminary results are encouraging [572].

Among the chiral borohydrides, ate-complexes generated from α-pinene **2.12** (R = Et, PhCH₂OCH₂CH₂) reduce 2-octanone or acetylcyclohexane at −100°C with good enantiomeric excesses [581, 582, 589]. A reagent formed from N-benzoylcysteine **6.5**, *tert*-BuOH and LiBH₄ reduces arylalkylketones to (R)-alcohols with a good selectivity (90%) [566]. Similar selectivities are observed with alkoxyborohydride **6.6** derived from diacetoneglucose **1.48** [566, 574].

Asymmetric reductions by boranes have been widely studied, and two classes of reagents are used: (1) those in which the reducing hydride is linked to boron, and (2) those in which the reducing hydride is linked to a carbon β to the boron atom.

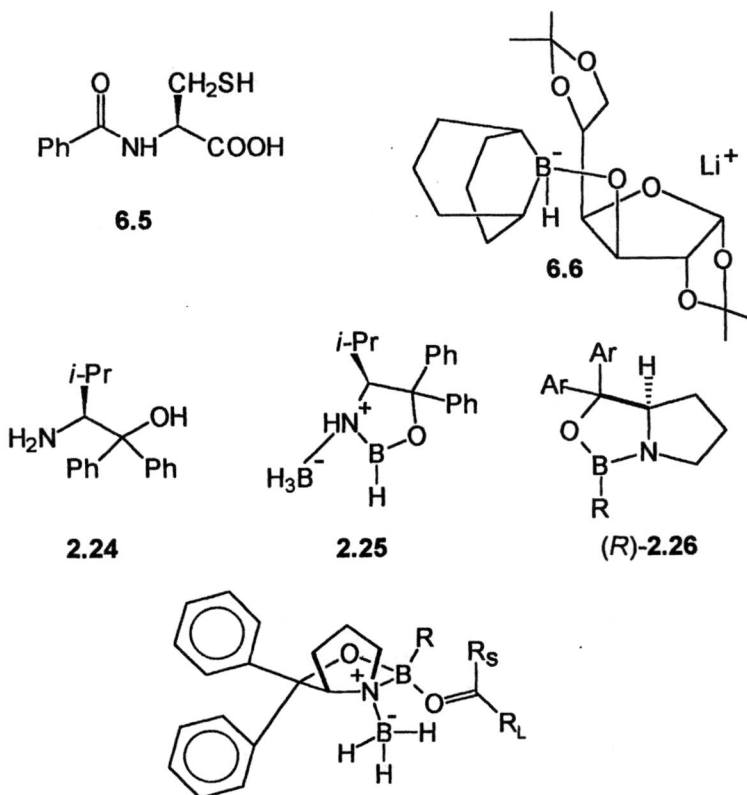

Figure 6.5

Among chiral boranes, pyridine-borane included in crystalline cyclodextrins reduces PhCOMe and $PhCH_2CH_2COMe$ with a high enantiomeric excess (90%) [579]. With other ketones low selectivities are obtained. Itsuno and coworkers proposed the use of BH_3 in the presence of aminoalcohol **2.24** to perform the asymmetric reduction of arylalkylketones to (R)-alcohols with an excellent selectivity (ee > 95%). These reductions occur near room temperature, but the selectivity is lower with dialkylketones (ee 55 - 73%) [566]. Corey and coworkers showed that the active reducing agent is an ate-complex of borane and a chiral 2,5-oxazaborolidine **2.25**. The asymmetric induction depends upon the geometry of the complex formed by coordination of the carbonyl oxygen to the Lewis acidic heterocyclic boron atom (Figure **6.5**). These authors have developed the use of stable oxazaborolidines (CBS reagents) **2.26** (Ar = Ph, 2-Np, R = Me, n-Bu, c-C_6H_{11}, tert-Bu) which, in stoichiometric or in catalytic amounts (§ 2.3 and Figure **2.3**), mediate the asymmetric reduction of aldehydes, ketones and α-enones by borane or catecholborane with excellent enantiomeric excesses [35, 566, 598, 599, 600, 601, 604, 606, 607, 608, 610, 611, 1089, 1091, 1092, 1093] (Figure **6.6**). Both enantiomeric oxazaborolidines **2.26** are available, so it is possible to obtain at will (R)- or (S)-alcohols. The reductions of ketones take place between –20°C and room temperature, and those of aldehydes occur at –126°C. Reductions of ketones that contain heteroatoms prone to coordinate borane, particularly nitrogen, can be highly enantioselective. For example, reductions of β-aminoacetophenones are highly selective [1094]. However, 2-, 3- and 4-acetylpyridines give poor selectivities [1094]. Some theoretical treatments of these reductions have been published [1095, 1096].

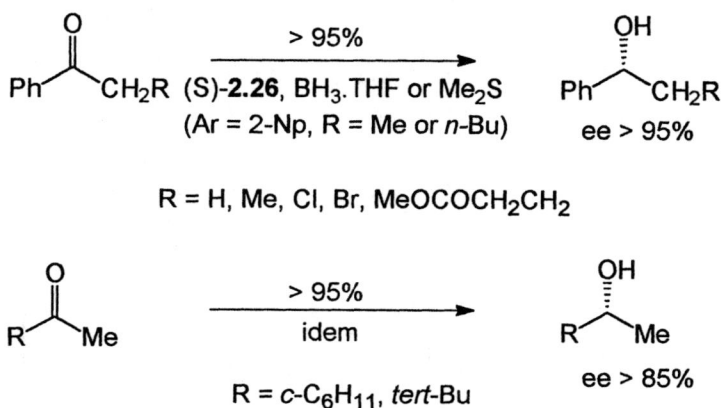

$$Ph-CO-CH_2R \xrightarrow[\text{(Ar = 2-Np, R = Me or } n\text{-Bu)}]{> 95\% \quad \text{(S)-2.26, BH}_3.\text{THF or Me}_2\text{S}} Ph-CH(OH)-CH_2R$$

ee > 95%

R = H, Me, Cl, Br, $MeOCOCH_2CH_2$

$$R-CO-Me \xrightarrow[\text{idem}]{> 95\%} R-CH(OH)-Me$$

R = c-C_6H_{11}, tert-Bu

ee > 85%

Figure 6.6

> 90%

(S)-2.26,
(Ar = 2-Np, R = n-Bu)

ee > 90%

R = n-C7H15, c-C6H11, Ph, 4-BrC6H4

89%

(S)-2.26, BH3.THF
(Ar = Ph, R=Me)

X = Br, I

ee > 90%

> 95%

(S)-2.26,
(Ar = Ph, R = n-Bu)

ee 90 - 98%

R = n-C5H11, PhCH2CH2, 2-NpCH2, c-C6H11, tert-Bu   X=F, Cl, Br

92%

(S)-2.26, BH3.THF
(Ar = Ph, R=Me)

ee 98 - 99%

n = 1,2,3

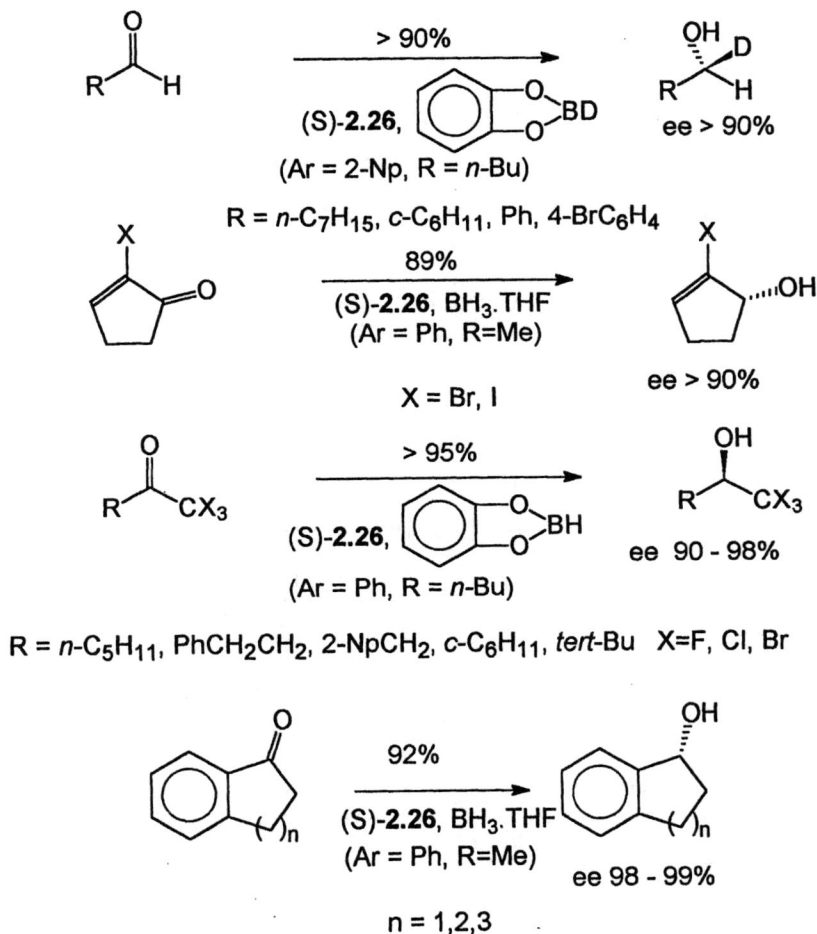

Figure 6.6 (continued)

These oxazaborolidine reagents have been broadly developed in a number of directions [601]. For example, reduction of trihaloalkylketones into the corresponding alcohols opens the way to the synthesis of nonracemic α-hydroxy- and α-aminoacids [609, 1091]. Addition of Et$_3$N can enhance the enantioselectivity of some reductions [1097]. Other aminoalcohols [601, 1096a] such as (S)-prolinol **1.64** (R = CH$_2$OH) [615], aminoalcohol **2.28** [616], indolinemethanols **2.29** (R = Ph) [613], methionine or cysteine-derived aminoalcohols [35, 617, 1097a], and (1S,2R)- and (1R,2S)-2-amino-1,2-diphenylethanols **1.58** [605, 614] are recommended to reduce ketones with enantioselectivities similar to **2.26**. However, the reagents formed *in situ* in THF from **1.58** [605] give a better enantioselectivity than **2.26** in the reduction of 3- and 4-acetylpyridine (ee 76 - 80%). Other systems [368, 611], including β-hydroxysulfoximines [620] and oxazaphospholidines [619], are less efficient.

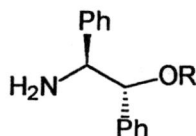

(1*S*,2*R*)-**1.58**

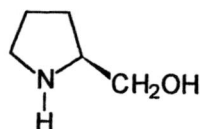

1.64(R = CH₂OH)

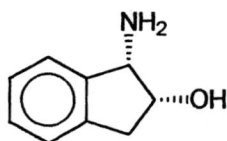

2.28

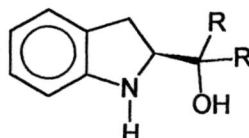

2.29

Masamune and coworkers reduced unsymmetrical dialkylketones RCOR' (R' = branched alkyl) with boranes **2.14** (R = H), and this reaction is highly enantioselective. Asymmetric induction occurs by coordination of the carbonyl group of the ketone to the mesylate **2.14** (R = OSO_2Me) which is present in catalytic amounts. This is followed by hydride transfer in a fashion so that steric interactions are minimized [85, 566, 1089, 1098, 1099] (Figure **6.7**). The availability of (*R*,*R*)- and (*S*,*S*)-**2.14** allows the formation of either enantiomeric secondary alcohol.

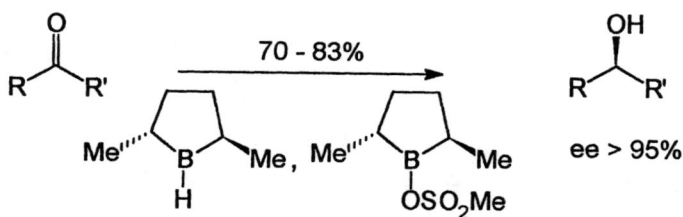

R = Me, *n*-C₅H₁₁, *n*-C₆H₁₃
R' = PhCH₂, *tert*-BuCH₂, *i*-Pr, *i*-Bu, *c*-C₆H₁₁, *tert*-Bu

Figure 6.7

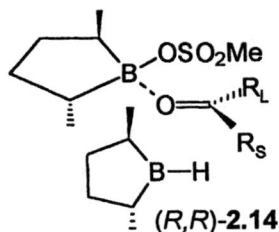

(R,R)-**2.14**

Figure 6.7 (continued)

Chiral boranes **2.15** and **2.16** are capable of transferring a hydride bound to a β-carbon, and they have been prepared by Brown, Midland and their coworkers from enantiomeric α-pinenes [584, 589, 591, 592, 593, 1100]. Asymmetric induction is effected by coordination of the lone pairs of the carbonyl oxygen to the boron atom in a boat geometry with minimization of steric interactions [589] (Figure **6.8**).

Alpine-Borane **2.15** (BR_2 = 9-boratabicyclononyl) reduces only reactive carbonyl compounds like aldehydes, α-haloketones, and α-ynones [584, 589] (Figure **6.9**). These reactions are faster when they are run under pressure, and arylalkylketones can thus be reduced with a good enantiomeric excess. B-Chlorodiisocamphenylborane (DIP chloride) **2.16** (R = Cl) is a better Lewis acid than Alpine-Borane, and it is more reactive. Between −25 and 0°C, DIP chloride reduces the following classes of substrates with an excellent enantioselectivity: arylalkylketones [589, 590, 1101, 1102] sufficiently dissymetrical dialkyl- and cyclic ketones [589], trifluoromethylketones [1103], α-ynones bearing bulky substituents [589], acylsilanes, and some acyclic α-enones [589] (Figure **6.9**). This reduction has been used in the synthesis of several drugs [1104, 1105], and it has occasionally displayed a better enantioselectivity than reduction with CBS reagents. 2-Aminoacetophenones are reduced at −78°C in Et_2O to β-aminoalcohols with a high enantioselectivity [1106].

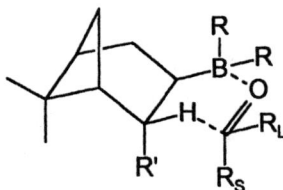

Figure 6.8

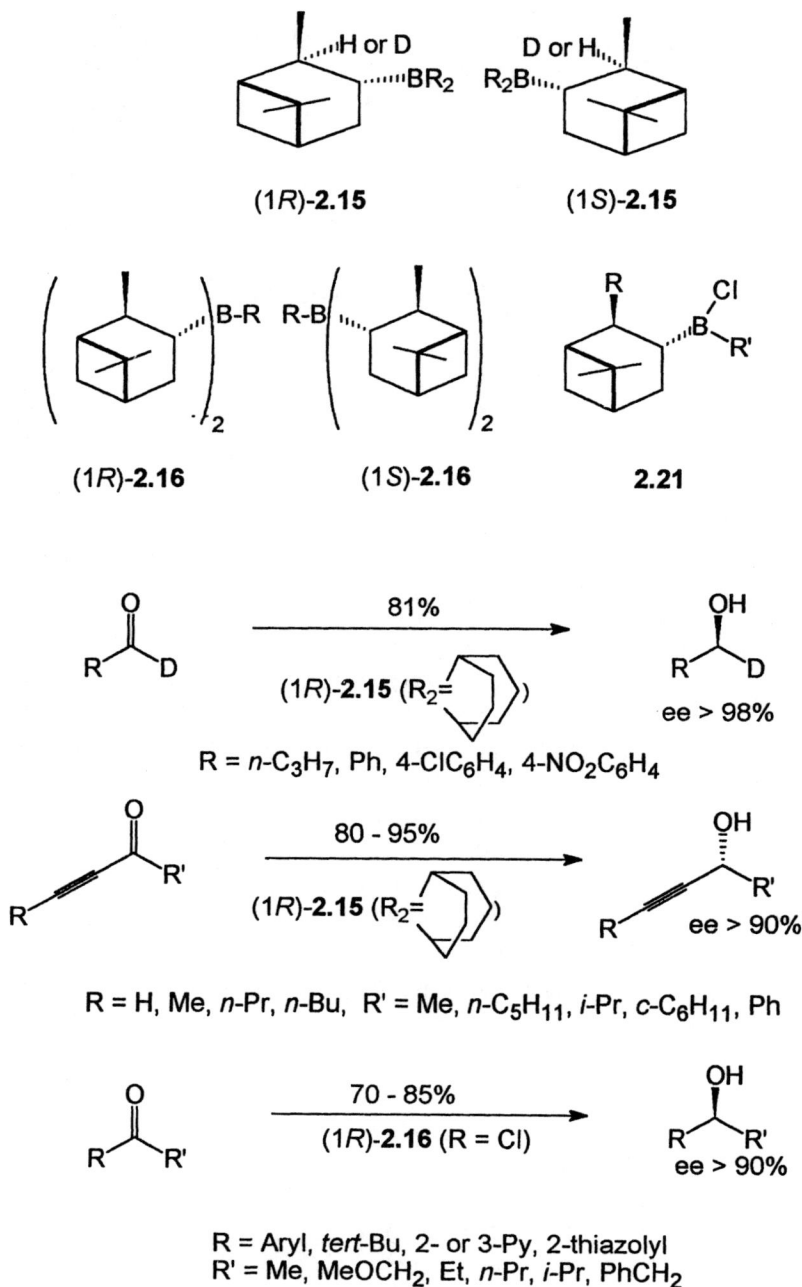

(1R)-2.15   (1S)-2.15

(1R)-2.16   (1S)-2.16   2.21

R = n-C₃H₇, Ph, 4-ClC₆H₄, 4-NO₂C₆H₄

$$R = n\text{-}C_3H_7,\ Ph,\ 4\text{-}ClC_6H_4,\ 4\text{-}NO_2C_6H_4$$

$$R = H,\ Me,\ n\text{-}Pr,\ n\text{-}Bu,\quad R' = Me,\ n\text{-}C_5H_{11},\ i\text{-}Pr,\ c\text{-}C_6H_{11},\ Ph$$

$$R = Aryl,\ \textit{tert}\text{-}Bu,\ 2\text{-}\ or\ 3\text{-}Py,\ 2\text{-}thiazolyl$$
$$R' = Me,\ MeOCH_2,\ Et,\ n\text{-}Pr,\ i\text{-}Pr,\ PhCH_2$$

Figure 6.9

70%
(1R)-**2.16** (R = Cl)

ee 98%

70 - 90%
(1R)-**2.16** (R = Cl)

$R \quad CF_3$

ee 78 - 96%

R = Me, n-C$_7$H$_{15}$, n-C$_8$H$_{17}$, c-C$_6$H$_{11}$, Ph, 1-Np, 2-Np

75 - 80%
(1R)-**2.16** (R = Cl)

ee > 95%

R = Ph, c-C$_5$H$_9$, n-C$_8$H$_{17}$ R' = *tert*-Bu, Et$_3$C, Me$_2$PhC

60 - 67%
(1R)-**2.16** (R = Cl)

$R \quad SiR'_3$

ee > 96%

R = Me, Et, i-Pr, R' = Me, i-Pr, n-Bu

65 %
(1R)-**2.16** (R = Cl)

Ph   Me

ee > 81%

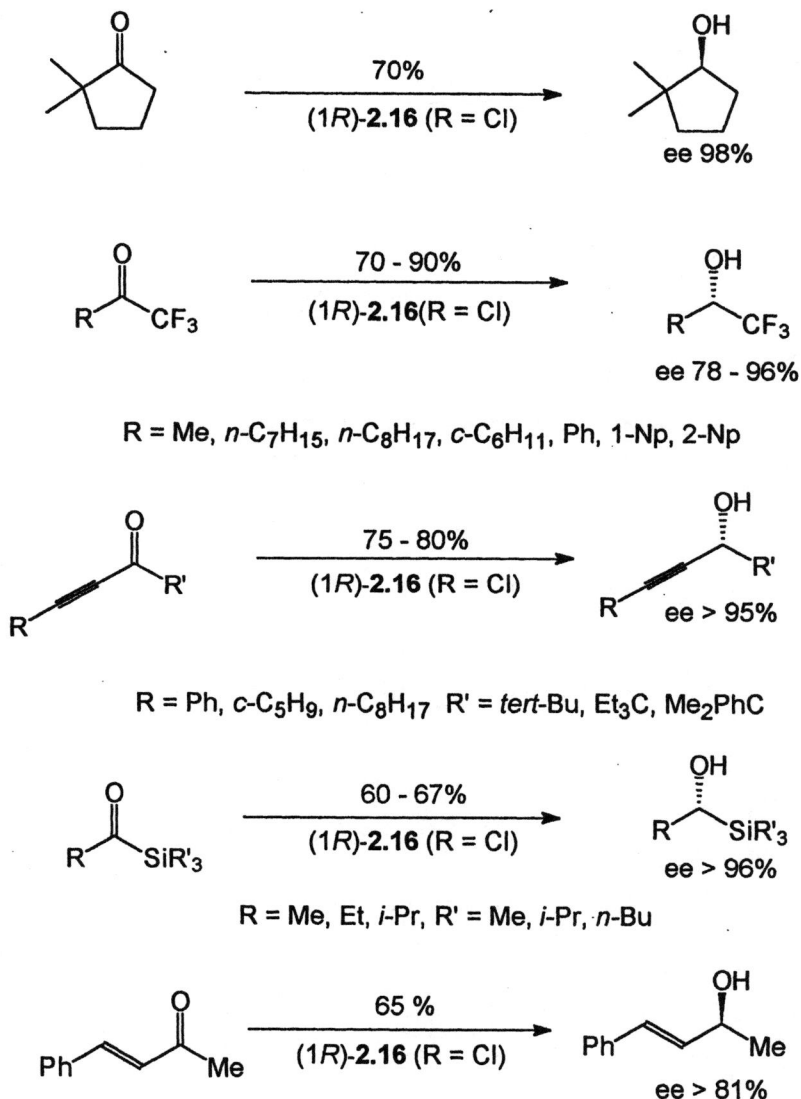

Figure 6.9 (continued)

Although less reactive, the analog of **2.16** in which the Me group is replaced by an Et group gives improved results in the reductions of unsymmetrical dialkylketones [592]. Chloroborane **2.21** (R' = *tert*-Bu, R = Et or PhCH$_2$CH$_2$-OCH$_2$CH$_2$) has also been used [593]. The observed selectivities with this reagent are opposite to those predicted by the previous model (Figure **6.8**).

Trifluoro- and trichloroacetophenones are reduced by 1,4-dihydronicotamides **2.31**, but the enantioselectivity of this process is modest [591, 627].

6.1.2 Reductions of Functionalized Aldehydes and Ketones

Functionalized carbonyl compounds may be reduced by chiral reagents or, if the functional group is chiral, by achiral reagents.

6.1.2.1 *Reductions by Chiral Reagents*

The keto group of α-ketoesters, -nitriles and -amides is more electrophilic than that of nonfunctionalized ketones, and relatively unreactive chiral reducing agents can be used for an enantioselective reduction of these activated carbonyl compounds. Alpine-Borane **2.15** (BR_2 = 9-boratabicyclononyl, R' = Me) reduces α-ketoesters and α-ketonitriles with a high enantioselectivity [584] (Figure **6.10**). α-Ketoesters and α- or β-hydroxy- or alkoxyketones are reduced by sodium borohydride/tartaric acid with good enantiomeric excesses [575].

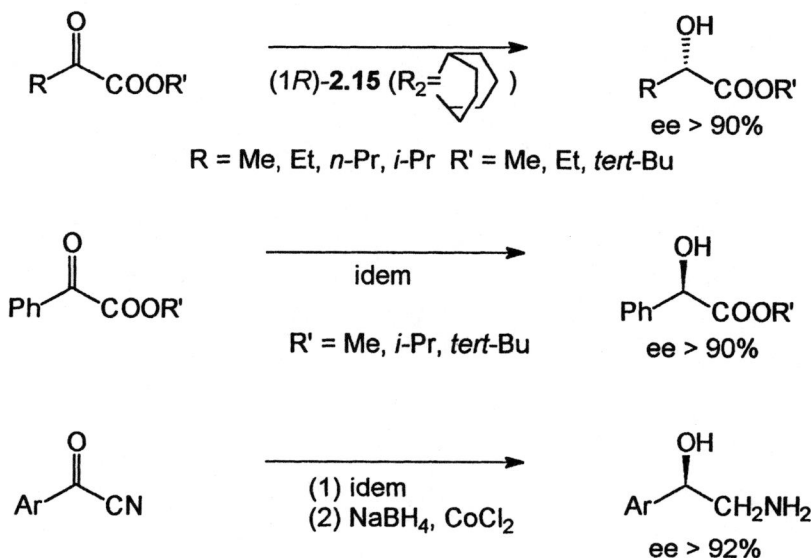

R = Me, Et, *n*-Pr, *i*-Pr   R' = Me, Et, *tert*-Bu

ee > 90%

R' = Me, *i*-Pr, *tert*-Bu

ee > 90%

(1) idem
(2) NaBH$_4$, CoCl$_2$

ee > 92%

Figure 6.10

Nature's reducing agent, NADPH, has inspired many biomimetic reductions of α-ketoesters with 1,4-dihydropyridines [591, 625, 627, 970, 1107]. Reagents **2.31** (R = R' = R" = R'" = Me or R = Me, R' = *n*-Pr, R" = H, R'" = PhCHMe), **2.33** and **2.34** reduce methyl or ethyl benzoylformate to the corresponding (*R*)-α-hydroxyester with an enantiomeric excess greater than 95%. The reaction time is typically less than 90 minutes, except with **2.31** (R = Me, R' = *n*-Pr, R" = H, R"' = PhCHMe). The other enantiomer is obtained with reagents **2.32** or **2.35** (R = Me, R' = PhCH$_2$, A = CH$_2$OH), but these reductions take 4-5 days (Figure **6.11**). Reductions of ketolactone **6.7** are less selective. With reagents **2.33** and **2.35**, (*R*)- or (*S*)-alcohols are formed in poor chemical yields (Figure **6.11**).

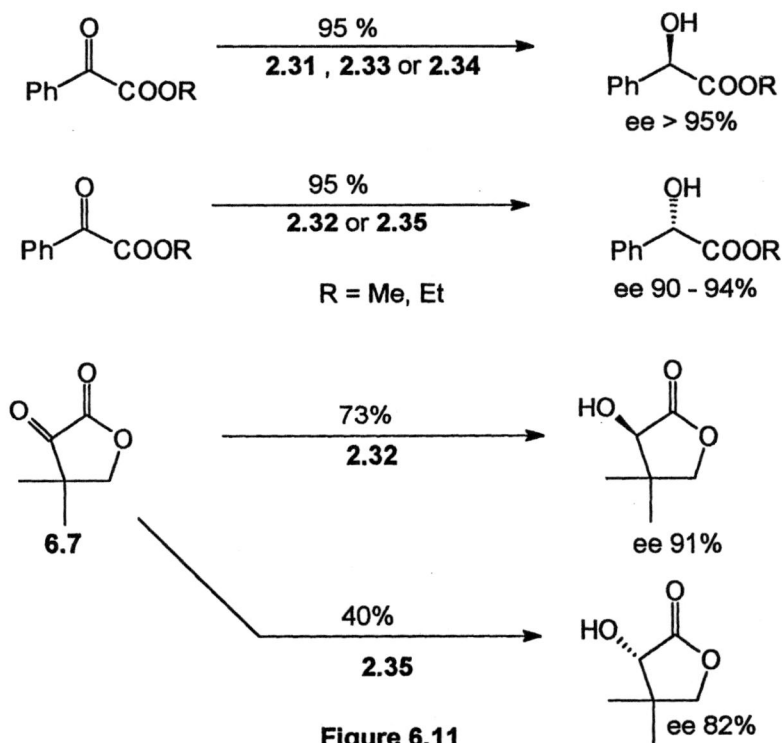

Figure 6.11

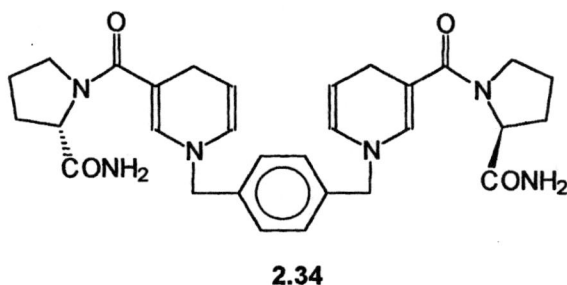

2.34

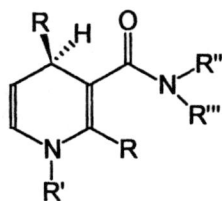

2.31

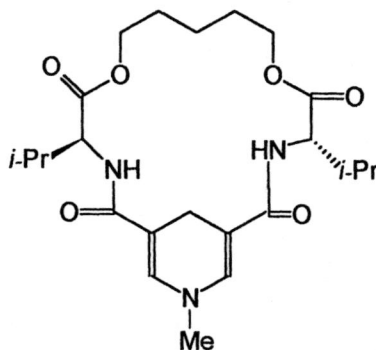

2.32

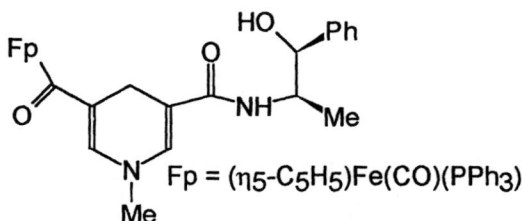

Fp = (η5-C5H5)Fe(CO)(PPh3)

2.33

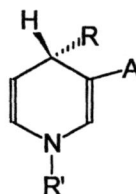

2.35

All these reductions are catalyzed by Mg(ClO$_4$)$_2$ and take place on a rigid complex. Two different models, **6.8** and **6.9**, have been proposed to account for the observed selectivities. In model **6.8**, the magnesium chelates between the dihydropyridine nitrogen and the carbonyl oxygen, and hydride transfer takes place as indicated to minimize steric interactions between the substituents of the carbonyl group and those of the dihydropyridine. In model **6.9**, the magnesium chelates between the amide and ketone carbonyls, and hydride transfer occurs by a Dunitz-Bürgi trajectory (§ I.2.1). In model **6.9**, polar effects determine the face of the carbonyl group to which the hydride is transferred [625] (Figure **6.12**).

6.1.2.2 *Reductions of Carbonyl Compounds Bearing Chiral Residues*

The presence of a chiral acetal, aminal or oxathiane in the vicinity of a carbonyl group can direct the reduction of a ketone toward a diastereoisomeric alcohol, whether or not chelation is operative. For example, the LiAlH$_4$ or Li s-Bu$_3$BH reductions shown in Figure **6.13** give predominantly the diastereoisomer predicted by the Felkin-Ahn model while reductions with DIBAL or LiAlH$_4$/MgBr$_2$ give predominantly the diastereoisomer predicted by the chelation model [87, 94, 213, 226].

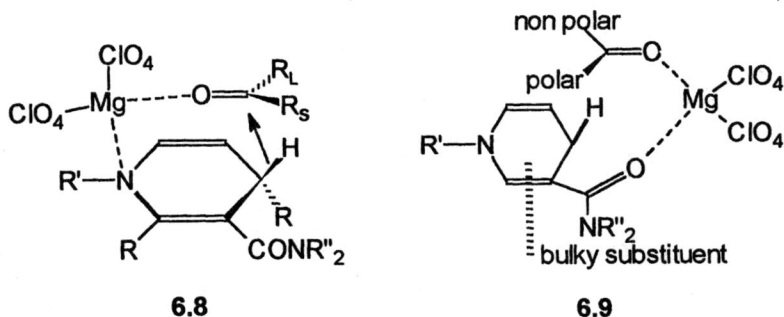

6.8   **6.9**

Figure 6.12

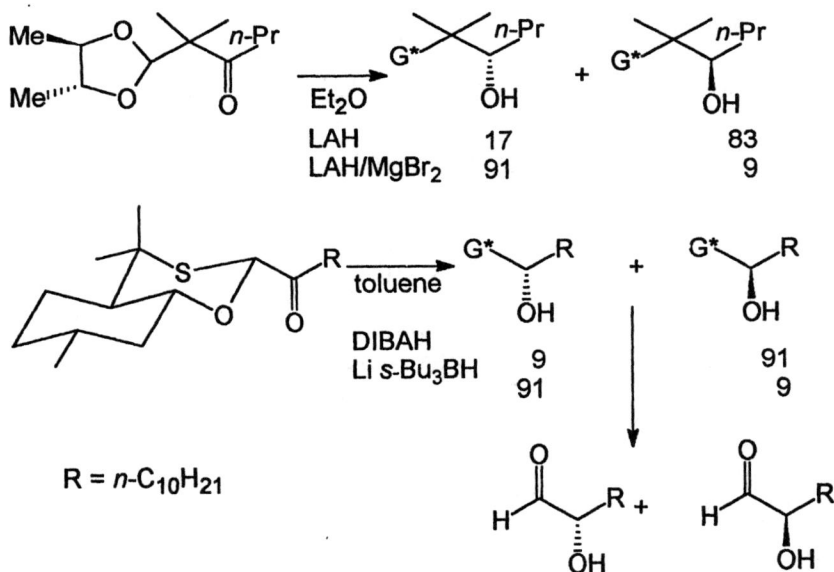

Figure 6.13

Reductions of α- or β-ketoesters of chiral alcohols also lead to either enantiomeric alcohol according to the reducing reagent. Reduction of phenmenthyl pyruvate **1.23** (R = Me) or phenylglyoxylate **1.23** (R = Ph) leads to either enantiomeric diol with a high selectivity, depending on whether the reaction takes place via a chelate or not [144, 180, 182] (Figure **6.14**). The use of a bulkier auxiliary can improve the selectivity [144, 155]. Similar results are obtained from β,γ-unsaturated α-oxoesters [1108]. Carbohydrate glyoxylates are reduced by K s-Bu$_3$BH, and either enantiomeric α-hydroxyacid is obtained, depending on whether the reduction is carried out in the presence or absence of 18-crown-6

[1109].   The reduction of β-ketoesters **6.10** with excess DIBAH or $Zn(BH_4)_2/ZnCl_2$ leads to diol (R)-**6.12** after $LiAlH_4$ cleavage of the auxiliary [181, 1110]. This reaction occurs by attack on the least hindered face of chelate **6.11**. On the other hand, reduction by DIBAH/BHT does not involve chelation, and leads to the enantiomer (S)-**6.12** after the same cleavage (Figure **6.14**). If the keto groups are located in a more remote position, rigidification of the system is more difficult to accomplish. Tamai and coworkers [236] performed the enantioselective reduction of γ-ketoester **1.48** by DIBAH/MgBr₂ (Figure **6.14**). Molander and Bobbitt [70] reduced δ-ketoboronate **6.13** by $BH_3 \bullet SMe_2$ at 0°C with a high selectivity (Figure **6.14**). This reaction probably occurs through a six-membered internal chelate, although this chelate could not be characterized by NMR spectroscopy [69].

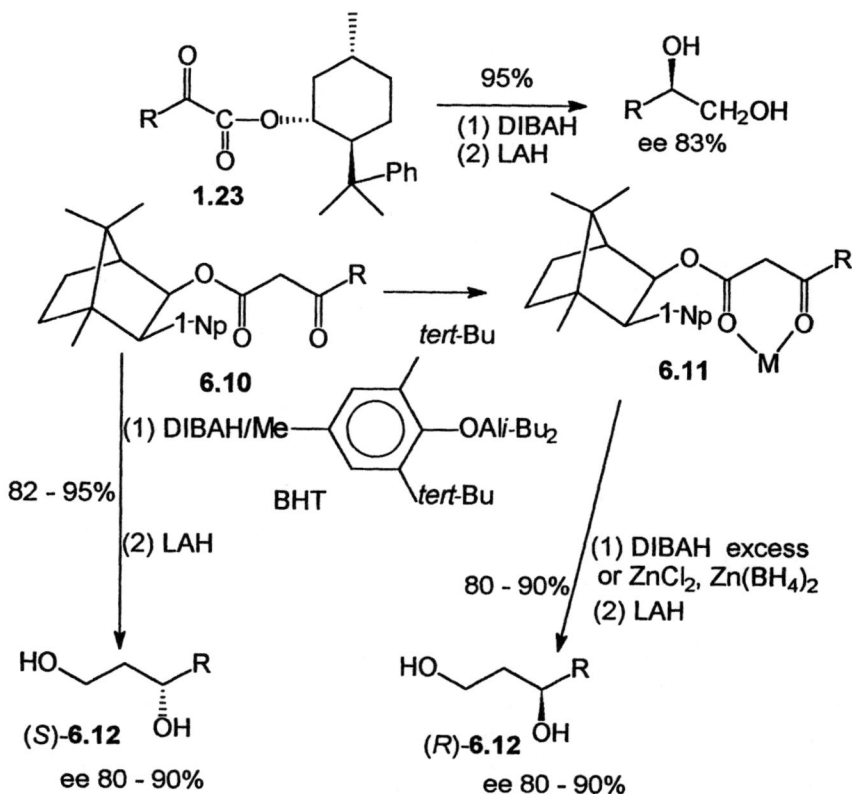

Figure 6.14

1.48 (n = 1)

ee 82%

6.13

87 - 95%

(1) BH₃.SMe₂

(2) H₂O₂, NaOH

ee 92 - 98%

R = n-C₅H₁₁, Cl(CH₂)₃, CN(CH₂)₁₀, MeOCO(CH₂)₄,c-C₆H₁₁, Ph

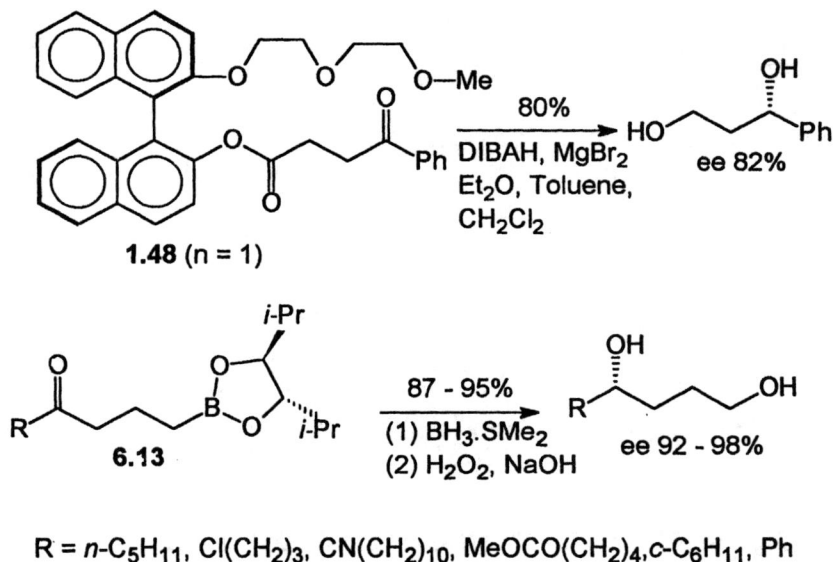

Figure 6.14 (continued)

The reductions of chiral α-ketoamides have also been frequently studied. To avoid racemizations when the chiral amide residue is hydrolyzed, easily cleavable amides of **1.65** (R = CH₂OMe) must be used [87] (Figure **6.15**). Reductions of β-keto-*N*-acyloxazolidinones **6.14** are highly stereoselective. After hydrolysis or transesterification, β-hydroxyacids or esters are obtained with excellent selectivities [87] (Figure **6.15**). However, the most efficient reducing reagent varies from case to case.

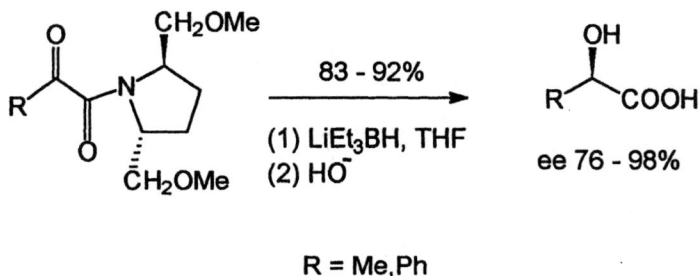

83 - 92%

(1) LiEt₃BH, THF

(2) HO⁻

ee 76 - 98%

R = Me,Ph

Figure 6.15

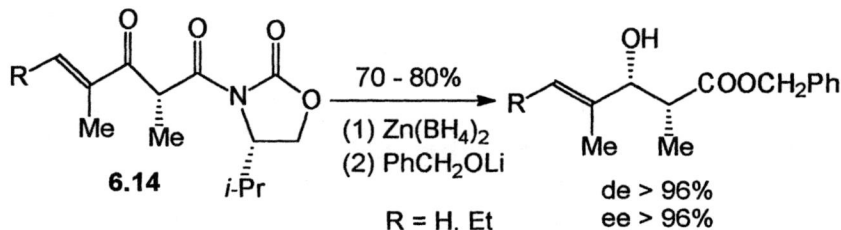

Figure 6.15 (continued)

The reductions of chiral ketosulfoxides have been studied by Solladié and coworkers [87, 473]. These reductions occur at −78°C, and either diastereoisomeric β-hydroxysulfoxide can be formed with a high selectivity, depending on whether DIBAH in THF is used alone or in the presence of a stoichiometric or catalytic [103] amount of $ZnCl_2$. After Raney nickel desulfurization, enantiomeric secondary alcohols are obtained [473, 1111] (Figure **6.16**). The observed selectivities are interpreted by models **6.16A** and **6.16B**. During the DIBAH reduction, the reagent is coordinated to the sulfoxide oxygen, and hydride transfer takes place so that steric repulsions are minimized. In the presence of $ZnCl_2$, a zinc chelate forms. In the transition state for reduction of this chelate, the aluminum atom of DIBAH is coordinated to the carbonyl oxygen of the ketone and to one chlorine atom of the zinc chloride, and hydride transfer takes place on the other face of the carbonyl group [103] (Figure **6.16**). The sulfoxide group of the secondary alcohol products can be reduced by $LiAlH_4$ to the corresponding sulfide, opening the way to nonracemic aminoalcohols with a high enantiomeric excess [474] (Figure **6.16**).

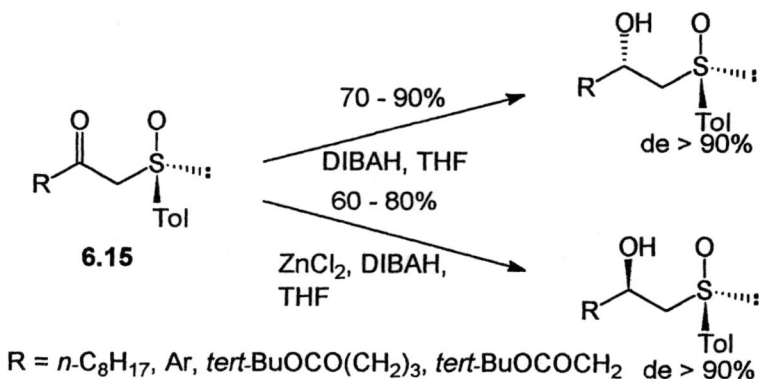

$R = n\text{-}C_8H_{17}$, Ar, tert-BuOCO(CH_2)_3, tert-BuOCOCH_2   de > 90%

Figure 6.16

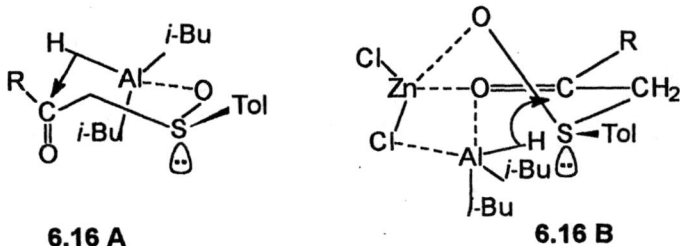

6.16 A   **6.16 B**

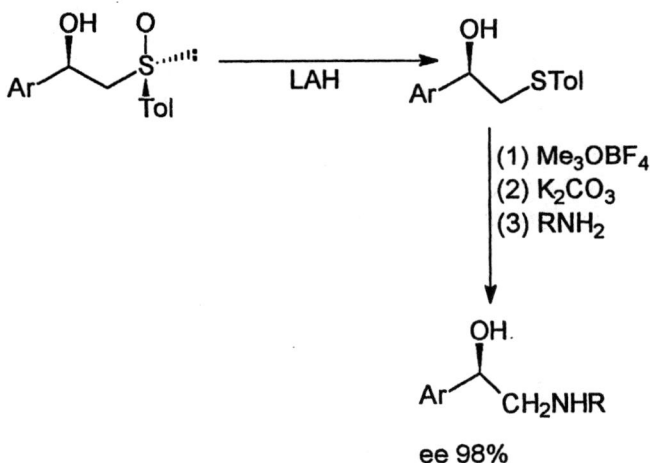

ee 98%

Figure 6.16 (continued)

The stereoselective reductions of α- or β-ketoaldehydes in which the aldehyde functionality is transformed into an *N*-tosyloxazolidine derived from ephedrine give rise to enantioenriched α- or β-hydroxyaldehydes after aldehyde unmasking [1112, 1113, 1114].

Chiral ironcarbonyl complexes bearing an acyl group such as **6.17** are stereoselectively reduced by LiAlH₄ or NaBH₄ [526, 537, 1115] (Figure 6.17). The hydride transfer takes place on the least hindered face of the complex with the ketone in the *s-cis* conformation. Ketones in chiral arenechromium carbonyl complexes are reduced on the face opposite to the organometallic residue. This reaction is highly stereoselective provided that the keto group has a favored conformation. Reductions of rigid indanone- and tetralonechromium carbonyl **6.18** and **6.19** (n = 1 or 2) show the utility of such an approach [87, 539] (Figure **6.17**).

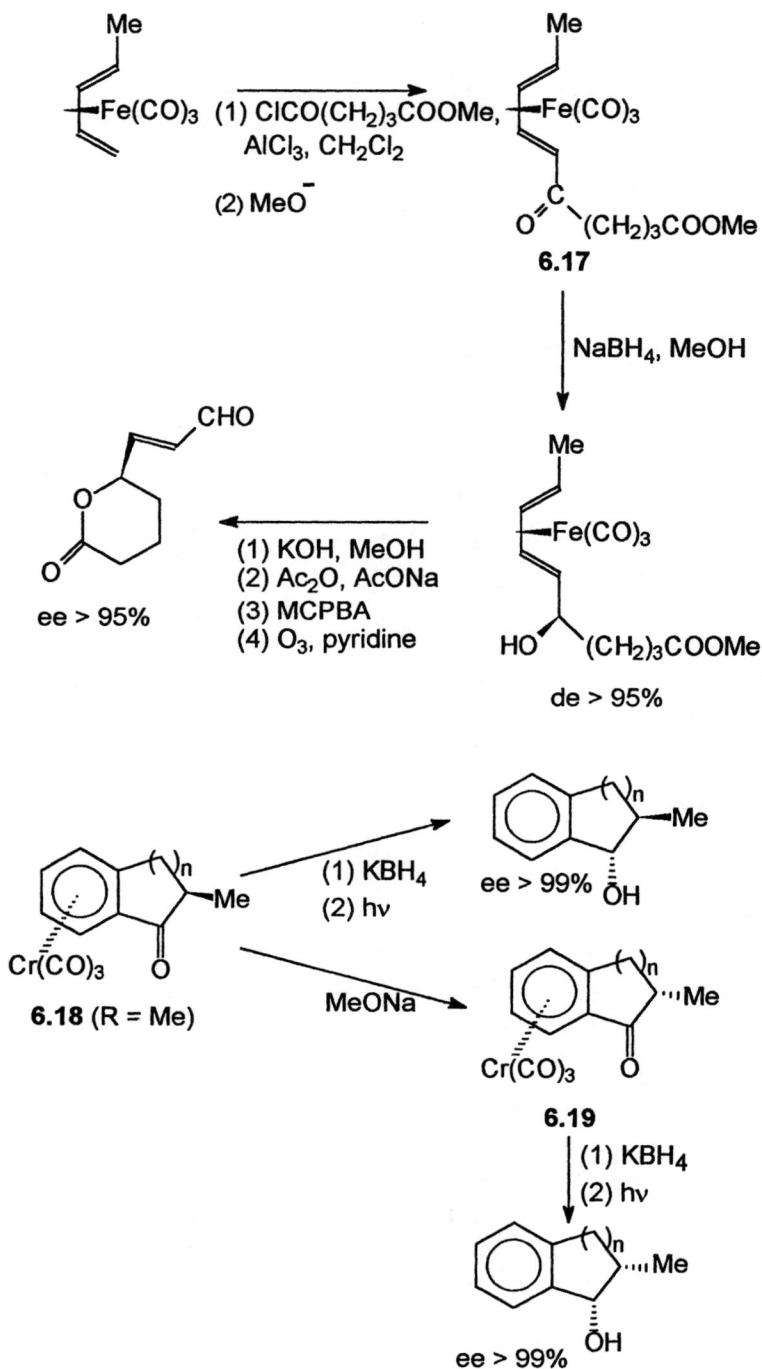

Figure 6.17

6.1.3 Reductions of Chiral Acetals

The reduction of ketals by Br_2AlH or $Et_3SiH/TiCl_4$ leads to secondary diol-monoethers, which are easily transformed into the corresponding alcohols [213]. Therefore, this method is equivalent to the reduction of ketones. When the ketals are formed from enantiopure diols, these reductions can be stereoselective, provided that predominant conformers are involved. Such is the case with 1,3-dioxanes **6.20** which prefer the conformation **6.21** in which the large group (R) is equatorial. Selective coordination of O-3 to the Lewis acid induces the cleavage of the C-2–O-3 bond [1116, 1117]. This is followed by hydride transfer either on the same face of the molecule ($AlBr_2H$) or on the opposite face (Et_3SiH). According to the reagent, either the (R)- or (S)-alcohol is produced with a good enantioselectivity after the removal of the diol residue (Figure **6.18**). Such a reductive cleavage can also be performed by using Et_2AlF and C_6F_5OH [222, 1118].

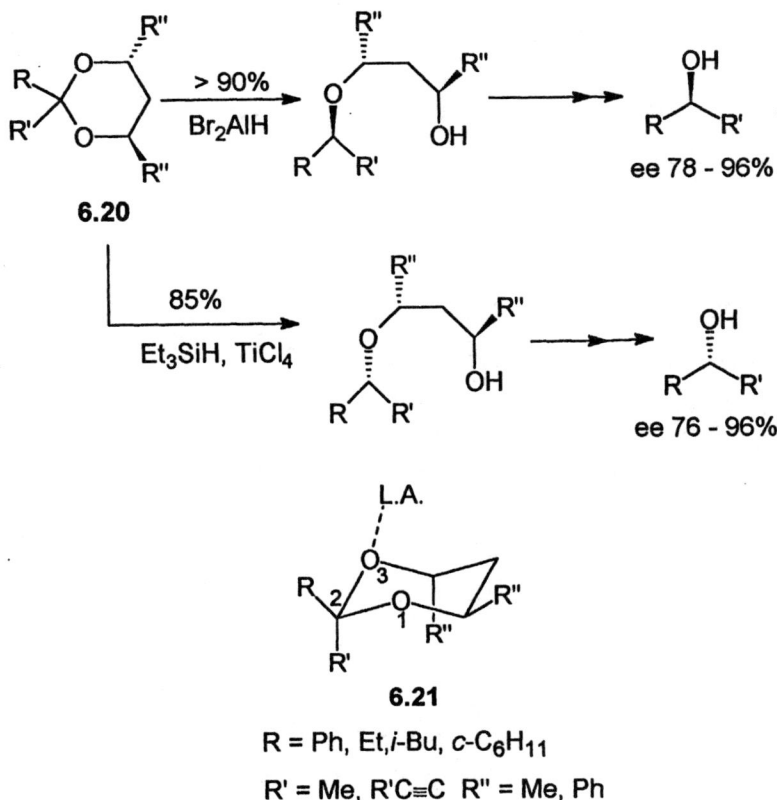

R = Ph, Et,i-Bu, c-C_6H_{11}

R' = Me, R'C≡C R" = Me, Ph

Figure 6.18

6.1.4 Reductions of Imines and Derivatives

6.1.4.1 *Reductions by Chiral Reagents*

Arylalkyl *N*-phenylketimines are reduced by borane in the presence of aminoalcohol **2.24** with a good enantioselectivity, but low selectivities are observed with *N*-alkyl analogs or with dialkyl *N*-phenylketimines [368, 601, 622] (Figure **6.19**). Sodium triacyloxyborohydride **2.9** generated from *N*-benzyloxycarbonyl-proline **2.10** (R = PhCH$_2$O) effects the enantioselective reduction of cyclic imines **6.22** to nonracemic tetrahydroquinolines [261, 566] (Figure **6.19**). Oxime ethers of arylalkylketones also suffer highly enantioselective reductions with BH$_3$ in the presence of aminoalcohols such as (1*R*,2*S*)-norephedrine **1.14** (R = H) or (1*S*,2*R*)-1-aminoindan-2-ol **2.28** [601, 616, 658] (Figure **6.19**). The configuration of the amine product depends upon the E- or Z-geometry of the oxime ether (Figure **6.19**). Less useful results are obtained with aminoalcohol **2.24**, except in the reduction of the oxime ether of acetophenone by NaBH$_4$/ZnCl$_4$, which leads to (*S*)-1-phenethylamine with an excellent enantiomeric excess [575, 601, 1102].

Some valuable asymmetric reductions of *N*-phosphinylimines (RR'=NPOPh$_2$) have been observed [1119]. Hydride transfer from nicotinamide **2.31** (R = R' = R" = Me) to activated imines takes place with a high enantiomeric excess, but with variable chemical yields [628, 629].

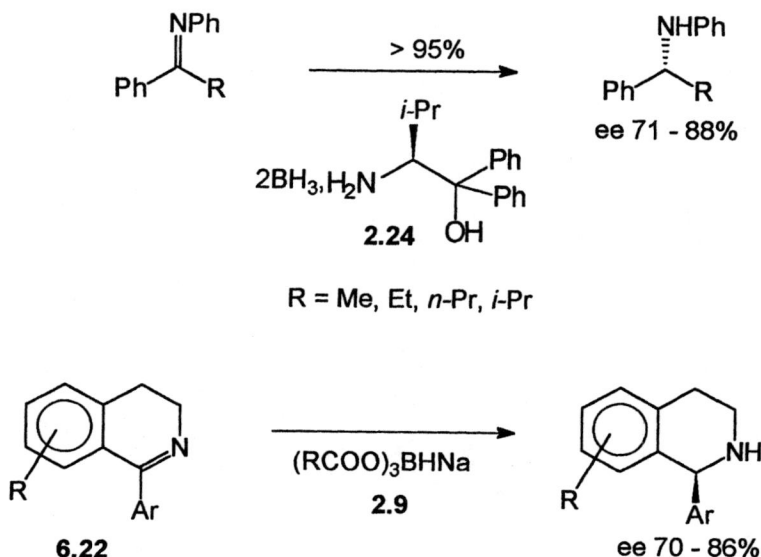

R = Me, Et, *n*-Pr, *i*-Pr

Figure 6.19

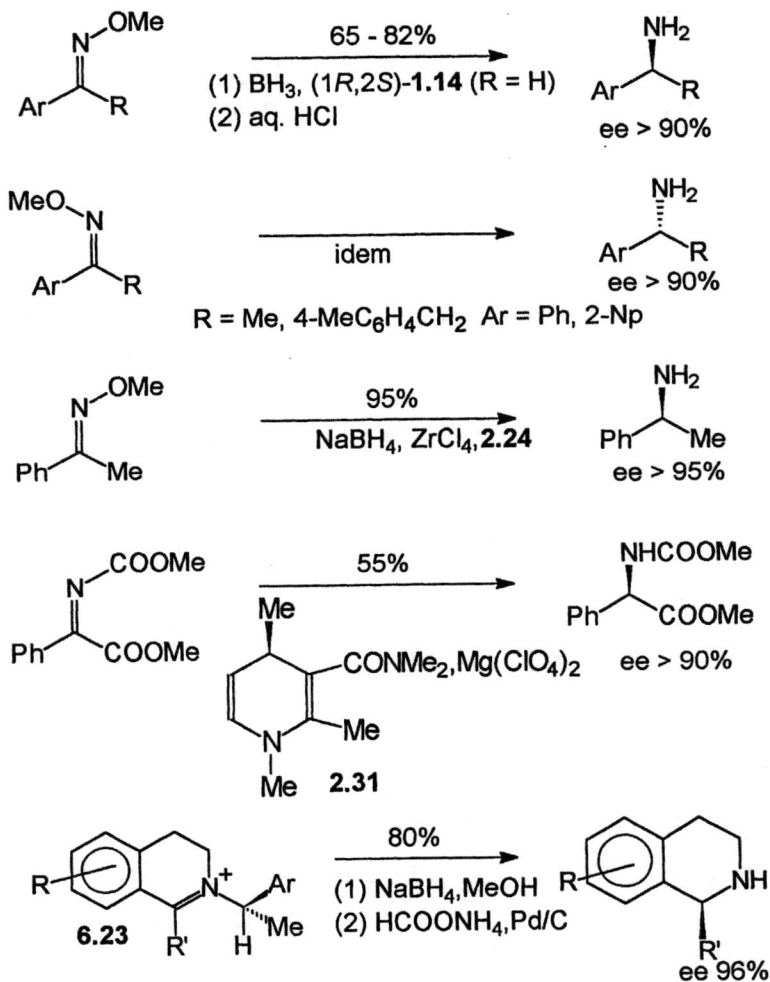

Figure 6.19 (continued)

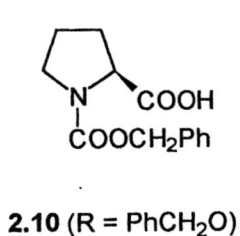

2.10 (R = PhCH₂O)

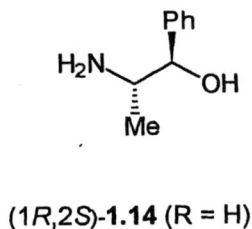

(1*R*,2*S*)-**1.14** (R = H)

6.1.4.2 *Reductions of Imines Bearing Chiral Residues*

There are few examples of enantioselective reductions of imines bearing chiral substituents. According to Polniaszek and Dillard [139], reduction of imminium salts **6.23** by NaBH$_4$, followed by hydrogenolysis, leads to nonracemic-substituted tetrahydroquinolines. To observe a high asymmetric induction, it is necessary to introduce a 2,6-dichlorophenyl group on the chiral substituent (Figure **6.19**).

The presence of a chiral sulfoxide group in the β-position allows the highly selective reduction of imines by Li s-Bu$_3$BH or LiBEt$_3$H [482, 1120] (Figure **6.20**). Reduction of β-sulfonylimines with DIBAH/ZnCl$_2$ at −78°C is also very efficient, and produces nonracemic amines after Raney nickel desulfurization [1121, 1122] (Figure **6.20**). A chelated complex that is similar to the one involved in the reduction of β-ketosulfoxides (§ 6.1.2.2) is proposed as a reaction intermediate.

Chiral sulfinimines **1.148** are selectively reduced by DIBAH at −30°C. After treatment of the products with CF$_3$COOH, chiral primary amines are obtained with an excellent enantiomeric excess [510, 512] (Figure **6.20**). The reduction of recyclable sulfinimides obtained from **1.142** with DIBAH has been proposed by Wills and coworkers [500].

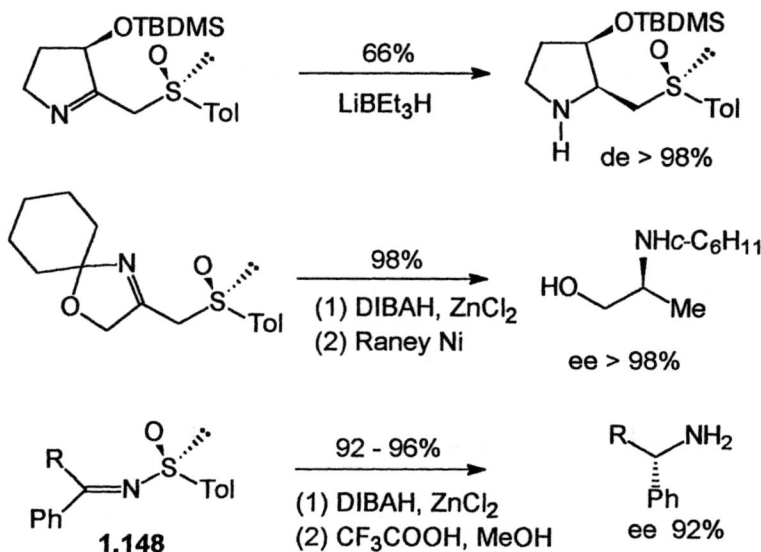

Figure 6.20

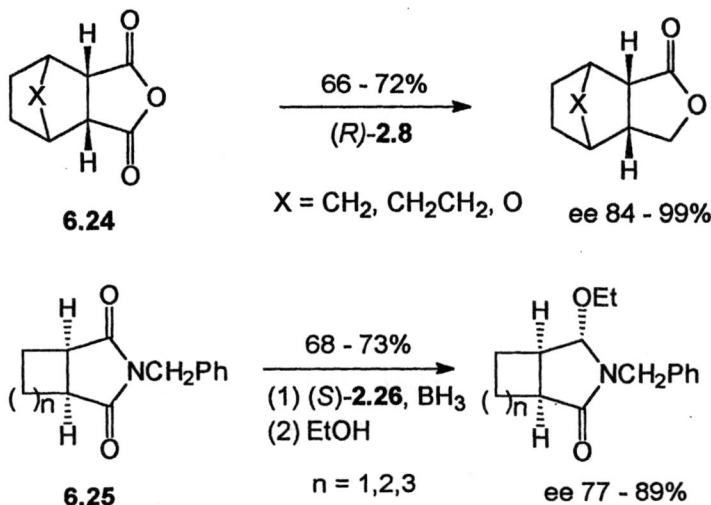

6.24   X = CH2, CH2CH2, O   ee 84 - 99%

66 - 72%
(R)-2.8

6.25   n = 1,2,3   ee 77 - 89%

68 - 73%
(1) (S)-2.26, BH3
(2) EtOH

Figure 6.20 (continued)

6.1.5 Reductions of *Meso* Anhydrides and Imides

Highly enantioselective reductions of *meso*-1,2-dicarboxylic anhydrides **6.24** with Binal **2.8** have been performed by Matsuki and coworkers at −78°C [1123] (Figure **6.20**). The model for Binal reductions proposed by Noyori and coworkers (§ 6.1.1) accounts for the observed selectivity. In the presence of BH3, CBS reagent **2.26** (R = H, Ar = Ph) selectively reduces *meso*-imides **6.25**. After EtOH treatment, ethoxyimides are obtained with a good enantioselectivity [1124] (Figure **6.20**).

6.2 CATALYTIC HYDROGENATION

6.2.1 Hydrogenation by Dihydrogen

Catalytic hydrogenation of the keto groups of α-ketoesters or -amides bearing a chiral residue usually leads to alcohols with a low stereoselectivity [1125]. Such is also the case with related imines. However, Miao and coworkers observed a stereoselective catalytic hydrogenation of the imine of 2-PyCOCH2c-C6H11 and (*R*)-phenylglycinol or (*R*)-valinol [1126]. In practice, most asymmetric hydrogenations require the use of catalysts bearing chiral ligands.

Functionalized ketones are hydrogenated under pressure in the presence of two types of homogeneous catalysts: binap-RuX2 [859, 881, 883, 893] and bcpm-Rh [114]. There are very few examples of efficient asymmetric hydrogenations of ketones that do not bear a second ligating functionality. Takaya and coworkers [901] recently devised a binap **3.43**-Ir(I)-aminophosphine system **6.26** that promotes the asymmetric hydrogenation of 1-tetralones and related compounds and β-thiacyclanones at 90°C under H_2 pressure (57 atm) with a high enantiomeric excess (Figure **6.21**). Other catalysts are far less efficient, as is the hydrogenation of other substrates.

Binap-$RuCl_2$ catalysts are obtained from the reaction of $RuCl_2$-$(C_6H_6)_2$ or $RuCl_2$-(cyclooctadiene)$_2$ and (R)- or (S)-binap **3.43** (Ar = Ph), occasionally in the presence of Et_3N or traces of a strong acid [859, 893, 894, 1089, 1127]. These catalysts are very efficient in enantioselective hydrogenations of ketones bearing dialkylamino, hydroxy, alkoxy, silyloxy, keto, ester, thioester or amide groups in α-, β- or γ-positions. According to the configuration of the binap, either enantiomeric alcohol is obtained with an excellent enantiomeric excess (Figure **6.21**). The proposed mechanism involves the coordination of both the carbonyl oxygen and the other Lewis basic site to the metal to generate a five-to-seven-membered chelate prior to hydrogen transfer.

The asymmetric hydrogenation of β-ketoesters has been widely used. This reaction can be carried out at room temperature under 100 atm H_2 or at 80-100°C under 4 atm [859, 899, 900, 1128], and it is compatible with the presence of isolated double bonds or chlorine substituents. This reduction can be applied to the synthesis of precursors of nonracemic β-substituted β-lactones [1129] as well as to the formation of β-hydroxy-α-aminoesters [812, 898] (Figure **6.22**) and α-hydroxy-β-aminoacids [1130]. When racemic α-substituted β-ketoesters or β-diketones **6.27** are used as substrates, one enantiomer is preferentially hydrogenated due to double diastereodifferentiation (§ I. 6). If the other enantiomer suffers epimerization faster than hydrogenation, it is possible to obtain a single diastereoisomeric alcohol with a very high enantiomeric excess and in a high chemical yield [881, 898, 1131, 1132, 1133] (Figure **6.22**). The enantioselective hydrogenation of prochiral γ-ketoesters **6.28**, followed by acidic treatment, is the simplest way to synthesize nonracemic γ-lactones [1134] (Figure **6.22**). For this purpose, the recommended catalyst is binap-Ru(OCOMe)$_2$ [1135], and the reaction is carried out at 25°C under 100 atm.. Biphemp **3.45** (R = Ph) is also an efficient ligand in Ru-catalyzed hydrogenation of ketones [1128] while bichep **3.45** (R = c-C_6H_{11})-Ru complex is recommended for hydrogenation of phenylglyoxylic derivatives [906].

The asymmetric hydrogenation of functionalized ketones may also be performed in the presence of neutral Rh(I) complexes bearing **3.36** (R = c-C_6H_{11}) as a ligand [114, 859, 1136-1139]. (R)-Pantolactone **1.16** is obtained from **6.7** with a high enantiomeric excess by this method. Methyl glyoxylate (MeCOCOOMe) is also transformed into the corresponding (R)-alcohol.

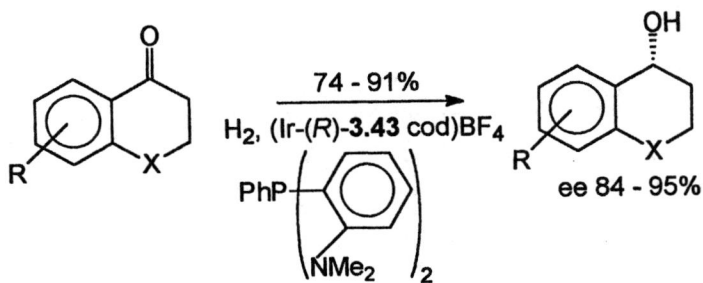

6.26

X = CH$_2$, O, S

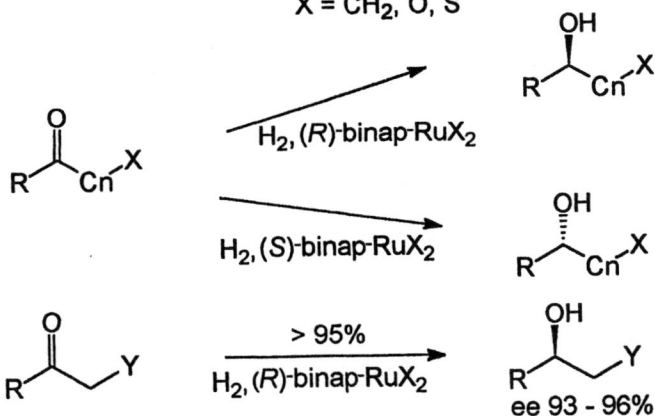

R = Me, i-Pr, tert-Bu, Ph Y = NMe$_2$, OH

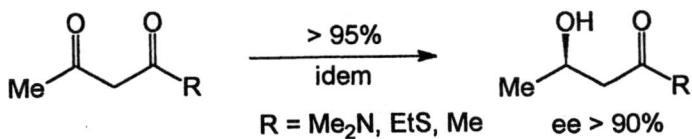

R = Me$_2$N, EtS, Me   ee > 90%

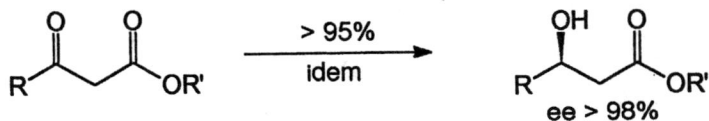

ee > 98%

R = Me, n-Pr, n-Bu, i-Pr, MeCH=CHCH$_2$CH$_2$
R' = Me, Et, i-Pr, tert-Bu

Figure 6.21

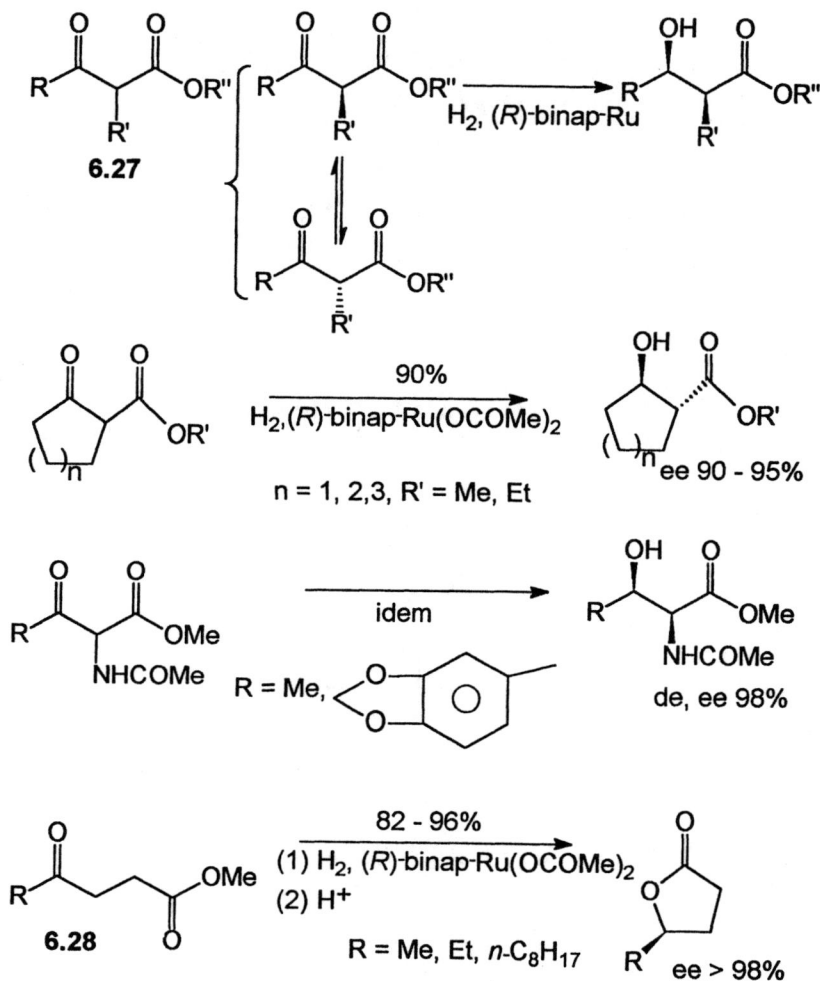

Figure 6.22

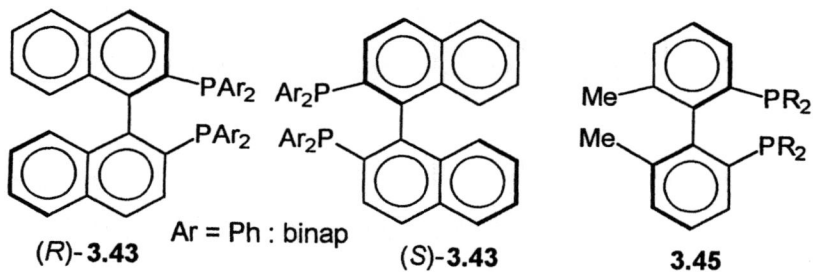

(R)-**3.43**   Ar = Ph : binap   (S)-**3.43**   **3.45**

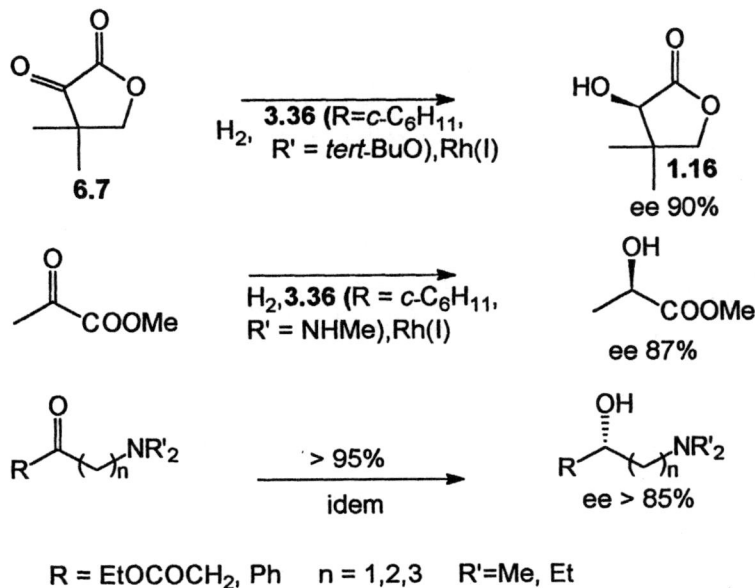

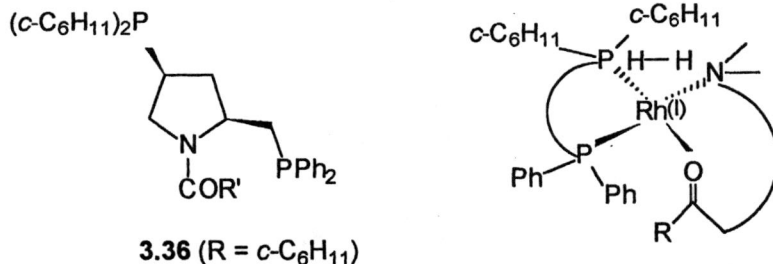

Figure 6.23

Under 50 atm of H_2 pressure, the asymmetric hydrogenation of α-, β- or γ-ketoamines gives high selectivities (Figure **6.23**). According to Achiwa and co-workers [114], the coordination of the keto and NR_2 groups to the rhodium is strengthened by the presence of the electron-rich dicyclohexylphosphine ligands in the *anti* position. The rigidification of the system enhances the influence of the chirality of the ligand (Figure **6.23**). Some arylalkylamidophosphinephosphinites have also been proposed as ligands of rhodium in asymmetric hydrogenation of α-ketoesters or -amides [120a].

The asymmetric hydrogenation of ketones over heterogeneous catalysts has also been studied [578, 1140]. Hydrogenation of α-ketoesters must be performed on Pt/Al_2O_3 in the presence of dihydrocinchonidine **2.87** (R' = H) [578, 812, 1141].

For β-ketoesters, ultrasound activation [1142] of Raney nickel modified by tartaric acid **2.2** (R = H) in the presence of NaBr [578] is the method of choice (Figure **6.24**). The hydrogenation of β-diketones under similar conditions leads to a mixture of *meso* and chiral diols. In the most favorable cases, the chiral diols are obtained in 60% chemical yield and with a good enantiomeric excess [1142] (Figure **6.24**). Under these conditions, hydrogenation of β-ketosulfones is less selective (ee 70%) [578].

The asymmetric hydrogenation of imines usually yields disappointing results. However, Willoughby and Buchwald have recently obtained high enantiomeric excesses in hydrogenations of imines in the presence of a titanocene catalyst. The most useful results are obtained from cyclic imines [830, 1143, 1144].

The asymmetric hydrogenation of *N*-aroylhydrazones of arylalkylketones and of α-ketoesters can be performed in the presence of Rh(I) cationic complexes bearing (*R,R*)-**3.47** as a ligand [1143, 1145]. After SmI$_2$ treatment, nonracemic amines are obtained with a high enantiomeric excess (Figure **6.25**).

Double diastereodifferentiation takes place in rhodium-catalyzed hydrogenation of chiral 1-phenethyl imines in the presence of bdpp **3.37** (n = 1) [295]: the matched system is (*R*)-imine–(2*S*,4*S*)-bdpp.

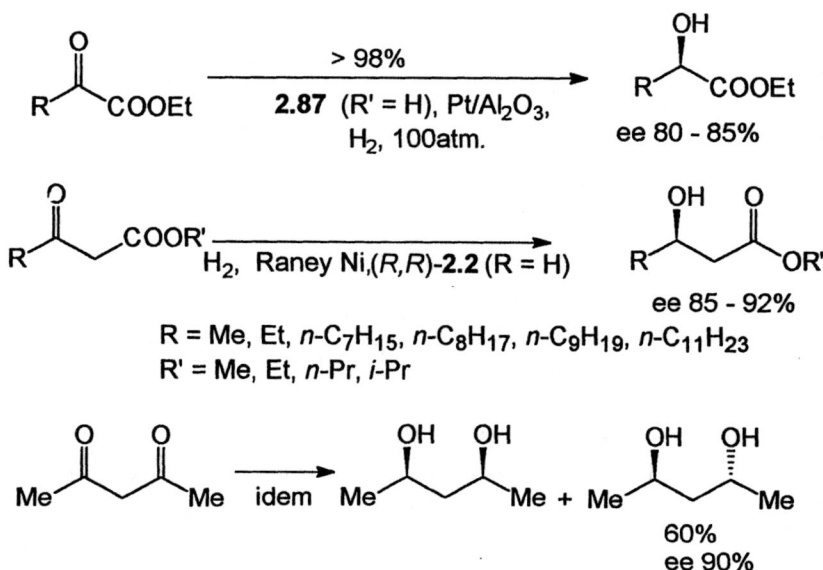

$$R = Me, Et, n\text{-}C_7H_{15}, n\text{-}C_8H_{17}, n\text{-}C_9H_{19}, n\text{-}C_{11}H_{23}$$
$$R' = Me, Et, n\text{-}Pr, i\text{-}Pr$$

Figure 6.24

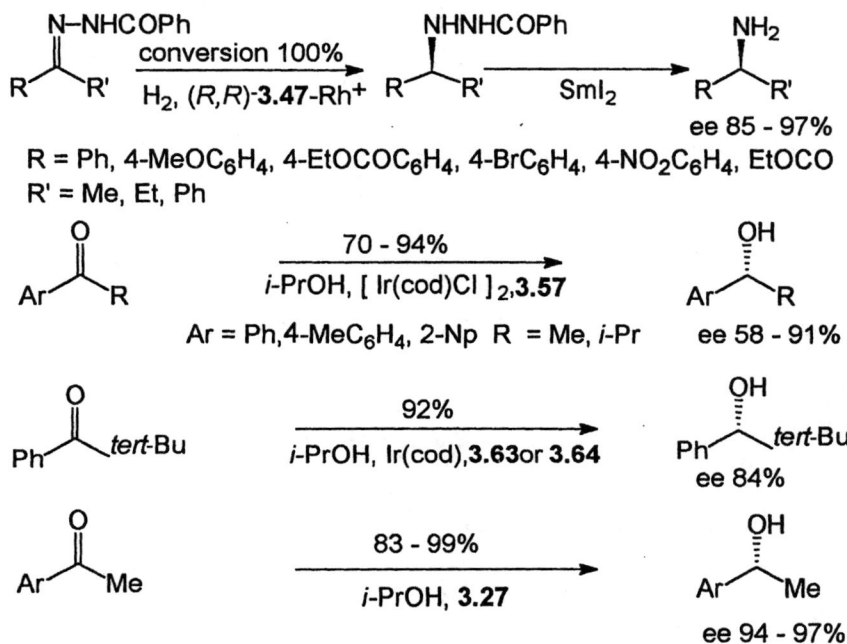

R = Ph, 4-MeOC₆H₄, 4-EtOCOC₆H₄, 4-BrC₆H₄, 4-NO₂C₆H₄, EtOCO

R = Ph, $4\text{-MeOC}_6\text{H}_4$, $4\text{-EtOCOC}_6\text{H}_4$, $4\text{-BrC}_6\text{H}_4$, $4\text{-NO}_2\text{C}_6\text{H}_4$, EtOCO

R' = Me, Et, Ph

Figure 6.25

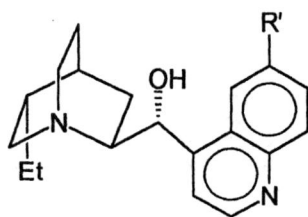

2.87

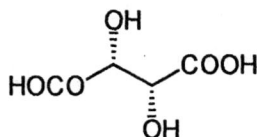

(R,R)-2.2

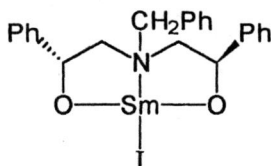

3.27

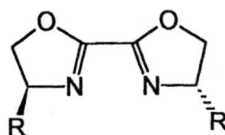

3.57

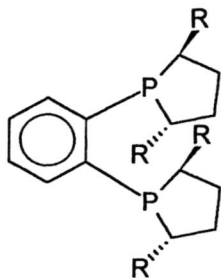   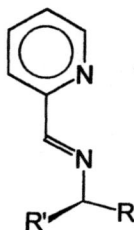   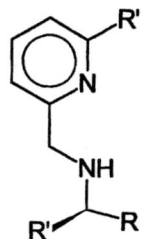

         3.47   **3.63**   **3.64**

6.2.2 Hydrogen Transfer Hydrogenation [873]

Catalyzed hydrogenation of ketones can be carried out by hydrogen transfer from alcohols such as *i*-PrOH or from ammonium formate. This process gives modest enantioselectivities with rhodium or ruthenium complexes bearing various chiral ligands as catalysts [873, 932, 1146]. Iridium complexes generated from [Ir(cod)Cl]$_2$ and bis-oxazolines **3.57** efficiently catalyze the asymmetric reduction of arylalkylketones [929, 942] (Figure **6.25**). Other iridium complexes bearing 2-(*N*-alkylimino)-pyridines **3.63** or pyridylmethylamines **3.64** as ligands also give useful results, especially when the ligands are grafted on polymers [873, 942] (Figure **6.25**). From dialkylketones, the selectivities are much lower.

Evans and coworkers [626] performed the asymmetric Meerwein-Pondorf-Verley reduction of arylmethylketones at room temperature with a high enantiomeric excess by using samarium complex **3.27** as a catalyst. (Figure **6.25**).

6.3 HYDROSILYLATION [868]

Addition of a silicon hydride (hydrosilylation) to a carbonyl group is an alternative to a metal hydride reduction because the silyl ether products can easily be hydrolyzed. Hydrosilylation can be catalyzed by acids or transition-metal complexes. Cationic rhodium (I) complexes, initially recommended by Ojima and coworkers [752, 855], are the most efficient transition metal catalysts. Koga and Morokuma have provided a theoretical treatment of this reaction [1147]. The use of diop **3.30** as a ligand allows the asymmetric hydrosilylation of α-ketoesters and -amides by H$_2$SiPh1-Np with a good enantiomeric excess [566] (Figure **6.26**). However, to obtain high enantiomeric excesses in hydrosilylation of arylalkylketones with H$_2$SiPh$_2$ requires the use of nitrogen ligands for rhodium such as

(R)-pythia **3.55** or (S)-pymox **3.56** recommended by Brunner [113, 566, 752] (Figure **6.26**). Complexes of bis-oxazolines **3.57** (R = PhCH$_2$), **3.58** and **3.59** (R = i-Pr)-RhCl$_3$ also give high selectivities [566, 928, 930, 931, 1148] (Figure **6.26**). Nevertheless, hydrosilylation of dialkylketones, α-enones and imines under similar conditions gives modest enantioselectivities [566, 752]. Seebach and co-workers [925] obtained some useful results with the phosphonite of **2.50** (R = R' = Me) in the rhodium-catalyzed enantioselective hydrosilylation of arylmethylke-tones with H$_2$SiPh$_2$. Intramolecular hydrosilylation of silylether **6.29** takes place with a high enantioselectivity when catalyzed by a cationic rhodium complex bear-ing diphosphine **3.47** (R = i-Pr) as a ligand [656]. If R = Ph, the selectivity is lower (Figure **6.26**).

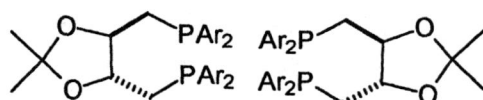

(R,R)-**3.30**   (S,S)-**3.30**

Ar = Ph:diop

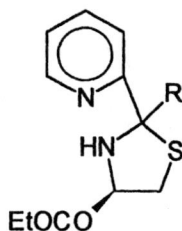

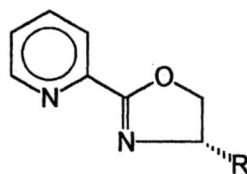

3.55 : (R)-pythia   **3.56** : (S)-pymox

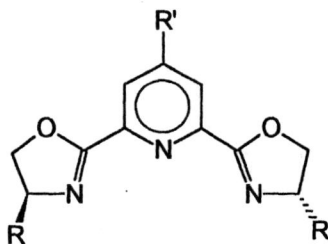

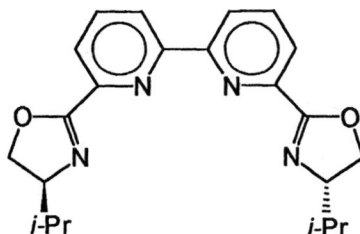

3.58   **3.59**

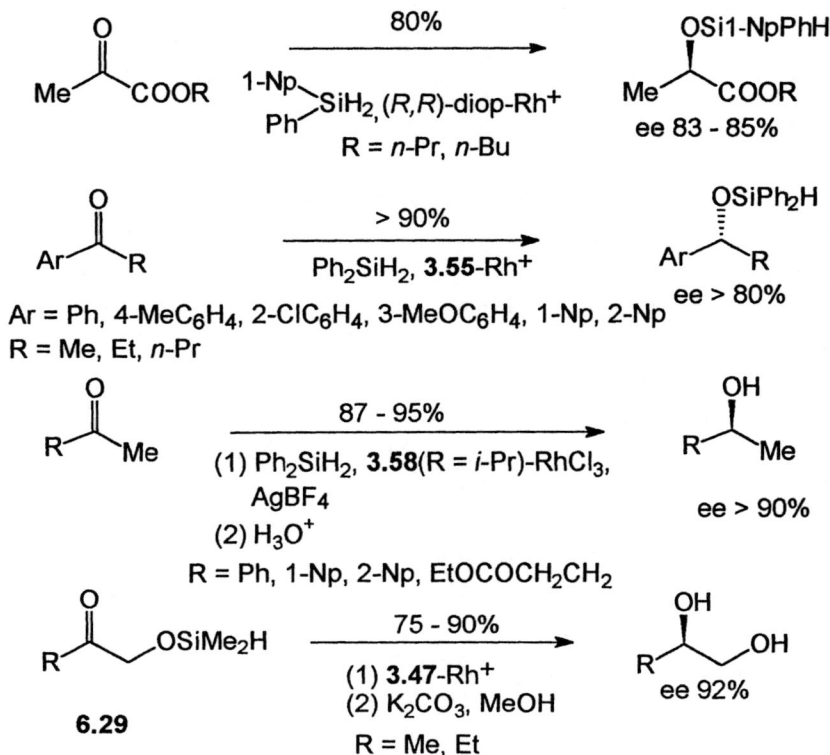

Figure 6.26

6.4 HYDROCYANATION, SILYLCYANATION, STRECKER REACTION

6.4.1 Hydrocyanation of Aldehydes [1149]

Asymmetric hydro- or silylcyanation of aldehydes can take place either with HCN or Me_3SiCN in the presence of a chiral catalyst. The reaction of HCN with aldehydes in the presence of a cyclic dipeptide, cyclo-[(S)-phenylalanino-(S)-histidine] **3.4** (or its (R,R)-enantiomer), promotes the highly enantioselective hydrocyanation of aromatic aldehydes [1150, 1151]. This reaction may occur through an aldehyde-peptide complex in which the carbonyl group is activated by hydrogen-bonding and steric hindrance is minimized (Figure 6.27). The approach of the CN anion is directed by the histidine residue. This reaction also occurs with a good selectivity in the presence of $Ti(OEt)_4$ and acyclic dipeptide **3.21** (R = i-Pr, R' = i-Pr or 3-indolyl) [825, 1152]. Similar hydrosilylation of aliphatic aldehydes occurs with modest enantiomeric selectivity.

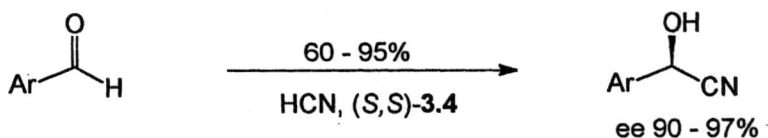

Ar = Ph, 3-MeOC$_6$H$_4$, 4-MeC$_6$H$_4$, 2-Np, 4-PhOC$_6$H$_4$

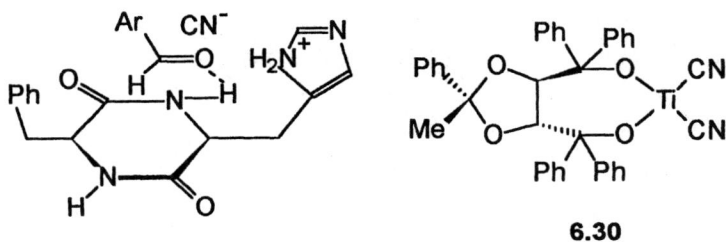

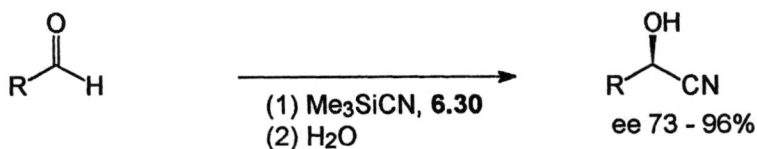

6.30

R =PhCH$_2$, PhCH$_2$CH$_2$, n-C$_8$H$_{17}$, c-C$_6$H$_{11}$, Ph

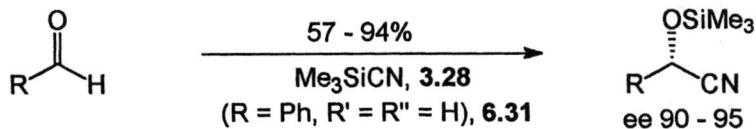

R = n-C$_6$H$_{13}$, Et$_2$CH, c-C$_6$H$_{11}$, tert-Bu

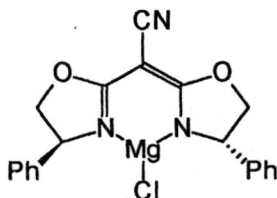

6.31

Figure 6.27

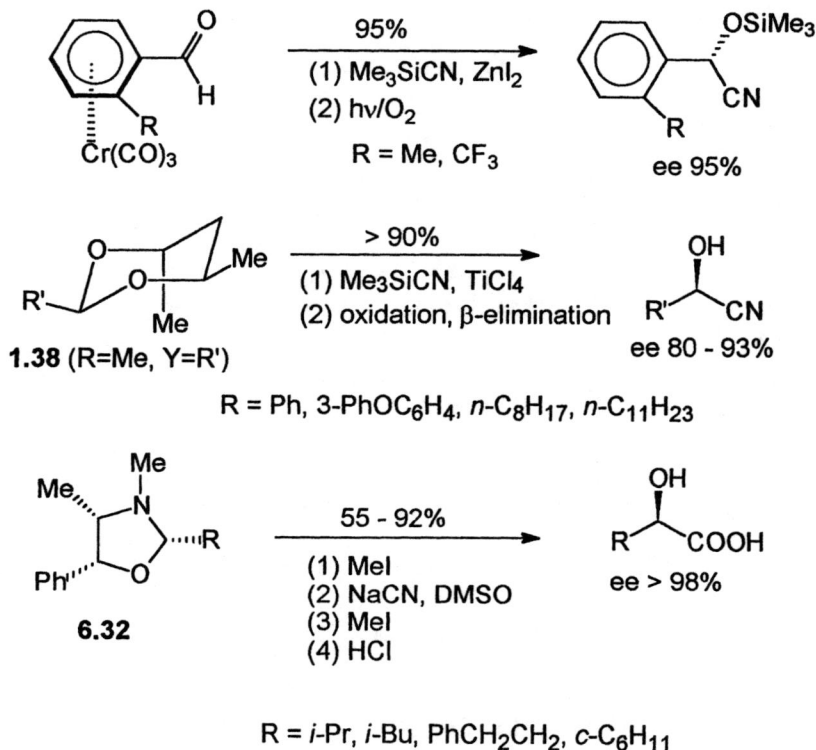

Figure 6.27(continued)

Asymmetric silylcyanation with Me_3SiCN is catalyzed by Lewis acids, including $Sn(OTf)_2$-cinchonine **3.2** (R = H) [1153] or $TiCl_4$-n-BuLi-binaphthol **3.7** (R = H) complexes [666, 778]. The complex $TiCl_2(Oi\text{-}Pr)_2$-diol **2.50** (R = Ar = Ph, R' = Me) is also a very efficient Lewis acid in this process. ^{1}H NMR studies have shown that complex **6.30** is preformed at room temperature, and that this is the reactive species. The reaction of **6.30** with aliphatic and aromatic aldehydes takes place at −78°C and can be highly enantioselective [778, 1154] (Figure **6.27**). Titanium complexes generated from diethyl- or diisopropyl tartrates can also give interesting selectivities [1155], as can chiral Schiff base complex **3.20** (R = i-Pr)-$Ti(Oi\text{-}Pr)_4$ [268]. Recently, Corey and Wang [1156] introduced asymmetric trimethylsilylcyanations in the presence of two oxazoline catalysts, **3.28** (R = Ph, R' = R" = H) and **6.31**. Magnesium amide **6.31** is prepared from **3.28**. High selectivities are observed in hydrogenations of aliphatic aldehydes at −78°C (Figure **6.27**), but unsaturated and aromatic aldehydes give lower selectivities. Aluminum complexes are less useful [804, 1157].

Arenechromium tricarbonyl benzaldehydes bearing a substituent in 2-position in order to fix the conformation of the carbonyl group also undergo highly selective silylcyanation with Me_3SiCN. After decomplexation, precursors of aminoalcohols are obtained with a high enantioselectivity [539] (Figure 6.27).

6.4.2 Hydrocyanation of Acetals

Chiral acetals derived from (R,R)-2,5-pentanediol 1.38 (R = Me, Y = R') react with Me_3SiCN in the presence of $TiCl_4$. After appropriate treatment, (R)-cyanohydrins are obtained in a good enantiomeric excess [213] (Figure 6.27). Following quaternization with MeI, cyclic N,O-acetals 6.32 derived from (1R, 2S)-ephedrine 1.61 (R = Me) suffer stereoselective ring opening by NaCN in DMSO at 130°C [1158]. (R)-α-Hydroxyacids are obtained with an excellent enantiomeric excess after quaternization and treatment with HCl (Figure 6.27).

6.4.3 Strecker Reaction [861]

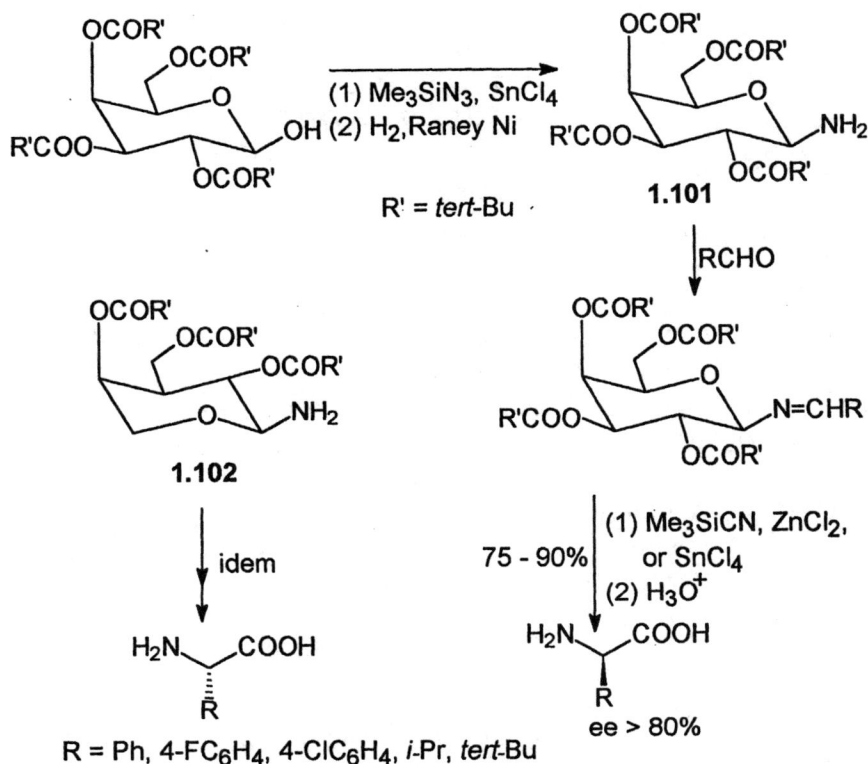

Figure 6.28

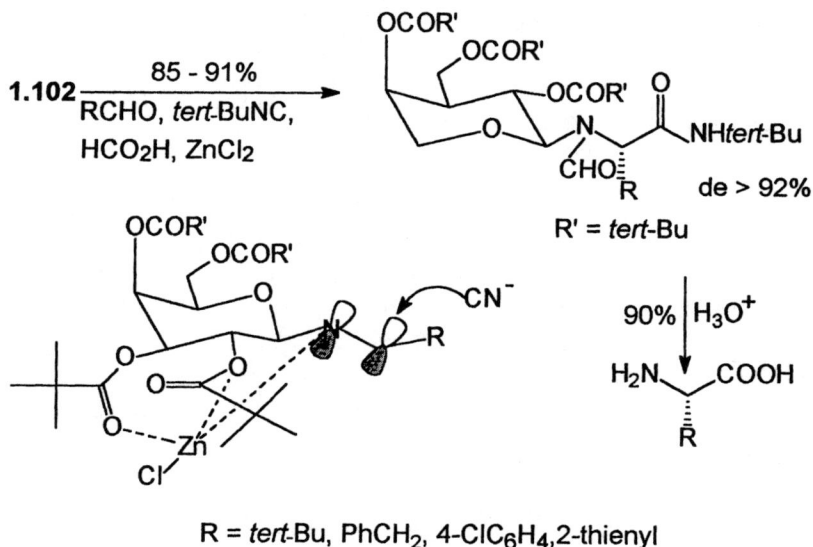

R = *tert*-Bu, PhCH$_2$, 4-ClC$_6$H$_4$,2-thienyl

Figure 6.28 (continued)

The Strecker reaction is a method used to prepare α-aminoacids by reaction of an aldehyde with HCN and NH$_3$ or an amine. The reaction occurs through the intermediacy of an imine, and it can also be performed with Me$_3$SiCN. Asymmetric Strecker reactions have been carried out with chiral amines such as 1-phenethylamine or α-phenylglycinol **1.60** (R = Ph, R' = H). After the reaction, the chiral nitrogen substituent is removed by hydrogenolysis. Amines **1.101** and **1.102** (derived from carbohydrates through the corresponding azides) have been transformed into aromatic or branched aliphatic imines. The reaction of these imines with Me$_3$SiCN/ZnCl$_2$, followed by acidic hydrolysis, leads to α-aminoacids with a good enantiomeric excess [154, 248, 360, 1159]. The attack of cyanide takes place on the least hindered face of a zinc chelate involving two of the carbohydrate esters groups (Figure **6.28**). The chiral auxiliary may be recycled. The Ugi variant of the Strecker reaction, which is the reaction of an aldehydo-imine with an isonitrile, also gives high selectivities [154] (Figure **6.28**).

6.5 REACTIONS OF ORGANOMETALLIC REAGENTS

The reactions of organolithium, -magnesium, -zinc, -copper, and -titanium reagents with aldehydes, ketones, acetals, imines and hydrazones will be described in this section. The reactions of enolates and enamines will be subsequently examined (§ 6.8 to 6.10).

6.5.1 Reactions with Nonfunctionalized and α-Unsaturated Aldehydes and Ketones

In most of the cases, the reactions of aldehydes with organolithium or organomagnesium reagents in the presence of chiral ligands lead to secondary alcohols with modest enantiomeric excesses [110, 253, 559, 634, 681]. Two exceptions merit citation, although these reactions must be run at −120°C. Mukaiyama, Cram and their coworkers performed the reaction of benzaldehyde with *n*-BuLi in the presence of either ligand **2.36** (R = H) or **2.37** leading, respectively, to (*S*)- or (*R*)-1-phenylpentanol with a high enantiomeric excess (Figure **6.29**). Similar results have been obtained by Johnson and coworkers from Me₃SiC≡CLi and an unsaturated aldehyde [1160] (Figure **6.29**). The reactions of aldehydes with functionalized organolithium reagents have given disappointing selectivities [1004, 1161].

Encouraging results were obtained with aromatic aldehydes by using organotitanium reagents bound to binaphthols **1.44** [131, 663, 666] or to diol **2.50** (R = *tert*-Bu, R' = H, Ar = Ph) [673], or with complex **2.48** (R = Tol) [664, 666] (Figure **6.29**).

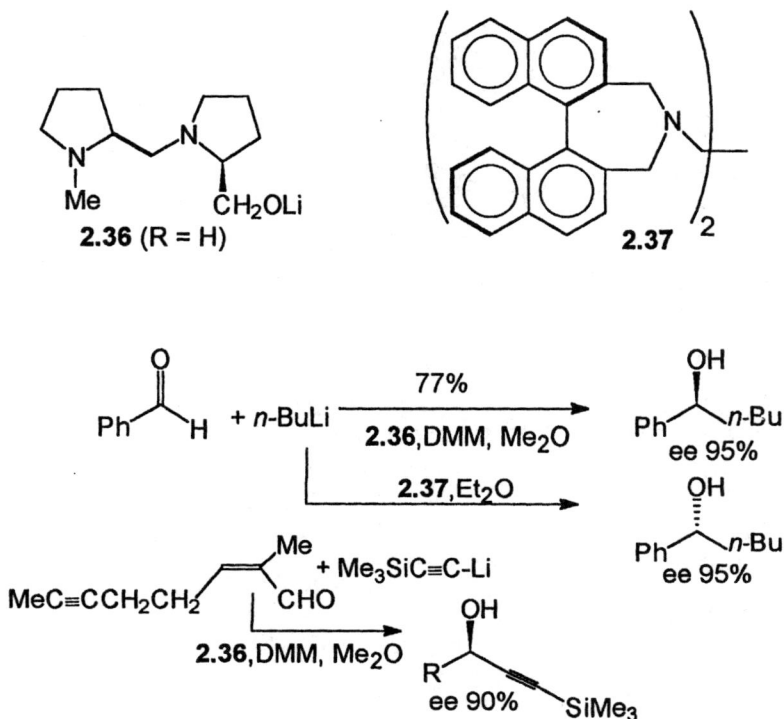

Figure 6.29

Ar = 2-MeC6H4, 4-MeC6H4, 4-MeOC6H4, 1-Np, 2-Np

2.48 (R = Tol)

Ar = Ph, 2-NO2C6H4, 2-MeC6H4, 1-Np

R = Me, n-Bu, i-Pr, CH2=C(Me), CH2=C(i-Pr)

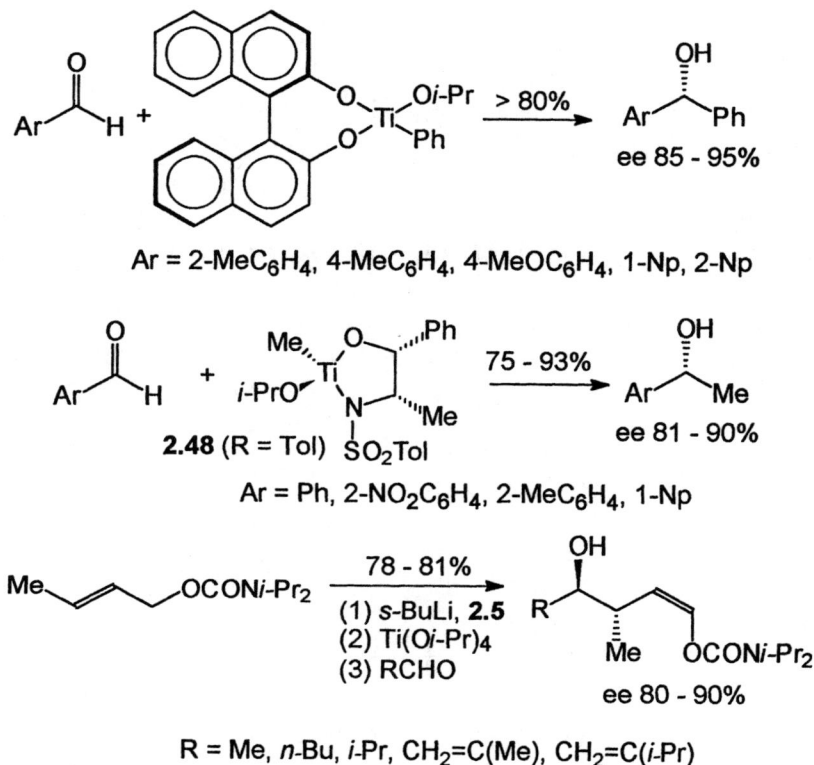

Figure 6.29 (continued)

The reaction of aldehydes with another titanium complex generated from E-crotyl N-diisopropyl carbamate, s-BuLi, sparteine 2.5 and Ti(Oi-Pr)4 gives *anti* homoallylic alcohols with a good enantioselectivity [109, 1162, 1163] (Figure 6.29). Addition of allylaluminum reagents to aromatic aldehydes in the presence of Sn(OTf)2 and a chiral amine also gives useful results [842].

The chemistry of chiral allyltitanium reagents took a new turn when Duthaler and coworkers prepared cyclopentadienyldialkoxyallyltitanium reagents from allylmagnesium chloride and 2.49 [665, 666, 667]. Two enantiopure alcohols, diacetoneglucose 1.48 and diol 2.50 (R = R' = Me, Ar = Ph), are the best precursors of titanium complex 2.49 for asymmetric allylations. Either the (R)- or (S)-homoallylic alcohol is obtained with high enantioselectivity from the appropriate allyltitanium complex and an aliphatic or aromatic aldehyde (Figure 6.30). The reaction can be carried out with crotyl or other substituted reagents; however, the complexes E-6.33 are formed regardless of the conditions of the reactions. The reaction of these complexes with PhCHO or n-C9H19CHO leads to the expected *anti* alcohols with a very high diastereo- and enantioselectivity (Figure **6.30**). Unfortunately, the reaction of complexes **6.33** with α-substituted aldehydes is not diastereoselective, and a mixture of *syn* and *anti* alcohols is obtained (except from glyceraldehyde acetonide and related aldehydes) [667].

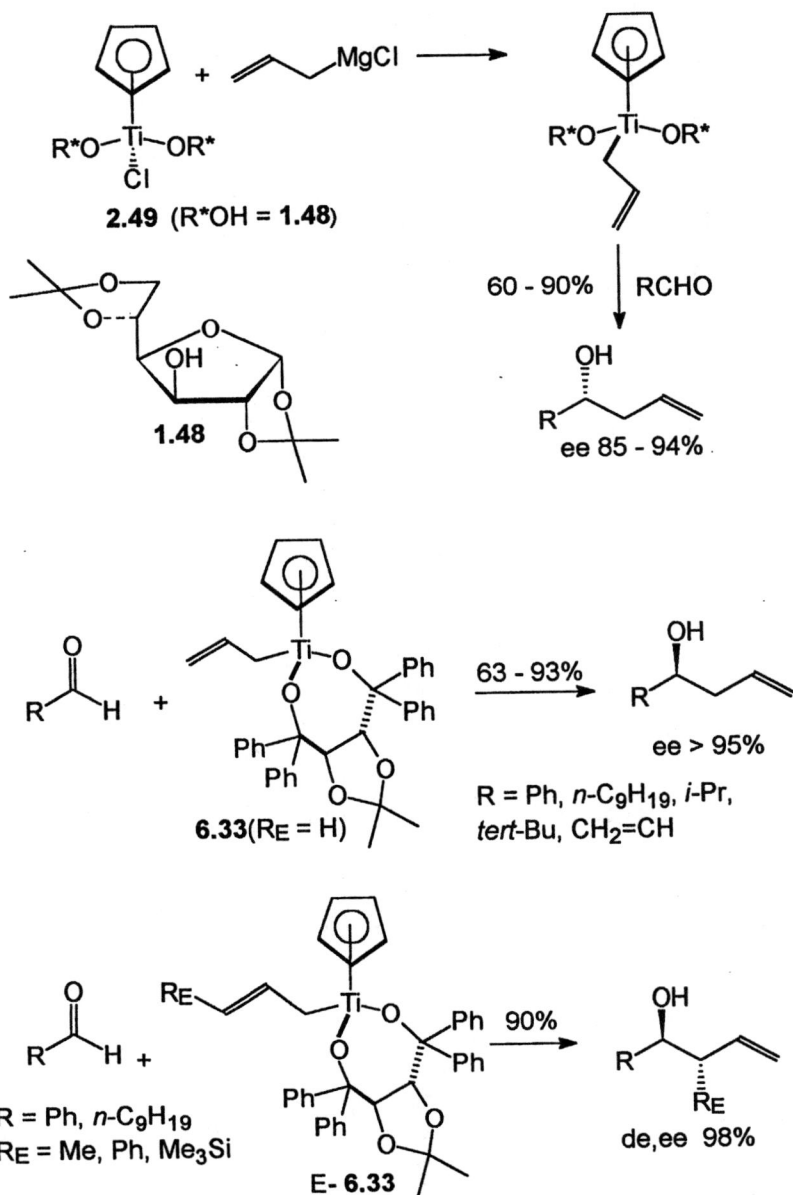

Figure 6.30

Enantioselective addition of dialkylzinc reagents to aldehydes can be catalyzed by enantiopure aminoalcohols. This class of reactions has rapidly been developed since the first observations of Oguni and Omi in 1984 and of Noyori and coworkers in 1986 [110]. The method was first restricted to the reactions of aromatic or α-unsaturated aldehydes with dialkylzincs conducted in the presence of catalytic amounts of bornane-**1.8** (R = NMe$_2$) or ephedrine-derived aminoalcohols **1.14** (R = *n*-Bu). Occasionally, the reagents were grafted onto polymers [110, 559, 640, 642, 643, 1164] (Figure **6.31**). This reaction has quickly been extended to aliphatic aldehydes [111, 640]. Depending on the absolute configuration of the catalyst, either enantiomeric secondary alcohol can be prepared with a high enantiomeric excess in reactions conducted near room temperature [1165] (Figure **6.31**). The reaction also works with PhMgBr/ZnCl$_2$, but the aminoalcohol must be used in stoichiometric amounts (Figure **6.31**). Aminoalcohol **2.45** also induces good selectivities, and it is recommended as a catalyst in the reaction of ferrocene- or ruthenocene carboxaldehydes with Et$_2$Zn or Me$_2$Zn [1166].

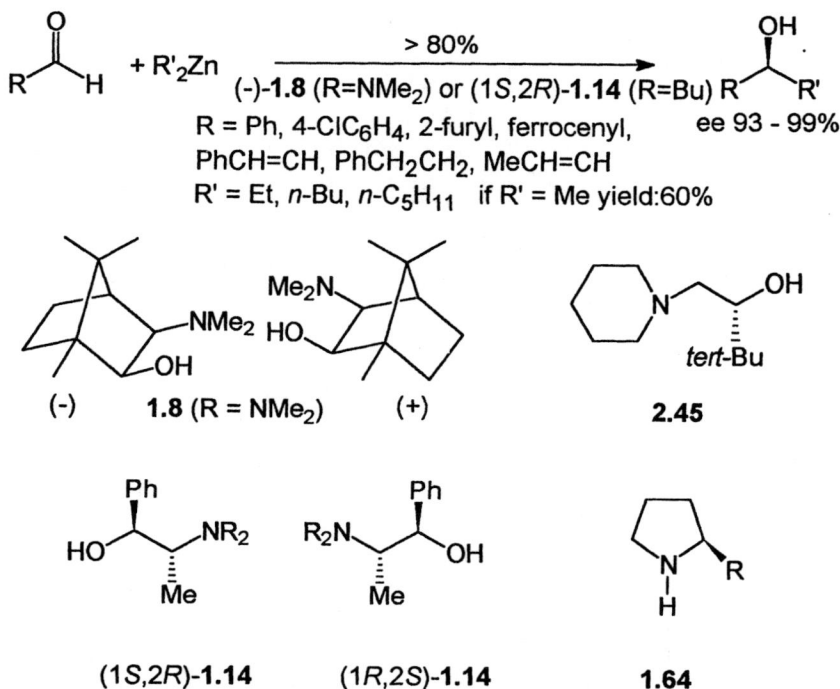

Figure 6.31

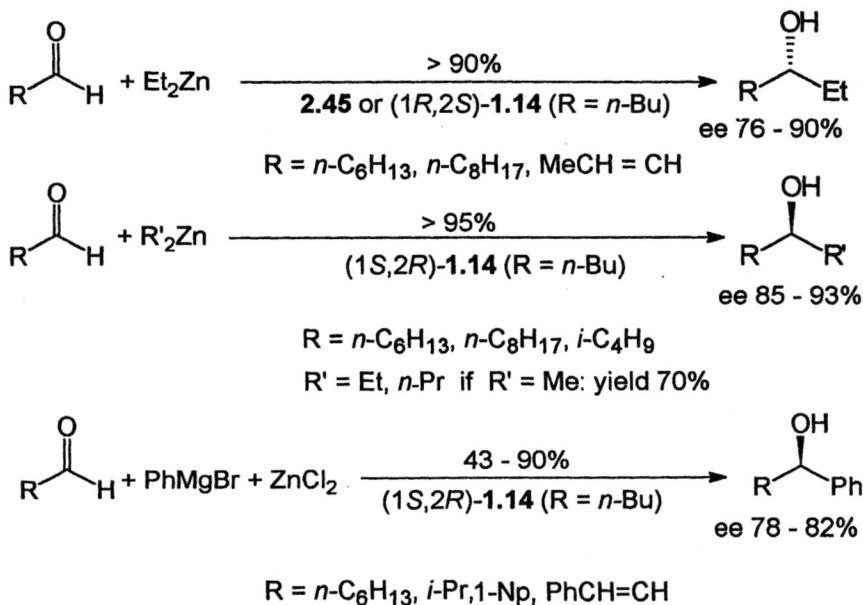

Figure 6.31 (continued)

In some cases, the use of lithium alcoholates of **1.14** (R = Me) gives improved results. For example, alkynyl- or alkenylzinc reagents can be used with these alcoholates [650, 1167] (Figure **6.32**), although alkenylzincs can react in the presence of **1.8** alone in catalytic amounts [1168, 1169] (Figure **6.32**).

Noyori and coworkers [110] have shown that these reactions involve a binuclear zinc complex (§ 2.5.1). Chirality transfer takes place from the Zn binuclear assembly **6.34**. The aldehyde is coordinated to Zn$_A$, and the repulsion between the R' group located on Zn$_A$ and the R substituent of the aldehyde is the key interaction in determining the structure of the complex. The transferred R' group is located on the other zinc atom (Zn$_B$) (Figure **6.32**). The approach depicted in **6.34** results in the enantioselective formation of an (S)-alcohol from the (–)-enantiomer of **1.8** [110, 253, 1170]. This process is subject to chirality amplification (§ 2.5.1).

Many other aminoalcohols and related compounds have been recommended as catalysts in enantioselective additions of organozinc reagents to aldehydes. Tricarbonyl chromium complexation of the aryl ring of **1.14** (R = Me,n-Bu) can lead to improved enantioselectivities [1171]. Other arenechromium carbonyl complexed aminoalcohols have also been proposed as Zn ligands [1172].

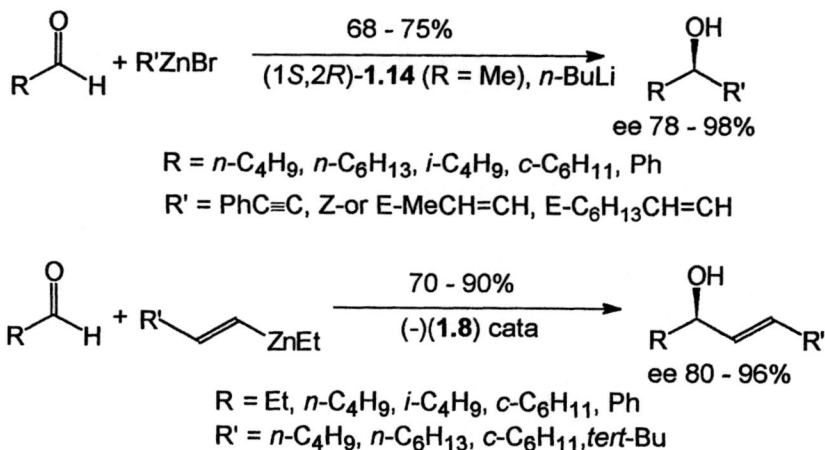

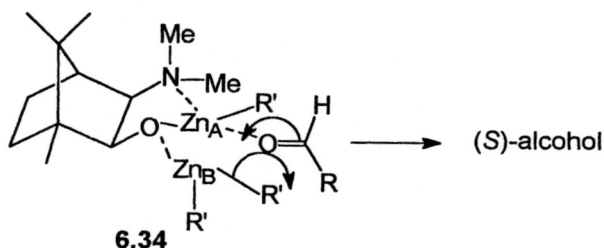

Figure 6.32

(1R,2S)-**1.61** (R = SO$_2$Tol)   **2.47**   **2.13** (R = CPh$_2$OH)

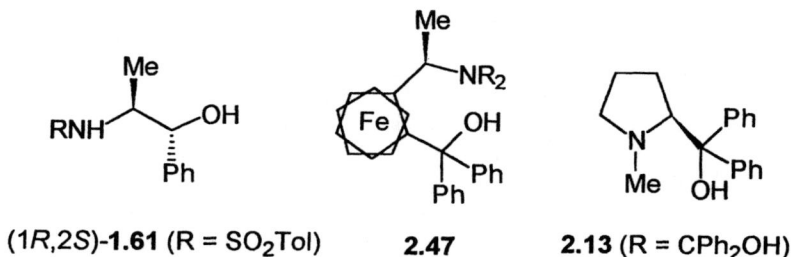

Additional catalysts have also been proposed for the reaction of Et$_2$Zn with aromatic aldehydes, including: (S)-proline derivative **2.13** (R = CPh$_2$OH) [110] and a four-membered analog [646], (1S,2R)-**1.61** (R = SO$_2$Tol), **2.47**, and pyridine-derived aminoalcohols [110, 644, 651, 1173]. Other catalysts include sulfur derivatives of ephedra alkaloids [645, 728], the Li diamide of piperazine **2.46**, diamines **1.64** (R = 2-Py) and other related 2-aminopyridines [367, 648, 649, 1174]. β-Hydroxysulfoximines have also been used as catalysts in these reactions [1175], as has an oxazaborolidine derived from ephedrine [1176].

Bimetallic Zn-Ti reagents [666, 1177] also give useful results. These reactions are run in the presence of complexes **2.52** or **2.53** [653, 654, 655, 666, 668, 670]. Functionalized dialkylzinc reagents may be used [672, 1177, 1178], and these form polyfunctional secondary saturated or allylic alcohols (Figure **6.33**). The functional dialkylzinc reagents can be prepared *in situ* from symmetrical dialkylmagnesiums [669, 1179] (Figure **6.33**). The bimetallic zinc-titanium reagents are generated at low temperatures and the reactions are run between −60°C and room temperature. However, the reaction of BrZnCH$_2$CN with aromatic aldehydes does not need the formation of a bimetallic reagent, and takes place in the presence of **2.13** (R = CPh$_2$OH) in stoichiometric amounts [1180].

Organocerium reagents modified by binaphthol **1.44** react at −100°C with aromatic aldehydes to give secondary alcohols with an enantiomeric excess in the range of 77 - 85% [680]. Because the addition of oxazolines **4.8**-derived organomagnesium reagents to PhCHO is poorly stereoselective, Gawley and Zhang designed a new auxiliary that promotes a diastereoselective reaction with **6.35** (de = 82%). After recrystallization of the major diastereoisomer and treatment with LiAlH$_4$, precursors of natural products are obtained with a high enantiomeric excess (Figure **6.34**).

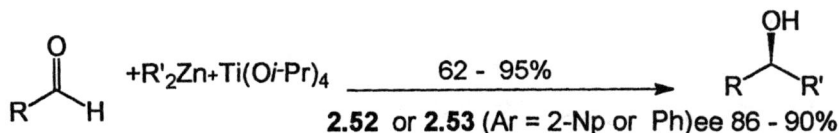

R = n-C$_5$H$_{11}$, n-C$_6$H$_{13}$, Ph, PhCH=CH, PhC≡C
R' = Et, n-Bu, n-C$_8$H$_{17}$, Cl(CH$_2$)$_4$, MeCOO(CH$_2$)$_{3, 4}$ or $_5$

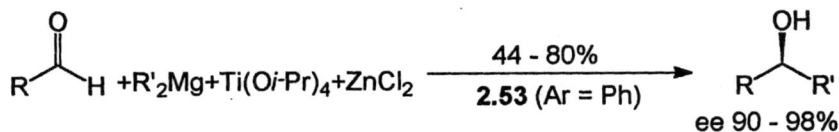

R = Ph, i-Pr, c-C$_6$H$_{11}$, CH$_2$=CH(CH$_2$)$_3$, PhCH$_2$CH$_2$
R' = Me, Et, n-Pr, n-C$_4$H$_9$, i-C$_4$H$_9$, Me$_2$CH(CH$_2$)$_2$, CH$_2$=CH(CH$_2$)$_2$
CH$_2$=CH(CH$_2$)$_3$, Me$_2$C=CH(CH$_2$)$_2$, PhCH$_2$CH$_2$

Figure 6.33

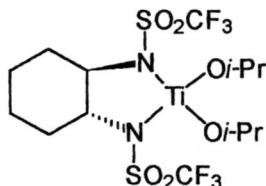

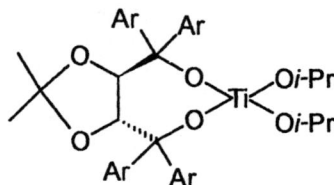

2.52   **2.53**

The reactions of carbonyl compounds with α-metalated sulfoxides are generally poorly stereoselective [481, 1181]. However, the reaction of lithiated sulfoxide (*R*)-**1.136** (R = 1-Np, Y = Me) with arylalkylketones leads preferentially to one diastereoisomer. After Raney nickel desulfurization, (*S*)-tertiary alcohols are obtained with a high enantioselectivity [479] (Figure **6.34**). The approach of the reagents involves chelation of lithium by the C=O and S–O groups, and a π-π through space stabilizing interaction is invoked in the transition state (Figure **6.34**). The enantioselective condensation of acetone with an allylic sulfone anion in the presence of a chiral ligand has also been described [632].

Hanessian and Beaudoin [310] have performed the diastereoselective condensation of lithiated phosphonamides **1.74** with prochiral or chiral cyclohexanones. Each enantiomer of the nonracemic alkene product is prepared from the appropriate reagent (Figure **6.34**). Denmark and Chen studied similar reactions of oxazaphosphorinanes **1.96** (R = Ph, SPh) [356]. These reactions are performed in a two-step fashion, and olefin formation requires activation with Ph$_3$COTf in the presence of 2,6-lutidine.

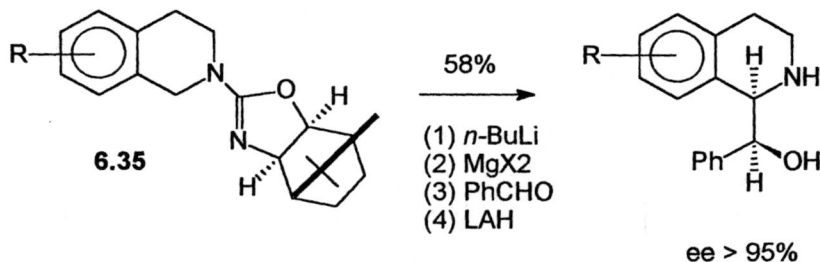

Figure 6.34

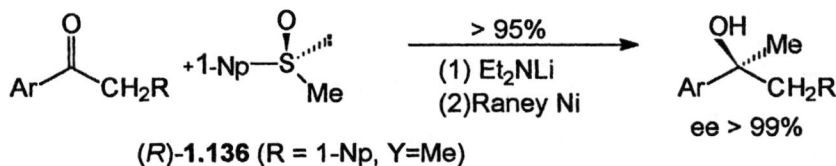

(*R*)-**1.136** (R = 1-Np, Y=Me)

R = H, Me, Et, *n*-Pr

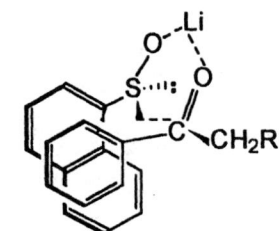

π-π interaction

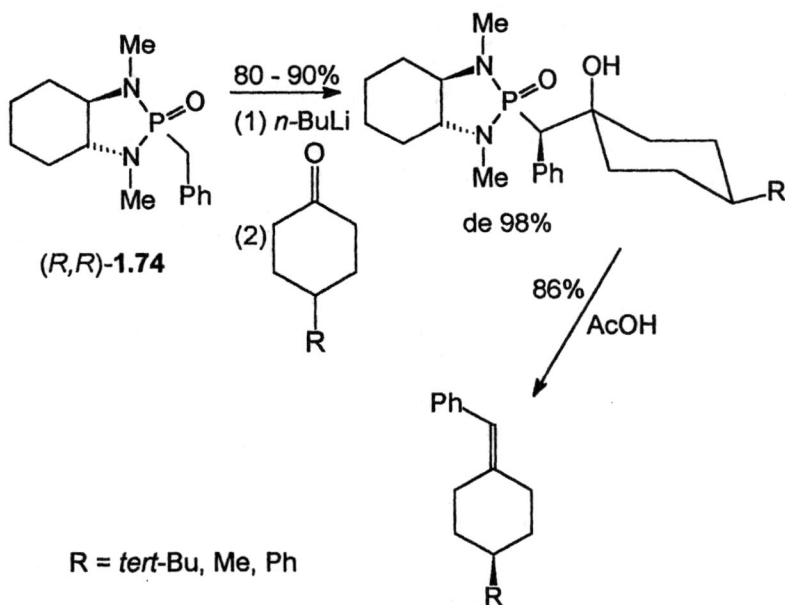

R = *tert*-Bu, Me, Ph

Figure 6.34 (continued)

6.5.2 Reactions with Functionalized Aldehydes and Ketones

Although the reaction of dialkylzincs with α-alkoxyaldehydes or α-keto-esters is catalyzed by chiral aminoalcohols, such reactions are usually poorly selective [110]. Recently, Soai and coworkers [1182] succeeded in promoting the enantioselective addition of Et$_2$Zn to (R)-3-benzyloxybutanal. Knochel and co-workers [1183] also realized the enantioselective addition of functionalized dialkylzinc reagents to β-silyloxypropanal by using **2.52** as a catalyst. Addition of Et$_2$Zn to racemic α-thio- and α-selenoaldehydes has also been examined [652]. However, most of the reactions of organometallic reagents have been performed with functionalized carbonyl compounds bearing chiral residues.

2-Substituted glycerol derivatives **6.36** have been obtained with a high enantiomeric excess by reaction of organomagnesium reagents with a cyclic acetal of menthone **6.37** displaying a locked conformation. Nucleophilic attack takes place from the axial direction [373] (Figure **6.35**). The formation of chiral monoacetals, -thioketals or aminals of α-ketoaldehydes or diketones promotes the stereoselective attack of one face of the remaining carbonyl group (Figure **6.35**). Useful reagents for such reactions include: chiral 1,2-diols **1.35** (R = CH$_2$OMe) [213], sulfur derivative **1.40** or still better aminoalcohol **1.80** (obtained from pulegone according to Eliel and coworkers [227, 329]), diamines **1.64** (R = CH$_2$NHPh) [253, 261], and (S)-prolinol **1.64** (R = CH$_2$OH). Interesting selectivities are obtained under chelation control (Figure **6.35**). By using Eliel's reagents **6.38** or aminals **6.39** generated from **1.64** (R = CH$_2$NHPh), either enantiomer of α-hydroxy-α,α'-dialkylaldehydes can be obtained by varying the sequence of addition of the two organomagnesium reagents (Figure **6.35**). 2-Acyloxazolidines **6.40** can be prepared from ketoaldehydes. Their reactions with organometals are stereoselective, and the direction of selectivity depends upon the reaction conditions. With organomagnesium or better yet alkyltrialkoxytitanium R'Ti(OR)$_3$ reagents, the reaction is chelation-controlled. In contrast, attack on the other face of the carbonyl group is favored in reactions with organolithium reagents conducted in the presence of HMPA or DMPU. These latter reactions typically occur with a lower selectivity [328] (Figure **6.35**). After hydrolysis, enantioenriched α-hydroxy-α,α'-dialkylaldehydes are obtained. Ephedrine-derived oxazolidines have been used in a similar fashion [1184].

The addition of organomagnesiums to α-ketoesters RCOCOOG* **1.23** is diastereoselective. This reaction is chelation-controlled, and one face of the carbonyl group of phenmenthyl **1.4** (R = Ph) esters is shielded by the aromatic ring [66, 144, 253] (Figure **6.36**). Similar results are obtained with other organometallic reagents [1185]. Esters of trans-2-phenylcyclohexanol **1.5** [66, 1186] or of L-quebrachitol **1.49** [243] are used for similar purposes. These reactions can lead to either enantiomeric α-hydroxyacid after hydrolysis. The direction of selectivity is controlled by the selection of organomagnesium or organolithium reagents, which in turn dictates whether the reaction is under chelation control or not (Figure **6.36**).

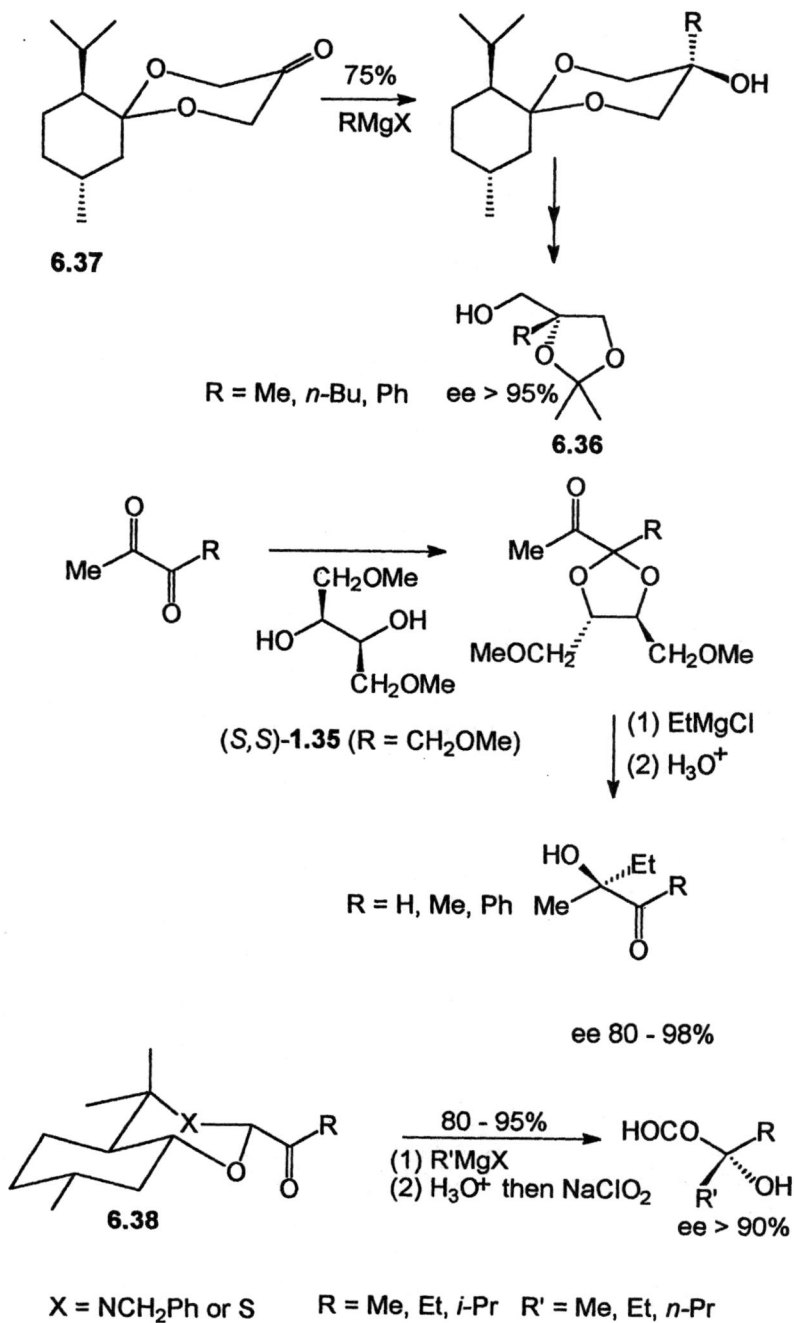

R = Me, *n*-Bu, Ph   ee > 95%

6.36

(S,S)-**1.35** (R = CH$_2$OMe)

(1) EtMgCl
(2) H$_3$O$^+$

R = H, Me, Ph

ee 80 - 98%

6.38

80 - 95%
(1) R'MgX
(2) H$_3$O$^+$ then NaClO$_2$

ee > 90%

X = NCH$_2$Ph or S R = Me, Et, *i*-Pr   R' = Me, Et, *n*-Pr

Figure 6.35

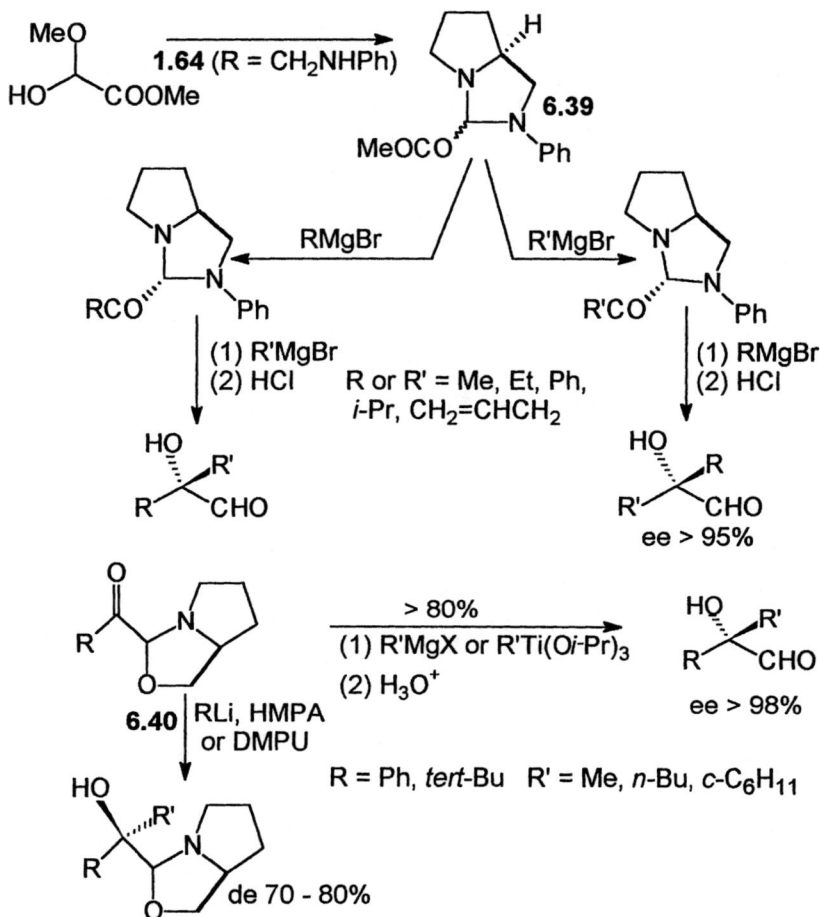

Figure 6.35 (continued)

α-Ketoamides RCOCONR'$_2$ derived from pyrrolidines having a C$_2$ axis of symmetry and bearing CH$_2$OR' substituents to facilitate hydrolysis of the products have received some applications [217a, 292] (Figure 6.36). The best addition selectivities are observed with organotitanium reagents, and the reactions are chelation-controlled. The reaction of ketoester 1.45 with R'MgX-MgBr$_2$ is stereoselective. A nonracemic gem-dialkylated lactone is formed *in situ* with a high enantiomeric excess provided that the polyether side chain has the proper length (n = 0 or 2) [1187] (Figure 6.36). The reaction of nonracemic β-ketosulfoxides with AlMe$_3$ in the presence of ZnCl$_2$ is not very stereoselective [483], except from a *tert*-butyl ketone (de 90%).

1.23, G*OH = (1R,2S,5R) **1.4** (R = Ph)

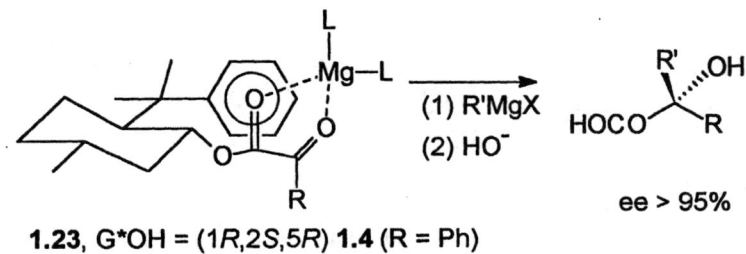

R = H, Me, Ph   R' = Me,n-C$_6$H$_{13}$, c-C$_6$H$_{11}$, Ph

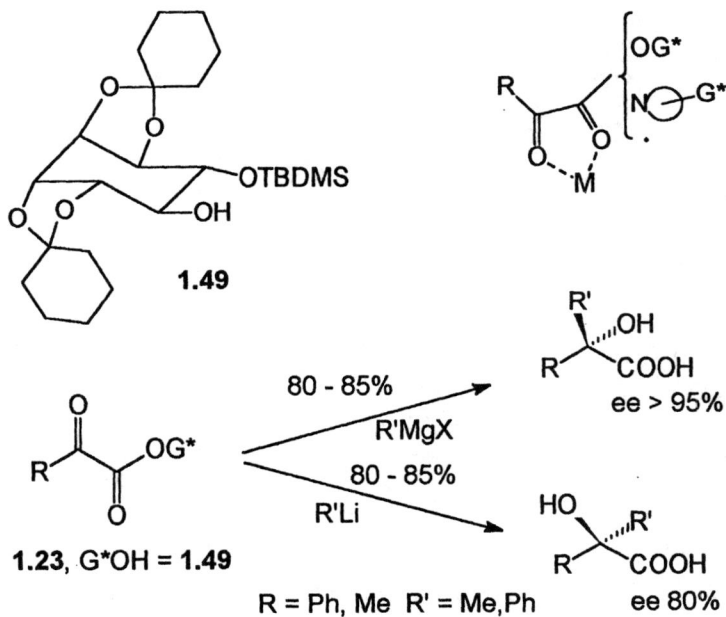

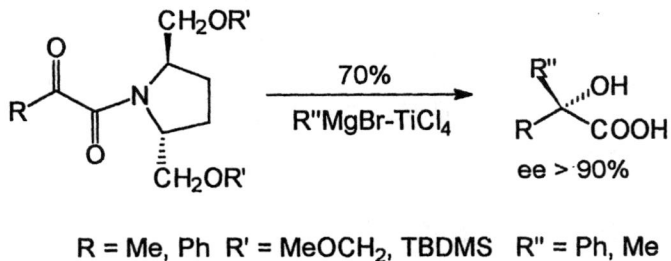

1.49

1.23, G*OH = **1.49**

R = Ph, Me   R' = Me,Ph   ee 80%

R = Me, Ph   R' = MeOCH$_2$, TBDMS   R" = Ph, Me

Figure 6.36

1.45   R = Ph, Me   R' = Me, Ph

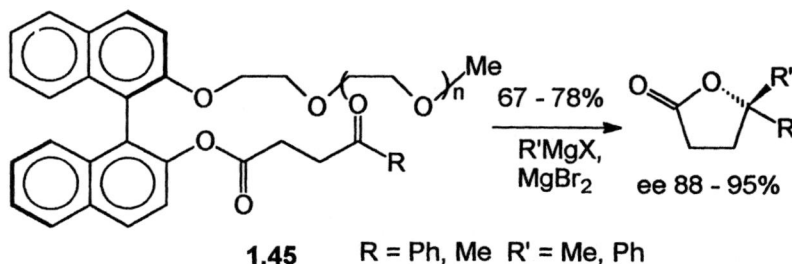

Figure 6.36 (continued)

The reaction of organometallic reagents with carbonyl compounds bearing a butadieneiron carbonyl residue can be highly stereoselective, provided that the conformation of the carbonyl group is fixed. Such is the case for the reaction of organolithium reagents with acylcomplexes **6.41** (R = alkyl) [526, 533] (Figure **6.37**), or for the reaction of allylzinc reagents with formyltrimethylenemethaneiron carbonyl [529]. However, similar reactions run with aldehydes **6.41** (R = H) are usually poorly stereoselective [528, 531]. Addition of organomagnesium compounds to rigidified chromium tricarbonyl complexed ketones **6.18** takes place on the face opposite to the organometallic residue. 2-Substituted aldehydes **6.42** react with various organomagnesium and -lithium reagents with a good stereoselectivity due to the restricted carbonyl conformation coupled with the steric hindrance induced by the chromium carbonyl group [539, 540, 1188] (Figure **6.38**). The nitromethane anion and lithioformamidine $LiCH_2N(Me)CH=N$-*tert*-Bu are precursors of a CH_2NHMe residue, and they react similarly [539, 540]. After decomplexation, secondary benzyl alcohols are obtained with an excellent enantiomeric excess (Figure **6.37**)

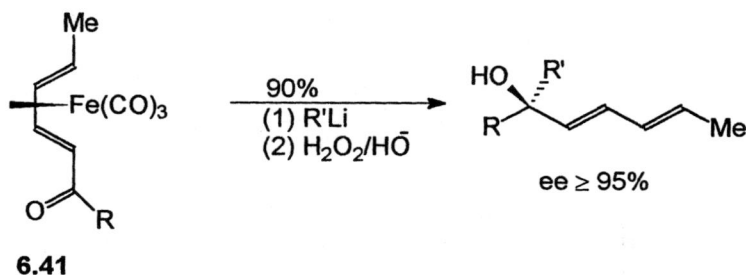

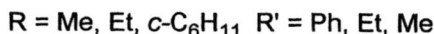

6.41

R = Me, Et, c-C_6H_{11} R' = Ph, Et, Me

Figure 6.37

6.42

R = Me, MeO, CF$_3$, Cl

R' M = MeMgX, PhMgX, C$_2$F$_5$Li, NaCH$_2$NO$_2$, LiCH$_2$

Me$_3$SiC≡CLi, HC≡CMgBr

Figure 6.37 (continued)

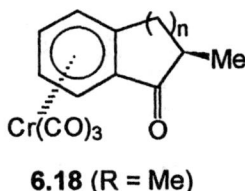

6.18 (R = Me)

6.5.3 Reactions with Chiral Acetals

Like the related reductions (§ 6.1.3), the reactions of organometallic reagents with chiral acetals **6.20** are often selective, and these reactions provide nonracemic secondary or tertiary alcohols [213, 1118]. As described for the reductions, the dioxane must adopt a preferred conformation (that is, the sizes of R and R' must be sufficiently different) to observe good selectivities. Furthermore, the stereochemistry of the reaction depends upon the nature of the organometal and the Lewis acid (if added). Frequently, the reactions of 1,3-dioxanes **6.20** (R' = H, R" = Me) prepared from aldehydes and (R,R)-2,4-pentanediol result in the formation of the new C–C bond with inversion of configuration. Inversion occurs with organocopper reagents under BF$_3$ catalysis, and with organomagnesium, organolithium, organozinc reagents in the presence of TiCl$_4$ [213, 1118] (Figure **6.38**). The reaction of these acetals with R$_3$Al is not stereoselective. However, Yamamoto and coworkers [222, 1118] discovered that the use of R$_2$AlOC$_6$F$_5$, generated *in situ* by action of C$_6$F$_5$OH on R$_3$Al, promotes the stereoselective reaction of **6.20** (R = alkyl, R' = H or Me, R" = Me) with retention of configuration (Figure **6.38**). Yamamoto and coworkers also performed related stereoselective ring openings from racemic 2-alkyl-4-methyl-1,3-dioxanes derived from 1,3-butanediol and excess Et$_3$Al [222].

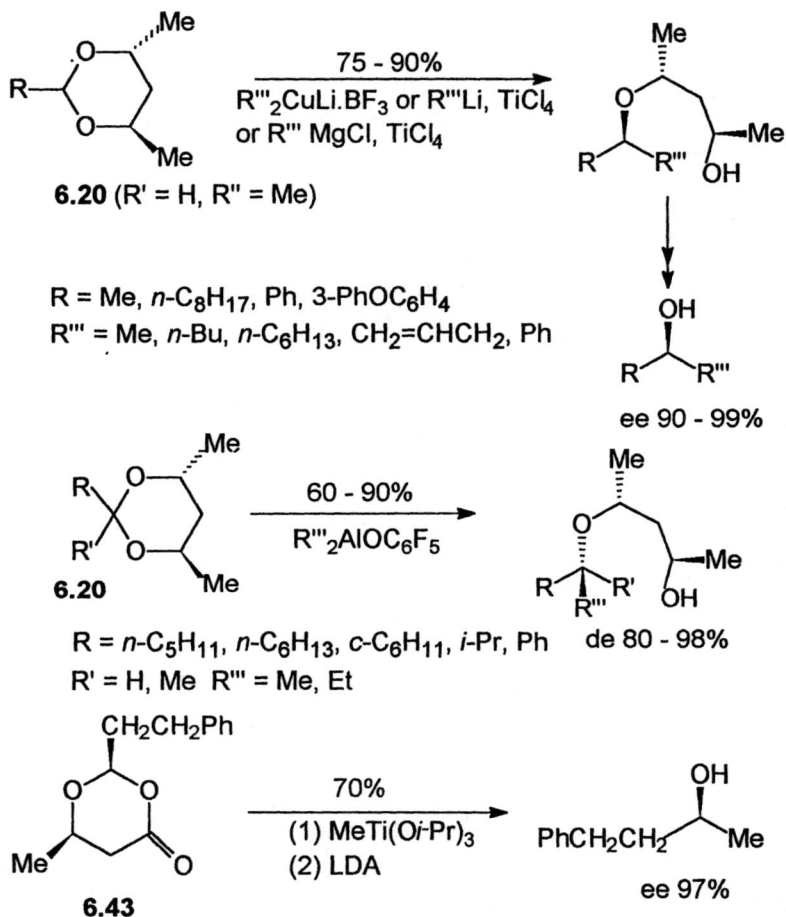

6.20 (R' = H, R" = Me)

R = Me, n-C$_8$H$_{17}$, Ph, 3-PhOC$_6$H$_4$
R''' = Me, n-Bu, n-C$_6$H$_{13}$, CH$_2$=CHCH$_2$, Ph

ee 90 - 99%

6.20

R = n-C$_5$H$_{11}$, n-C$_6$H$_{13}$, c-C$_6$H$_{11}$, i-Pr, Ph   de 80 - 98%
R' = H, Me R''' = Me, Et

6.43

ee 97%

Figure 6.38

A related transformation comes from Seebach and coworkers [386], who treated dioxanone **6.43** with MeTiO(i-Pr)$_3$. After cleavage of the resulting ether with LDA, a nonracemic secondary alcohol was obtained with a high enantiomeric excess (Figure **6.38**). Organolithium or -magnesium reagents attack the carbonyl carbon of **6.43**, while cuprates do not react at all.

6.5.4 Reactions with Imines and Derivatives

Asymmetric addition of organolithium reagents to prochiral diarylimines **6.44** has been performed by Tomioka, Koga and their coworkers with **2.40** as chiral ligand at −100°C in toluene [636, 637, 638]. In these reactions, the presence of a 4-methoxy group and a 2-substituent on the nitrogen-substituted aryl ring increase the enantioselectivity. If the nitrogen substituent of the cinnamyl imine is c-C$_6$H$_{11}$, then conjugate addition takes place instead of C=N addition. Dialkylzincs do not react with N-silyl- or N-phenylimines, even in the presence of aminoalcohols. However, this reaction does occur with N-diphenylphosphinylimines **6.45**, and when it is carried out in the presence of aminoalcohol (1S,2R)-**1.14** (R = n-Bu), nonracemic branched primary amines are obtained after hydrolysis with a high enantiomeric excess [1189] (Figure **6.39**). The reactions of the lithiated anion of (R)-methyl tolylsulfoxide with imines PhCH=NR are stereoselective, but the diastereoisomeric excesses are lower than 80% [480]. The same reaction performed with nitrones is also poorly stereoselective, unless cinchona alkaloids are added [1190].

Ukaji and coworkers [1191] performed the enantioselective addition of organomagnesiums to 3,4-dihydroquinoline N-oxide in the presence of aminoalcohol **2.6** at −78°C . With one exception, the enantioselectivity is in the 80% range.

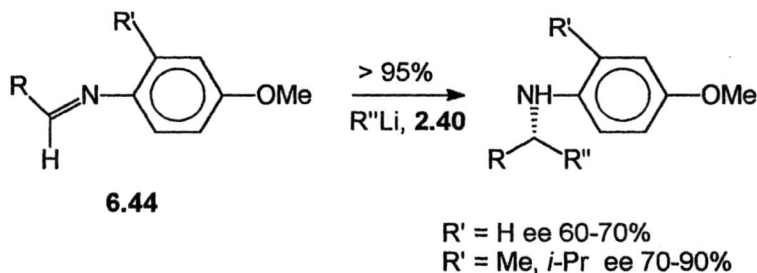

R' = H ee 60-70%
R' = Me, i-Pr ee 70-90%

R = Ph, 1-Np, 2-Np, PhCH=CH R''=Me, n-Bu, CH$_2$=CH

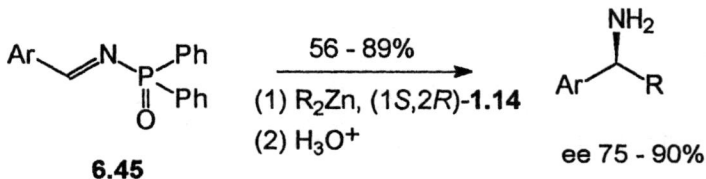

ee 75 - 90%

Ar = Ph, 2-Np, 4-MeC$_6$H$_4$ R = Me, Et, n-Bu

Figure 6.39

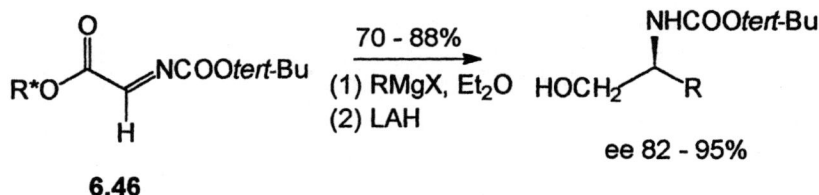

6.46

R*OH = (1R,2S,5R)-**1.4** (R = Ph) R = Me, i-Pr, i-Bu, Ph

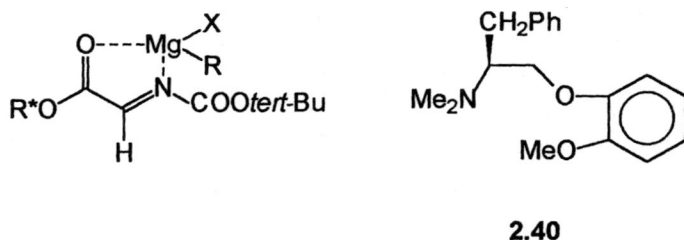

2.40

Figure 6.39 (continued)

More interesting results have been obtained from imines, imminium salts or hydrazones bearing chiral residues. The reaction of organomagnesium compounds with imines **6.46** bearing an ester functionality derived from phenmenthol **1.4** (R = Ph) is highly stereoselective. After treatment of the products with LiAlH$_4$, aminoalcohols are obtained with a high enantiomeric excess [1192, 1193]. Alternatively careful hydrolysis leads to nonracemic α-aminoacids. The stereochemistry of this reaction is under chelation control (Figure **6.39**). Addition of allylzinc reagents to (S)-valine imines **6.47** at room temperature is highly stereoselective. If R in **6.47** is aromatic, then the reaction must be conducted in the presence of CeCl$_3$ to avoid equilibration [1194]. After appropriate treatment, (S)-homoallylamines are obtained (Figure **6.40**).

Reactions of organomagnesium reagents with cyclic imminium salts **6.27** (R' = H) give, after hydrogenolysis, 2-alkyltetrahydroquinolines with a good selectivity [139] (Figure **6.40**). The enantiomer that is obtained is the same as the one formed by NaBH$_4$ reduction of **6.27** (R' = alkyl). These results have been interpreted by the intervention of different reactive conformers in each case (Figure **6.40**). Reaction of MeMgBr with chiral imine oxide **6.48** selectively leads to nonracemic hydroxylamines, from which the corresponding amines are obtained with a high enantioselectivity [1195]. Attack of MeMgBr takes place from the least hindered side of a chelate (Figure **6.40**).

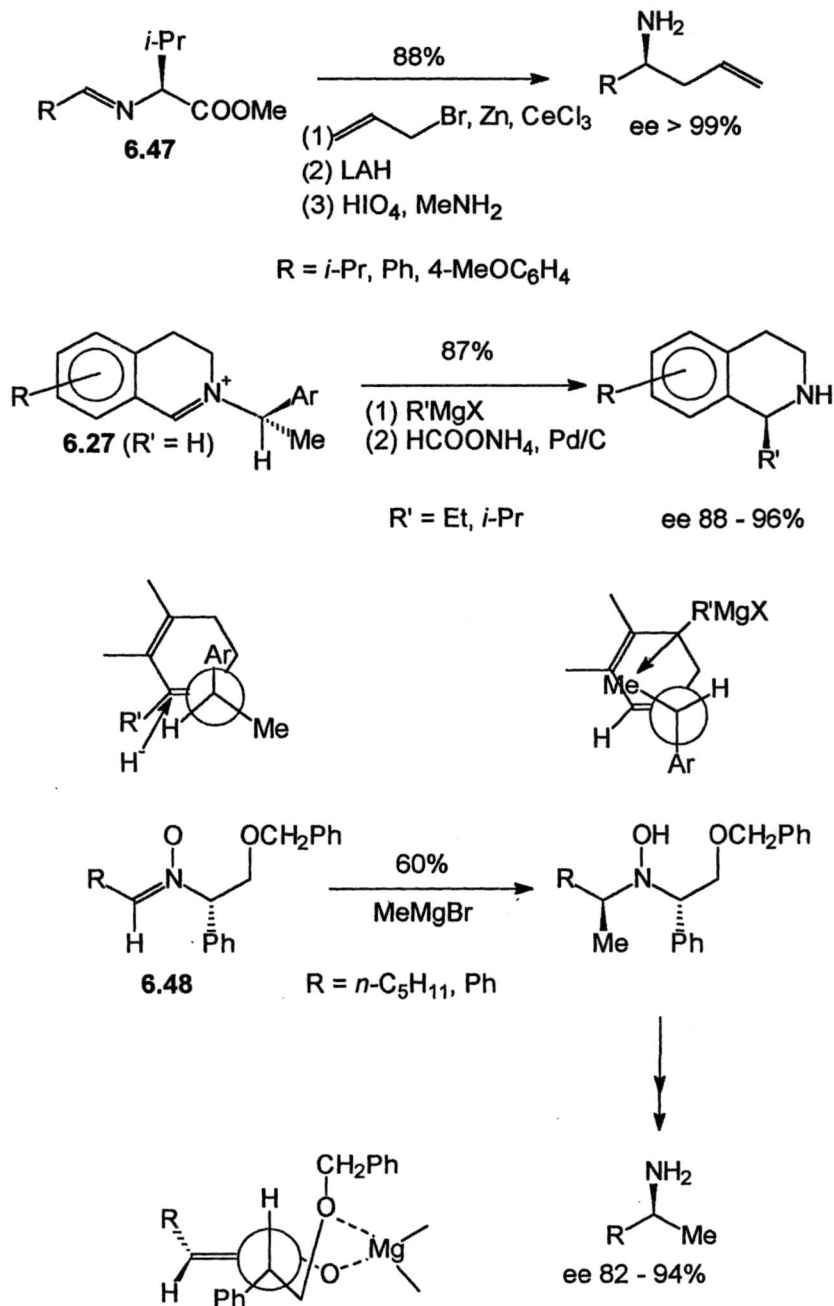

R = i-Pr, Ph, 4-MeOC₆H₄

R' = Et, i-Pr   ee 88 - 96%

6.48   R = n-C₅H₁₁, Ph

ee 82 - 94%

Figure 6.40

Addition of allylmagnesium bromide to chiral sulfimines **1.148** is stereoselective. After treatment of the products with CF_3COOH, homoallylamines are obtained with a high enantiomeric excess [512] (Figure **6.41**).

Useful results were simultaneously obtained by Denmark [316], Enders [317a, 1195a] and their coworkers in the reactions of organolithium reagents with chiral hydrazones **1.77**. These reactions, which are sometimes conducted in the presence of $CeCl_3$, are highly stereoselective, and further treatment of the products with Raney nickel leads to (R) or (S) nonracemic amines according to the configuration of the auxiliary **1.76** (Samp or Ramp) (Figure **6.41**). The auxiliary can be easily recycled [317a]. In a few cases, the selectivity is improved by using a Samp or Ramp reagent in which the CH_2OMe group is replaced by CH_2OCH_2OMe [318].

The reaction of chiral aminals of the monohydrazone of glyoxal **6.49** with organolithium reagents is stereoselective. After Raney nickel treatment, followed by protection of the amine and hydrolysis, α-aminoaldehydes are obtained with an excellent enantiomeric excess [302]. Reaction of organomagnesium reagents with **6.49** in toluene, followed by the same treatment, gives the other enantiomer [1196] (Figure **6.41**).

A few examples of stereoselective additions of organomagnesium or -cerium reagents to imines bearing a chiral arenechromium tricarbonyl or dieneiron tricarbonyl residue have been described [539, 1197]. Reactions of allylmetals with an oxime bearing a chiral ether functionality can be stereoselective provided that the oxime displays the E configuration [1198].

Comins and coworkers examined the reactions of organomagnesium compounds with 4-methoxy-N-carboalkoxypyridinium salts bearing a chiral ester residue **6.50** [186, 187, 188]. The best selectivities are observed when X = $Si(i\text{-}Pr)_3$ and when the chiral auxiliary is (1R,2S,5R)-phenmenthol **1.4** (R = Ph, $C_6H_4OC_6H_{11}$-4) (Figure **6.41**). The $Si(i\text{-}Pr)_3$ group may subsequently be removed with oxalic acid. When the reaction is carried out with $Ph_3SiMgBr$, the other diastereoisomer is formed [1199]. Similar reactions with other pyridinium salts have been used in alkaloid synthesis [769].

Nonracemic amines can be prepared from chiral perhydrooxazines **6.51** generated from aminoalcohol **1.80** and aldehydes [332]. The reaction of **6.51** with RMgX at room temperature mainly gives one diastereoisomer, from which nonracemic amines are obtained by a convenient procedure. However, the chiral auxiliary cannot be recovered (Figure **6.41**).

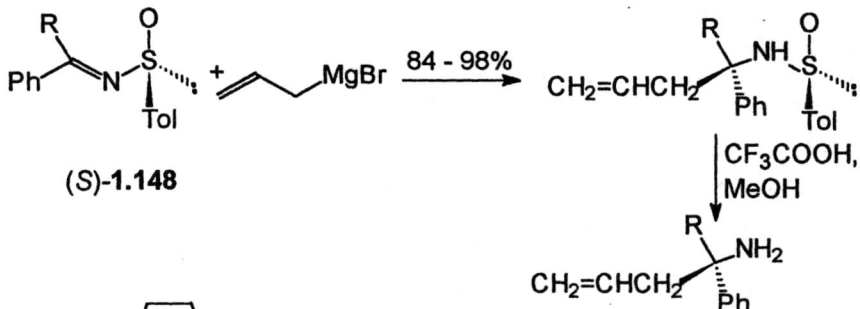

(S)-**1.148**

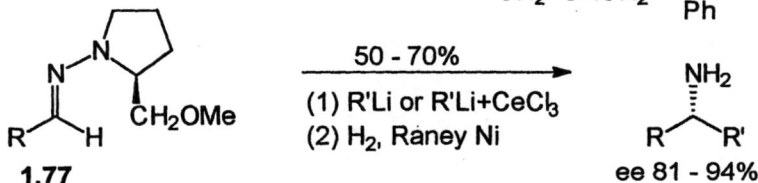

1.77

ee 81 - 94%

R = i-Pr, tert-Bu, c-C₆H₁₁, Ph, PhCH₂, PhCH₂CH₂, (EtO)₂CH
R' = Me, n-Bu, i-Pr, tert-Bu, Ph

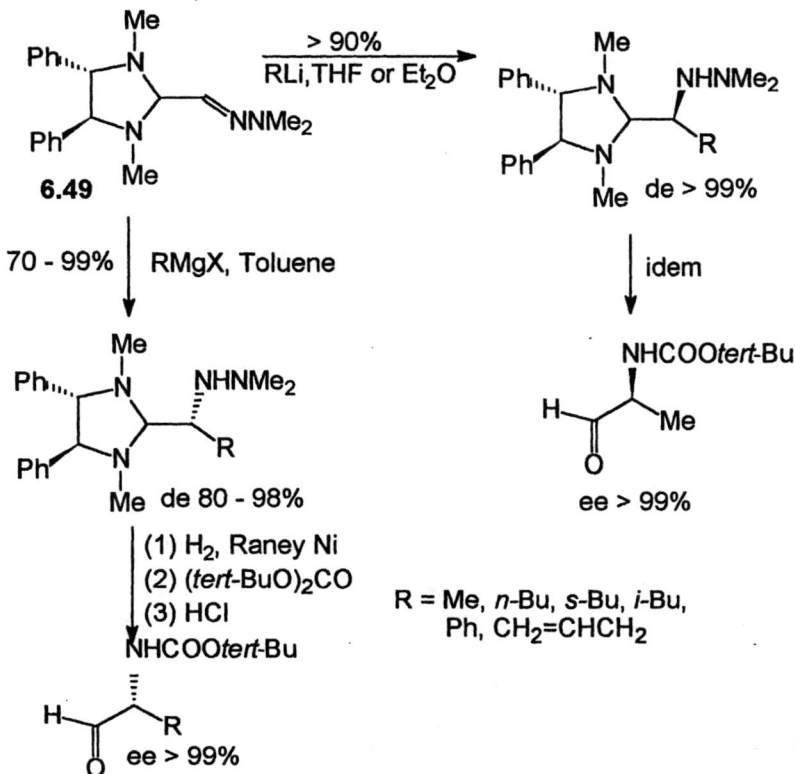

R = Me, n-Bu, s-Bu, i-Bu,
Ph, CH₂=CHCH₂

Figure 6.41

6.50

ee 80 - 95%

R = Me, *i*-Bu, *c*-C$_6$H$_{11}$, EtOCH$_2$CH$_2$OCH$_2$CH$_2$, Ph, 4-MeC$_6$H$_4$, 4-ClC$_6$H$_4$
G*OH = (1R,2S,5R)-**1.4** (R = 4-C$_6$H$_5$OC$_6$H$_4$)

6.51

de 86 - 99%

R = Me, Et, *n*-Pr
R' = Ph, *i*-Pr, *c*-C$_6$H$_{11}$

(1) P$_2$O$_5$, toluene reflux
(2) hydrogenolysis

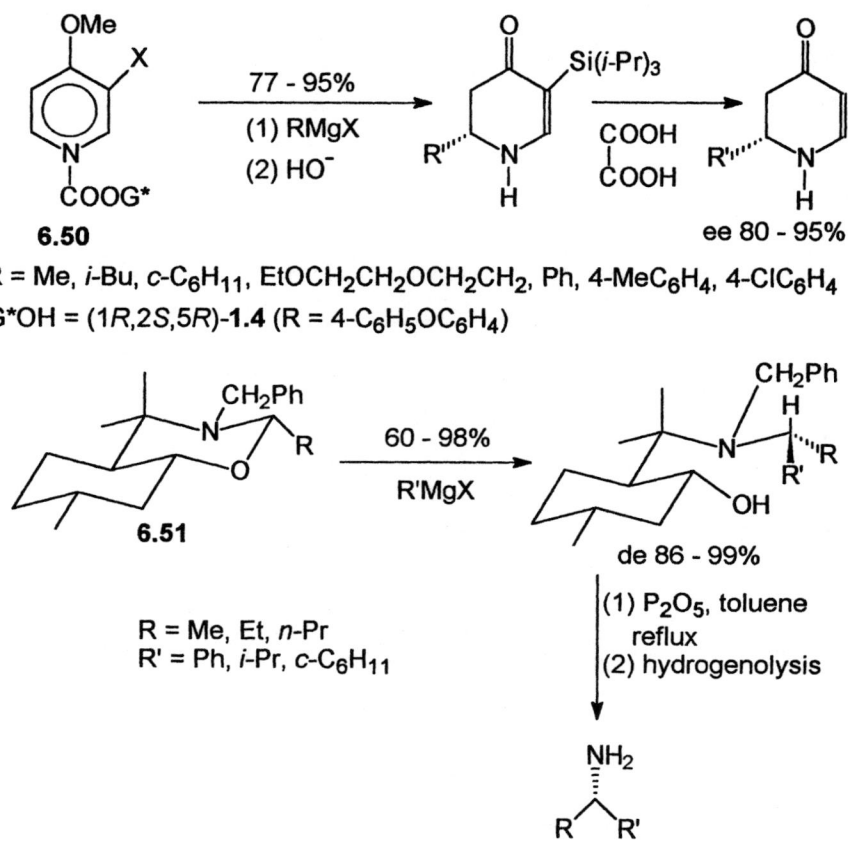

Figure 6.41 (continued)

6.6 REACTIONS OF ALLYLBORANES, -BORONATES, -SILA-NES, -STANNANES AND UNSATURATED ANALOGS [698]

The reactions of allylboranes, -silanes and -stannanes with carbonyl compounds and imines always take place with double-bond migration, and the structural stability of the reagent is a common problem that is encountered (§ 2.7). The reactions of allylboranes and -boronates occur without catalysis, while those of allylsilanes and -stannanes usually require the presence of a Lewis acid [253]. The mechanism of the reactions of allylboron derivatives is concerted, and the addition occurs via a six-membered cyclic transition state. A slightly distorted chair transition state model in which the oxygen of the carbonyl group is coordinated to the boron atom is usually invoked (Figure **6.42**). Various steric and polar interactions dictate whether the *Si* or *Re* face of the prochiral aldehyde is attacked (models **C$_1$**

to C_4). In contrast to the uncatalyzed reactions, Lewis acid-catalyzed reactions involve open transition states (models A_1 to A_6) in which the oxygen of the carbonyl group is coordinated to the Lewis acid. Repulsive interactions mainly determine which face of the carbonyl will be attacked (Figure 6.42). Similar reactions take place with alkynyl-, alkenyl- and propargylboranes, -silanes and -stannanes.

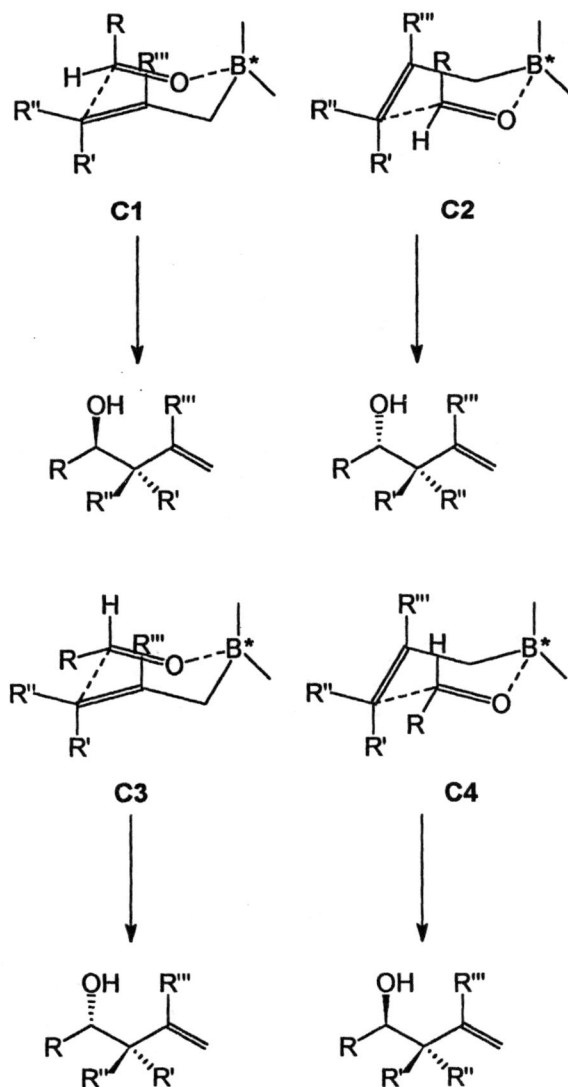

Figure 6.42

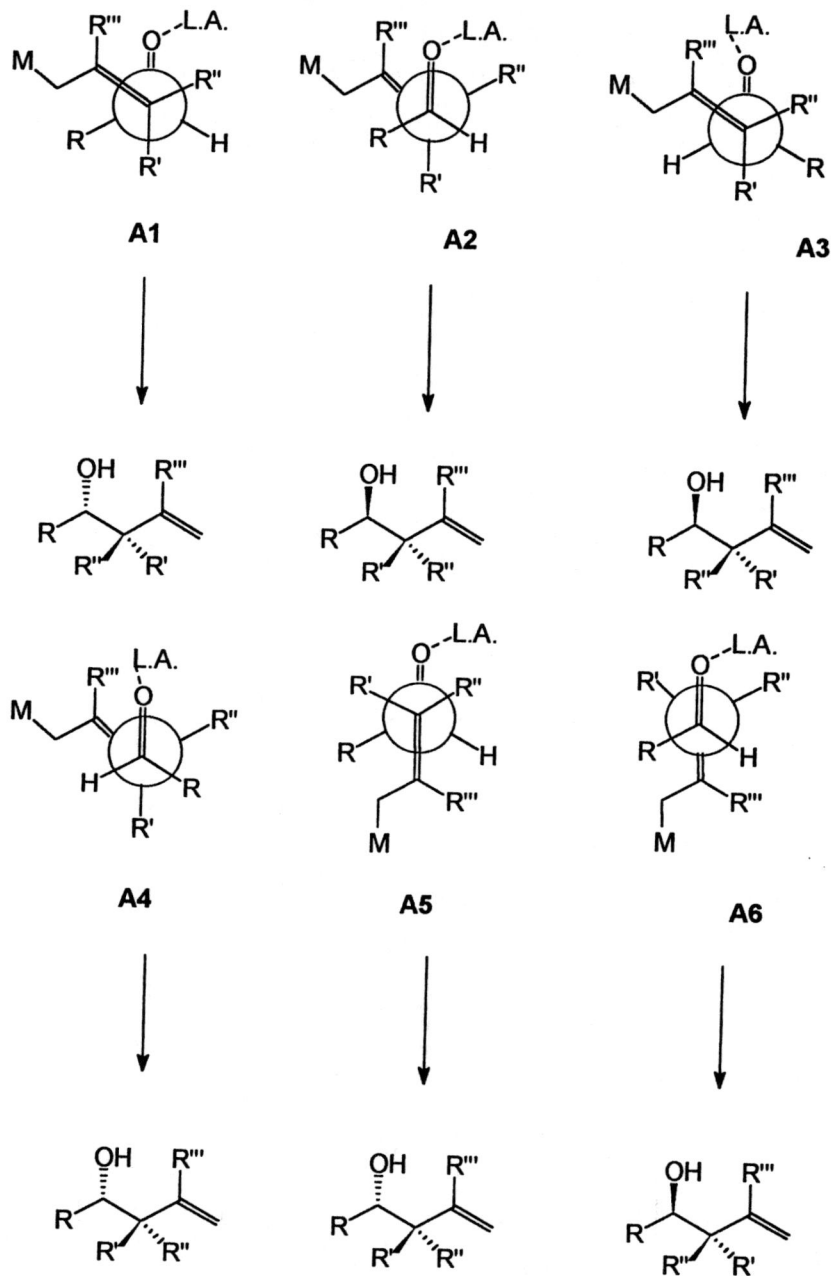

Figure 6.42 (continued)

6.6.1 Reactions with Nonfunctionalized Aldehydes

6.6.1.1 *Boranes and Boronates* [601]

The stereoselectivity of the reactions of aldehydes with allylboranes and -boronates bearing chiral substituents on the boron atom depends upon the terminal substituent of the allylic double bond. If the C-3 carbon is symmetrically substituted, then the products are enantiomers, and if it is unsymmetrically substituted, then diastereoisomers will be formed (Figure **6.42**).

Reactions of allylboranes **2.16** (R = CH_2=$CHCH_2$, Me_2C=$CHCH_2$, CH_2=$C(Me)CH_2$) [706, 707, 1200, 1201], **2.66** (R = CH_2=$CHCH_2$) [703, 707, 1200, 1200a], and **2.67** (R = CH_2=$CHCH_2$, CH_2=$C(Me)CH_2$) [714] with aliphatic, aromatic and α,β-unsaturated aldehydes at −78°C lead to homoallyl alcohols with an excellent enantioselectivity. The initial addition step is followed by treatment of the products with aminoalcohols or oxidizing agents to remove the boron (Figure **6.43**).

Similar selectivities are observed with allylboronate **6.52** prepared from tartramide **2.75** [1202], with boronamides **2.70** (R = H, X = NSO_2Me) [725] and with **2.62** (X = CH_2=$CHCH_2$, CH_2=$CClCH_2$, CH_2=$CBrCH_2$) [734]. In most cases, either enantiomeric homoallylic alcohol can be prepared from the proper reagent with an enantiomeric excess higher than 90%.

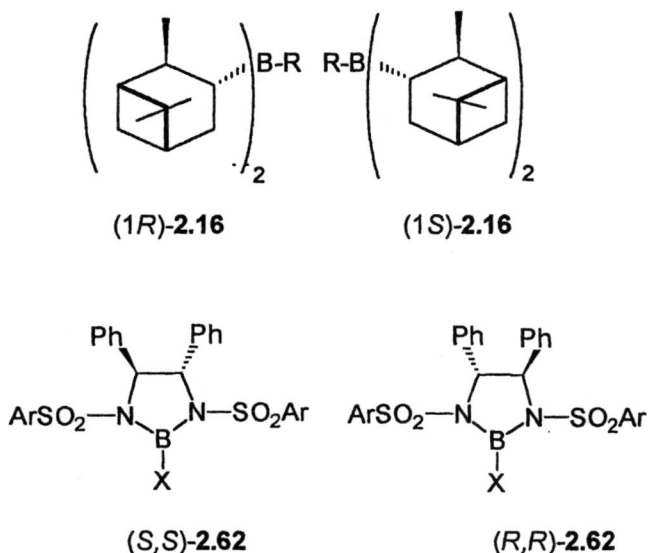

(1*R*)-**2.16**   (1*S*)-**2.16**

(*S*,*S*)-**2.62**   (*R*,*R*)-**2.62**

2.66   **2.67**

6.52   **2.70 (R = H, X = NSO₂Me)**

BR′₂ : (1R)-**2.16**, (S)-**2.67**

BR′₂ : (1S)-**2.16**, **2.66**, (R,R)-**2.62**, (R)-**2.67**

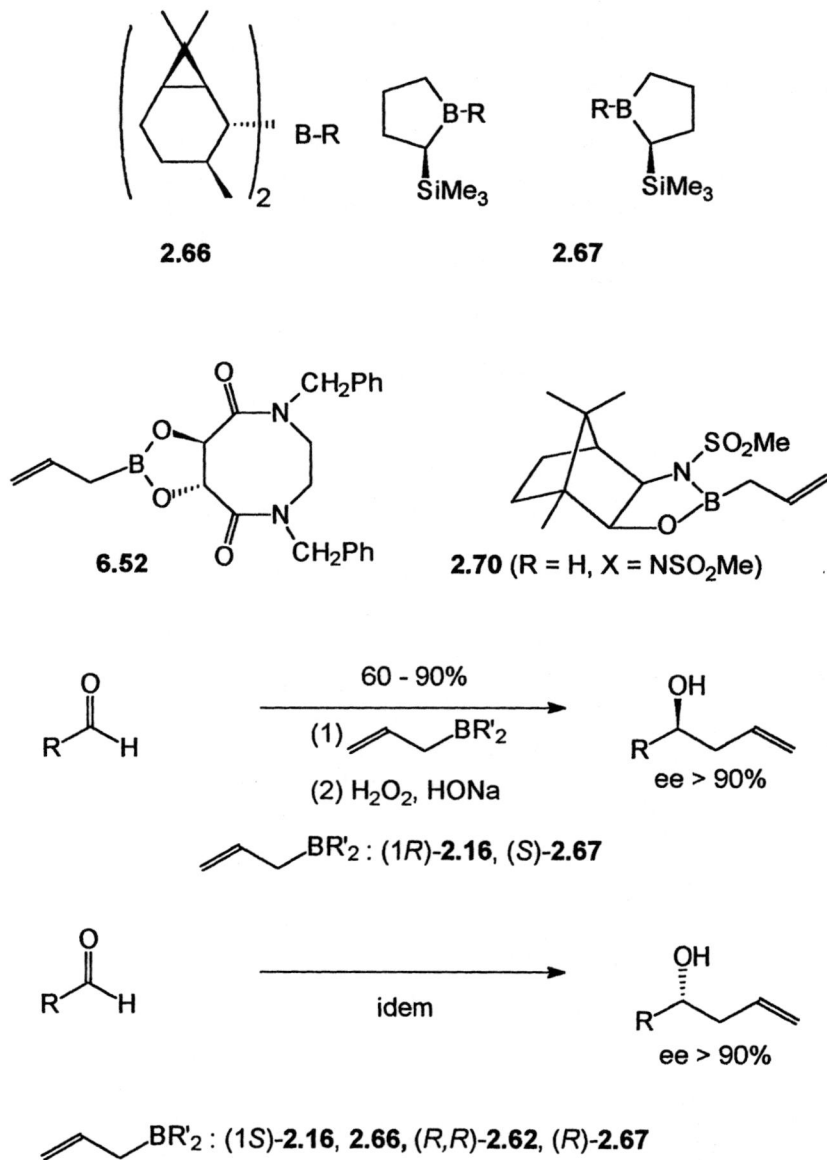

R = Me, Et, n-Pr, i-Pr, tert-.Bu, MeCH=CH,
c-C₆H₁₁, Ph,2,3,4-Py, 2-furyl, 2-thienyl

Figure 6.43

O
R⎓H

55%

(1) Me / ‖ / BR'₂ (1S)-**2.16**

(2) H₂O₂, HONa   R = Me, Et, n-Pr, i-Pr, CH₂=CH

OH   Me
R⤳⤳
ee > 90%

O
R⎓H

85%

(1) Me / Me / BR'₂ (1S)-**2.16**

(2) H₂O₂, HONa   R = Me₂C=CH

OH
R⤳
Me   Me
ee 96%

O
R⎓H

73 - 85%

(S,S)-**2.62** (X = CH₂=CClCH₂ or CH₂=CBrCH₂)

OH   X
R⤳⤳
ee 80 - 99%

X = Cl, Br R=n-C₅H₁₁, Ph, PhCH=CH

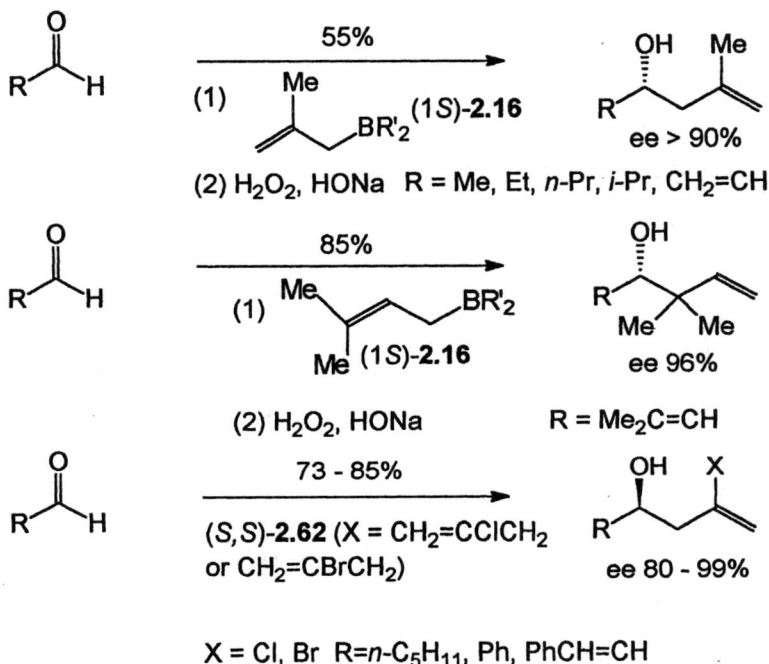

Figure 6.43 (continued)

The reactions of boronates **2.68** (R$_E$ = R$_Z$ = H, R = i-Pr) are somewhat less selective [698]. However, by using the arenechromium tricarbonyl complex of benzaldehyde or dicobalt hexacarbonyl complexes of α-alkynylaldehydes, homoallyl alcohols are obtained with a high selectivity after decomplexation [722, 1203] (Figure **6.44**). These selectivities are interpreted by distorted chair transition states (Figure **6.44**). In the reactions of allylboranes, the approach of the aldehyde minimizes both the steric interactions with the boron substituents and the eclipsing 1,3-interactions of the aldehyde C–R bond with the B–C bond. In the case of boronates **2.68**, repulsive interactions between the oxygen lone pairs are also avoided [698, 1204] (Figure **6.44**).

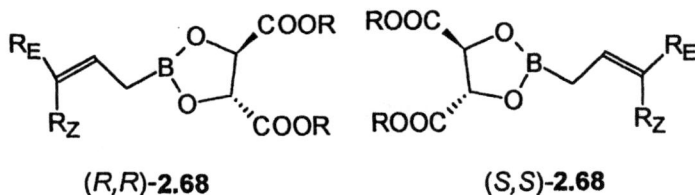

R$_E$ / COOR / B⟨O,O⟩ / COOR / R$_Z$ (R,R)-**2.68**

ROOC / O,O⟩B / ROOC / R$_E$ / R$_Z$ (S,S)-**2.68**

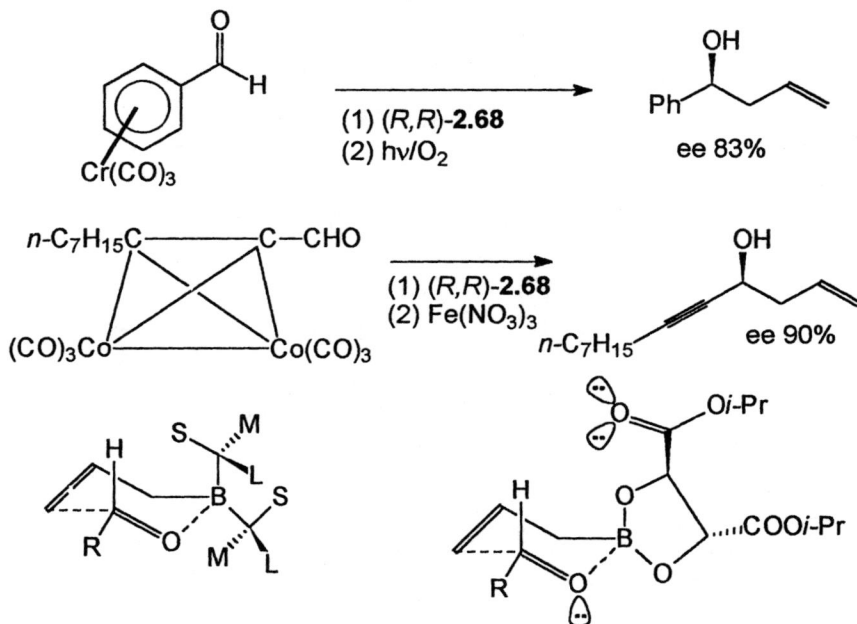

Figure 6.44

If the aldehyde is chiral, double diastereodifferentiation (§ I.6) is expected, and the stereoselectivity depends upon matching or mismatching of the partners. Brown [702], Corey [734], Roush [718, 720, 721] and their coworkers as well as Ganesh and Nicholas [1203] have prepared homoallylic alcohols with excellent diastereo- and enantiomeric excesses (Figure 6.45).

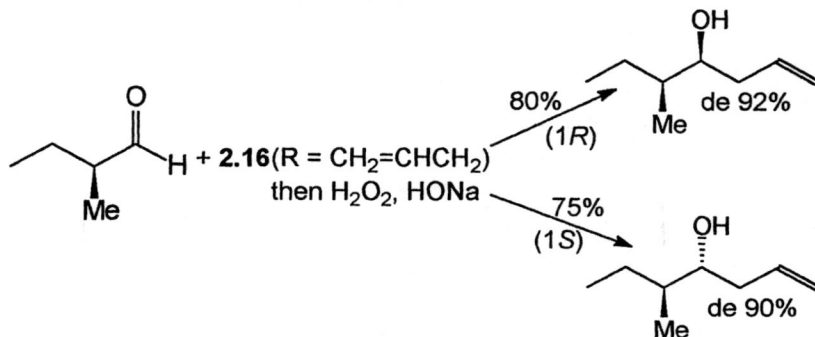

Figure 6.45

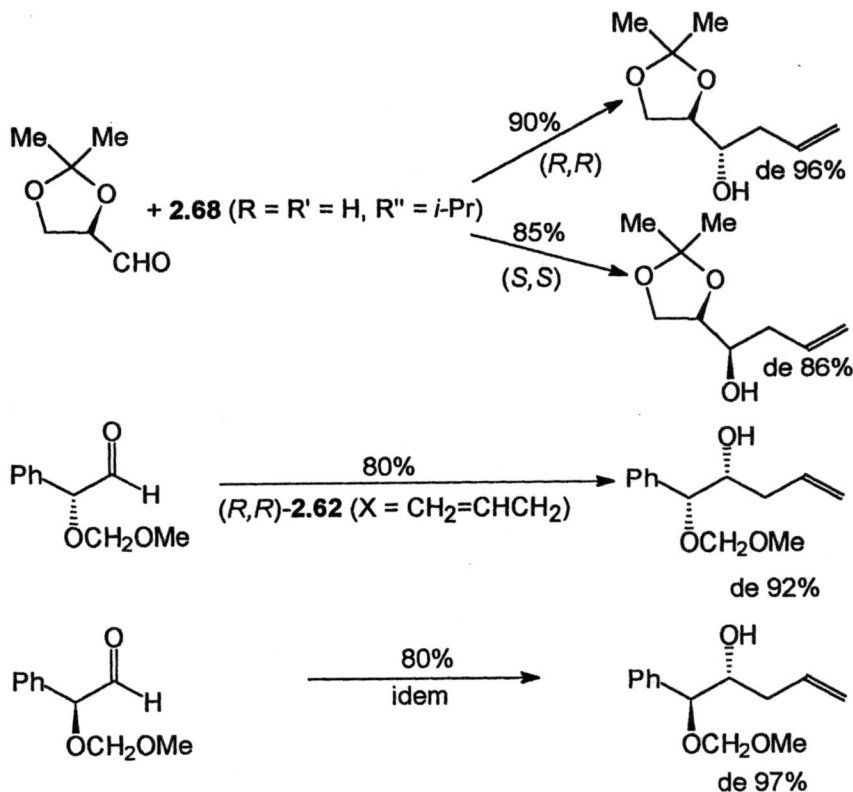

Figure 6.45 (continued)

Reactions of dissymmetrical allylboranes and -boronates ($R_E \neq R_Z$) with pro-chiral aldehydes also lead to diastereoisomers. The reactions of Z- and E-crotyl boranes **2.14**, **2.16**, **2.66** (R = Z- or E-MeCH=CHCH₂) and **2.68** (R_Z, R_E = H, Me or Me, H, R = *i*-Pr) have been examined. Because the Z- and E-crotylboranes interconvert at room temperature (§ 2.7), the borane reagents must be generated *in situ* and the reactions run at –78°C. In contrast, boronates **2.68** are stable at room temperature [253, 704, 717], and their reactions with aldehydes are stereoselective. *Syn* homoallylic alcohols **6.53** are obtained from Z-isomers while *anti* alcohols **6.54** are generated from E-isomers (Figure 6.46). The absolute configuration of these products depends upon that of the starting reagent [700, 701, 704, 713, 717, 722, 1205, 1206]. As mentioned above, boronates **2.68** give less selective reactions than boranes, but complexation of the aldehydes again improves the selectivity [722, 1207]. The best selectivities are observed with the carene-derived boranes **2.66** [1208] or the tartramide-derived boronamide, although the reactions are very slow in the later case [1202].

6.53
de > 95%
ee 85 - 99%

6.54
de > 90%
ee 85 - 95%

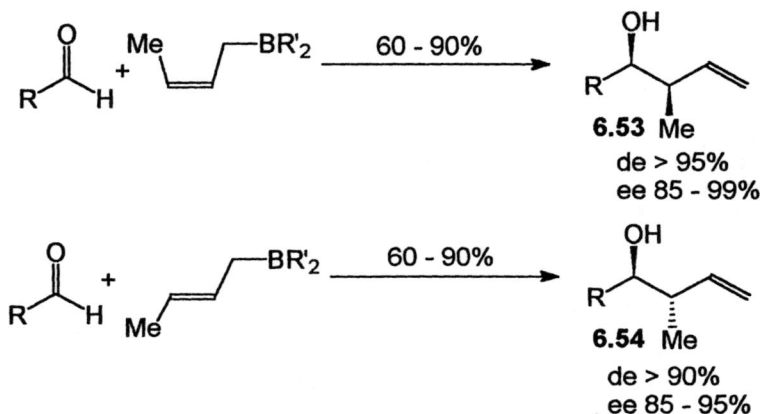

MeCH=CHCH$_2$BR'$_2$:(R,R)-**2.14**, (1R)-**2.16**, (R,R)-**2.68**
R = Me, Et, i-Pr, tert-Bu, Ph, CH$_2$=CH,
    n-C$_9$H$_{19}$, n-C$_5$H$_{11}$, PhCH$_2$CH$_2$

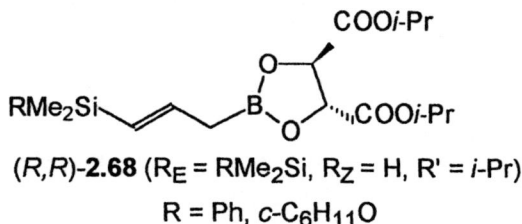

(R,R)-**2.68** (R$_E$ = RMe$_2$Si, R$_Z$ = H, R' = i-Pr)
R = Ph, c-C$_6$H$_{11}$O

Figure 6.46

Other substituents, such as cycloalkylidene [699], R$_Z$ = MeO [709], MeOCH$_2$O [1209], Z-Me$_2$i-PrNSi [710] have been introduced on the double-bond of allylboranes **2.16** (R = R$_Z$CH=CHCH$_2$). Silicon substituents, precursors of alcohols after H$_2$O$_2$ oxidation, have also been introduced on boronates **2.68** [723, 724] (Figure **6.46**). The reactions of these vinyl silanes with aldehydes give selectivities as high as those obtained from crotyl analogs. In a similar fashion, *trans* aminosubstituted allylboranes **2.16** (R = E-Ph$_2$NCH=CHCH$_2$) give *anti*-β-diphenylaminoalcohols with a high stereo- and enantioselectivity, albeit with a poor chemical yield [711, 712]. From **2.68** (R$_E$ = Cl(CH$_2$)$_n$, R$_Z$ = H, R = i-Pr), Brown and Phadke [719] prepared enantioenriched *cis*-α,β-disubstituted tetrahydrofurans (n = 2) or -tetrahydropyrans (n = 3) with 85 - 98% ee. Homoallyl alcohols bearing at least three stereocenters can be prepared from chiral aldehydes, provided that the reagents are matched. A few examples are given in Figure **6.47**. Since the product homoallyl alcohols can lead to aldehydes **6.55** after ozonolysis, this method is an alternative to aldol reactions (§ 6.8) (Figure **6.47**).

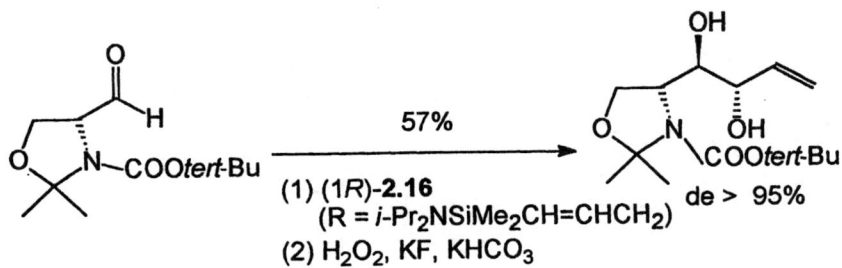

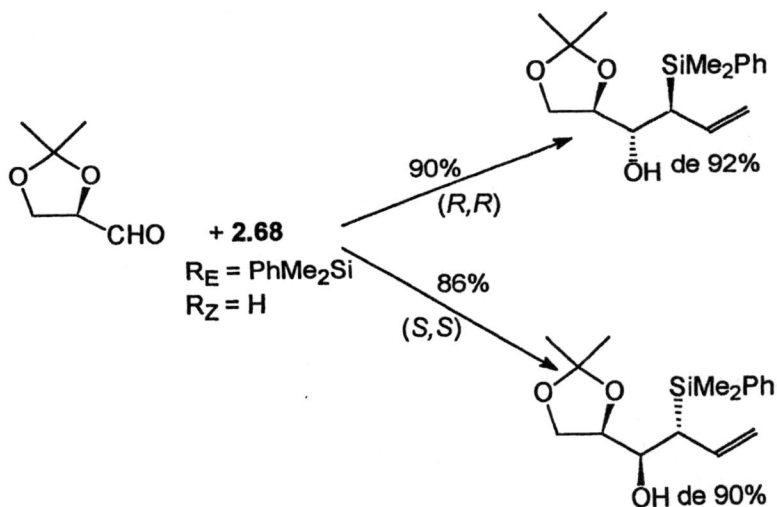

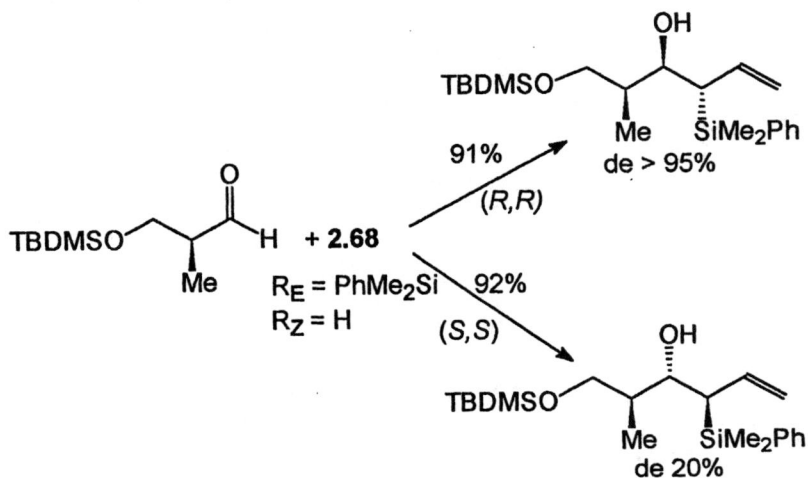

Figure 6.47

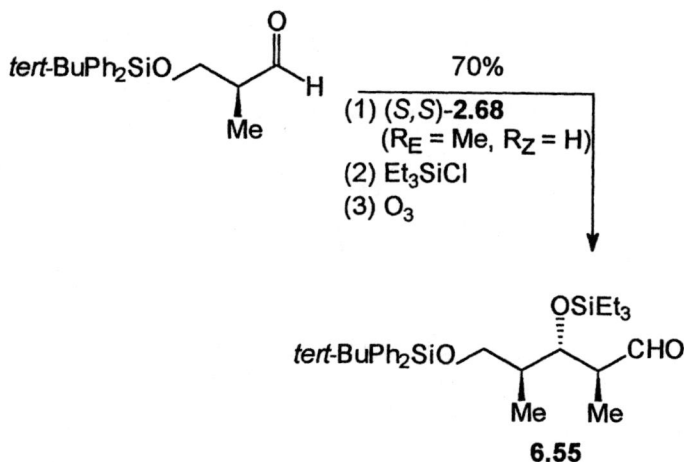

6.55

Figure 6.47 (continued)

Cyclic transition states (models C_3 and C_4) are favored for these reactions. In these transition states, 1,3 eclipsing interactions between the aldehyde C–R and boron B–C bonds are avoided [1210]. Other interactions with the aldehyde substituents must also be taken into account. The transition-state models proposed by Roush [721] in the reactions of (S)-α-methylaldehydes with (R,R)- and (S,S)-crotylboronates E-**2.68** (R_E = Me, R_Z = H, R = i-Pr) are shown as representative examples in Figure **6.48**. In the first case, model C_4 is favored and the reagents are matched. In the second case, model C_3 is lowest in energy, and the reagents are mismatched.

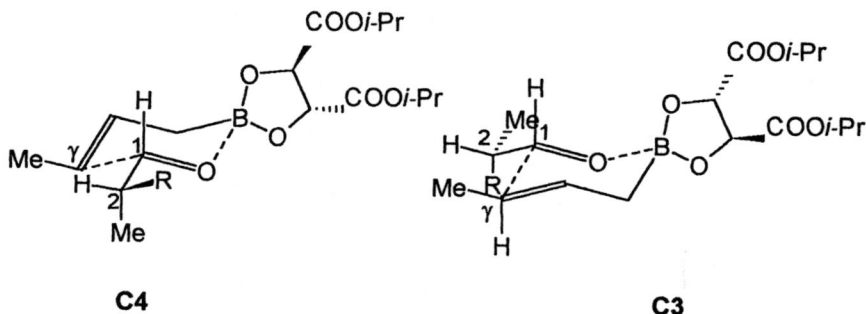

C4   C3

Figure 6.48

Reactions of aldehydes with chiral α-substituted allylboronates **2.71**, **2.73** and **2.74** (X = Cl, Me, MeO) have been examined by Hoffmann and coworkers [715, 716, 728, 729, 731, 733, 1211]. Chirality transfer is very efficient with these reagents (Figure **6.49**). Enantiomeric *anti* Z-substituted homoallyl alcohols **6.56** are obtained with a high diastereo- and enantioselectivity from E-substituted boronates, while Z-boronates **2.71** (R_E = H, R_Z = Me) lead to *syn* isomers **6.57**. Disubstituted reagents **2.71** (R_E, R_Z = alkyl, X = Me) give the corresponding dialkylated alcohols **6.57** (R_E, $R_Z \neq$ H) (Figure **6.50**).

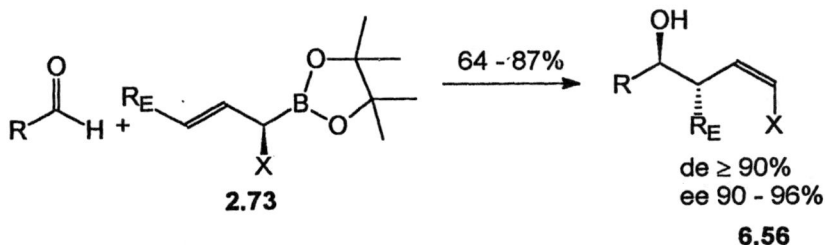

R = Me, Et, *i*-Pr, *n*-Bu, Ph R_E = H, Me X = Cl, MeO

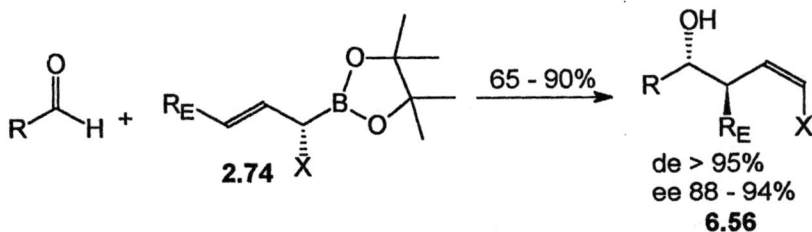

R = Me, Et, *i*-Pr, *n*-Pr, *n*-Bu, Ph X = Cl, MeO

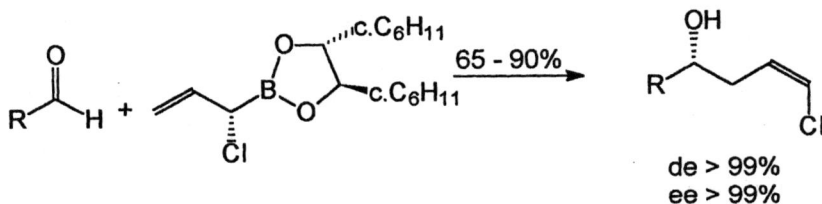

(*R,R*)-**2.71** (R_E = R_Z = H, X = Cl)

R = Me, Et, *i*-Pr, Ph, Me$_2$C=CH

Figure 6.49

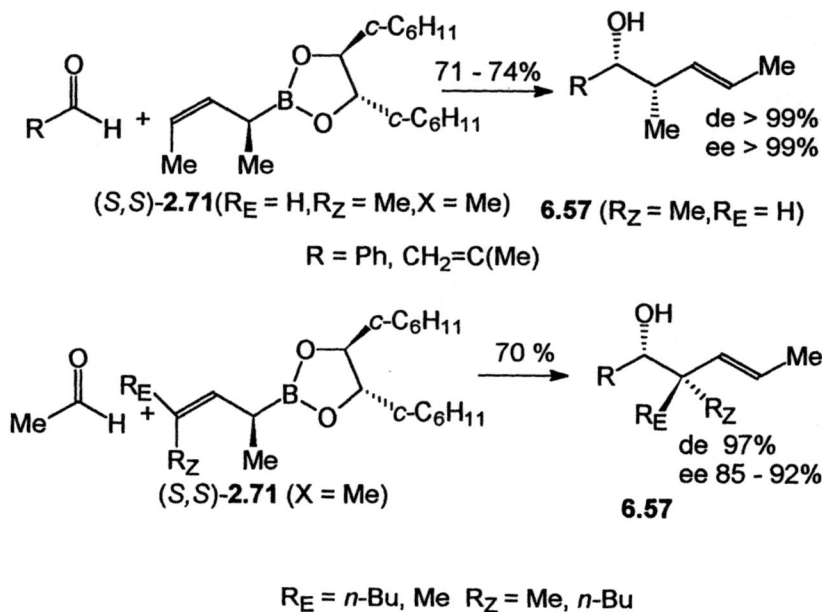

(S,S)-**2.71**(R_E = H,R_Z = Me,X = Me)   **6.57** (R_Z = Me,R_E = H)

R = Ph, CH$_2$=C(Me)

(S,S)-**2.71** (X = Me)

R_E = n-Bu, Me R_Z = Me, n-Bu

Figure 6.50

Unfortunately, the reaction of PhCHO with **2.71** (R_E = H, R_Z = Me) is not enantioselective [730]. The results with α-substituted reagents can be interpreted by cyclic transition states in which dipolar repulsions between the C–X (X = MeO, Cl) and B–O bonds are crucial. With E-substituted reagents, transition state **6.58** is preferred. However, for 2-substituted boronates, A(1,3) strain disfavors geometries with the C–X bond axial so that **6.59** is preferred when X = Me (Figure **6.51**).

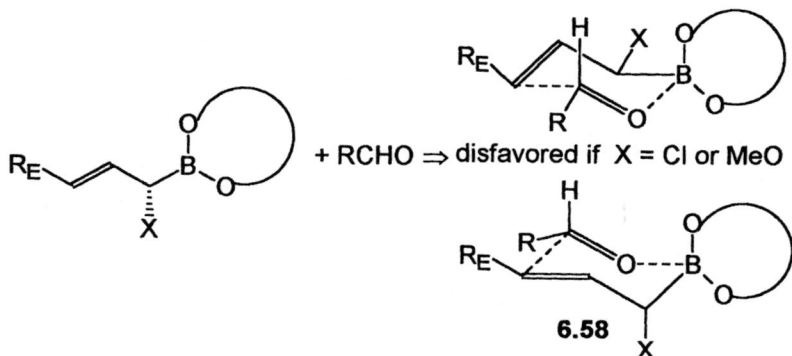

+ RCHO ⇒ disfavored if X = Cl or MeO

6.58

Figure 6.51

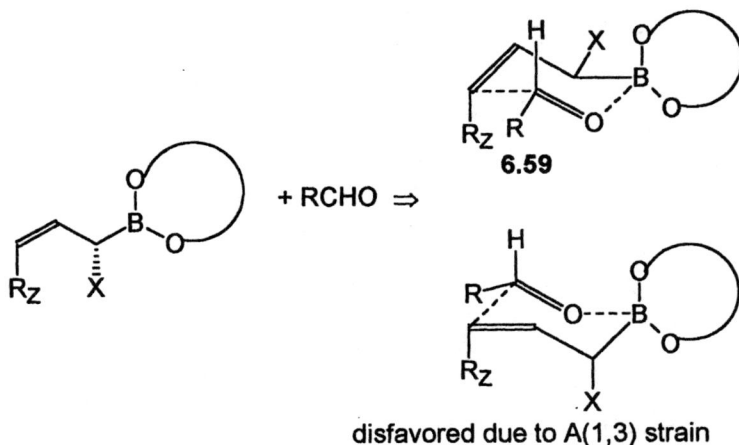

Figure 6.51 (continued)

When the allylboranes react with a chiral aldehyde, products with three stereocenters are formed, and two examples of such transformations are given in Figure 6.52 [1211].

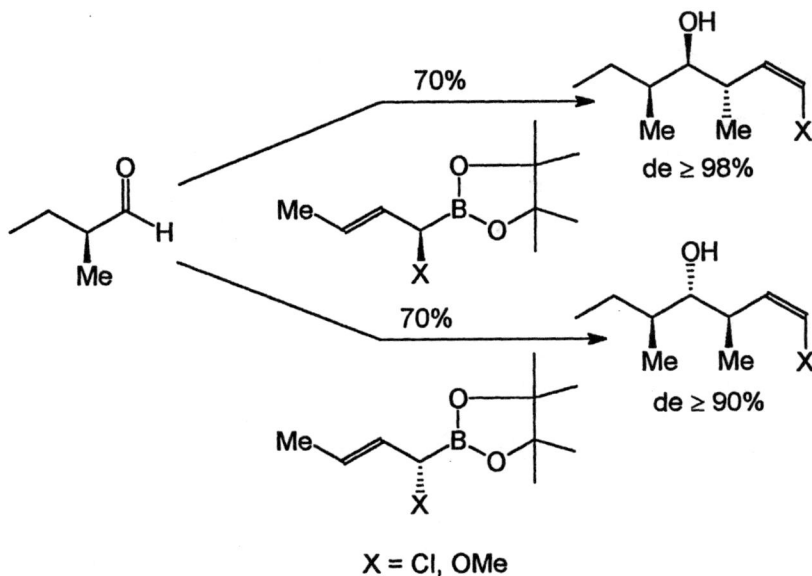

X = Cl, OMe

Figure 6.52

6.6.1.2 *Silanes and Stannanes* [1212]

The additions of allylsilanes and -stannanes to carbonyl groups are usually catalyzed by a Lewis acid. Therefore, chirality has been introduced either on the Lewis acid [786, 816] or on the reagent [569, 698, 737, 739, 740, 1213-1217]. The use of a chiral Lewis acid has been proposed by Yamamoto and coworkers [777a , 786, 791]. Acyloxyboranes (*R,R*)- or (*S,S*)-**3.9** (R' = *i*-Pr, R = H or 3,5-(CF$_3$)$_2$C$_6$H$_3$) catalyze the asymmetric addition of allylsilanes and -stannanes to aldehydes at –78°C. Provided that R" ≠ H, the chemical yields and the selectivities of the reactions of allylsilanes are good only with PhCHO or α,β-unsaturated aldehydes [791]. Better results are obtained from allylstannanes and any aldehyde [816] (Figure **6.53**).

The use of titanium catalysts formed from (*S*)- or (*R*)-binaphthol **1.44** and Ti(O*i*-Pr)$_4$ or Ti(O*i*-Pr)$_2$Cl$_2$ has been proposed by Keck, Umani-Ronchi and their coworkers [1218-1221] for the asymmetric allylation of aldehydes with CH$_2$=C(R)CH$_2$SnBu$_3$ (R = H,Me). These reactions occur near room temperature in the presence of molecular sieves, and excellent yields and enantiomeric excesses are obtained (Figure **6.53**).

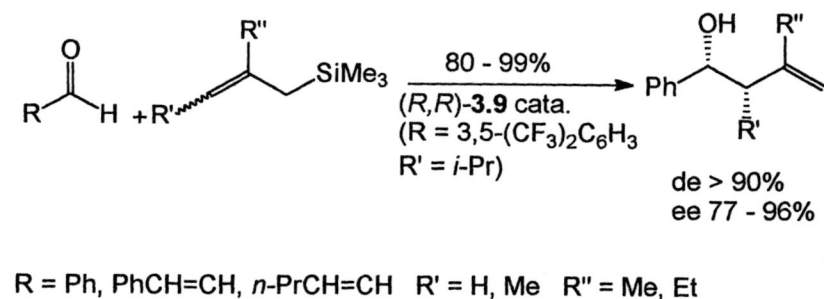

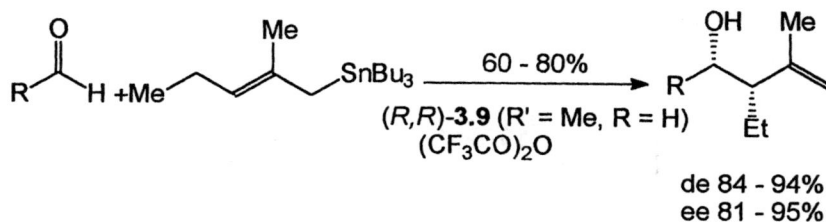

R = Ph, PhCH=CH, *n*-PrCH=CH   R' = H, Me   R" = Me, Et

R = *n*-C$_3$H$_7$, *i*-C$_3$H$_7$, E-MeCH=CH, MeCH=C(Me)

Figure 6.53

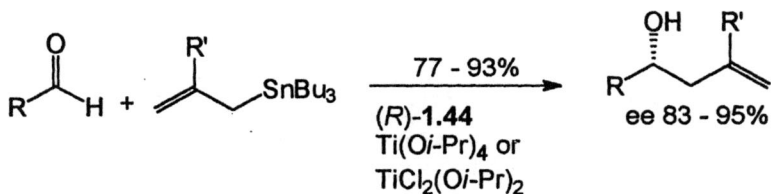

R = n-C$_5$H$_{11}$, n-C$_8$H$_{17}$ R' = H
R = PhCH$_2$CH$_2$, c-C$_6$H$_{11}$, Ph, PhCH=CH, 2-furyl   R' = H, Me

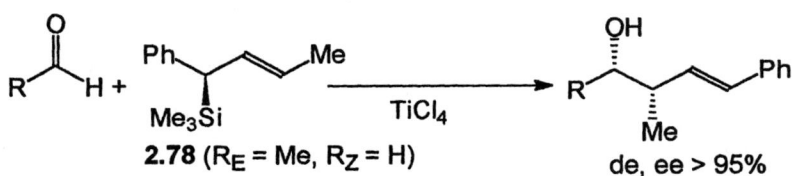

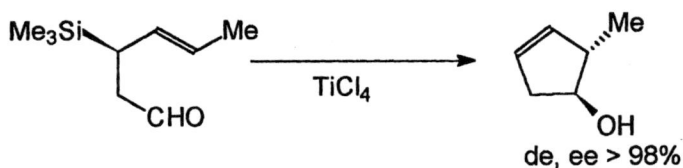

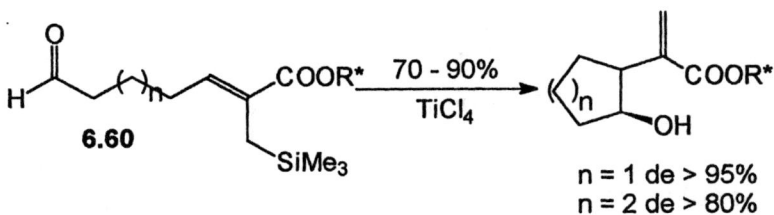

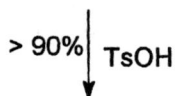

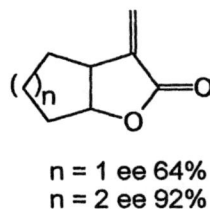

2.78 (R$_E$ = Me, R$_Z$ = H)

6.60

Figure 6.53 (continued)

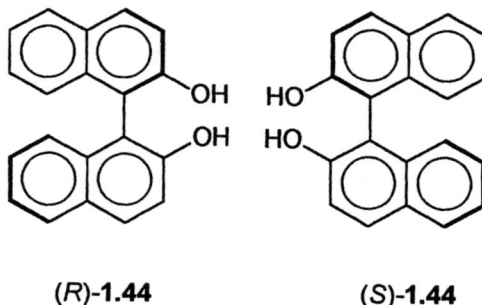

(R)-**1.44**   (S)-**1.44**

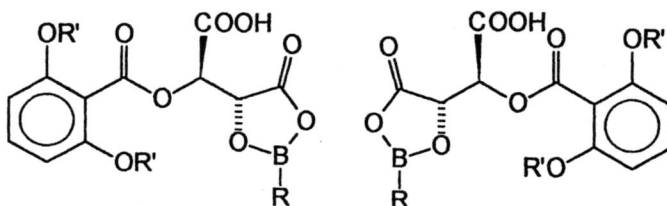

3.9

If the silicon atom of allylsilane is stereogenic, disappointing selectivities are obtained [1156]. On the other hand, if the silicon substituents are chiral, a chirality transfer may be observed if the allylic group itself bears the asymmetry [54, 736] or asymmetric induction may be observed if another substituent of silicon is chiral [737, 1014]. Chirality transfer in inter- or intramolecular reactions of allylsilanes **2.78** with aldehydes is very efficient under TiCl$_4$ catalysis; however, from allylsilanes **2.79**, a poor selectivity is observed [737, 1014] (Figure **6.53**). More useful results were obtained by Nishitani and Yamakawa [1222] in the cyclization of **6.60**, whose chiral ester is derived from (1R,2S,5R)-phenmenthol **1.4** (R = Ph). The chiral auxiliary is recovered after lactonization, but the absolute configuration of the *cis* lactones thus formed has not been determined (Figure **6.53**).

Enantioenriched α-alkoxystannanes **2.80** are easily prepared from the corresponding alcohols (§ 6.1.1) [1213]. The condensation of **2.80** with aldehydes is catalyzed by BF$_3$·Et$_2$O and leads to *syn*-homoallyl alcohols **6.61**, often as Z-, E-mixtures [739, 1213] (Figure **6.54**). A few reactions conducted with ethers in which R" = PhCH$_2$OCH$_2$ have given good selectivities [739, 740] (Figure **6.54**), as have some intramolecular reactions [1215]. The reactions of α-substituted aldehydes may also be highly selective [1216], as shown by the example in Figure **6.54**.

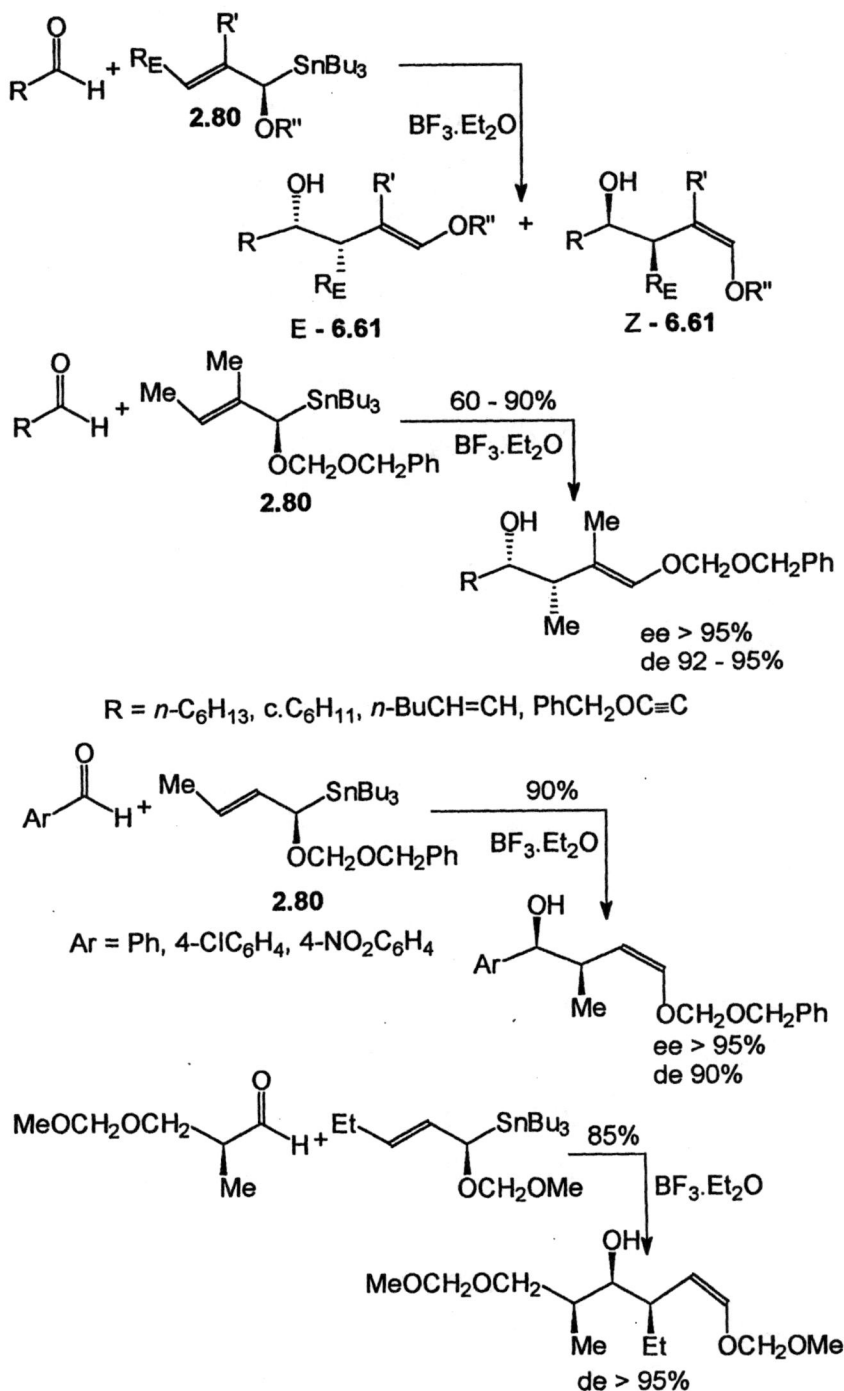

E - 6.61 Z - 6.61

60 - 90%
BF₃.Et₂O

ee > 95%
de 92 - 95%

R = n-C₆H₁₃, c.C₆H₁₁, n-BuCH=CH, PhCH₂OC≡C

90%
BF₃.Et₂O

Ar = Ph, 4-ClC₆H₄, 4-NO₂C₆H₄

ee > 95%
de 90%

85%
BF₃.Et₂O

de > 95%

Figure 6.54

The results with α-oxyallylstannanes are interpreted by open transition-state models (Figure **6.55**). In the reactive conformer of the allylstannane, the C–Sn bond should be orthogonal to the allylic system and *anti* to the incoming electrophile. The approach of the aldehyde coordinated to the Lewis acid should be either antiperiplanar (models **A5** and **A6**) [253, 258, 739, 1213] or synclinal (model **A4**) [740, 1212]. In the synclinal approach, the formation of *syn* Z-alcohol involves the inside alkoxy effect as suggested by Houk [740] (Figure **6.55**). When the antiperiplanar approach is considered, the Z- or E-geometry of the allylstannanes does not influence the direction of stereoselectivity of the process, which is stereoconvergent toward *syn* alcohols [253, 258].

The reactions of aldehydes with γ-alkoxystannanes **2.81** are catalyzed by BF$_3$·Et$_2$O, and they are more selective toward *syn* E-homoallyl alcohols **6.62** than the previous examples [1214, 1217] (Figure **6.56**). This holds for the reactions with prochiral or α-chiral aldehydes provided that the reagents are matched [569]. Transition-state model **A5** accounts for this selectivity. This method has been applied by Yamamoto and coworkers [1223] to the reaction of aromatic aldehydes with γ-(tetrahydropyranyloxy)-allylstannanes under AlCl$_3$ catalysis.

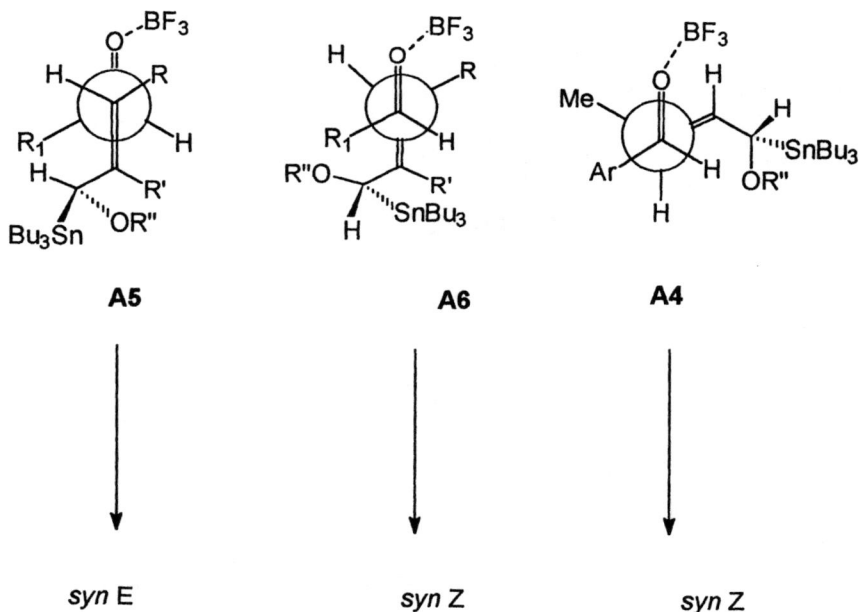

Figure 6.55

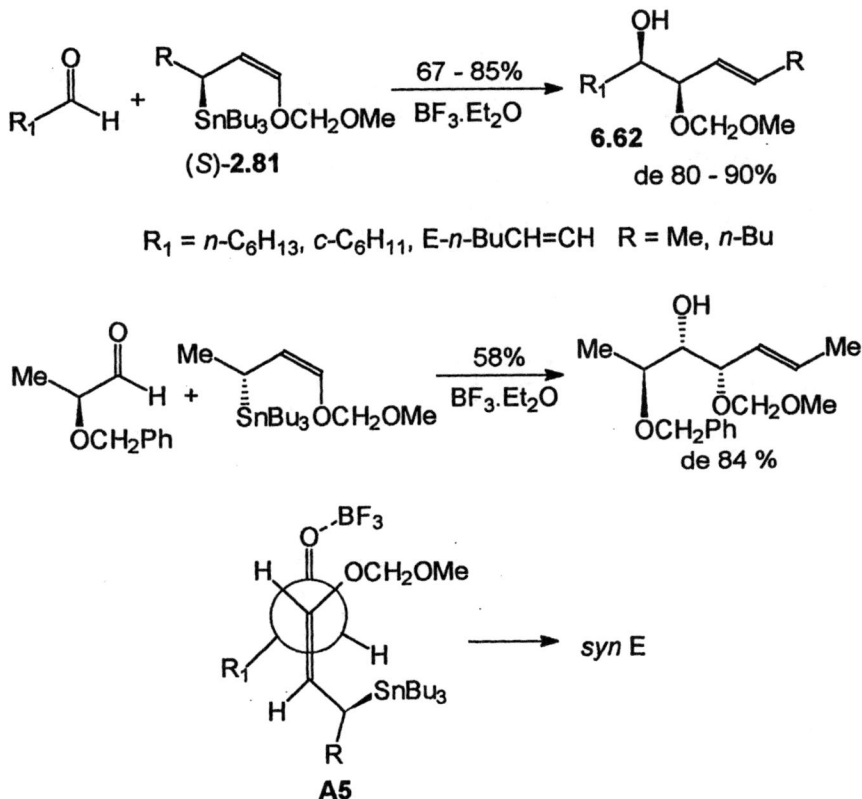

Figure 6.56

6.6.2 Reactions with Functionalized Aldehydes and Ketones

The reaction of allylstannanes with α-alkoxyaldehydes bearing a modified carbohydrate residue **1.103** (R = PhCH$_2$, R' = CH$_2$CHO) is highly stereoselective when mediated by MgBr$_2$ at −60°C. After exposure of the ether products to allyl alcohol in an acidic medium, Charette and coworkers [366, 1224] obtained the related allyl alcohols with a high enantiomeric excess (Figure **6.57**). By using the triisopropylsilyloxy ether **1.103** (R = i-Pr$_3$Si, R' = CH$_2$CHO), the isomeric alcohol is obtained with a lower selectivity. It is thought that a chelation-controlled reaction takes place in the first case, but not in the second one. From the related α-isomer **1.104** (R' = CH$_2$CHO), a high stereoselectivity is only observed if R = PhCH$_2$O [366]. The reaction of prochiral benzyloxyacetaldehyde with enantioenriched allylsilanes **6.63** can lead either to *anti* or to *syn* homoallyl alcohols, depending on whether the Lewis acid catalyst favors chelation control (MgBr$_2$) or not (BF$_3$•Et$_2$O) [1225].

The TiCl$_4$ or SnCl$_4$ catalyzed reactions of allylsilanes with glyoxylates of chiral alcohols **1.23** (R = H, Ph) or with α-ketoamides derived from C$_2$ symmetrical amines **1.65** are stereoselective [66, 147, 243, 292, 1226]. The most selective auxiliaries are (1R,2S,5R)-phenmenthol **1.4** (R = Ph), *trans*-(1R,2S)-2-phenylcyclohexanol **1.5** (R = Ph) [66], alcohol **1.49** [243], and **1.65** (R = TBDMSOCH$_2$) [292] (Figure 6.57). The reagent attacks the *Re* face of the ketone held in a cyclic, bidentate chelate with the other carbonyl group and the Lewis acid. Similar reactions are observed with (1R,2S,5R)-phenmenthylglyoxylate **1.23** (R = H) and vinylsilanes [1227] or E-crotyltributylstannane [253, 258] (Figure 6.57).

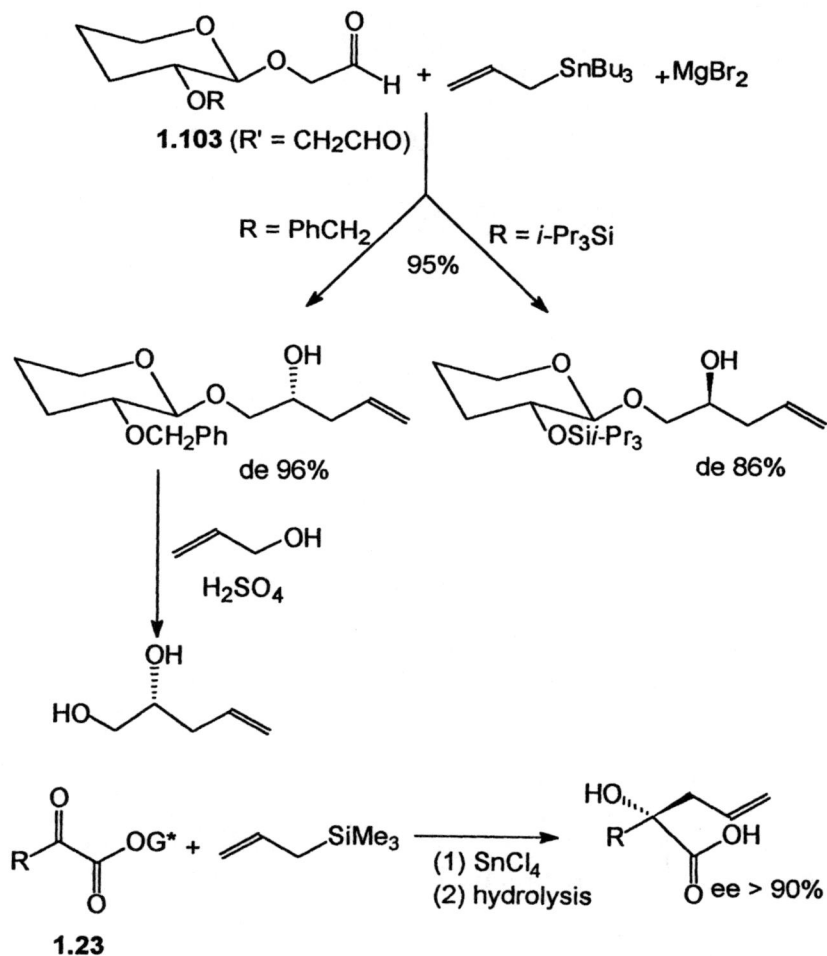

Figure 6.57

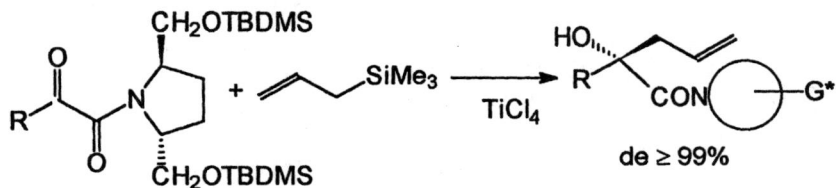

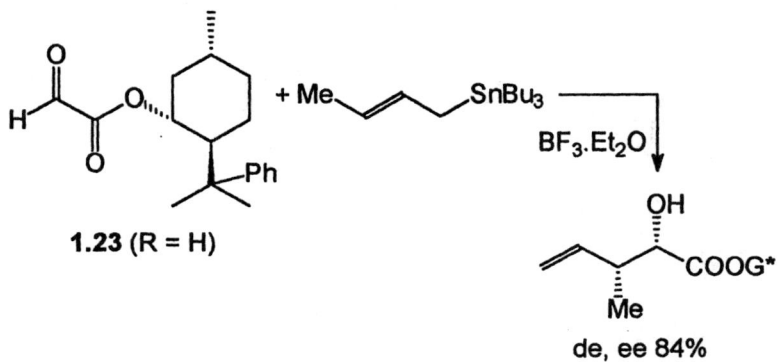

1.23 (R = H)

de, ee 84%

Figure 6.57 (continued)

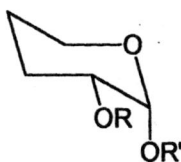

1.104

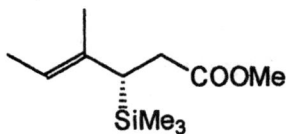

6.63

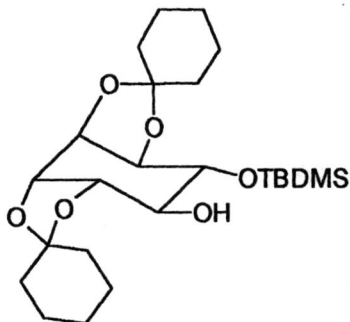

1.49

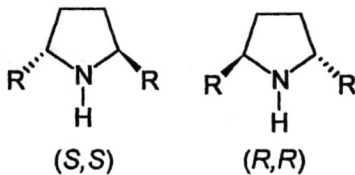

1.65

6.6.3 Reactions with Acetals and Related Compounds

As shown in racemic series [220], the stereoselectivity of the allylation of chiral 1,3-dioxanes **6.20** depends upon the reaction conditions and upon the nature of the reagent (silane or stannane). The most selective reagent is $CH_2=CHCH_2SnBu_3$. Most of the reactions have been carried out with trimethylallylsilane in the presence of $TiCl_4/Ti(Oi\text{-}Pr)_4$ under the conditions recommended by Johnson and coworkers [213]. After oxidation of the products and treatment with a base, nonracemic homoallylic alcohols are obtained with excellent enantiomeric excesses (Figure **6.58**). From $CH_2=C=CHCH_2SiMe_3$, dienyl alcohols are similarly obtained [1228, 1229] (Figure **6.58**). The $TiCl_4$-catalyzed reactions of dioxanes **6.20** with alkynylsilanes also lead to nonracemic α-alkynyl alcohols [223].

When the reactions are performed with α-chiral aldehydes, notably in the steroid series [213, 224], a single homoallylic alcohol is obtained from either an allylsilane or -stannane when the aldehyde and the dioxane substituents are matched, for example from (R,R)-**6.20**. In contrast, if the substituents are mismatched, the allylsilane gives one diastereoisomer and the allylstannane gives the other, although with modest selectivities. The interpretation of these results hinges upon the mechanism of dioxane ring cleavage [1230]. When the nucleophile is allylsilane, the Lewis acid induces the cleavage of the C-2–O-3 dioxane bond prior to the formation of the new C–C bond. Thus, stereoselection depends only upon the steroid structure. When the reagent is an allylstannane, C-2–O-3 bond breaking and C–C formation are concerted, and therefore the dioxane configuration also influences the stereoselection. Similar results are observed with alkynylsilanes and -stannanes [213] (Figure **6.58**).

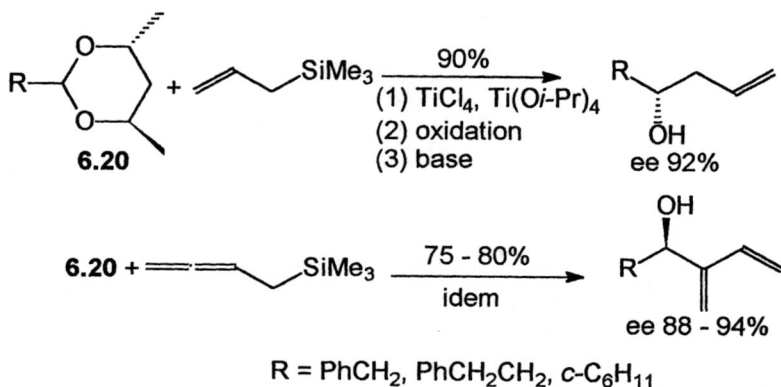

R = $PhCH_2$, $PhCH_2CH_2$, $c\text{-}C_6H_{11}$

Figure 6.58

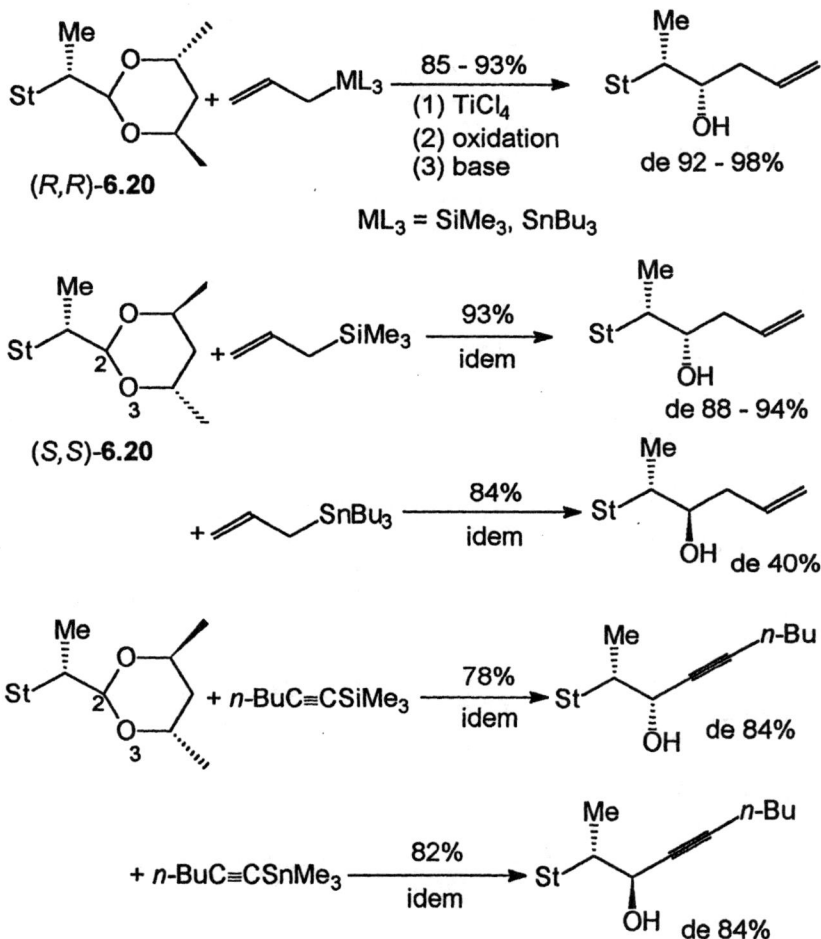

Figure 6.58 (continued)

Seebach and coworkers [386] have also prepared nonracemic homoallyl alcohols with a high selectivity by reacting allylsilane with dioxanone **1.110** under TiCl$_4$ or TiCl$_3$Oi-Pr catalysis, followed by treatment with LDA. Meyers and Burgess [349, 350] used chiral bicyclic lactam **1.90** (R = Ph, n = 1). After suitable treatment, a 5-allylpyrolidin-2-one is obtained with a good selectivity (Figure **6.59**). Less selective reactions are observed with **1.90** (R ≠ H). Tietze and coworkers [1231] performed the asymmetric allylation of aldehydes in the presence of the trimethylsilyl ether of (1R,2R)-N-trifluoroacetylnorpseudoephedrine at −78°C, under Me$_3$SiOTf catalysis (Figure **6.59**). After treatment of the products with Na/NH$_3$, homoallyl alcohols were obtained with a high enantioselectivity. These reaction probably occur through the formation of an intermediate oxazolidinium ion **6.64**. However, with PhCHO or electron-poor aromatic aldehydes, the selectivity is lower.

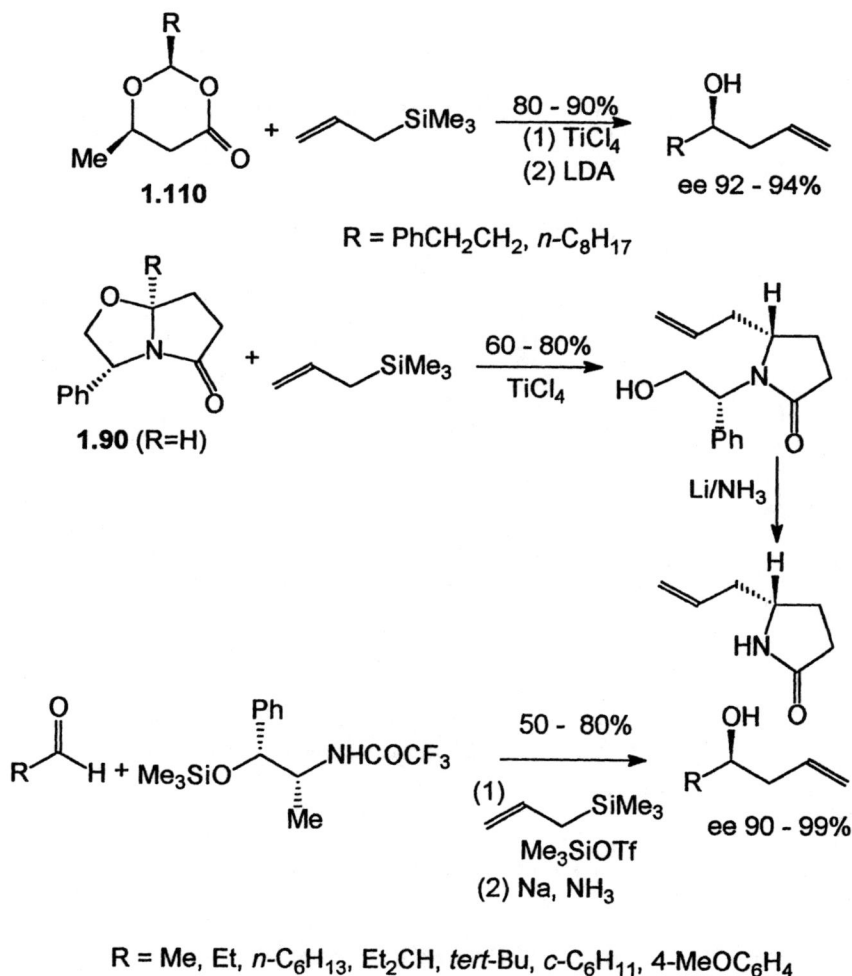

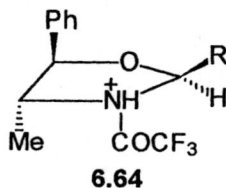

R = PhCH$_2$CH$_2$, n-C$_8$H$_{17}$

R = Me, Et, n-C$_6$H$_{13}$, Et$_2$CH, tert-Bu, c-C$_6$H$_{11}$, 4-MeOC$_6$H$_4$

6.64

Figure 6.59

Johnson and coworkers conducted cyclizations of trimethylsilyl-substituted polyenes bearing chiral 1,3-dioxolane or 1,3-dioxane groups under $SnCl_4$ catalysis [213]. The highest selectivities occurred with propargyltrialkylsilanes systems (§ 6.6.5).

6.6.4 Reactions with Imines and Acylimminium Ions [1232]

The reaction of chiral imines derived from 1-phenethylamine with allyl-9-BBN **2.19** (R = CH_2=CHCH$_2$) or CH_2=CHCH$_2$SnBu$_3$ under $TiCl_4$ catalysis is *syn* selective [258], in agreement with the Felkin-Anh model (§ I.3). After hydrogenolysis, nonracemic amines are obtained with a high enantiomeric excess [258] (Figure **6.60**). In contrast, the *anti* isomer is formed from methallyl-9-BBN **2.19** (R = CH_2=C(Me)-CH$_2$) (Figure **6.60**). These results have been interpreted by quasi-chair or quasi-boat transition-state models, respectively (§ I.6). Kunz and Laschat [248, 361] recommended *N*-arylimines bearing carbohydrate residues **1.101** and **1.102**, and the reactions of these imines with CH_2=CHCH$_2$SiMe$_3$ require a Lewis acid catalyst ($SnCl_4$ is preferred). According to the auxiliary used, (*R*)- or (*S*)-amines are obtained with a good selectivity, unless the aryl group is 2-substituted. The approach of the reagent occurs from the least hindered side of a chelated $SnCl_4$ complex **6.65** (Figure **6.60**).

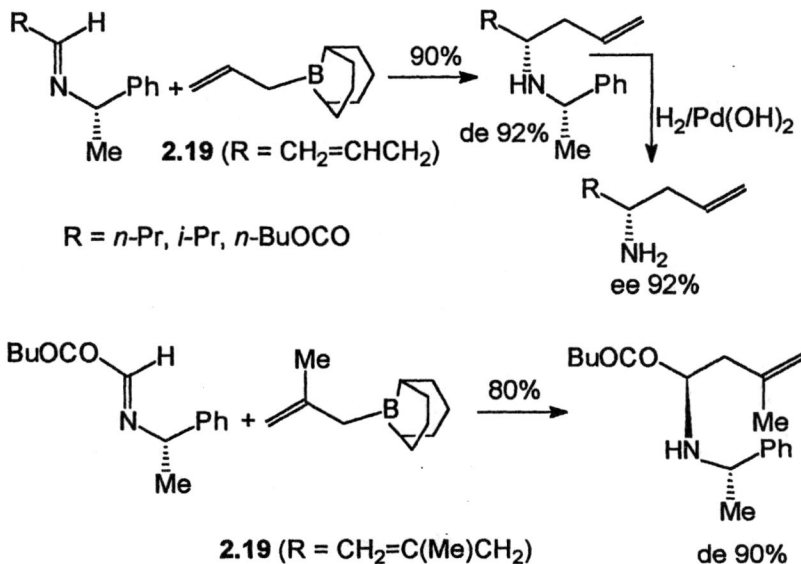

Figure 6.60

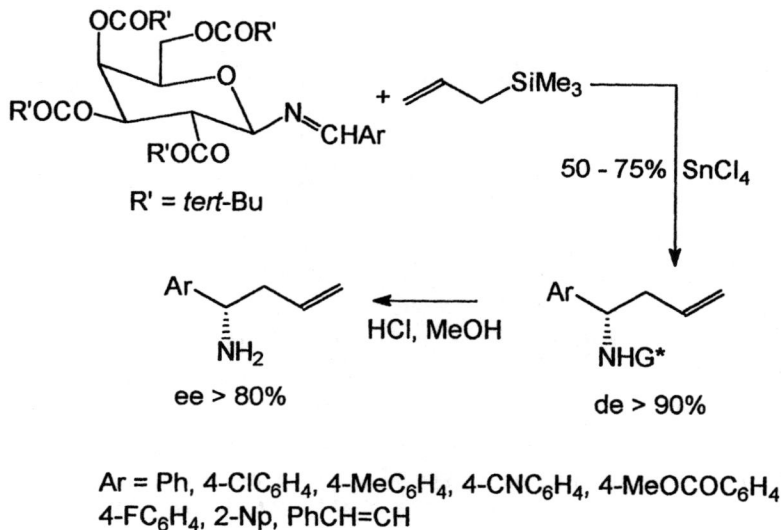

Ar = Ph, 4-ClC$_6$H$_4$, 4-MeC$_6$H$_4$, 4-CNC$_6$H$_4$, 4-MeOCOC$_6$H$_4$
4-FC$_6$H$_4$, 2-Np, PhCH=CH

Figure 6.60 (continued)

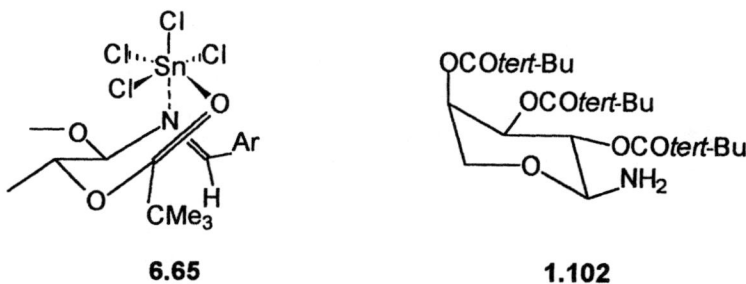

Imines generated from aliphatic aldehydes do not react under these conditions. The reactions of these imines with CH$_2$=CHCH$_2$SnBu$_3$/SnCl$_4$ lead to α- and β-epimeric mixtures in low chemical yields [361].

Additions of trimethylallylsilane to cyclic imminium salts **6.66** (R = Me or MeO) generated from α-hydroxylactams in the presence of SnCl$_4$ have been studied by Polniaszek [254], Fujisawa and their coworkers [1233]. Starting from **6.66** (Ar = 2,6-Cl$_2$C$_6$H$_3$ or Cl$_5$C$_6$, R = Me), the reaction leading to **6.67** is highly stereoselective. After treatment of **6.67** with HCOONH$_4$ in the presence of Pd/C, nonracemic 2-propylactams are obtained with a high enantioselectivity [254]. When Ar = C$_6$H$_5$ and R = Me, the other diastereoisomer is formed with a lower selectivity (Figure **6.61**). Treatment of the intermediate product with Na/NH$_3$ generates 2-allyllactam **6.68**. From **6.66** (Ar = Ph, R = MeO, n = 1) Fujisawa and co-workers [1233] obtained **6.67** (Ar = Ph, R = MeO, n = 1) by treating **6.66** with allylsilane and SnCl4 in CHCl3 at –60°C, but the selectivity was lower (de = 84%). By using a different Lewis acid, the other isomer predominated, but the stereoselectivity was lower still (60%).

Massy-Westropp and coworkers [164] described the reactions of allylstannanes with (1*R*,2*S*,5*R*)-phenmenthyl α-bromoaminoacetate **6.69**. These reactions, which seem to take place via a radical pathway, are highly selective. After hydrolysis, *N*-protected α-aminoacids are obtained with excellent enantiomeric excesses (Figure **6.61**).

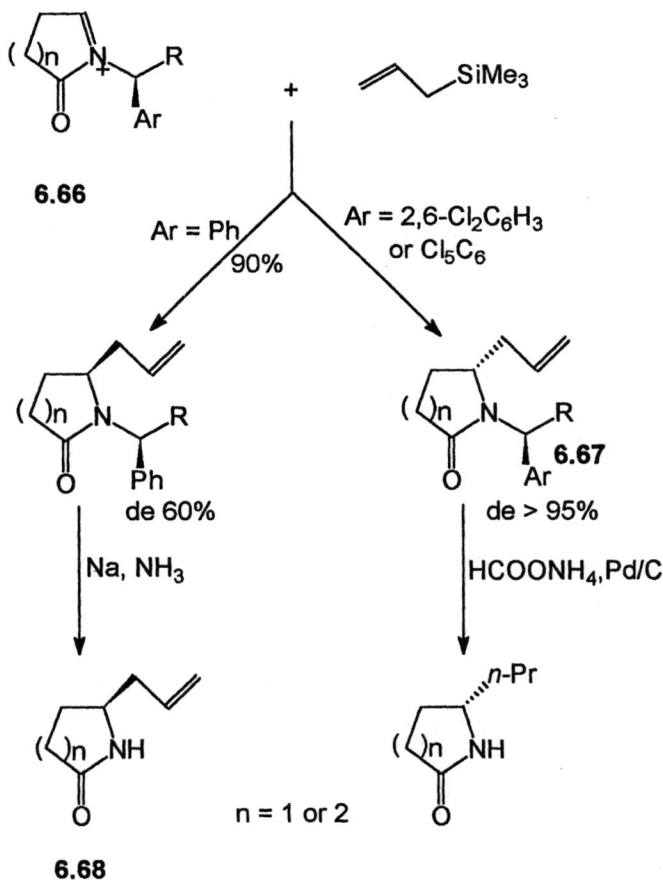

Figure 6.61

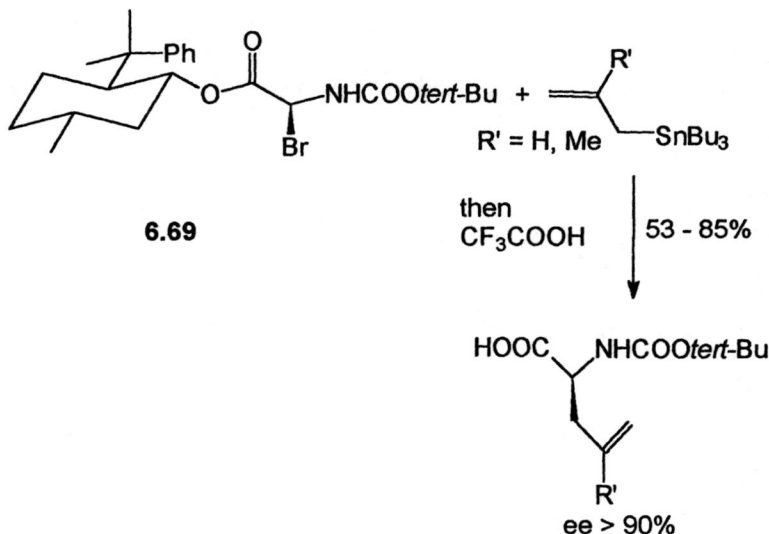

Figure 6.61(continued)

6.6.5 Reactions of Allenyl- and Propargylboranes, -silanes and -stannanes

The reactions of allenyl and propargyl derivatives of boron, silicon or tin with carbonyl compounds or acetals take place with double-bond migration ($S_{E'}$ reactions). Therefore, allenyl derivatives will lead to homopropargyl alcohols and propargyl analogs to allenyl alcohols.

Yamamoto and coworkers [169] generated allenylboronates **6.70** *in situ* from dialkyl tartrates. The reactions of **6.70** with aliphatic aldehydes lead to homopropargyl alcohols **2.77** with an excellent enantioselectivity provided that R' is bulky enough (R' = $(Me_2CH)_2CH$) (Figure **6.62**). Aromatic aldehydes, however, give less satisfactory results. Corey and coworkers [423, 735] also prepared homopropargyl- or homoallenylalcohols **2.77** and **2.76** with a very high enantioselectivity from reagents **2.62** (X = $CH_2=C=CH$ or $HC\equiv CCH_2$) obtained from the bromoborane precursors **2.62** (X = Br) (Figure **6.62**). Both enantiomers of **2.62** (X = Br) are available, so (R)- or (S)-alcohols can be prepared. The chiral diamine precursor of **2.62** (R = Br) is easily recycled. As in the related reactions described above, these selectivities are interpreted by cyclic transition-state models.

The reactions of allenyl- and propargylsilanes with functionalized aldehydes bearing chiral residues have also been studied. Grée and coworkers observed high diastereoselectivity in the reaction of aldehyde **6.71** bearing a butadieneiron carbonyl residue with 1-pentyl-1-trimethylsilylallene under TiCl$_4$ catalysis (Figure **6.63**). After decomplexation with Ce(NH$_4$)$_2$(NO$_3$)$_6$ and partial reduction of the triple bond of the resulting homopropargyl alcohol, compound **6.72** (a precursor of leukotrienes) was obtained with an excellent enantiomeric excess [536] (Figure **6.63**).

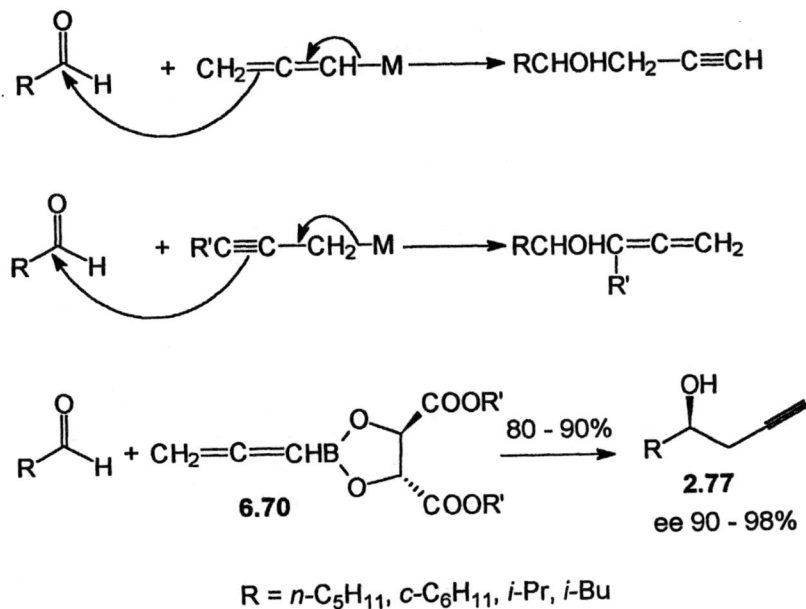

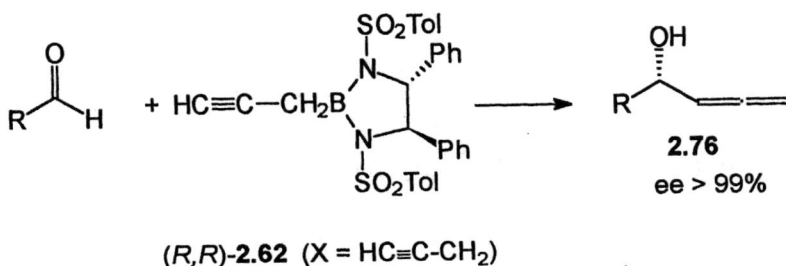

R = n-C$_5$H$_{11}$, c-C$_6$H$_{11}$, i-Pr, i-Bu

(R,R)-**2.62** (X = HC≡C-CH$_2$)

R = n-C$_5$H$_{11}$, Ph, PhCH=CH, tert-Bu

Figure 6.62

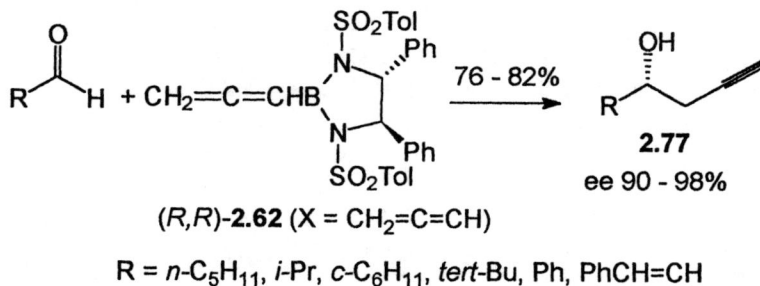

(R,R)-**2.62** (X = CH$_2$=C=CH)

R = n-C$_5$H$_{11}$, i-Pr, c-C$_6$H$_{11}$, $tert$-Bu, Ph, PhCH=CH

Figure 6.62 (continued)

Marshall and coworkers [1234] examined the reactions of chiral allenylstannanes **6.73** with prochiral or α-chiral aldehydes. These reagents are prepared from nonracemic propargyl alcohols, which are in turn easily obtained by asymmetric reduction of alkynylketones (§ 6.1.1). The most efficient catalyst for addition to prochiral aldehydes is BF$_3$•Et$_2$O (Figure **6.63**). An acyclic transition state analogous to model **A$_5$** proposed for allylsilanes (Figure **6.55**) accounts for the observed selectivity. From α-chiral aldehydes bearing an ether functionality, the use of MgBr$_2$ as a catalyst may lead to high selectivities (Figure **6.63**). In this case, a chelated transition state **6.74** is proposed to interpret the results [1234]

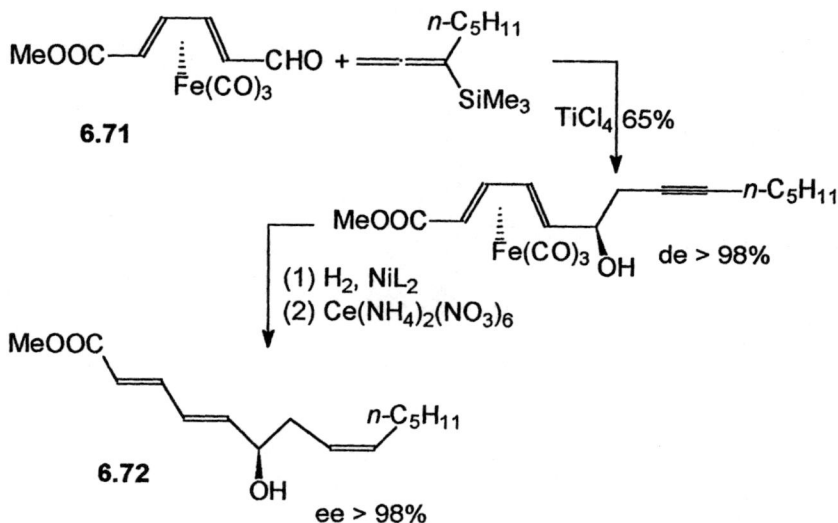

Figure 6.63

$$R = i\text{-Pr, } tert\text{-Bu} \quad R' = n\text{-}C_7H_{15}$$

6.73

6.73 (R' = Et)

de > 98%

89%

BF$_3$·Et$_2$O

95%

MgBr$_2$

de > 99%

de 98%

6.74

6.69

then H$_3$O$^+$ 53%

ee 93%

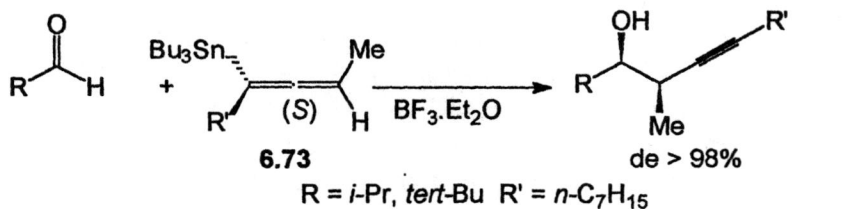

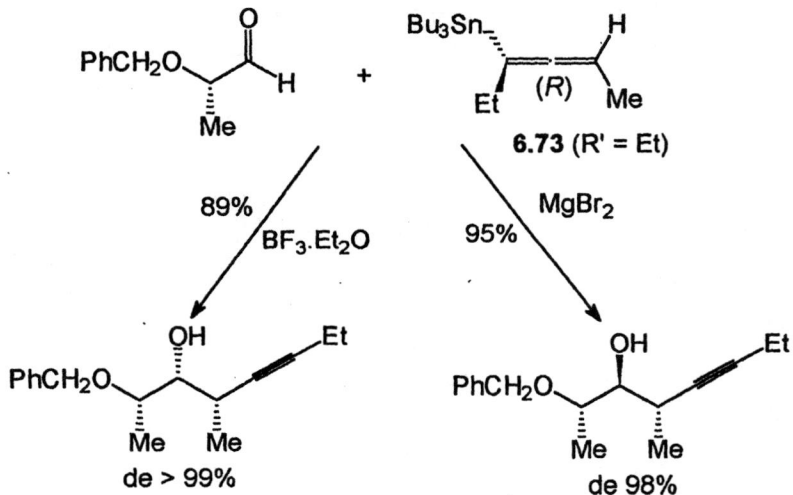

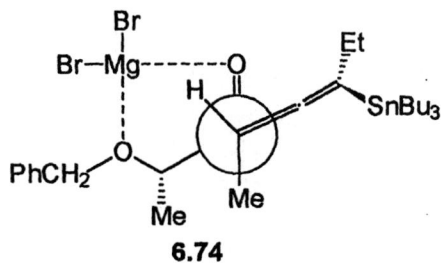

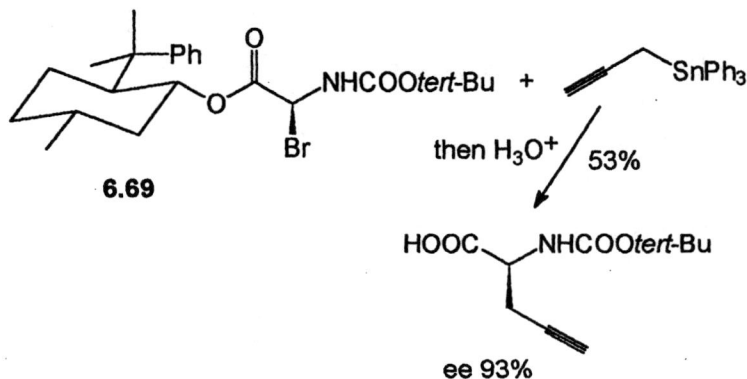

Figure 6.63 (continued)

The reaction of chiral bromoester **6.69** with HC≡CCH₂SnPh₃ takes place without transposition of the propargyl group [164]. After hydrolysis, the N-protected α-aminoacid is obtained with a high enantiomeric excess (Figure **6.63**). One of the most elegant applications of the reaction of propargylsilanes with chiral acetals is due to Johnson and coworkers [213, 1235], who performed Lewis-acid-catalyzed bi- and tetracyclizations of polyenes with a high stereoselectivity as a route to making steroids (Figure **6.64**).

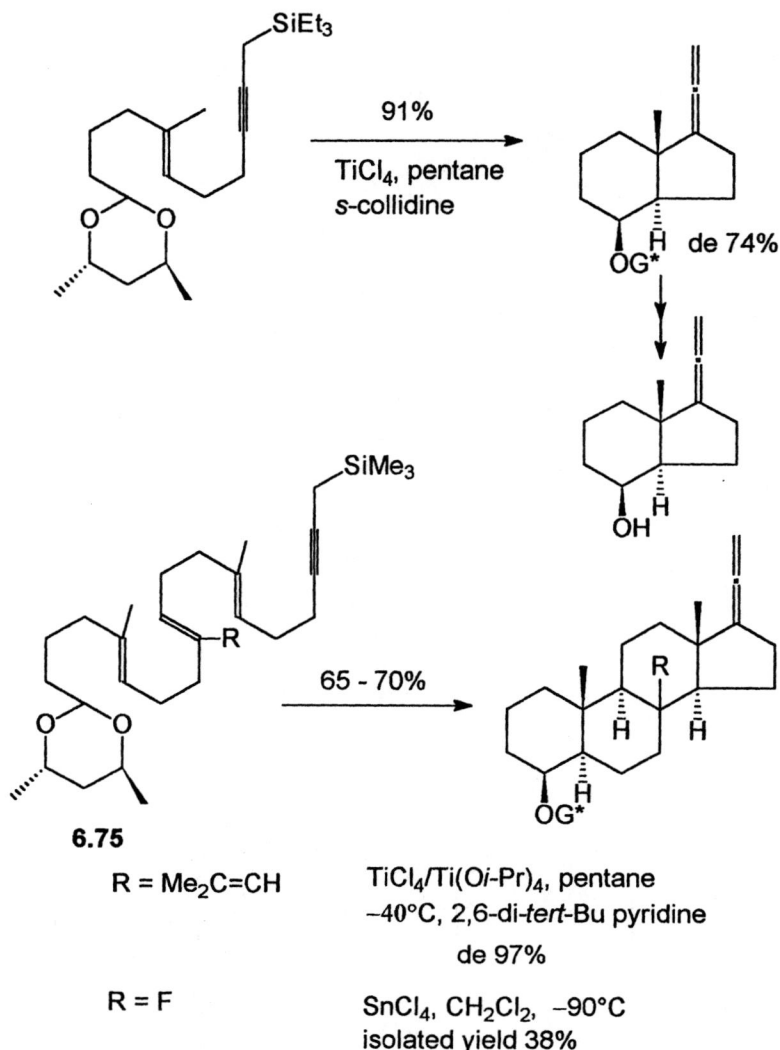

Figure 6.64

To obtain high chemical yields and diastereoselectivities, it is necessary to introduce a removable cation-stabilizing group (R = Me$_2$C=CH or F) in the proper position on the polyenic skeleton of **6.75** (Figure **6.64**). As usual, the acetal auxiliary is cleaved by oxidation and β-elimination (§ 1.1.2). These cyclizations are carried out at low temperature (−40 to −90°C), and the Lewis acid is either TiCl$_4$/Ti(Oi-Pr)$_4$ or SnCl$_4$.

6.7 ENE REACTIONS [1236, 1237]

The ene reaction provides another route to homoallylic alcohols. The additions of highly reactive aldehydes to olefins bearing a C–H bond in the allylic position are catalyzed by Lewis acids. Intramolecular ene reactions do not require a highly activated aldehyde. Asymmetry can be induced either by the use of a chiral catalyst or by introduction of a chiral auxiliary on the aldehyde.

6.7.1 Chiral Lewis Acid Catalyzed Reactions [819, 1236]

Yamamoto and coworkers [1238] recommended the use of the chiral organoaluminum complex derived from 2,2'-triphenylsilylbinaphthol **3.7** (R = Ph$_3$Si) to catalyze the reaction of haloaldehydes with 2-phenylthio- or 2-methylpropene. When this reaction is run in the presence of molecular sieves, the Lewis acid may be used in catalytic amounts. Homoallyl alcohols are obtained with high yield and selectivity (Figure **6.65**). Mikami, Nakai and coworkers [816, 820, 1239, 1240] proposed dihalotitanates prepared *in situ* from (*R*)- or (*S*)-binaphthol **3.7** (R = H) and Ti(Oi-Pr)$_2$X$_2$ (X = Br or Cl) as catalysts. At −30°C in the presence of molecular sieves, these reagents promote asymmetric ene reactions from fluoral or alkylglyoxylates and terminal olefins with a high enantioselectivity (Figure **6.65**). Some isomerization to the corresponding allylic alcohols takes place when the reaction is run with CF$_3$CHO, while the reaction of chloral gives poor enantioselectivities [1239]. If the olefin is dissymmetrical or if Z- and E-isomers can be generated, mixtures are usually formed, as illustrated in Figure **6.65** [819].

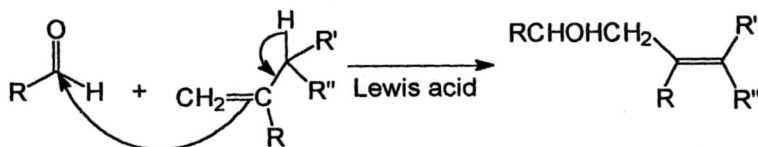

Figure 6.65

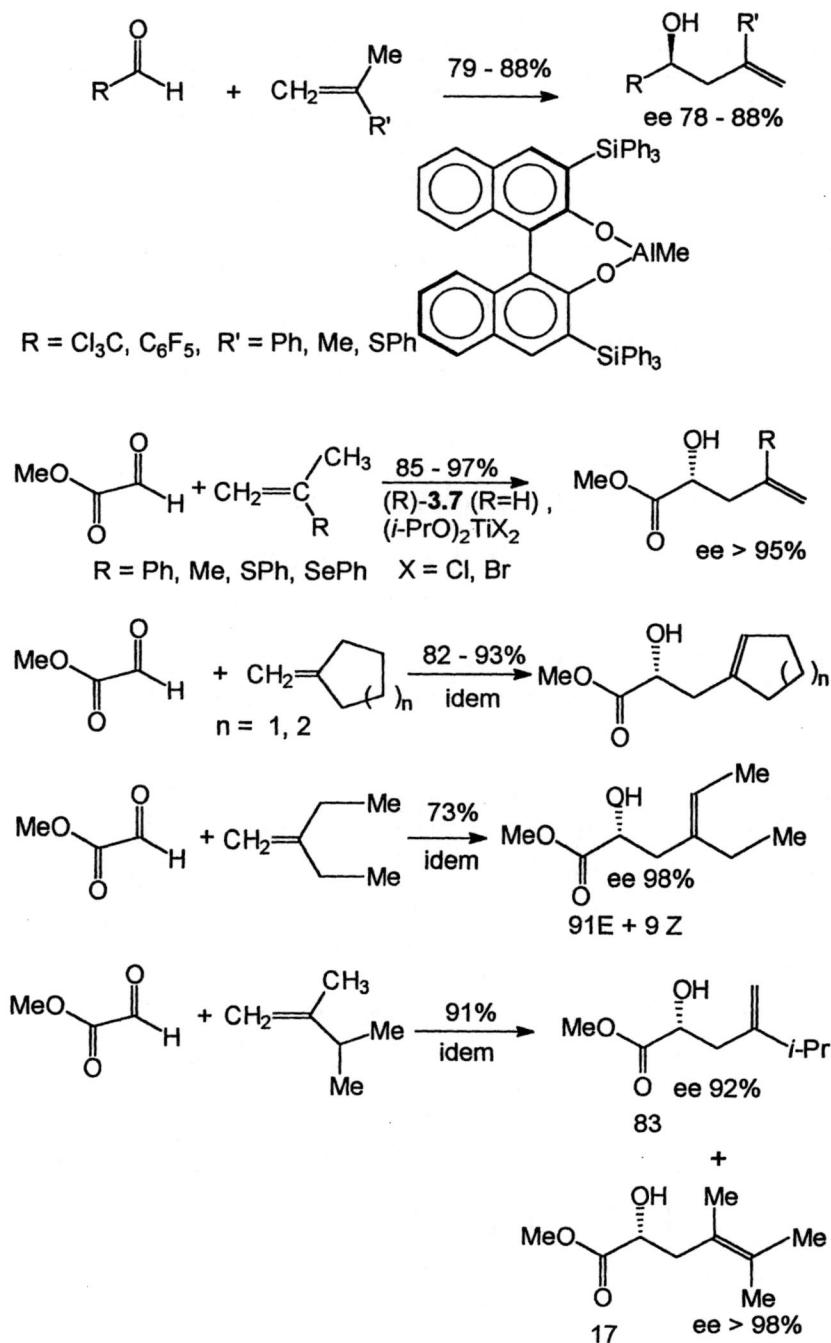

R = Cl₃C, C₆F₅, R' = Ph, Me, SPh

R = Ph, Me, SPh, SePh   X = Cl, Br

n = 1, 2

ee 78 - 88%

ee > 95%

82 - 93%
idem

73%
idem

ee 98%
91E + 9 Z

91%
idem

ee 92%
83

+

17   ee > 98%

Figure 6.65 (continued)

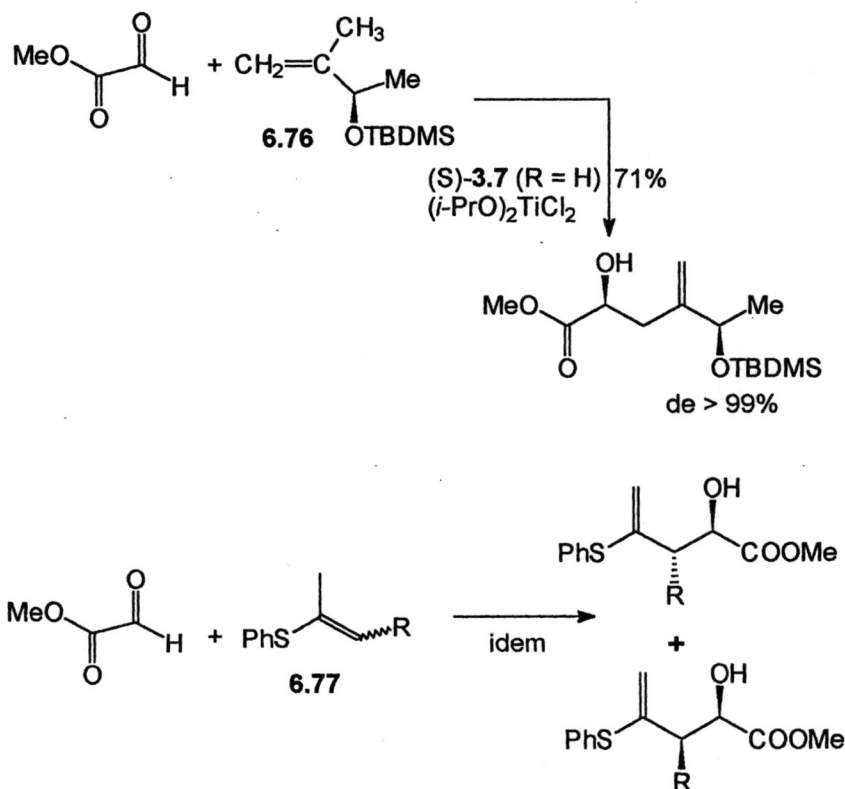

Figure 6.66

Double diastereodifferentiation (§ I.6) can take place when an α-chiral olefin such as **6.76** is used, and a single regio- and stereoisomer is obtained if the reagent and the Lewis acid are matched (Figure **6.66**). In contrast, when the (*R*)-binaphthol derived catalyst is used instead of the (*S*)-enantiomer, a mixture of stereoisomers is formed. Mikami and coworkers performed the reactions of Z- or E-vinylsulfides **6.77** with methyl glyoxylate, and they usually obtained mixtures of *syn*- and *anti*-α-hydroxy-β-alkyl esters [1240]. This method has been extended by Van der Meer and Feringa to the reactions of exocyclic olefins [1241]. Ene cyclizations are also catalyzed by other binaphthol-derived catalysts. The use of excess (*R*)-binaphthol zinc alcoholate promotes the cyclization of terpene aldehydes **6.78** provided that the R groups are methyl [819] (Figure **6.67**). Mikami and Nakai's catalyst can also give good results in such cyclizations if AgClO$_4$ or AgOTf is added as a cocatalyst [819, 822, 1236, 1242] although *cis / trans* mixtures are occasionally obtained (Figure **6.67**).

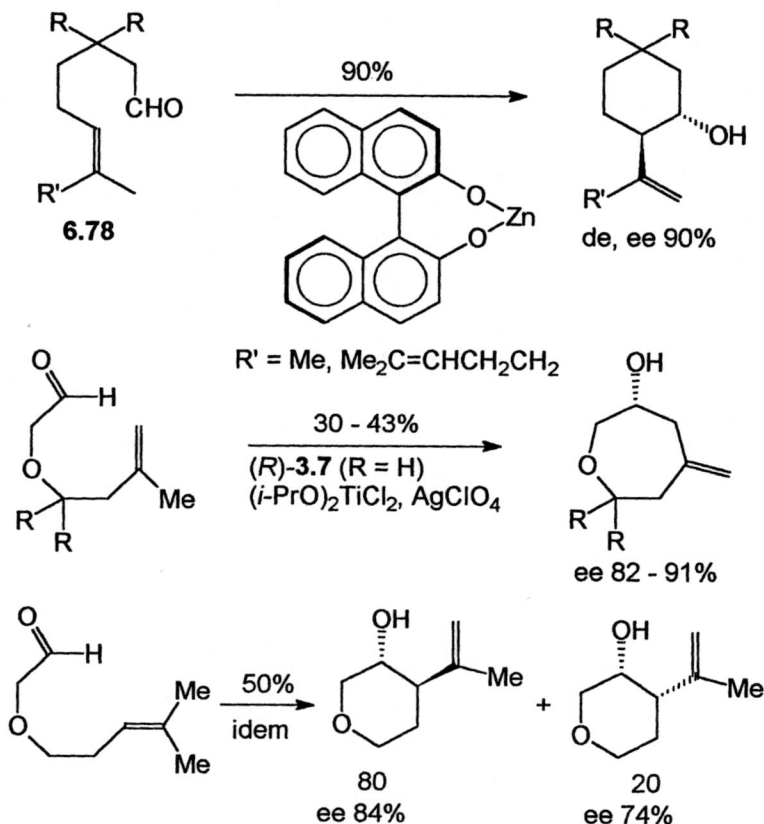

Figure 6.67

6.7.2 Reactions of Chiral Glyoxylates and Derivatives [66, 147]

The reactions of terminal olefins with chiral glyoxylates **1.23** (R = H) derived from phenmenthol **1.4** (R = Ph), *trans*-2-phenylcyclohexanol **1.5**, or alcohol **1.7** (R = *tert*-BuCH₂O) take place in the presence of stoichiometric amounts of SnCl₄ or TiCl₄. These reactions are highly stereoselective (de > 99%) (Figure **6.68**). Most surprisingly, esters of **1.4** (R = Ph) and **1.5** that have comparable configurations lead to opposite diastereoisomers. These divergent results are not yet understood [66]. Similar reactions occur with (1*R*,2*S*,5*R*)-phenmenthyl glyoxylate *N*-benzylimine [1236] or better yet with its tosylimine [1243] (Figure **6.68**), and lead to precursors of aminoacids. In most of the cases, the stereoselectivity of these reactions is interpreted by chelation control when SnCl₄ or TiCl₄ are the Lewis acids [1236].

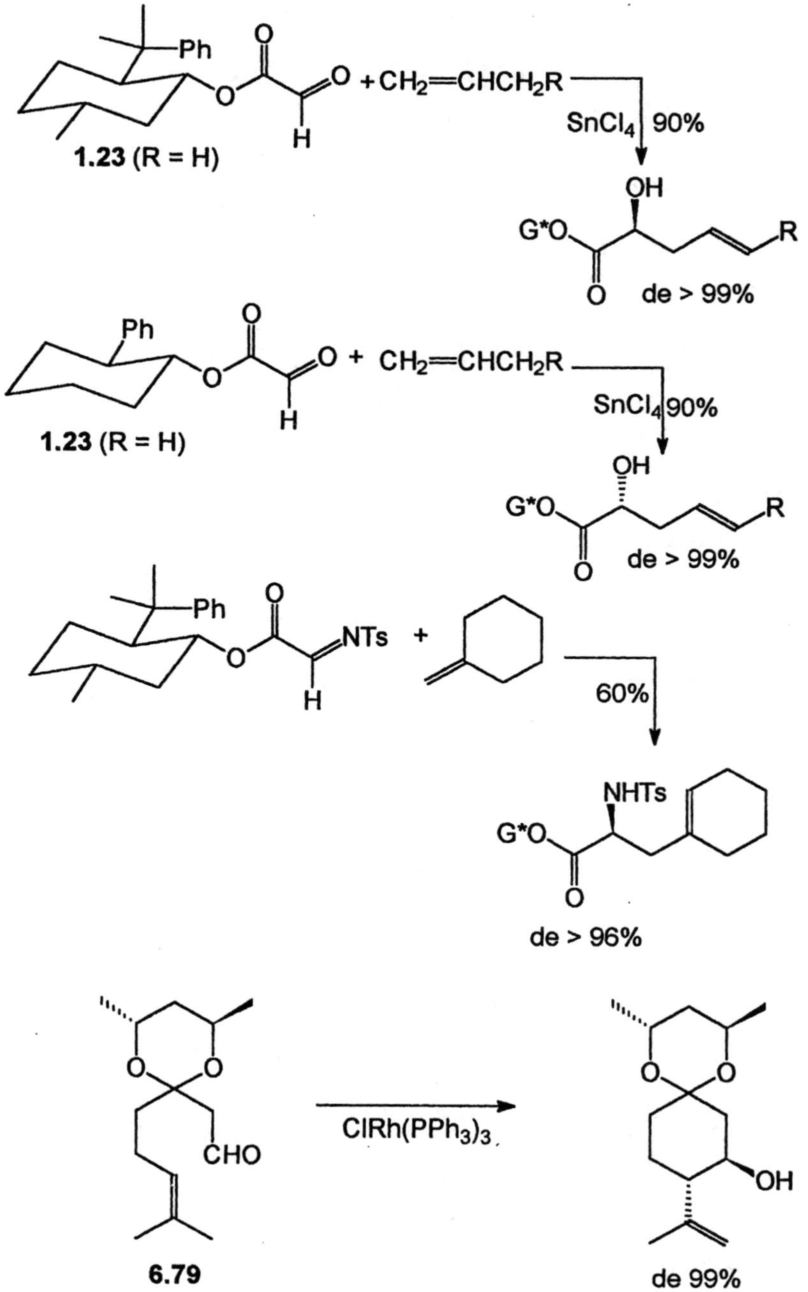

1.23 (R = H)

de > 99%

1.23 (R = H)

de > 99%

de > 96%

6.79

de 99%

Figure 6.68

Rhodium (I)-catalyzed cyclizations of 6-octen-1-als bearing a (4R,6R)-di-methyl-1,3-dioxane group at different positions do not lead to hydroacylation (§ 7.5.3) products, but lead instead to the products of formal ene reaction reactions [1244]. 3-Substituted aldehyde **6.79** (R,R = 4R,6R-1,3-dioxolanyl) gives *trans*-cyclohexanols with a high selectivity (Figure **6.68**), while 2-substituted analogs cyclize to *cis*-cyclohexanols in lower selectivities.

6.8 REACTIONS OF BORON AND METAL ENOLATES [160, 209, 253, 407, 408]

The reactions of enolates with aldehydes (aldol reactions) or with imines have been widely developed since the 1970s. Asymmetric aldol-type reactions are very important in the multistep synthesis of complex molecules such as ionophores or β-lactam antibiotics. Chirality has been introduced either on the substituents of boron, on the metal ligands or on the carbon skeleton of the enolate. Aldol reactions are usually run at low temperatures, and when metal enolates are used, the reactions are sometimes easily reversible [160, 209].

6.8.1 Reactions with Nonfunctionalized and α-Unsaturated Aldehydes and Ketones

Depending on the enolate substituents, one or two new stereocenters can be created during an aldol reaction. A single new stereocenter is created when a 2-substituted enolate E_1 or E_2 is reacted with CH_2O or a symmetrical ketone R_2CO, or when a 2-unsubstituted enolate E_3 or E_4 is reacted with a prochiral aldehyde. When substituted enolates E_1 or E_2 react with prochiral aldehydes, two stereocenters are created and the *syn/anti* selectivity depends upon the Z- or E-geometry of the enolate [160, 209, 407, 408, 1016], the associated cation, and the reaction conditions [160, 209, 253, 408] (Figure **6.69**). As already mentioned (§ 5.3.2), priority is given to OM in Z, E descriptors regardless of the nature of R' or G*.

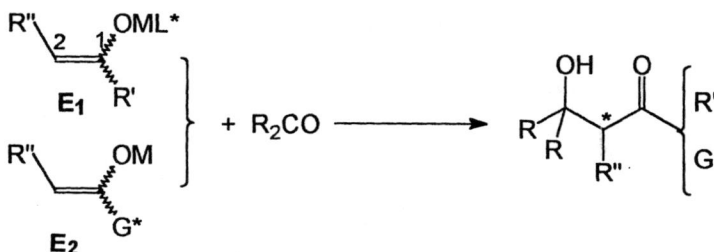

Figure 6.69

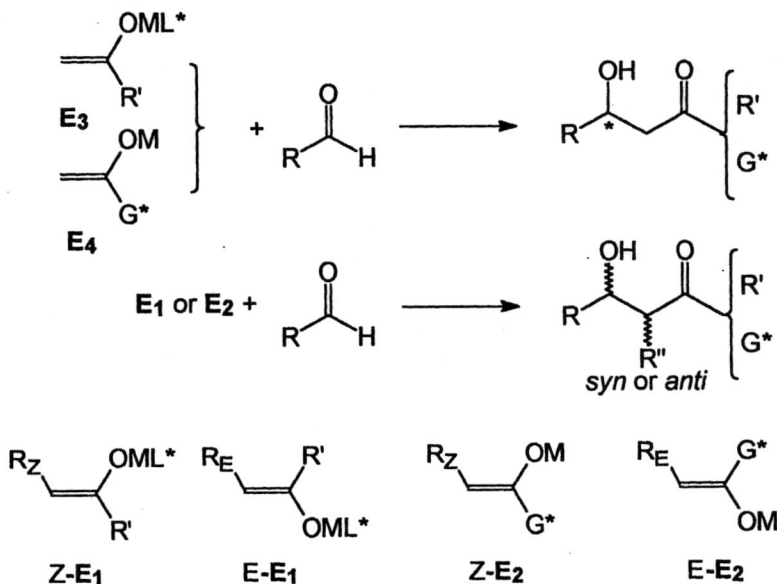

Figure 6.69 (continued)

Most metal enolates are generated by transmetalation from Li enolates. However, Ti-enolates can be formed by action of $TiCl_4/i\text{-}Pr_2NEt$ on carbonyl compounds [404, 1042] and Zr-enolates can be generated by similar reactions with $Zr(O\text{-}tert\text{-}Bu)_4$ [1245]. Lithium E-enolates are obtained by deprotonation of ketones or esters with a branched Li-amide (LDA, LICA, LOBA, LTMP) in a weakly polar medium (THF or THF-hexane), while Z-enolates are formed by using LDA or LHMDS in the presence of HMPA or DPMU [1016]. Tertiary amides always give Z-enolates, and difunctionalized derivatives such as Evans's oxazolidinones **5.30** and **5.31** are chelated to the metal prior to enolization.

Boron or tin (II) Z-enolates are generated by reaction with the corresponding triflates with a carbonyl compound in the presence of tertiary amines like $i\text{-}Pr_2NEt$ or N-ethylpiperidine (except when using dicyclopentylboron triflate [407]). E-Enolates are prepared by using dicyclohexyl- or other cyclic chloroboranes in the presence of Et_3N or Me_2NEt [407, 685, 686, 1246, 1247, 1248]. Because enolization does not take place under such conditions with esters or aliphatic tertiary amides, thiophenyl esters RCH_2COSPh have been used as ester/amide substitutes. Furthermore, Z-boron enolates of ketones can be prepared by conjugate addition of acid derivatives of dialkylboranes to α-enones [687].

Under kinetic control, the reactions of prochiral aldehydes with Z-enolates generally lead to *syn* aldols, while E-enolates lead to *anti* aldols. The presence of bulky R' groups on the enolates, however, may alter these selectivities. The highest diastereoselectivities are observed with boron or titanium enolates. These selectivity trends are interpreted by a concerted cyclic mechanism. The favored transition state resembles a distorted chair, in line with the Zimmermann-Traxler proposals [57, 160, 253] (Figure **6.70**). This model has been supported by theoretical studies [9, 40, 41, 125, 1249]. Transition states analogous to C_2 and C_4 (Figure **6.70**) are destabilized by 1,3-eclipsing interactions between the C–R, M–L and C–R' bonds, so that models C_1 and C_3 are more favorable. For the sake of simplification, only the reaction on one face of the enolates is shown in these models, but enolate face selectivity will be discussed later. In some cases, boatlike transition-state models are invoked to interpret selectivity inversions [401, 402, 666]. Moreover, Heathcock and coworkers [105] obtained evidence for the influence of an excess of n-Bu$_2$BOTf on the stereoselectivity of the aldol reactions of Z-enolates. In such reactions, *anti* aldols can be formed preferentially (see below).

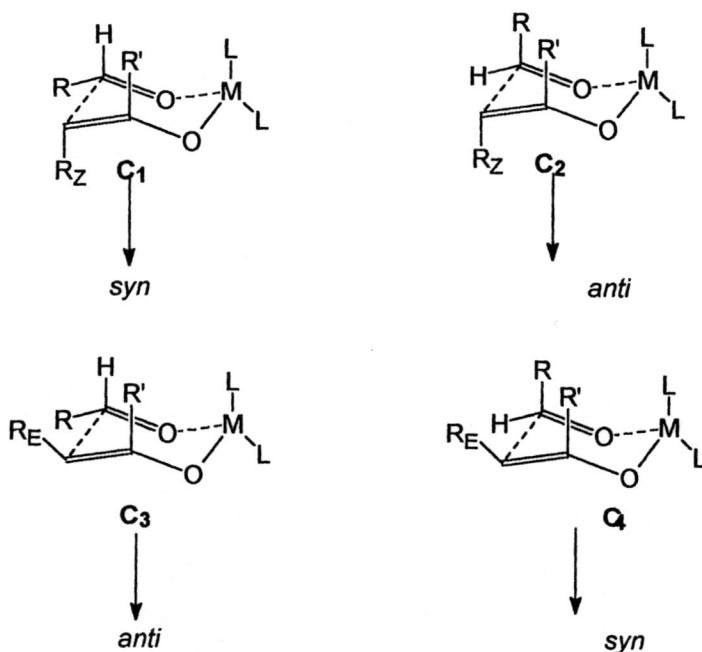

Figure 6.70

6.8.1.1 *Reactions of Enolates Bearing Chiral Ligands*

The reactions of lithium enolates generated by deprotonations of ketones with lithium amides of chiral diamines can be enantioselective [209, 557, 558]. Aggregates are often involved in these processes [558], but in most cases the observed selectivities are not very high. However, useful results were obtained by Koga and coworkers [561]. These authors prepared aldol **6.80** with a high ee, but poorer selectivities were observed with other aldehydes and other enolates (Figure **6.71**). Sparteine **2.5** was used as a zinc ligand as early as 1973 by Guetté and coworkers in the Reformatsky reaction [160, 1249a]. A highly enantioselective reaction was observed (Figure **6.71**) with benzaldehyde, but not with ketones. Soai and coworkers [1250] performed an enantioselective Reformatsky reaction of *tert*-Bu bromoacetate and PhCOMe at –15°C by using *N,N*-diallylnorephedrine **1.14** (R = CH₂=CHCH₂) as the zinc ligand (ee 74%). Tin (II) enolates, generated in the presence of enantiopure amines bearing the (*S*)-proline skeleton **2.13** (R = CH₂N(CH₂)₅ or CH₂NHAr), also promote highly enantioselective aldol reactions [253, 261, 559] (Figure **6.71**).

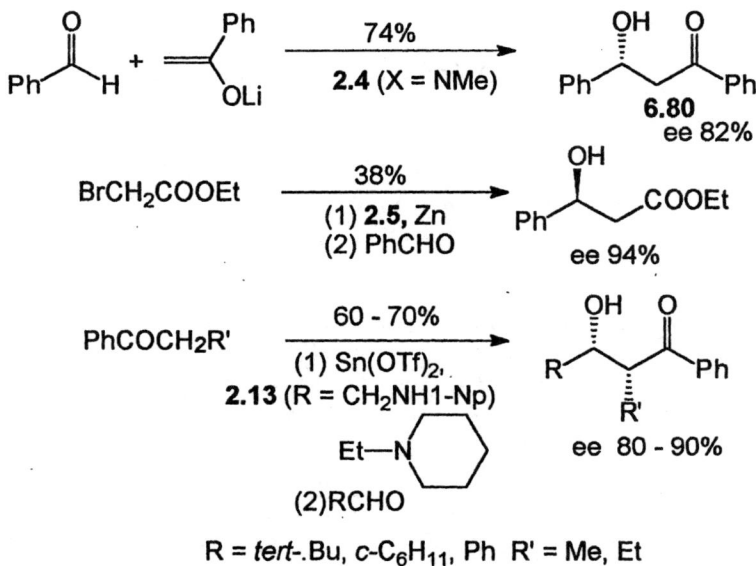

Figure 6.71

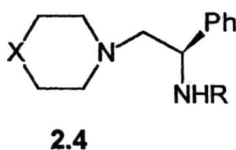

2.4

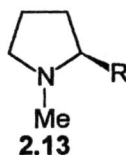

2.13

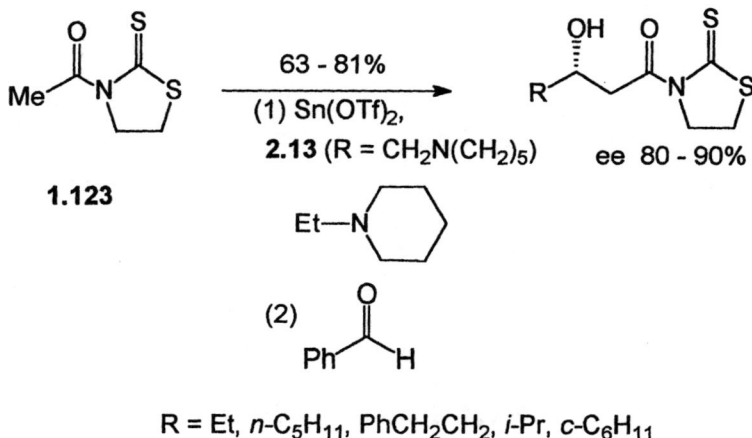

63 - 81%

(1) Sn(OTf)$_2$,

2.13 (R = CH$_2$N(CH$_2$)$_5$)   ee 80 - 90%

1.123

Et—N⟨piperidine⟩

(2) Ph—CHO

R = Et, n-C$_5$H$_{11}$, PhCH$_2$CH$_2$, i-Pr, c-C$_6$H$_{11}$

Figure 6.71 (continued)

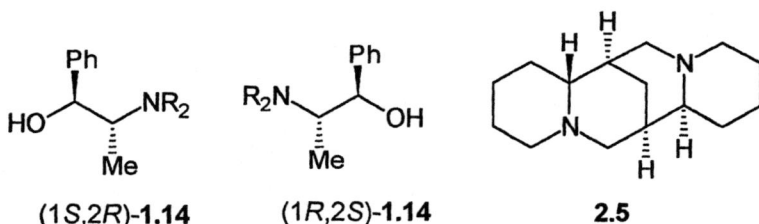

(1S,2R)-**1.14**   (1R,2S)-**1.14**   **2.5**

Titanium enolates bearing chiral ligands have received numerous applications. Titanium enolates of *tert*-butyl acetate bearing diacetoneglucose **1.48** or diol **2.50** (R = R' = Me, Ar = Ph) as ligands are prepared by exchange of the lithium enolate with the appropriate complexes **2.49** at −78°C. Aldol reactions of these enolates are highly enantioselective, and lead to enantiomeric β-hydroxyesters **6.81** [408, 665, 666] (Figure **6.72**). Similar aldol condensations with the enolate **6.82** of an aminoacetate afford enantiomeric *syn*-α-amino-β-hydroxyesters in a highly selective fashion [408, 665, 666, 861] (Figure **6.72**). To perfom such reactions with propionates, 2,6-dimethylphenyl esters must be used. E- or Z-titanium enolates are generated under carefully controlled conditions (Figure **6.72**) from the Li ester enolates and **2.49**. Unexpectedly, the E-enolate yields the *syn* aldol, and the Z-enolate yields the *anti* aldol. In both cases, the *Re* face of the aldehyde is attacked [683] (Figure **6.72**). These selectivities are rationalized by a boat transition state in which the steric interactions are minimized. Similar aldol reactions cannot be carried out with titanium enolates of ketones or hydrazones, but the Z-enolate of *N*-propionyloxazolidinone **5.30** (R$_E$ = Me, R = H) does react with high selectivities [665, 666].

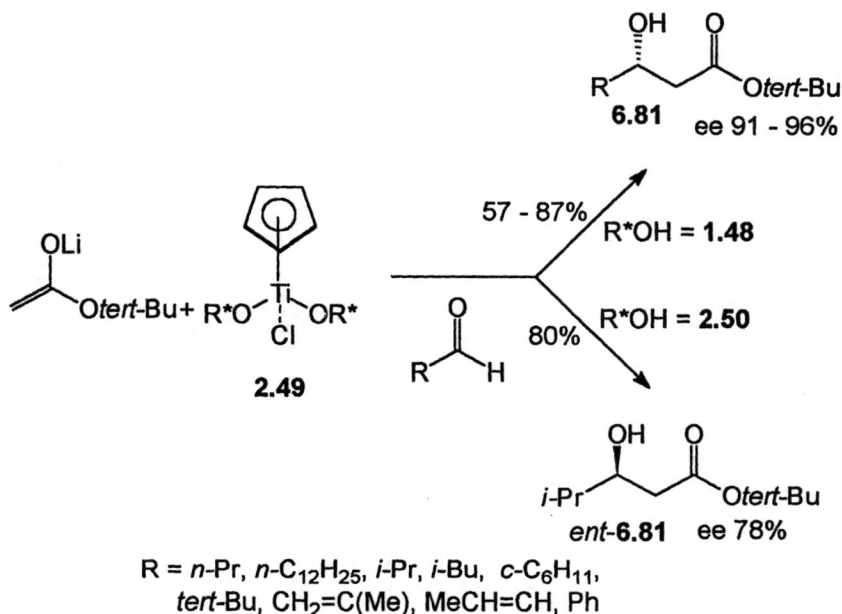

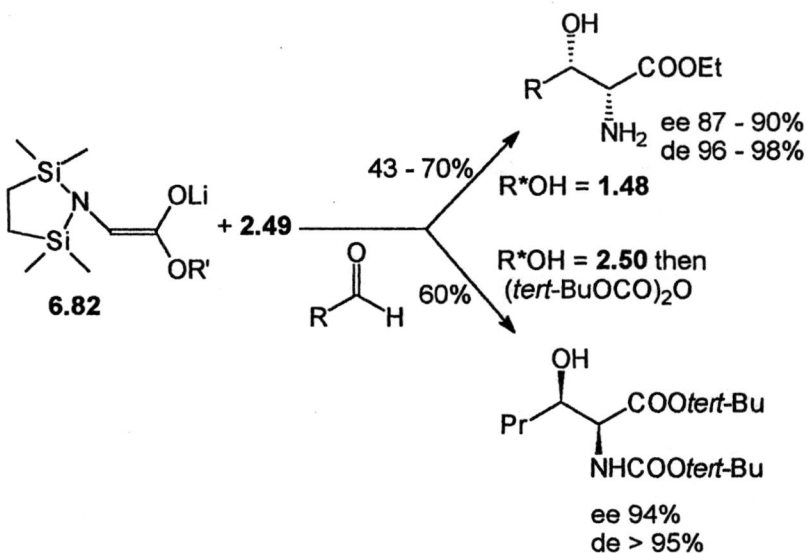

R = *n*-Pr, *n*-C₁₂H₂₅, *i*-Pr, *i*-Bu, *c*-C₆H₁₁,
tert-Bu, CH₂=C(Me), MeCH=CH, Ph

R = Me, *n*-Pr, *tert*-Bu, CH₂=CH, CH₂=C(Me), Ph

Figure 6.72

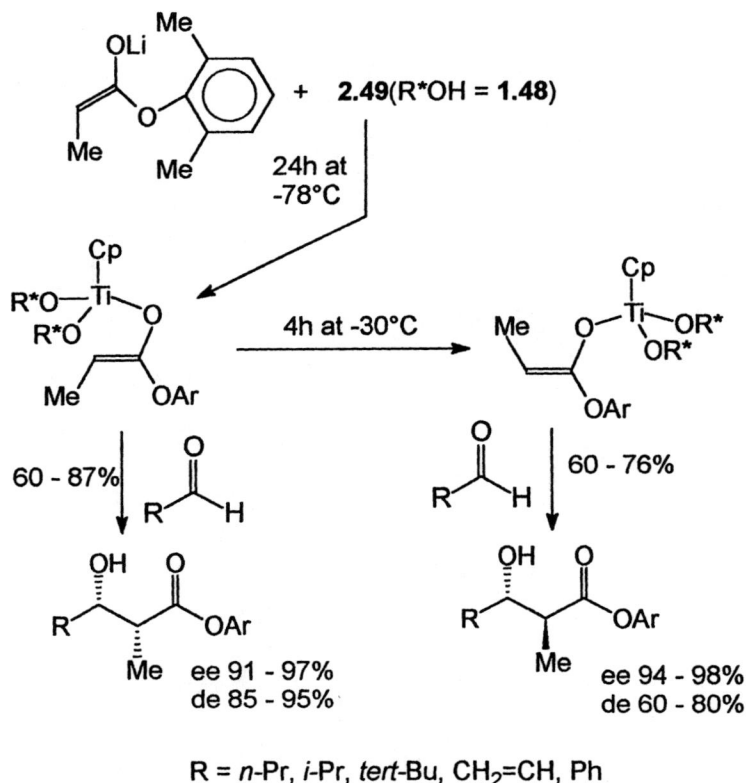

Figure 6.72 (continued)

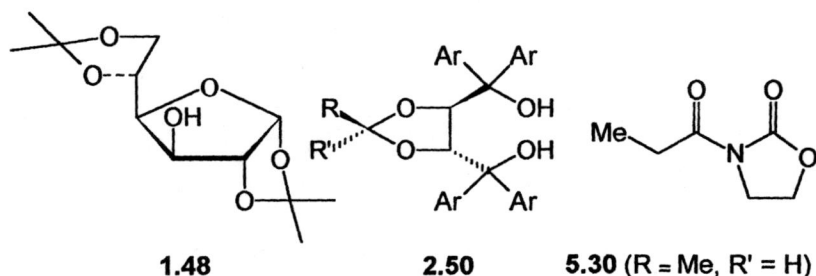

Chiral boron enolates **2.60** are generated by the reaction of chiral boron tri-flates or haloboranes **2.14** (R = OTf or Cl), **2.16** (R = OTf or Cl), **2.61** or **2.62** (X = Br) with carbonyl compounds in the presence of tertiary amines [82, 407, 684, 690, 693, 694, 1251], or by conjugate addition of **2.16** (R = H) with α-enones [687, 688]. Recently, Gennari, Paterson and coworkers [1252] proposed the use of haloboranes **6.83** (X = Cl, Br) prepared from menthone. Enol borinates derived from bulky thiolacetates lead to β-hydroxy-thiolesters with an excellent enantiomeric excess at −78°C by using boron triflate **2.14** (R = OTf), bromoborane **6.83** (X = Br) or chloroborane **2.61** as precursors [407, 415, 689, 1248]. Corey's bromosulfonamides **2.62** are useful as well, starting from phenylthiol acetate [693] (Figure **6.73**). Aldol reactions of α-chiral aldehydes have also been carried out and the expected aldol adducts **6.84** are obtained with a high diastereo- and enantioselectivity [690] (Figure **6.73**). However, the reactions of the enolborinates of methyl ketones (RCOMe) are less selective, and enantiomeric excesses are about 60 - 70% [407, 684, 1252].

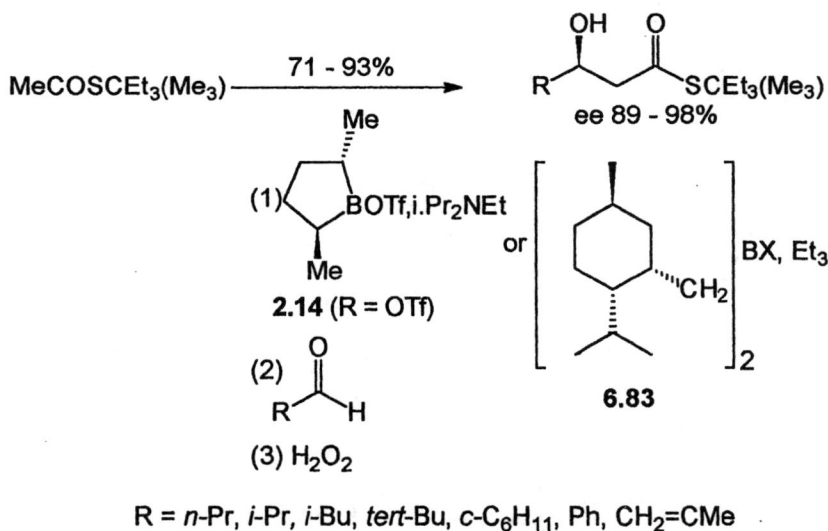

R = n-Pr, i-Pr, i-Bu, tert-Bu, c-C₆H₁₁, Ph, CH₂=CMe

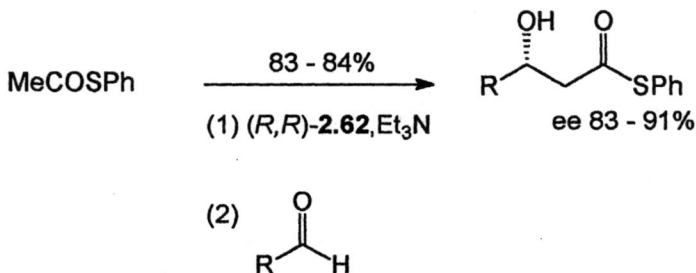

R = i-Pr, Ph

Figure 6.73

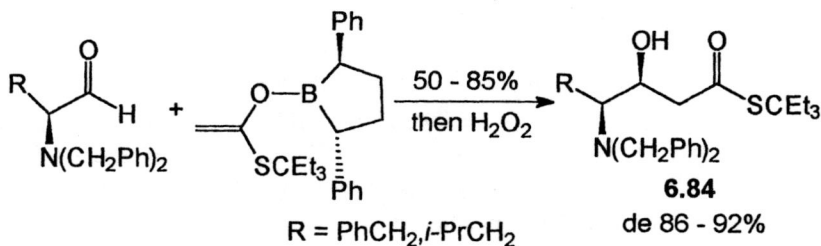

Figure 6.73 (continued)

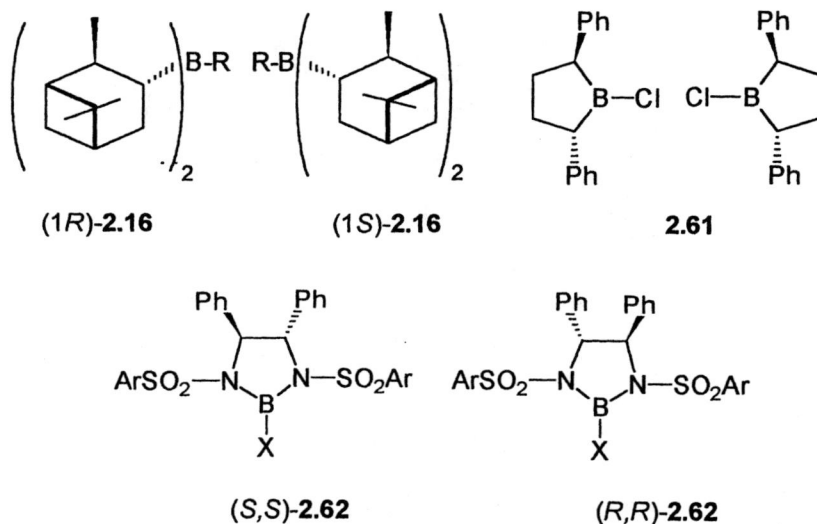

(1R)-**2.16**   (1S)-**2.16**   **2.61**

(S,S)-**2.62**   (R,R)-**2.62**

    The method of generation of Z- or E-enolborinates **2.60** determines their geometry. From ethylketones (RCOCH$_2$Me), Z-enolborinates are formed by using enantiomeric diisopinocampheylborane triflates **2.16** (R = OTf) in the presence of i-Pr$_2$NEt. Subsequent additions to aldehydes are both diastereoselective (*syn* isomers favored) and enantioselective [407, 684, 693] (Figure **6.74**). The Zimmermann-Traxler model accounts for these observed selectivities [1249]. The *Si* face of the aldehyde is attacked when the enolborinate exhibits the (1R) configuration in order to minimize the interaction between the enolborinate R' substituent and the methyl group located on the 2-position of the axial boron ligand. Transition-state model **C$_1$** *Si* is thus favored over **C$_1$** *Re* [125, 1253] (Figure **6.74**). Differently substituted Z-enolborinates **2.60** (R ≠ Me) can be prepared by conjugate additions of **2.16** (R = H) to α-enones, and these have been reacted with PhCHO. Useful selectivities are observed when R = PhCH$_2$ and R' = Ph or Me [687, 688].

E-Enolborinates of ethylketones are prepared with **2.16** (R = Cl) or **6.83** (R = Br) in the presence of Et$_3$N. These are precursors of *anti* aldols, but the enantioselectivity of these reactions is not extremely high (≤ 75%) [684, 1248, 1279, 1252].

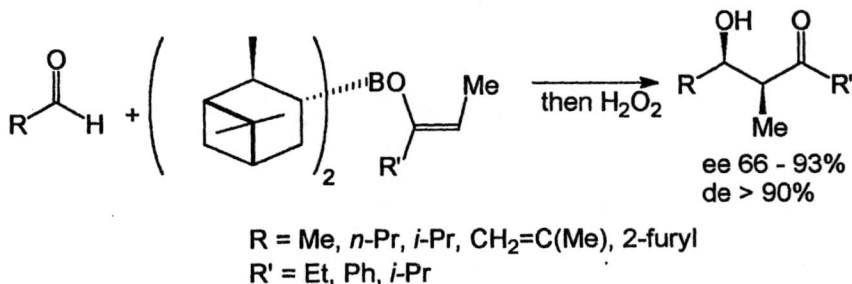

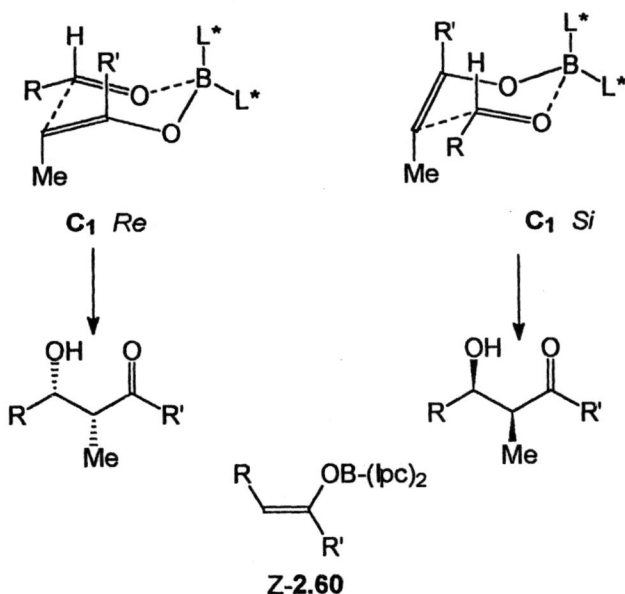

R = Me, *n*-Pr, *i*-Pr, CH$_2$=C(Me), 2-furyl
R' = Et, Ph, *i*-Pr

Figure 6.74

The geometry of enolborinates of thiolesters depends mainly upon the nature of the sulfur substituent. Masamune and coworkers [691] have shown that the reaction of thiophenyl propionate (MeCH$_2$COSPh) with **2.14** (R = OTf) leads to E-**6.85**, while triethylmethylthiol propionate (MeCH$_2$COSCEt$_3$) leads to Z-**6.85** (Figure **6.75**). A similar trend is observed in the reaction of MeCH$_2$COS-*tert*-Bu with **6.83** (X = Br) [1248]. The reaction of Z-enolborinates Z-**6.85** generated from **2.14** (R = OTf), **2.61** (R = OTf) or **6.83** (X = Br) with aldehydes is highly diastereo- and enantioselective [407, 691, 1204, 1248]. *Anti*-β-hydroxy-thiolesters are easily obtained at −78°C (Figure **6.75**). This reaction is also useful with α-chiral aldehydes provided that the reagents are matched (§ I.6) [407, 690]. Corey's reagents **2.62** (X = Br, Ar = 4-NO$_2$C$_6$H$_4$ or 3,5-(CF$_3$)$_2$C$_6$H$_3$) can also be used. Starting from thiophenyl propionate, enantiomeric *syn* aldols are obtained with a high enantioselectivity [693] (Figure **6.75**). In contrast, starting from *tert*-Bu propionate (R' = Me) or bromoacetate (R' = Br) and **2.62** (X = OTf, Ar = 3,5-(CF$_3$)$_2$C$_6$H$_3$), enolborinates **6.86** are generated. These enol borinates are precursors of *anti* aldol adducts, which are obtained with high diastereo- and enantioselectivity [694, 1254, 1255] (Figure **6.75**). Cyclic transition-state models minimizing steric interactions account for these results [1255]. The nonsubstituted or the transoid (*R,R*)-**6.86** reagents attacks the *Re* face of an aldehyde, while the cisoid reagent (*R,R*)-E-**6.85** attacks the *Si* face. This reversal is due to different repulsive interactions involved in the transition states [1255]. Each type of reagent is available in either enantiomeric configuration, so both enantiomeric aldols can be made. However, the reactions run with boranes require a further treatment with H$_2$O$_2$ to give aldol products, while the use of Corey's reagents require only a simple hydrolysis. Additionally, the recycling of the Corey auxiliary is easy. Meyers and Yamamoto [407, 697] reacted boron azaenolates **6.87** with aldehydes, and they obtained *anti* aldols with a high selectivity but a poor chemical yield (Figure **6.75**). A catalytic asymmetric nitroaldol reaction has been described by Shibasaki and coworkers [847, 848, 1256]. In the presence of La binaphtholates prepared from La$_3$(O-*tert*-Bu)$_9$ or LaCl$_3$ and Li (*R*)- or (*S*)-binaphtholates, nonracemic 2-nitroalcohols are obtained from MeNO$_2$ with a good enantiomeric excess at −40°C provided that lithium salts and water are present in the reaction medium [849, 1257] (Figure **6.75**).

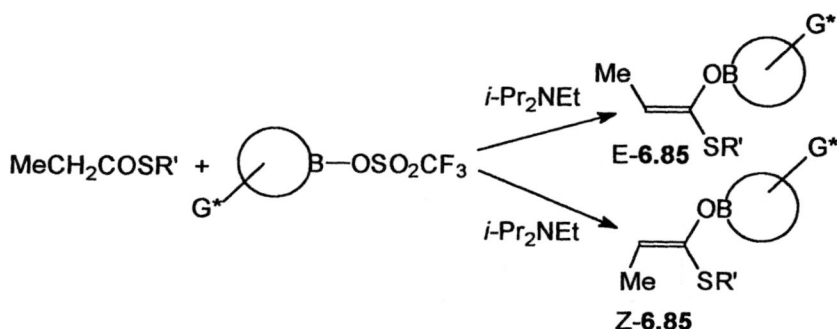

Figure 6.75

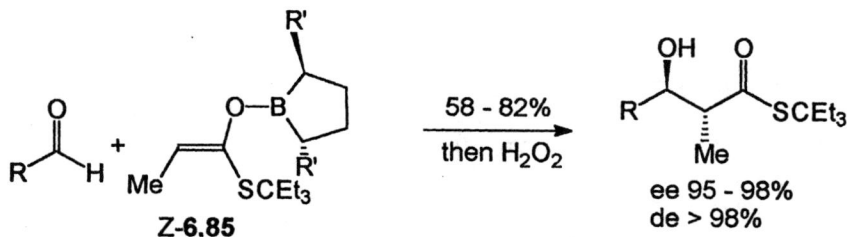

R = Pr, *i*-Pr, *tert*-Bu, c-C$_6$H$_{11}$, Ph R' = Me, Ph

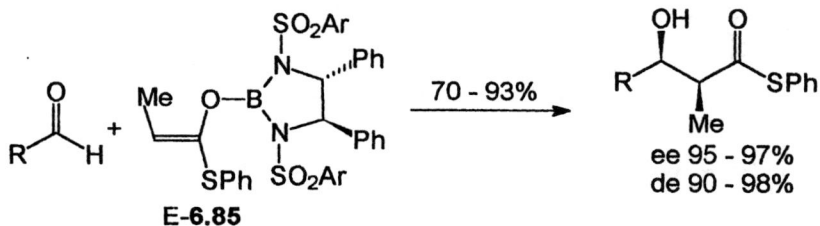

R = Ph, *i*-Pr, c-C$_6$H$_{11}$, PhCH$_2$CH$_2$

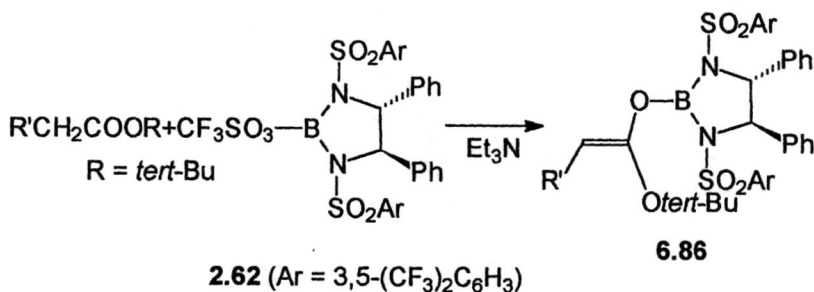

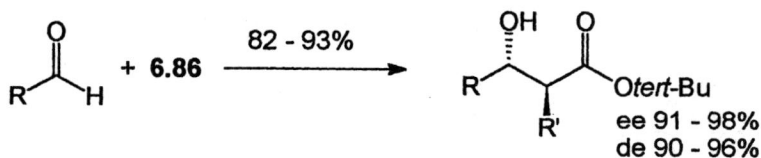

R = PhCH$_2$CH$_2$, c-C$_6$H$_{11}$, Ph, PhCH=CH R' = Me, Br

Figure 6.75 (continued)

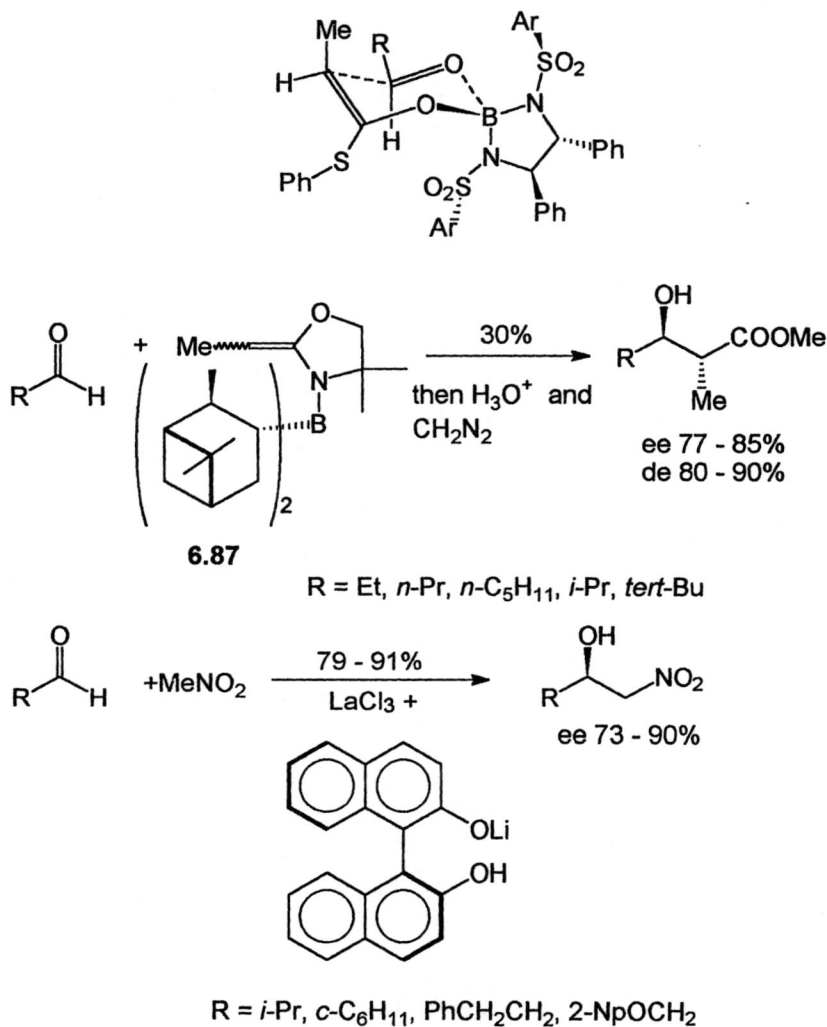

6.87

R = Et, n-Pr, n-C₅H₁₁, i-Pr, tert-Bu

ee 77 - 85%
de 80 - 90%

ee 73 - 90%

R = i-Pr, c-C₆H₁₁, PhCH₂CH₂, 2-NpOCH₂

Figure 6.75 (continued)

6.8.1.2 *Reactions of Enolates Bearing Chiral Residues: Generation of a Single New Stereocenter*

The reactions of aldehydes with enolates of acetic esters (MeCOOG*) **1.18** (R = H) are often poorly selective [66, 147, 209, 408, 1186, 1258]. For this reason, the asymmetric synthesis of acetate aldols is usually performed by reduction of α-bromo analogs with Bu₃SnH [1254] or from sulfoxides [1049] (see below). However, the reactions of lithium or magnesium enolates of monoacetates of diols

(*R*)- and (*S*)-**1.37** (R = Ph) with RCHO are highly selective at −100°C [204, 207, 209, 210] (Figure **6.76**). Similar reactions with α-chiral aldehydes can also be selective [208] (Figure **6.76**). The tin (II) enolate of *N*-acetylthiazolidinone **1.123** (R = *i*-Pr, R' = H) reacts highly selectively with α,β-unsaturated aldehydes at −78°C. After removal of the chiral auxiliary with PhCH$_2$ONH$_2$, β-hydroxyacylhydroxylamines are obtained with an excellent enantiomeric excess [438] (Figure **6.76**). Disappointing results have been obtained from enolates of amides (MeCONG*$_2$) [265, 408] and methylketones (MeCOR') [407]. Enders and coworkers [1259] performed enantioselective aldol reactions from hydrazones of pyruvic esters at −78°C. To obtain good selectivities, it is necessary to use a bulky hydrazone together with a 2,6-di-*tert*-Bu-4-methoxyphenyl ester **6.88**. Samp **1.76** hydrazone gives low selectivities (Figure **6.76**).

Highly selective aldol reactions have been carried out by Davies and coworkers [408, 522, 1260] with acetyl cyclopentadienyliron carbonyl complexes **1.151** (R = Me). Reactions of aldehydes with aluminum enolates under controlled conditions lead, after oxidative cleavage of the iron complex, to nonracemic β-hydroxyesters (Figure **6.76**). If one of the phenyl groups bound to phosphorus is replaced by a pentafluorophenyl group, then lithium enolates give similar selectivities [408]. Tin (II) enolates promote the formation of the other enantiomer (Figure **6.76**). The reactions with α-chiral aldehydes are also highly selective [522, 1260]. One of the drawbacks of this method is the low temperature required (−100°C). The interpretation of the observed selectivities from aluminum and lithium enolates is similar to that proposed for alkylation reactions (see Figure **5.29** which shows an attack on the enolate side that is not shielded by a phenyl ring).

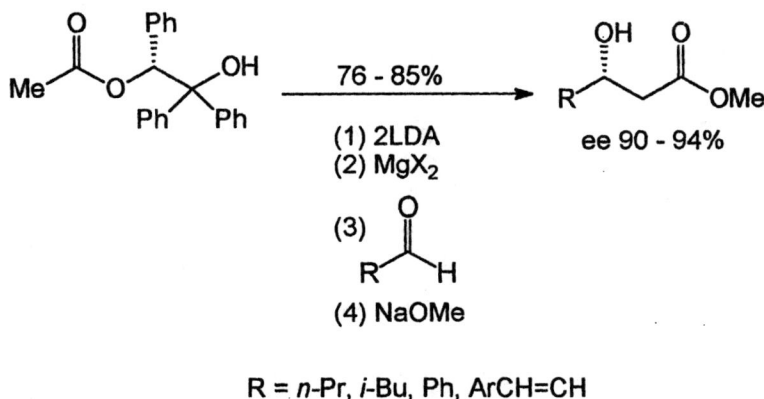

R = *n*-Pr, *i*-Bu, Ph, ArCH=CH

Figure 6.76

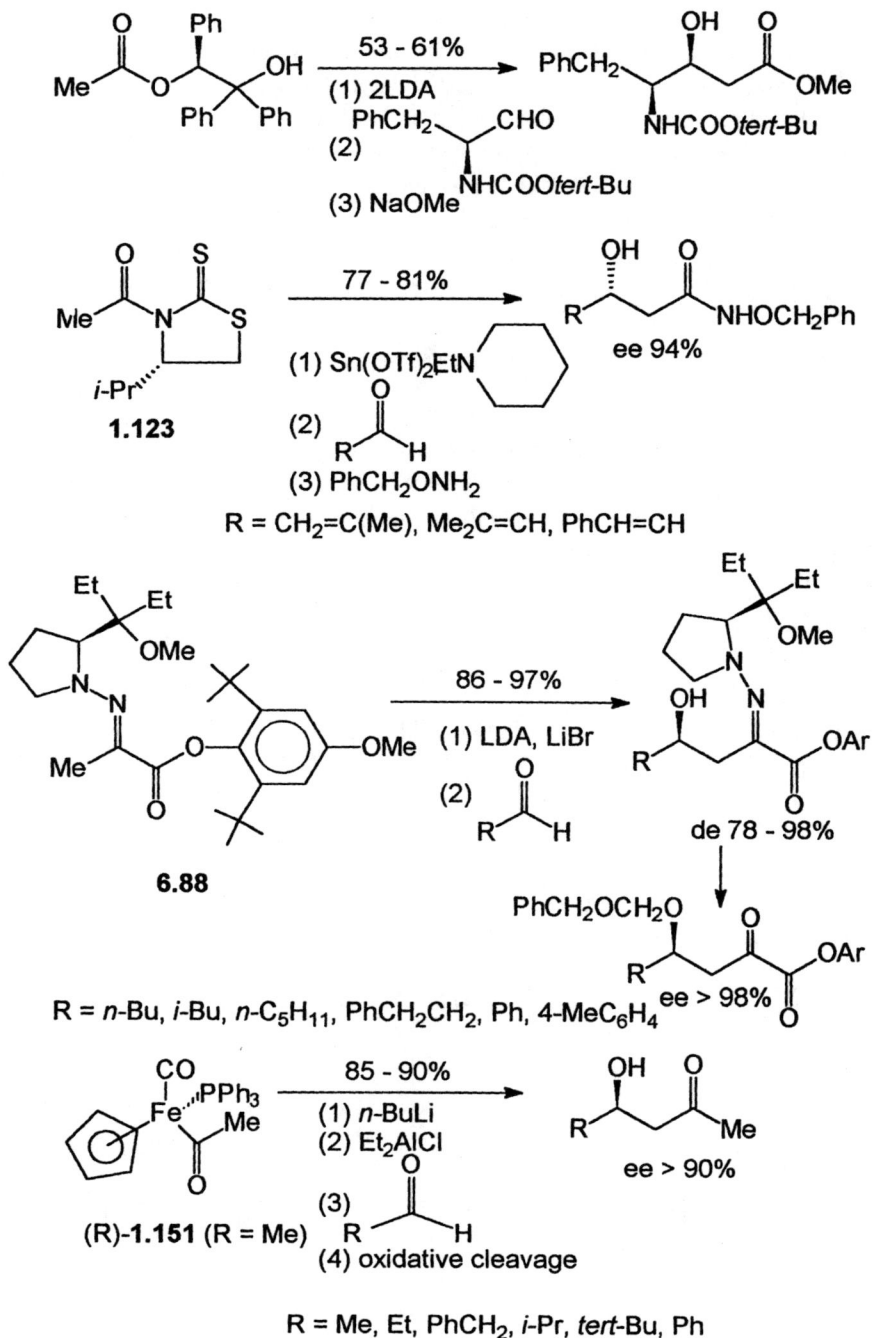

Figure 6.76 (continued)

(R)-**1.151** $\xrightarrow[\substack{(1)\ n\text{-BuLi}\\(2)\ SnCl_2\\(3)\ i\text{-PrCHO}\\(4)\ \text{oxidative cleavage}}]{74\%}$

i-Pr with OH and O groups, OMe

ee 84%

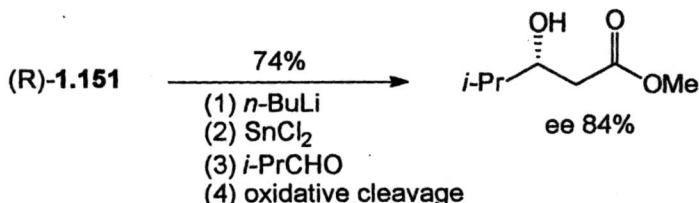

Figure 6.76 (continued)

The reactions of enolates bearing chiral auxiliaries with formaldehyde or symmetrical ketones can be stereoselective. After removal of the auxiliary, nonracemic primary or tertiary alcohols are obtained. The reaction of lithium enolates of Schöllkopf's lactim ethers **1.114** with symmetrical carbonyl compounds are highly stereoselective, as are the reactions of enolates of Seebach's imidazolidinone **5.39** (R = Ph). In both cases, the enolate reacts from its least hindered face [154, 261] (Figure 6.77). After acidic hydrolysis, β-hydroxy-α-aminoacids are obtained with a high enantiomeric excess. However, when R' = H, some unwanted epimerization can take place.

The titanium enolate of *N*-propionyloxazolidinone **5.30** (R = Me) reacts highly selectively with *s*-trioxane at −78°C. The least hindered face of the chelate **6.89** is attacked [666, 1042] (Figure 6.77). Unexpectedly, the reaction of the boron enolate of **5.30** (R = Me, *n*-Bu, PhCH₂) with hexafluoroacetone leads to compound **6.90**, probably through an open transition state [1261] (Figure 6.77).

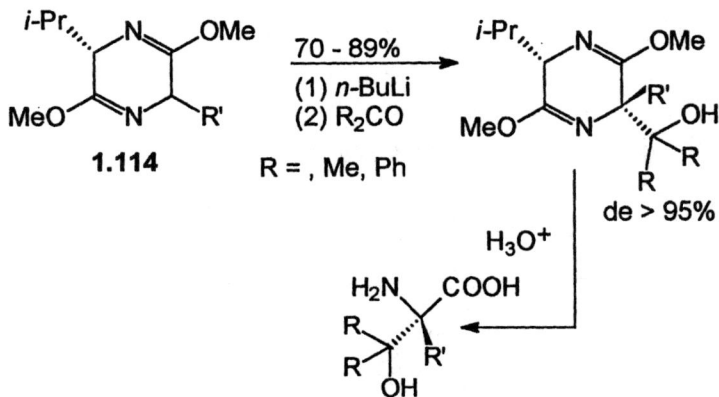

1.114 → 70 - 89% (1) *n*-BuLi (2) R₂CO, R = , Me, Ph

de > 95%

H₃O⁺ → H₂N, COOH amino acid

Figure 6.77

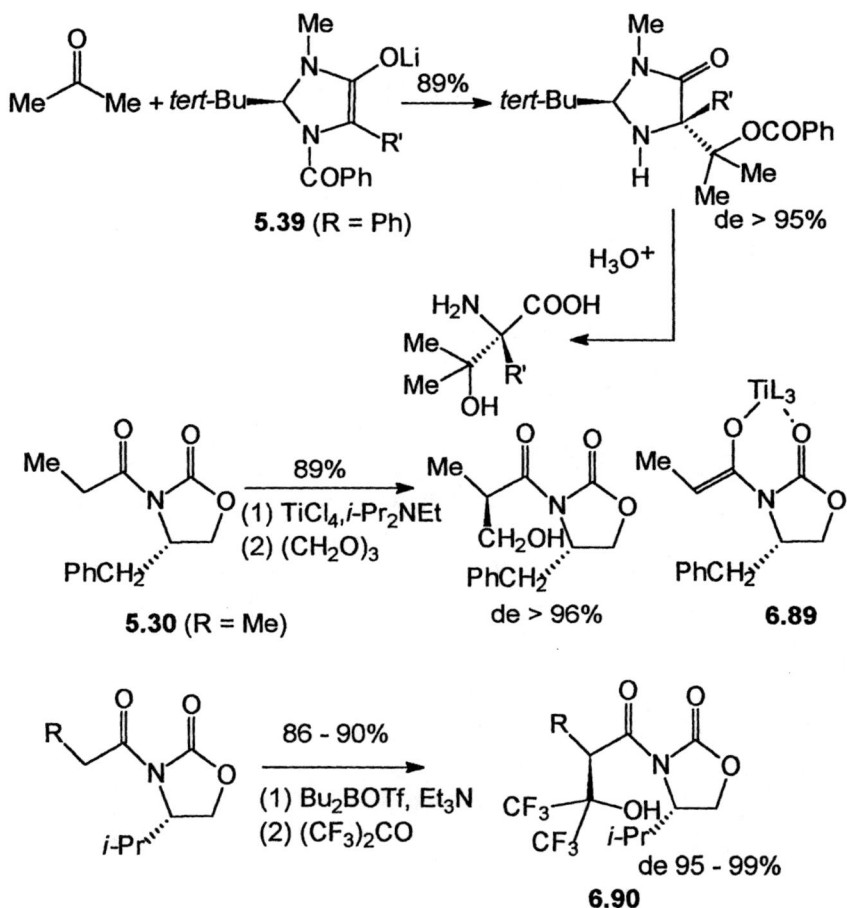

5.39 (R = Ph)

5.30 (R = Me)

de > 96%

6.89

86 - 90%

de 95 - 99%

6.90

Figure 6.77 (continued)

6.8.1.3 *Generation of Two New Stereocenters*

Due to the importance of polypropionate antibiotics, mainy chiral auxiliaries have been introduced on propionic acid derivatives in order to perform asymmetric aldol reactions. The use of esters of chiral alcohols usually gives disappointing results [147, 209]. In an important exception, Braun and Sacha [149] recommended the propionate of a trimethylsilyloxyalcohol **1.12**. The reaction of the derived dicyclopentylchlorozirconium enolate with aliphatic aldehydes at −105°C leads to *anti* aldols with an excellent facial stereoselectivity (Figure **6.78**). The selectivity is lower with benzaldehyde. The use of titanium enolates of *N*-tosyl-aminoephedrine **1.61** (R = Ts) propionate has recently been advocated [1262].

The reaction of aldehydes with zirconium enolates of chiral propionamides derived from amines bearing a C_2 axis of symmetry also leads to useful selectivities [408] (Figure 6.78).

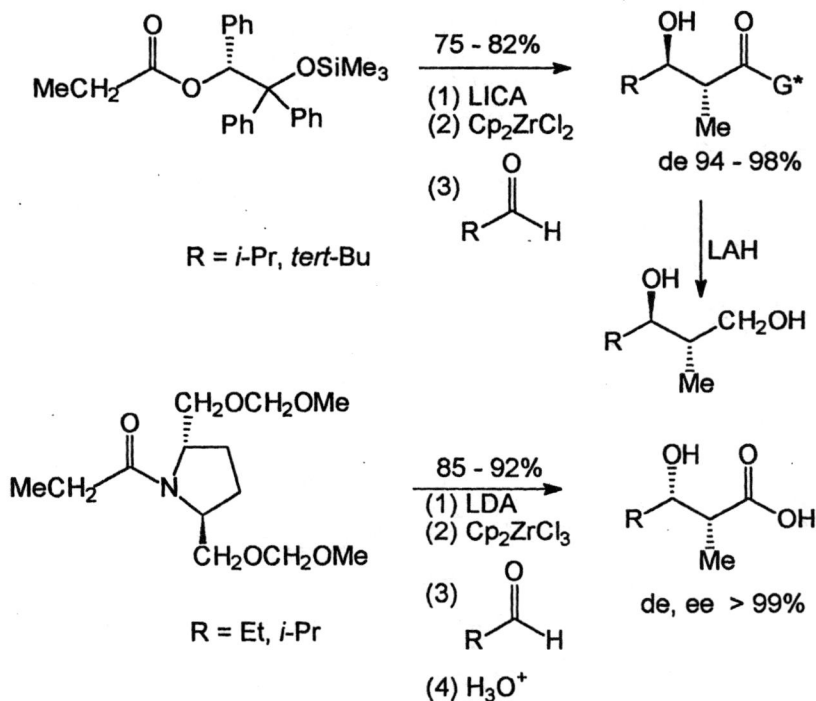

Figure 6.78

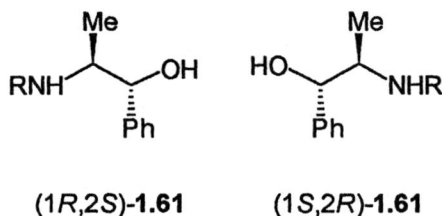

(1R,2S)-**1.61**   (1S,2R)-**1.61**

Derivatives of Evans's oxazolidinones **1.116** and **1.117** have been broadly developed for use in aldol reactions [160, 167, 407, 408]. The reactions of aldehydes with lithium enolates are usually poorly stereoselective, but remarkable results have been obtained from boron, tin(II) and titanium enolates. The boron enolate **6.91** is generated by reaction of *N*-propionoyloxazolidinone with stoichiometric amounts of *n*-Bu$_2$BOTf in the presence of Et$_3$N or *i*-Pr$_2$NEt. After addition of the aldehyde at −78°C, followed by treatment with H$_2$O$_2$ and removal of the chiral auxiliary, *syn*-β-hydroxyacids **6.92** or their derivatives are formed with an excellent diastereo- and enantioselectivity [406] (Figure **6.79**). This method can be applied on a large scale for natural product synthesis [1263]. The other *syn* enantiomer can be prepared by using **1.117** as chiral auxiliary instead of **1.116**. From some disubstituted aromatic aldehydes, the nature of the amine used to generate the boron enolate can influence the selectivity of the aldol reaction [1264]. However, the reaction of CF$_3$CHO with **6.91** is poorly stereoselective, and non-Evans *syn* and *anti* aldols are obtained [1261]. When the reaction of aldehydes with **6.91** is carried out either in the presence of excess *n*-Bu$_2$BOTf [105, 1265] or another added Lewis acid [106], either *anti* **6.93** or *syn* aldols **6.94** with the opposite configuration are obtained. The selectivity of the reaction depends upon the aldehyde, the Lewis acid and the experimental protocol (Figure **6.79**).

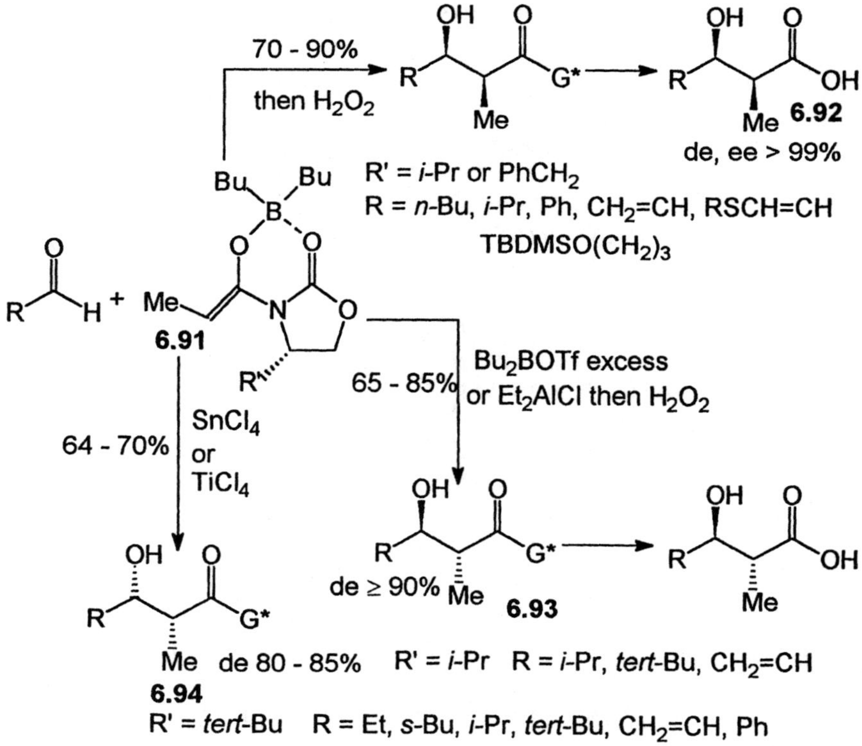

Figure 6.79

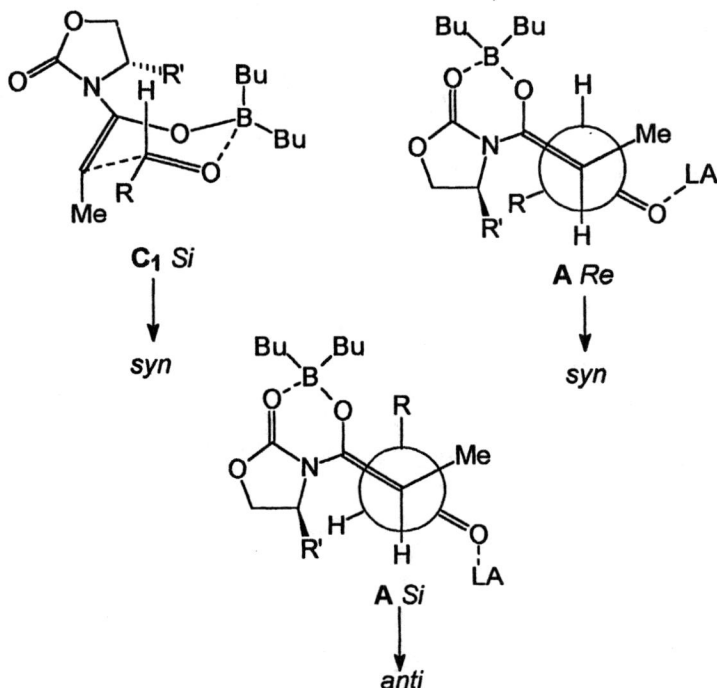

Figure 6.79 (continued)

These results are interpreted as shown in Figure **6.79**. The formation of **6.92** takes place via a cyclic transition state C_1 *Si*. Chelate **6.91** is disrupted in order to allow the coordination of the aldehyde to the boron atom. As usual, steric interactions are minimized in the favored transition state. In the presence of excess boron triflate or of another Lewis acid which can activate the aldehyde carbonyl group, the boron chelate is no longer disrupted. Two acyclic transition-state models, **A** *Re* and **A** *Si*, can be envisioned according to the nature of the Lewis acid [106]. Open transition states are also proposed for the reactions of **6.91** with CF_3CHO, which does not require electrophilic assistance [1261].

Outstanding selectivities are also obtained with α-chiral aldehydes, and three new stereocenters are thereby generated. These aldol reactions have been applied in synthesis of natural products [1040, 1266, 1267]. A few examples are shown in Figure **6.80** [407, 1265]. Modeling of related transition states has been performed by Gennari and coworkers [1268]. The stereoselectivity of aldol reactions of the titanium enolates of *N*-propionyloxazolidinone **5.30** (R = Me) also depends upon the reaction conditions [408, 666].

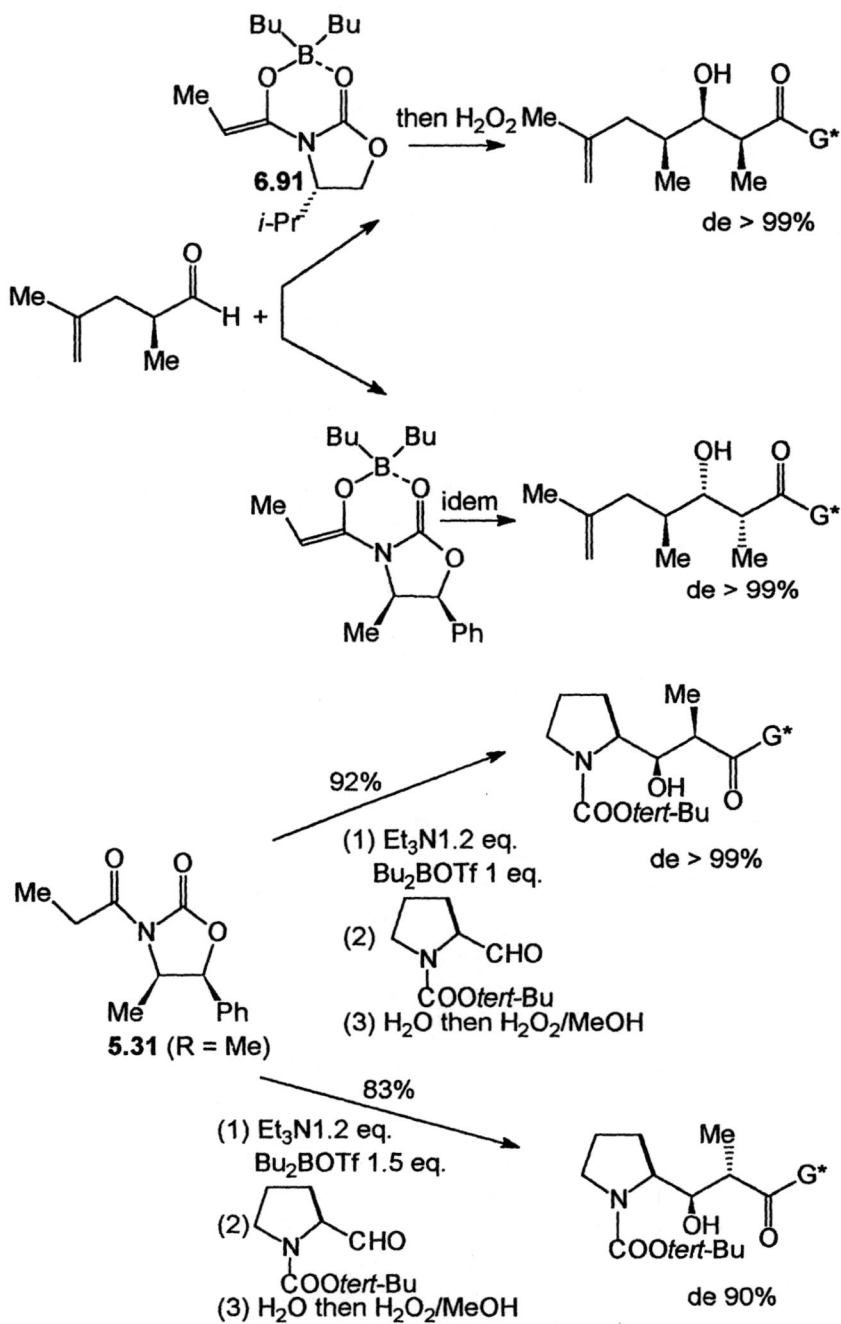

Figure 6.80

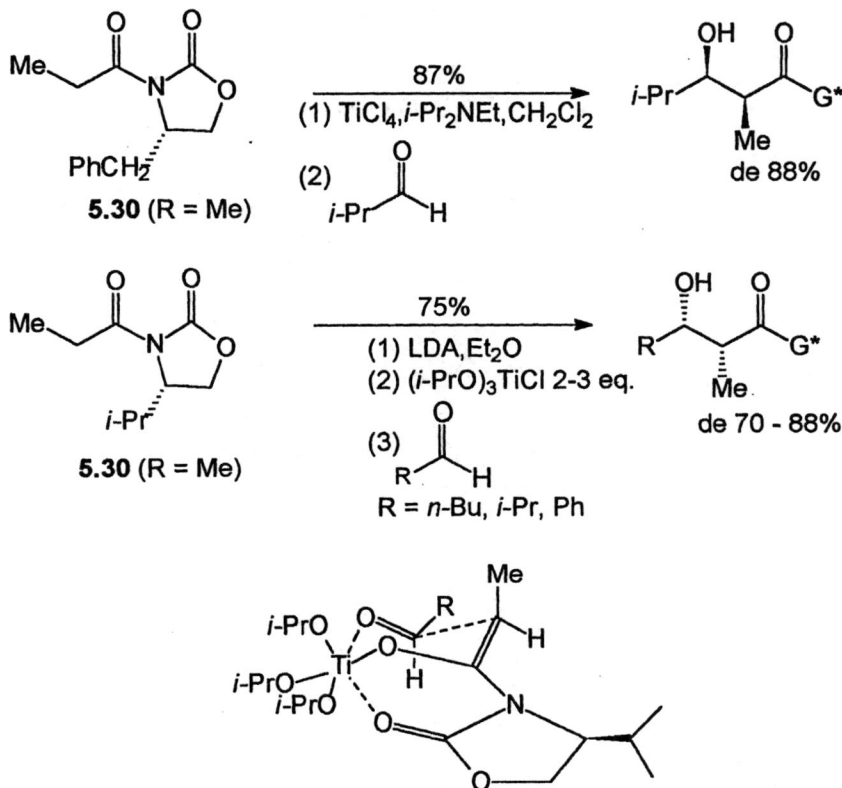

Figure 6.81

When the titanium enolate was generated by reaction with 1 equiv of $TiCl_4$ and i-Pr_2NEt in CH_2Cl_2, Evans and coworkers obtained the same *syn* aldol with i-PrCHO as obtained from boron enolate [404], although with a lower selectivity. In contrast, Thornton and coworkers [401, 403] showed that exchange between the lithium enolate and 2-3 equiv of $(i$-$PrO)_3TiCl$ generates another species which leads to the enantiomeric *syn* aldol after reaction with aldehydes in Et_2O (Figure **6.81**). This last result is interpreted by formation of a hexacoordinated titanium chelate (Figure **6.81**). Similar selectivity changes have been noted by Pridgen and coworkers [123].

Extension of this method to *N*-acyloxazolidinones **5.30** or **5.31** ($R = PhCH_2$, Br, Cl,F, NCS) [123, 154, 398, 402, 405, 409, 1269], ($R = MeCH=CH$) [414], ($R = $ 1,3-dithianylidene) [1044] or **6.95** [403a] has been accomplished. As expected, Z-boron enolates lead to *syn* aldols **6.96** with an excellent selectivity

(Figure **6.82**). Tin (II) enolates, generated with Sn(OTf)$_2$/Et$_3$N [123, 154, 403a], also usually give the same *syn* aldols [123, 154, 398, 402, 405]. However, when using 2-substituted benzaldehydes and **5.30** (R = Br), *anti* aldols are preferentially formed, probably through a boat transition state [123]. *Anti* aldols are also predominantly obtained, though with a modest selectivity, from benzaldehyde and tin (IV), Li and zinc enolates [123] of **5.30** (R = Br, Cl), and a twist boat transition state has been proposed. The selectivity is higher from the tin (IV) enolate of **5.30** (R = F). The synthesis of *anti*-β-hydroxy-α-aminoacids by reaction with NaN$_3$ followed by the usual treatment [154, 399, 405, 861] is a nice application of the use of haloacetyloxazolidinones **5.30** (R = Br, Cl).

Enolization of **6.95** with Sn(OTf)$_2$/Et$_3$N or TiCl$_4$/*i*-Pr$_2$NEt is regioselective. The reaction of the enolates formed in this way with aldehydes leads to either enantiomeric *syn* aldol with a high selectivity [403a, 666] (Figure **6.82**). Evans has proposed two cyclic transition states **6.97** and **6.98** to interpret these results (Figure **6.82**).

Other oxazolidinones have been used as chiral auxiliaries in asymmetric aldol reactions. Bornane derivatives **1.121** (X = O or S) and **1.122** are readily transformed into *N*-acyl derivatives. The reactions of their boron or titanium enolates with aldehydes give the same selectivities as Evans's reagents [426, 428, 429, 431, 436]. *N*-Acylimidazolidinones **1.131** and **1.132** [449, 1270] lead to similar results, but the selectivities observed are somewhat lower.

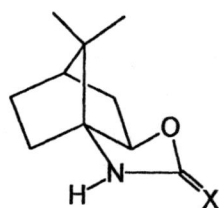

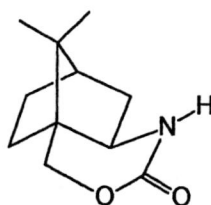

1.121 (X = O or S)   **1.122**

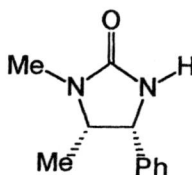

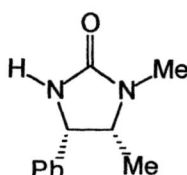

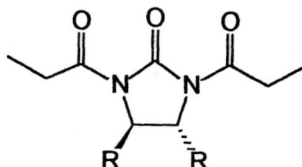

1.131   **1.132**

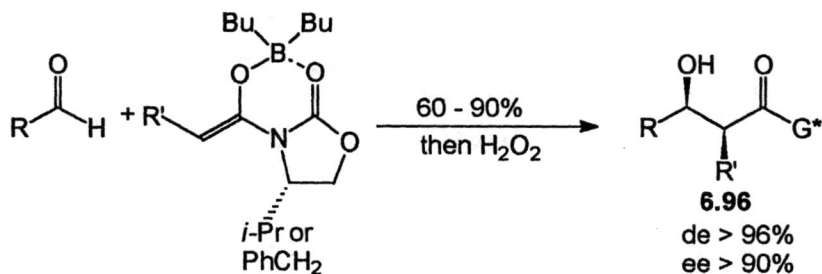

R = Me, Et, n-C$_5$H$_{11}$, i-Pr, Ph
R' = Cl, Br, CH$_2$=CH, CH$_2$=C(Me), PhCH$_2$

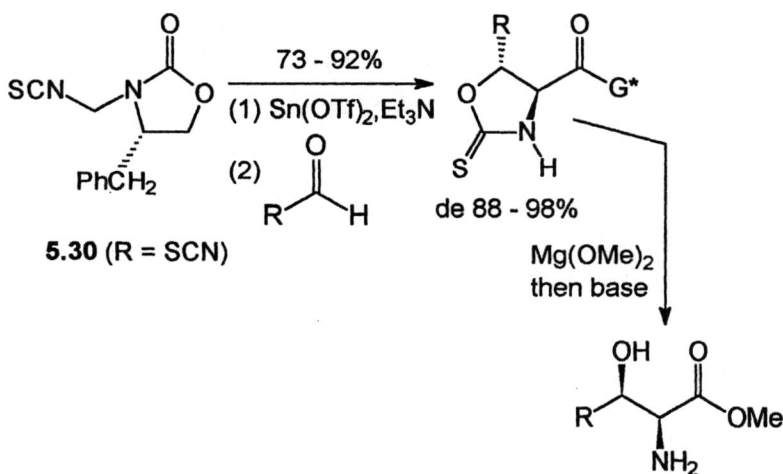

R = Me, i-Pr, Ph, MeCH=CHCH$_2$CH$_2$,

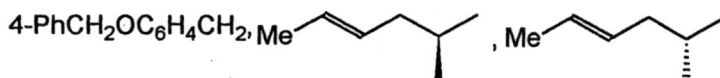

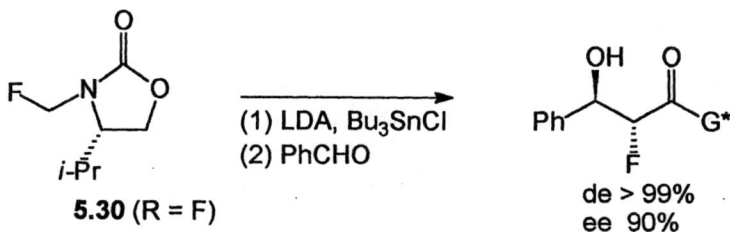

4-PhCH$_2$OC$_6$H$_4$CH$_2$, Me~~~~, Me~~~~

Figure 6.82

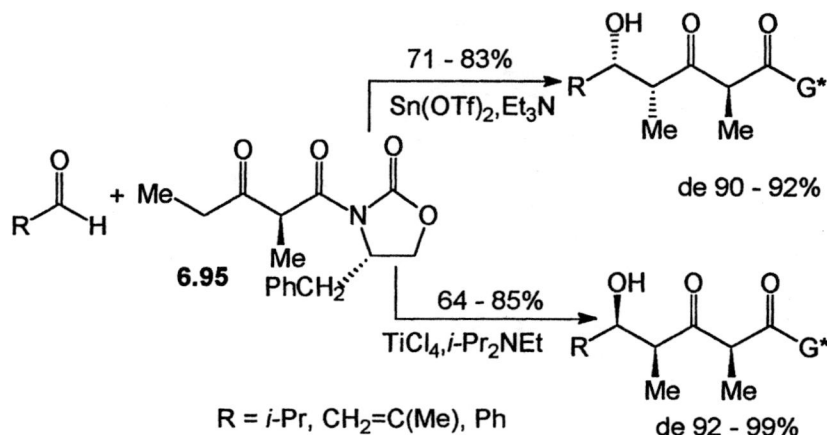

R = *i*-Pr, CH₂=C(Me), Ph

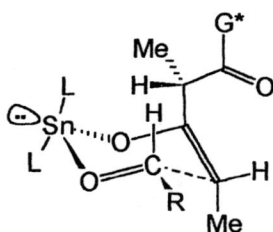

6.97

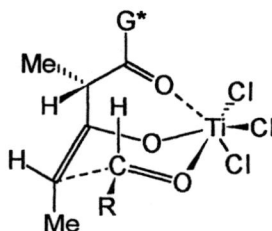

6.98

Figure 6.82 (continued)

N-Acylthiazolidinones or -thiazolidinethiones **1.123** (R = COOMe, X = O or S), initially devised by Nagao and coworkers [261], have received new applications in asymmetric aldol reactions. With these reagents, a simple hydrolysis liberates the aldols generated from the corresponding boron enolates [434]. This is in contrast to oxazolidinones, which require H_2O_2 treatment. Boron or tin (II) enolates lead to *syn* aldols with a high selectivity, and either enantiomer is obtained according to the configuration of the reagent [261, 434, 435, 641] (Figure **6.83**).

Oppolzer's sultams **1.133** are also efficient auxiliaries in asymmetric aldol reactions [209, 404, 407, 457, 1271]. Boron, titanium or Sn (IV) enolates of *N*-propionoylsultams lead stereoselectively to either enantiomeric *syn* aldol at −78°C. These products are easily purified by fractional crystallization (Figure **6.83**). After treatment with LiOH/H_2O_2 and CH_2N_2, *syn*-β-hydroxyesters are obtained with an excellent enantiomeric excess. The drawback of this method is the need to use an excess of aldehyde to obtain good chemical yields. As in the case of oxazolidinones (see above), *anti* aldols are prepared from boron enolates and various alde-hydes in the presence of TiCl$_4$ [1271]. *Anti* aldols are also obtained in a highly selective fashion by using the titanium enolate of *N*-alkylideneglycinamide **6.99** bearing a 2,2-dimethyloxazolidine (*S*)-**1.85** (R' = H, R" = Me) chiral controller [1272]. These compounds are precursors of β-hydroxy-α-aminoacids (Figure **6.83**). A twist boat transition state again accounts for the observed stereoselection (Figure **6.83**).

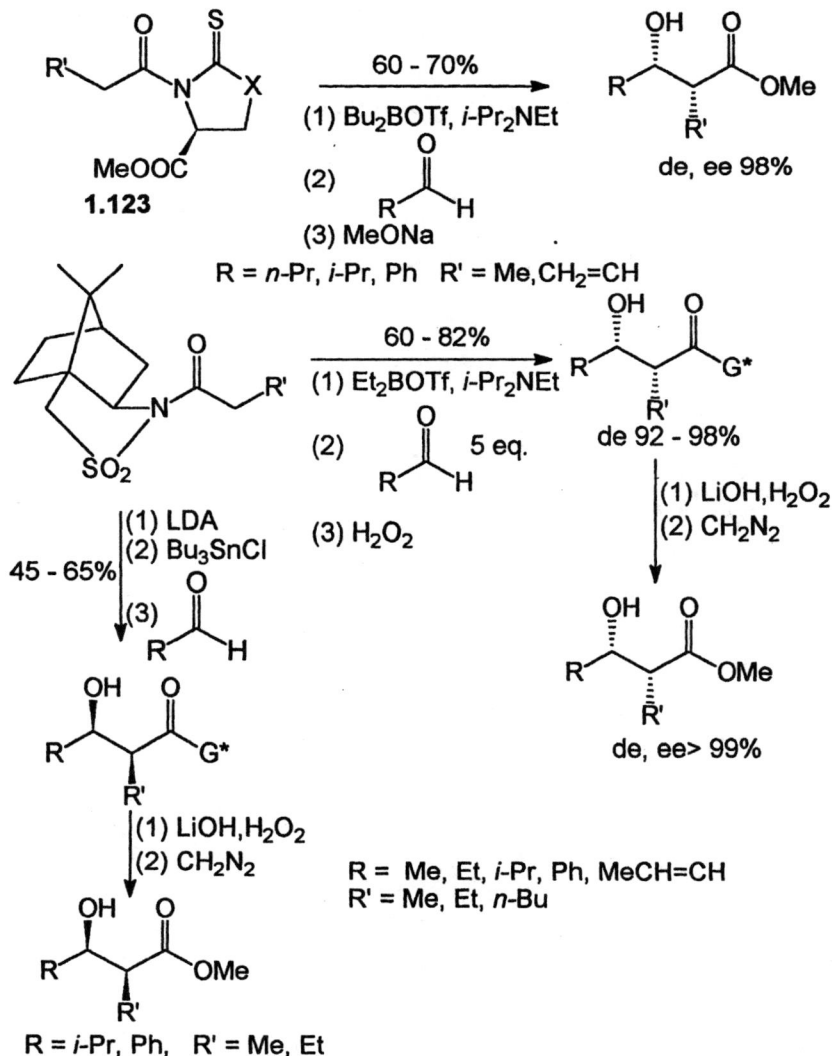

Figure 6.83

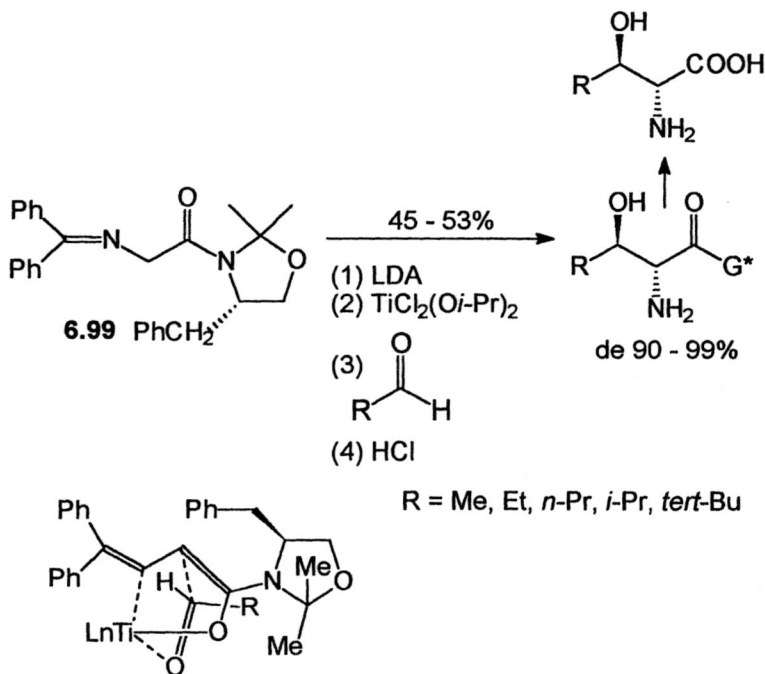

Figure 6.83 (continued)

α-Functionalized β-hydroxyesters have been prepared from dioxolanone **1.106** [209, 375], and the selectivity is useful with *i*-PrCHO. The asymmetric synthesis of nonracemic β-hydroxy-α-aminoacids has been performed by using several chiral auxiliaries. From aldehydes or ketones and titanium enolates generated from Schöllkopf's reagents **1.114**, *syn* β-hydroxy-α-aminoesters are obtained with a high selectivity after treatment of the initial products with CF_3COOH [154, 261, 861, 1273, 1274, 1275]. The reagent approach takes place on the least hindered face of the enolate, so that steric hindrance is minimized (Figure **6.84**). Seebach's heterocyclic reagents **1.125**, **1.126** and **1.127** are also very useful. Among these, **1.127** is preferred due to the easy removal of the chiral auxiliary [154, 261, 439, 441, 443, 861, 1052]. The lithium enolates of **1.127** are stable only at low temperatures, and the aldol condensation has been carried out at −100°C. The aldols, formed with a very high diastereoselectivity, are transformed into *syn*-β-hydroxy-α-aminoacids by hydrogenolysis and hydrolysis under mild conditions (Figure **6.84**). According to the configuration of the reagent **1.127**, either *syn* enantiomer of the product can be obtained. Model **6.100**, in which an unfavorable interaction between the nitrogen substituent of the enolate and the aldehyde C–R bond is avoided, accounts for the observed stereoselectivity [443]. The reactions of α-chiral aldehydes are also highly selective, as shown Figure **6.84** [154].

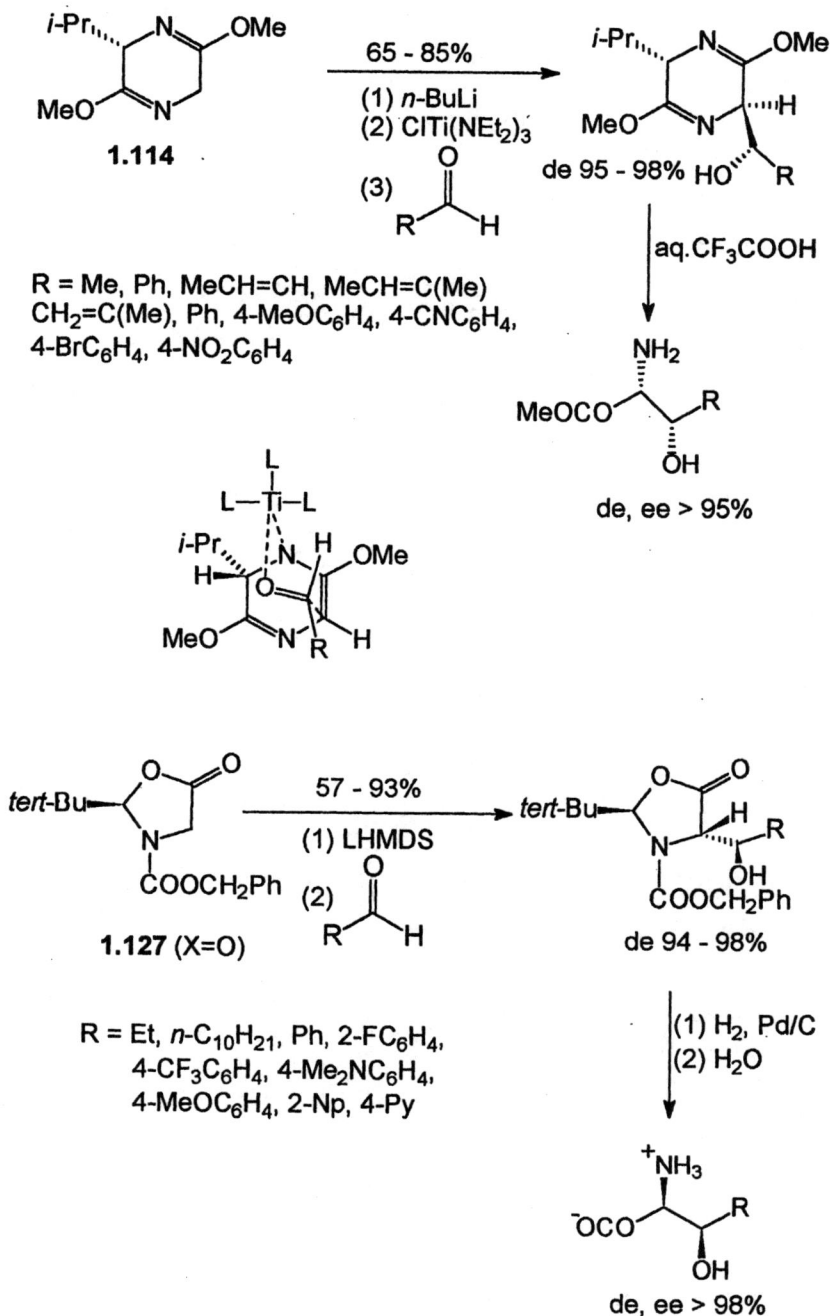

R = Me, Ph, MeCH=CH, MeCH=C(Me)
CH₂=C(Me), Ph, 4-MeOC₆H₄, 4-CNC₆H₄,
4-BrC₆H₄, 4-NO₂C₆H₄

R = Et, n-C₁₀H₂₁, Ph, 2-FC₆H₄,
4-CF₃C₆H₄, 4-Me₂NC₆H₄,
4-MeOC₆H₄, 2-Np, 4-Py

Figure 6.84

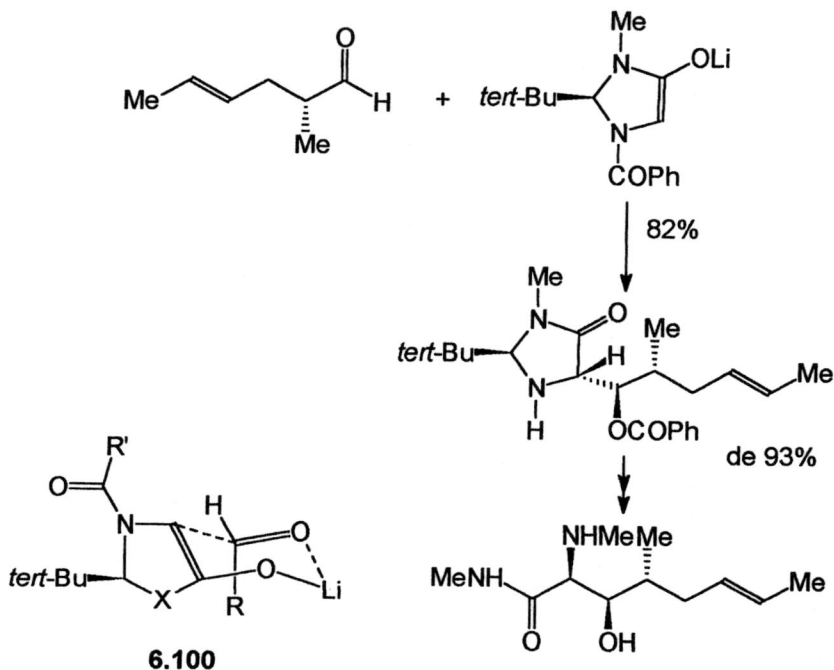

Figure 6.84 (continued)

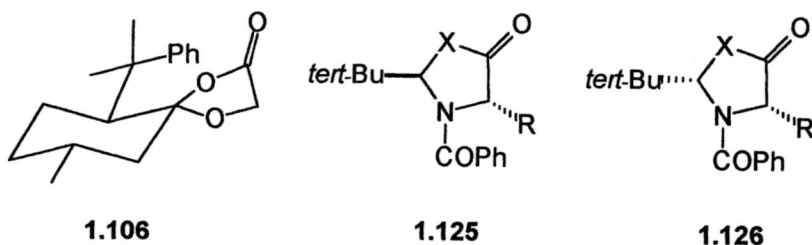

1.106   **1.125**   **1.126**

Belokon and coworkers used the copper or nickel complex of the imine generated from **1.109** (R = H, Ph) and glycine as a chiral reagent [261, 384, 1276]. When reactions with this reagent are carried out at room temperature in the presence of MeONa, *syn*-(1*S*,2*R*)-β-hydroxy-α-aminoacids are selectively obtained from MeCHO or aromatic aldehydes. When they are run in the presence of Et₃N, a high selectivity for the (1*S*,2*S*)-isomer is only obtained with MeCHO (Figure **6.85**).

1.109 (R = H or Ph)+H₂NCH₂COOH

Ni(NO₃)₂(R = Ph)
or CuSO₄(R = H)

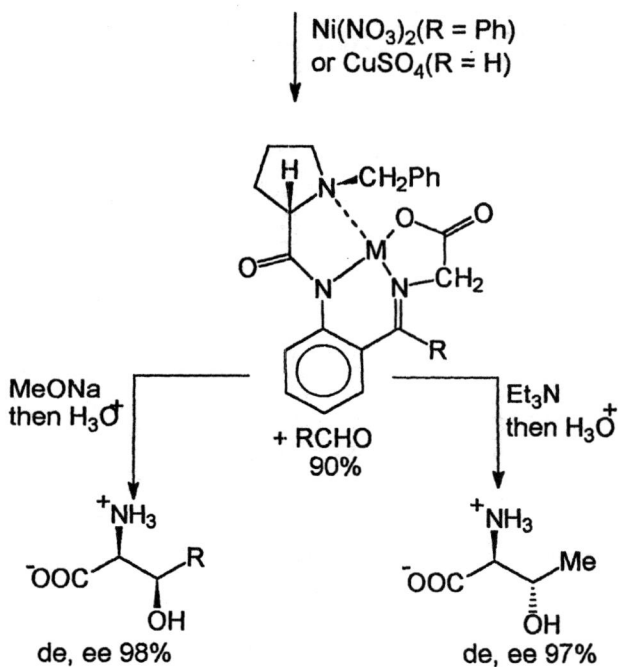

MeONa
then H₃O⁺

+ RCHO
90%

Et₃N
then H₃O⁺

de, ee 98%

de, ee 97%

R = Me, Ph, 2- or 4-FC₆H₄

R = Me

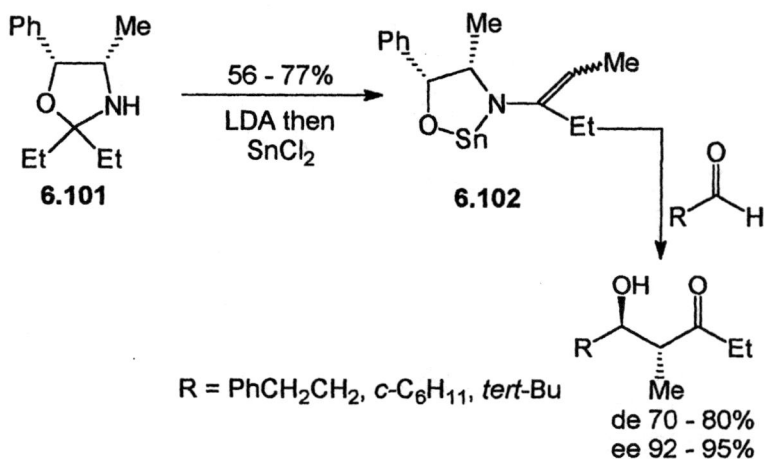

56 - 77%

LDA then
SnCl₂

6.101

6.102

R = PhCH₂CH₂, c-C₆H₁₁, tert-Bu

de 70 - 80%
ee 92 - 95%

Figure 6.85

Boron azaenolates derived from chiral oxazolines have been reacted with aldehydes [154, 697], and *anti* aldols are formed with high diastereoselectivity (> 95%) but with modest enantiomeric excess. More satisfactory results were obtained by Narasaka and Niwa [1277] by using tin azaenolate **6.102** (Figure **6.85**).

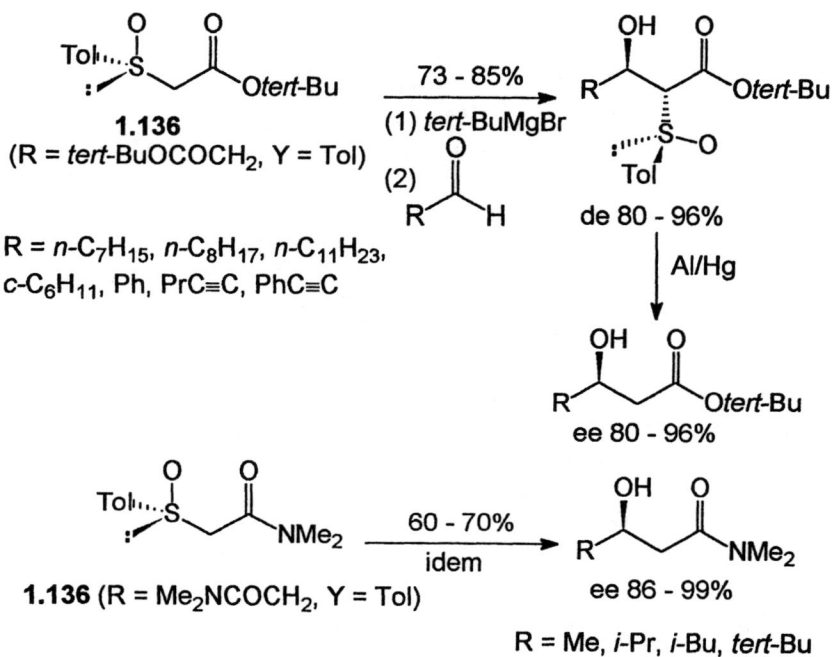

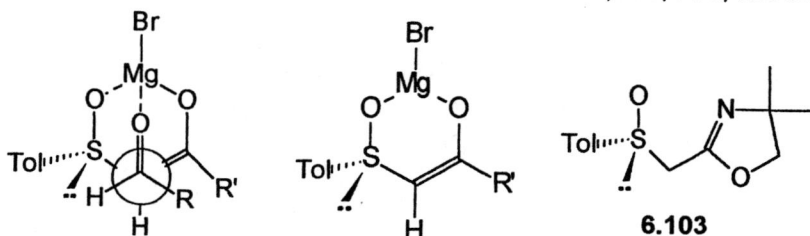

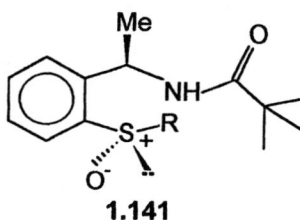

Figure 6.86

1.141

The reactions of aldehydes with enolates of esters and amides bearing a chiral sulfoxide group have been studied by Solladié and coworkers [209, 481, 681, 1278].   Magnesium chelates generated from **1.136** (R = *tert*-BuOCOCH$_2$, Me$_2$NCOCH$_2$) react very selectively with prochiral saturated and unsaturated aldehydes [481] (Figure **6.86**).   The aldehyde approaches from the same side of the chelate as the sulfoxide lone pair, so that steric hindrance between the sulfoxide and aldehyde R group is minimized (Figure **6.86**).   Aluminum amalgam reduces the sulfoxide residue and completes a highly enantioselective synthesis of β-hydroxyesters or -amides (Figure **6.86**).   Recoverable chiral sulfoxide **1.141** also has been successfully used in asymmetric aldol reactions [499, 501].   Cinquini and coworkers [209, 475, 475a, 481, 1278] extended the sulfoxide strategy to thioamides **1.136** (R = Me$_2$NCSCH$_2$) and dihydrooxazoles **6.103**, but the observed selectivities are often lower.   Lower selectivities are also observed in reactions run with ketones **1.136** (R = MeCOCH$_2$) [209] and 2-triarylsilylethylsulfoxides **1.136** (R = Ph$_3$SiCH$_2$CH$_2$) [1279].

Propionylcyclopentadienyliron carbonyl complexes **1.151** (R = MeCH$_2$) form enolates whose aldol condensations are highly selective.   Depending on the associated metal, either *anti* (aluminum) or *syn* (copper) aldols are predominantly formed at −100°C.   The absolute configuration of these aldols depends upon that of the starting complex [408, 522] (Figure **6.87**).

These results are interpreted by the attack of the aldehyde on the least hindered side of the enolate **6.104**, through a chair or twist-boat-like transition state [408] (see above).   Both (*S*)- and (*R*)-chromium tricarbonyl complexes of 2-methoxyacetophenone have been prepared, and their boron enolates react selectively with MeCH=CHCHO, although further decomplexation was not carried out [1280].   Chiral aminonitrile **1.71** bearing the ephedrine skeleton [301] or menthone-derived ketals [374] have been used as chiral auxiliaries in aldol reactions, but modest selectivities were obtained.

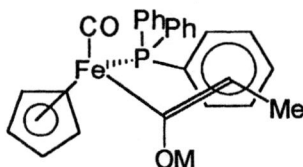

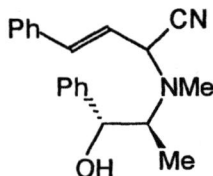

6.104   **1.71**

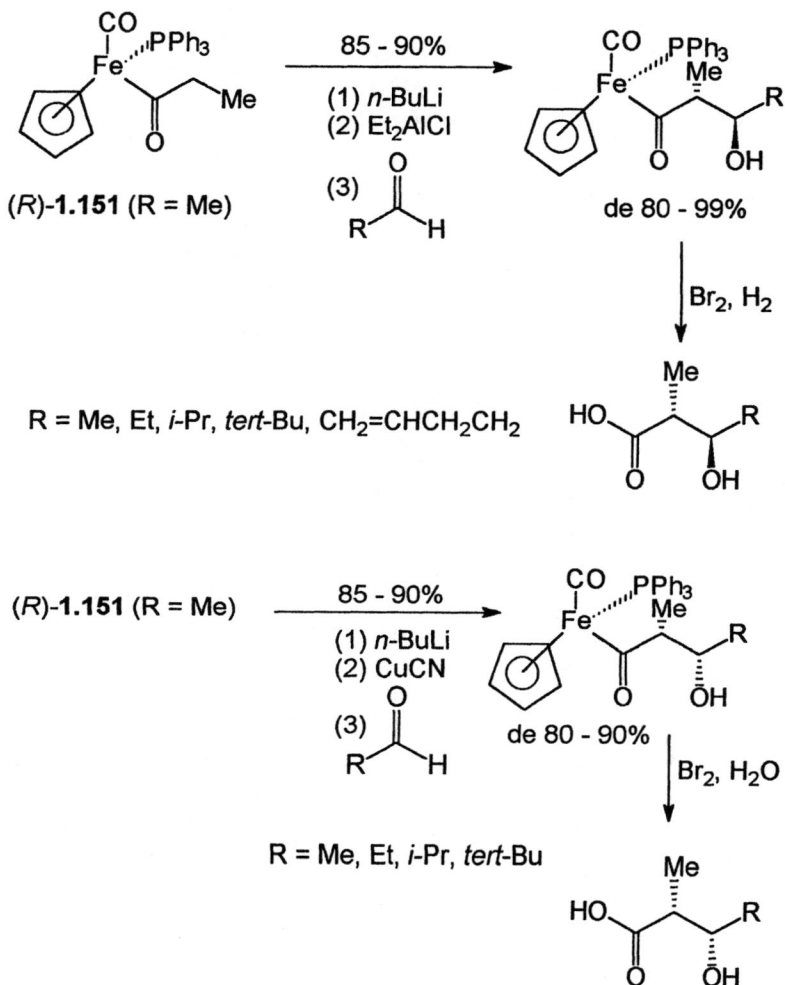

Figure 6.87

6.8.2 Reactions with Functionalized Aldehydes and Ketones

The reaction of α-ketoesters with the tin (II) enolate of N-acetyl-thiazolidinethione **1.123** (R = R' = H, X = S) has been carried out with an excellent enantioselectivity in the presence of chiral amine **2.13** (Figure **6.88**). The reactions of α- or β-alkoxyaldehydes with metal enolates may or not be under chelation control (Cram cyclic model, § I.4.2). Reetz has shown that triisopropoxy- or tris-diethylaminotitanium enolates display a sufficiently weak Lewis acidity so as to avoid chelation control [408].

Chiral 2-substituted benzaldehyde chromium tricarbonyl complexes have been reacted with chloroacetophenone in the presence of KO-*tert*-Bu [544]. After decomplexation, the E-epoxyketone is obtained with a high selectivity (Figure **6.88**). This Darzens reaction with ClCH$_2$COO-*tert*-Bu is poorly stereoselective. Condensation of the same aldehydes with methyl acrylate or acrylonitrile in the presence of DABCO, followed by decomplexation, also leads highly selectively to β-hydroxyesters or -nitriles **6.105** (Y = COOMe or CN) [547] (Figure **6.88**). An *anti* aldol product is also obtained with a high selectivity from a chromium complex and the titanium enolate of PhCH$_2$OCH$_2$COS-*tert*-Bu at −78°C [1281, 1282]. Chiral aminals of α-ketoaldehydes react with lithium or sodium enolates of ethyl acetate. After treatment with acid, compounds **6.106** are obtained with a high enantiomeric excess (Figure **6.88**).

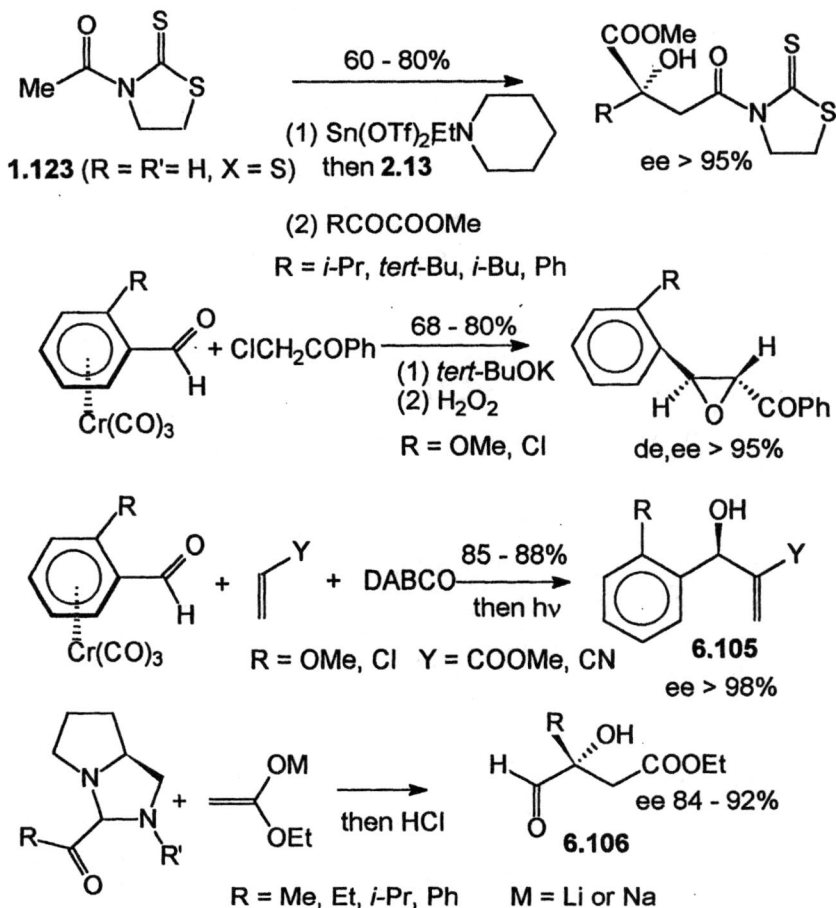

Figure 6.88

6.8.3 Reactions with Imines and Derivatives

The reactions of imines with ester enolates provide a route to β-lactams. Because of the importance of this class of antibiotics, much study has been devoted to this type of reaction [116a, 316a]. Recently published theoretical treatments [1283, 1284] suggest a mechanistic pathway different from that of the aldol reaction.

Prochiral imines and the boron enolates of thiolesters Z-**6.85** designed by Corey and coworkers [692] react at –78°C, leading highly selectively to *anti*-β-aminothiolesters. 3,4-Disubstituted nonracemic *trans*-β-lactams are formed by treating these products with *tert*-BuMgCl (Figure **6.89**). A twist-boat-like transition state model is proposed to account for the observed selectivity. As in the case of aldehydes, the coordination of the imine nitrogen by the boron atom of the reagent is invoked (Figure **6.89**).

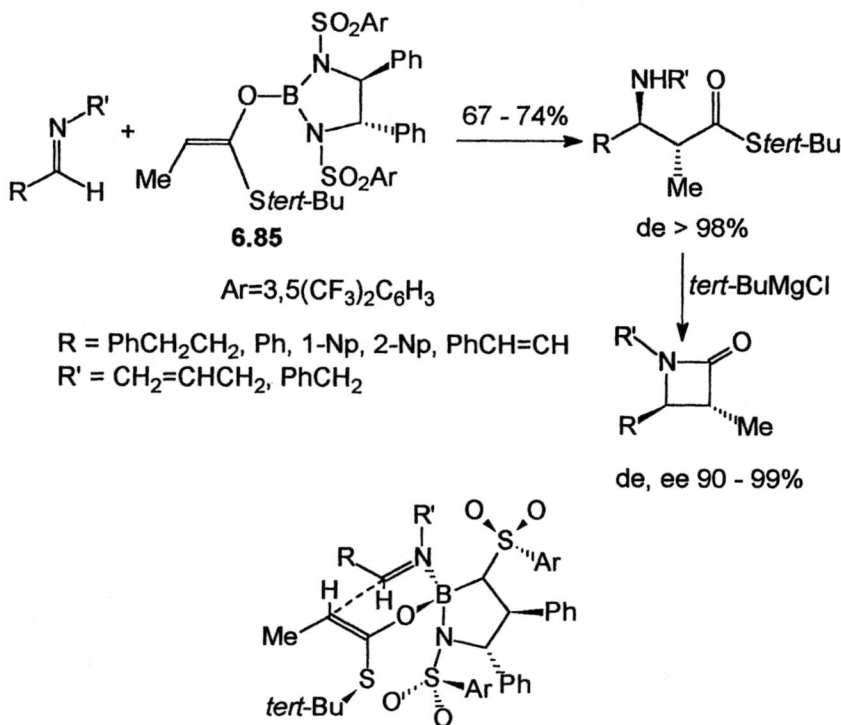

Ar=3,5(CF$_3$)$_2$C$_6$H$_3$

R = PhCH$_2$CH$_2$, Ph, 1-Np, 2-Np, PhCH=CH
R' = CH$_2$=CHCH$_2$, PhCH$_2$

Figure 6.89

Hart [147], Ojima and their coworkers [131, 150, 1285] performed related reactions with ester enolates of chiral alcohols at $-78°C$. *N*-Arylimines react selectively with the lithium E-enolate of propionate ester **6.107** having a bornane skeleton. After treatment of the product with $Ce(NO_3)_6(NH_4)_2$, an *N*-unsubstituted *cis*-β-lactam is obtained with a high diastereo- and enantioselectivity (Figure 6.90). Similarily, lithium E-enolates of *trans*-2-phenylcyclohexyl triisopropylsilyloxyacetates **6.108** also react highly selectively with *N*-aryl or *N*-silylimines, giving *cis*-β-lactams (Figure 6.90). The reactions of other alkoxyacetates are less satisfactory, but good results were obtained from bis-silylaminoacetates of chiral alcohols **6.109** [131]. Ito and coworkers prepared chiral β-lactams by a Reformatsky reaction between imines and *N*-bromopropionoyloxazolidinone **6.110**, and the selectivities were around 80% [1285a]. Tin (II) enolates of chiral acylthiazolidinethiones **1.123** react selectively with acylimminium salts generated *in situ* from α-acetoxylactams. After hydrolysis, nonracemic substituted lactam-acids or their derivatives are obtained [437, 1286, 1286a, 1287] (Figure 6.90). This method has been applied in the synthesis of alkaloids and functionalized β-lactams. Greene and coworkers used Oppolzer's sultam as chiral auxiliary in similar reactions [1288].

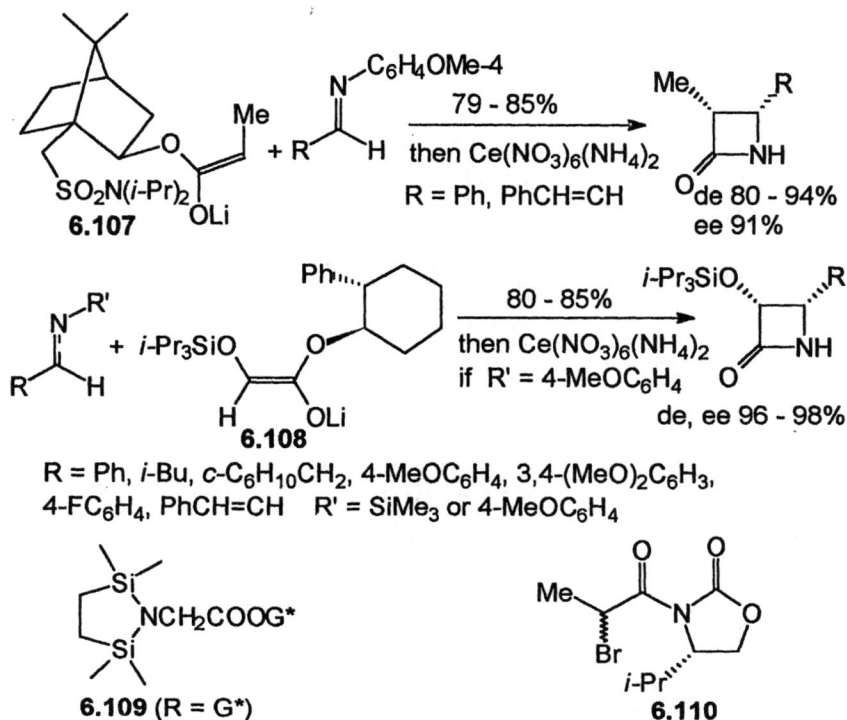

$R = Ph, i\text{-}Bu, c\text{-}C_6H_{10}CH_2, 4\text{-}MeOC_6H_4, 3,4\text{-}(MeO)_2C_6H_3,$
$4\text{-}FC_6H_4, PhCH=CH \quad R' = SiMe_3 \text{ or } 4\text{-}MeOC_6H_4$

Figure 6.90

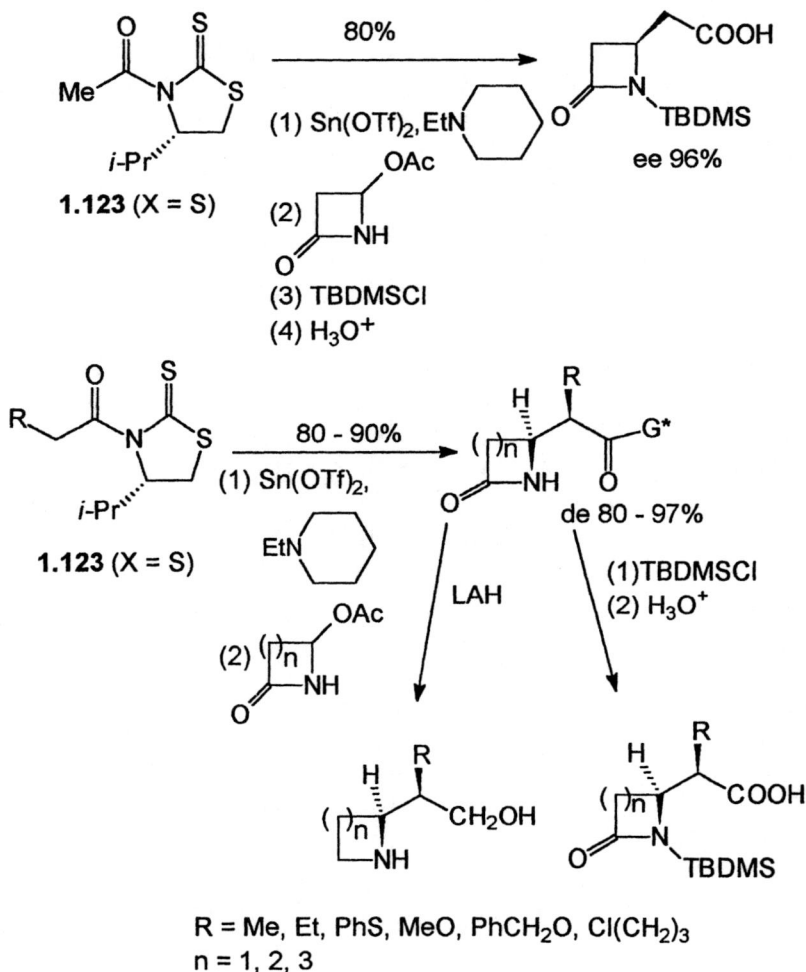

Figure 6.90 (continued)

Chirality also has been introduced on the imine. After treatment with Na/NH$_3$, 3-amino-2-azetidinones are obtained from reactions of (R)- or (S)-1-phenethylamine imines and zinc enolates of ethyl bis-silylaminoacetate **6.109** (R = Et) [131, 1289]. In Et$_2$O, *trans* isomers are formed, while in THF-HMPA, *cis* isomers predominate. The facial selectivity depends upon the (R)- or (S)-configuration of the imine N-substituent (Figure **6.91**). If the imine C-substituent R is Me$_3$C≡C, then the selectivity is lower. Cyclic transition-state models account for the observed selectivities. From the titanium enolate of 2-pyridylthioester Me$_2$CHCOS-2-Py and benzaldehyde (S)-1-phenethylimines **6.111**, Cinquini, Cozzi and coworkers obtained nonracemic β-lactams with a high selectivity at −78°C [1290] (Figure 6.91). However, the selectivity was lower when starting from other thioesters (RCH_2COS-2-Py, R = Et, i-Pr, i-Pr_3Si).

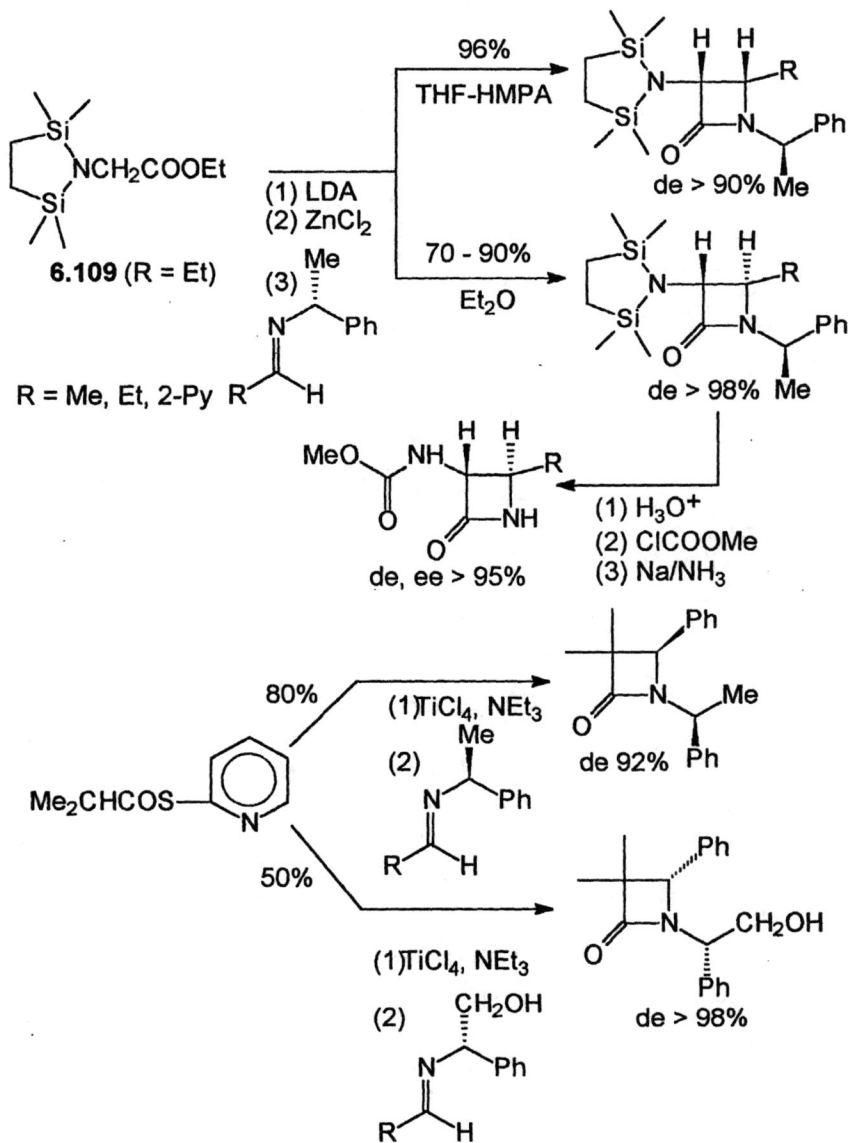

Figure 6.91

The use of chiral sulfoximines **1.136** (R = Ph, Tol, Y = ArCH=N) has allowed the enantioselective synthesis of β-aminoesters after cleavage of the S–N bond by CF₃COOH [510]. Preliminary studies showed that the reaction of C-arenechromium tricarbonyl imines and the lithium enolate of Me₂CHCOOEt gave chiral β-lactams after decomplexation with an excellent enantiomeric excess [549, 1291].

(R)-**1.136**   (S)-**1.136**

A route to functionalized β-lactams has been developed by Fujisawa and co-workers [218, 1292-1294a] by using imine **6.111** bearing a chiral acetal residue. Condensation of the imine **6.111** (R = Me) with an excess of a symmetrically substituted ester enolate at −78°C leads highly selectively to either diastereoisomeric β-lactam, depending on the associated cation (Li or Zn on the one hand, Ti on the other hand) (Figure **6.92**). Removal of the chiral acetal can be accomplished under acidic conditions. The reactions of titanium enolates of monosubstituted *tert*-Bu esters in the presence of HMPA are highly diastereoselective. Reactions of zinc enolates are also selective, and generate a different isomer. From the lithium enolate, a third isomer is predominantly obtained, although with a poorer selectivity (70%) [1190] (Figure **6.92**).

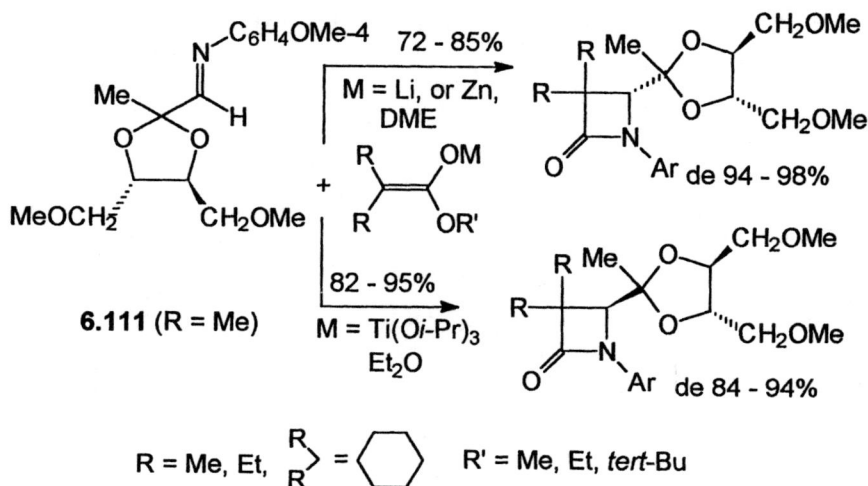

Figure 6.92

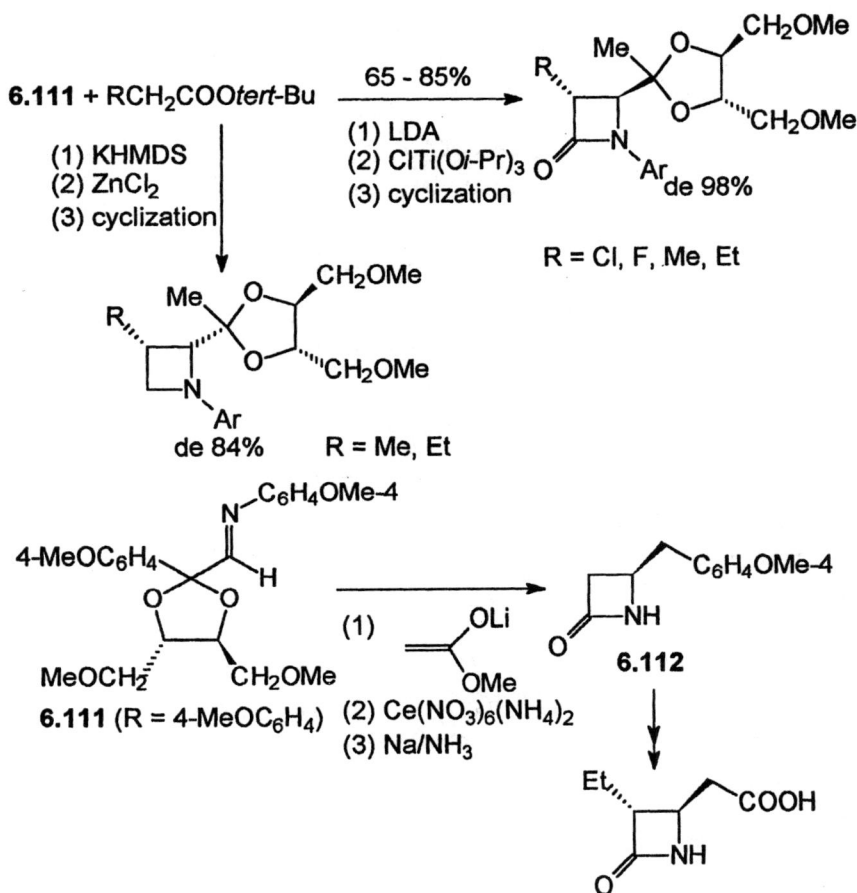

Figure 6.92 (continued)

A precursor **6.112** of carbapenems [1294] is selectively obtained by the reaction of **6.111** (R = 4-MeOC₆H₄) with the lithium enolate of methyl acetate, followed by treatment of the product with Ce(NO₃)₆(NH₄)₂ and then with Na/NH₃ (Figure **6.92**).

6.9 REACTIONS OF ENOXYSILANES [1295]

Lewis-acid-catalyzed reactions of enoxysilanes with aldehydes and ketones have been developed by Mukaiyama and coworkers [832]. The Lewis acid activates the C=O group of the electrophile. The O–Si bond of the reaction product is usually cleaved by a nucleophile or during hydrolytic workup. However, aldol silylethers are obtained under carefully controlled conditions [1296]. It has been shown that metal enolates are not involved in this process (Figure **6.93**). However, like metal enolates, the method of generation of the enoxysilanes determines their E- or Z-geometry [1297, 1298].

6.9.1. Reactions with Nonfunctionalized and α-Unsaturated Aldehydes

The reactions of aldehydes with enoxysilanes of ketones usually exhibit poor stereoselectivity. Enoxysilanes derived from esters and thiolesters, on the other hand, give good results. Chirality has been introduced either on the Lewis acid or by using an ester of an enantiopure alcohol. As in the case of enolate reactions, one or two new stereocenters may be created. All these reactions take place at low temperatures, and they are sometimes limited by the instability of certain ketene silylacetals.

6.9.1.1 *Lewis Acid Bearing Chiral Ligands*

Mukaiyama and coworkers have performed the reaction of aldehydes and ketene or thioketene silylacetals under $Sn(OTf)_2$ catalysis in the presence of a chiral diamine **2.13** (R = $CH_2N(CH_2)_5$ or CH_2NH1-Np) and Bu_3SnF, SnO or $Bu_2Sn(OAc)_2$ [551, 833, 835-838, 841, 1299, 1300, 1301]. The reactions with enoxysilanes derived from benzyl or thioethyl acetates are highly enantioselective at −95°C (Figure **6.93**). The reactions of ketene silylacetals derived from thioethyl or phenyl propionate or benzyloxyacetate are diastereo- and enantioselective at −78°C, and the most efficient chiral amines are **2.13** (R = CH_2NH1-Np or CH_2NH1-tetrahydro,5,6,7,8-Np) [835, 839, 840, 1299, 1301] (Figure **6.93**). In reactions of α-alkynyl aldehydes or methacrolein, the selectivities are somewhat lower (de 90%, ee 80 - 92%) [836, 1299]. Depending on the reaction partners, $Sn(OTf)_2$, the chiral amine, and the Sn(IV) additive are used in catalytic or in stoichiometric amounts.

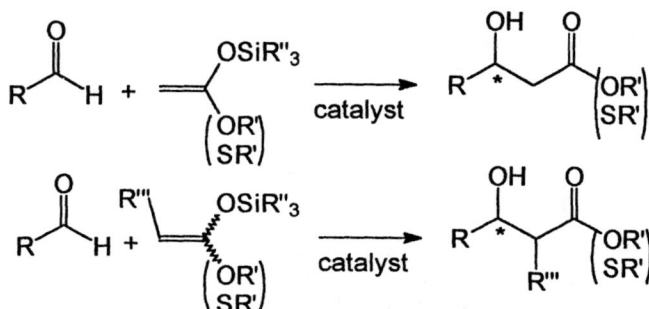

Figure 6.93

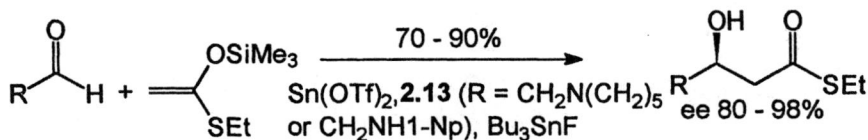

R = PhCH₂CH₂, *i*-Pr, *tert*-Bu, Ph

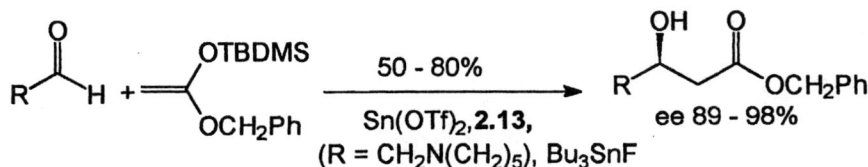

R = *i*-Pr, *c*-C₆H₁₁, MeCH=CH, Ph, 4-MeC₆H₄, 4-ClC₆H₄, *n*-BuC≡C

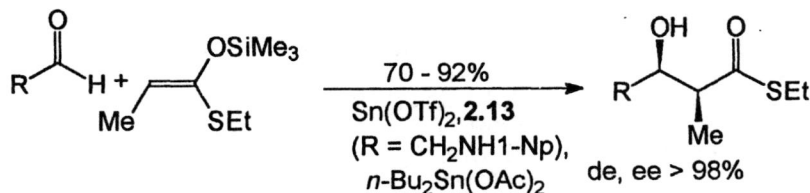

R = *n*-C₇H₁₅, *c*-C₆H₁₁, *i*-Pr, Ph, 4-ClC₆H₄, 4-MeOC₆H₄
MeCH=CH, PhCH=CH, *i*-Bu, 2-furyl, 3-thienyl

SnO, **2.13** (R=CH₂NH—⟨naphthyl⟩)

R = *n*-C₅H₁₁, MeCH=CH,
CH₂=CH, *n*-PrCH=CH

Figure 6.93 (continued)

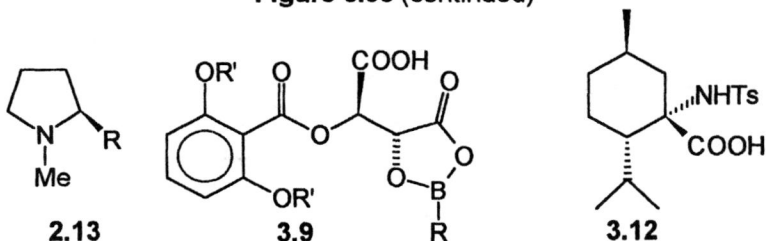

2.13   **3.9**   R   **3.12**

Chiral boranes have been recommended as Lewis acids catalysts by Reetz [689], Yamamoto [787, 788], Kiyooka [795, 1302], Masamune and their coworkers [796, 797]. These groups used, respectively, boranes **2.61**, **3.9** (R = H, R' = *i*-Pr), **3.10** (R = *i*-Pr or *tert*-Bu, R' = H) and derivatives of **3.12** and **3.13**. These boranes are very efficient catalysts in asymmetric additions of symmetrically substituted ketene silylacetals **6.113** to aldehydes (Figure 6.94). Similar reactions can also be conducted with enoxysilanes derived from methylketones or from *tert*-Bu thiolacetate [787, 794, 796]. Oxazaborolidine **3.10** derived from tryptophan **3.11** is also a very potent catalyst [794].

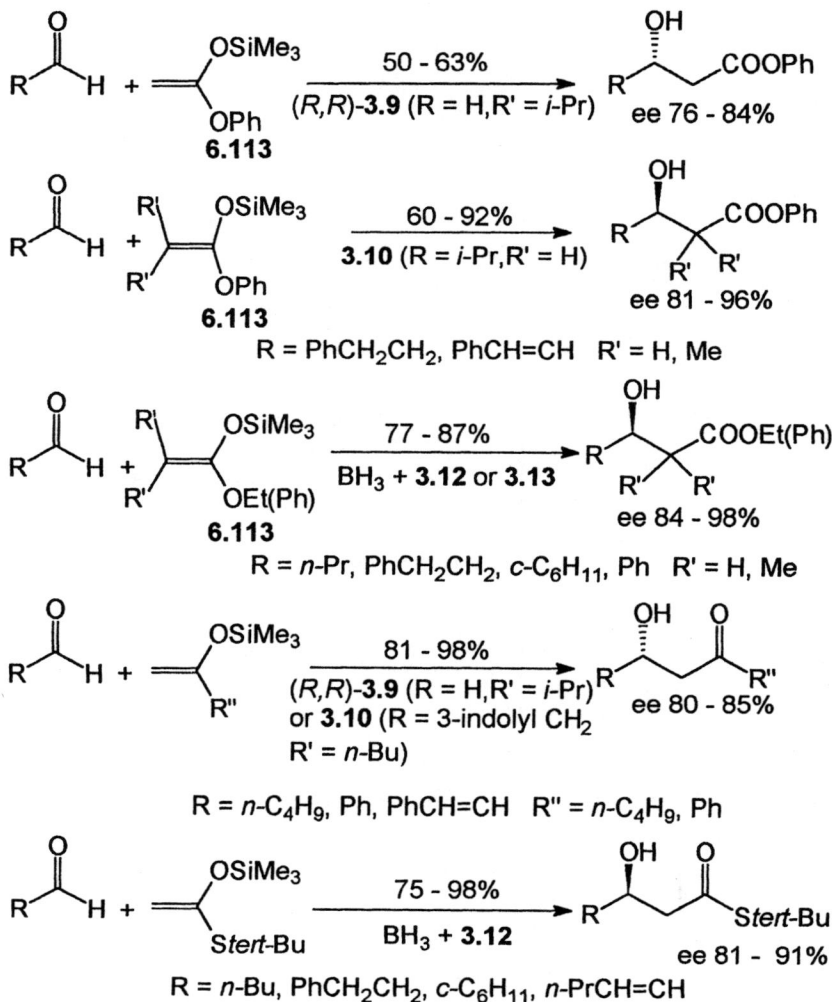

Figure 6.94

With the exception of **3.9**, these borane catalysts give lower selectivities with enoxysilanes of propionic esters or ethylketones (< 80%) [796, 1302]. Using **3.9** as a catalyst, high diastereo- and enantioselectivities are observed in the reactions of the E-ketene silylacetal of phenyl propionate with α,β-unsaturated aldehydes [788] and in the reaction of the enoxysilane of diethylketone with PhCHO [787] (Figure 6.95). All these results are interpreted by acyclic transition state models in which steric repulsions are minimized (Figure 6.95).

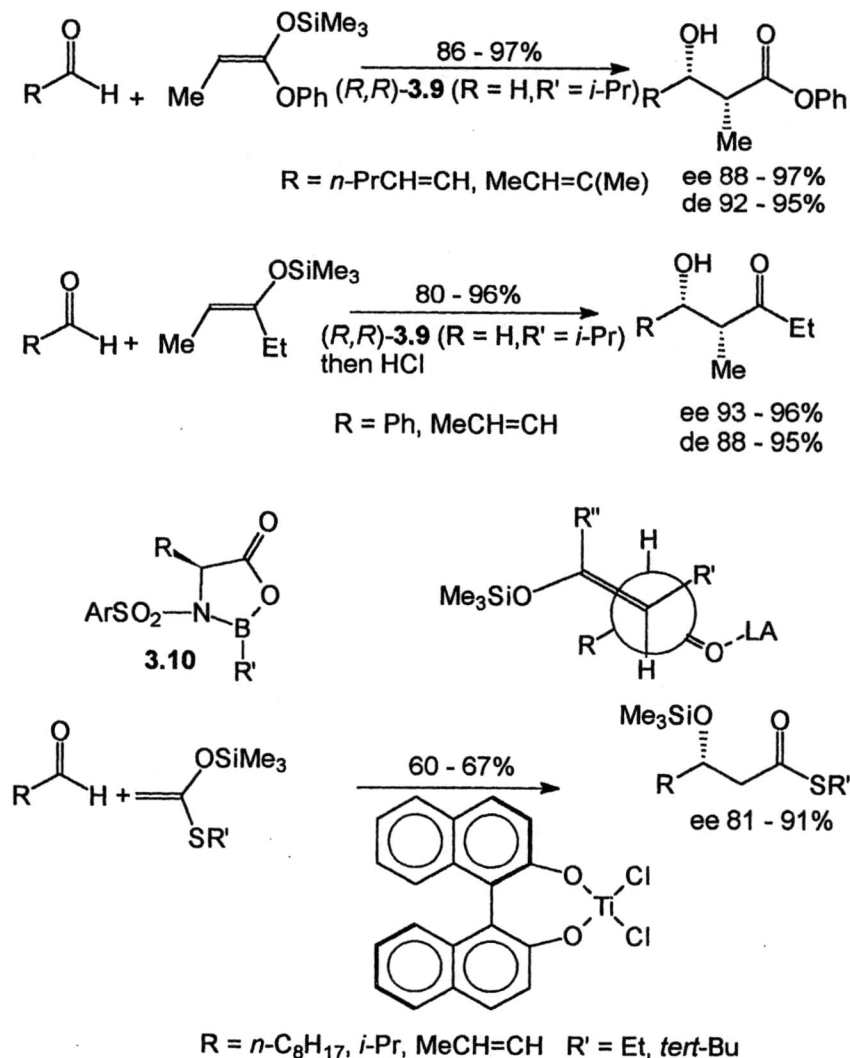

Figure 6.95

Lewis acid complexes formed by the reactions of various aminoalcohols with Et$_2$AlCl [778, 824] or by the reaction of Et$_2$Zn with a chiral sulfamide [806] have displayed a low efficiency in the asymmetric condensations of ketene and thioketene silylacetals derived from acetic acid with aldehydes. Disappointing selectivities have also been observed with some binaphtol-titanium complexes [778]. However, Mikami and Matsukawa [1296] recently performed the enantioselective condensation of various aldehydes with acetic acid derivatives in the presence of a chiral binaphtol-titanium complex. Good selectivities were observed when the reaction was performed at 0°C in toluene (Figure **6.95**). Quaternary ammonium fluorides derived from cinchona alkaloids have been proposed as catalysts to perform additions of enoxysilanes derived from ketones to PhCHO, but the observed selectivities are modest [1303].

6.9.1.2 *Ketene Acetals and Analogs Bearing Chiral Residues*

Esters of chiral alcohols having a bornane skeleton such as **1.8**, **1.10** [147, 1295] or esters of *N*-methylephedrine **1.14** (R = Me) [151, 152, 1295] have been transformed into the corresponding ketene silylacetals. These react with aldehydes in the presence of TiCl$_4$ or BF$_3$•Et$_2$O. From acetic acid derivatives, β-hydroxyacids or their precursors are obtained with a high enantioselectivity (Figure **6.96**). Lower selectivities are obtained from acetate derivatives of Oppolzer's sultam [1304]. The reactions of ketene silylacetals derived from propionates with prochiral aldehydes generate two new stereocenters. The formation of *anti* aldols is always favored, but the preferred face of attack of the aldehyde varies according to the E- or Z-geometry of the reagent, the nature of the silicon substituents and the Lewis acid. The reactions of ketene acetals **6.114** with PhCHO are poorly selective. From *i*-PrCHO, either *anti* aldol adduct can be obtained depending on whether E- or Z-**6.114** is used in the presence of TiCl$_4$ or of BF$_3$•Et$_2$O [147] (Figure **6.96**). From *N*-methylephedrine derivatives E-**6.115**, Gennari and coworkers [152, 1295] obtained *anti*-β-hydroxyesters with high selectivity by conducting reactions in the presence of TiCl$_4$ and PPh$_3$ (Figure **6.96**).

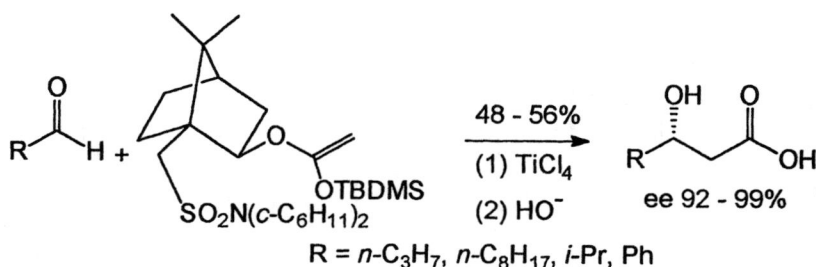

R = n-C$_3$H$_7$, n-C$_8$H$_{17}$, i-Pr, Ph

Figure 6.96

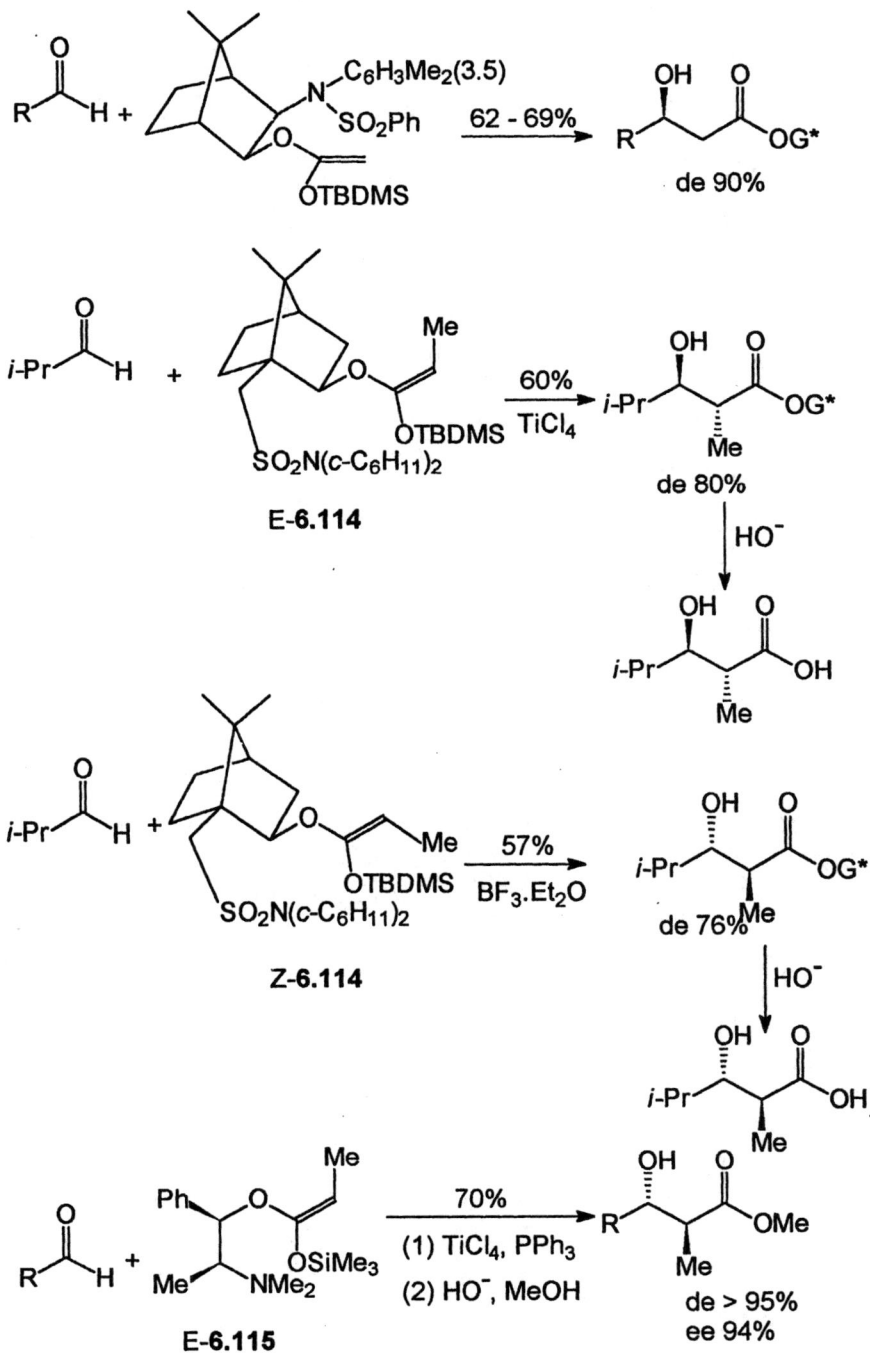

Figure 6.96 (continued)

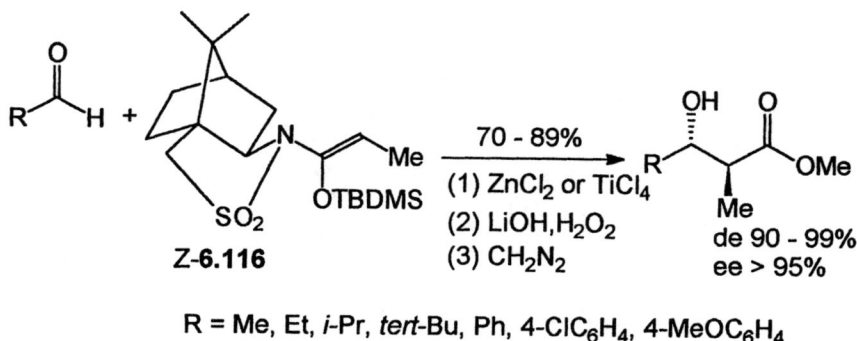

R = Me, Et, *i*-Pr, *tert*-Bu, Ph, 4-ClC$_6$H$_4$, 4-MeOC$_6$H$_4$

Figure 6.96 (continued)

    Oppolzer and coworkers also synthetized *anti*-β-hydroxyesters with a high selectivity starting from Z-**6.116** [459] (Figure **6.96**). When the silicon atom is sterically hindered, the results are consistent with acyclic transition states A *Re* from Z-ketene silylacetals and A *Si* from E-isomers [147]. The approach should take place on a Lewis-acid-aldehyde complex having an *anti* conformation similar to that favored in the ground state [85]. Repulsive interactions between the chiral residue and the aldehyde R group (A *Re* model) or between the bulky Lewis acid and the OTBDMS group (A *Si* model) are minimized in each case (Figure **6.97**). If the silicon substituents are not bulky (Me$_3$Si), TiCl$_4$ may coordinate to the silicon atom and induce nucleophilic assistance [147, 1305]. The attack of the *Re* face of the aldehyde observed by Gennari in the presence of PPh$_3$ [1295] can be understood by such a six-coordinate titanium complex **6.117** (Figure **6.97**). However, a cyclic mechanism has been advocated in other cases (see below).

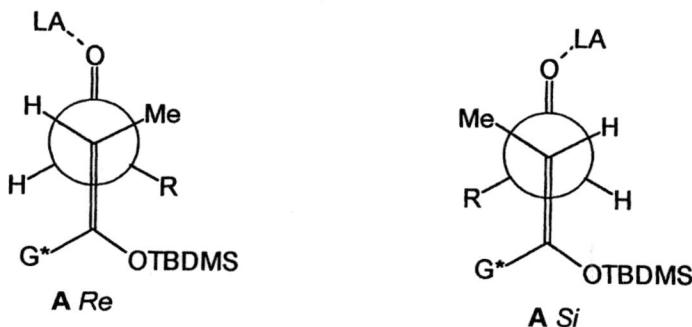

Figure 6.97

6.117

60 - 77%

1.94

R = Et, *i*-Pr, Ph

de 96 - 99%

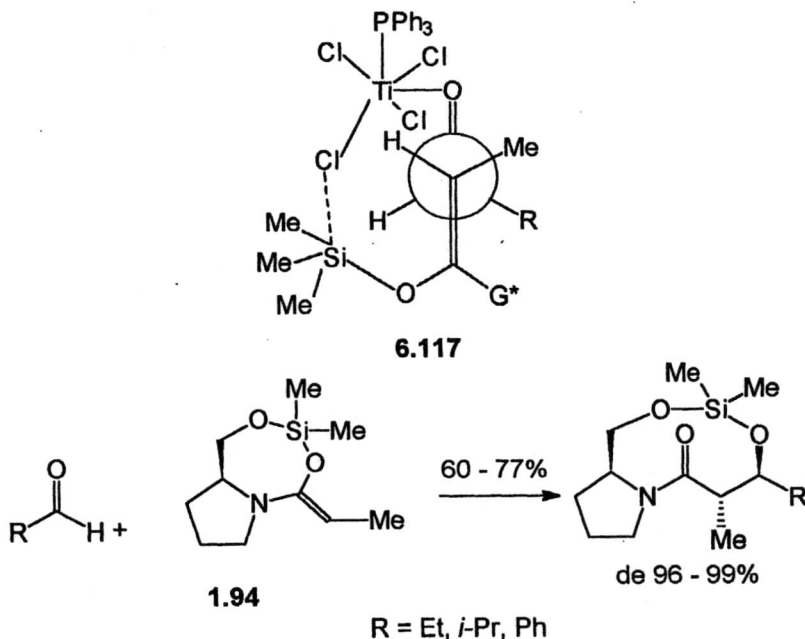

Figure 6.97 (continued)

The reactions of enoxysilanes derived from ketones bearing a chiral butadieneiron carbonyl group with aldehydes are poorly stereoselective [1306]. The reactions of α-chiral aldehydes occur with double diastereodifferentiation, and high
selectivities are obtained when the reagents are matched [1295]. The reactions of
aldehydes with chiral ketene aminal **1.94** do not require Lewis acid activation, and
they are highly stereoselective [353]. Although the chiral auxiliary has not yet been
removed, this method is quite promising (Figure **6.97**).

6.9.2 Reactions with Functionalized Aldehydes and Ketones

α-Ketoesters have been reacted with ketene silylacetals in the presence of
chiral Lewis acids at −78°C. Reaction of thioethyl acetal derivative **6.118** (R' = H)
with $Sn(OTf)_2$, chiral amine **2.13** (R = $CH_2N(CH_2)_5$) and Bu_3SnF [1300] forms
the expected hydroxydiester **6.119** in a high enantiomeric excess (Figure **6.98**).
Similar reactions were conducted with alkoxyacetic acid analogs **6.118** [1307].
Anti isomers predominated when R' = $PhCH_2O$, and *syn* isomers predominated
when R' = TBDMSO. However, the structure of the chiral diamine must be modified to obtain satisfactory selectivities (de 70 - 80%, ee 90%). Reactions of methyl
or butyl glyoxylate with enoxysilanes or with **6.118** (R = H) are catalyzed by an

(R)-binaphthol-titanium complex, and they occur with very high selectivity at 0°C
[1296, 1308] (Figure 6.98). Enoxysilanes lead to syn-2-hydroxy-3-alkyl Z-enol-
ethers 6.120 independent of the E- or Z-geometry of the reagent. A cyclic ene-
mechanism has been proposed to interpret the formation of 6.120 [1308].
Glyoxylates of chiral alcohols have also been used as electrophiles in these reactions. The reactions of phenmenthyl pyruvate MeCOCOOG* 1.23 (R = Me,
G*OH = 1.4, R = Ph) with enoxysilanes or ketene silylacetals are not highly
stereoselective [1226]. On the contrary, reactions of 1.23 (R = Me, Ph) with
similar reagents at −78°C are highly stereoselective when G* is quebrachitol derivative 1.49 [1309] (Figure 6.98).

Ketene silylacetals also react with α- or β-alkoxy- or -aminoaldehydes. Chelation control may take place in the reaction of ephedrine-derived ketene silylacetal
6.121 with β-benzyloxyaldehydes. Under TiCl$_4$ catalysis, syn isomers are favored
(Figure 6.98), but the reaction is highly selective only if the aldehyde is α-alkylated
and the reagents are matched [1295]. The results are interpreted through the intervention of a six-coordinate titanium complex 6.122 (Figure 6.98).

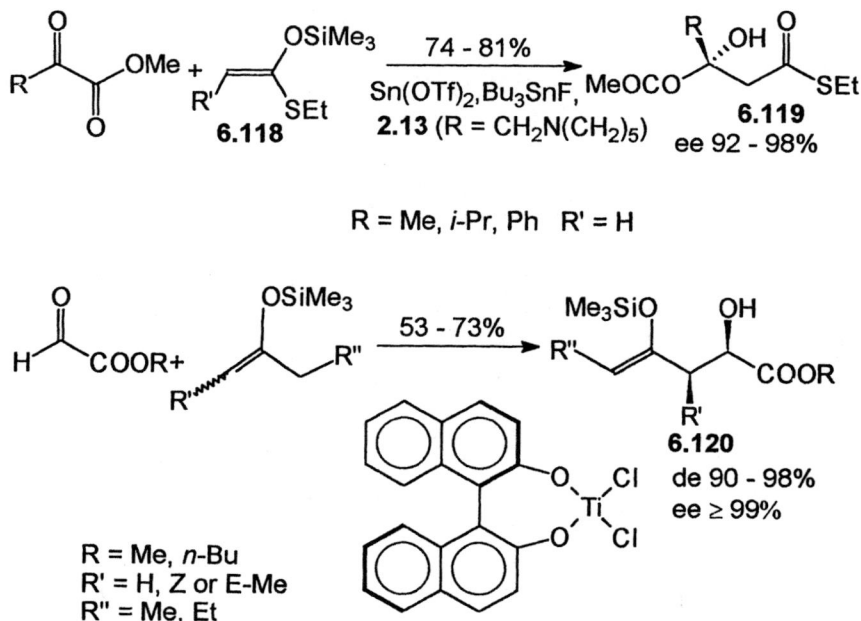

R = Me, i-Pr, Ph   R' = H

R = Me, n-Bu
R' = H, Z or E-Me
R" = Me, Et

Figure 6.98

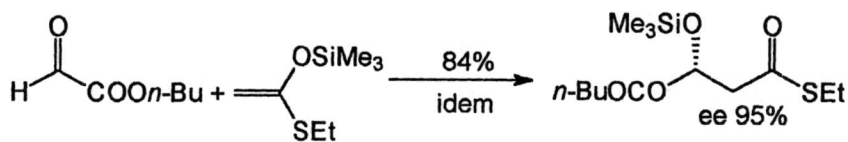

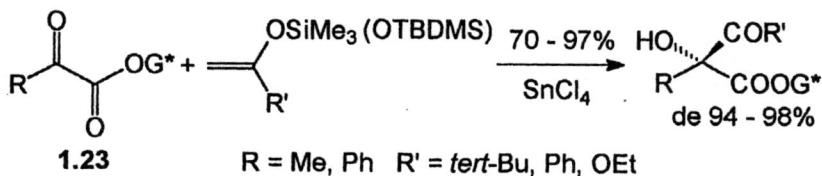

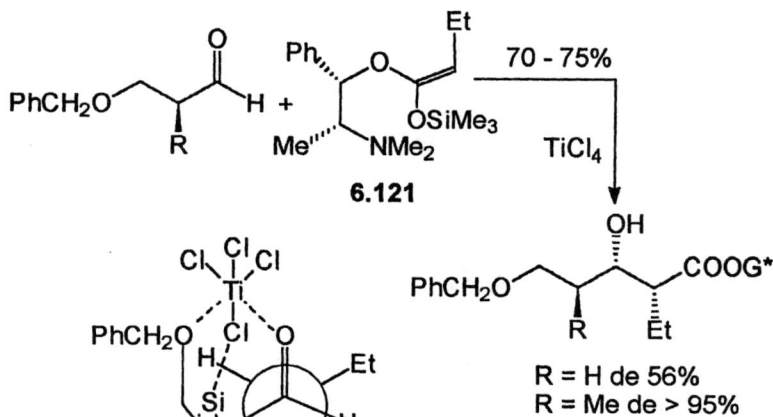

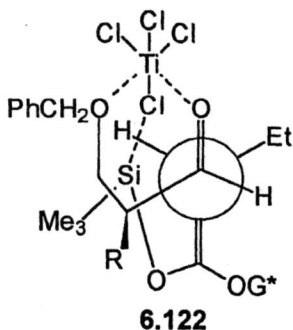

Figure 6.98 (continued)

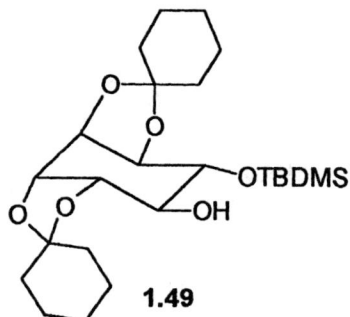

1.49

Reactions of α-benzyloxy-, α-chloro- and α-BOCNH acetaldehyde with **6.118** (R' = H) are catalyzed by a binaphthol-titanium complex, they are highly enantioselective when conducted in toluene at 0°C [1296]. A cyclic silatropic ene pathway has been proposed to interpret these results, with possible chelation of titanium to the heteroatom substituent (Figure **6.99**).

2-Substituted arenechromium tricarbonyl aldehydes have been reacted with enoxysilanes derived from cyclanones [545] or thiol propionates [546]. After de-complexation, the expected aldols are obtained with excellent selectivities (Figure **6.99**).

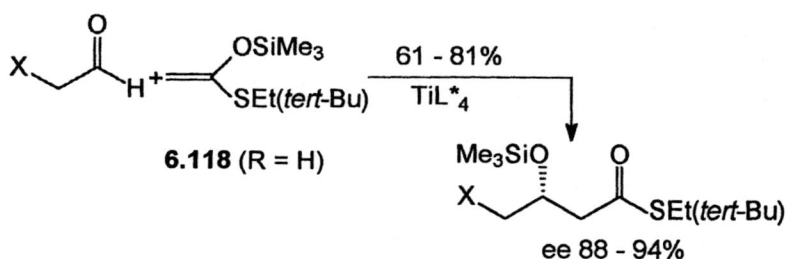

X = PhCH$_2$O, Cl, BOCNH

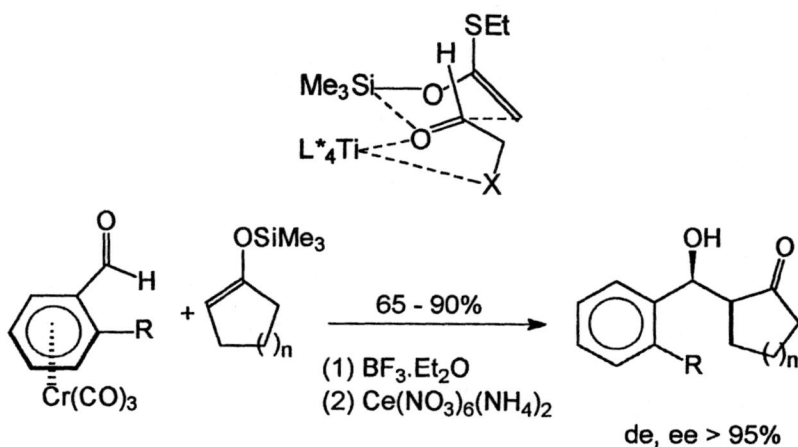

R = Me$_3$Si, MeO, Me, Et

Figure 6.99

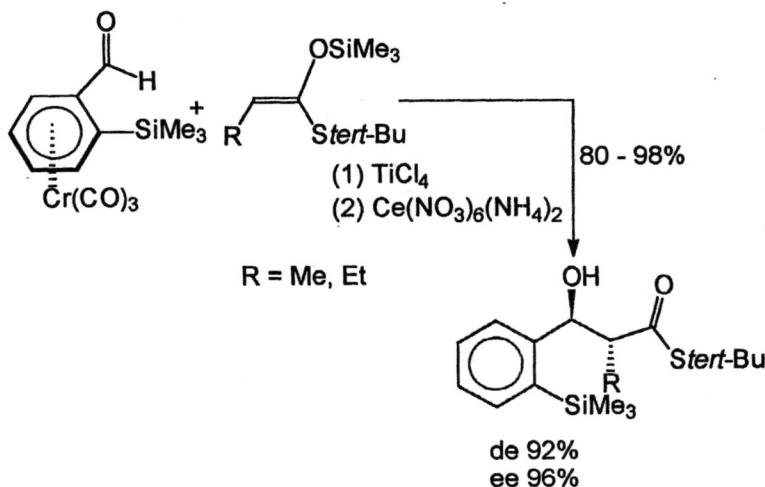

Figure 6.99 (continued)

6.9.3 Reactions with Chiral Acetals and Analogs

Chiral 1,3-dioxanes **6.20** (R' = H, R" = Me) react highly selectively with enoxysilanes and ketene silylacetals under $TiCl_4$ catalysis [678, 1295]. The removal of the chiral auxiliary is easy for the ester products, and nonracemic β-hydroxyesters can be obtained (Figure **6.100**). Because the selective cleavage of the diketones formed by oxidation of **6.123** is delicate, the use of (R)-1,3-butanediol instead of 2,3-pentanediol has been proposed [213, 1310]. Scolastico and coworkers [322, 333, 1311] used chiral oxazolidines **1.84** (EWG = $COOCH_2Ph$ or Ts), which react with **6.118** (R' = Me, $PhCH_2O$) in the presence of $TiCl_4$. After proper treatment, nonracemic aldehydes or methylketones are obtained with a high enantiomeric excess (Figure **6.100**).

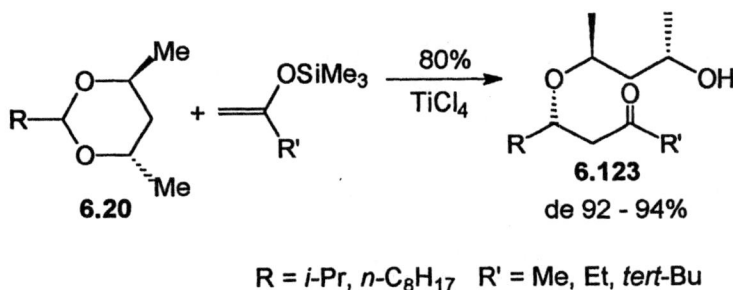

R = i-Pr, n-C₈H₁₇   R' = Me, Et, tert-Bu

Figure 6.100

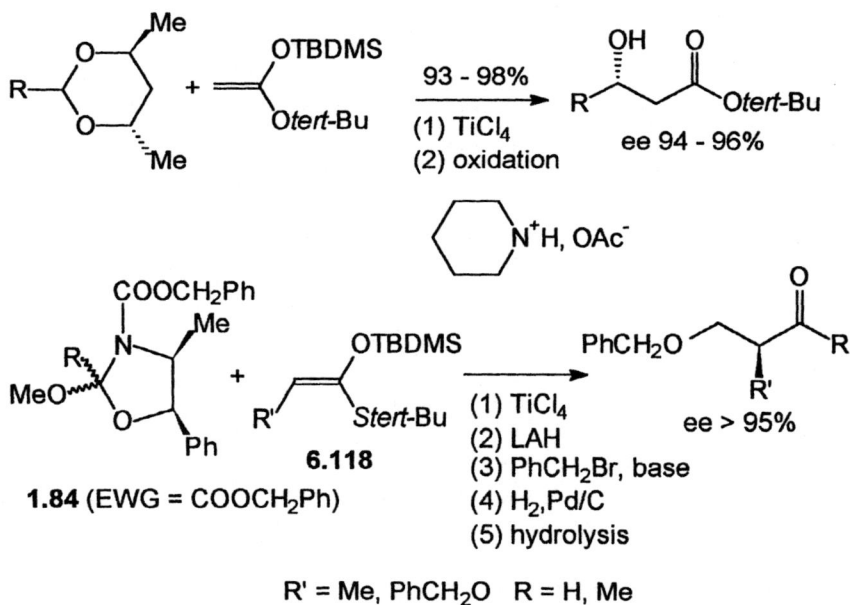

Figure 6.100 (continued)

6.9.4 Reactions with Imines and Derivatives

The reactions of ketene silylacetals with imines have been widely studied as a route to precursors of β-lactam antibiotics. The TiCl$_4$-catalyzed reactions of prochiral imines with ketene silylacetals derived from (1S,2R)-N-methylephedrine have been performed by Gennari and coworkers [1295] (Figure **6.101**). After treatment of the adducts with LHMDS, *trans*-β-lactams are obtained with a good selectivity when starting from benzaldehyde N-phenylimine. From other imines, the results are less satisfactory [1295]. Better selectivities were obtained by Ojima and coworkers [1295] by introducing chiral substituents on the nitrogen atom. Ketene silylacetals **6.113** (R' = Me) react with imines derived from nonracemic 1-phenethylamine or α-aminoesters and yield chiral β-lactams with a high selectivity under TiCl$_4$ catalysis (Figure **6.101**). This reaction is chelation controlled, and **6.124** reacts from its least hindered side (Figure **6.101**). The chiral auxiliaries can be removed by hydrogenolysis or hydrolysis. Kunz and coworkers [360] have also obtained useful selectivities from the same ketene silylacetal and imines bearing a chiral residue derived from a carbohydrate.

Yamamoto and coworkers [782, 1312] applied the double diastereodifferentiation concept (§ I.6) to the reaction of enantiopure (S)-1-phenethylimines with ketene silylacetals at −78°C in the presence of a chiral boron catalyst generated *in situ* from binaphthol **1.44** and B(OPh)$_3$ (Figure **6.101**).

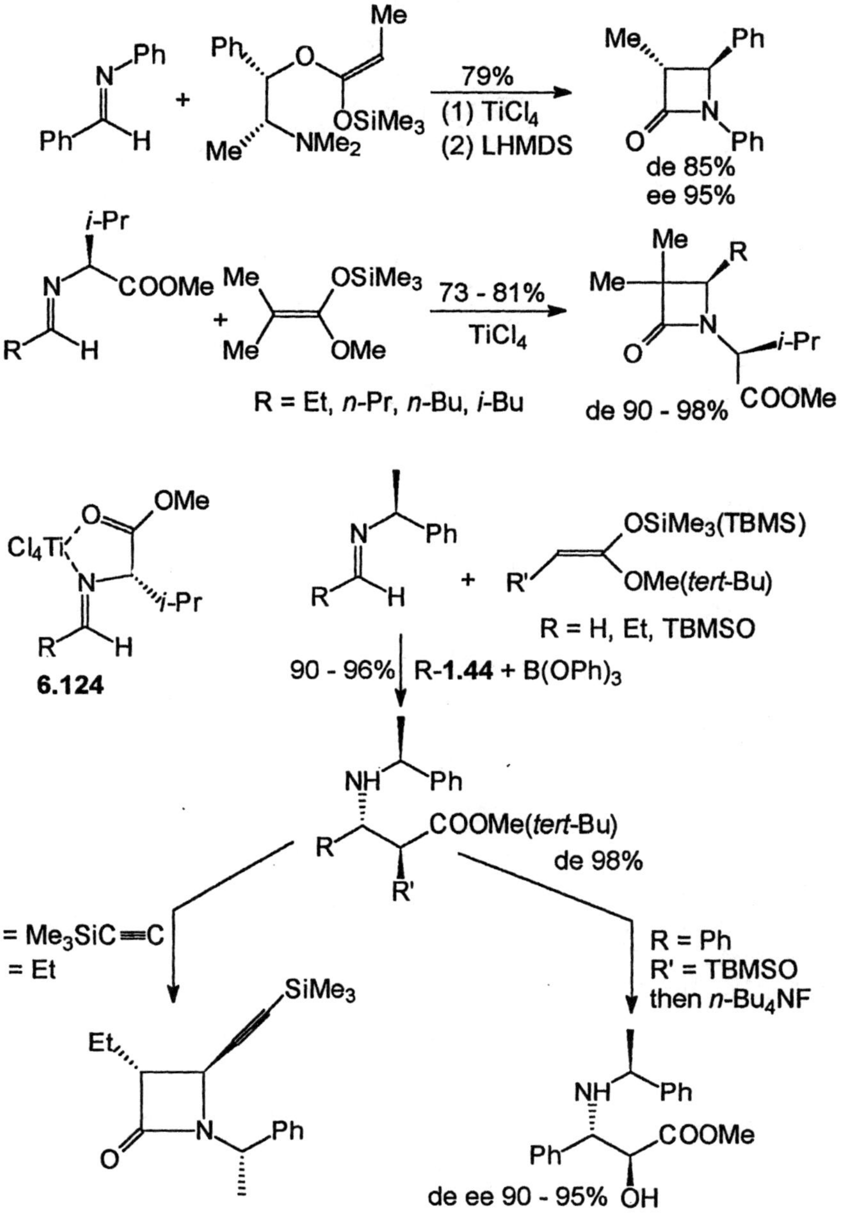

Figure 6.101

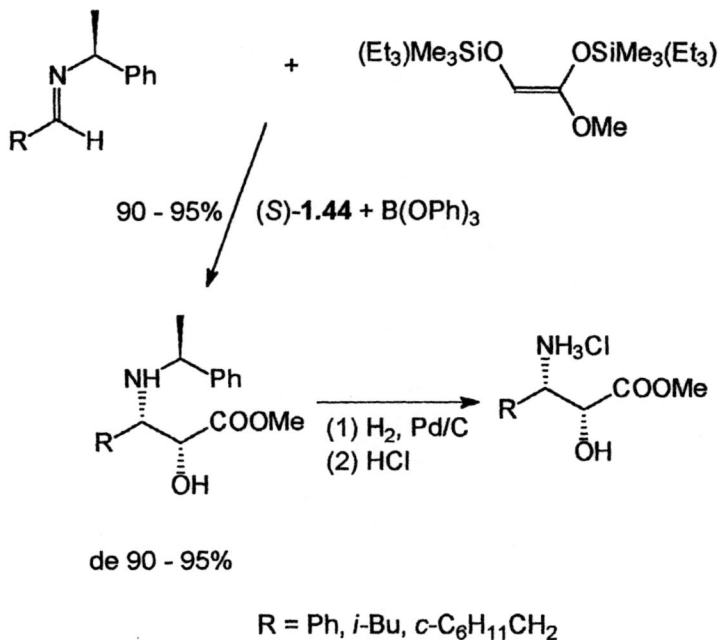

R = Ph, *i*-Bu, *c*-C$_6$H$_{11}$CH$_2$

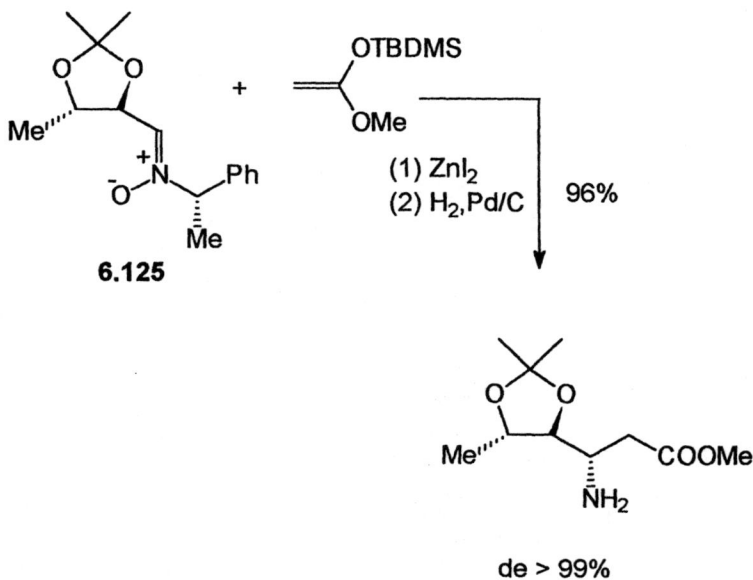

Figure 6.101 (continued)

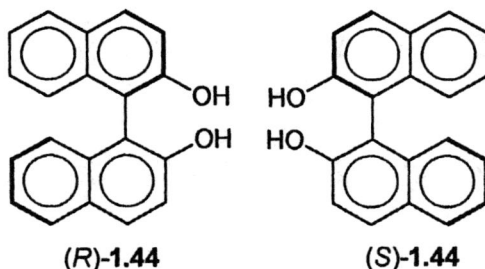

(R)-1.44   (S)-1.44

According to the E- or Z-geometry of the ketene silylacetal, the (R)-bi-naphthol catalyst or the (S)-enantiomer gives the best selectivity. The E-ketene silylacetal and the (R)-binaphthol-derived catalyst are the matched pair and vice-versa. High selectivities are obtained from alkyl- or silyloxysubstituted E-ketene silylacetals and aryl- or alkynylimines. This provides a route to nonracemic β-lac-tams or to α-hydroxy-β-aminoesters. The reactions of silyloxysubstituted Z-ke-tene acetals and aliphatic or arylimines are also highly selective (Figure **6.101**). Double diastereodifferentiation is also operative in the reaction of nitrone **6.125** with a ketene silylacetal derived from methyl acetate. If the 1-phenethyl group has the (S)-configuration, a single stereoisomer is obtained under ZnI_2 catalysis. If the (R)-enantiomer is used, the de is only 88% (Figure **6.101**). The chiral nitrogen substituent is removed by hydrogenolysis, and the stereochemistry of the attack is an agreement with the Felkin-Anh model.

6.10 REACTIONS OF ENAMINES

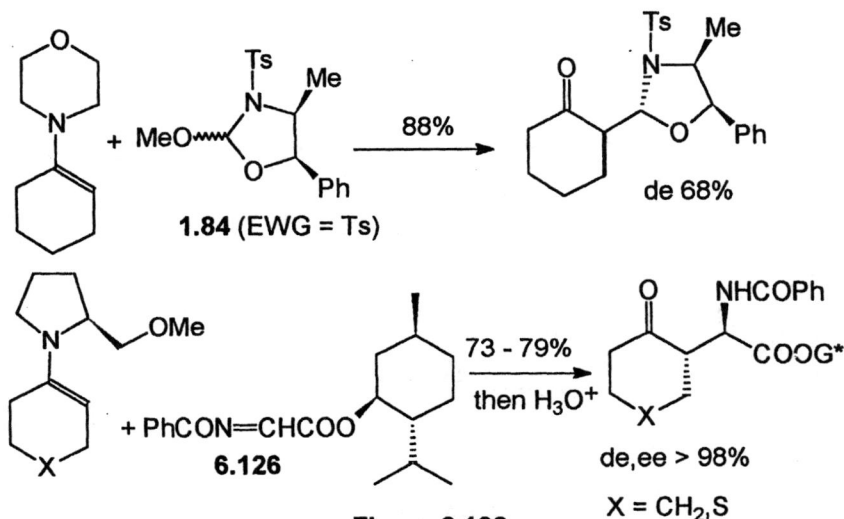

Figure 6.102

Stork and coworkers [624e] have introduced enamines as a nucleophilic substitute of enols, and a few asymmetric aldol reactions have been performed with enamines. Scolastico and coworkers [1311] have reacted morpholine enamines with chiral oxazolidine **1.84** (EWG = Ts), and in some cases they obtained higher selectivities than those obtained from enoxysilanes (§ 6.9.3) (Figure **6.102**). Chiral enamines derived from pyrrolidine **1.64** (R = MeOCH$_2$) react with acyliminoesters of chiral alcohols at −100°C [1313]. Double diastereodifferentiation is at work so that from matched reagents, for example the pyrrolidine enamine and iminoester **6.126** shown in Figure **6.102**, β-keto-α-aminoesters are obtained with a high diastereo- and enantioselectivity. The esters of either enantiomer of menthol or of achiral alcohols give mediocre asymmetric induction.

The Robinson annelation catalyzed by (S)-proline is an important asymmetric reaction involving chiral enamines as intermediates. This reaction was devised simultaneously by Hajos, Wiechert and their coworkers [261, 853, 1008, 1060]. The cyclization of triketone **6.127** in the presence of 3% (S)-proline **1.64** (R = COOH) at room temperature leads to aldol **6.128**. When this reaction is carried out under acidic conditions, α-enone **6.129** is obtained. In both cases, the bicyclic diketones are formed with a remarkable enantioselectivity (Figure **6.103**). Agami and coworkers [775] have determined the mechanism of the reaction. They showed that an (S)-proline enamine, which exists as carboxylate salt, is involved as an intermediate. This chiral enamine attacks one of the enantiotopic carbonyl groups of the cyclopentanedione, which is itself activated by hydrogen bonding to another (S)-proline molecule. The most favorable reactive conformation of this assembly is the one indicated in Figure **6.103**, and the *Si* face of one of the carbonyl groups is attacked. Under kinetic control, this attack provides an intermediate imminium salt **6.130**, which is the precursor of **6.128** [775]. Agami, Kagan and their coworkers [1314] have shown that this reaction responds to chirality amplification (§ 2.5.1). This reaction is the key step in a total synthesis of optically active steroids [853]. Other enantiopure aminoacids may be used as catalysts; (S)-amino acids lead to (S)-cyclenones and vice-versa, with a similar selectivity [261]. Other acyclic or cyclic di- and triketones bearing diverse substituents also undergo similar (S)-proline catalyzed asymmetric cyclizations [775].

6.11 ALDOL REACTIONS CATALYZED BY TRANSITION METAL COMPLEXES [1063]

Hayashi and Ito have shown that the reaction of methyl isocyanoacetate (MeOCOCH$_2$NC) with aldehydes is catalyzed by gold [339, 408, 752, 1060] or silver [951] complexes. In the presence of ferrocenyl bisphosphines **3.41** (R = CH$_2$CH$_2$NMe$_2$ or CH$_2$CH$_2$N(CH$_2$)$_5$) the reaction is diastereo- and enantio-selective, and provides *trans* oxazolines **6.131**. These oxazolines are precursors of nonracemic β-hydroxy-α-aminoacids (Figure **6.104**). The reaction has been extended to TsCH$_2$NC in the presence of silver complexes [952] (Figure **6.104**) or to amides or phosphonates under gold complex catalysis [952].

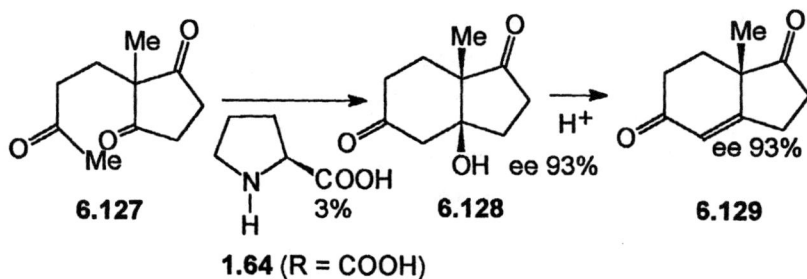

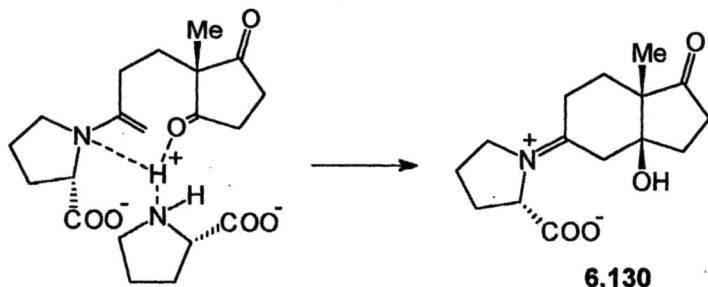

Figure 6.103

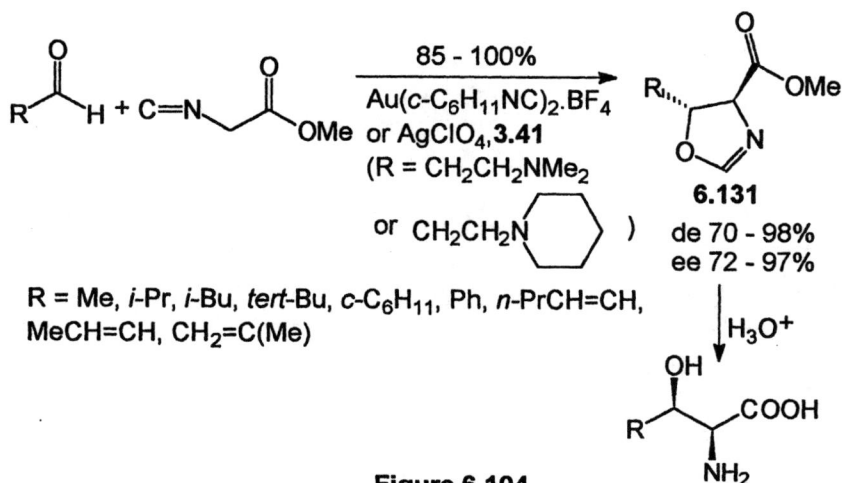

R = Me, *i*-Pr, *i*-Bu, *tert*-Bu, *c*-C$_6$H$_{11}$, Ph, *n*-PrCH=CH,
MeCH=CH, CH$_2$=C(Me)

Figure 6.104

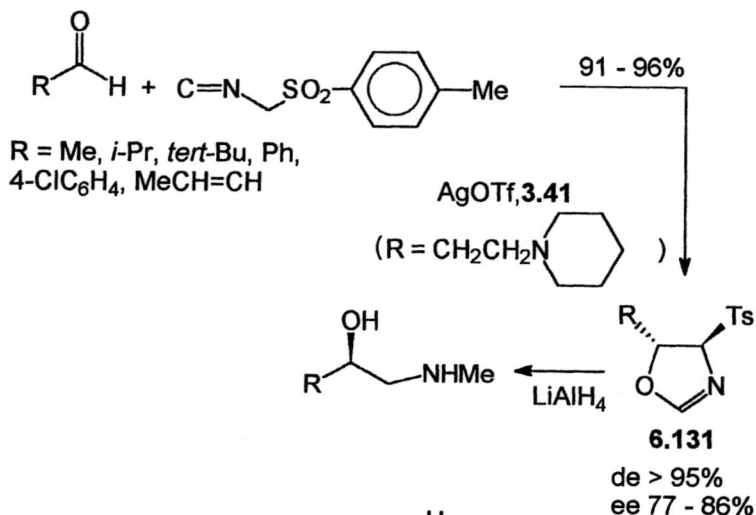

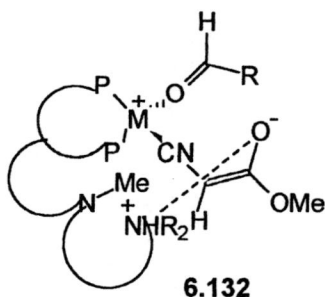

Figure 6.104 (continued)

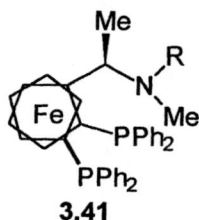

Tricoordinated metal complexes should be involved in order to observe a high selectivity [951]. The transition-state assembly is envisioned as having an ammonium enolate coordinated to the transition metal through the isocyanate residue **6.132** (Figure **6.104**).

6.12 REACTIONS OF CHIRAL ALCOHOLATES WITH *MESO* ANHYDRIDES

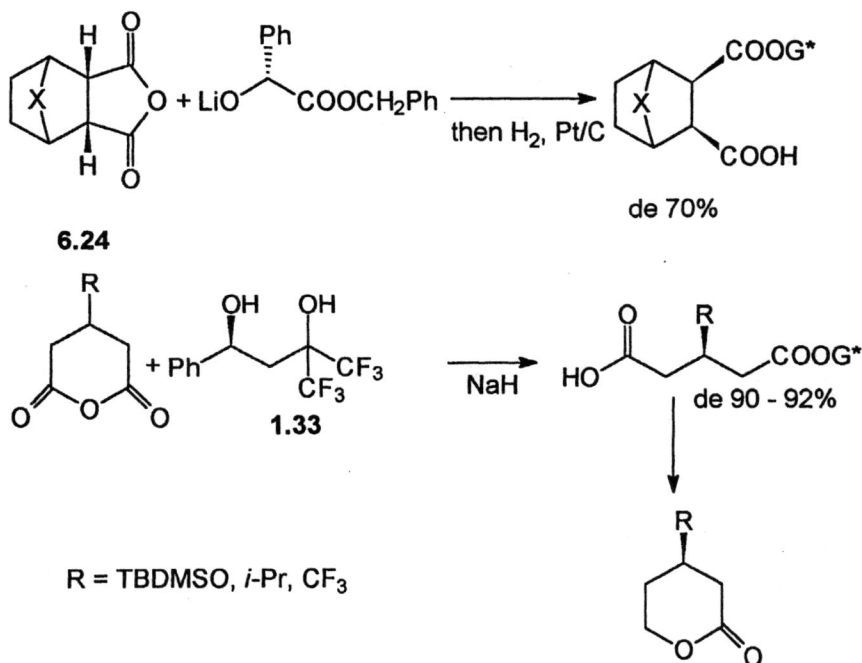

Figure 6.105

As already discussed for reductions (§ 6.1.5), *meso* anhydrides **6.24** undergo enantioselective addition to one of their carbonyl groups in reactions with alkali alkoxides of enantiopure alcohols at −78°C. The lithium alcoholate of the benzyl ester of (*R*)-mandelic acid (*R*)-**1.1** (Ar = Ph, R = COOCH$_2$C$_6$H$_4$OMe-4) has been used for such a purpose [132]. Although the selectivity of this process is not very high (70%), hydrogenolysis of the mixture followed by fractional crystallization of the half-ester formed this way provides the enantioenriched major isomer (Figure **6.105**). However, the use of the sodium alcoholate of (*S*)-1-phenyl-3,3-bis-trifluoromethylpropane-1,3-diol **1.33** leads to higher selectivities (92%) [211]. This holds for **6.24** as well as for *cis*-cyclohexane- or cyclohexene dicarboxylic anhydrides or else for 3-substituted glutaric anhydride, provided that the R substituent is bulky enough (Figure **6.105**). When R = PhCH$_2$, Ph or Me, the selectivity is lower (de 65%).

CHAPTER 7

ADDITIONS TO CARBON-CARBON DOUBLE BONDS

Olefins are very important industrial raw materials, and much effort has been devoted toward using them as substrates in asymmetric synthesis [811, 812, 853]. The industrial synthesis of nonracemic α-aminoacids by catalytic hydrogenation was one of the first important uses of olefins in asymmetric synthesis [859]. Today, the Sharpless epoxidation of allylic alcohols [807, 808, 809] is one of the most popular methods in asymmetric synthesis. The importance of pyrethrinoid pesticides, bearing a cyclopropane skeleton, justifies the efforts devoted to the asymmetric synthesis of cyclopropanes from alkenes [811, 812, 937].

In this chapter, all addition reactions to activated and unactivated carbon-carbon double bonds will be described. Because the mechanisms of these various reactions are different, the step in which asymmetric induction occurs will be discussed case by case.

7.1 CATALYTIC HYDROGENATION

7.1.1 Hydrogenation with Catalysts Bearing Chiral Ligands [812, 859, 860]

Catalytic hydrogenation of prochiral carbon-carbon double bonds with dihydrogen in the heterogeneous phase on supports modified by chiral additives has typically provided mediocre levels of asymmetric induction [578]. In contrast, catalytic asymmetric hydrogenation by H_2 under homogeneous conditions with rhodium or ruthenium complexes has been highly successful over the last two decades. Both soluble complexes and those grafted onto polymer supports have been used. To date, numerous prochiral functionalized olefins have been hydrogenated with an enantiomeric excess higher than 90%. In a few cases, the hydrogen source can be an alcohol or ammonium formate [873].

7.1.1.1 Rhodium Catalysts [114, 752, 861, 1061]

The modification of Wilkinson's catalyst [92] by the replacement of triphenylphosphine by chiral phosphines has been a popular approach to asymmetric hydrogenation. The first chiral phosphine ligands, diop **3.30** (Ar = Ph) and dipamp **3.31**, were designed by Kagan, Knowles and their coworkers. These ligands were used for the Rh-catalyzed asymmetric hydrogenation of dehydro-α-aminoacids

3.32 (R' = H). Since then, many other diphosphine ligands have been prepared and used in such hydrogenations [113, 860]. The catalysts are cationic Rh(I) complexes formed from $Rh(cod)_2ClO_4$ and one equivalent of a chiral diphosphine. Analogous rhodium complexes can also be used. The mechanism of the hydrogenation of Z-N-acetamidocinnamic acid and of its methyl ester Z-**3.32** (R_Z = Ph, Z = COMe, R' = H or Me) has been studied by Halpern and coworkers [2, 114, 752, 1061] (Figure **7.1**). Insight also comes from molecular mechanics calculations and from a study by NOE and ^{103}Rh NMR spectroscopy of the proposed intermediates in solution [1315, 1316]. Two diastereoisomeric cationic rhodium (I) complexes, **7.1** and **7.2**, are in rapid equilibrium. The lower energy complex **7.1** is the precursor of (R)-α-aminoacid **7.3** and the higher energy complex **7.2** is the precursor of the (S)-enantiomer. Coordination of the substrate **3.32** to the metal involves the carbon-carbon double bond and the carbonyl group, but these complexes cannot be considered as rigid chiral templates [1316]. The stereoselection is dictated by the oxidative addition of dihydrogen to these complexes. This addition leads to two octahedral rhodium (III) complexes **7.4** and **7.5** from which each enantiomer of **7.3** is irreversibly formed. The less populated complex **7.2** has been shown to be more reactive, and thus (S)-**7.3** is formed preferentially (§ I.1.2). The experimental conditions are very important to observe a high enantioselectivity. If the H_2 pressure is increased, then the rate of oxidative addition is accelerated so that the equilibration between **7.1** and **7.2** is no longer fast enough. This erodes the selectivity. A low reaction temperature also decreases the rate of this equilibration. A good compromise is to operate at room temperature under a hydrogen pressure of 1 atm.

It has been proposed that the energy levels of the *dyz* orbitals of the rhodium complexes **7.1** and **7.2** are crucial. The higher the energy level of this orbital (in other words, the more electron-rich the rhodium atom), the faster the oxidative addition of dihydrogen and the higher the level of enantioselection. Koenig and coworkers have shown that the presence of electron-withdrawing groups on the olefin is essential [114]. These substituents stabilize the rhodium chelates by d-π^*-donation, as do the carbonyl groups of the amide substituents.

Rhodium-catalyzed asymmetric hydrogenations have been performed with precursors of α-aminoacids, -esters, -nitriles or -amides [154, 169, 752, 858, 869, 948, 1317]. In these reactions, the chiral ligands of rhodium include tetraphenyldiphosphines **3.30** (Ar = Ph), **3.31**, **3.35** (X = O, NCH$_2$Ph, NCOO-*tert*-Bu), **3.36** (R = Ph), **3.37** and **3.38**, as well as the tetraphenylphosphinate **3.33** (Ar = Ph). Binap **3.43** is, however, less efficient as a ligand of rhodium. Some significant hydrogenation results are given in Figures **7.2** and **7.3**.

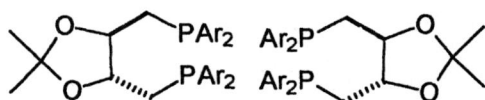
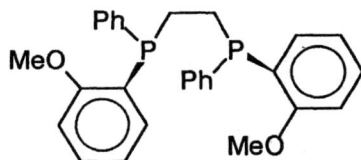

(R,R)-**3.30**   (S,S)-**3.30**   **3.31**

Ar = Ph:diop   dipamp

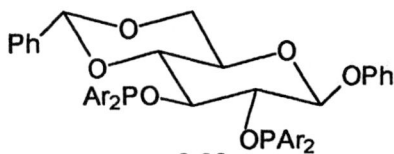
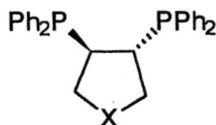

3.33   **3.35**

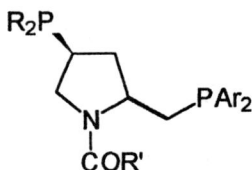
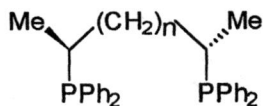

3.36   **3.37**

R = Ph: bppm   R = c-C$_6$H$_{11}$: bcpm   n = 0:chiraphos   n = 1: bdpp

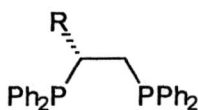

3.38

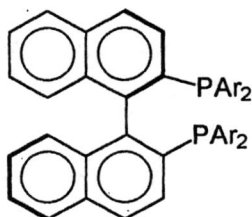
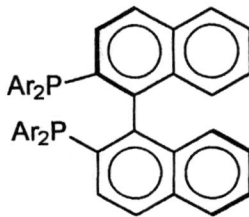

(R)-**3.43**   (S)-**3.43**

Ar = Ph: binap

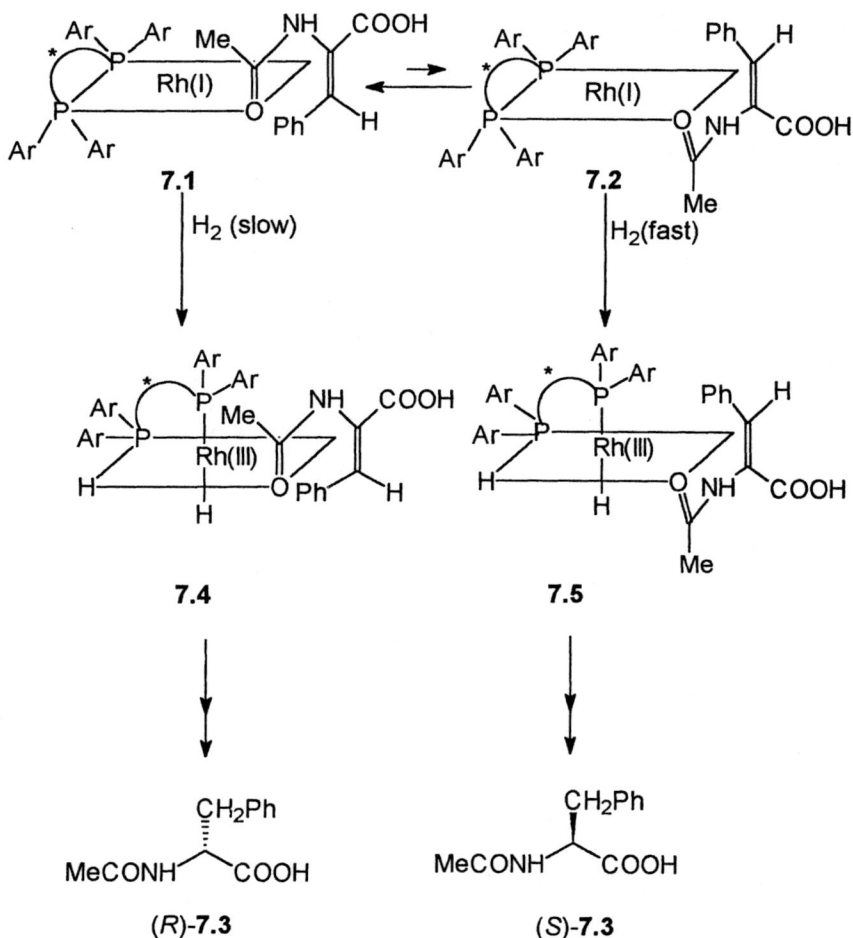

Figure 7.1

As a rule, an asymmetric hydrogenation of Z-**3.32** gives a higher enantiose-lectivity than that of E-**3.32** bearing the same substituents. The influence of various amide protective groups on the rhodium-catalyzed asymmetric hydrogenation of Z-acylaminocinnamic acid methyl esters Z-**3.32** (R_Z = Ph, Z = COMe, R' = Me) has been examined [1318]. The replacement of the NHCOMe or COPh residue by a more readily hydrolyzed NCOO-*tert*-Bu group can increase or decrease the enantioselectivity of the reaction, depending on the nature of the chiral diphosphine ligand.

$$
\underset{\text{Z-3.32 (R=H)}}{\overset{\displaystyle R\diagup\!\!\diagdown^{NHZ}}{\diagdown_{COOH}}} \quad\xrightarrow[\text{H}_2,\ \text{Rh}^+\ (I),\ L^*]{\text{conversion 100\%}}\quad \text{RCH}_2^*\text{CH}\overset{\displaystyle NHZ}{\diagdown_{COOH}}
$$

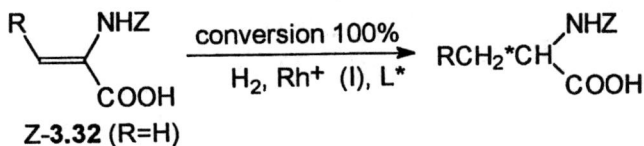

R	Z	L*	configuration	ee%
H	COMe	(R,R)-**3.30** (Ar = Ph)	(R)	73
		(R,R)-**3.31**	(S)	94
		(S,S)-**3.37** (n = 0)	(R)	91
		3.35 (X = O)	(S)	96
Ph	COMe	(S,S)-**3.30** (Ar = Ph)	(S)	82
		(R,R)-**3.31**	(S)	94
		(S,S)-**3.37** (n = 0)	(R)	99
		(S)-**3.43** (Ar = Ph)	(R)	84
		3.36 (R = Ar = Ph, R' = tert-BuO)	(R)	78
		(R)-**3.38** (R = Me)	(S)	90
		(R)-**3.38** (R = c-C$_6$H$_{11}$)	(S)	88
		3.35 (X = O)	(S)	97
		3.35 (X = NCH$_2$Ph)	(S)	99
		3.35 (X = NCOO-tert-Bu)	(S)	90
		3.33 (Ar = Ph)	(S)	94
		3.33 (Ar = 3,5-Me$_2$C$_6$H$_3$)	(S)	99
Ph	COPh	(R,R)-**3.31**	(S)	96
		(S,S)-**3.30**	(S)	64
		(S,S)-**3.37** (n = 0)	(R)	95
i-Pr	COMe	(S,S)-**3.37** (n = 0)	(R)	>99

Figure 7.2

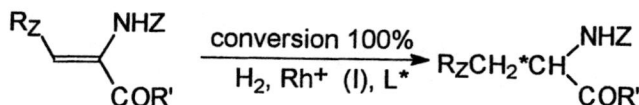

R_Z	R'	Z	L*	configuration	ee%
Ph	OMe	COMe	(R,R)-**3.31**	(S)	97
n-Pr	OMe	COMe	(R,R)-**3.31**	(S)	96
Ph	NH_2	COPh	(R,R)-**3.31**	(S)	94
i-Pr	NH_2	COPh	(R,R)-**3.31**	(S)	95

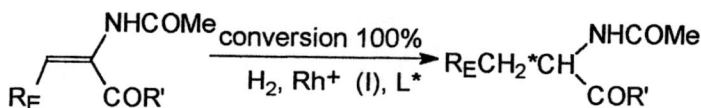

R_E	R'	L*	configuration	ee%
Ph	OMe	(S)-**3.43** (Ar = Ph)	(R)	87
Ph	OMe	**3.35** (X = NCOO-*tert*-Bu)	(S)	91
i-Pr	OMe	(R,R)-**3.31**	(S)	78
n-Pr	OMe	(R,R)-**3.31**	(S)	95

Figure 7.3

    Asymmetric hydrogenation of precursors of dipeptides has mainly been studied by Kagan, Ojima and coworkers [154, 752]. Double diastereodifferentiation comes into play (§ I.6), and excellent selectivities are obtained when the ligands of rhodium are diop **3.31** (Ar = Ph), dipamp **3.31** or bppm **3.36** (R = Ph, R' = NHAr). A few examples of dipeptide hydrogenations are shown in Figure **7.4**.

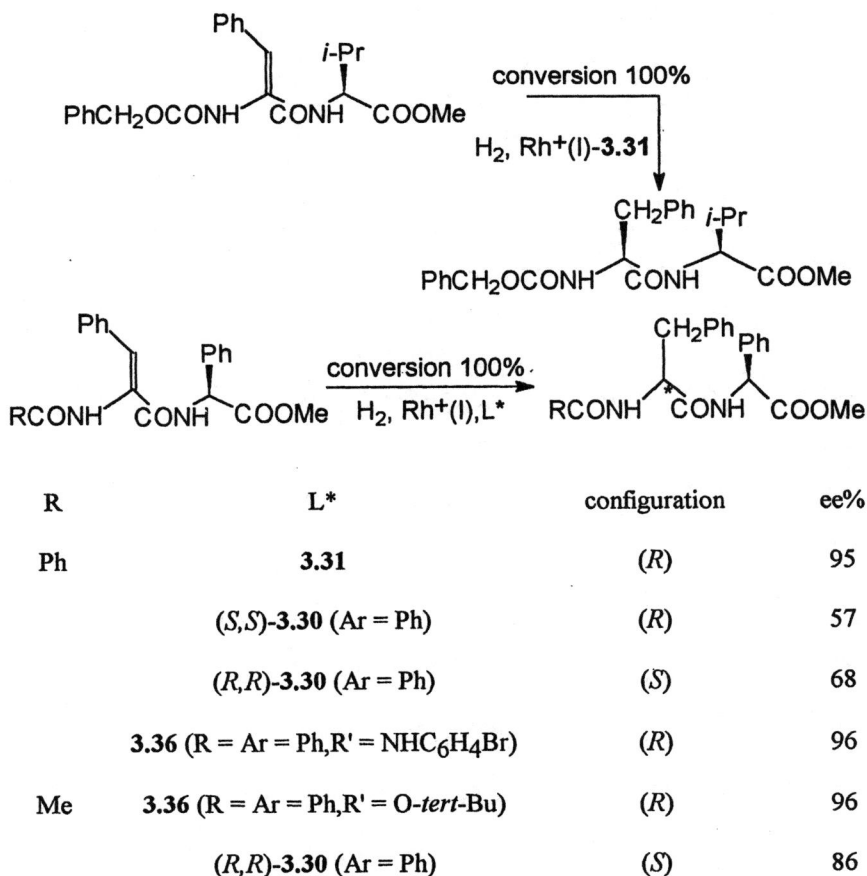

Ph

i-Pr

PhCH$_2$OCONH CONH COOMe

conversion 100%

H$_2$, Rh$^+$(I)-**3.31**

CH$_2$Ph i-Pr

PhCH$_2$OCONH CONH   COOMe

Ph

Ph

RCONH CONH COOMe   conversion 100%.

H$_2$, Rh$^+$(I),L*

CH$_2$Ph Ph

RCONH CONH COOMe

R	L*	configuration	ee%
Ph	**3.31**	(R)	95
	(S,S)-**3.30** (Ar = Ph)	(R)	57
	(R,R)-**3.30** (Ar = Ph)	(S)	68
	3.36 (R = Ar = Ph,R' = NHC$_6$H$_4$Br)	(R)	96
Me	**3.36** (R = Ar = Ph,R' = O-*tert*-Bu)	(R)	96
	(R,R)-**3.30** (Ar = Ph)	(S)	86

Figure 7.4

Grafting the ligands onto silica or polymeric supports often decreases the efficiency of the catalyst. However, the use of **3.35** (X = N-polymer) gives useful results. For example, the enantiomeric excess in asymmetric hydrogenation of Z-**3.32** (R = Ph, Z = COMe) is still 95%. The introduction of water-solubilizing substituents on the ligands [865] allows asymmetric hydrogenation to be conducted in aqueous solution with a high enantioselectivity. Sulfonate and amino substituents have been introduced onto the aromatic rings of diop **3.30**, chiraphos **3.37** (n = 0), bddp **3.37** (n = 1) [752, 866], and binap **3.43** [884]. Polyether groups have also been introduced on the acetal functionality of diop [752]. Another way to perform asymmetric hydrogenations in water under Rh-bppm **3.36** (R = Ar = Ph, R' = O-*tert*-Bu) catalysis is to add a surfactant to the solution [1319].

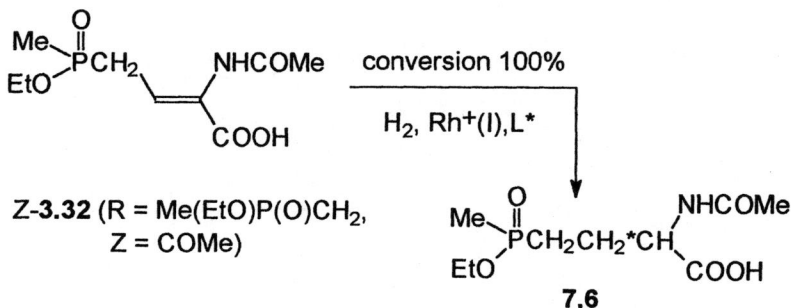

$$Z\text{-}3.32 \ (R = Me(EtO)P(O)CH_2,$$
$$Z = COMe)$$

7.6

L*	configuration	ee%
(R,R)-**3.39**	(S)	91
(S,S)-**3.37** (n = 0)	(R)	91

Figure 7.5

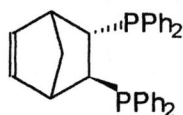

3.39 : norphos

A number of different functional groups on the R substituent of **3.32** are tolerated in asymmetric hydrogenation. For instance, Z-**3.32** (R = Me(EtO)P(O)CH$_2$, Z = COMe) is hydrogenated to (S)- or (R)-phosphinothricin **7.6** with a high ee when the ligand of rhodium is (R,R)-**3.39** or (S,S)-**3.37** (n = 0) [870] (Figure 7.5).

To broaden the range of available catalysts, more basic diphosphines have been designed to increase the electron density on the rhodium atom of the complex [114]. Electron-donating substituents have been introduced in the 4-position of the aryl groups of the tetraarylphosphines. Mod-diop **3.30** (Ar = 3,5-Me$_2$-4-MeOC$_6$H$_2$) and xyl-diop **3.30** (Ar = 3,5-Me$_2$-4-Me$_2$NC$_6$H$_2$) are useful ligands of rhodium in asymmetric hydrogenation of itaconic acid **7.7** (R = H) or its methyl ester **7.7** (R = Me). When diop is used as the ligand, the ee of these reductions is significantly lower [114] (Figure 7.6). A similar modification has been made to bppm. The use of **3.36** (Ar = 3,5-Me$_2$-4-MeOC$_6$H$_2$, R' = O-*tert*-Bu) as a rhodium ligand improves the enantioselectivity of the asymmetric hydrogenation of Z-**3.32** (R = Ph, R' = H) compared to the unsubstituted ligand (Figure 7.7). A similar improvement is observed with **3.37** (n = 1). However, the catalysts bearing **3.35** (X = O or NCH$_2$Ph) as ligand are efficient enough so that no modification is needed to obtain an excellent asymmetric induction in their reactions [752, 948] (Figure 7.2)

Phosphinite **3.33** is a useful ligand that has been modified by RajanBabu and coworkers [115]. Replacement of the phenyl groups (Ar = Ph) by 3,5-Me$_2$C$_6$H$_3$ increases the enantioselectivity of the asymmetric hydrogenation of Z-**3.32** (R = Ph, 3-MeOC$_6$H$_4$, 4-FC$_6$H$_4$, Z = COMe, R' = H or Me), while introduction of electron-withdrawing substituents decreases the ee (Figure 7.2). In this study, these authors discovered that the use of D-glucose-derived phosphinite **7.8** (Ar = 3,5-Me$_2$C$_6$H$_3$) as a ligand of rhodium promoted the highly enantioselective hydrogenation of Z-**3.32** (R = Ph, 2-Np, 3,5-F$_2$C$_6$H$_3$, 3-BrC$_6$H$_4$, Z = COMe, R' = H) to form unnatural (*R*)-aminoacids (ee 96 - 98%) [115].

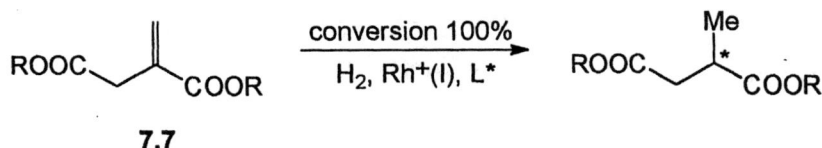

7.7

R	L*	configuration	ee%
H	(*R,R*)-**3.30** (Ar = Ph)	(*S*)	62
H	(*R,R*)-**3.30** (Ar = 3,5-Me$_2$-4-MeOC$_6$H$_2$)	(*S*)	91
H	**3.36** (R = *c*-C$_6$H$_{11}$, Ar = Ph, R' = O-*tert*-Bu)	(*S*)	92
Me	(*R,R*)-**3.30** (Ar = Ph)	(*S*)	< 10
Me	(*R,R*)-**3.37** (n = 0)	(*S*)	98
Me	(*R,R*)-**3.30** (Ar = 3,5-Me$_2$-4-Me$_2$NC$_6$H$_2$)	(*S*)	84
Me	(*R*)-**3.45** (R = *c*-C$_6$H$_{11}$)	(*R*)	96
Me	(*R,R*)-**3.47** (R = Et)	(*R*)	> 95
Me	(*S,S*)-**3.47** (R = Me)	(*S*)	90

Figure 7.6

$$R_Z\diagdown_{\diagdown}\!\!\stackrel{NHCOMe}{\underset{COOR'}{|}} \quad \xrightarrow[\substack{H_2,\ Rh^+(I),\ L^*}]{\text{conversion 100\%}} \quad R_ZCH_2{}^*CH\stackrel{NHCOMe}{\underset{COOR'}{\diagdown}}$$

Z-3.32

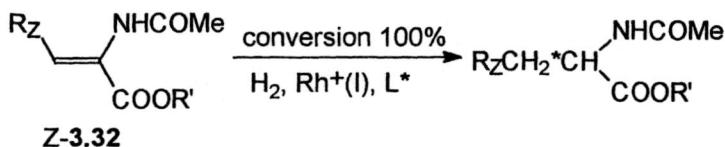

R	R'	L*	configuration	ee%
Ph	H	**3.36** (R = Ar = Ph, R' = O-*tert*-Bu)	(R)	78
Ph	H	**3.36** (R = 3,5-Me$_2$-4-MeOC$_6$H$_2$, R' = O-*tert*-Bu)	(R)	98
Ph	Me	(R,R)-**3.46** (n = 2)	(R)	85 - 93
Ph	Me	(S,S)-**3.46** (n = 2)	(S)	85 - 93
Ph	Me	(R,R)-**3.47**	(R)	87 - 98
i-Pr	Me	(R,R)-**3.46** (n = 2, R = i-Pr)	(R)	99
i-Pr	Me	(R,R)-**3.47**	(R)	95 - 99
H	Me	(R,R)-**3.46** (n = 2)	(R)	91 - 98
H	Me	(R,R)-**3.47**	(R)	95 - 99
Ph	Et	(R)-**3.45** (R = c-C$_6$H$_{11}$)	(S)	97 - 99

Figure 7.7

Another way to obtain more basic phosphines is to replace the phenyl groups by cycloalkyl groups. Such a modification has been made to bppm **3.36** (R = Ar = Ph, R' = O-*tert*-Bu). Although bcpm **3.36** (R = c-C$_6$H$_{11}$, Ar = Ph, R' = O-*tert*-Bu) is not more efficient in asymmetric hydrogenation of Z-**3.32** (R = Ph, Z = COMe) than bppm, it behaves as a very useful ligand of rhodium in asymmetric hydrogenation of itaconic acid **7.7** (R = H) (Figure 7.6).

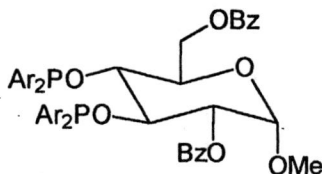

OBz

Ar₂PO
Ar₂PO

BzO OMe

7.8

A new class of diphosphines, bisphospholanes **3.46** (n = 2, R = Me, Et, *i*-Pr) and **3.47** (R = Me, Et, *i*-Pr), has been proposed by Burk and coworkers [907, 908, 909]. These bisphospholanes are very efficient ligands of rhodium in asymmetric hydrogenation of acetamidoacrylates Z-**3.32** (Z = COMe) (Figure 7.7), of methyl itaconate **7.7** (R = Me) (Figure 7.6), and of enol acetates **7.9** (Figure 7.8). Moreover, the reactions with these ligands are faster than reactions with other ligands, although the mechanism seems to be the same [1320]. Ligands **3.45** (R = *c*-C₆H₁₁) are also useful for asymmetric hydrogenation of acetamidoacrylates (Figure 7.7) and methyl itaconate (Figure 7.6)

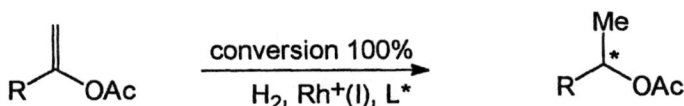

$$R \overset{O}{\underset{OAc}{||}} \quad \xrightarrow[H_2, Rh^+(I), L^*]{conversion\ 100\%} \quad R \overset{Me}{\underset{OAc}{|_*}}$$

7.9

R	L*	configuration	ee%
Ph	(*S,S*)-**3.47** (R = Me)	(*S*)	89
1-Np	(*R,R*)-**3.46** (n = 2, R = Et)	(*R*)	94
1-Np	(*S,S*)-**3.47** (R = Me)	(*S*)	93
COOEt	(*S,S*)-**3.47** (R = Me)	(*S*)	99
CF₃	(*S,S*)-**3.47** (R = Me)	(*S*)	94

Figure 7.8

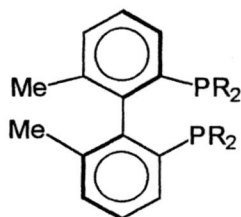

3.45

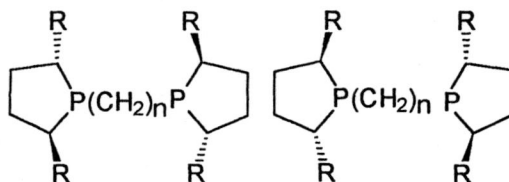

3.46 : duphos

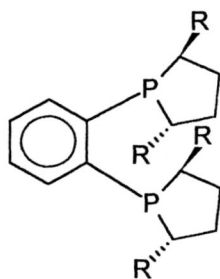

3.47

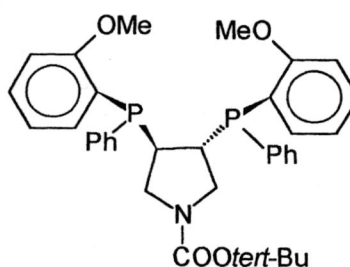

7.10

The chiral poisoning strategy (§ 3.3) has been applied by Faller and Parr to Rh(I) catalyzed asymmetric hydrogenation of dimethyl itaconate **7.7** (R = Me). Racemic chiraphos **3.37** (n = 0) was used as a ligand in the presence of (S)-(Ph$_2$POCH$_2$CH(NMe$_2$)CH$_2$CH$_2$SMe). The chiral poison binds selectively to the (S,S)-chiraphos-Rh(I) complex. The unbound (R,R)-chiraphos complex produces (S)-methylsuccinate, but the ee is at best 49% [856]. A hybrid ligand **7.10** displaying partial structures of both (R)-pyrphos **3.35** (X = NCOO-tert-Bu) and dipamp **3.31** has shown a high efficiency in asymmetric hydrogenation of Z-**3.32** (R = Ph, Z = COMe, R' = H) to form the (S)-aminoacid (ee 99%) [1321].

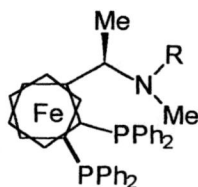

3.41

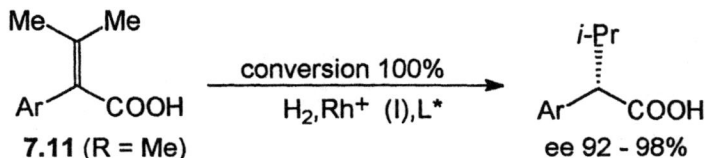

L*: **3.41** R = CH$_2$CH$_2$NR'$_2$
Ar = Ph, 4-ClC$_6$H$_4$, 4-MeOC$_6$H$_4$, 2-Np

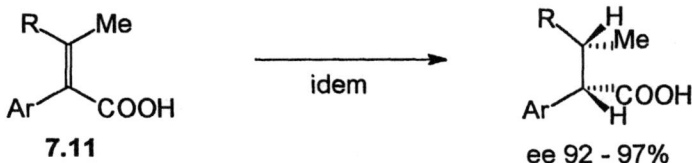

R = Ph, Et

Figure 7.9

Asymmetric hydrogenation of trisubstituted α,β-unsaturated acids **7.11** is catalyzed by cationic Rh(I) complexes, and can be performed with ferrocenyl-diphosphines **3.41** as ligands [169, 752, 1063]. These reactions are run under elevated pressure and are highly enantioselective (Figure **7.9**).

A new, highly efficient family of ferrocenyldiphosphines has been recently described [1322]. Neutral rhodium complexes, prepared with chiral diphosphines and (RhCl(nbd))$_2$ or (RhCl(cod))$_2$, have also been used as catalysts in asymmetric hydrogenation, though they do not give better selectivities than cationic complexes [904]. Exceptions to this generalization are the asymmetric hydrogenations of itaconic acid **7.7** (R = H) and acid **7.12** (R = Ph) by using **3.36** (R = Ar = Ph, R' = NHPh) as ligand. These reactions lead to (S)-diacids with enantiomeric excesses higher than 95% [1323] (Figure **7.10**). Similarly good selectivities are obtained with diester **7.7** (R = Me) when the ligand is **3.36** (R = 4-Me$_2$NC$_6$H$_4$, R' = O-*tert*-Bu) [114] (Figure **7.10**).

Asymmetric hydrogenation of itaconic acid **7.7** (R = H) with trimethylammonium or (S)-1-phenethylammonium formate in the presence of Rh(I)-bppm **3.36** (R = Ar = Ph, R' = O-*tert*-Bu) complex has been carried out [873]. This reaction takes place near room temperature, and it is highly enantioselective [1324]. With other ligands or other substrates, the results are less useful.

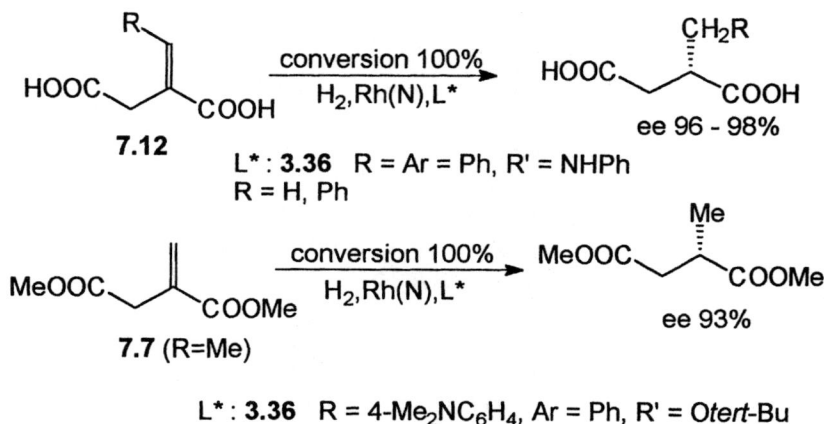

R
HOOC ⟍ ═ ⟋ COOH
7.12

conversion 100%
H₂,Rh(N),L*
→

CH₂R
HOOC ⟍ ⟋ COOH
ee 96 - 98%

L* : **3.36** R = Ar = Ph, R' = NHPh
R = H, Ph

MeOOC ⟍ ═ ⟋ COOMe
7.7 (R=Me)

conversion 100%
H₂,Rh(N),L*
→

Me
MeOOC ⟍ ⟋ COOMe
ee 93%

L* : **3.36** R = 4-Me₂NC₆H₄, Ar = Ph, R' = O*tert*-Bu

Figure 7.10

7.1.1.2 *Ruthenium Catalysts* [559, 752, 881, 1061]

Ruthenium catalysts bearing chiral diphosphine ligands display a higher efficiency in asymmetric hydrogenation than their rhodium counterparts. Complexes of binap **3.43** (Ar = Ph)-Ru(II)(OCOR)₂ are the most popular [1135]. The precursors of aminoacids Z-**3.32** (R = Ph) are hydrogenated with an excellent enantiomeric excess (Figure **7.11**). However, the configuration of the aminoacid derivative is opposite to that obtained when using the Rh(I) catalyst bearing the same ligand. Asymmetric hydrogenation of Z-**3.32** (R = Ph) catalyzed by (*R*)-binap-Ru(OCOMe)₂ leads predominantly to the (*R*)-aminoacid, while in the presence of (*R*)-binap-Rh(I) complex, the (*S*)-enantiomer is formed (Figure **7.11**). While substituted acrylates E-**7.13** and **7.14** suffer rhodium-catalyzed asymmetric hydrogenations with poor selectivities, the formation of β-aminoesters in the presence of binap-Ru(OCOMe)₂ complex takes place with a high enantiomeric excess [895, 1325] (Figure **7.11**). However, low ee's are obtained in hydrogenation of Z-**7.13** with this Ru-complex. Ruthenium-catalyzed asymmetric hydrogenations of α,β- or β,γ-unsaturated acids give excellent results (ee 85 - 97%), even when the double bond is di- or trisubstituted [571, 752, 881]. For example, the anti-inflammatory drug naproxen (**7.15**) is obtained by Ru-binap catalyzed hydrogenation in 97% ee [811, 812, 881, 923] (Figure **7.11**). Equally good results are obtained from 2-alkylidenecyclopentanones **7.16** or from allylic alcohols such as geraniol (E-**7.17**) or nerol (Z-**7.17**) [752] (Figure **7.11**). Nevertheless, these hydrogenations usually require high H₂ pressure (100 atm) when conducted at room temperature. The use of complex **3.43** (Ar = 4-MeC₆H₄)-Ru(OCOCF₃)₂ allows reactions to occur at lower pressure (30 atm) [752].

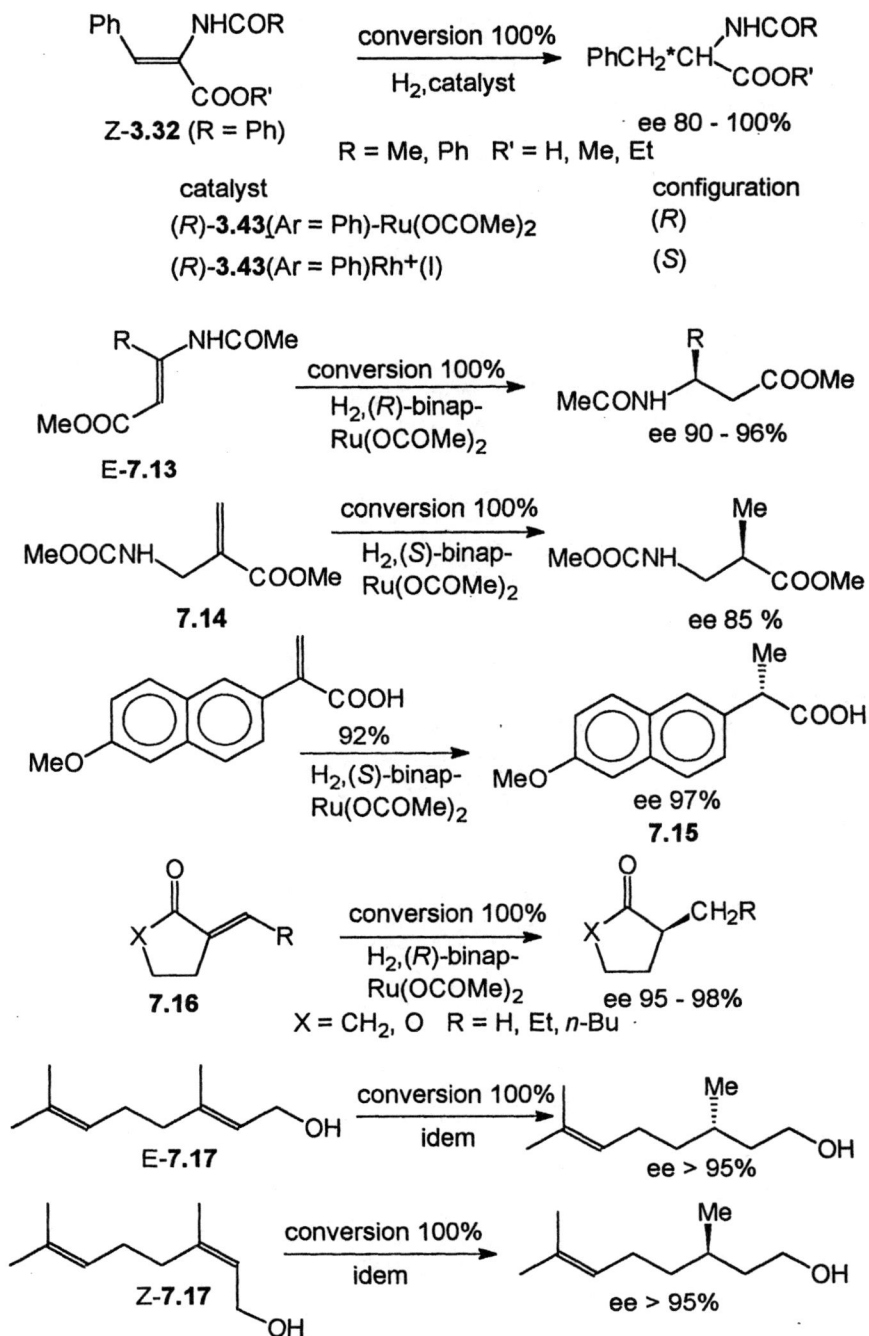

Figure 7.11

Cyclic Z-enamides **7.18** are precursors of alkaloids. The asymmetric hydrogenation of these enamides takes place under 1-4 atm of H_2 pressure at room temperature with a high enantioselectivity when the catalyst is **3.43** (Ar = Ph)-Ru(OCOMe)$_2$ [752, 881, 883, 1326] (Figure **7.12**). The use of (*R*)-binap as a ligand gives the (*R*)-enantiomer and vice-versa. However, the related E-alkenes do not undergo hydrogenation under such conditions. Useful results have also been obtained when using other atropisomeric diphosphines **3.45** (R = Ph) as ligands of ruthenium [892]. For instance, asymmetric hydrogenation of **7.19** is more rapid and more enantioselective with the catalyst **3.45** (R = Ph)-Ru(OCOCF$_3$)$_2$ than with the binap analog (Figure **7.12**).

Other ruthenium complexes, including binap-Ru$_2$X$_4$•NEt$_3$(X = Cl, Br), (binap/C$_6$H$_6$-RuCl)Cl and binap-Ru(2-methallyl)$_2$, have also been proposed as catalysts for asymmetric hydrogenation [752, 1326, 1327, 1328]. In some cases, it is possible to work at lower H_2 pressure [1327]. Excellent enantioselectivities have been obtained in hydrogenations of cyclic Z-enamides **7.18** [752, 1326], alkylidene butyrolactones **7.20** and **7.21** (n = 1), diketene **7.21** (n = 0) and α-fluoro- or E-α-methyl-α,β-unsaturated acids [1327-1330] (Figure **7.12**). Water-soluble catalysts have been prepared by using sulfonated binap ligand **3.43** (Ar = 4-NaOSO$_2$C$_6$H$_4$) [896]. Other atropisomeric chiral diphosphines such as **3.45** (Ar = Ph) are also good ligands for ruthenium in these complexes [892, 1327].

The resolution of racemic secondary allyl alcohols can be performed in the presence of certain ruthenium chiral catalysts through enantioselective asymmetric hydrogenation [811, 881]. Chiral poisoning also works in such kinetic resolutions. For example, hydrogenation of 2-cyclohexenol under (±)-binap-Ru catalysis in the presence of (1*R*, 2*S*)-ephedrine **1.61** (10 equiv) provides unreacted (*R*)-2-cyclohexenol in 95% ee after 60% conversion [857].

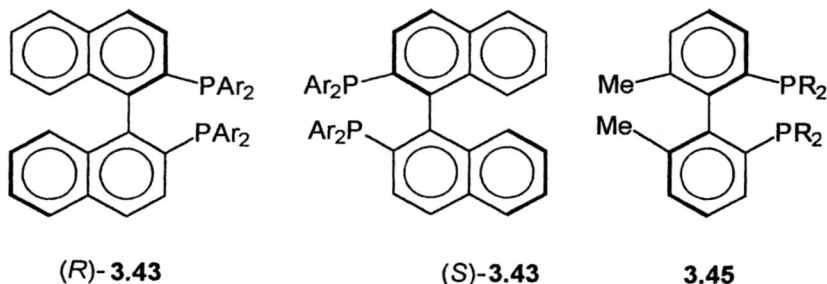

(*R*)- **3.43**   (*S*)-**3.43**   **3.45**

Ar = Ph: binap

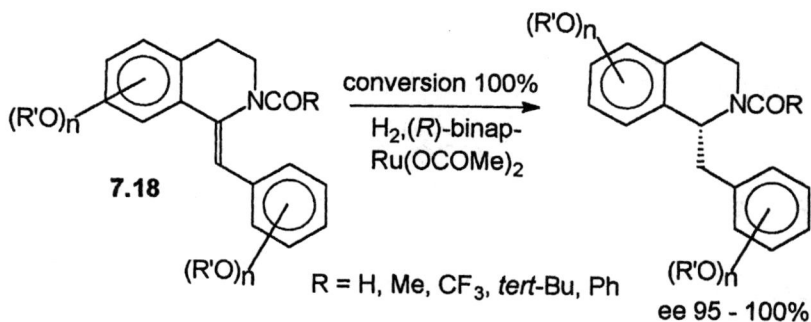

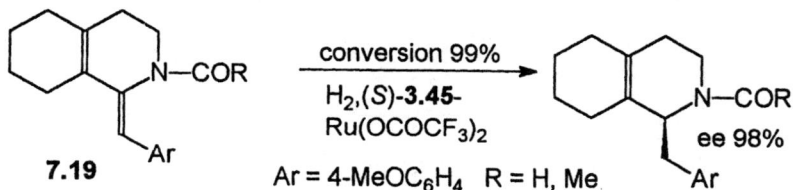

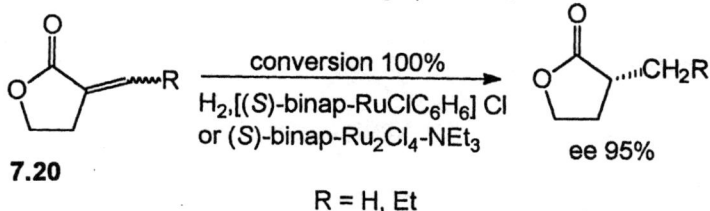

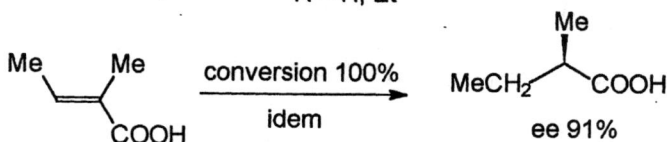

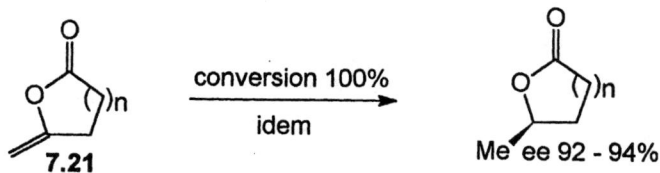

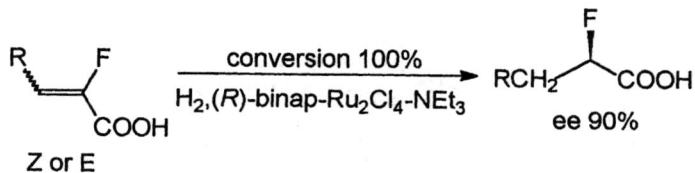

Figure 7.12

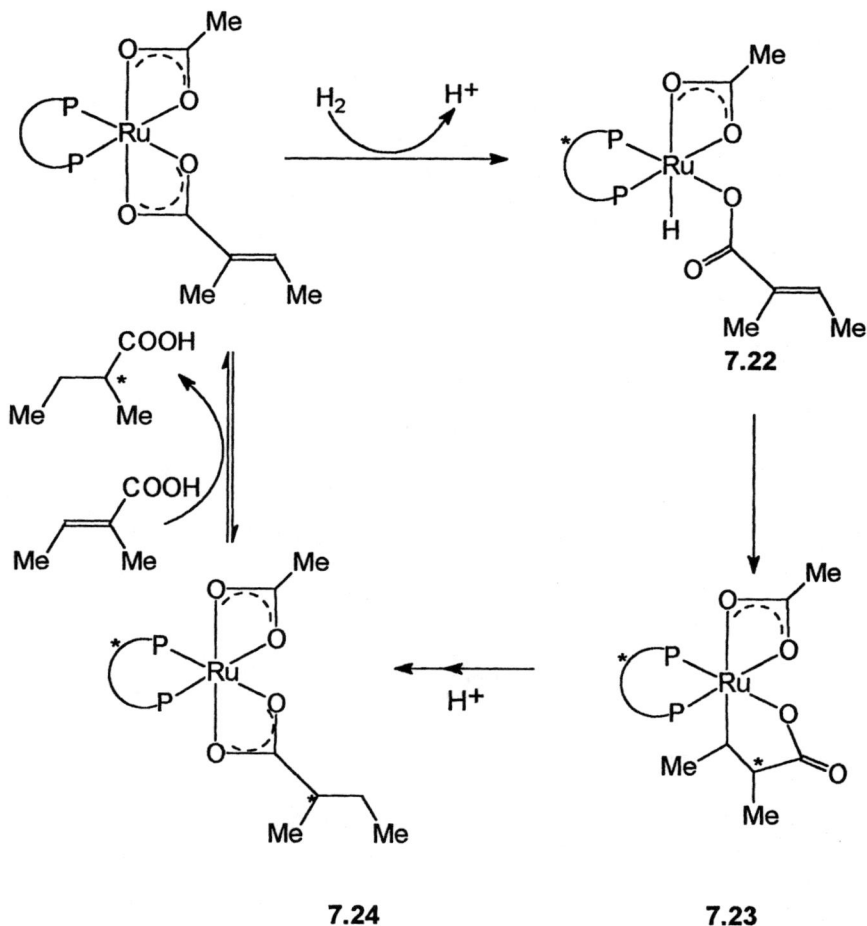

Figure 7.13

According to Takaya, Noyori [1331] and Halpern [1332], the mechanism of the ruthenium-catalyzed hydrogenation (Figure **7.13**) is different from its rhodium counterpart. Indeed, during the hydrogenation of α,β-unsaturated acids with ruthenium complexes, a single hydrogen atom of dihydrogen is transferred to the substrate. These authors proposed the formation of an intermediate monohydride complex **7.22** in which the unsaturated acid is coordinated to the metal through its carboxylic group. The key step that dictates the stereoselection is a hydride transfer to generate complex **7.23**. Protonation leads to complex **7.24**, which bears the saturated acid as a ligand (Figure **7.13**). Ligand exchange between two acids then completes the catalytic cycle.

Asymmetric hydrogen transfer from EtOH, *i*-PrOH or triethylammonium formate to Z-α-acetamidocinnamic acid Z-**3.32** (R = Ph, Z = COMe, R' = H) or to itaconic acid **7.7** (R = H) is highly enantioselective when the catalyst is a binap-Ru complex [873, 1333] (Figure **7.14**). The relationship between the absolute configuration of the saturated acid and the binap ligand is the same as in the catalytic hydrogenation. However, hydrogen transfer to other α,β-unsaturated acids, such as the precursor of naproxen (**7.15**), is less enantioselective [1333].

7.1.1.3 *Titanocene Catalysts*

A titanocene catalyst bearing a chiral fused cyclopentadienyl ligand has been proposed by Vollhardt and coworkers [1334] for asymmetric hydrogenation of 2-phenyl-1-butene: high ee are obtained at −75°C.

A binaphthol titanocene catalyst has been proposed by Buchwald and Broene for use in asymmetric hydrogenation of nonfunctionalized trisubstituted olefins [829]. These reactions take place under H_2 pressure at 65°C. Acyclic or cyclic olefins lead to saturated, branched alkanes with high chemical yield and enantioselectivity (ee 83 - 99%). Unfortunately, in most of the cases, the absolute configuration of the product has not yet been determined.

7.1.1.4 *Cobalt Catalyst* [173]

Pfaltz and coworkers have performed the asymmetric reduction of prochiral β,β-disubstituted α,β-unsaturated esters and amides with $NaBH_4$ in the presence of catalytic amounts of $CoCl_2$ and semicorrin **2.11**. These reactions are run at room temperature in protic solvents. Depending on the Z- or E-geometry of the alkene, (*S*)- or (*R*)-β-branched esters or amides are obtained with an excellent selectivity [576, 577, 929]. When R = Ph, the enantiomeric excess is lower (Figure **7.14**).

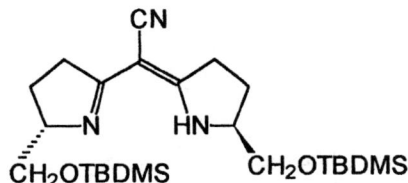

CN

$CH_2OTBDMS$ $CH_2OTBDMS$

2.11 (R = $CH_2OTBDMS$)

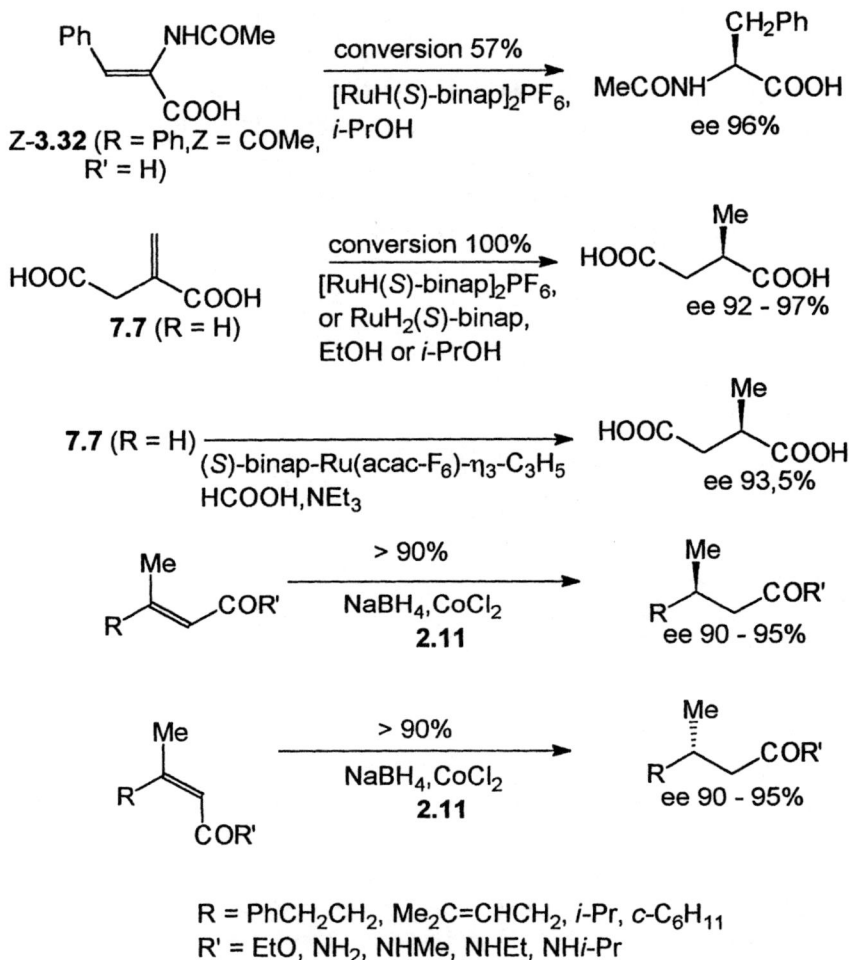

Figure 7.14

7.1.2 Hydrogenation of Functionalized Olefins Bearing Chiral Residues

The double bond of Z-*N*-tosyloxazolidine **7.25** (Z = Ts) derived from an α,β-unsaturated aldehydoester can be selectively hydrogenated with H_2 under Pd/C catalysis or with $NaBH_4/CoCl_2$ [1335] (Figure **7.15**). If Z = COOMe or if the double bond has the E-geometry, the selectivity is lower. These trends have been interpreted by H_2 transfer to the less sterically hindered face of the substrate through a transition state that resembles the ground-state-favored conformation.

Hydrogenation of prochiral β,β-disubstituted enoyl sultams **1.134** (R = CH=CRR') with H_2 in the presence of Pd/C takes place with a high diastereofacial discrimination [147, 454, 464, 465]. The interpretation of this selectivity invokes reaction through an *s-cis* conformation in which the carbonyl and SO_2 groups are coordinated to the surface of the metal catalyst. Consistent with this, if a substituent is introduced in the α-position of **1.134**, then the *s-cis* conformer is destabilized by interaction with the skeleton of the chiral auxiliary, and the selectivity of the hydrogenation is lower [147]. Raney-nickel-mediated catalytic hydrogenation of chiral aminoester **7.26** is also highly selective. After hydrogenolysis of the 1-phenethyl residue, a *cis*-2,3-disubstituted pyrrolidine is obtained with a good enantiomeric excess [1336] (Figure **7.15**). Enantioenriched α-aminoacids may also be prepared by catalytic hydrogenation of chiral imidazolidinones **7.27**, which takes place from the least hindered face. Mild hydrolysis is required to liberate the final products [442, 861].

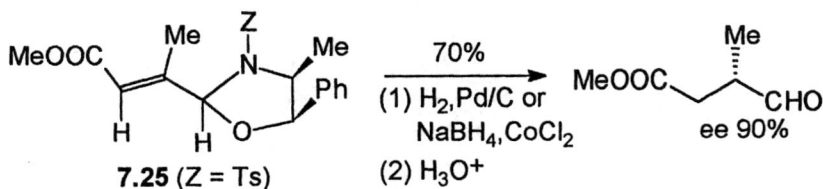

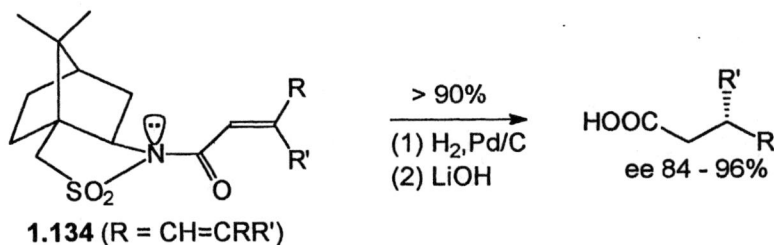

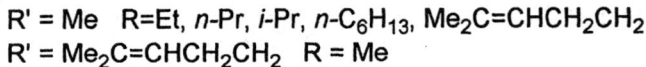

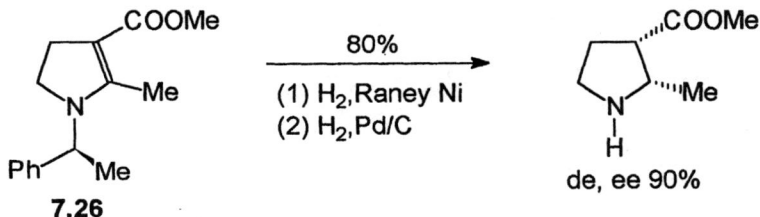

Figure 7.15

Me
|
N ,O
tert-Bu''''⟨ ∥
N CHR
|
COOtert-Bu
7.27

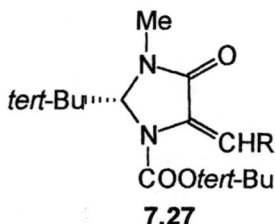

7.2 REDUCTIONS BY HYDRIDES

Conjugate addition of Li s-Bu$_3$BH to prochiral β,β-disubstituted sultams **1.134** (R = CH=CRR') in toluene, followed by aqueous NH$_4$Cl treatment, leads to imides that are precursors of acids having the opposite configuration of those obtained by catalytic hydrogenation (§ 7.1.2) [147, 464]. In the reactive s-cis conformer of **1.134**, the SO$_2$ and C=O dipoles are opposed, in agreement with X-ray crystal structures. The hydride approach takes place on the face of the double bond that is opposite to the pyramidalized nitrogen atom (Figure **7.16**). Trapping of the intermediate 1,4-adducts by MeI, instead of a proton donor, leads to α,β-disubstituted carboxylic acids with high diastereo- and enantiomeric excesses after removal of the chiral auxiliary [147].

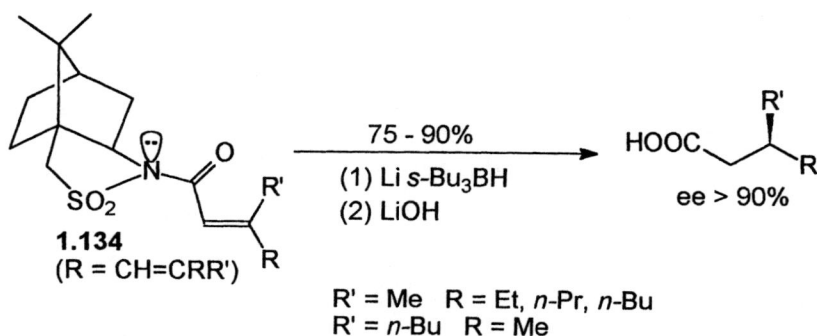

75 - 90%

(1) Li s-Bu$_3$BH
(2) LiOH

1.134
(R = CH=CRR')

HOOC

ee > 90%

R' = Me   R = Et, n-Pr, n-Bu
R' = n-Bu   R = Me

Figure 7.16

7.3 HYDROBORATION [451]

Asymmetric hydroboration of prochiral olefins can be performed either with a chiral borane in the absence of a catalyst or with catecholborane and a chiral rhodium catalyst [549]. The rich chemistry of the boranes generated by these methods opens the way to a variety of enantioenriched molecules bearing different functionalities [583].

7.3.1 Noncatalyzed Hydroboration [583]

Noncatalyzed asymmetric hydroborations require control of both stereo- and regioselectivity. The mechanism of hydroboration is an electrophilic attack of the borane on the olefin. Therefore, the predominant frontier orbital interaction involves the HOMO of the olefin and the LUMO of the borane. The most favorable overlap occurs when the electrophilic borane approaches the center of the double bond with minimization of steric interactions [9]. Mono- and diisopinocamphenyl-boranes 2.15 (R = H) and 2.16 (R = H) designed by Brown, and 2,5-dimethylborolane 2.14 proposed by Masamune [580, 583] are the most popular reagents. Monoisopinocamphenylborane 2.15 (R = H) promotes the stereoselective hydroboration of E-1,2-disubstituted olefins with a good facial discrimination, provided that one of the substituents is bulky enough (R = tert-Bu). This is also a useful reagent for hydroboration of trisubstituted arylolefins (Figure 7.17). Diisopinocamphenylborane is the reagent of choice for the hydroboration of symmetrical or unsymmetrical Z-1,2-disubstituted olefins [169]. In the later case, the differences between the two alkene substituents must be sufficient to induce a regioselective addition. Reagent 2.14 (R = H) hydroborates Z- and E-1,2-disubstituted and trisubstituted olefins with a high selectivity [583] (Figure 7.17). However, disappointing results are obtained with terminal olefins. Typically, the chiral borane adducts are oxidized by H_2O_2 in basic media to provide the corresponding alcohols with retention of configuration [169, 583, 588]. According to the configuration of the borane reagent, either enantiomeric secondary alcohol can be prepared with a high enantiomeric excess [583]. This method has been extended to hydroboration of acyclic pyrrolidino- or morpholino-E-enamines [1337], opening the way to enantioenriched β-aminoalcohols. Better enantioselectivities are obtained at 0°C with morpholine derivatives (Figure 7.17). Unfortunately, cyclohexanone-derived enamines give disappointing results [1337].

The treatment of di- or trialkylborane products with MeCHO regenerates α-pinene and delivers chiral boronates (§ 2.3), which are precursors of many enantioenriched, branched alkyl-substituted compounds including olefins, ketones, aldehydes, acids, nitriles or diamines [580, 583, 585, 587, 595].

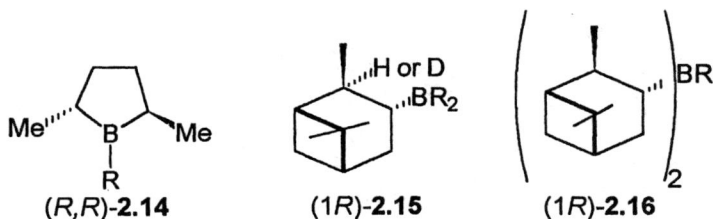

(R,R)-**2.14** $\qquad$ $(1R)$-**2.15** $\qquad$ $(1R)$-**2.16**

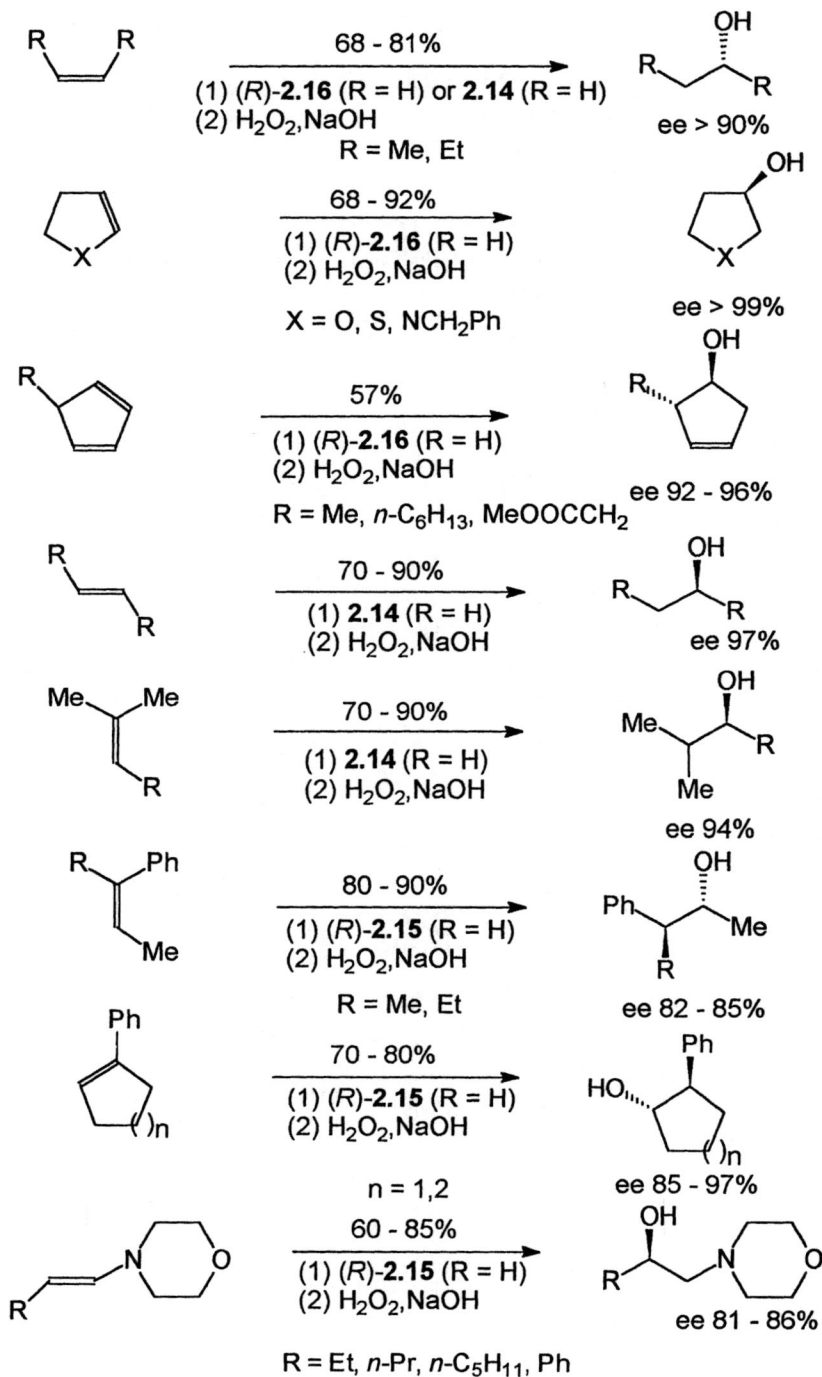

Figure 7.17

7.3.2 Rhodium-catalyzed Hydroboration

Catecholborane is less electrophilic than alkylboranes, and its use requires catalysis by a cationic rhodium (I) complex. Asymmetric hydroboration of styrenes is carried out in the presence of chiral ligands such as diop **3.30** (Ar = Ph) or binap **3.43** (Ar = Ph) at −78°C [886, 1177, 1338] (Figure **7.18**). This reaction occurs at room temperature with a phosphine ligand bearing a quinoline residue **3.53** [922]. Following hydroboration, secondary alcohols are obtained with a high regio- and enantioselectivity by oxidation with H_2O_2 (Figure **7.18**). Hydroboration of 2-naphthylstyrene or 1-phenylbutadiene are somewhat less selective [886, 887]. Asymmetric hydroboration of norbornene is also enantioselective, provided that the ligand of rhodium is either (S,S)-bdpp **3.37** or (R,R)-2-MeO-diop **3.30** (Ar = 2-MeOC$_6$H$_4$) [864] (Figure **7.18**). The use of cyclic boranes derived from norephedrine has been proposed, but these hydroborations are poorly regioselective [888]. The mechanism of the catalyzed hydroboration involves the formation of a rhodium hydridoborane complex **7.28,** which then coordinates with the olefin [1338a]. Insertion of hydride is followed by reductive elimination to form boronate **7.29**, which is the primary product of the reaction (Figure **7.18**).

A palladium catalyst was proposed to perform asymmetric hydroboration of but-1-en-3-ynes to chiral allenylboranes; unfortunately, low enantiomeric excesses were obtained [1339].

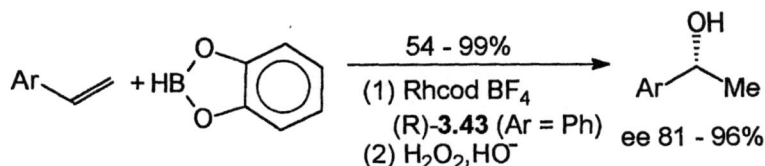

Ar = Ph, 4-ClC$_6$H$_4$, 4-MeC$_6$H$_4$, 2- and 4-MeOC$_6$H$_4$, 3-ClC$_6$H$_4$

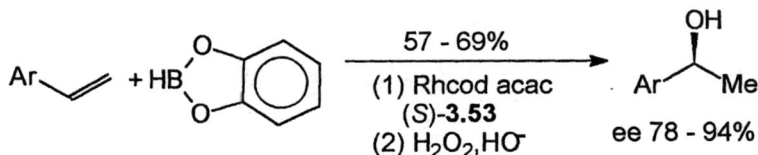

Ar = Ph, 4-ClC$_6$H$_4$, 4-MeOC$_6$H$_4$

Figure 7.18

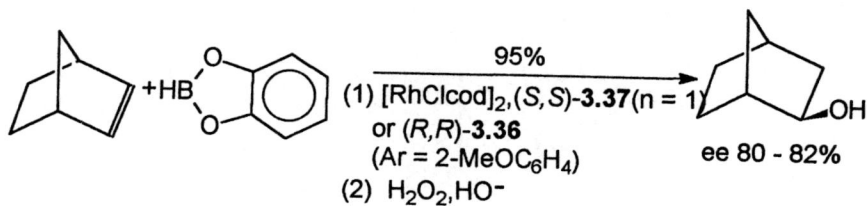

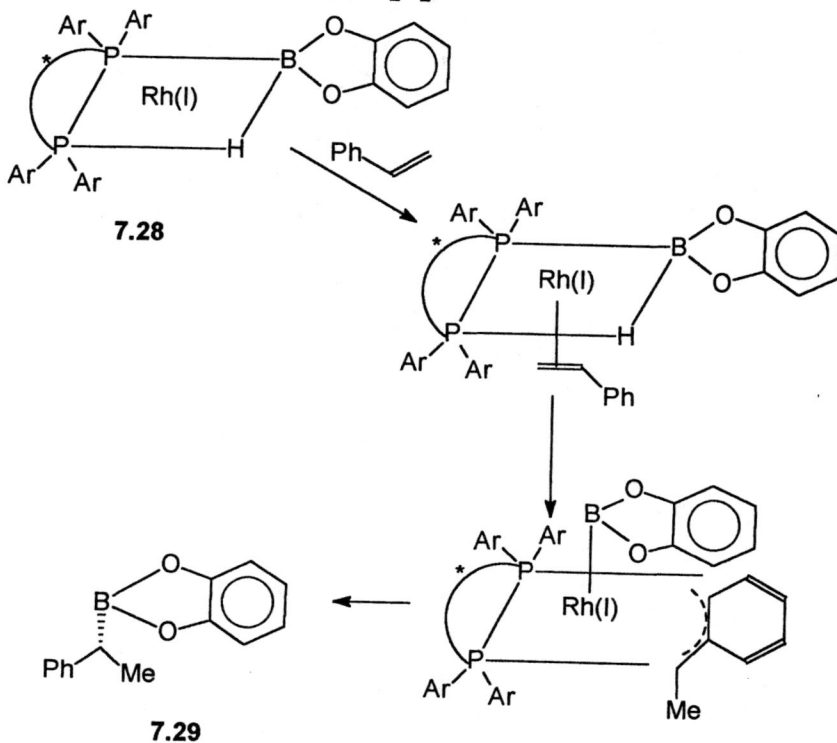

Figure 7.18 (continued)

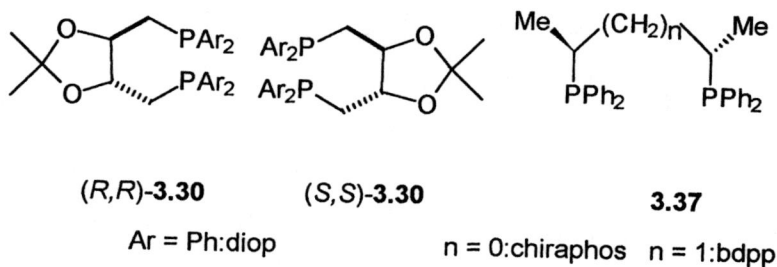

(R,R)-**3.30**   (S,S)-**3.30**   **3.37**

Ar = Ph:diop   n = 0:chiraphos   n = 1:bdpp

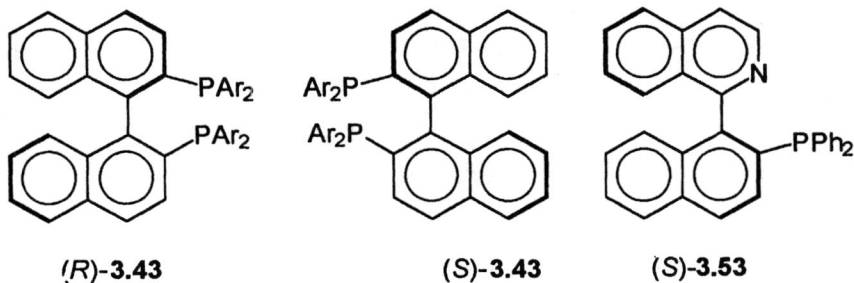

(R)-**3.43**   (S)-**3.43**   (S)-**3.53**

7.4 HYDROSILYLATION [868, 1339a]

Asymmetric hydrosilylation of C=C bonds is not as generally applicable as hydrosilylation of C=O bonds [752]. The regioselectivity of alkene hydrosilylation is often difficult to control, and cleavage of a C–Si bond is more difficult than cleavage of an O–Si bond. However, Bosnich and coworkers [885] described the highly selective intramolecular asymmetric hydrosilylation of silylether **7.30**. This reaction is catalyzed by cationic Rh(I) complexes of (S,S)-chiraphos **3.37** (n = 0) or (S)-binap **3.43** (Ar = Ph). Exposure of siloxane **7.31** to basic H$_2$O$_2$ provides primary-secondary 1,3-diols with an excellent enantiomeric excess, provided that **7.30** bears a terminal aryl substituent (Figure **7.19**).

Asymmetric hydrosilylations of terminal alkenes, 1-arylalkenes, norbornenes and dihydrofurans with HSiCl$_3$ have been successfully performed by Hayashi and coworkers [914, 915, 916, 1340, 1341]. These reactions take place at 40°C when catalyzed by chiral palladium complexes, and the most efficient ligand is monophosphine **3.51** (R = Me) (Figure **7.19**). The regioselectivity of the hydrosilylation of terminal olefins is opposite to that usually observed; after treatment with H$_2$O$_2$/KF, secondary alcohols are obtained as major products [752, 855, 1340]. The regioisomeric primary alcohols are typically formed in only about 10% yield in these reactions.

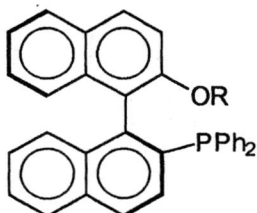

3.51 : (S)-mop

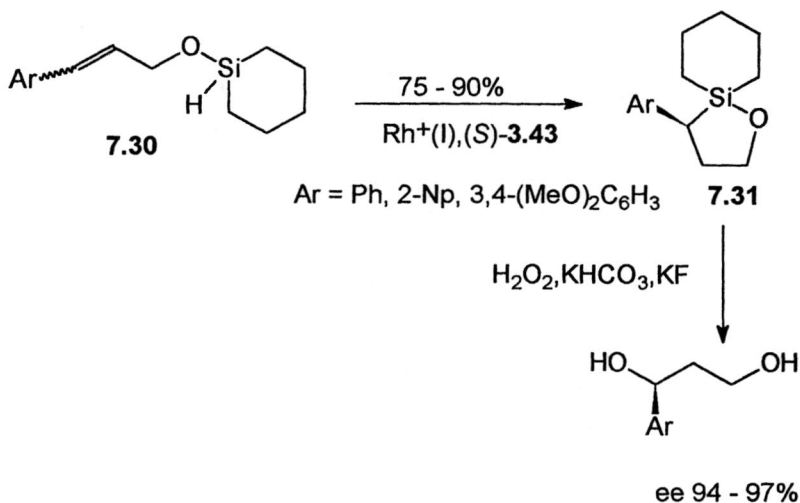

Ar = Ph, 2-Np, 3,4-(MeO)$_2$C$_6$H$_3$ **7.31**

ee 94 - 97%

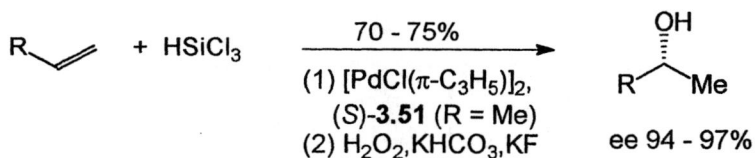

R = n-C$_4$H$_9$, n-C$_6$H$_{13}$, n-C$_{10}$H$_{21}$

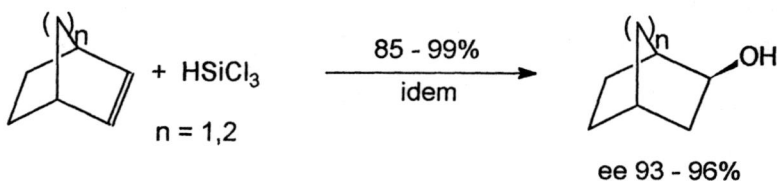

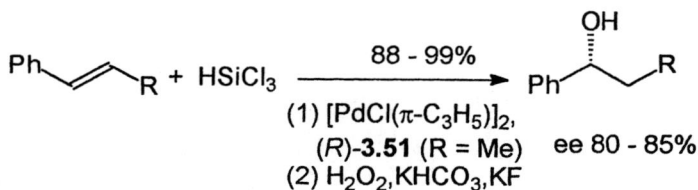

R = H, Me, MeOCH$_2$, PhCH$_2$OCH$_2$

Figure 7.19

7.5 HYDROCARBONYLATION, HYDROCARBOXYLATION, HYDROACYLATION, HYDROCYANATION

7.5.1 Hydrocarbonylation [752, 812, 853, 858, 874, 875]

The metal-catalyzed reaction of olefins with carbon monoxide under dihydrogen pressure is an industrial method for the synthesis of aldehydes, and rhodium or platinum catalysts are preferred. When applied to prochiral terminal olefins, asymmetric hydrocarbonylation poses two problems. First, regioselectivity in these hydrocarbonylations is often low, and second, the chiral aldehyde products are easily epimerized. Therefore, many disappointing results have been obtained [752, 891]. However, asymmetric hydroformylation of styrenes has been performed with a good enantiomeric excess over a catalyst formed from $PtCl_2$, $SnCl_2$ and 3.36 (R = Ph, R' = tert-BuO) (Figure 7.20). These reactions are slower when run in the presence of $HC(OEt)_3$, but the aldehyde is transformed in situ into the corresponding acetal. Epimerization is thereby prevented, and the enantiomeric excesses are higher [752, 923]. The aldehyde is recovered after transketalization with acetone in acidic media. However, the chemical yields of chiral aldehydes are low due to the poor regioselectivity of the process. This method has been applied to the synthesis of 2-arylacetic acids, which are anti-inflammatory drugs [923]. Recently, Takaya and coworkers [912, 1342] advocated the use of phosphine-phosphinite Rh(I) complexes as catalysts in asymmetric hydroformylation of styrenes and vinylacetates. These reactions are highly enantioselective, but display a modest regioselectivity ($\leq 75\%$). However, useful results are obtained with Z-but-2-ene, E-1-phenylprop-1-ene or with cyclic olefins (Figure 7.20). The recommended ligand is (R,S)-binaphos 3.51 (R = (S)-1,1-binaphthalen-2,2'-diyl phosphinite).

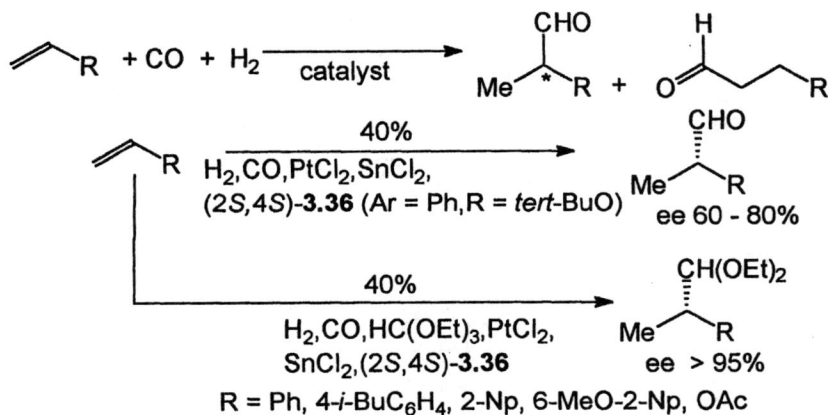

Figure 7.20

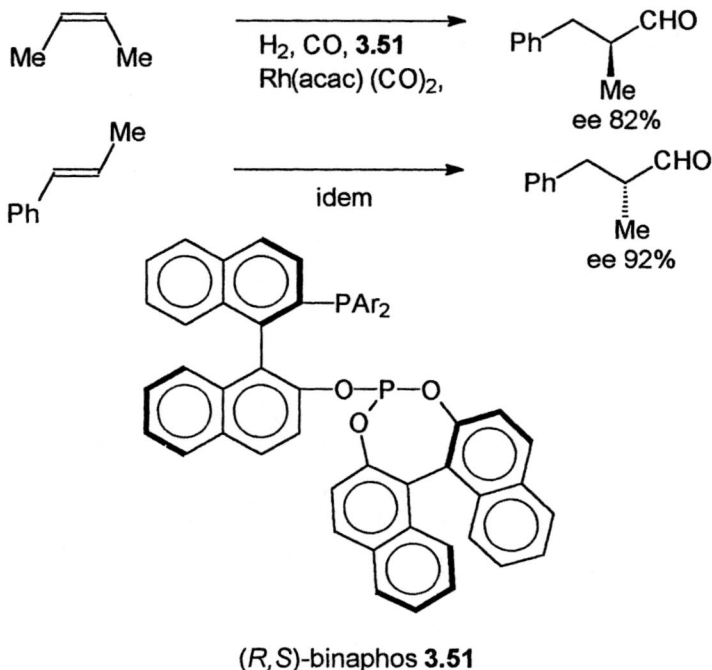

(R,S)-binaphos **3.51**

Figure 7.20 (continued)

7.5.2 Hydrocarboxylation [875]

When the palladium-catalyzed reaction of olefins with carbon monoxide is carried out in a protic solvent, acids or esters are obtained. The problem of the regioselectivity in the reaction of terminal olefins is the same as in hydrocarbonylation (§ 7.5.1). The use of chiral phosphines as ligands of palladium has given disappointing results [752]. However, in hydroorganic media, Alper and coworkers succeeded in performing a regio- and enantioselective synthesis of 1-arylacetic acids in the presence of $PdCl_2$, $CuCl_2$ and binaphthol phosphoric ester **3.54** [923] (Figure 7.21).

7.5.3 Hydroacylation

Intramolecular hydroacylation of 4-substituted pent-4-enals is catalyzed by Rh (I) complexes and leads to 3-substituted cyclopentanones. When the ligand of rhodium is (1S,2S)-**3.35** (X = CH_2CH_2) or even better (R)- or (S)-binap **3.43** (Ar = Ph), cycloalkanones are obtained with an excellent enantiomeric excess

[1197, 1343] (Figure **7.21**). *Trans*-3,4-disubstituted cyclopentanones can be obtained with a high diastereo- and enantioselectivity by a similar process [275].

7.5.4 Hydrocyanation

Asymmetric catalyzed hydrocyanation of prochiral olefins usually gives disappointing selectivities. However, RajanBabu and Casalnuovo [1344] have recently performed asymmetric hydrocyanation of 2-vinylnaphthalenes under $Ni(cod)_2$ catalysis in the presence of phosphinite **3.33** (Ar = 3,5-$(CF_3)_2C_6H_3$). These hydrocyanations occur with a high regio- and enantioselectivity (Figure **7.21**), and the major product is obtained in enantiopure form by fractional crystallization.

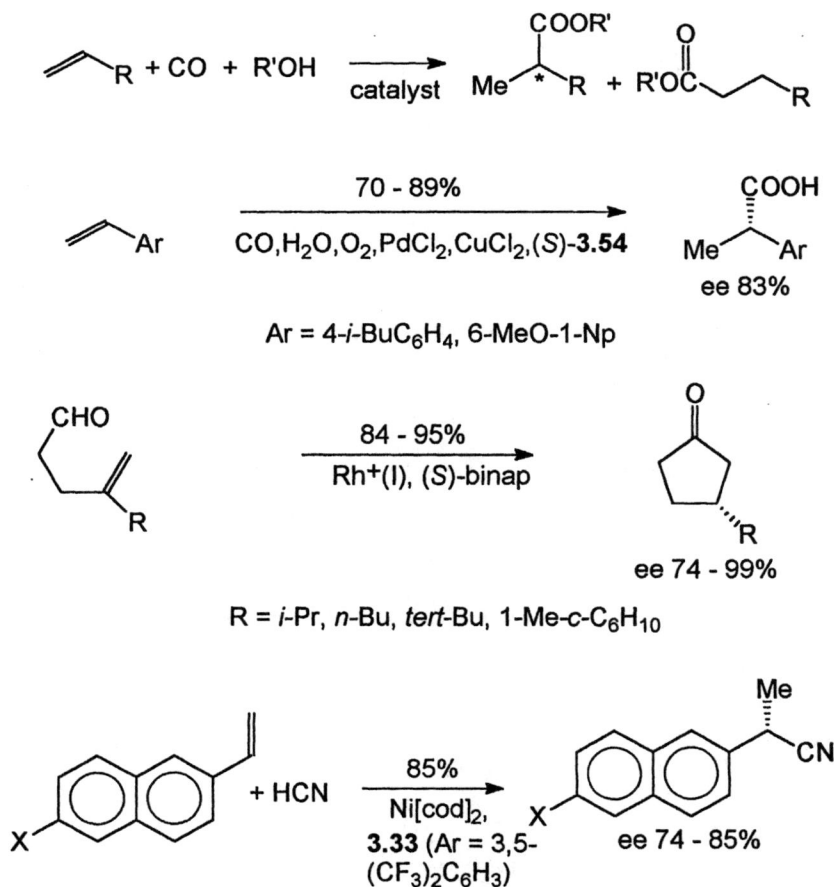

Figure 7.21

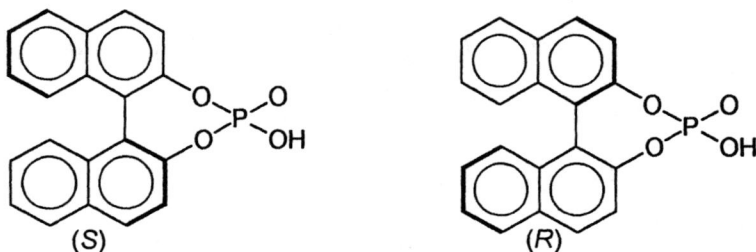

(S)   (R)

3.54

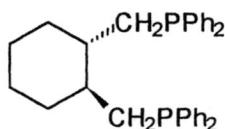

(1S,2S)-**3.35**(X = CH$_2$CH$_2$)

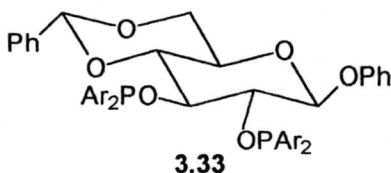

3.33

7.6 DIHYDROXYLATION

Syn-dihydroxylation of olefins is the method of choice for synthesis of
1,2-diols. Dihydroxylation reactions are usually conducted with OsO$_4$ either in
stoichiometric amounts or catalytic amounts in the presence of cooxidants such as
NMO. Osmic esters are intermediates in these dihydroxylations **2.89** (§ 2.9).
Asymmetric dihydroxylations can be performed in the presence of chiral ligands of
osmium [559, 750, 751, 754] or by reaction of OsO$_4$ or other reagents with olefins
bearing a chiral residue.

7.6.1 Osmylation in the Presence of Chiral Ligands [751, 753]

Two classes of ligands have been successfully used in asymmetric osmylation
of prochiral olefins: chiral diamines and cinchona alkaloids. The diamines have
been developed mainly by Corey, Koga and Tomioka, Hirama and their coworkers
[559, 750, 754, 755, 756], and cinchona alkaloids have been designed and imple-
mented by Sharpless and coworkers [750, 753-756].

7.6.1.1 *Chiral Diamines*

Hydroxylation reactions of alkenes in the presence of chiral diamines **2.38**, **2.84**, **2.85** or *trans*-1,2-diaminocyclohexane derivative **7.32** [559, 750, 1345, 1346, 1347] are usually carried out at −90°C. The reagent is used in stoichiometric amounts. These reactions take place through hexacoordinated osmium complexes **7.33**. Evidence for the structures of these complexes comes from X-ray crystallography. After reductive hydrolysis of the osmic ester **2.89**, 1,2-diols are obtained with excellent enantiomeric excesses from terminal olefins and from symmetrical or unsymmetrical disubstituted E-olefins. By using **2.38** and **7.32** as ligands, methyl Z-cinnamate and some trisubstituted olefins also give diols in good ee's [1345] (Figure 7.22). A chiral bispiperazine has also been used as ligand [1347a]. This method has been applied to the synthesis of anthracycline antibiotics [1348] and chiral alkoxysubstituted α,β-unsaturated esters **7.34** [1349] (Figure 7.22). In the reactions of **7.34**, the use of OsO_4 in catalytic amounts in the presence of NMO as reoxidant decreases the ee [1349].

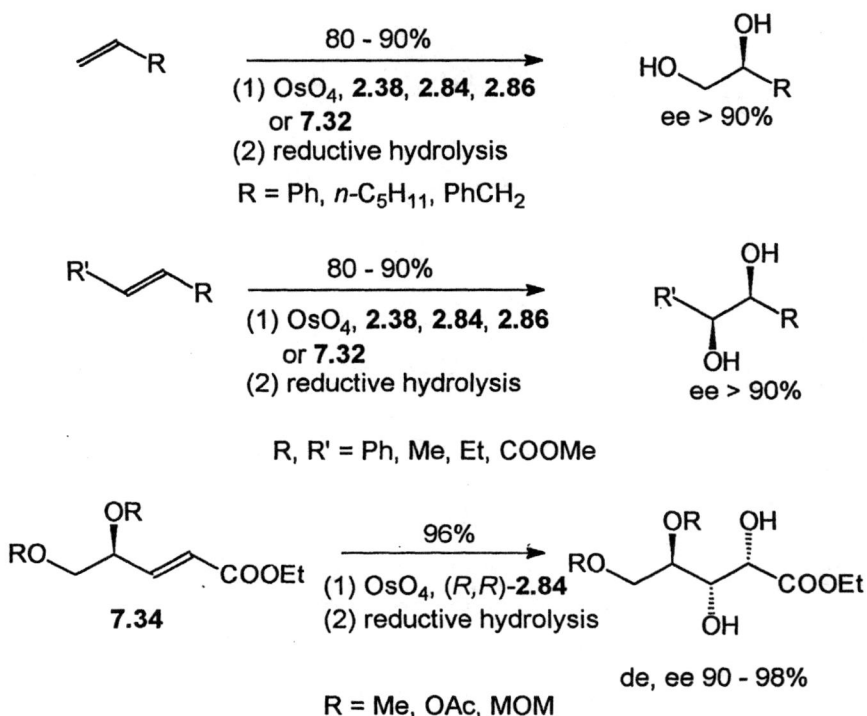

$$\text{80 - 90\%}$$
(1) OsO_4, **2.38**, **2.84**, **2.86**
or **7.32**
(2) reductive hydrolysis

R = Ph, n-C_5H_{11}, $PhCH_2$

ee > 90%

$$\text{80 - 90\%}$$
(1) OsO_4, **2.38**, **2.84**, **2.86**
or **7.32**
(2) reductive hydrolysis

R, R' = Ph, Me, Et, COOMe

ee > 90%

7.34

$$\text{96\%}$$
(1) OsO_4, (*R,R*)-**2.84**
(2) reductive hydrolysis

R = Me, OAc, MOM

de, ee 90 - 98%

Figure 7.22

2.85   **7.32**

2.38   **2.84**

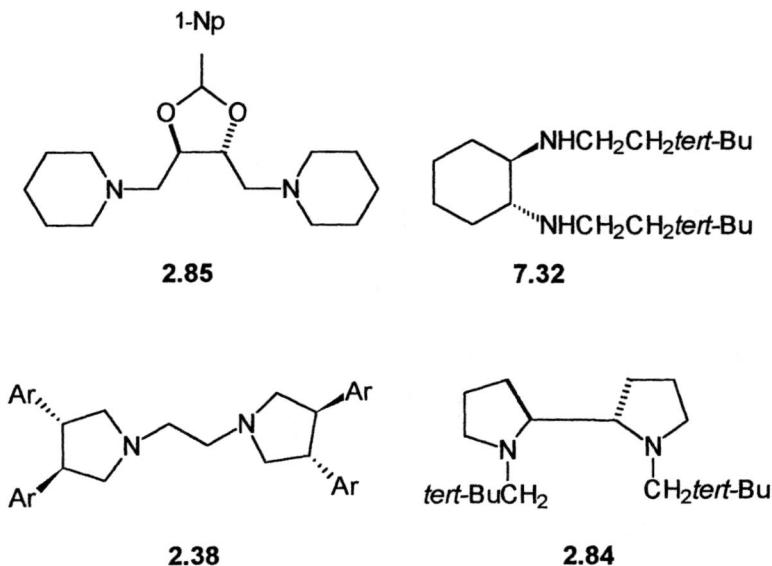

Two reaction mechanisms have been proposed for these dihydroxylations (pathway a or b, Figure **7.23**), either a concerted [3+2] cycloaddition of the olefins on osmium-diamine complex **7.33** or a stepwise reversible [2+2] cycloaddition followed by a rearrangement [559, 1350]. An X-ray crystal structure of the resulting osmic ester **2.89A** shows its symmetrical structure. Houk's calculations [1351] are in favor of a concerted reaction, and his transition state model is reactant-like, with steric interactions dictating the face selectivity of osmylation.

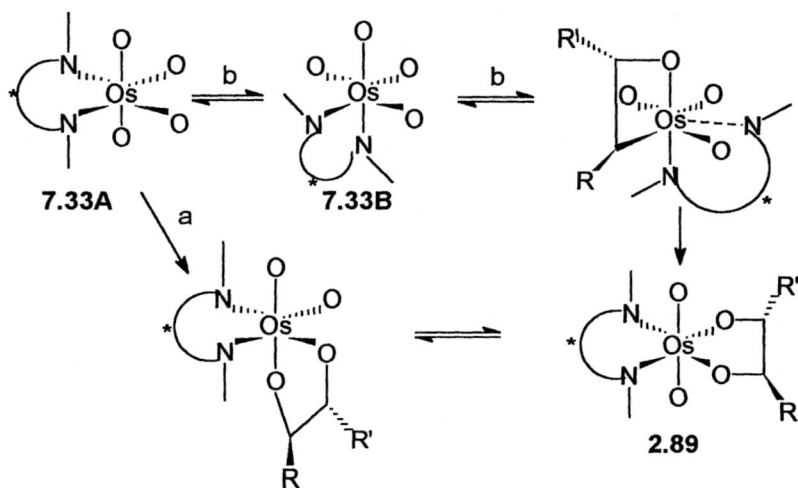

Figure 7.23

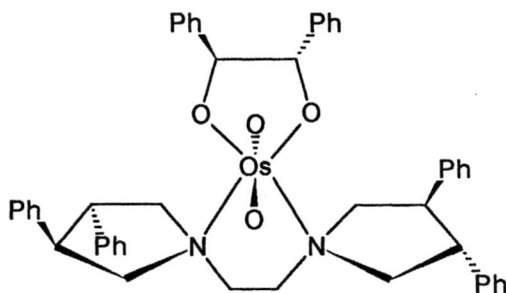

2.89A

Figure 7.23 (continued)

7.6.1.2 Cinchona Alkaloid Derivatives [750, 753-756]

The cinchona-alkaloid-based asymmetric dihydroxylation is a very popular method that has been devised and continuously improved by Sharpless and co-workers. The original method required stoichiometric amounts of OsO$_4$, but ex-perimental conditions were soon developed to allow the use of the toxic and ex-pensive OsO$_4$ in catalytic amounts in the presence of a reoxidant. Only small amounts of the chiral ligands are required. The mechanism of this asymmetric os-mylation has been studied both with monoligands such as **2.91** and bis-symmetrical phthalazine ligands **2.92** or analogs [758, 759, 760, 763, 1352]. From the various studies, it appears that a single ligand is linked to a single OsO$_4$ molecule in the intermediate pentacoordinate osmium complex **7.34**. The nature of the complexes formed from the mono- and bis-phthalazine ligands is essentially the same. The osmium-ligand complex probably reacts with the olefin via a stepwise [2+2] proc-ess to produce osmic ester **2.89B** (Figure 7.24). The significant rate increase that is observed with the most efficient ligands is likely due to π stacking of the olefin and the ligand substituents in the transition state [1352]. A drawback of the cata-lytic process is that the cooxidant can transform ester **2.89B** into osmate ester **7.35**. This compound can be hydrolyzed *in situ* to the 1,2-diol and OsO$_4$, or it can oxidize a second molecule of an alkene to form osmate ester **7.36**. This second osmylation must be prevented because it is not stereoselective. Experimental conditions must be designed to ensure that hydrolysis of **2.89B** or **7.35** is fast enough. Indeed, the replacement of NMO as the reoxidant by K$_3$Fe(CN)$_6$ in aqueous alcohol at 0°C fulfills this goal. The resulting reagent combination, AD-mix (OsO$_4$ 0,2 mol.%, ligand, 1 mol % K$_3$Fe(CN)$_6$, K$_2$CO$_3$) (§ 2.9), has found many applications. Among the numerous ligands tested by Sharpless's group, those derived from dihydroquinine **2.87** (AD-mix-α) and from dihydro-quinidine **2.88** (AD-mix-β) often give rise to enantiomeric diols with a high facial discrimination. Steric interactions are the most important ones, as shown by the structure of osmic ester **2.89B** in which the ligand is **2.87** 4-chlorobenzoate (Figure **7.24**). As an empirical rule, AD-mix-α induces *syn*-dihydroxylation of the bottom face of the olefin as drawn in Figure **7.25**, while AD-mix-β induces dihydroxylation on the top face.

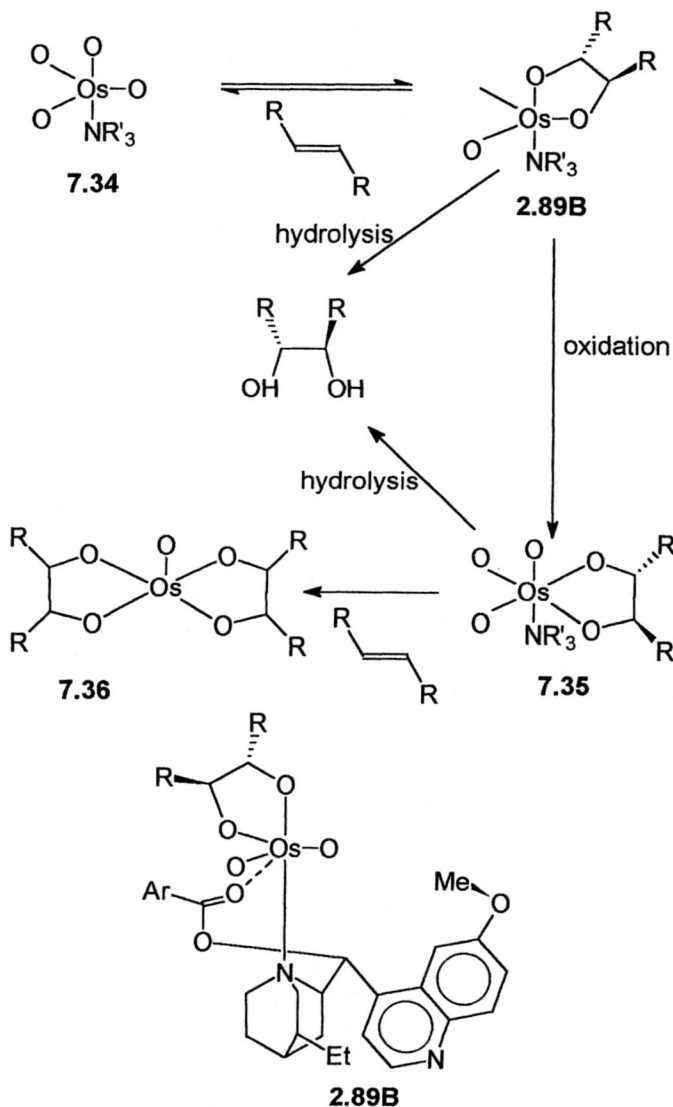

Figure 7.24

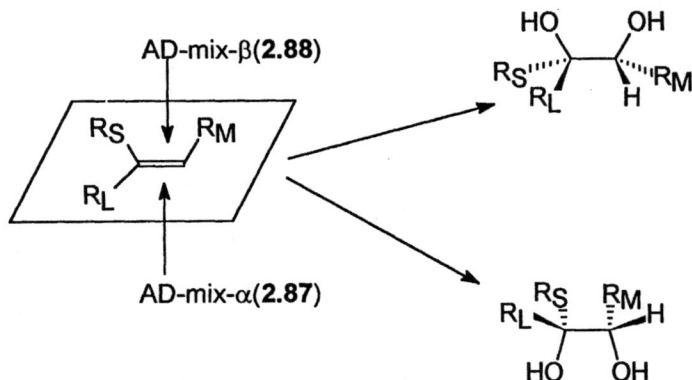

Figure 7.25

Some of the most significant results are summarized in Figure **7.26**. Depending on the substitution pattern of prochiral olefin, the most efficient ligands derived from **2.87** or **2.88** may be phenanthrol ether **2.90**, quinoline **2.91**, or 1,4-dihydroxyphthalazine **2.92** or pyrimidine **2.93** derivatives [1353, 1354]. In nearly all the cases, the predominant isomer follows from the empirical rule given in Figure **7.25**. With terminal olefins, the best ligand is **2.93** [1354]. Substitution of the MeO group on the alkaloid skeleton of **2.92** by an i-C_5H_{11} group brings an improvement in the enantioselectivity with this type of ligand [33]. E-1,2-Disubstituted olefins bearing various substituents give *syn*-diols with excellent enantioselectivities, whatever the ligand [1355]. Vinyl and allylsilanes also give useful results [1356, 1357], as do 1-substituted propenes [1354], some 1-substituted styrenes [1358], and trisubstituted olefins. Arylallyl ethers also give high selectivities unless the phenyl group is *o*-substituted [1359]. A limitation of the method is the low selectivity usually obtained in the asymmetric dihydroxylation of Z-1,2-disubstituted olefins [753]. This drawback was surmounted by Wang and Sharpless [1360], who used as ligands **2.87** (R = *N*-indolylcarbamoyl) (Figure **7.26**). Tetrasubstituted olefins generate osmate esters whose *in situ* hydrolysis is difficult. However, when the reaction is run with a higher amount of catalyst in the presence of $MeSO_2NH_2$, useful selectivities can sometimes (but not always) be obtained with **2.92** and **2.93** as ligands [1361]. Enol ethers are transformed into α-hydroxyketones with high enantiomeric excesses [1362]. Conjugated or nonconjugated dienes can also be dihydroxylated, and under AD-mix conditions with **2.92** as ligand, it is possible to perform the selective dihydroxylation of one of the two double bonds [1363]. For example, asymmetric dihydroxylation of geranyl acetate **7.37** or of isomeric neryl acetate is highly regio- and stereoselective [1364] (Figure **7.27**).

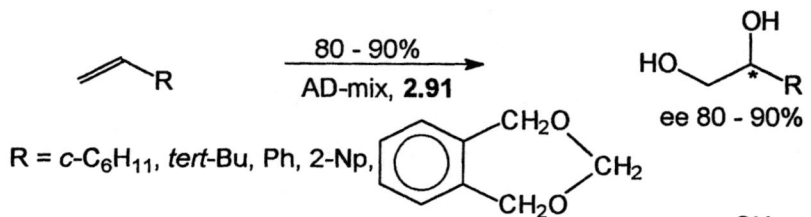

R = c-C$_6$H$_{11}$, tert-Bu, Ph, 2-Np,

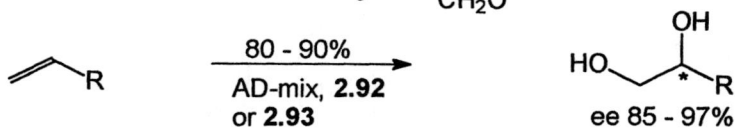

R = n-C$_8$H$_{17}$, Ph, PhOCH$_2$, 2-NpOCH$_2$, c-C$_5$H$_9$, c-C$_6$H$_{11}$, c-C$_7$H$_{13}$

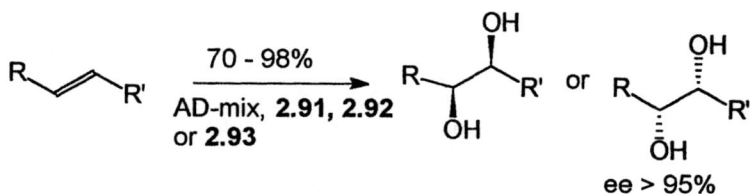

R,R' = Et, n-Bu, n-C$_5$H$_{11}$, Ph, PhCH=CH, COOMe, COOEt, PhC≡C,
Me$_3$SiC≡C, CH$_2$OAc, CH$_2$OH   R = SiR$_3$ or CH$_2$SiR$_3$, R' = alkyl

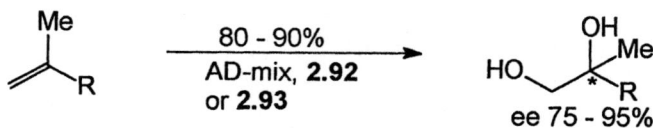

R = n-C$_5$H$_{11}$, Ph, PhC≡C

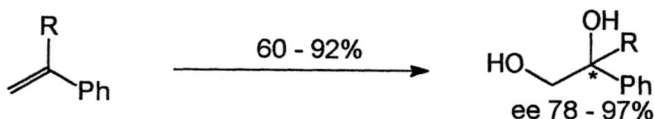

R = CH$_2$Cl, CH$_2$Br, CH$_2$OMe, CH$_2$OCH$_2$Ph

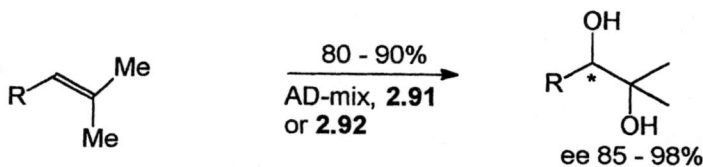

R = n-C$_5$H$_{11}$, Ph

Figure 7.26

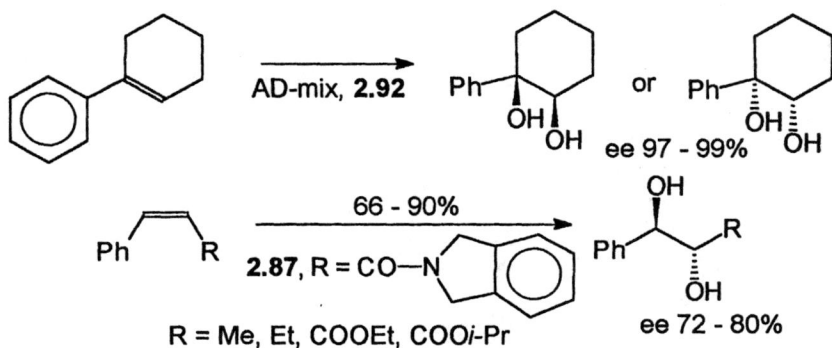

Figure 7.26 (continued)

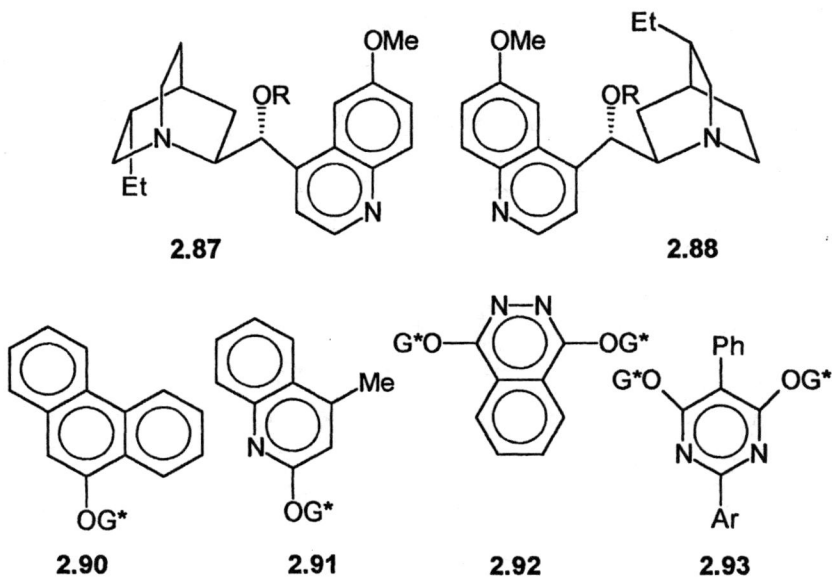

2.87   2.88

2.90   2.91   2.92   2.93

Asymmetric dihydroxylations of certain functionalized compounds require both a higher amount of OsO_4 reagent and the presence of $MeSO_2NH_2$. These compounds include tertiary allylic alcohols [1365], α,β-unsaturated esters and amides [756, 1366], and β,γ-unsaturated amides [1194] (Figure 7.27). With α,β-unsaturated ketones, it is necessary to operate in buffered conditions to avoid epimerization or retro-aldol fragmentation of the products [1367]. Allylic and homoallylic amines are also good substrates for asymmetric dihydroxylation, provided that the nitrogen atom is protected as a BOC-carbamate [1368] (Figure 7.27).

However, an anomalous enantioselectivity has been observed in the asymmetric dihydroxylation of 1,1-disubstituted allylic alcohols [1369].

Asymmetric dihydroxylation of olefins bearing chiral substituents can take place with double diastereodifferentiation (§ I.6), so that matched and mismatched pairs will be observed [498, 753, 762].

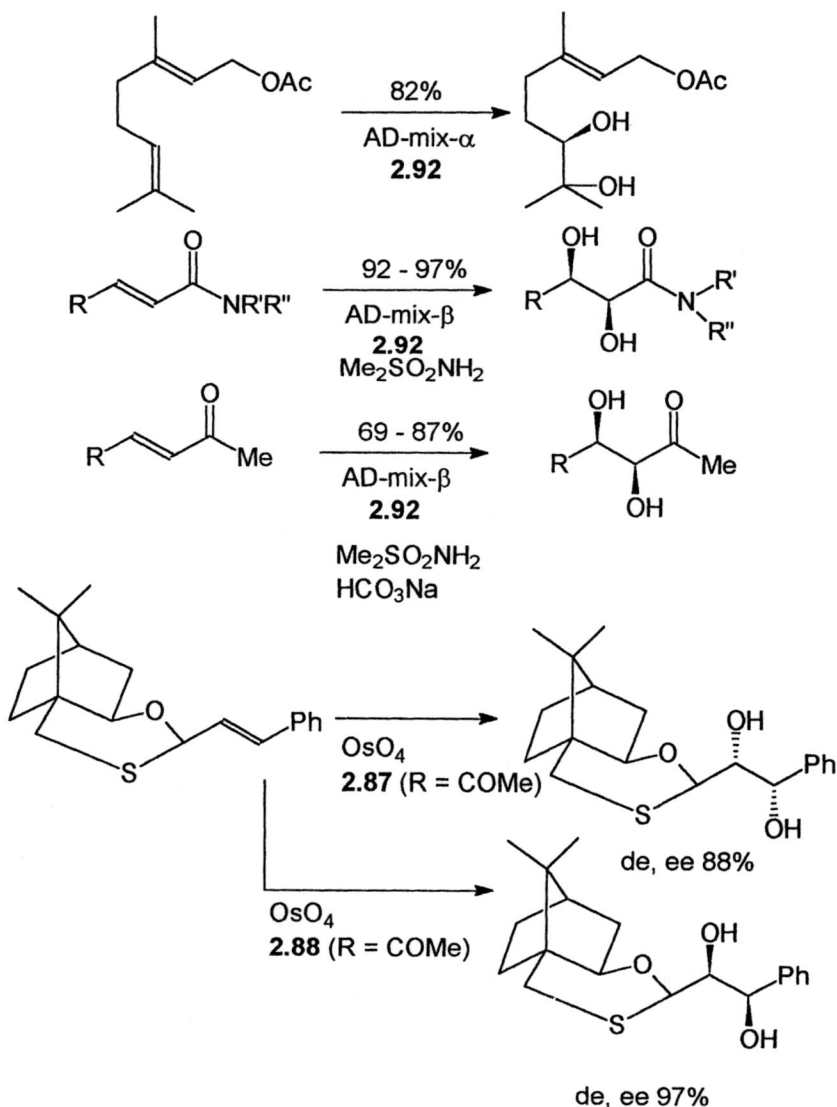

Figure 7.27

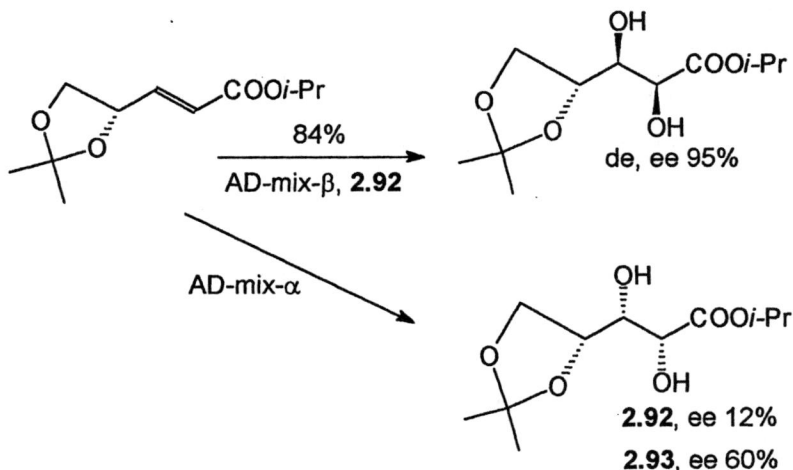

Figure 7.27 (continued)

Two examples are given in Figure 7.27. In the first example, OsO$_4$ and ligands **2.87** and **2.88** (R = COMe) are used in stoichiometric amounts and, in both cases, the enantioselectivity is good. In the second example, the use of catalytic conditions gives very good results with the matched pair but poor results with the mismatched pair. However, the selectivity can be improved by changing the ligand in the mismatched pair. An application of this concept to kinetic resolution of racemic olefins and allylic acetates has recently been published [1370, 1371].

Asymmetric dihydroxylation is a powerful and popular tool in drug and natural product synthesis [812, 1359, 1366, 1372].

7.6.2 Osmylation of Functionalized Olefins Bearing Chiral Residues

The OsO$_4$-promoted dihydroxylations of α,β-unsaturated chiral acetals or of α,β-unsaturated esters of chiral alcohols have given disappointing selectivities [147, 750]. The N-enoylsultams **1.134** (R = CH=CRR') are valuable reagents for OsO$_4$ dihydroxylations, and 1,2-diols are obtained with a high selectivity. After acetalization, the auxiliary is cleaved, and the corresponding acids are formed [147, 454, 463] (Figure **7.28**). The approach of the OsO$_4$ takes place on the least hindered face of the substrate, which is in the s-cis conformation. Osmylation of chiral butadieneiron carbonyl complexes also takes place with a high selectivity on the face opposite to the metal. HETES precursors have been prepared in this fashion [538, 1373].

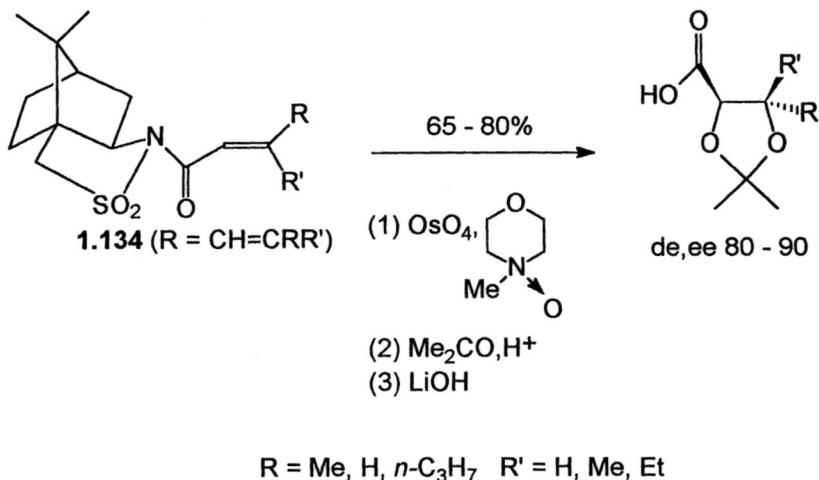

R = Me, H, *n*-C₃H₇ R' = H, Me, Et

Figure 7.28

7.7 EPOXIDATION

7.7.1 Nonfunctionalized Olefins [957]

    The development of an asymmetric epoxidation of nonfunctionalized olefins with an enantiomeric excess higher than 80% is a challenge that was taken up in 1990 by Jacobsen [954, 958, 960, 1374], Katsuki [596, 954, 955, 1375, 1376, 1377] and their coworkers. These epoxidations were performed at room temperature with sodium hypochlorite or iodosylbenzene in the presence of manganese(III) complexes **7.38** and **7.39** bearing chiral ligands derived from salen **3.70** or cyclohexyldiimine **3.29**. Recently, Katsuki used (Me₃Si)₂O₂ or H₂O₂ as the oxidant in the presence of *N*-methylimidazole and an ammonium salt [1376]. High enantiomeric excesses are obtained from Z-arylpropenes, methyl Z-cinnamate, and dihydronaphthalenes **7.40** (X = CH₂) or chromenes **7.40** (X = O) (Figure **7.29**). Enantiofacial selection occurs during the approach of the olefin to the least hindered side of an *oxo*-manganese(V) complex [1375]. The approach of *cis*-olefins probably takes place along the nitrogen-metal bond axis.

    Mukaiyama and coworkers [1378] have performed such oxidations with molecular oxygen in the presence of pivaldehyde under catalysis with related salen **3.70** Mn(III) complexes; however, lower enantioselectivities were obtained. The use of various manganese or iron porphyrins as catalysts also gives lower enantioselectivities [971, 1379, 1379a], as do the oxidations with chiral oxaziridines in stoichiometric amounts [741].

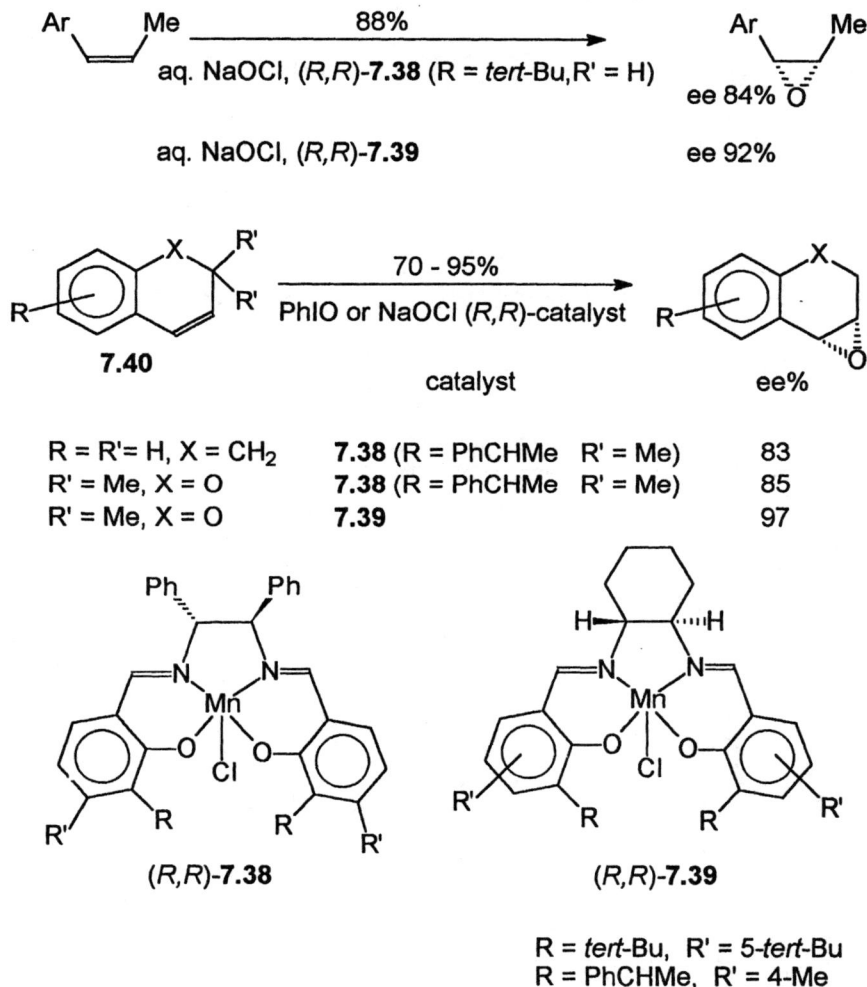

R = R'= H, X = CH$_2$ 7.38 (R = PhCHMe   R' = Me)   83

R' = Me, X = O   7.38 (R = PhCHMe   R' = Me)   85

R' = Me, X = O   7.39   97

(R,R)-7.38   (R,R)-7.39

R = tert-Bu, R' = 5-tert-Bu
R = PhCHMe, R' = 4-Me

Figure 7.29

7.7.2 Allyl Alcohols [169, 559, 752, 807, 808, 810, 858]

The Sharpless epoxidation of allyl alcohols **3.16** by tert-butyl hydroperoxide under catalysis with chiral titanium complexes is a very popular method that has frequently been used in industry [811, 812, 853]. This epoxidation was initially developed with stoichiometric amounts of tartrate catalysts. Today, it is usually performed in the presence of catalytic amounts of Ti(Oi-Pr)$_4$ and diethyl or diisopropyl tartrate (2R,3R)- or (2S,3S)-**2.69** (R = Et or i-Pr). The reactions are conducted at or near room temperature in the presence of molecular sieves. Several titanium complexes are generated, and they are in fast equilibrium in the reaction medium [813]. From structural and kinetic studies, Sharpless proposed that the major complex is also the catalytic species [813, 1380, 1381]. This complex is a dimer with one tartaric ester residue per titanium atom. Ligand exchange with the allyl alcohol and *tert*-BuOOH on the same Ti atom should generate an intermediate complex **3.18**. In this complex, coordination of the oxidant with titanium activates the peroxide. The oxygen transfer to the double bond takes place along the O–O bond of **3.18**, and the two carbon-oxygen bonds are probably formed simultaneously. Steric hindrance around the metallic site favors a single conformer of the transition state. As a rule, when using (*R,R*)-tartaric esters, attack of the double bond occurs from the upper face of the allyl alcohol as shown in Figure **7.30**, while the lower face is attacked when using the (*S,S*)-tartrates.

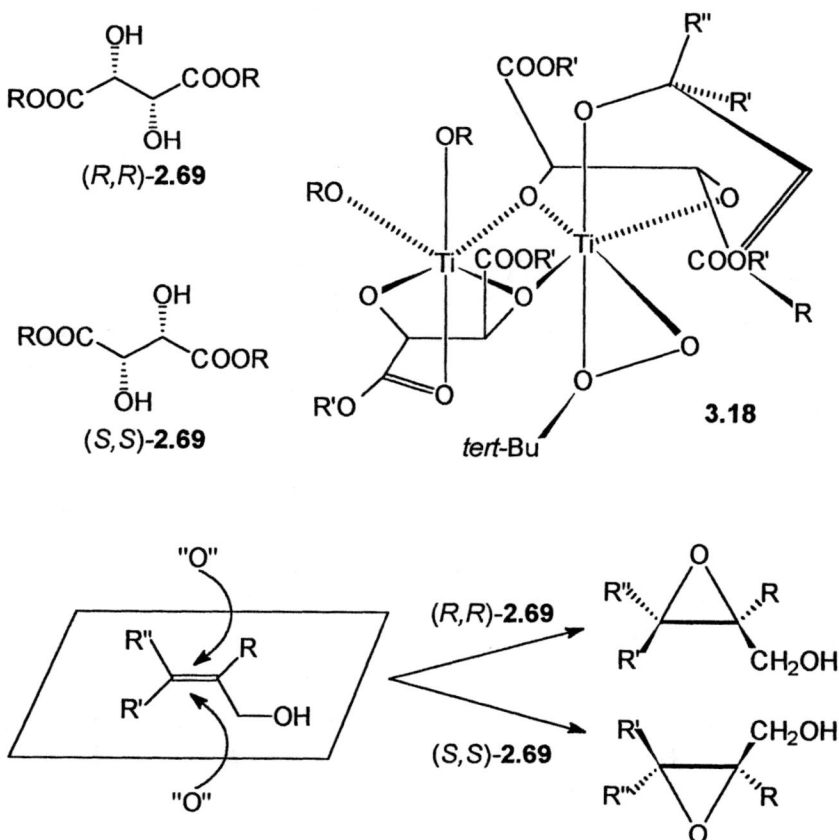

Figure 7.30

Asymmetric epoxidation of many primary allyl alcohols can be performed under catalytic conditions with an excellent enantiomeric excess [808] (Figure **7.31**). The reactions of low molecular weight allylic alcohols **3.16** ($R_Z = R_E = H$, R' = H or Me) do not give high isolated yields due to the volatility of the epoxyalcohol products. However, if these epoxyalcohols are esterified *in situ*, then the yields are significantly improved. Though selectivities are frequently good, a few Z-allylic alcohols Z-**3.16** (R = *i*-Pr or *tert*-Bu) are epoxidized with mediocre enantioselectivities. When dealing with chiral allyl alcohols, double diastereodifferentiation (§ I.6) occurs. The stereoselectivity of this process depends upon matching or mismatching of the partners, as shown by the example in Figure **7.32**.

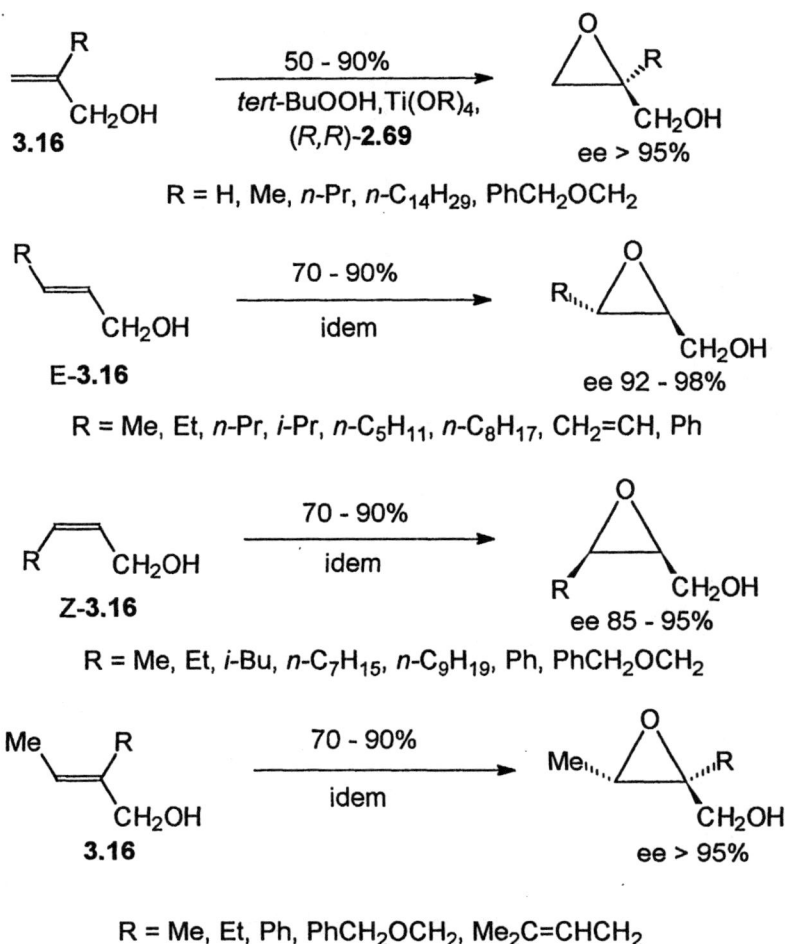

R = H, Me, *n*-Pr, *n*-C$_{14}$H$_{29}$, PhCH$_2$OCH$_2$

R = Me, Et, *n*-Pr, *i*-Pr, *n*-C$_5$H$_{11}$, *n*-C$_8$H$_{17}$, CH$_2$=CH, Ph

R = Me, Et, *i*-Bu, *n*-C$_7$H$_{15}$, *n*-C$_9$H$_{19}$, Ph, PhCH$_2$OCH$_2$

R = Me, Et, Ph, PhCH$_2$OCH$_2$, Me$_2$C=CHCH$_2$

Figure 7.31

$$R = Me_2C=CHCH_2 \quad Me_2C=CHCH_2CH_2$$

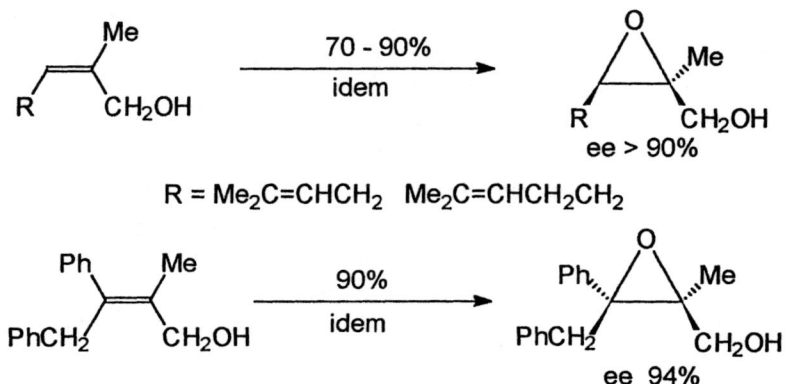

Figure 7.31 (continued)

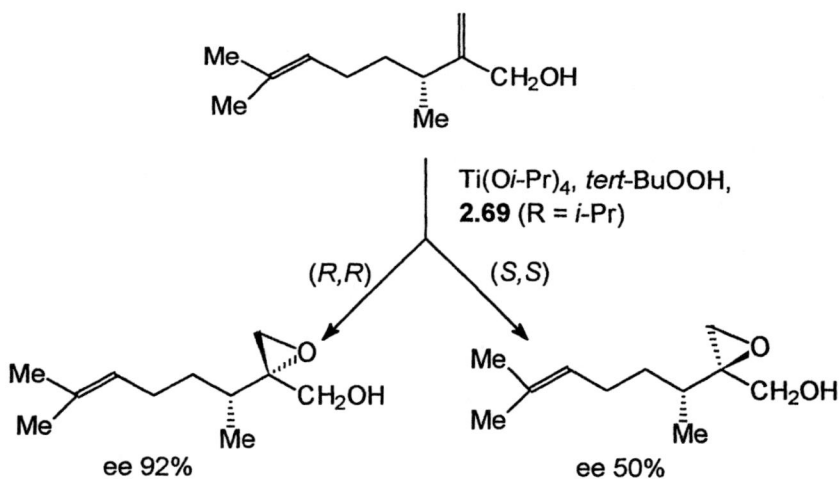

Figure 7.32

    Asymmetric epoxidation of racemic secondary allyl alcohols **3.17** takes place with kinetic resolution [127]. The presence of a substituent on the same face as the reagent at transition state induces a decrease in rate due to steric hindrance. Therefore, according to the (R)- or (S)-absolute configuration of the substrate, the rate of epoxidation with a given catalyst will be different (Figure **7.33**). The ratio of rates in a kinetic resolution depends upon the nature of the R substituent, the temperature, and the structure of the tartrate **2.69** (R = Me, Et, i-Pr). Cyclohexyl tartrates have been recommended for kinetic resolutions because bulkier esters give higher relative rate ratios [808]. A few examples of resolutions are shown in Figure **7.34**. Useful results are obtained with a number of different substrates, including **3.17** (R_E = Me$_3$Si, I, R' = H). The process is less satisfactory with intracyclic secondary allyl alcohols. Furthermore, tertiary allyl alcohols are not reactive enough. The asymmetric epoxidation of homoallyl alcohols is less enantioselective than that of allyl alcohols, and moreover, the facial selectivity is reversed. The Sharpless epoxidation is compatible with many functional groups (acetals, alkynes, aldehydes, alcohols, amides, esters, ketones, nitriles, isolated double bonds, sulfones, sulfoxides, ureas, carbamates) [808], and it has been applied in the synthesis of natural products [810].

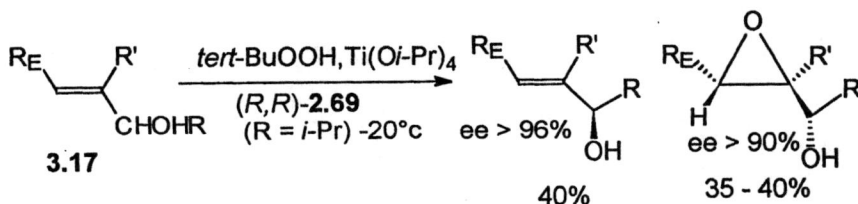

R	R'	R_E	k rel.(-20°C)
n-C$_6$H$_{13}$	H	H	83
n-C$_4$H$_9$	Me	H	138
c-C$_6$H$_{11}$	H	Me	104

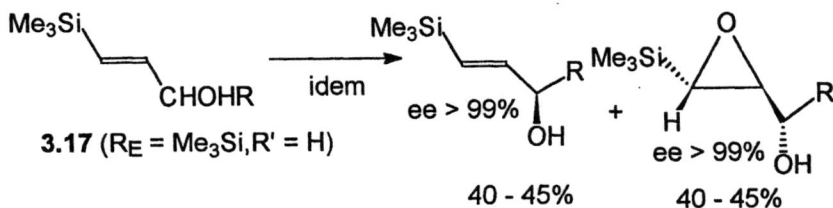

$R = n$-C$_5$H$_{11}$, i-Pr, Ph, PhOCH$_2$, C$_5$H$_{11}$CH=CHCH$_2$

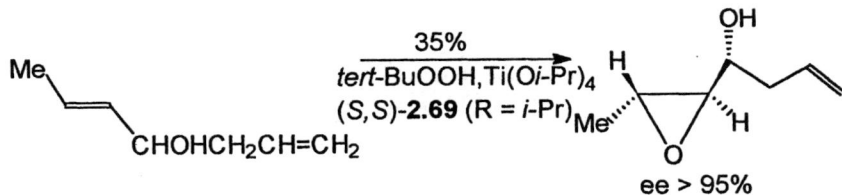

Figure 7.34

The facial discrimination of the asymmetric epoxidation of allyl alcohols follows the empirical rule given in Figure 7.30 in nearly all the cases examined. However, exceptions have been observed with dialkenyl glycols [1382, 1383]. Due to the efficiency of the Sharpless epoxidation, the use of chiral auxiliaries for the epoxidation of allylic alcohols is not generally required. However, Charette and Côté performed the epoxidation of carbohydrate ethers of allyl alcohols with MCPBA and observed a good stereoselectivity (80%) [369].

7.7.3 α,β-Unsaturated Carbonyl Compounds

Double bonds substituted by electron-withdrawing groups easily suffer epoxidation with H_2O_2 or alkali hypochlorites in basic media. These reactions can be performed under phase-transfer catalysis, and chiral catalysts have been introduced [767]. Quaternary ammonium salts of cinchona alkaloids have been studied, but enantioselectivities are less than 55% [771, 1058]. Polyalanines and analogs give better results, but only with diaryl enones ArCH=CHCOAr' (ee 90%). These reactions are run with aqueous H_2O_2 in the presence of the chiral base and toluene [776, 777, 1384].

In an application of the chiral auxiliary technique, succindialdehyde has been transformed into mono-N-tosyloxazolidine **1.83** (n = 0, Y = CH=CHCOOH) derived from ephedrine. Treatment of **1.83** with KOCl leads to an epoxyacid with high diastereoselectivity [1385]. After nucleophilic ring opening of epoxide, the chiral auxiliary is recovered by treatment with ethanedithiol (Figure 7.35). Epoxidation of the (2R,4R)-pentane-2,4-diol monoketal of a prochiral quinone with tritylhydroperoxide has been performed with a moderate selectivity by Corey and Wu en route to a natural product (70%) [1386].

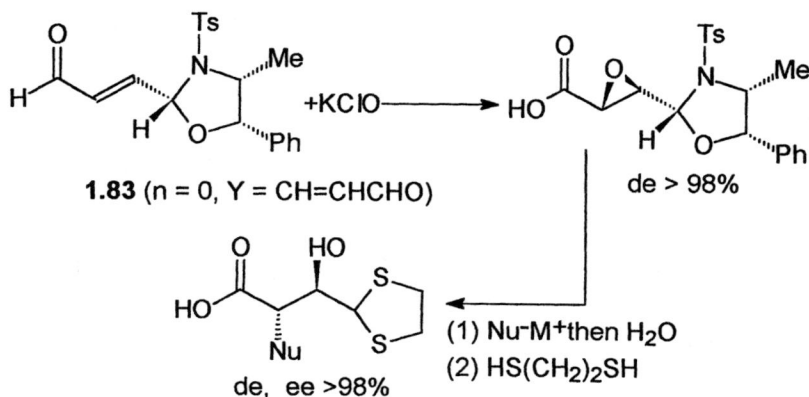

Figure 7.35

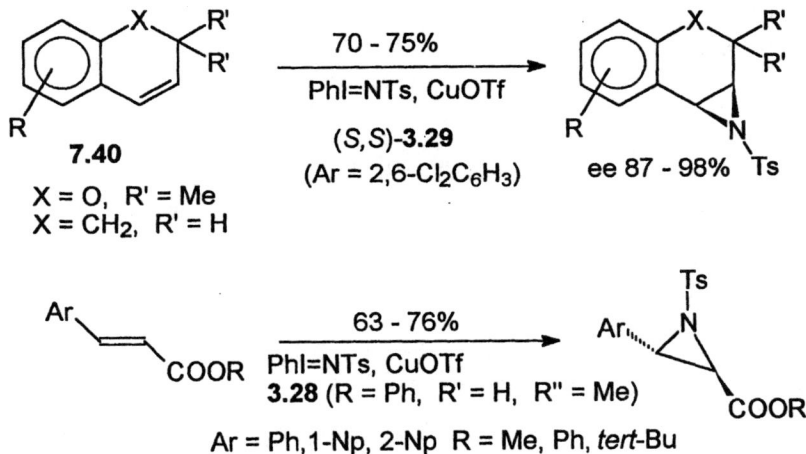

Figure 7.35 (continued)

7.8 AZIRIDINATION

Nitrene-group transfer bears several features in common with oxygen-transfer, and indeed PhI=NTs can be used as an aziridination reagent in the presence of manganese(III) catalysts **7.38** [961]. These reactions are poorly enantioselective, but copper catalysts showed higher selectivity [1387]. Recently, two types of copper ligands have been used for asymmetric aziridination of olefins. Jacobsen and coworkers [962] proposed ligand **3.29** (R = 2,6-Cl$_2$C$_6$H$_3$). With this ligand, aziridination of chromene **7.40** (X = O, R' = Me, R = 4-CN) takes place at −78°C with an ee superior to 98% and aziridination of **7.40** (X = CH$_2$, R = R' = H) takes place with 87% ee. From other olefins, disappointing results are observed. Evans and coworkers recommended bis-oxazolines **3.28** (R = Ph, R' = H, R'' = Me) as copper ligands for asymmetric aziridination of cinnamic esters [965] (Figure 7.35). These reactions take place in benzene at room temperature in the presence of molecular sieves. However, other olefinic substrates again do not give high selectivities.

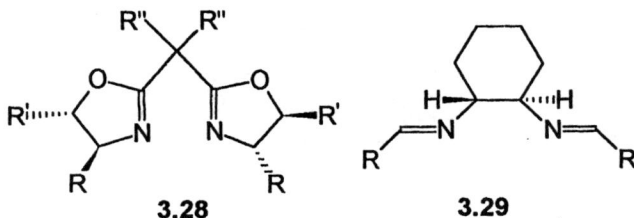

3.28   3.29

7.9 ADDITIONS OF HETEROATOMIC NUCLEOPHILES TO ELECTROPHILIC DOUBLE BONDS

The double bonds of electron-deficient olefins (carbonyl compounds, nitriles, sulfones, sulfoxides, nitro derivatives, etc.) have a low-lying LUMO that can allow the attack of various nucleophiles. The nucleophiles can be neutral or negatively charged heteroatomic species, or they can be carbon species such as organometallic reagents or enolates. In the case of heteroatomic nucleophiles, asymmetric additions can be performed in the presence of chiral catalysts, with chiral reagents, or with substrates bearing chiral residues.

7.9.1 Chiral Catalysts

Conjugate additions of thiols or thioacetates to α-enones, α,β-unsaturated esters or nitrostyrenes are catalyzed by cinchona alkaloids [173, 1058]. Even under pressure [1388], the adducts are obtained with an enantioselectivity lower than 75%. However, the use of **7.41** as catalyst induces higher selectivities [173] (Figure **7.36**).

7.9.2 Additions of Chiral Lithium or Magnesium Amides

Conjugate additions of lithium or magnesium amides derived from enantiopure amines such as (R)- or (S)-(α-methylbenzyl)-benzylamine [1389-1394] or (S)-binaphthylmethylamine [563] to α,β-unsaturated esters are highly stereoselective at −78°C (Figure **7.36**). After hydrogenolysis, β-aminoacids derivatives are obtained with a high enantiomeric excess. α-Alkyl substituted α,β-unsaturated esters provide *syn*-α-alkyl-β-aminoesters with a high selectivity while trapping the intermediate magnesium enolates with MeI generates *anti* isomers [1391, 1392, 1393] (Figure **7.36**). Trapping these enolates can also be accomplished with camphoroxaziridines **2.82**, and this provides *anti*-β-amino-α-hydroxyacids with an excellent diastereo- and enantioselectivity [1081, 1082, 1395] (Figure **7.36**). To make the subsequent hydrogenolysis easier, Bovy and coworkers [1394] replaced (R)-(α-methylbenzyl)-benzylamine by N-trimethylsilyl-(R)-1-phenethylamine.

7.9.3 Additions to α,β-Unsaturated Compounds Bearing Chiral Residues

Conjugate additions of thiols to N-enoyloxazolidinones E- or Z-**7.42** derived from **1.116** [415] or to lactone derivative **1.31** [66] take place with a good selectivity. After removal of the chiral auxiliary, the corresponding adducts are readily obtained. Conjugate additions of amines to lactone **1.36** (G* = menthyl) [66, 1396] are also highly selective. Reduction of the resulting adducts with LAH provides chiral aminodiols, while treatment of the adducts with MeOH in the presence of TsOH leads to chiral β-aminoesters [1396] (Figure 7.36).

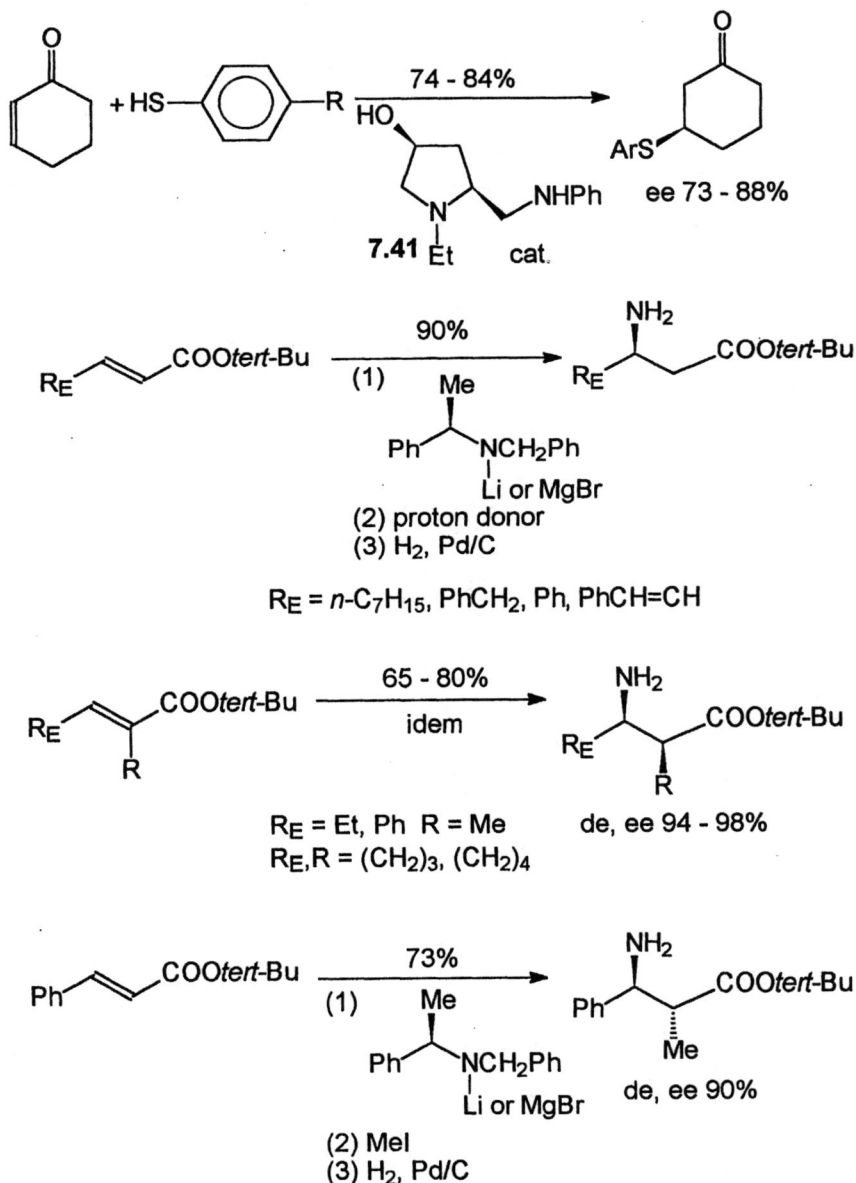

Figure 7.36

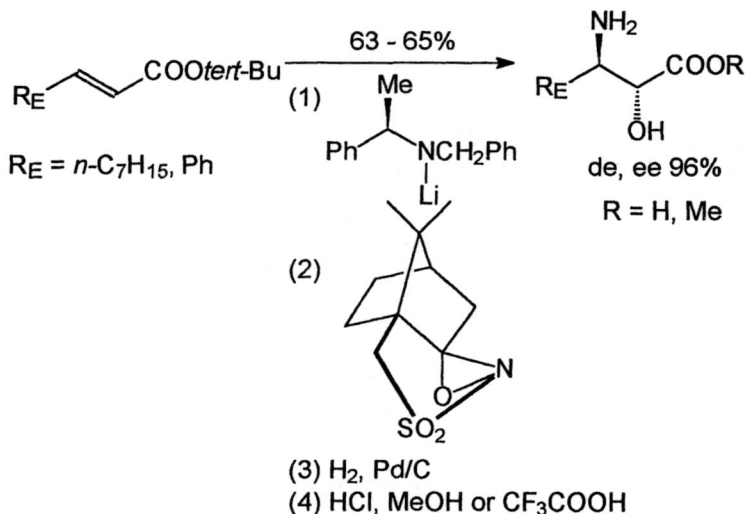

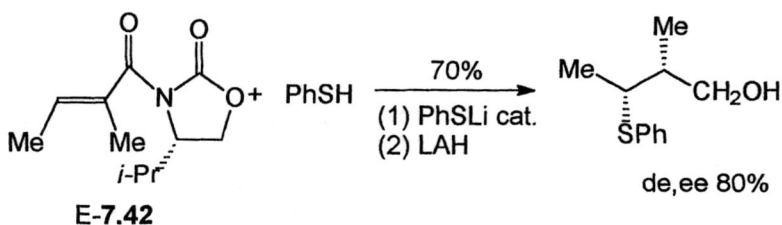

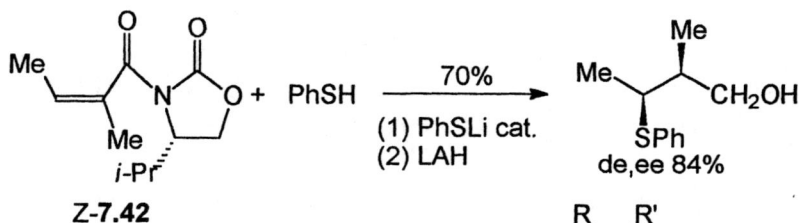

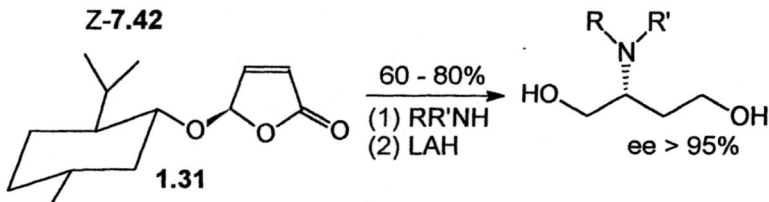

$R_E = n\text{-}C_7H_{15}, Ph$

63 - 65%

(1) Me / Ph–CH–NCH$_2$Ph / Li

(2)

(3) H$_2$, Pd/C
(4) HCl, MeOH or CF$_3$COOH

de, ee 96%
R = H, Me

E-**7.42**

+ PhSH

70%
(1) PhSLi cat.
(2) LAH

de,ee 80%

Z-**7.42**

+ PhSH

70%
(1) PhSLi cat.
(2) LAH

de,ee 84%

1.31

60 - 80%
(1) RR'NH
(2) LAH

ee > 95%

RR' = (CH$_2$)$_4$, (CH$_2$)$_5$, (CH$_2$)$_2$O(CH$_2$)$_2$, Et$_2$, n-C$_4$H$_9$ and H, PhCH$_2$ and H

1.31

45%
(1) PhCH$_2$NH$_2$
(2) MeOH,TsOH

ee 92%

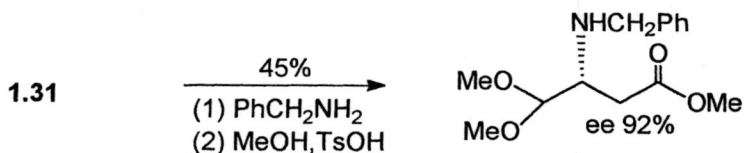

Figure 7.36 (continued)

Conjugate additions of primary amines to crotonates of chiral alcohols require high pressure [134, 173], and esters **7.43** of (1*R*,2*S*,5*R*)-menthols substituted in the 8-position **1.4** (R = Ph, 4-$C_6H_5OC_6H_4$, 2-Np) give highly selective reactions. Facial discrimination is due to the shielding of one face of the double bond by the aromatic ring. Proton NMR and X-ray crystallography studies have shown that the ground-state conformation of crotonates **7.43** is *s-cis, syn*. However, in polar media, the reaction is thought to proceed through the *s-trans, syn* conformation (§ I.4) [91] (Figure 7.37). 1,4-Addition of lithiocopper amides to unsaturated esters of chiral alcohols **7.44** (G* = OR*) or to the corresponding sultams **7.44** is highly stereoselective, and the face selectivity is reversed under pressure [1397]. Lithium amides can also be used, provided that the chiral auxiliary phenmenthol **1.4** (R = Ph) is replaced by the conformationally biased alcohol 2,2,6,6-tetramethyl-3,5-heptanediol monobenzoate [1398]. Sequential conjugate additions to 2,7-diene-1,9-dioate derivatives of chiral alcohols result in asymmetric cyclizations [1399]. Crotonimides derived from imidazolidinones **1.131** react with $PhCH_2ONH_2$ in the presence of Me_2AlCl at −78°C with a good selectivity [1400]. Lithiated amides also react highly selectively with E-α,β-unsaturated cyclopentadieneiron carbonyl complexes **1.151** (R = E-MeCH=CH). After oxidative cleavage of the chiral residue with bromine, β-lactams are obtained with excellent diastereo- and enantiomeric excesses [173, 522] (Figure 7.37). These results are interpreted by reaction of the least hindered face of the double bond of the complex through its *s-cis* conformation (Figure 7.37). Additions of amines or metal amides to chiral α,β-unsaturated sulfoxides or sulfoximines are usually poorly selective, except in the case of hydroxybornane-derived sulfoxide **7.45** [173, 1401].

Additions of various alcohols to *N*-enoyloxazolidines **1.83** (Y = RCH=CH, n = 1) or to some *N*-enoyl-1,3-oxazolidin-2-ones have also given useful results [321, 1402].

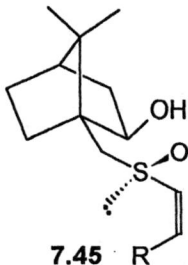

7.45

R = Ph, TBDMSO

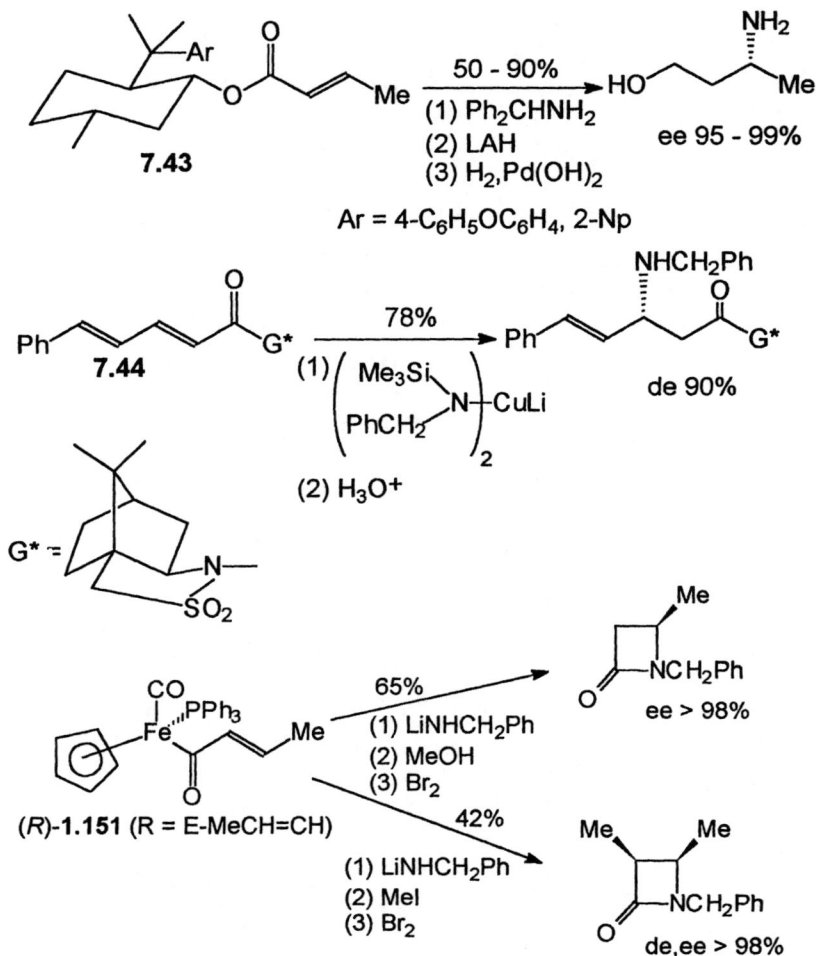

Figure 7.37

7.10 ADDITIONS OF ORGANOMETALLIC REAGENTS TO ELECTROPHILIC DOUBLE BONDS

Because nucleophilic additions of organometallic reagents to electrophilic double bonds are very popular in organic synthesis [624e], extensions to asymmetric synthesis have been widely studied. Chirality is introduced either on the organometallic reagent, which can bear chiral ligands, or on the electrophile. As in Chapter 6, the reactions of enolates, enamines and related reagents will be described later. A theoretical description of the conjugate addition of MeCu to α,β-unsaturated aldehydes has been reported by Dorigo and Morokuma [26].

7.10.1 Reactions of Organometallic Reagents Bearing Chiral Ligands

Koga and coworkers [112, 262, 559, 635] have described the conjugate addition reactions of organolithiums to α,β-unsaturated imines. Chiral diethers are used as metal ligands, and diether **2.39** has shown the greatest efficiency for the asymmetric addition of n-BuLi or PhLi to unsaturated N-cyclohexylimines. After hydrolysis, α-branched aldehydes are obtained with an excellent selectivity (Figure **7.38**). The observed facial discrimination has been interpreted through formation of an intermediate complex **7.46**. The R group is then transferred to the less hindered face of the unsaturated imine. These authors used the same ligand for lithium to prepare chiral binaphthyls from hindered 1-X-substituted imine **7.47** and 1-NpLi. These substitution reactions occur through an addition-elimination mechanism, and the substituted binaphthyl is obtained with a good enantioselectivity [1403] (Figure **7.38**). The same ligand for lithium has been used to perform asymmetric addition reactions of aryllithiums to hindered esters of 1- or 2-naphthalene carboxylic acid [1404].

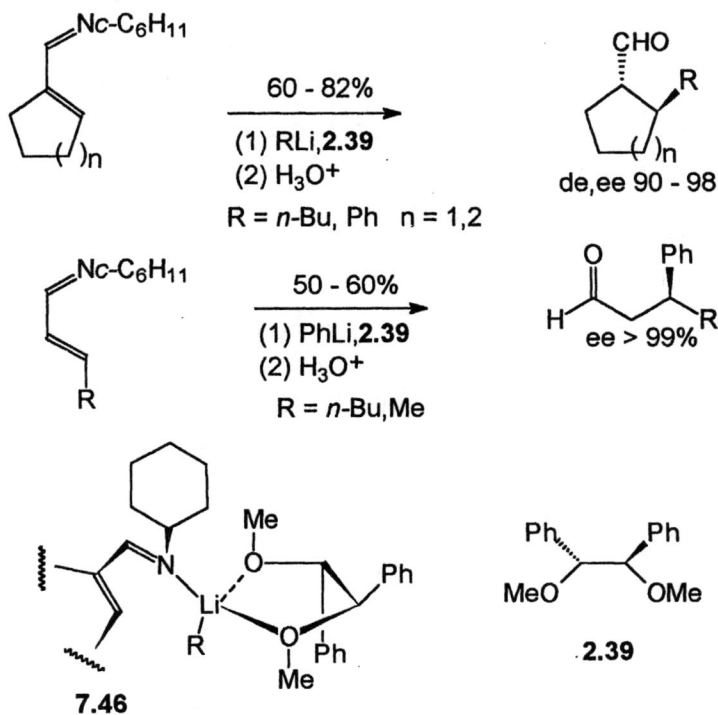

Figure 7.38

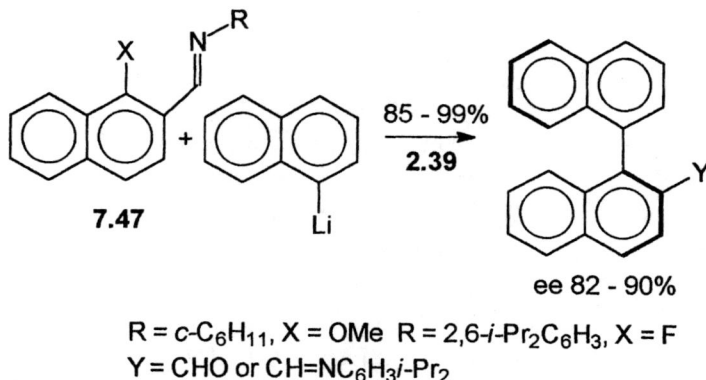

85 - 99%
2.39

ee 82 - 90%

7.47

$R = c\text{-}C_6H_{11}$, $X = OMe$ $R = 2,6\text{-}i\text{-}Pr_2C_6H_3$, $X = F$
$Y = CHO$ or $CH=NC_6H_3i\text{-}Pr_2$

Figure 7.38 (continued)

Asymmetric conjugate additions of organozincs to α-enones in the presence of chiral α-aminoalcohols usually give mediocre selectivities, even in the presence of a nickel cocatalyst [173, 657, 659, 1405]. However, Soai and coworkers [112] succeeded in performing the asymmetric conjugate addition of Et_2Zn to phenyl substituted α-enones by using stoichiometric amounts of (1*S*,2*R*)-**1.14** (NR_2 = $N(CH_2)_5$) as a Zn ligand. The same reaction can be performed with chalcone (PhCH = CHCOPh) in the presence of a nickel cocatalyst. The ligands for this reaction are aminoalcohols (*R*)-**3.72** or (1*S*,2*R*)-**1.14** (R = *n*-Bu), and the 1,4 adducts are obtained with a high enantioselectivity [112, 559, 972]. Even better enantioselectivity is obtained by using as a cocatalyst nickel acetylacetonate coordinated to a proline-derived amide grafted onto a polymeric support [1406] (Figure **7.39**).

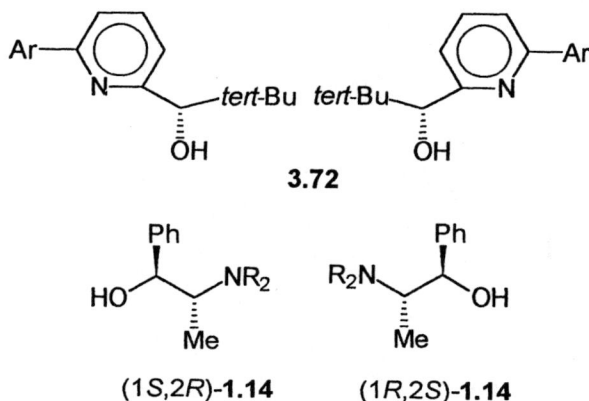

3.72

(1*S*,2*R*)-**1.14**   (1*R*,2*S*)-**1.14**

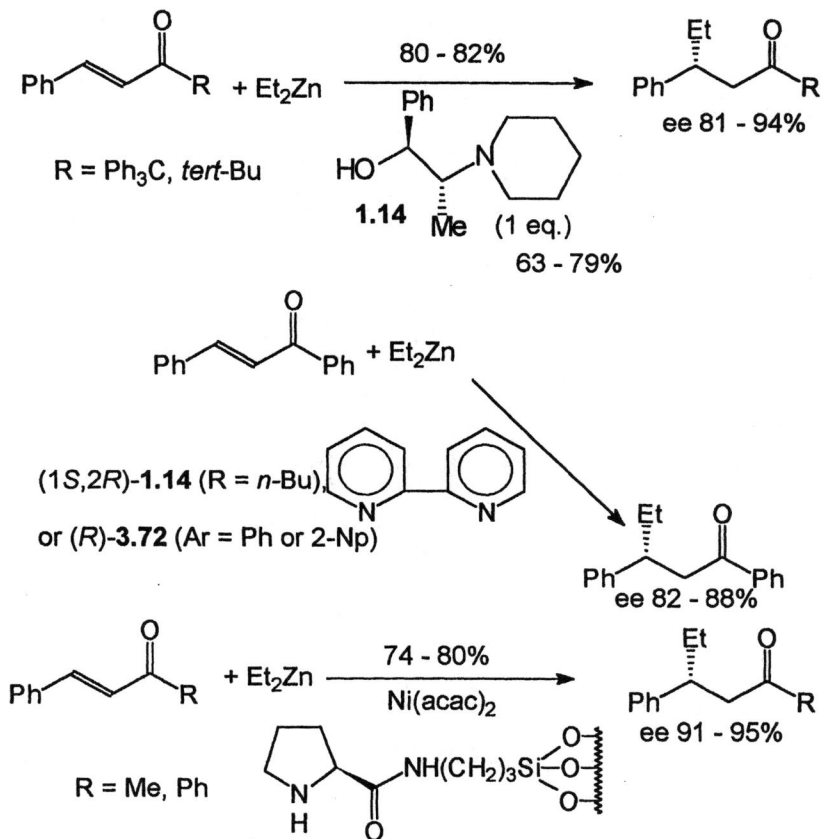

Figure 7.39

Organocuprates are the most popular reagents for conjugate additions [112, 173, 559] and a number of chiral cuprates have been developed. Chiral heterocuprates can be generated *in situ* from RLi or RMgX and CuI in the presence of a Li or Mg alcoholate of a chiral aminoalcohol. Heterocuprates can also be generated from a chiral lithium amide and an organocopper (RCu) reagent. Both types of reagents transfer an R group to α-enones at −78°C, but in most cases the enantioselectivity is low. However, Leyendecker and coworkers and Corey and coworkers [112, 559] recommended (*S*)-*N*-methylprolinol **2.13** (R = CH$_2$OH) and **2.55** as chiral aminoalcohols. With these additives, conjugate additions to chalcone (PhCH=CHCOPh) and 2-cyclohexenone occur with very good enantioselectivities (Figure **7.40**). By using (*R*)- or (*S*)-aminoalcohols **2.56** (X = S or NMe) having a bornane skeleton, Tanaka and coworkers generated methylheterocuprates which reacted with E-cyclopentadecen-2-one to give muscone (or its enantiomer) with an excellent enantioselectivity [674, 675, 1407]. This method has been extended to other Z- and E-cycloalkenones [1408]. Corey has proposed model **7.48** to interpret the observed selectivities (Figure **7.40**).

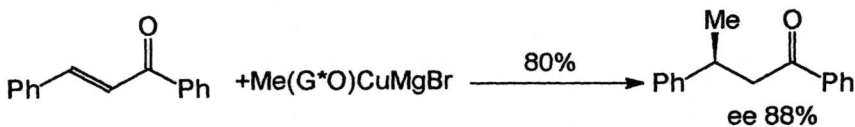

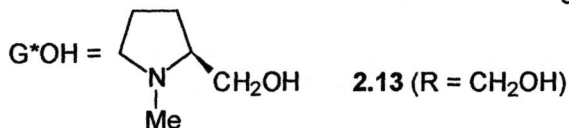

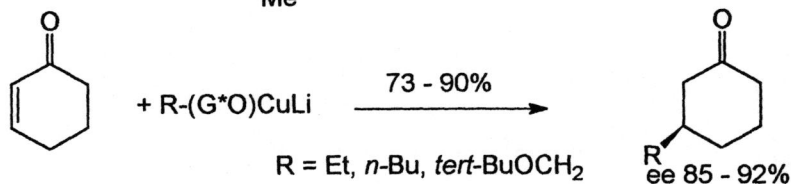

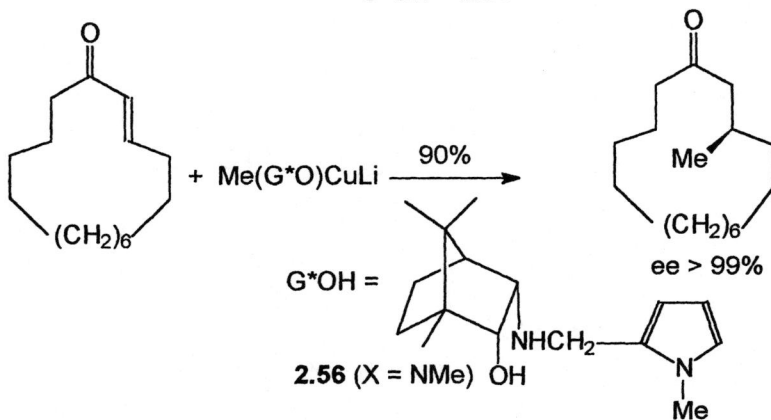

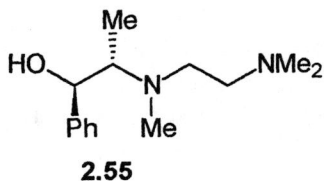

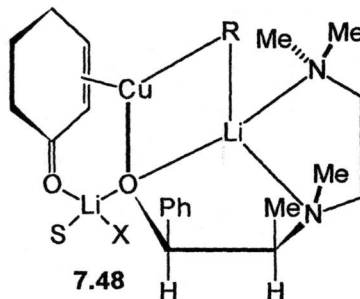

Figure 7.40

Among the amide ligands of heterocuprates, (S)-proline derivatives **1.64** (R = CH$_2$OMe or CH$_2$OEt) have been used to perform the asymmetric conjugate addition of methyl- or isopropenylcuprates to 2-cyclohexenone or 2-methyl-2-cyclopentenone [112, 173, 1409]. However, the use of CuSCN as copper salt under very strict experimental conditions is required to obtain good selectivities (Figure **7.41**). Rossiter and coworkers [112, 676] recommended a linear diamine **2.57** as ligand in similar reactions; a heterocuprate is generated in the presence of CuI and the observed enantioselectivity is useful (Figure **7.41**).

Homocuprates bearing noncovalently bound copper ligands have also been used in asymmetric conjugate additions. 1,3-Oxazoline **3.52** (X = SH) has been used by Zhou and Pfaltz for copper catalyzed 1,4-addition of n-BuMgBr and i-PrMgBr to 2-cycloalkenones; however, useful selectivities are observed only with 2-cycloheptenone [969]. Proline derivative **7.49** is also useful as a copper ligand. Tomioka and coworkers reported the use of stoichiometric amounts of aminophosphine **2.59** bearing the (S)-proline skeleton as a copper ligand to promote enantioselective addition of Me$_2$CuLi to chalcone at −20°C. Additions of cyanocuprates to 2-cycloalkenones at −78°C in the presence of LiBr can also be conducted [679, 1410, 1411] (Figure **7.41**).

Alexakis and coworkers [677, 678] used an excess of aminophosphine **2.58** (R = i-Pr) as a copper ligand. The best enantioselectivities were observed in the conjugate addition of medium order cuprates R$_5$Cu$_3$Li$_2$ to 2-cyclohexenone (Figure **7.41**).

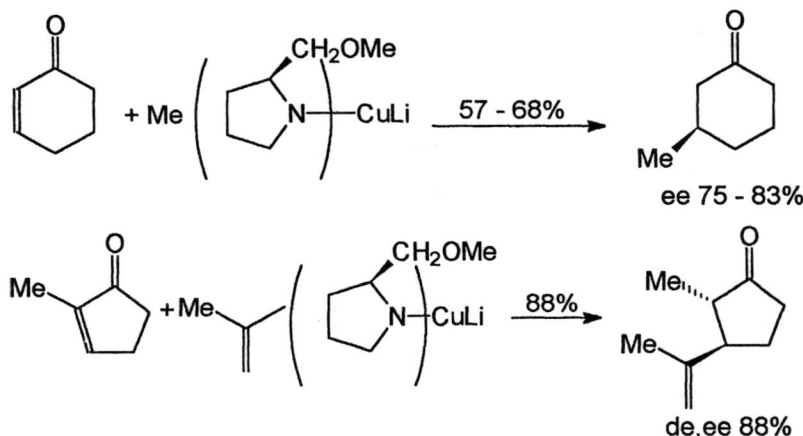

Figure 7.41

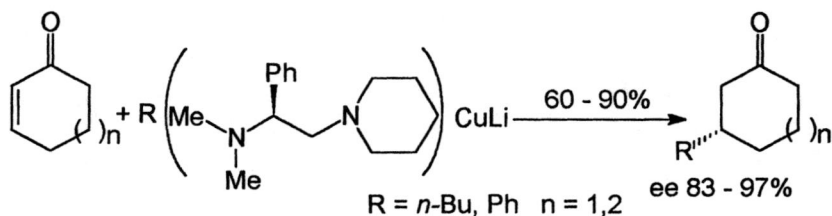

R = *n*-Bu, Ph   n = 1,2

ee 83 - 97%

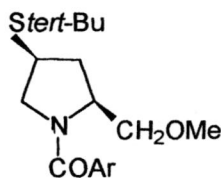

7.49

Ar = Ph, 1-Np

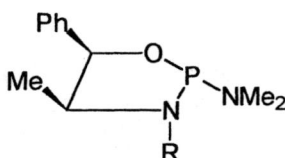

2.58

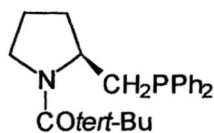

2.59

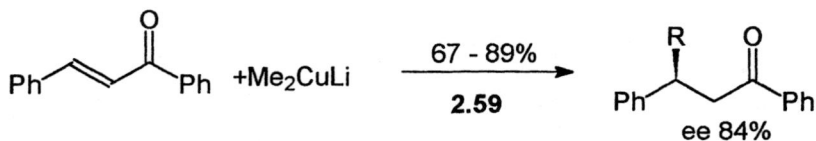

ee 84%

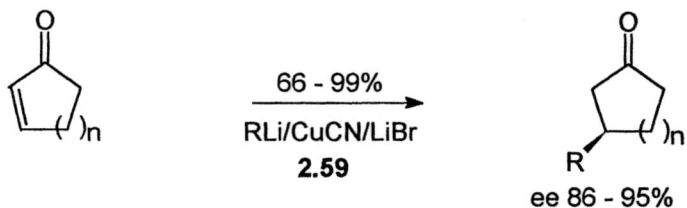

R = Me, Et, *n*-Bu   n = 1,2

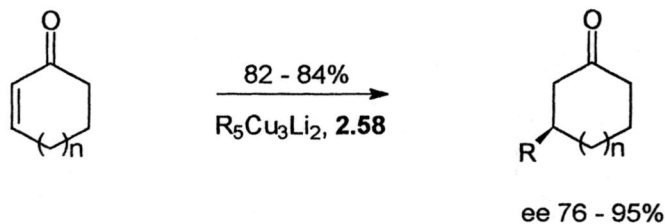

R = Et, *n*-Bu, *tert*-BuO(CH₂)₄

n = 1,2

Figure 7.41 (continued)

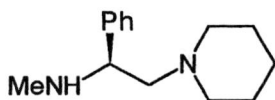

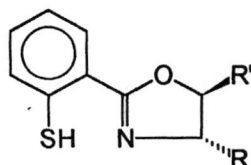

2.57   **3.52** (X = SH)

7.10.2 Reactions with Unsaturated Compounds Bearing Chiral Residues

Many examples of selective conjugate additions of organometallic reagents to chiral substrates have been described. The chiral residue can be introduced either on the electron-withdrawing group or on the carbon skeleton of the acceptor.

7.10.2.1 Reactions with α,β-Unsaturated Aldehydes and Ketones and Derivatives

Koga and coworkers [112, 173, 261, 262, 1008] have studied the asymmetric conjugate additions of organomagnesium reagents to α,β-unsaturated aldimines derived from chiral aminoesters. The best results were obtained with imines of *tert*-leucine *tert*-Bu esters **7.50** and **7.51** (Figure 7.42). An intermediate chelate **7.52** is involved, and the R' group of the organomagnesium reagent is transferred to the least hindered face of the double bond. The resulting magnesium enolate can be trapped by an alkylating agent. After removal of the imine functionality, α,β-disubstituted aldehydes can be obtained with a high selectivity [112]. Similar reactions can be carried out with chiral 1-naphthylimines **7.53** [112] (Figure **7.42**). Enders and coworkers [1412] performed the conjugate addition of trialkylstannyllithiums to SAMP-hydrazones of 2-cyclohexenones at −100°C. Protonation of the resulting lithium enolates leads to the 1,4-adducts with a low selectivity. However, trapping the enolates with alkyl halides at low temperature gives, after ozone treatment to remove the chiral auxiliary, 2,3-disubstituted cyclohexanones with a high diastereo- and enantioselectivity (Figure **7.42**).

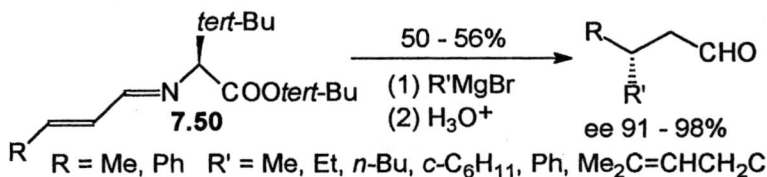

Figure 7.42

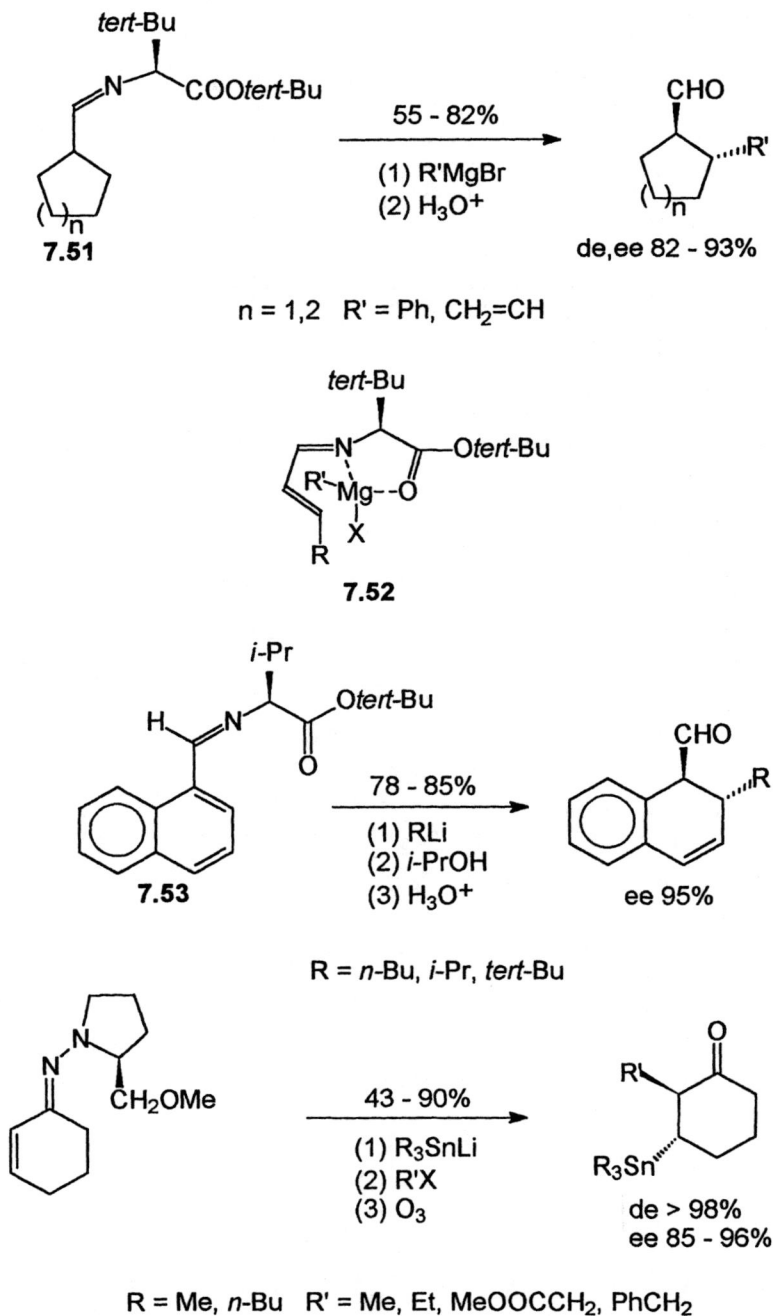

Figure 7.42

At −78°C, organolithium reagents add in a 1,4 fashion to E-α,β-unsaturated acylironcarbonyl complexes **1.148** (R = E-RCH=CH) [112, 522]. If an α-hydrogen is available, then the corresponding Z-isomers suffer deprotonation. When conjugate addition does take place, the resulting enolate can either be protonated or be trapped by an alkyl halide. The various traps are introduced to the complex on the face of the double bond that is opposite to one of the phenyl rings of the PPh₃ substituent (Figure **7.43**). After oxidative cleavage of the chiral residue, α,β-dialkyl-substituted acid derivatives are obtained with an excellent diastereo- and enantioselectivity (Figure **7.43**).

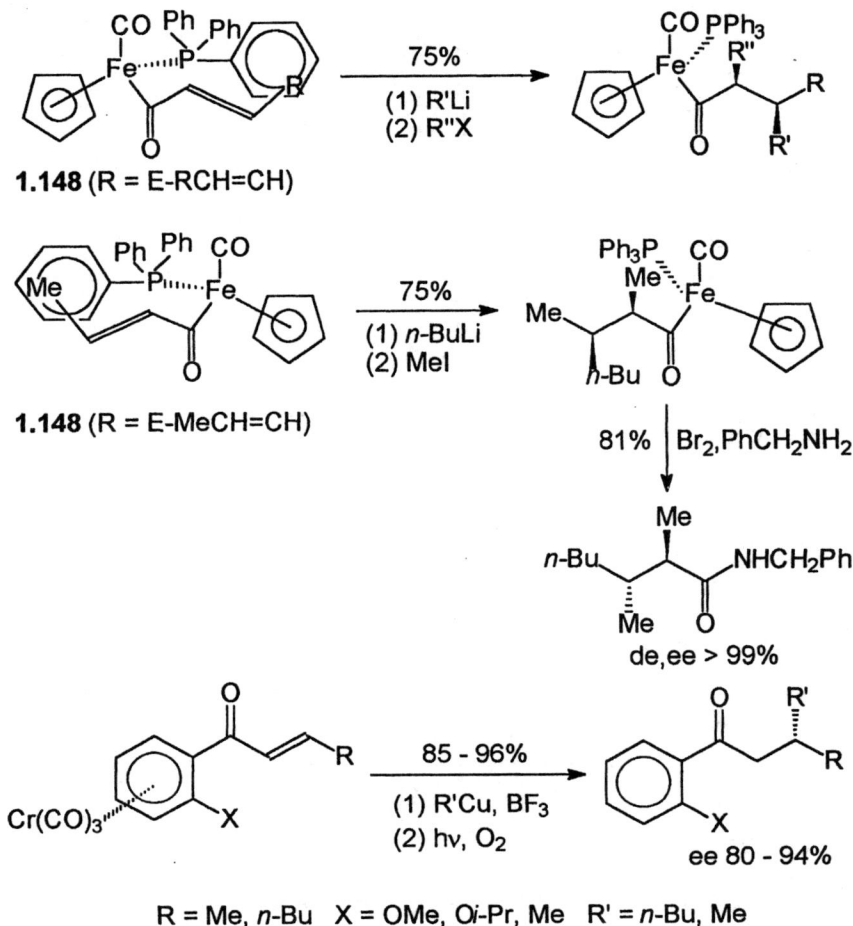

1.148 (R = E-RCH=CH)

1.148 (R = E-MeCH=CH)

75%
(1) R'Li
(2) R"X

75%
(1) n-BuLi
(2) MeI

81% | Br₂,PhCH₂NH₂

de,ee > 99%

85 - 96%
(1) R'Cu, BF₃
(2) hv, O₂

ee 80 - 94%

R = Me, n-Bu   X = OMe, Oi-Pr, Me   R' = n-Bu, Me

Figure 7.43

Complexation of the aryl ring of α-enones RCH=CHCOAr (Ar = 2-MeO-C_6H_4, 2-i-PrOC$_6$H$_4$) provides chiral chromium complexes that undergo 1,4-addition of R$_2$CuLi or of RCu•BF$_3$. In the reactions with RCu•BF$_3$, β-branched aryl ketones are obtained with a high enantiomeric excess after decomplexation [1413] (Figure 7.43). Interestingly, the conjugate addition of R$_2$CuLi gives predominantly the opposite diastereoisomer, albeit with a lower selectivity. With chiral copper reagents, double diastereodifferentiation (§ I.6) takes place and an excellent selectivity can be observed when the reagents are matched [1413].

α,β-Unsaturated acetals and ketals react with organoaluminum or organo-copper reagents in the presence of BF$_3$•Et$_2$O. The reactions of acetals with Me$_3$Al or alkylcopper reagents are not regioselective. However, the reactions of aryl- or alkenylcopper reagents with α,β-unsaturated acetals result in the attack of the double bond (S$_N$2$'$ addition). Likewise, the reactions of all those organometallic reagents with α,β-unsaturated ketals occur in an S$_N$2$'$ fashion. When reactions are performed in the presence of n-Bu$_3$P, enol ethers are formed. After hydrolysis, β-branched aldehydes or ketones are produced with an excellent regio- and enantioselectivity. The most efficient acetals are those of 2,3-butanediols and N,N'-tetramethyltartaramide [112, 173, 213, 1062] (Figure 7.44). The mechanism of these reactions is comparable to previous cases (§ 6.1.3, 6.6.3). The acetal resides in its preferred conformation, and the oxygen that is vicinal to the axial substituent of the acetal is coordinated to the Lewis acid. According to model 7.54, the organometallic residue approaches from the opposite face in the case of organocopper reagents. When using Me$_3$Al, the methyl group is transferred on the same face, in agreement with model 7.55 (Figure 7.44). The reactions of cuprates with α,β-unsaturated oxazolidines have been examined, but disappointing results were obtained in most cases [112, 320].

The presence of a chiral acetal group on the carbon skeleton of a prochiral α-enone does not influence the stereoselectivity of the conjugate additions. However, the presence of a chiral oxazolidine in the β-position allows the stereoselective addition of lithiocuprates to compounds 7.56. The chiral auxiliary is removed by transketalization with HS(CH$_2$)$_3$SH followed by MeI/CaCO$_3$ treatment [323] (Figure 7.44). Addition-elimination reactions of dialkylcuprates to cycloalkenones 7.57 bearing a chiral aminomethyl substituent occur at −90°C in the presence of excess LiBr, and lead to β-alkyl-2-methylenecyclanones with a high enantioselectivity [491] (Figure 7.44). Less useful results are obtained with acyclic acceptors. The interpretation of the facial selectivity of these reactions relies upon the formation of a chelate 7.58, which reacts from its least hindered side (Figure 7.44).

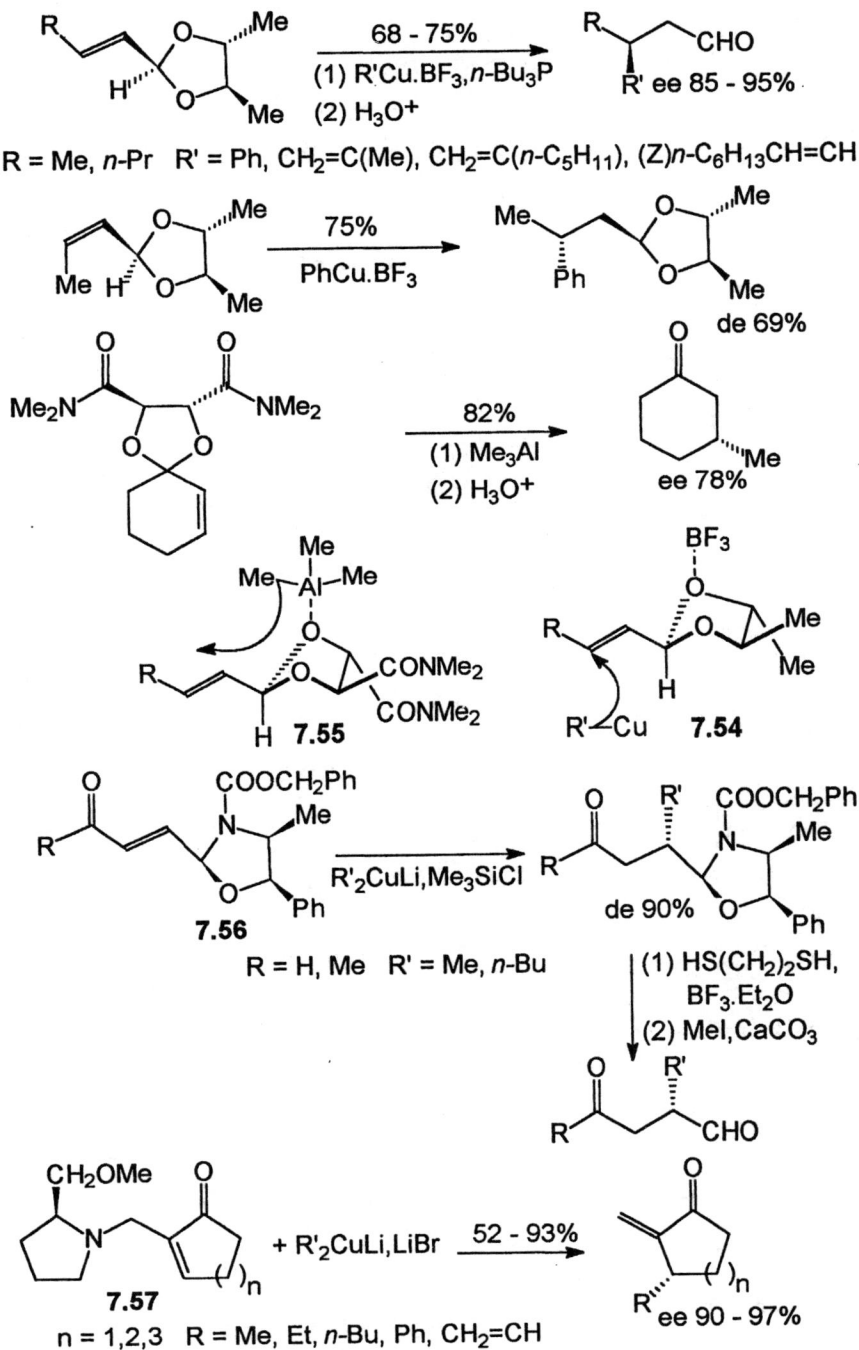

R = Me, n-Pr   R' = Ph, CH$_2$=C(Me), CH$_2$=C(n-C$_5$H$_{11}$), (Z)n-C$_6$H$_{13}$CH=CH

de 69%

ee 78%

7.55

7.54

7.56

R = H, Me   R' = Me, n-Bu

de 90%

(1) HS(CH$_2$)$_2$SH,
    BF$_3$.Et$_2$O
(2) MeI, CaCO$_3$

7.57
n = 1,2,3   R = Me, Et, n-Bu, Ph, CH$_2$=CH

ee 90 - 97%

Figure 7.44

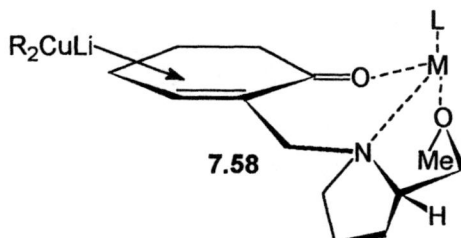

Figure 7.44 (continued)

7.10.2.2 Reactions with α,β-Unsaturated Carboxylic Acids Derivatives

Conjugate addition reactions of organometallic reagents to E-α,β-unsat-urated esters of chiral alcohols are usually performed with RCu•BF$_3$ or RCNLiCu•BF$_3$. The presence of n-Bu$_3$P can improve the stability of the reagent. These reactions allow the highly enantioselective synthesis of β-branched acids after hydrolysis of the chiral residue. The facial discrimination takes place by at-tack of the reagent on the least hindered side of an ester-BF$_3$ complex in the *s-trans* conformation [66, 83, 85, 147, 173]. By choosing the appropriate absolute configuration of the chiral auxiliary, it is possible to obtain either the (R)- or (S)-enantiomer of the acid, as shown by the first two examples in Figure 7.45. In these examples, the chiral auxiliaries are the two enantiomers of alcohol **1.13** (R = CH$_2$-*tert*-Bu). Equally good results are obtained with α,β-unsaturated esters E-RCH=CHCOOG* (R = Me, Et, *i*-Pr, Ph) of chiral alcohols **1.4** (R = Ph), **1.5** (R = Ph), **1.8** (R = 1-Np), **1.10**, **1.15**, **1.16**, and the conformationally locked 2,2,6,6-tetramethyl-3,5-heptanediol monobenzoate [1398]. The reactions of Z-α,β-unsaturated esters usually give lower selectivities.

Monoesters of chiral diols have also been recommended as acceptors. Lithiocuprates or magnesiocuprates provide 1,4-adducts with a high selectivity with α,β-unsaturated monoesters of (1R,2R)-1,2-cyclohexanediol **1.34** [112, 157] or of (R)- or (S)-binaphthol **1.44** [112, 237]. In the case of binaphthol, only Me$_2$CuLi gives useful results, and a β-branched methylketone is formed with a high enantiomeric excess (Figure 7.45). Deoxysugars have recently been proposed as chiral auxiliaries [247]. The results with these hydroxyesters are interpreted through the formation of an alcoholate from a cuprate cluster which is rigidified by chelation. The cluster **7.59** derived from cyclohexane diol monoester is a represen-tative example. Transfer of the alkyl group that is closest to the double bond of the ester takes place as directed by the structure of the chelate (Figure 7.45).

$R = Me, Et, n\text{-}Bu, n\text{-}C_8H_{17}, Me_2C=CHCH_2CH_2$
$R' = Me, i\text{-}Pr, CH_2=CH, CH_2=C(Me), Ph$

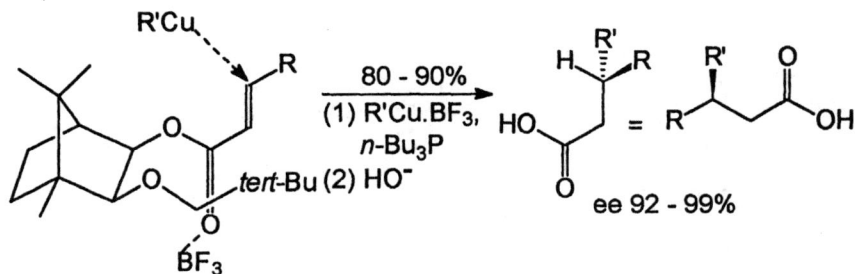

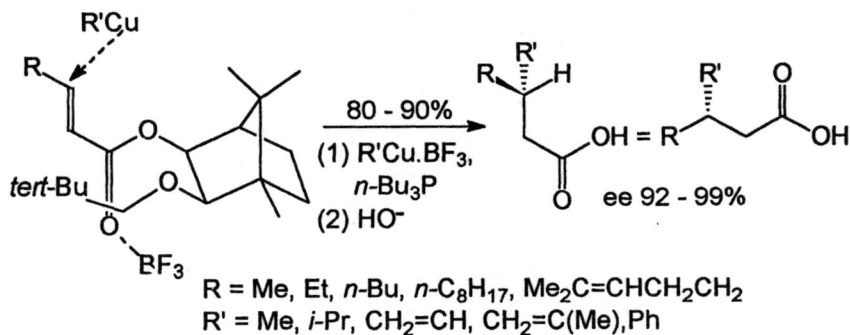

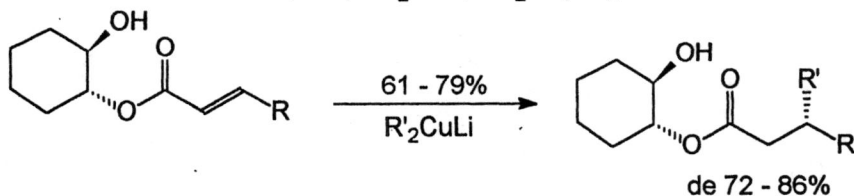

$R = Me, Et, Ph$ $R' = Me, n\text{-}Bu, Ph$

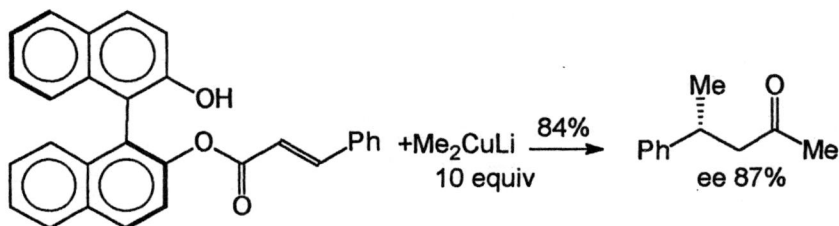

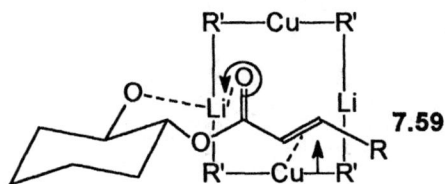

7.59

Figure 7.45

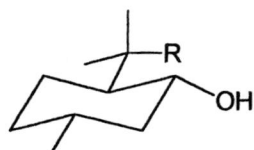

(1R,2S,5R)-**1.4**

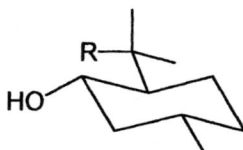

(1S,2R,5S)-**1.4**

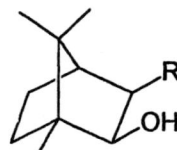

1.8

(1S,2R)-**1.5**

(1R,2S)-**1.5**

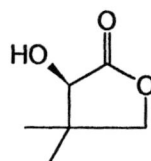

1.16

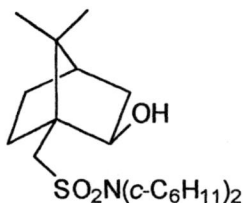

SO₂N(c-C₆H₁₁)₂

1.10

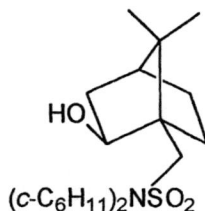

(c-C₆H₁₁)₂NSO₂

1.15

The chiral group does not always reside in the alcohol group of the ester. Alexakis and coworkers [112, 173, 213, 303, 306] have performed diastereoselective conjugate additions of lithiocuprates to chiral aldehyde derivatives of methyl 2-formylcinnamate. Imidazolidine **7.60** is the most efficient reagent discovered (Figure **7.46**). Mukaiyama and Asami [112, 173, 253] reacted magnesiocuprates with chiral aminal **7.61** and obtained substituted aldehydo-esters with high enantiomeric excess after hydrolysis. A similar reaction was performed by Scolastico and coworkers [112, 173, 321] who introduced an ephedrine-derived aldehyde-protecting group **7.62** (Figure **7.46**). The recovery of the chiral auxiliary from the products is effected by transketalization. Sakai and coworkers [216] used (S,S)-1,2-cycloheptanediol as the facial discriminating auxiliary in 1,4-additions of magnesiocuprates to 2-carbomethoxy-2-cyclopentanone. In all these reactions, the chiral auxiliary induces the facial selectivity through either a chelated or nonchelated intermediate complex, depending on the basicity and accessibility of the heteroatoms of the chiral residue [303].

Seebach and coworkers prepared chiral α,β-unsaturated heterocyclic reagents **7.63** from 1,3-dioxan-4-ones **1.110** (R = *tert*-Bu). Conjugate addition of lithiocuprates to these substrates is highly stereoselective. After acidic treatment, nonracemic β-hydroxyacids are obtained with a high selectivity [81, 173, 1414, 1414a] (Figure **7.46**). Similar methods were developed by Handke and Krause for 1,6-additions of organocuprates to 5-alkynylidene-1,3-dioxan-4-ones [1415].

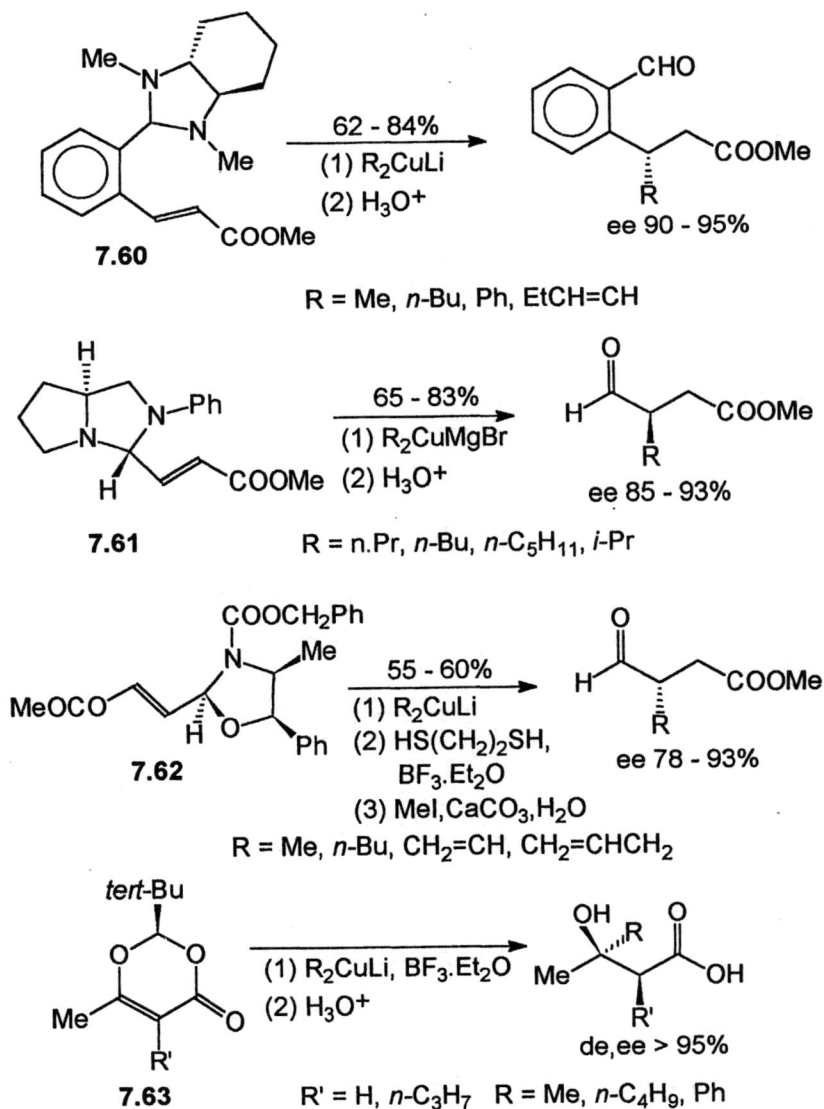

Figure 7.46

α,β-Unsaturated amides or imides bearing chiral residues can also undergo conjugate additions of organometallic reagents, and amide or imide derivatives of aminoalcohols are the most interesting reagents. Mukaiyama and coworkers [112, 173, 253] have performed conjugate additions of organomagnesium reagents to α,β-unsaturated amides derived from ephedrine **7.64**. They performed similar reactions on oxazepanes Z- and E-**1.93** generated from malonates and ephedrine, under $NiCl_2$ catalysis. After acid hydrolysis, both types of adducts give β-branched acids with an excellent enantiomeric excess (Figure **7.47**). The oxazepane method is limited by the formation of two diastereoisomeric oxazepanes, which must be separated before reaction. Cinnamamides of (*R*)- or (*S*)-*N*-alkyl-2-aminobutanol or better of the *N*-fluoroalkyl analogs are useful substrates [1416, 1417] on the way to nonracemic β-phenylalkanoic acids. The (*S*)-prolinol-derived amide **7.65** is similarly useful, but the alcohol functionality must be tertiary and the reaction must be run in the presence of a tertiary amine [112, 173, 261] (Figure **7.47**). The observed selectivities in these types of reactions are interpreted by formation of a rigid template **7.66** which transfers the R' group of R'MgX to the least hindered face of the alkene (Figure **7.47**). The conjugate addition of saturated or vinylic organomagnesium reagents to chiral amide **7.67** in the presence of $ZnBr_2$ is highly stereoselective [112, 1418]. After reaction of the products with NH_2OH/NaOAc, oximes of 3-alkylcyclohexanones are obtained with a good enantiomeric excess (Figure **7.47**). If PhMgBr or $CH_2=CHCH_2MgBr$ is used as a reagent, the reaction is no longer regioselective.

N-Enoyloxazolidinones **7.68** derived from (*S*)- or (*R*)-**1.116** (R = Ph) undergo highly stereoselective 1,4-additions of organomagnesium reagents in the presence of $CuBr/Me_2S$. Reaction temperatures vary between −78°C and −20°C according to the reagents [1071, 1419-1423]. The intermediate enolates can be trapped by NBS or tritylazide according to Evans (§ 5.6.1), opening the way to various nonracemic natural and nonnatural aminoacids (Figure **7.47**). The use of (*S*)-**1.116** (R = $PhCH_2$) as an auxiliary gives lower selectivities [1071]. Kunz and coworkers examined the photochemically induced addition of R_2AlCl (R = Me, Et, Pr, *i*-Bu) to *N*-enoyloxazolidinones **7.68** (R = $PhCH_2$) and observed an unsatisfactory facial discrimination [416]. However, with **1.119** as the chiral auxiliary, 1,4-additions of R_2AlCl to the related *N*-crotonoyl- or cinnamoyl derivatives RCH=CHCOG* (R = Me or Ph) take place with a high diastereoselectivity (90 - 96%). A radical process is probably involved in these reactions.

Stephan and coworkers have performed similar 1,4-additions of magnesium cuprates to *N*-enoylimidazolidinones derived from **1.131**. After hydrolysis, β-branched acids were obtained with a high enantioselectivity [1424].

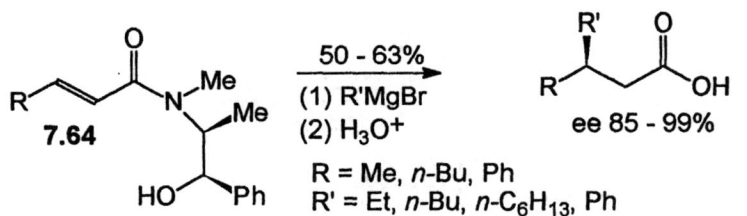

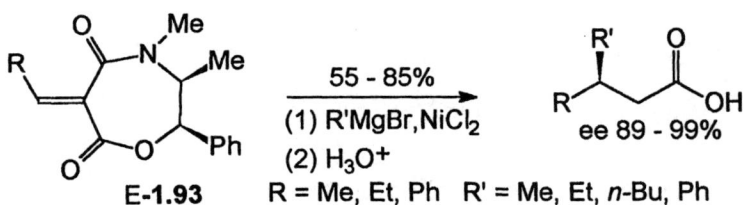

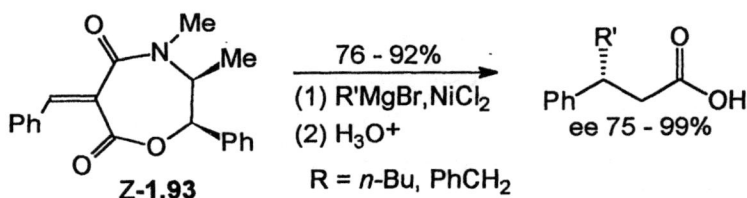

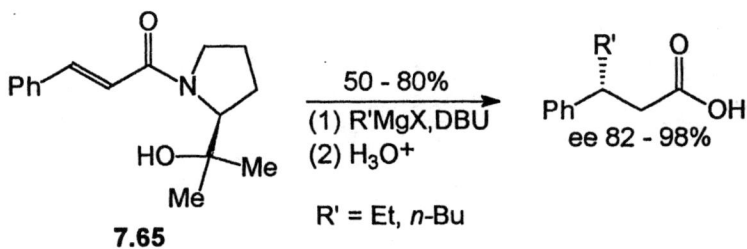

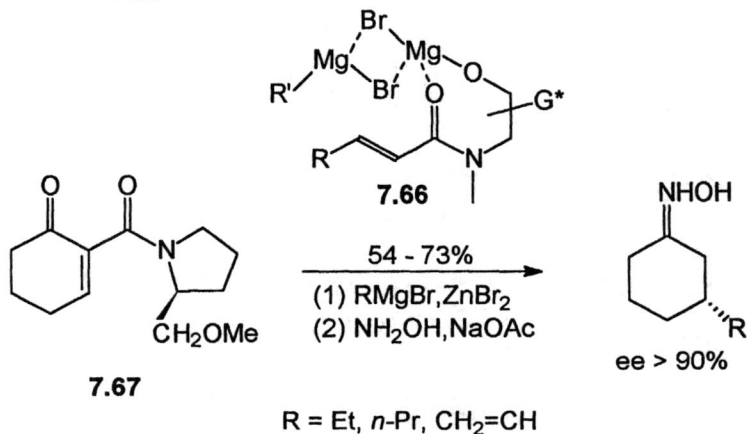

Figure 7.47

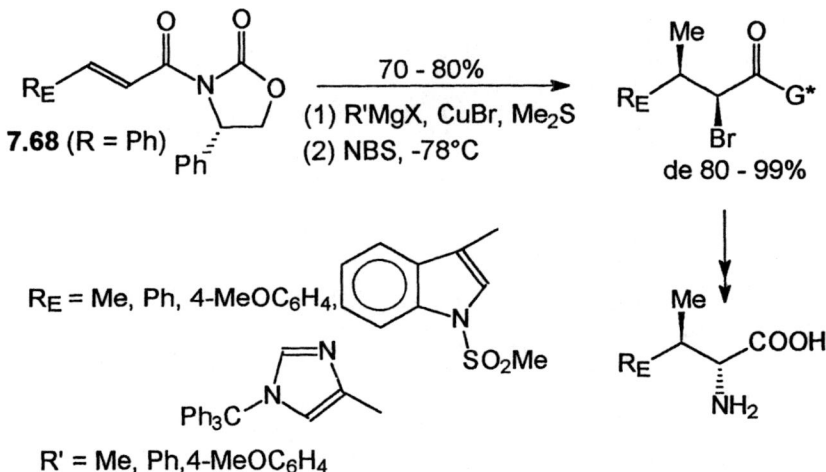

Figure 7.47 (continued)

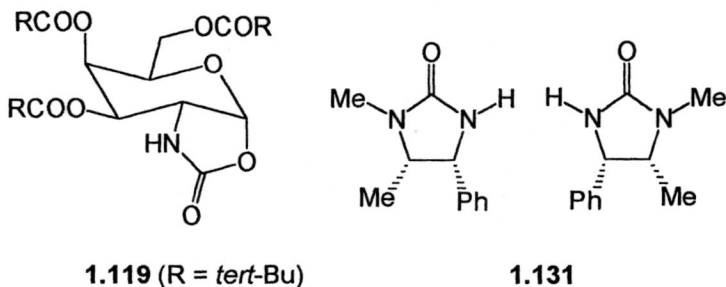

1.119 (R = *tert*-Bu)   **1.131**

Chiral α,β-unsaturated sultams **1.134** (R = R$_E$CH=CH) also undergo the conjugate addition of organometallic reagents with a very high facial discrimination [112, 147, 173]. By varying the substituents of the double bond and the nature of the organometallic reagent, the R' group can be introduced on either face of the prochiral double bond. Organomagnesium reagents transfer their R' groups from the least hindered side of a preformed template **7.69**. After fractional crystallization of the predominant isomer and hydrolysis, β-branched acids are obtained with an excellent ee (Figure 7.48). The intermediate chelated enolates can be trapped with MeI, and this leads stereoselectively to α-methyl-β-branched acid derivatives. The reactions of organocopper reagents in the presence of *n*-Bu$_3$P and EtAlCl$_2$ lead to the same diastereoisomers via an intermediate chelate [147]. Higher order silylcuprates (PhMe$_2$Si)$_2$CuCNLi$_2$ give 1,4-adducts with an excellent stereoselectivity (86 - 98%); by an appropriate treatment, the products are transformed into the corresponding alcohols with retention of configuration [147, 173, 1425]. The face selectivity in reactions of organometallic reagents with α,β-unsaturated α-alkylsultams 1.134 (R = R_ECH=CR") depends upon the nature of the metal. Organomagnesium reagents react through a chelated transition state similar to 7.69 (Figure 7.48). Lithio- or magnesiocuprates lead, after hydrolysis, to the enantiomeric *syn* acids 7.70, resulting from the attack of the opposite face of the double bond [1425a]. These results are interpreted as arising from A(1.3) allylic strain. This A-strain is not strong enough to disrupt the magnesium chelate. However, when the reaction is run with a lithiocuprate, the A-strain favors the conformer 7.71, which reacts via its least hindered face [173] (Figure 7.48).

Vinyloxazolines 1.86 (Y = RCH=CH) or 1.87 (R = RCH=CH, R' = H or Me) designed by Meyers and coworkers [112, 173, 324, 339] undergo conjugate additions of nearly all organolithiums at −78°C with a high selectivity. After hydrolysis, β-branched acids are obtained with an excellent enantiomeric excess (Figure 7.49). Trapping of the intermediate enolates with MeI is also quite efficient [112].

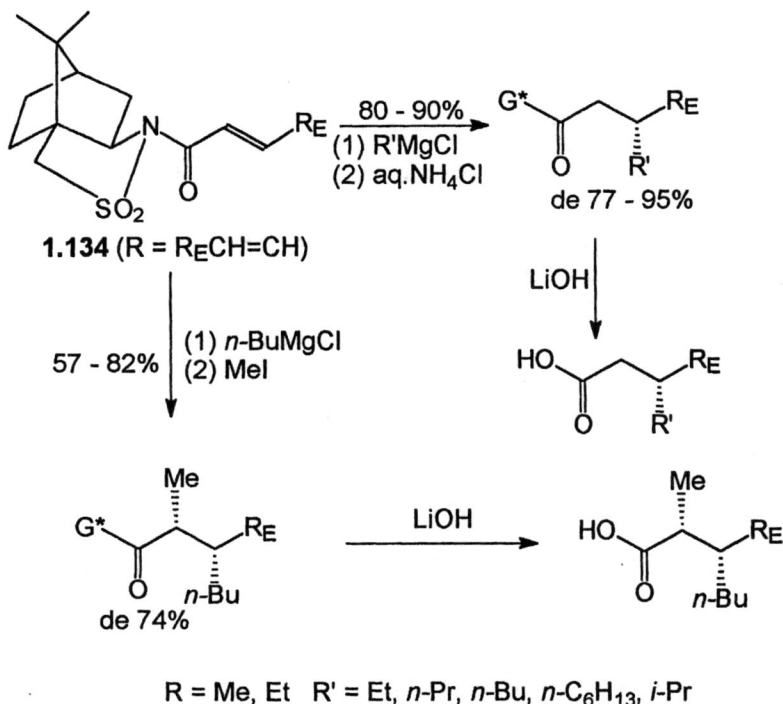

R = Me, Et   R' = Et, *n*-Pr, *n*-Bu, *n*-C$_6$H$_{13}$, *i*-Pr

Figure 7.48

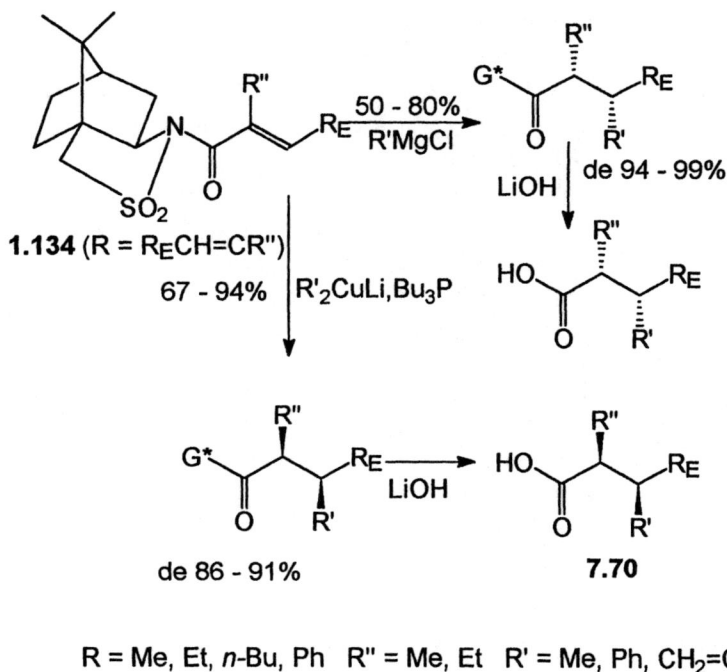

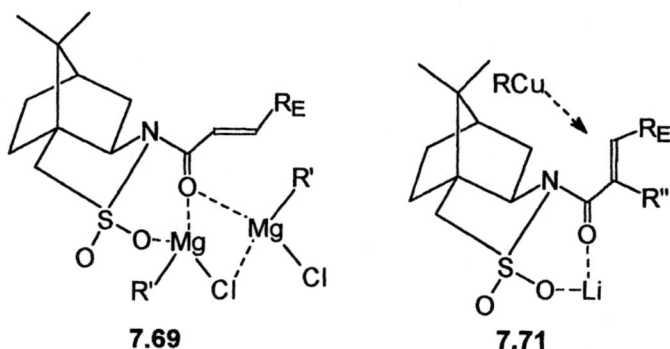

Figure 7.48 (continued)

The observed selectivities in these additions are interpreted by the formation of a chelate **7.72**, which reacts on its least hindered face (Figure **7.49**). A milder method to remove the chiral auxiliary from the products is a sequential treatment with MeOTf and NaBH₄. By using these conditions, β-branched aldehydes are obtained from **1.87** (R = RCH=CH, R' = Me) with an excellent enantioselectivity [339, 342] (Figure **7.49**).

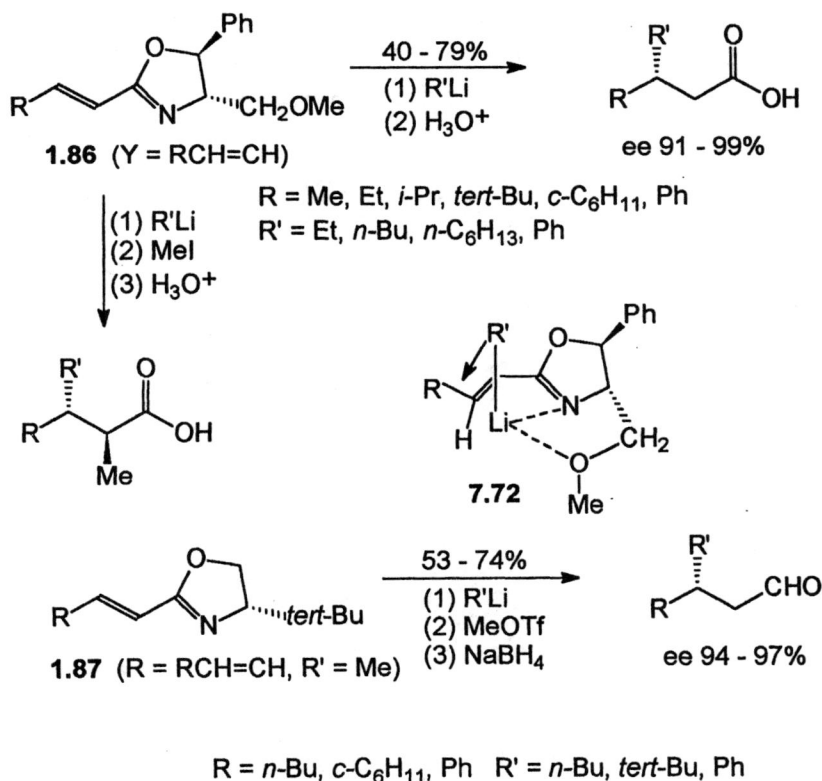

Figure 7.49

Chiral 1- and 2-naphthyloxazolines **1.86** (Y = 1- or 2-Np) also undergo 1,4-additions of RLi and R_3SiLi at −78°C. The intermediate azaenolates can be protonated or trapped with an electrophile. After treatment of the initial products with MeOTf and $NaBH_4$, substituted aldehydes are obtained with high selectivity [112, 173, 339, 1008, 1426] (Figure 7.50). This methodology has been applied to the synthesis of many alkaloids [112, 340]. The organolithium R group is introduced on the face opposite to the CH_2OMe group of the chiral residue. Protonation takes place on the same face, while alkylation occurs on the other face, so the relative stereochemistry of the R and CHO groups can be either *trans* or *cis* (Figure 7.50). Similar results are obtained with oxazolines **1.87** (R = 1-Np). In order to improve this method, Meyers and coworkers prepared two enantiomeric precursors of 1-naphthyloxazolines **7.73** from natural (*S*)-serine. The reactions of **7.73** with organolithium reagents are as useful as those of **1.86** [1427].

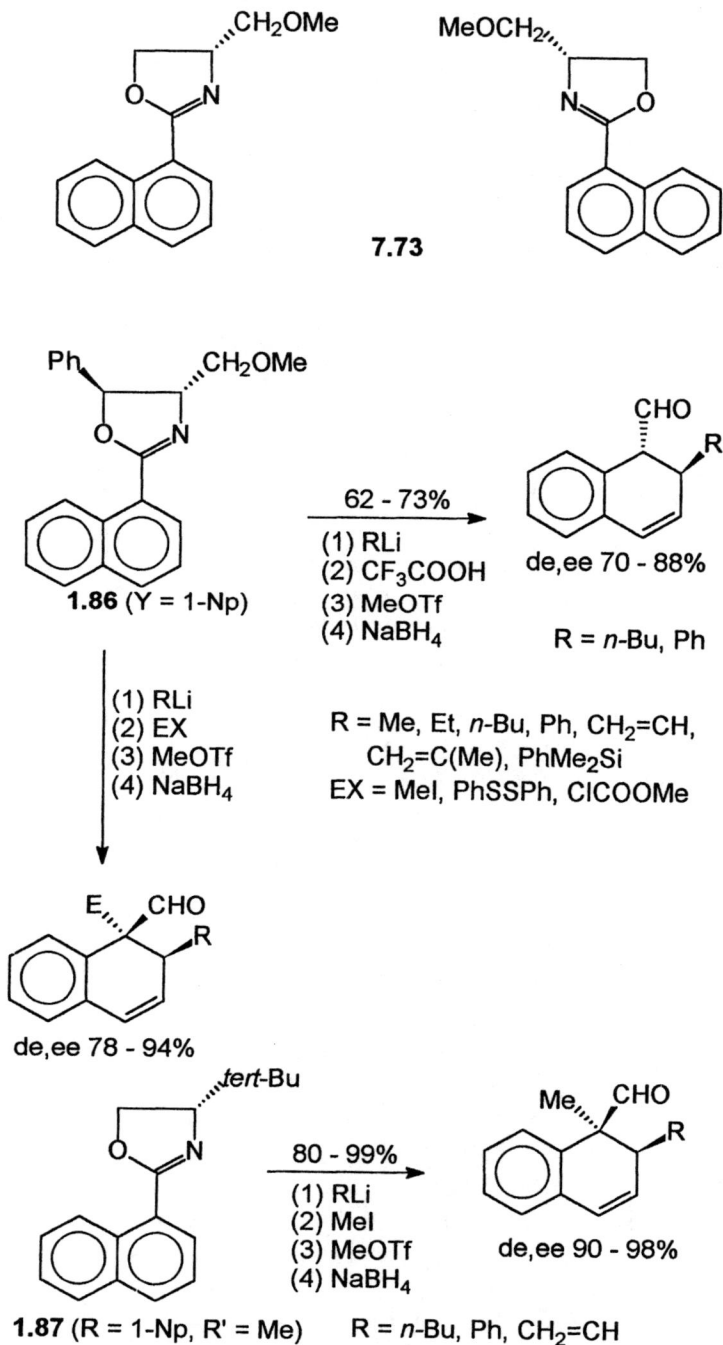

7.73

1.86 (Y = 1-Np)

62 - 73%

(1) RLi
(2) CF₃COOH
(3) MeOTf
(4) NaBH₄

de,ee 70 - 88%

R = n-Bu, Ph

(1) RLi
(2) EX
(3) MeOTf
(4) NaBH₄

R = Me, Et, n-Bu, Ph, CH₂=CH,
    CH₂=C(Me), PhMe₂Si
EX = MeI, PhSSPh, ClCOOMe

de,ee 78 - 94%

1.87 (R = 1-Np, R' = Me)

80 - 99%

(1) RLi
(2) MeI
(3) MeOTf
(4) NaBH₄

de,ee 90 - 98%

R = n-Bu, Ph, CH₂=CH

Figure 7.50

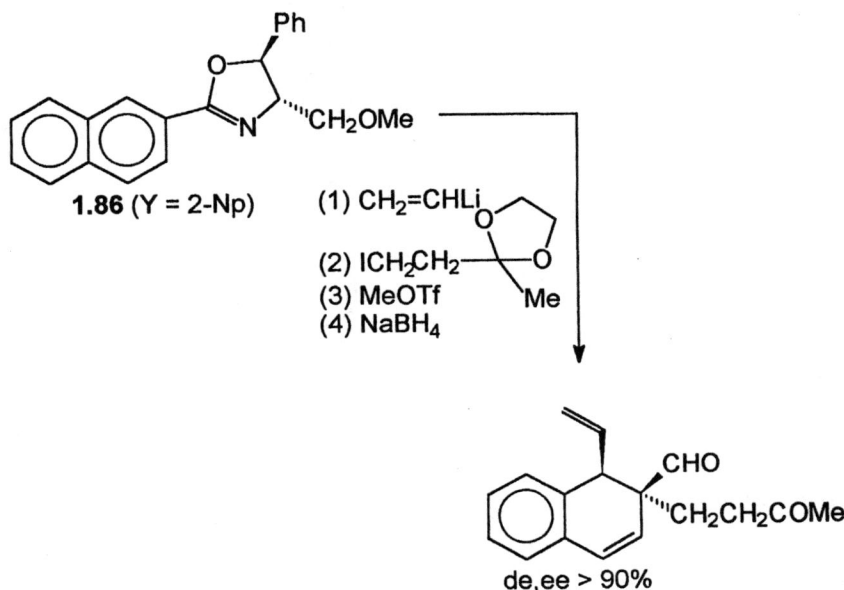

Figure 7.50 (continued)

Addition of aromatic organomagnesium reagents to chiral 1-methoxy-2-naphthyloxazolines **1.86** (Y = 1-MeO-2Np) or to monocyclic aryloxazolines **1.87** (R = 2,3,4,5-(MeO)$_4$C$_6$H) is a useful method for preparing chiral atropisomeric binaphthyls or biphenyls with high selectivity. The level of asymmetric induction depends upon the ability of the substituents of the organomagnesium reagent to promote chelation-controlled processes [112, 324] (Figure **7.51**). These reactions occur by an addition-elimination mechanism and the presence of several MeO groups on either reagent is mandatory to observe a good selectivity in asymmetric biphenyl synthesis. The best results are observed when the organometallic reagent bears MeO and CH$_2$OTBDMS or CH$_2$OSii-Pr$_3$ groups in the 2- and 6-positions [1428, 1429, 1430]. In all these reactions, the chiral auxiliary must be removed under mild conditions to avoid racemization [112]. Similar reactions were performed by Cram and Wilson [112], who introduced chirality on the alkoxy leaving group, starting from compounds **1.46** derived from cinchona alkaloids. Chiral biaryls were obtained with a good enantioselectivity but a low chemical yield.

Chiral sulfoxides have recently been proposed as leaving groups in the synthesis of chiral 2-isopropoxycarbonyl or dimethylamido-1,1'-binaphthyls [495].

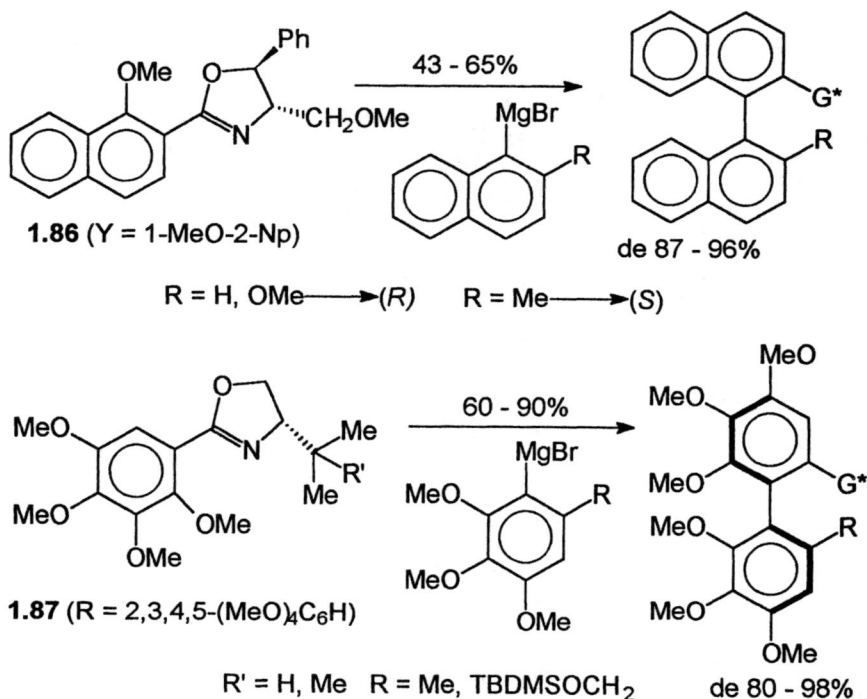

1.86 (Y = 1-MeO-2-Np)

43 - 65%   MgBr

de 87 - 96%

R = H, OMe ——▶(R)   R = Me ——▶(S)

1.87 (R = 2,3,4,5-(MeO)₄C₆H)

60 - 90%   MgBr

R' = H, Me   R = Me, TBDMSOCH₂   de 80 - 98%

Figure 7.51

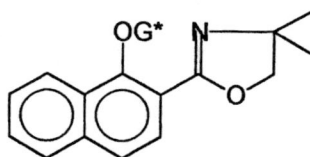

1.46

Meyers and coworkers [1431, 1432] have performed the conjugate additions of cyanocuprates to unsaturated chiral lactams **1.92** (R = Ph, R" = COOCH₂Ph). When R' ≠ H, the cyanocuprate R" group is introduced on the lower face of the substrate with a high diastereoselectivity (Figure **7.52**). The reverse selectivity is observed when R' = H, and the level is almost as good. A suitable treatment of the products gives access to enantioenriched 2,3-disubstituted or 3-monosubstituted pyrrolidines.

Imidazolidinones **7.27** undergo the stereoselective addition of R'Cu•BF₃ on the face opposite to the bulky *tert*-Bu group. Protonation takes place on the same side as conjugate addition leading to **7.74**. After acidic treatment, β-branched α-aminoacids are obtained with an excellent selectivity [442] (Figure **7.52**).

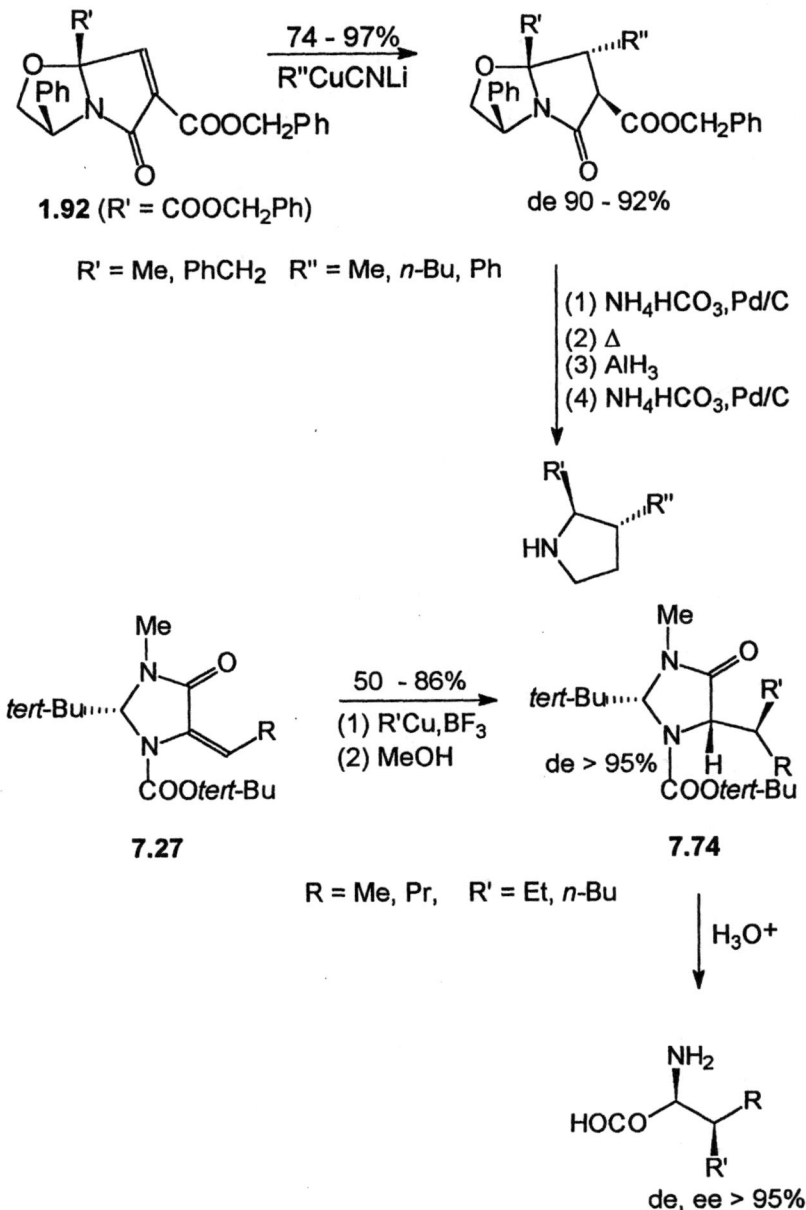

Figure 7.52

7.10.2.3 Reactions with α,β-Unsaturated Sulfoxides and Sulfones

The reactions of organometals with α,β-unsaturated sulfoxides pose the problem of competitive deprotonation in the α-position. Posner and coworkers have circumvented this problem by using chiral sulfoxides bearing a carbonyl group in the α-position. These compounds are easily prepared from menthyl p-toluene-sulfinate 1.137 [101, 102, 112, 173]. Disappointing results were obtained from linear sulfoxides 1.138 [101]; however, impressive selectivities were obtained with cycloalkenones 1.139 (X = CH_2) or unsaturated lactones 1.139 (X = O). The conjugate addition of triisopropoxyalkyltitanium reagents or organomagnesium reagents in the presence of $ZnBr_2$ to cyclic ketosulfoxides 1.139 (X = CH_2) is highly stereoselective. After desulfurization with aluminum amalgam or Raney nickel, nonracemic 3-alkylcyclanones are obtained with a high enantiomeric excess [101, 102] (Figure 7.53). The same reaction with a dialkylmagnesium reagent also is highly selective, but leads to the other enantiomer [101, 102, 112, 173] (Figure 7.53). These differing face discriminations are interpreted according to the reactive conformation of the substrate. In the presence of $ZnBr_2$ or titanium reagents, which are strong Lewis acids, a chelate 7.75 is formed. The R group of the organometallic reagent adds to the least hindered face of this chelate. On the other hand, when using R_2Mg, the sulfoxide reacts through nonchelated conformation 7.76, in which the two dipoles are opposed. To increase the face discrimination in the chelation-controlled additions, Posner and coworkers conducted the conjugate additions in 2,5-dimethyltetrahydrofuran, which is less basic than THF [102, 112]. Nonracemic 3-alkyllactones are also obtained with a high enantiomeric excess starting from 1.139 (X = O) (Figure 7.53). Lithiocuprates successfully add to 3-alkylsubstituted analogs 7.77, providing nonracemic 3,3-disubstituted cycloalkanones with a high enantiomeric excess [102, 253] (Figure 7.53). Conjugate addition of $PhCH_2OCH_2OCH_2Li$ is a useful technique for introduction of a protected hydroxymethyl group. Trapping of the intermediate enolates by various electrophiles has also been performed, and this method has been applied to the synthesis of optically active natural products such as estrone [102, 112]. Conjugate additions of organometallic reagents to a chiral α,β-unsaturated sulfoximine 1.149 have also been described [515].

Chirality may be introduced in another part of the molecule: Isobe and coworkers have performed the addition of organometallic reagents to chiral sulfones 7.78 [173]. After sequential treatment of the adducts with tetrabutylammonium fluoride, aluminum amalgam and $HgCl_2$, α-branched aldehydes were isolated with a high enantiomeric excess (Figure 7.54). Analogs formed from (S)-valinol have been used for similar purposes, but preparation of the substrates requires tedious purification [325].

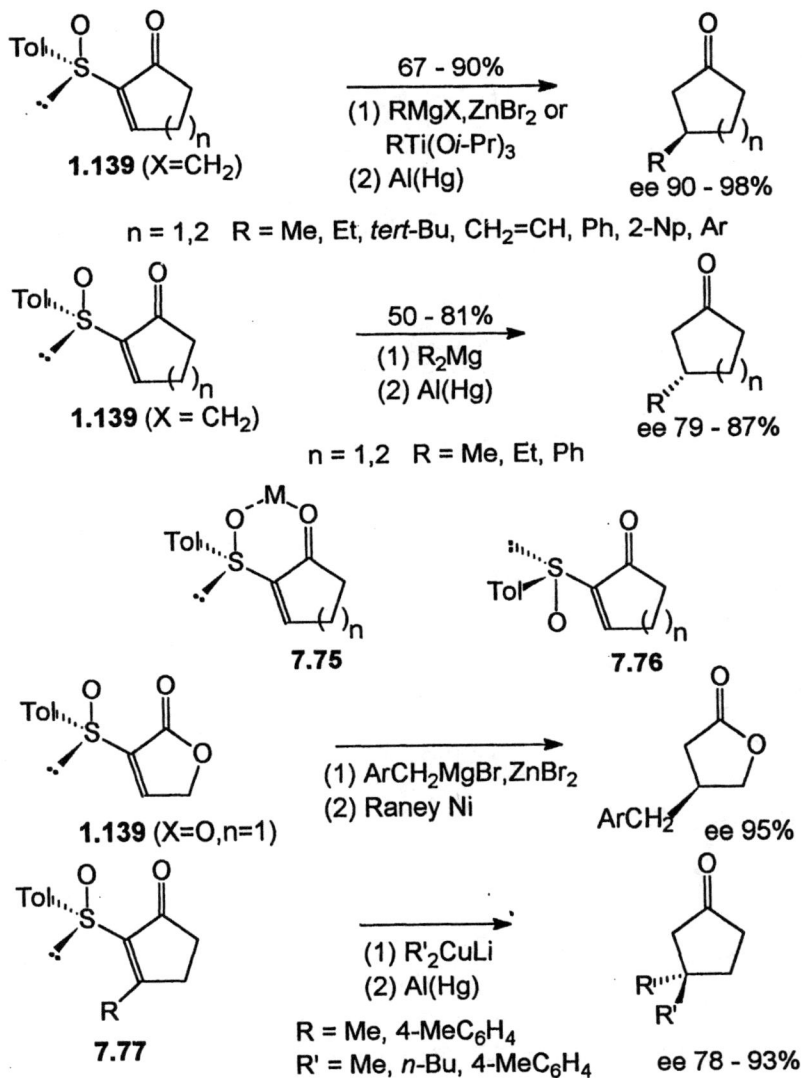

Figure 7.53

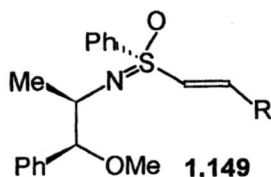

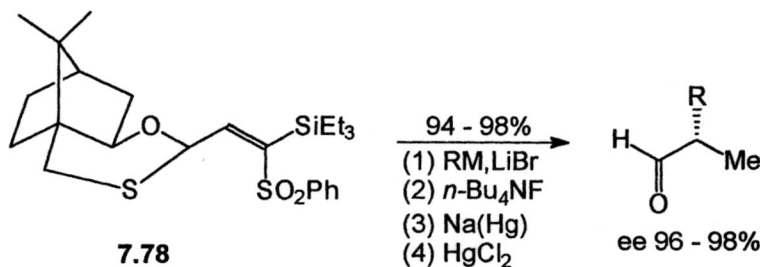

RM = MeLi, EtMgBr, *tert*-BuLi, 2-furylLi

Figure 7.54

7.10.2.4 *Reactions with Other Electrophilic Double Bonds*

The reactions of organometallic reagents with some *in situ* generated *N*-acylpyridinium salts lead to 4-substituted-*N*-acyl-1,4-dihydropyridines. Meyers [112, 324] and Mangeney and coworkers [112, 307, 1433] have introduced chiral residues in the 3-position of the pyridine ring to induce asymmetry. The reactions of organolithiums or -magnesiums with oxazoline **1.86** (Y = 3-Py) in the presence of ClCOOMe are highly regio- and stereoselective. After treatment of the products with FSO_3Me, and then $NaBH_4$, 4-substituted-3-formyl-1,4-dihydropyridines are obtained with a good enantiomeric excess (Figure **7.55**). The aminals **7.79** proposed by Mangeney and coworkers react with magnesiocuprates in THF at −70°C with a high regio- and stereoselectivity. After removal of the chiral auxiliary by acidic treatment, 4-substituted dihydropyridines are obtained with a high enantioselectivity (Figure **7.55**). The reaction of Et_2CuLi also gives useful results. Other acyl chlorides besides ClCOOMe can be employed to activate the pyridine functionality [307, 1433].

Recently, Kündig and coworkers [1434] have activated an aromatic ring bearing a chiral oxazoline group by complexation with a $Cr(CO)_3$ residue. Addition of RLi to the resulting substrates **1.87** (R = $Cr(CO)_3C_6H_5$, R' = Me) is highly regio- and stereoselective (Figure **7.55**), and the intermediates can be trapped with an alkyl halide. However, to date, the chiral residue has not been removed.

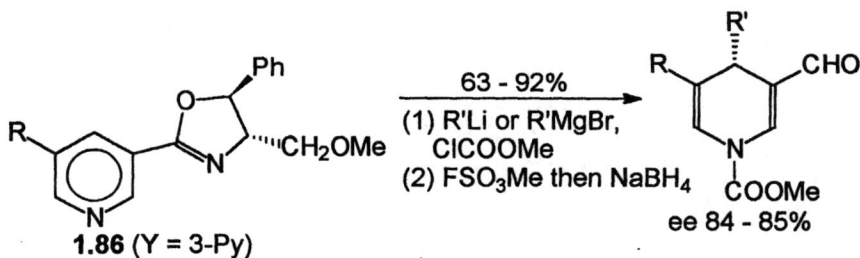

1.86 (Y = 3-Py)

63 - 92%
(1) R'Li or R'MgBr, ClCOOMe
(2) FSO₃Me then NaBH₄
ee 84 - 85%

R = H, COOEt, CONH₂   R' = Me, Et, *n*-Bu, Ph

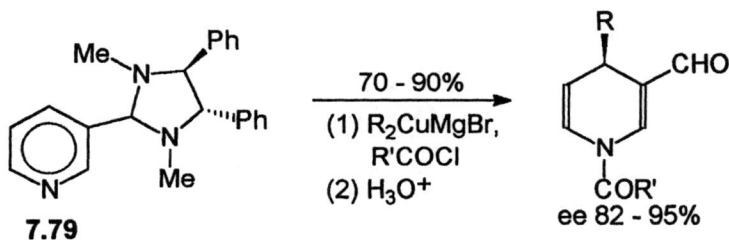

7.79

70 - 90%
(1) R₂CuMgBr, R'COCl
(2) H₃O⁺
ee 82 - 95%

R = Me, Et, *n*-Bu, Ph, CH₂=CH, 3-indolyl
R' = MeO, Me, 3-indolylmethyl

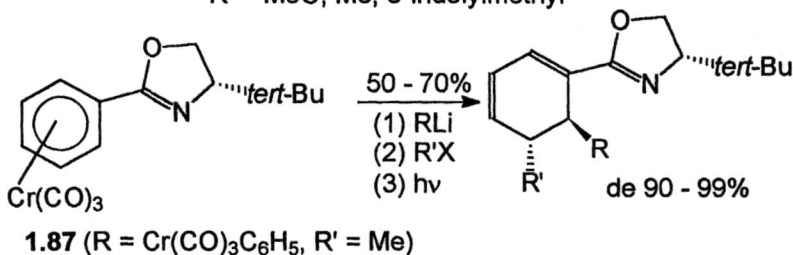

1.87 (R = Cr(CO)₃C₆H₅, R' = Me)

50 - 70%
(1) RLi
(2) R'X
(3) hν
de 90 - 99%

R = Me, *n*-Bu, Ph   R'X = MeI, CH₂=CHCH₂Br

Figure 7.55

7.11 ADDITIONS OF ALLYLSILANES TO ELECTROPHILIC DOUBLE BONDS

Under Lewis acid catalysis, the reaction of allylsilanes takes place easily with α-enones but not with α,β-unsaturated esters [162]. Because disappointing results were obtained in reactions of allylmetals with α,β-unsaturated carbonyl compounds, reactions of allylsilanes with electrophiles bearing chiral auxiliaries have been examined. Schultz and Lee [1435] performed the TiCl₄ catalyzed addition of trimethylallylsilane to compound **7.67** at −78°C. After treatment with methylhydroxylamine in acidic media, the 1,4-adduct **7.80** was obtained with high ee (Figure **7.56**). Under similar conditions, *N*-enoyloxazolidinone **7.68** (R = Ph) or sultam

1.134 (R = R$_E$CH=CH) react with trimethylallylsilanes, but the stereoselectivity of the reaction is lower than 80% if R' = H [419, 1436]. However, with **7.68** (R = Ph, R$_E$ = H, R' = Me) a higher selectivity (88%) is observed [1437]. Better results are obtained by reacting E-crotylsilane MeCH=CHCH$_2$SiMePh$_2$ with 2-cyclopentenones **7.81** bearing a chiral substituent [490] (Figure 7.56). After protection of the keto group as a dioxolane, the chiral auxiliary is removed by reduction with LiAlH$_4$.

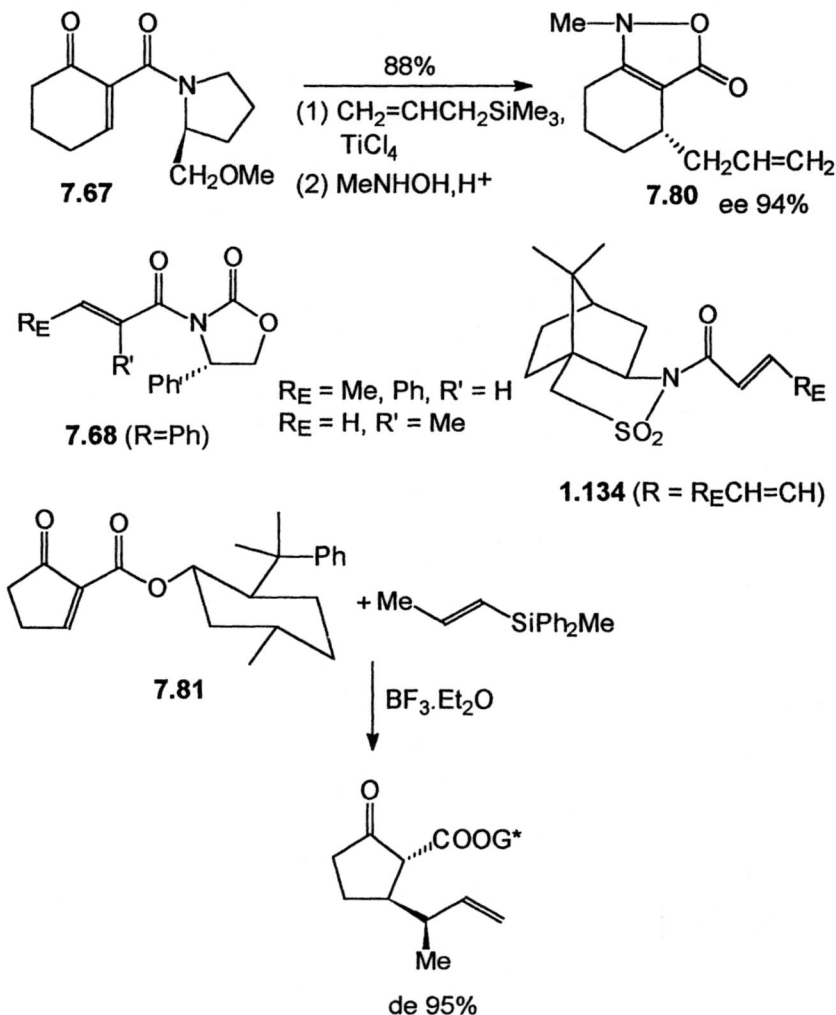

Figure 7.56

7.12 ENE REACTIONS [819, 820]

The only type of ene reaction that takes place easily under $EtAlCl_2$ catalysis is that of β-unsubstituted alkenes and acrylates. Therefore to induce asymmetry, it is necessary to use olefins whose double bonds are activated by more efficient groups. Narasaka and coworkers [819] and Snider and Zhang [419] performed ene reactions with N-enoyloxazolidinones. Achiral oxazolidinone **7.82** undergoes ene cyclization under catalysis of a chiral titanium dichloride derived from **2.52** with an excellent selectivity [819] (Figure **7.57**). Intermolecular ene reactions of chiral N-crotonoyloxazolidinones **7.68** (R_E = Me, R = i-Pr, $PhCH_2$) with different olefins under Me_2AlCl catalysis lead in the best cases to products with 80% diastereoisomeric excess [419]. The double bond of **7.83** is activated by both a chiral sulfoxide and a cyano group, and it undergoes a highly stereoselective intramolecular ene reaction [1438]. After oxidation of the sulfoxide to a sulfone and pyrolysis, the α,β-unsaturated nitrile **7.84** is obtained with an excellent selectivity (Figure **7.57**). The transition-state model proposed for this reaction involves chelation of the S–O and CN groups by the Lewis acid, and cyclization with minimization of steric interactions (Figure **7.57**).

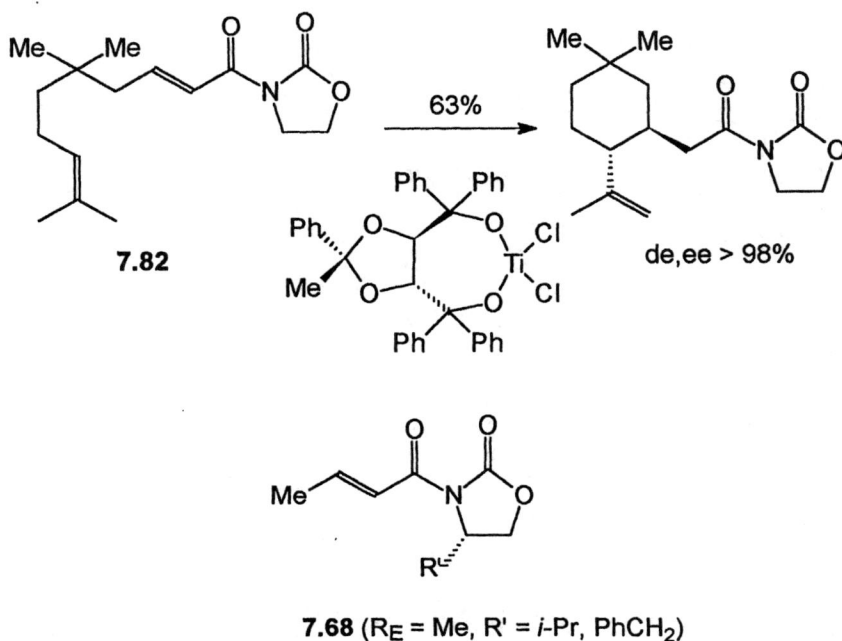

Figure 7.57

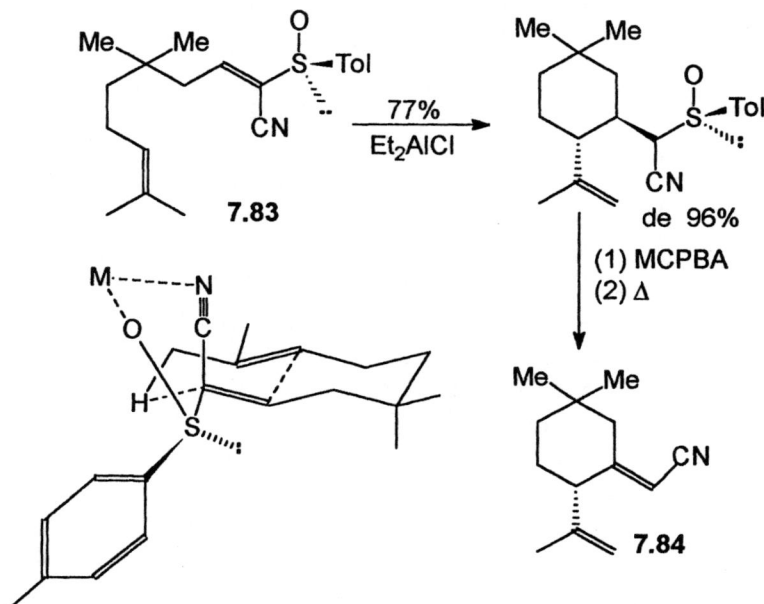

Figure 7.57 (continued)

7.13 ADDITIONS OF METAL ENOLATES AND ANALOGS TO ELECTROPHILIC DOUBLE BONDS [161]

Michael and related reactions are very popular methods to generate C–C bonds. When α-enones are used as acceptors, the competition between 1,2- and 1,4-addition and the reversibility of these reactions can pose problems. To induce asymmetry, chirality can be introduced on the ligands of the metal, on the enolate backbone, or on the Michael acceptor. Transition-state models that can interpret the diastereoselection of these reactions have been proposed. Just as in aldol reactions, *syn* and *anti* diastereoisomers can be generated when two new stereocenters are created in a Michael addition (Figure 7.58). The E- or Z-geometry of the enolate influences the facial discrimination, as does the *s-cis* or *s-trans* reactive conformation of the substrate. In some types of reactions, eight-centered chelates are thought to intervene. Seebach and Golinski [42] and Heathcock and coworkers [161] have proposed transition-state models in which the single and double bonds are staggered in order to minimize nonbonded interactions. Heathcock considered the reactions of Li enolates with α,β-unsaturated compounds. In order to favor chelation control, the carbonyl compound adopts the *s-cis* conformation. Chelates C1 and C2 lead to the products from E-enolates, and analogous chelates C3 and C4 are derived from Z-enolates.

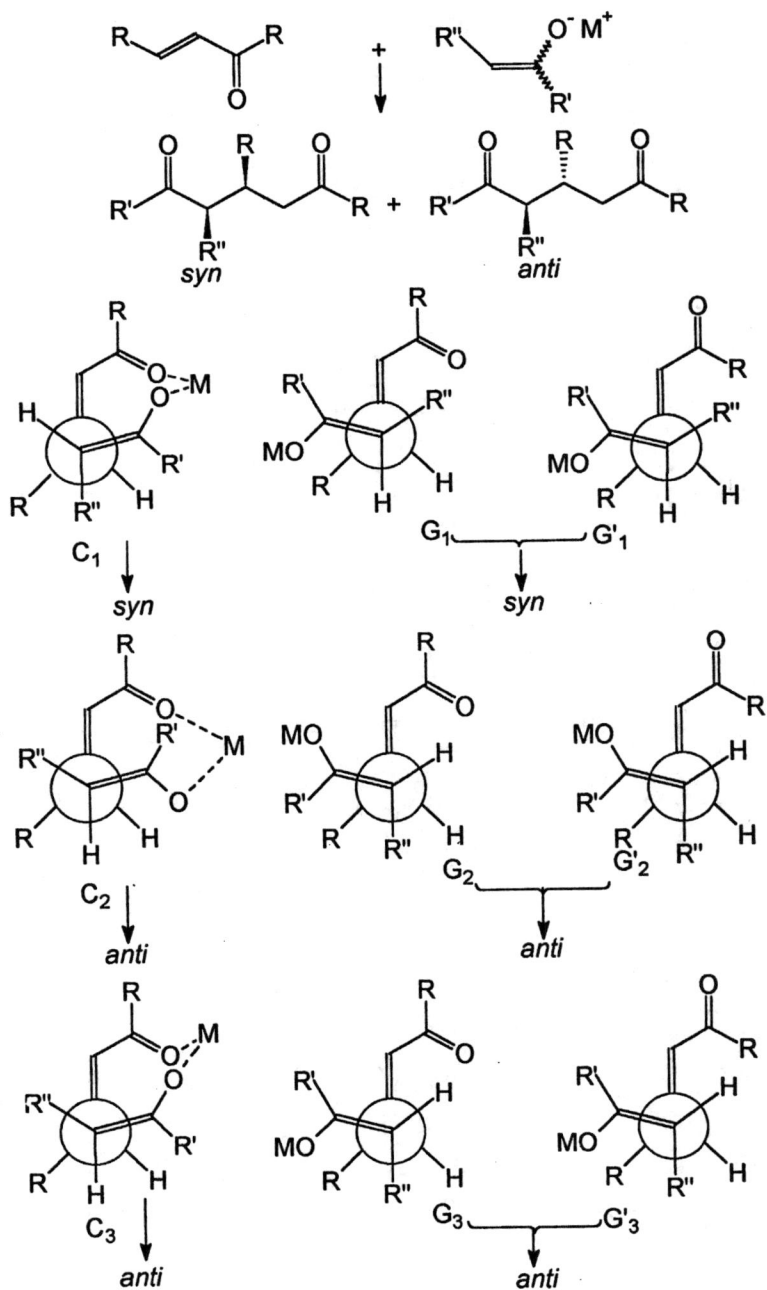

Figure 7.58

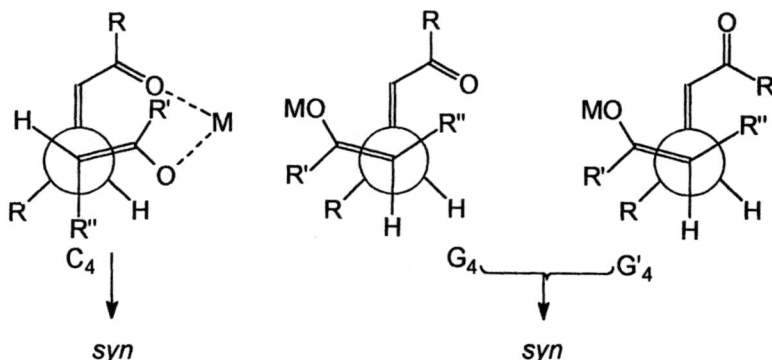

Figure 7.58 (continued)

These transition state models have been supported by theoretical calculations [126]. Seebach and Golinski's models do not constrain the reactants to approach in a *syn* fashion. Therefore, in addition to the previous models, approaches G_1, G_1', G_2, G_2', G_3, G_3', G_4 and G_4' which involve *s-trans* conformers are also considered. Stereoselectivity is interpreted by estimating the various steric interactions (Figure **7.58**).

7.13.1 Reactions of Enolates Bearing Chiral Ligands

Reactions of the lithium enolate of cyclohexanone with E-1-nitroalkenes in the presence of chiral lithium amides have been studied by Seebach and coworkers [558], and good diastereo- and enantioselectivities are obtained in a few cases. The tin enolate of *N*-propionoyloxazolidinone **6.83** undergoes diastereo- and enantioselective Michael reactions when coordinated to chiral amine **2.13** (R = NH-1-Np) [682] (Figure **7.59**). Similar reactions show low enantiomeric excesses (≤ 70%); however, some Michael additions catalyzed by chiral catalysts have shown high selectivities (§ 7.16).

7.13.2 Reactions of Enolates and Carbanions Bearing Chiral Residues

Metal enolates used in Michael reactions are generated from mono- or bifunctional compounds. For the sake of simplicity, bifunctional reagents will be classified under the heading corresponding to the functionality that bears the chiral auxiliary. The reactions of delocalized carbanions will also be considered in this chapter. As already discussed for aldol reactions, one or two new stereocenters can be created in the reactions of these reagents [161].

7.13.2.1 *Enolates of Aldehydes, Ketones and Derivatives*

Both asymmetric Michael reactions and alkylations (§ 5.3.1) are performed with anions of derivatives of aldehydes or ketones such as chiral imines, enamines or hydrazones. Azaenolates of acetone imines can be formed from chiral amines bearing ether groups, and these azaenolates undergo 1,4-additions to α-enones. Among the imines studied by Tsuji and coworkers [260], **7.85** is the most efficient. Conjugate additions of the derived lithiozincate or -cuprate to 2-cycloalkenones are highly stereoselective (Figure 7.59). Chemical yields are somewhat lower with cyclohexanone (n = 2) than with cyclopentanone (n = 1), and the 3-acetonylcyclanone is readily obtained with a good ee after hydrolysis. Koga and coworkers used lithium chelate **7.86** formed from a valine *tert*-Bu ester **1.59** (R = *i*-Pr, R' = *tert*-Bu) enaminoester as a chiral Michael donor. At −78°C, or better at −95°C in THF, high selectivities are observed with methylvinylketone and gem-diesters [161, 1439] (Figure 7.59). When the reaction is carried out in the presence of HMPA, the face selectivity is reversed. Ethyl acrylate gives a poor selectivity. From *tert*-Bu acetoacetate and dimethyl benzylidenemalonate, it is possible to prepare, after acid treatment and decarboxylation followed by exposure to CH_2N_2, 3-phenyl-1,5-ketoester **7.87** with a high enantiomeric excess (Figure 7.59). The selectivity is lower when alkylidene malonates are used. Enders and Karl performed similar reactions with lithioenamines [1440].

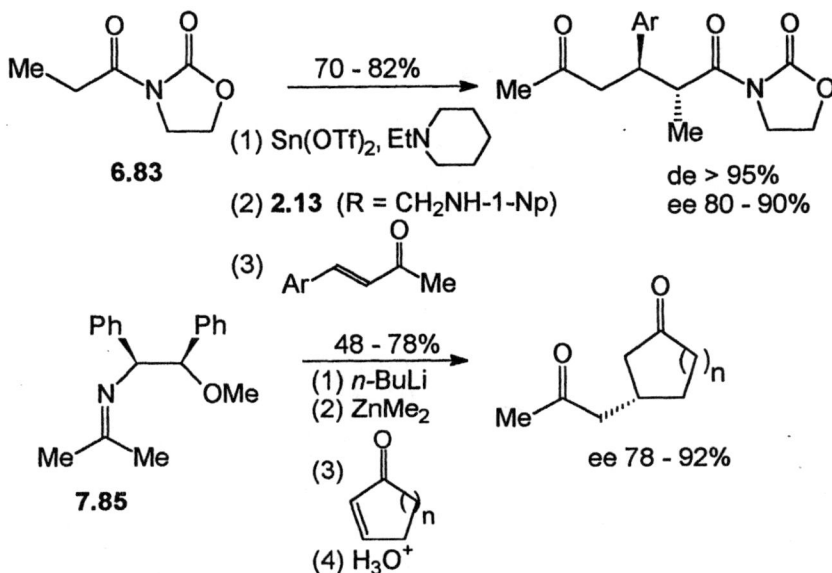

Figure 7.59

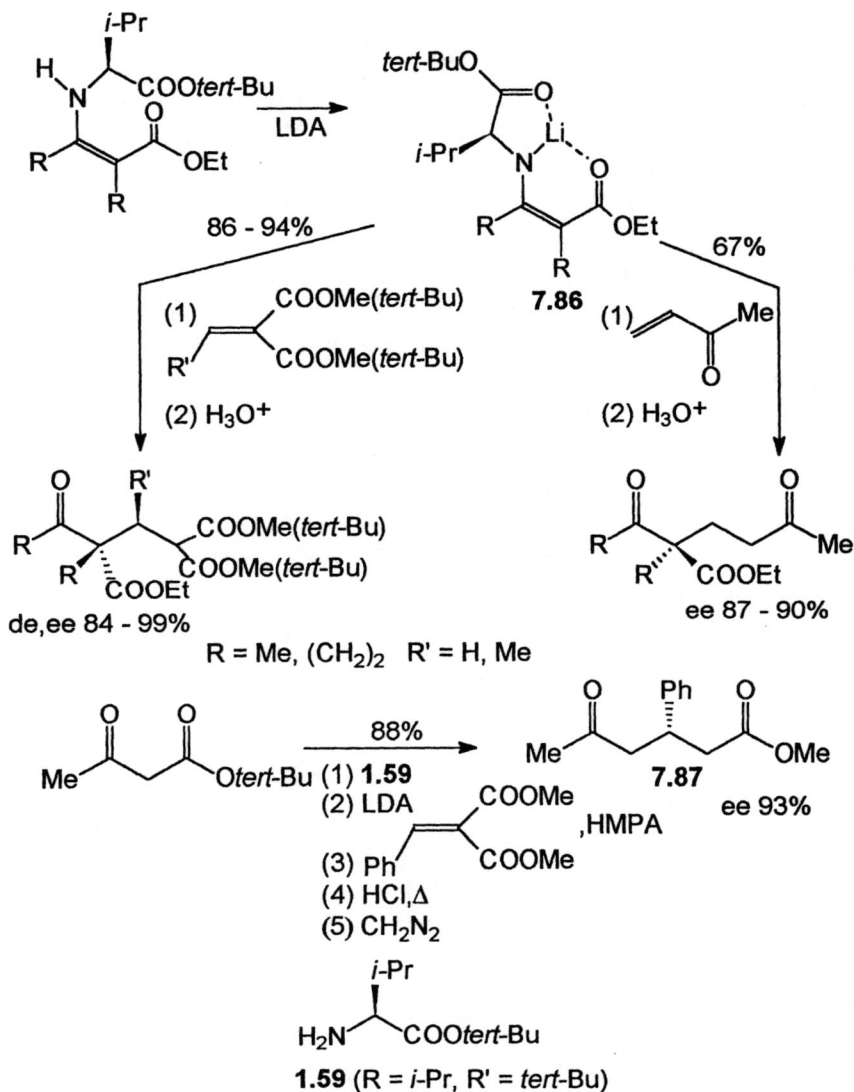

Figure 7.59 (continued)

The reactions of lithiated anions of chiral hydrazones **1.76** (Samp, Ramp) with α,β-unsaturated esters or sulfones are highly stereoselective at $-78°C$. By a sequence of 1,4-addition and removal of the chiral auxiliary with O_3, Enders and coworkers [161, 1441, 1442, 1443] prepared 5-keto-3,4-dialkylesters and 5-keto-3-alkylsulfones with high selectivities (Figure **7.60**). When ω-bromo-α,β-unsaturated esters or sulfones are used as Michael acceptors, cyclic ketoesters or sulfones such as **7.88** are obtained with high diastereo- and enantioselectivities (Figure **7.60**). Equally high selectivities are observed with benzylidene gem-ketoesters provided that a more hindered hydrazine Sadp **7.89** is used as the chiral auxiliary [1444]. An example of this last reaction is given in Figure **7.60**. After removal of the chiral residue and cyclization, nonracemic 1,4-dihydropyridines **7.91** are obtained with a high enantiomeric excess. Reactions of formaldehyde Samp hydrazone with nitroolefins have been studied, and the stereoselectivities of these reactions are at best 80% [1445]. All these reactions proceed through preformed lithium chelates which react with minimization of steric repulsions.

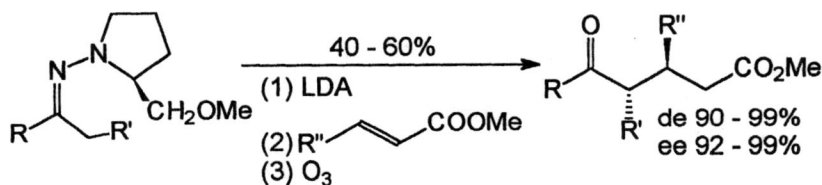

$R = H, Et, Ph \quad R' = Me, Et, n\text{-}C_5H_{11}, i\text{-}Pr, Ph \quad R'' = Me, Et, Ph$

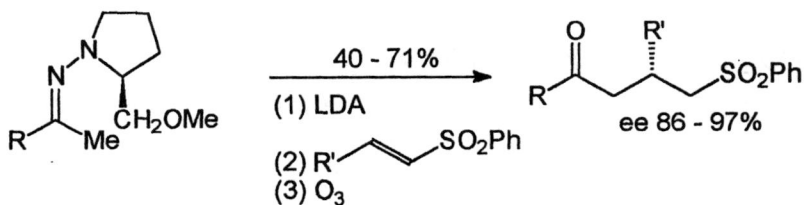

$R = Me, Et, i\text{-}Pr, Ph \quad R' = tert\text{-}Bu, Ph, 4\text{-}MeC_6H_4, 4\text{-}MeOC_6H_4$

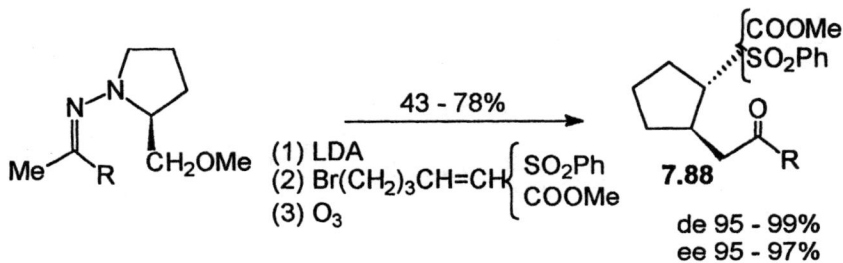

$R = Me, n\text{-}Bu, i\text{-}Bu, Ph$

Figure 7.60

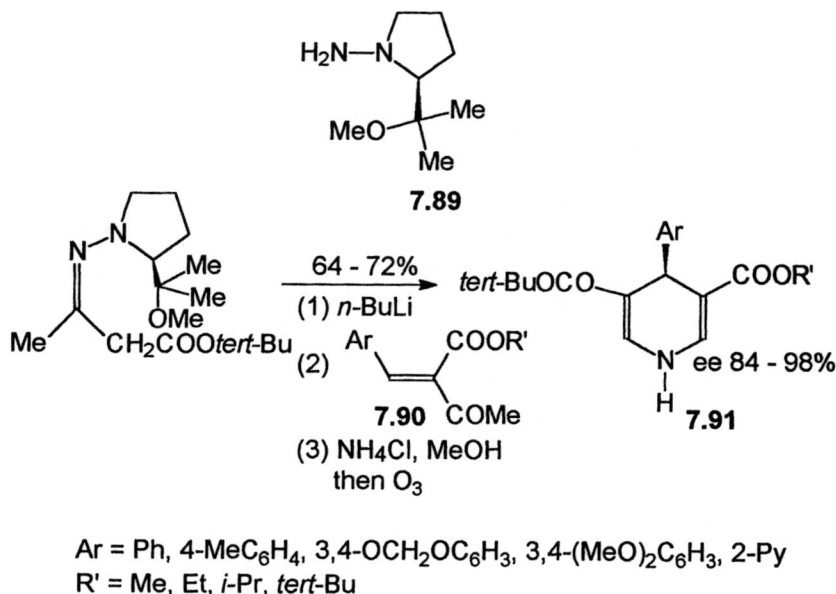

Figure 7.60 (continued)

7.13.2.2 *Enolates of Carboxylic Acid Derivatives* [161, 173]

Esters enolates of chiral alcohols can undergo stereoselective Michael reactions. Indeed, by adding the Li enolate of the propionate of (1*R*,2*S*,5*R*)-8-phenmenthol **1.4** (R = Ph) to methyl E-crotonate, Corey and Peterson obtained, the *syn* adduct with a good stereoselectivity at –100°C. Addition to the corresponding Z-crotonate is not very selective. The reaction of a phenmenthyl ester enolate with methacrolein is a key step in the synthesis of an inhibitor of acetylcholinesterase [1446]. Ester enolates derived from alcohols bearing the bornane skeleton do not generally give useful results. However, an intramolecular diastereoselective Michael reaction of malonate **7.92** has been realized by Stork and Saccomano (Figure **7.66**). Taber and coworkers [1447] reacted chiral aldehydo-ester **7.93** with methylvinylketone. After ketalization of both carbonyl groups with HS(CH$_2$)$_2$SH and LiAlH$_4$ reduction, compound **7.94** was obtained with a high enantiomeric excess (Figure **7.61**).

Lithium enolates of propionamides of chiral amines undergo stereoselective 1,4-addition reactions to α,β-unsaturated esters. The highest selectivities are obtained with C$_2$-symmetric amides derived from **1.65** (R = CH$_2$OCH$_2$OMe) bearing a substituent that is capable of metal chelation (Figure **7.61**). After hydrolysis, diacid **7.95** is obtained with a good selectivity [161]. Some nonracemic natural products can be prepared with a high enantioselectivity by reactions of these amide enolates with $EtOOCCH=CHCH_2CH_2COOEt$ as Michael acceptor [161, 173]. Titanium enolates of N-acyloxazolidinones **5.30** react at 0°C with terminal olefins $CH_2=CH$-EWG to give Michael adducts resulting from addition to the least hindered face of the corresponding titanium chelate **6.89** [413, 1042] (Figure **7.62**). After removal of the chiral auxiliary with H_2O_2/LiOH, followed by treatment with CH_2N_2, α-branched methyl esters are obtained with an excellent selectivity.

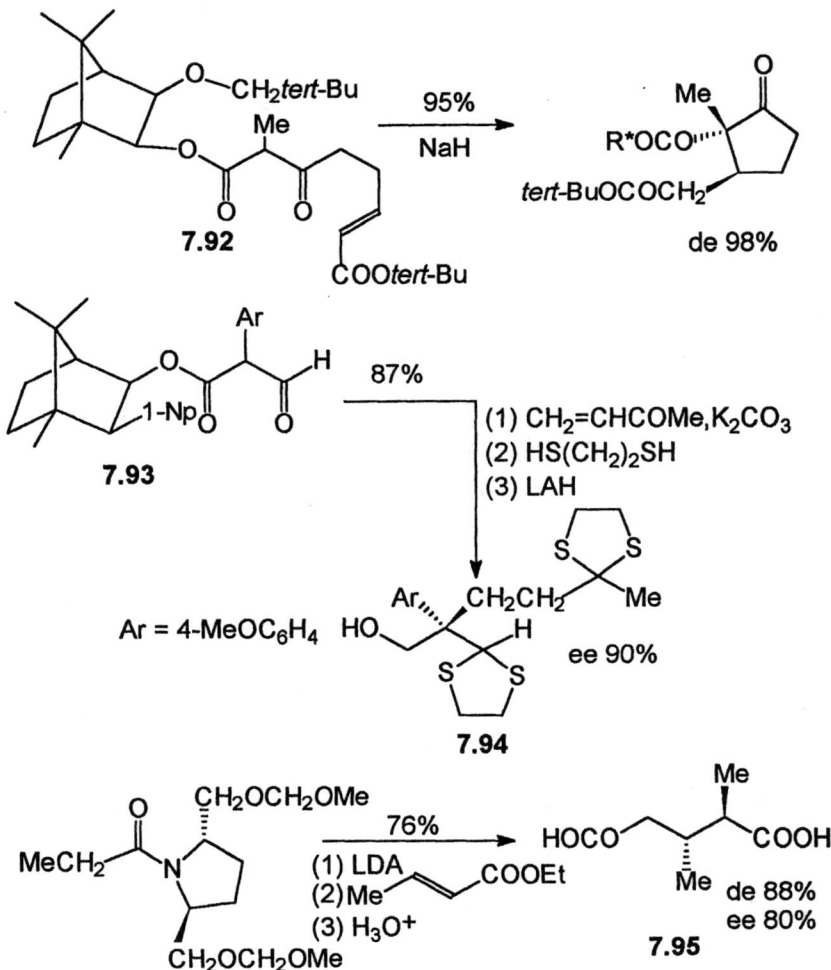

Figure 7.61

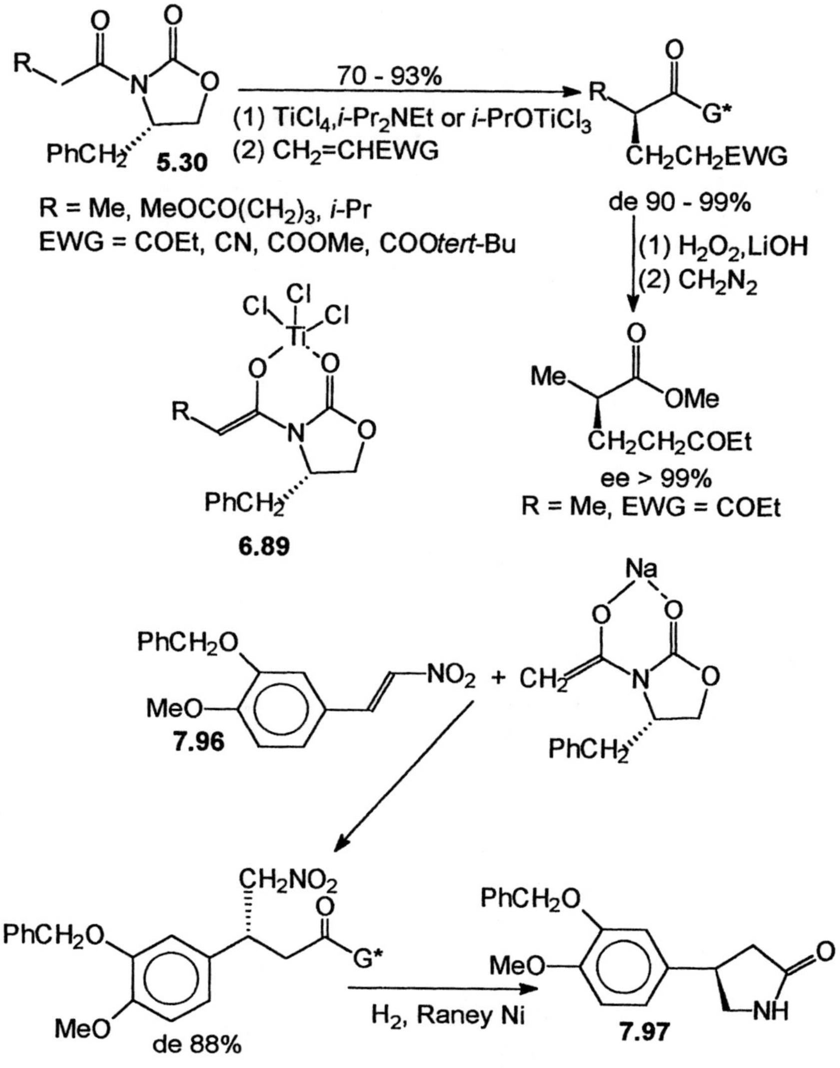

Figure 7.62

    Limitations of this method are the poor selectivity observed with 2-cyclo-hexenone and the lack of reactivity of these enolates toward β-alkyl-α,β-un-saturated esters [413]. Conjugate addition of the sodium enolate of **5.30** (R = H) to a substituted nitrostyrene **7.96** is the first step of the synthesis of an antidepres-sive drug [1448]. The chiral auxiliary is excised by lactam formation induced by hydrogenation of the nitro group. This leads to pyrrolidine **7.97** (Figure 7.62).

The reaction of the titanium enolate of *N*-propionoylsultam **1.134** (R = MeCH$_2$) with CH$_2$=CHCOEt gives a poor chemical yield [413], and the reaction of an analog **1.134** (R = MeCOCH$_2$) with arylidenemalononitriles is not highly stereoselective (de ≤ 70%) [1449].

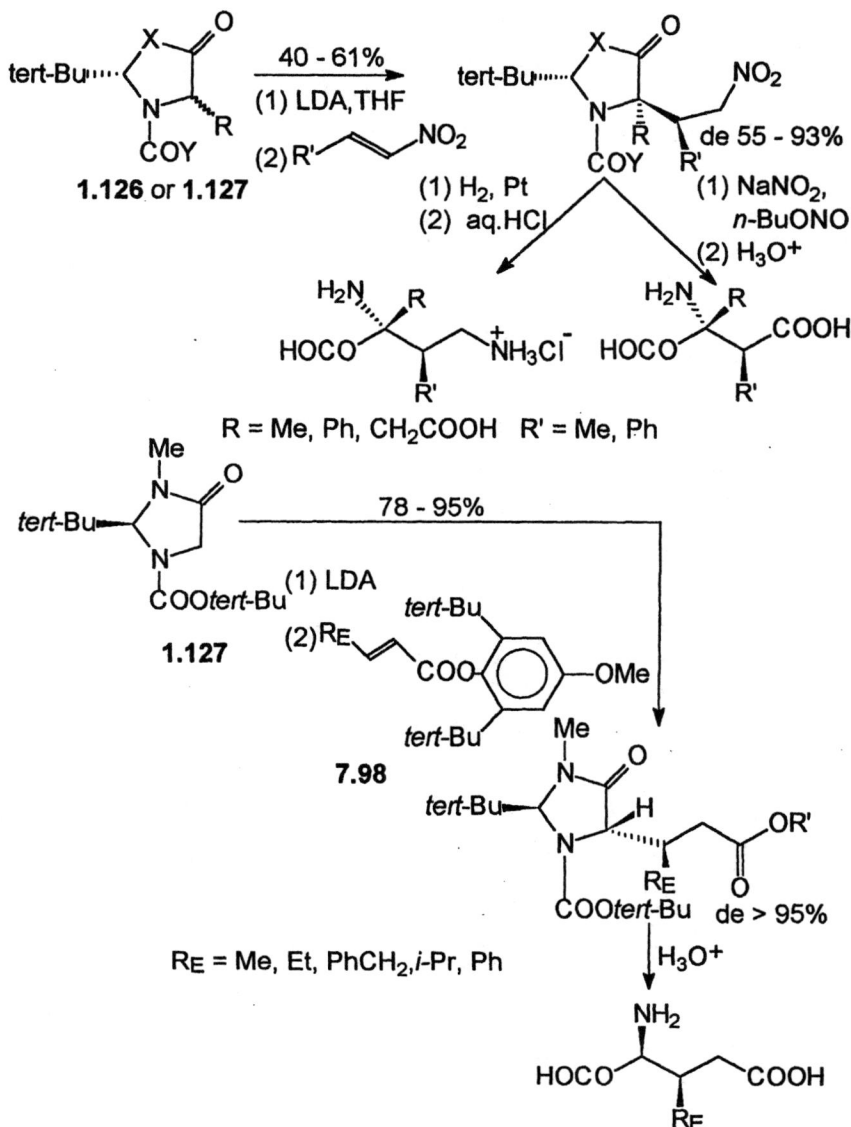

R = Me, Ph, CH$_2$COOH R' = Me, Ph

R$_E$ = Me, Et, PhCH$_2$,*i*-Pr, Ph

Figure 7.63

Lithium enolates of the heterocyclic reagents **1.125**, **1.126** and **1.127** developed by Seebach and coworkers [173, 1450] add selectively to α,β-unsaturated nitro derivatives at −100°C. Catalytic hydrogenation of the adducts gives α,γ-diaminoacids, while Nef reaction generates α-aminoacids with a high selectivity (Figure **7.63**). The reactions of the enolates of **1.127** with α,β-unsaturated esters at −78°C are also highly selective, provided that bulky esters such as **7.98** are employed [443]. The interpretation of the selectivity of these reactions of nitroalkenes is similar to that of aldehydes (see Figure **6.84**). The additions to α,β-unsaturated esters display an opposite topicity. Although no interpretation has yet been given for this, the reactive conformation of the esters certainly plays an important role.

Ester enolates of chiral imines of glycine **1.107** undergo conjugate addition to α,β-unsaturated esters, and the most useful results are obtained with *tert*-Bu esters. After removal of the chiral auxiliary with NH₂OH, 5-oxopyrrolidine-2-carboxylates **7.99** are formed with a high selectivity [376, 861]. Methacrylates give poorly selective reactions, but the reactions of gem-diesters are highly selective (Figure **7.64**). Double asymmetric induction is also observed [1451].

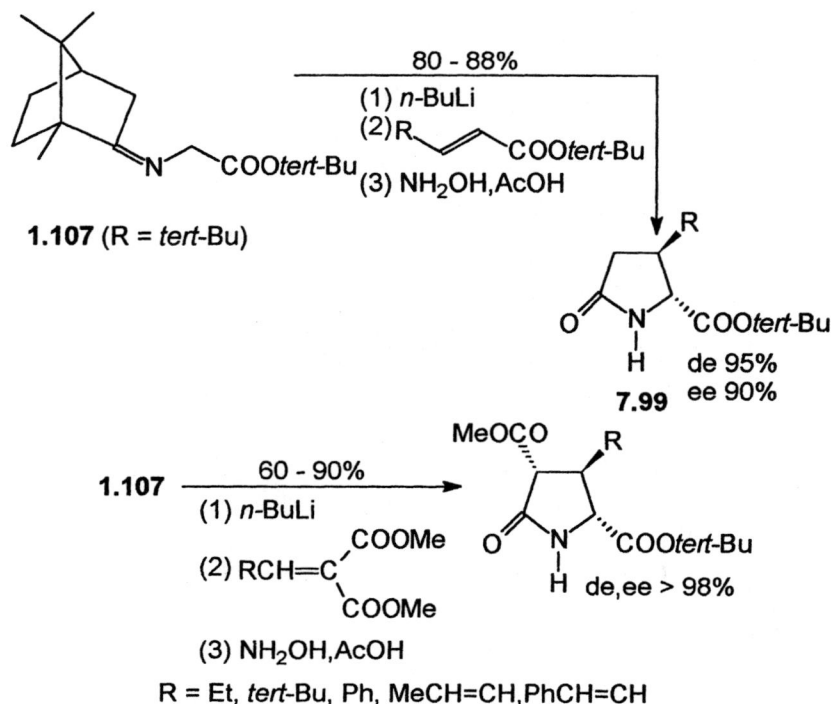

Figure 7.64

The enolates of chiral lactim ethers **1.114** undergo asymmetric conjugate additions to α,β-unsaturated ketones, esters and nitro derivatives [154, 161, 173, 861, 1452]. Lithium enolates give excellent selectivities with α,β-unsaturated esters (Figure **7.65**). Lithiocuprates are the reagents of choice for selective additions to 2-cycloalkenones, but with linear α-enones, mixtures of stereoisomers are obtained. Titanium enolates are recommended in reactions with nitroolefins, and aminotitanates give better selectivities than alkoxytitanates [1452]. After acid hydrolysis, functionalized α-aminoesters are obtained with a high selectivity (Figure **7.65**). Like the related alkylation reactions, the face selectivity of these 1,4-additions is interpreted by attack of the electrophile on the least hindered side of the organometallic reagent [154, 1452] (Figure **7.65**).

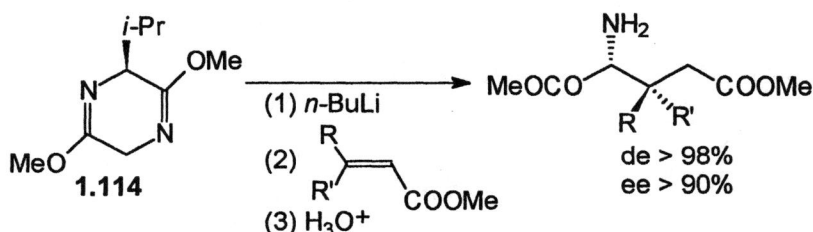

R = H, Me, Ph   R' = H, Ph   R = R' = H, Me

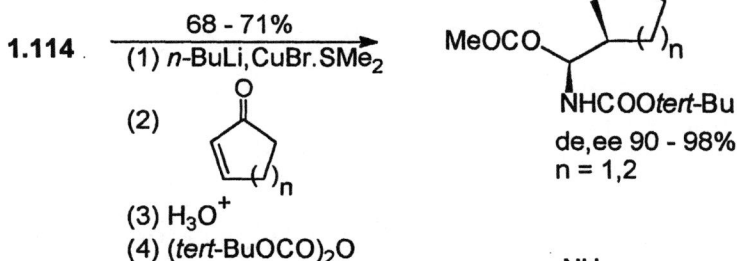

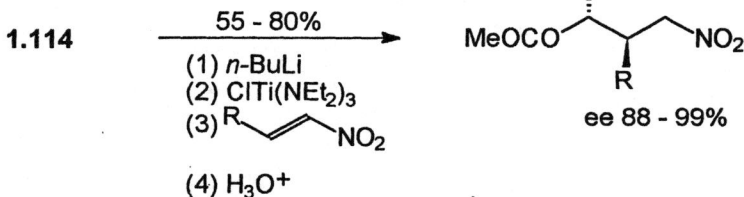

R = Me, Ph, 4-MeC₆H₄, 4-Br.C₆H₄, 1-Np

Figure 7.65

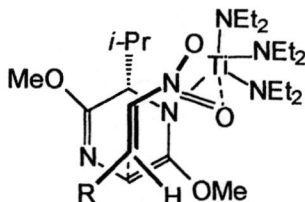

Figure 7.65 (continued)

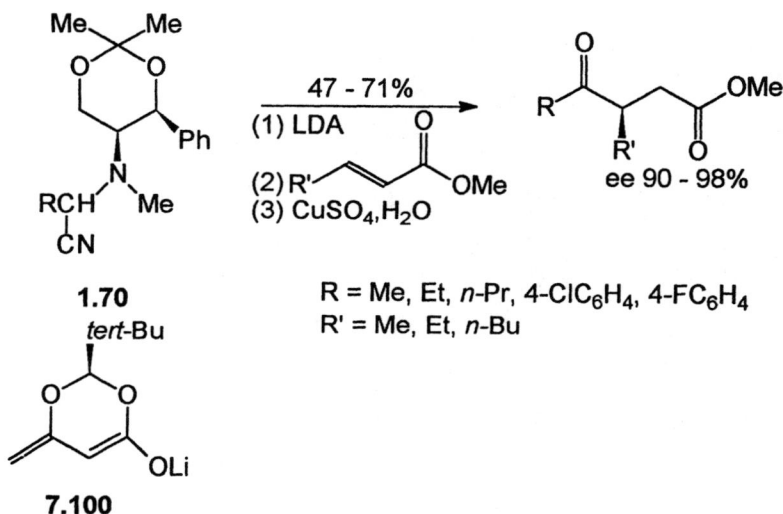

Figure 7.66

Enders and coworkers [300] used the lithiated carbanion of the chiral aminonitrile **1.70** as a Michael donor. Additions of this species to α,β-unsaturated
esters are performed at $-100°C$. After removal of the chiral residue with aqueous
$CuSO_4$, 3-branched-4-ketoesters are obtained with a high enantiomeric excess
(Figure 7.66). 2-Cyclohexenone can also be used as an acceptor with equally good
results [1453]. Dienolate **7.100** has also been proposed as a chiral Michael donor
[1454], but its reactions with α,β-unsaturated aldehydes give low chemical yields.

7.13.2.3 *Carbanions α to Sulfoxides and Phosphonamides* [161, 173, 1008,
         1278]

Conjugate additions of chiral lithiated sulfoxides **1.136** ($Y = CH_2=CHCH_2$,
$R = Tol$) are only highly selective with five-membered α,β-unsaturated cyclic ke-

tones and lactones (Figure 7.67). These reactions must be carried out at low temperature to prevent the epimerization of the reagent. The reactions of 4-substituted-2-cyclopentenones are subject to double diastereodifferentiation, and kinetic resolution occurs when the reactions are conducted with racemates (§ I.6). An example of a kinetic resolution is given in Figure 7.67 [161].

Hua and coworkers [477, 482] used the anion of chiral α-sulfinylketimine **1.136** (R = Tol, Y = CH$_2$(2-pyrrolinyl)) as a Michael donor toward α,β-unsaturated esters. High stereoselectivities are obtained with methyl acrylate or alkyl cycloalkenecarboxylates (Figure 7.67). The conjugate addition is followed by rapid intramolecular amidification. After reduction of the double bond and Raney nickel desulfurization, tricyclic lactams are obtained with a high stereo- and enantioselectivity (Figure 7.67). This reaction works with CH$_2$=C(NHBOC)COOMe, but a mixture of diastereoisomers is produced [482]. The observed face discrimination in this reaction is interpreted through a chelated transition state (Figure 7.67). The reactions of the anions of Solladié's chiral sulfinylacetates **1.136** (R = Tol, Y = COO-*tert*-Bu) are slow, and do not give useful selectivities [161].

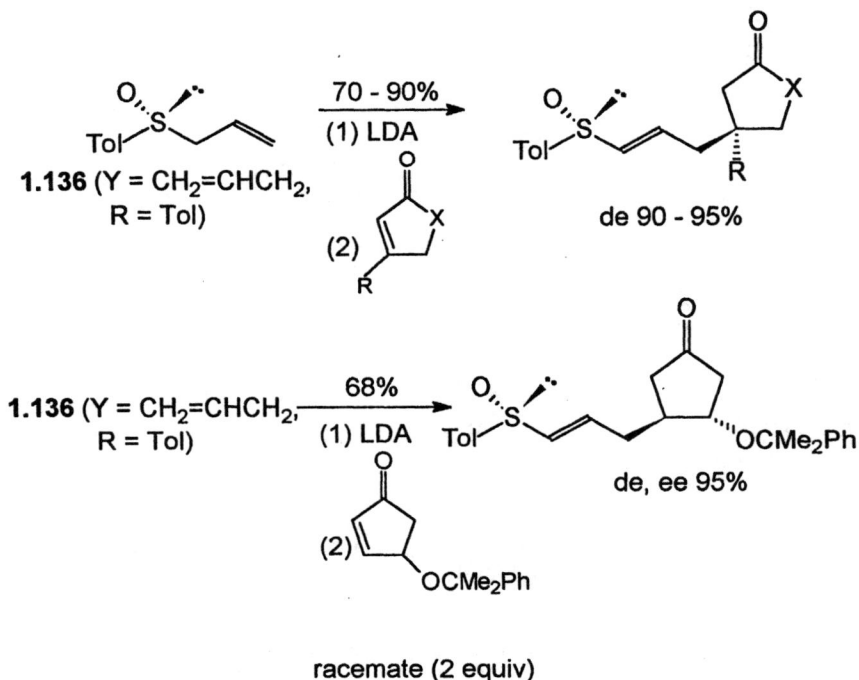

Figure 7.67

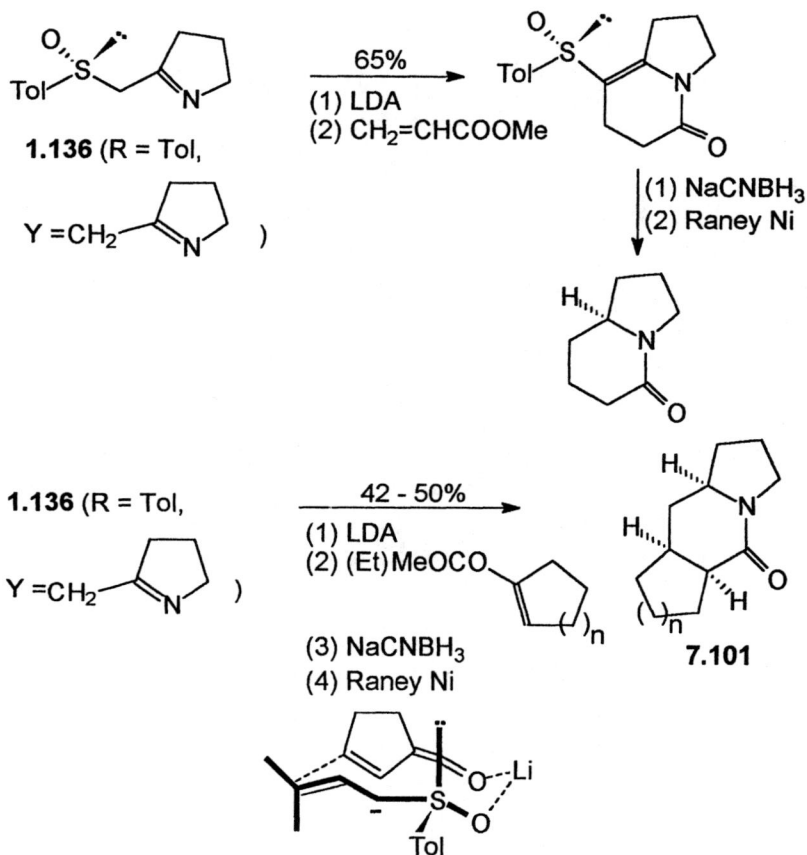

Figure 7.67 (continued)

Asymmetric conjugate additions of the anions of chiral allyl- and crotylphosphonamides **1.74** (Y = RCH=CH) to α,β-unsaturated cyclic ketones, lactones and lactams, and to *tert*-Bu cinnamate are highly stereoselective at −78°C [313] (Figure **7.68**). Hanessian and coworkers obtained 1,4-adducts with a high stereoselectivity. The level of selectivity was maintained even when the 3-position of the acceptor was substituted, so quarternary stereocenters can be generated. Both enantiomers of **1.74** are available, so it is possible to obtain either enantiomeric 1,4-adduct.

Hua and coworkers [173] previously performed conjugate additions of the anions of **7.102** obtained from the (1S,2R)-ephedrine derivative with 2-cycloalkenones. 1,4-Adducts were obtained with high selectivity, but the preparation of the reagent results in a mixture of stereoisomers that must be separated (Figure

7.68). Chelated transition states similar to those proposed with allylsulfoxides account for the observed selectivities.

7.13.2.4 Carbanions Derived from Chromium Carbonyl Complexes

Carbanions derived from indanone and tetralonechromium carbonyl complexes **6.18** (R = H) add to methylvinylketone [161]. The Michael acceptor is introduced on the face opposite to the organometallic residue with a good selectivity (80%).

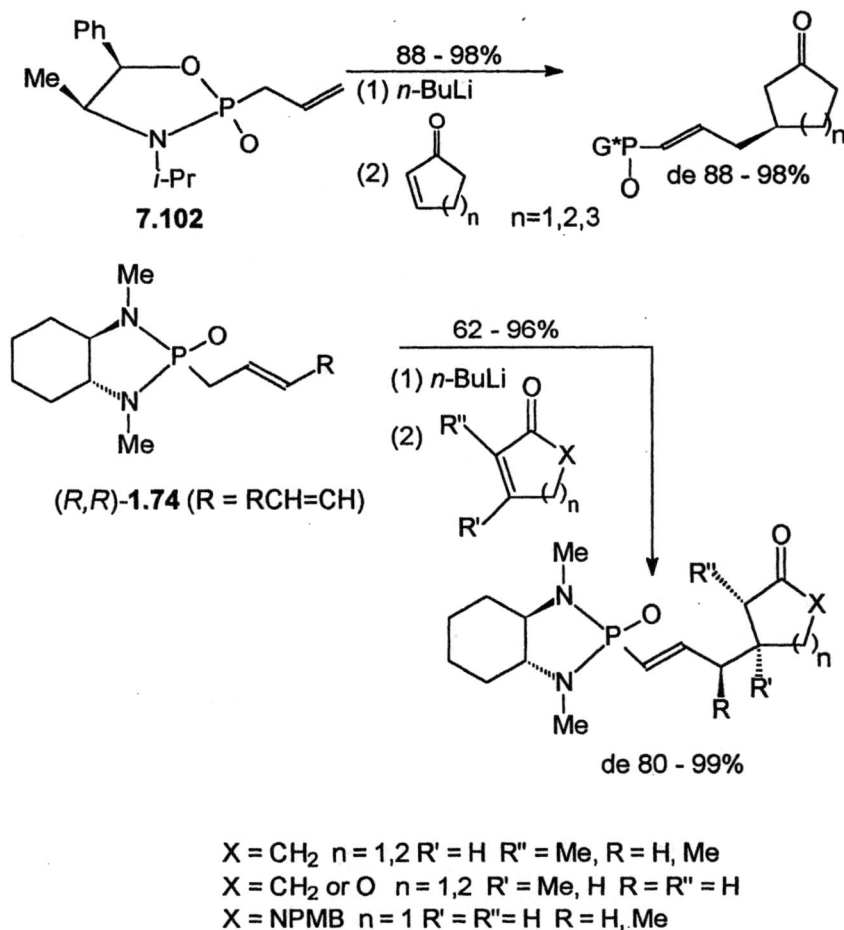

$X = CH_2$ n = 1,2 R' = H R'' = Me, R = H, Me
$X = CH_2$ or O n = 1,2 R' = Me, H R = R'' = H
X = NPMB n = 1 R' = R'' = H R = H, Me

Figure 7.68

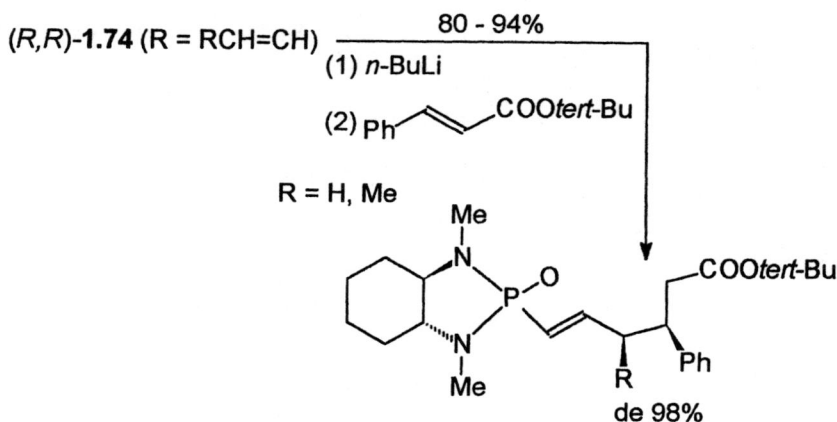

Figure 7.68 continued)

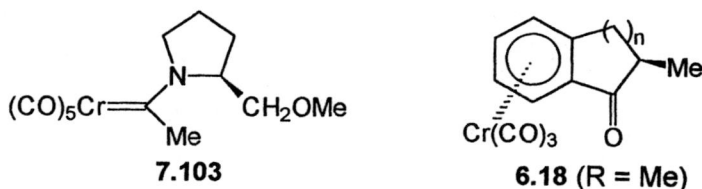

7.103

6.18 (R = Me)

The anions of chiral chromium aminocarbene complexes bearing a proline residue **7.103** have been used as Michael donors in reactions with 2-cycloalkenones at −78°C. After acid treatment, 3-acetaldehydosubstituted cycloalkanones are obtained. The enantioselectivities are not very high (60 - 70%), except with 4,4-dimethyl-2-cyclohexenone [297].

7.13.3 Reactions with α,β-Unsaturated Compounds Bearing Chiral Residues

There are few data in the literature on 1,4-additions of enolates or analogs to α,β-unsaturated aldehydes or ketones or to their derivatives bearing chiral residues. Yamada and coworkers [161] have added diethyl malonate to α,β-unsaturated imines of aminoacids **7.104**. When R = *tert*-Bu, the expected aldehyde is obtained with a high enantiomeric excess (Figure **7.69**). An asymmetric spirocyclization has been promoted by BF$_3$·Et$_2$O in the presence of (*S,S*)-1,2-cyclohexanediol [1455]. Additions of anions of 1,3-dithianes or dithioketals to menthyloxybutenolide **1.31** are highly stereoselective [137, 1456, 1457]. After trapping of the 1,4-adduct with an alkyl halide and desulfurization by

$NaBH_4/NiCl_2$ or Raney nickel, the menthyloxy group is removed with $NaBH_4/KOH$ to give 3,4-disubstituted butyrolactones with a high diastereo- and enantioselectivity (Figure **7.69**). Corey and Houpis [1458] have described asymmetric Michael reactions of ketone enolates with a 2-thiophenyl crotonate of 8-phenmenthol. Chirality has also been introduced on the amino group of 2-aminomethylacrylates to perform the asymmetric addition of the anion of the *tert*-Bu ester of cyclopentanecarboxylate [1459]. More important developments have been reported with chiral α,β-unsaturated sulfoxides and nitro compounds as Michael acceptors (see below).

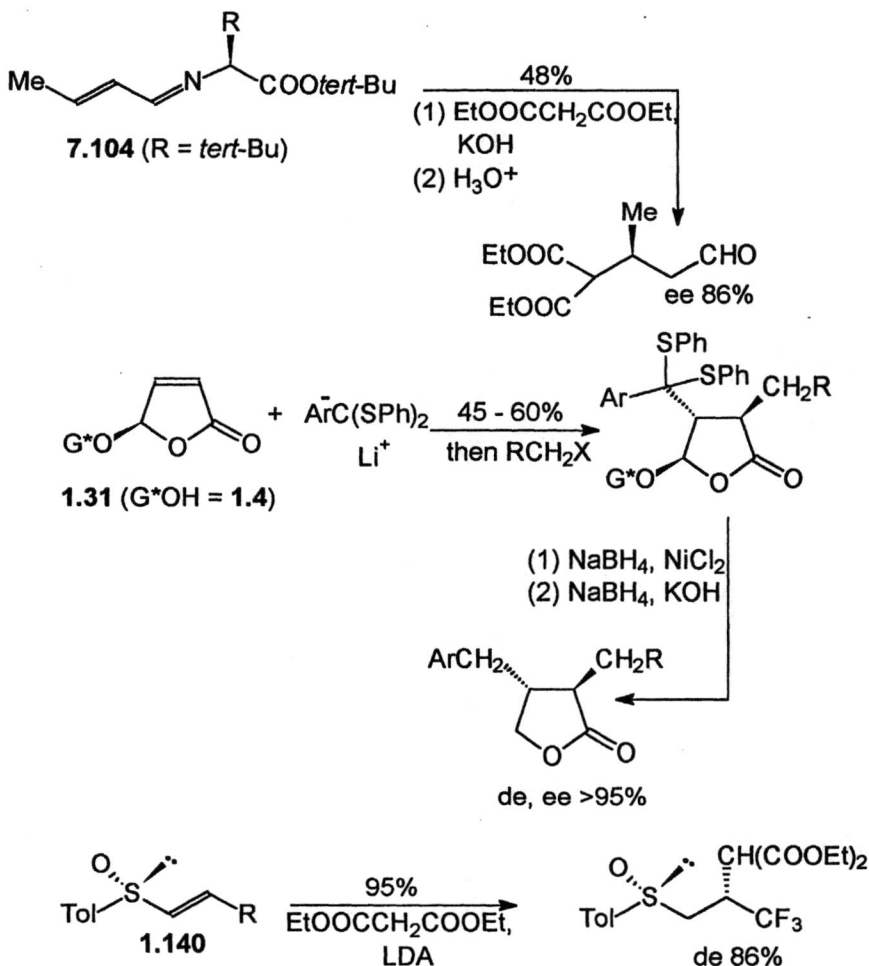

Figure 7.69

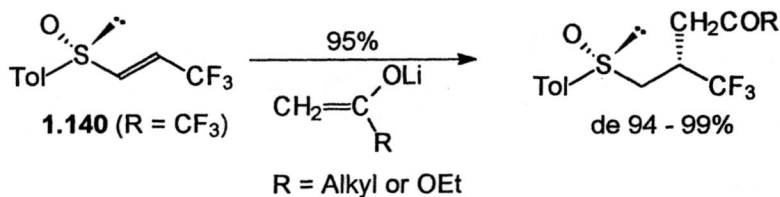

1.140 (R = CF$_3$)

95%

R = Alkyl or OEt

de 94 - 99%

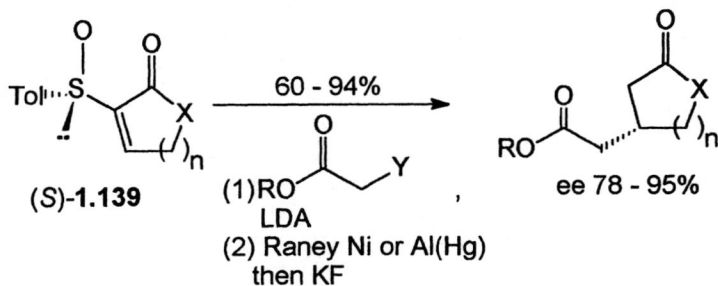

(S)-**1.139**

60 - 94%

(1) RO—CH$_2$—C(=O)—Y
LDA
(2) Raney Ni or Al(Hg)
    then KF

ee 78 - 95%

R = Me, MeOCH$_2$ X = CH$_2$,O   Y = PhS, Me$_3$Si   n = 1,2

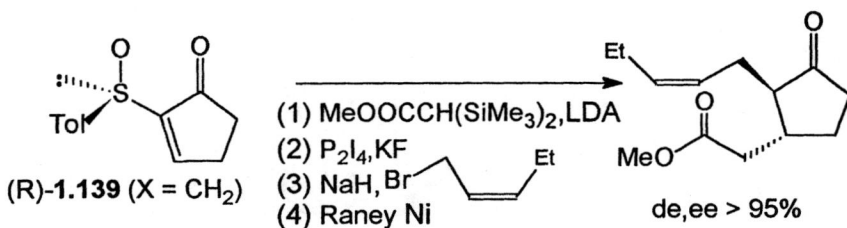

(R)-**1.139** (X = CH$_2$)

(1) MeOOCCH(SiMe$_3$)$_2$, LDA
(2) P$_2$I$_4$, KF
(3) NaH, Br
(4) Raney Ni

de,ee > 95%

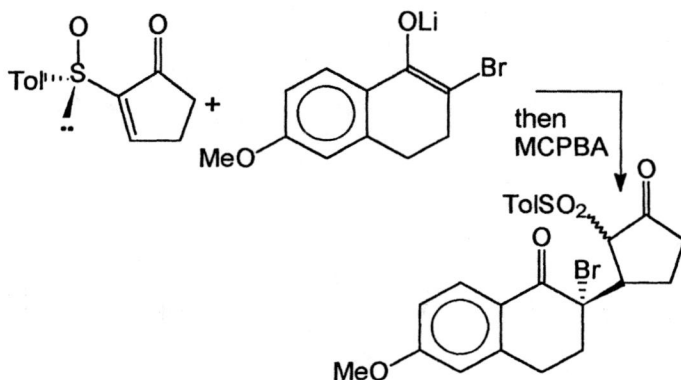

then
MCPBA

Figure 7.69 (continued)

7.13.3.1 *Reactions with α,β-Unsaturated Sulfoxides* [102, 161, 173]

In contrast to 1,4-additions of organometallic reagents (§ 7.10.2.3), the 1,4-additions of malonate anions to α,β-unsaturated sulfoxides **1.140** do not require the presence of another electron-withdrawing group in the α-position. However, good stereoselectivities are only observed in malonate additions to **1.140** (R = CF$_3$). Useful results are also obtained with **1.140** (R = CF$_3$) and lithium enolates of ketones or ethyl acetate (Figure **7.69**). The most remarkable conjugate additions of ketone or ester enolates to chiral α,β-unsaturated sulfoxides have been carried out with cyclic substrates **1.139**. For example, the enolates derived from phenylthio- or trimethylsilylacetate give 1,4-adducts with α,β-unsaturated ketosulfoxides **1.139** (X = CH$_2$) or lactones **1.139** (X = O) with excellent stereoselectivities. After treatment of the products with Raney nickel, or when silicon-substituted derivatives are used with aluminum amalgam, then KF, functionalized 3-substituted cyclanones or lactones are obtained with a high enantiomeric excess [102](Figure **7.69**). The 1,4-additions of disubstituted enolates occur on the other face of the sulfoxide double bond. Posner and coworkers applied this method to the synthesis of nonracemic methyl jasmonate and estrone. The key steps of these syntheses are in Figure **7.69**. The interpretation of the face discrimination in these additions is similar to that of the cuprate additions (Figure **7.53**). Monosubstituted enolates react with the conformer of the sulfoxide in which the C=O and S–O dipoles are opposed, while the addition of less reactive disubstituted enolates requires electrophilic assistance through a chelated conformer. In both cases, the nucleophile is introduced on the face opposite the S-tolyl group.

7.13.3.2 *Reactions with Nitroolefins* [1008, 1460]

Fuji and coworkers have studied the reactions of nitroolefins substituted in the β-position with chiral leaving groups. With these reagents, substitution reactions occur by an addition-elimination mechanism. Enamines **7.105** derived from (S)-2-methoxymethylpyrrolidine and sulfoxide **7.106** have received some applications. Zinc enolates of valerolactones **7.107** (X = O) react with **7.105** with a good selectivity, and functionalized nitroolefins bearing a quarternary stereocenter are formed with a high enantiomeric excess (Figure **7.70**). This reaction has been applied to the synthesis of alkaloids. Sulfoxide **7.106** reacts with many enolates. Zinc lactam enolates **7.107** (X = NR') react with useful selectivities, although the configuration of the resulting products was not determined (Figure **7.70**).

7.14 ADDITIONS OF ENOXYSILANES TO ELECTROPHILIC DOUBLE BONDS [162]

There are only a few highly selective examples of conjugate additions of enoxysilanes or ketene silylacetals bearing chiral residues to electrophilic double bonds. Ephedrine-derived ketene silylacetals add to methyl- or ethylvinylketone to give ketoesters with 72 - 75% selectivity [162]. Lewis-acid-promoted reactions of the trimethylsilylenolether of 1-acetylcyclohexene with acrylates of chiral alcohols take place with 70% diastereoselection [1461]. Following activation with $(CF_3CO)_2O$ the chiral oxazoline **1.89** reacts with enoxysilanes derived from acetophenone or 2-methylcyclohexanone. After hydrolysis, nonracemic δ-ketoacids are produced [347, 1462] (Figure **7.70**).

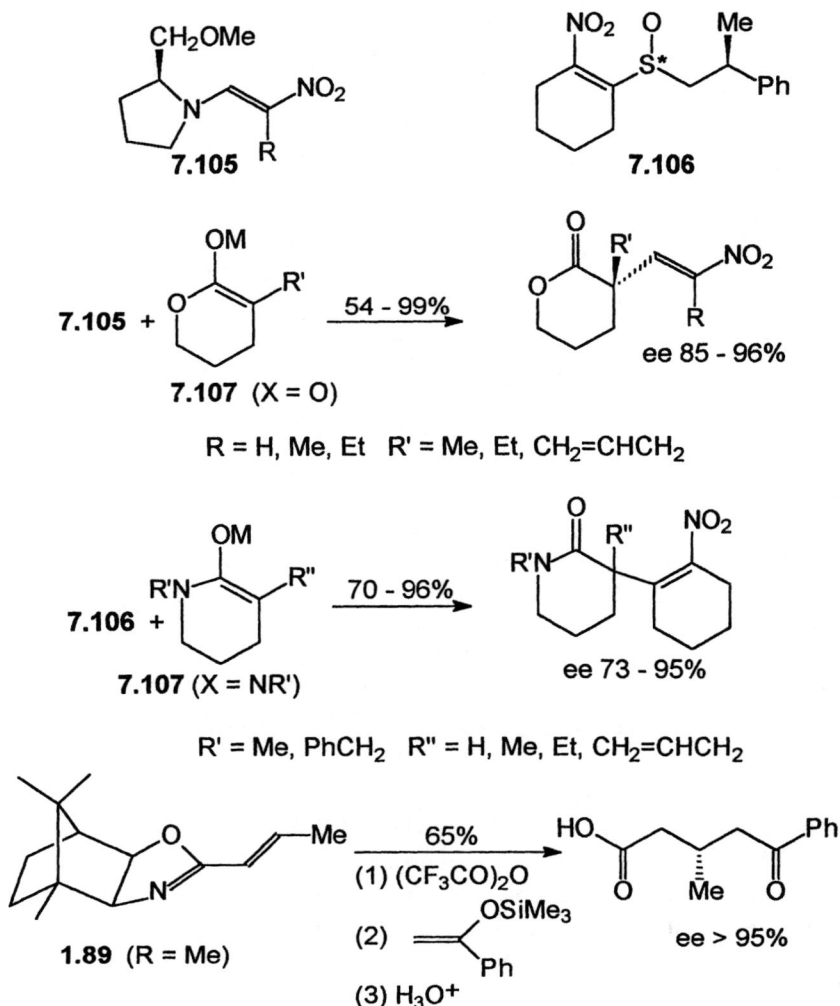

R = H, Me, Et   R' = Me, Et, $CH_2=CHCH_2$

R' = Me, $PhCH_2$ R" = H, Me, Et, $CH_2=CHCH_2$

Figure 7.70

7.15 ADDITIONS OF ENAMINES TO ELECTROPHILIC DOUBLE BONDS [162]

The nucleophilic properties of enamines uncovered by Stork have found a wide application in Michael additions. Secondary enamines are usually in equilibrium with the corresponding imines. These imines are generally more stable, unless the tautomeric enamine is stabilized by conjugation (Figure 7.71). The primary product of the reaction of an enamine with an α,β-unsaturated carbonyl compound is a dipolar intermediate 7.108. This intermediate is converted to a 1,5-dicarbonyl compound on exposure to aqueous acid. Proton transfers can take place before hydroysis to the ketone occurs, and the stereoselectivity of the process may be determined by such steps. Moreover, the enamine addition reaction can be reversible. These problems notwithstanding, the use of chiral amines to generate imines or enamines for use as Michael donors has been widely developed. The chiral imine/enamine can be preformed or, especially in the case of intramolecular reactions, the amine can be added to the reaction medium in stoichiometric amounts.

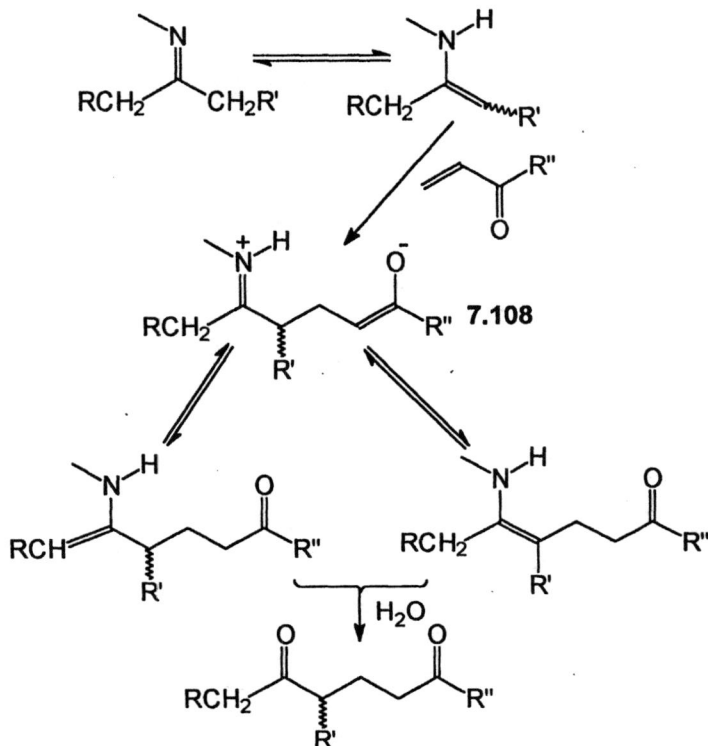

Figure 7.71

7.15.1 Tertiary Enamines [1008]

Most reactions of tertiary enamines have been performed with (S)-proline derivatives. The first examples of 1,4-additions to α,β-unsaturated aldehydes and ketones exhibited mediocre selectivities (ee < 60%) [173]. However, the reactions of two enamines of cyclohexanone, **7.109** and **7.110**, with methylacrylate lead to the expected adducts with a high enantiomeric excess but a poor chemical yield [173] (Figure 7.72). The addition of an (S)-prolinol-derived enamine to an α-trimethylsilyl-α,β-unsaturated ketone also gives useful selectivity [162]. Seebach and coworkers reacted enamine **7.110** (R = Me) with α,β-unsaturated gem-diesters or 2-aryl-1-nitroethylenes and obtained the expected 1,4-adducts with a high selectivity [162] (Figure 7.72). Martens and Lubben [294] proposed the use of enamine **7.111** for similar purposes, but the selectivity was not as high as with **7.110** (R = Me) (Figure 7.72).

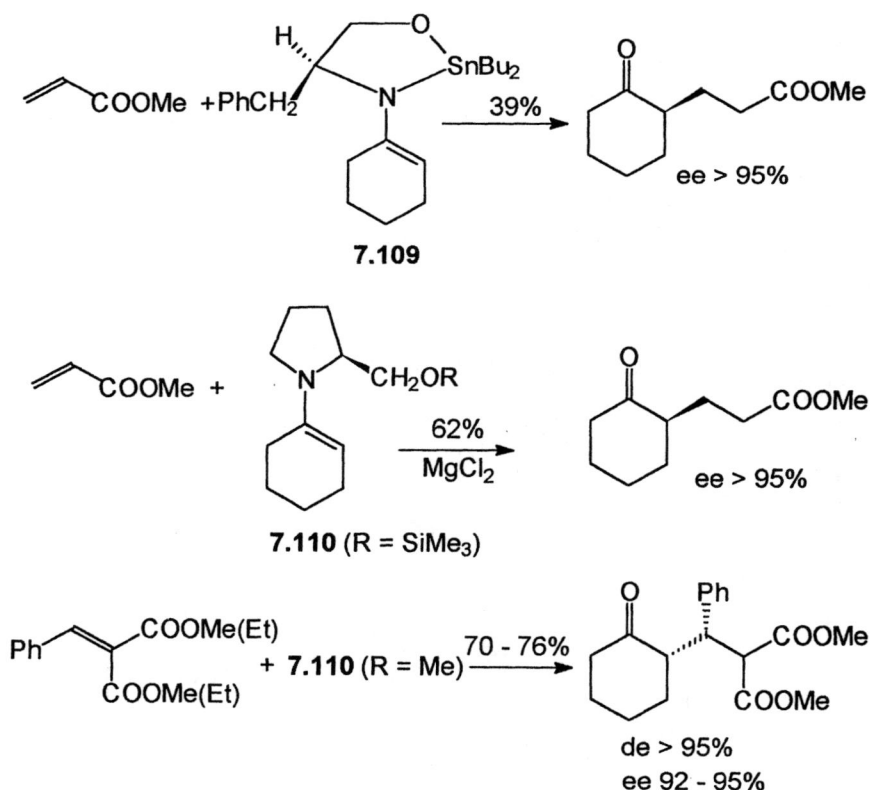

7.109

7.110 (R = SiMe₃)

de > 95%
ee 92 - 95%

Figure 7.72

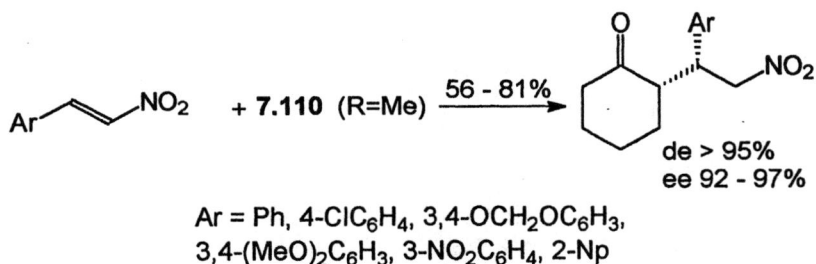

$$Ar\diagup\!\!\diagdown NO_2 \quad + 7.110 \ (R=Me) \xrightarrow{56 - 81\%}$$

de > 95%
ee 92 - 97%

Ar = Ph, 4-ClC$_6$H$_4$, 3,4-OCH$_2$OC$_6$H$_3$,
3,4-(MeO)$_2$C$_6$H$_3$, 3-NO$_2$C$_6$H$_4$, 2-Np

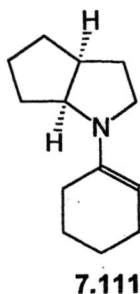

7.111

Figure 7.72 (continued)

7.15.2 Secondary Enamines [267]

1-Phenethylamine **1.56** (Ar = Ph) is the most popular auxiliary that is used to generate chiral enamines. Hirai and coworkers [1463] have performed the enantio-selective intramolecular cyclization of ketoesters **7.112** (n = 1,2) in the presence of (R)- or (S)-**1.56** (Ar = Ph), and they obtained functionalized pyrrolidines or piperidines with a high enantiomeric excess. In contrast, with (S)-proline enamines the ee is only about 34%. Important studies by Pfau, d'Angelo and coworkers [162, 267, 727] have shown that imines of 2-substituted cyclanones and (R)- or (S)-1-phenethylamine **1.56** (Ar = Ph) react with β-unsubstituted-α,β-unsaturated ketones, esters and sulfones. After hydrolysis, 2,2-disubstituted cyclanones are obtained with a high enantiomeric excess (Figure **7.73**). Methyl methacrylate and nitroethylene are polymerized under these reaction conditions. If the substituent of the cyclanone stabilizes the enamine by conjugation (COOR, for example), the re-action requires activation by high pressure or by a Lewis acid. Acrylonitrile is a good Michael acceptor under Lewis acid conditions (Figure **7.73**). Acceptors like CH$_2$=C(SPh)COOMe and CH$_2$=C(COOEt)$_2$ are also valuable substrates, but *tert*-Bu acrylate gives low chemical yields. The reactions of long chain vinylke-tones CH$_2$=CHCOR are less enantioselective.

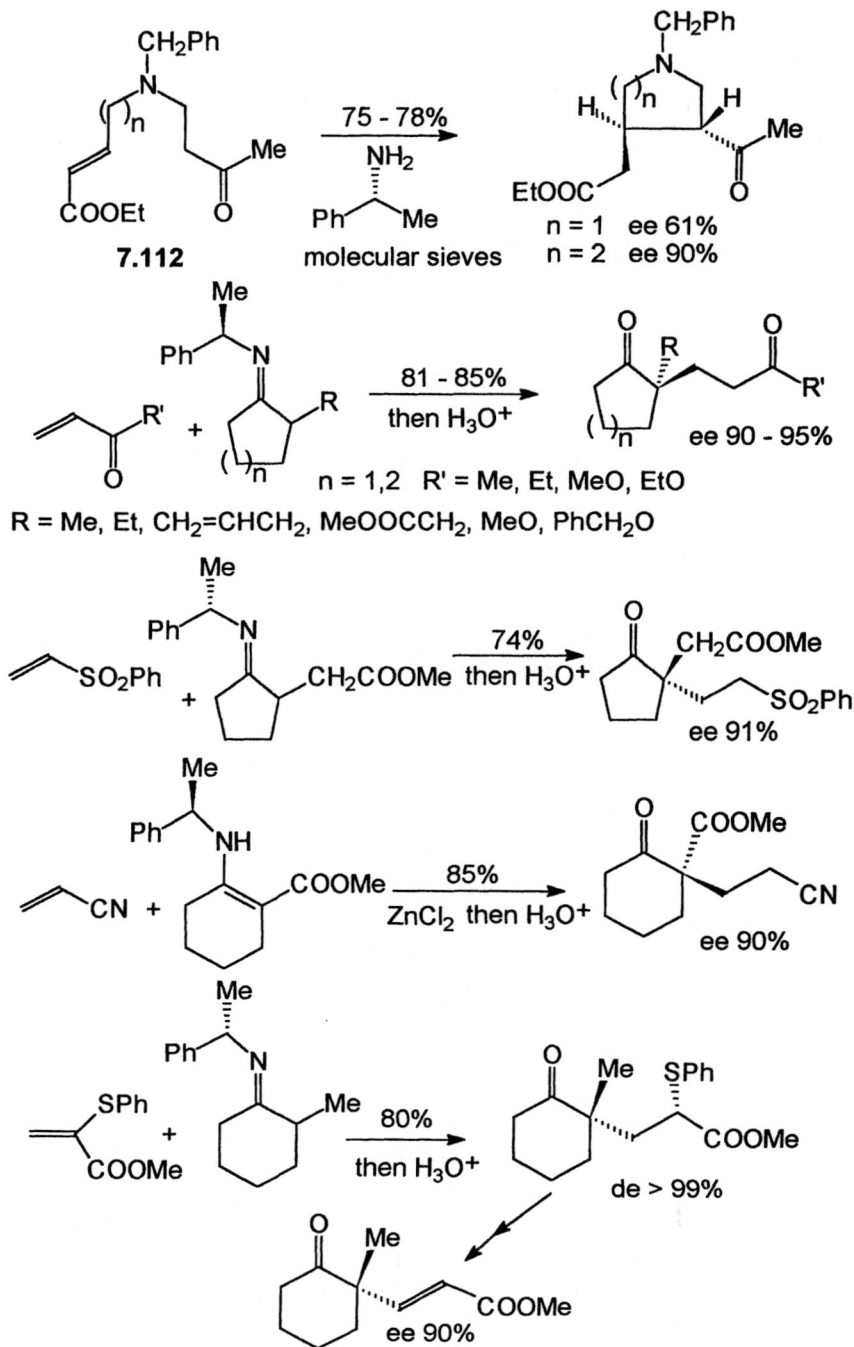

.Figure 7.73

Heteroatomic (O or S) or other substituents may be introduced on the cyclanone, although regioselectivity problems can arise in some cases. For these types of substrates, two regioisomeric enamines, E_1 and E_2, are in equilibrium. The preferred conformations of the enamines are those that minimize the effects of the A-strain. Addition to the Michael acceptor is performed in an apolar aprotic medium to minimize proton transfers. The equilibrium between the two primary dipolar adducts **7.108A** and **7.108B** is displaced by an intramolecular proton transfer. This transfer is easier in the case of **7.108A** due to proximity effects (Figure 7.74). The importance of this step is evidenced by the lack of regioselectivity of reactions performed in methanol. The facial selection is determined by the approach of the enamine to the Michael acceptor. In agreement with theoretical calculations, a chairlike transition-state model has been proposed. One of the faces of the enamine is hindered by the bulky phenyl group (Figure 7.74). This method has been applied to the synthesis of nonracemic natural products bearing quaternary stereocenters.

Recently, Gaidarova and Grishina [1464] have shown that the reactions of these enamines with methyl acrylate and acrylonitrile can be carried out on neutral alumina. Hydrolysis is done by elution with aqueous solvent.

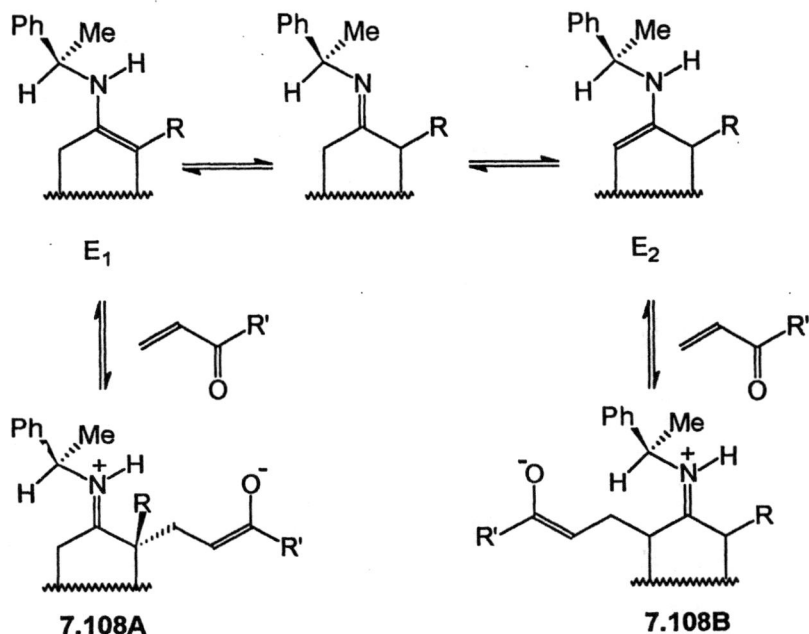

Figure 7.74

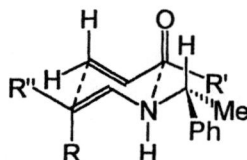

Figure 7.74 (continued)

7.16 CATALYZED ADDITIONS TO ELECTROPHILIC DOUBLE BONDS [161, 559, 1008, 1058]

When the Michael donors have a sufficiently low pKa (for example, malonates, β-ketoesters, phenylacetates), the Michael addition can be carried out in the presence of a catalytic amount of base. The conjugate additions of such donors to methyl vinylketone have been performed in the presence of chiral amines or transition metal complexes of (R,R)- or (S,S)-diamine **1.72** (R = Ph, R' = H). The most interesting results are those of Wynberg and coworkers, who performed the reaction of **7.113** with methyl vinylketone in the presence of quinine **3.1** (R = MeO) or quinidine **3.2** (R = MeO) and obtained either enantiomer of the Michael adduct. The reactions of other ketoesters or of nitromethane are less enantioselective (Figure 7.75). Asymmetric conjugate additions of diisopropylmalonate to E-enones RCH=CHCOR'(R = Me, n-C_5H_{11}, R' = Me, n-C_3H_7) take place with a moderate enantiomeric excess (74 - 77%) when catalyzed by a rubidium (S)-prolinate [1465]. With other (S)-proline salts or other Michael acceptors, the enantioselectivity is lower. Alkali-free lanthanum binaphtholate is an efficient asymmetric catalyst in conjugate additions of benzyl malonate to cycloalkenones at −20°C; methyl or ethyl malonates give lower enantiomeric excesses [1466] (Figure **7.75**).

Michael additions can also be performed under phase transfer conditions with an achiral base in the presence of a chiral quarternary salt as a phase transfer agent. Weinstock and coworkers [161, 1467] conducted the Michael addition of indanone **7.114** to methyl vinylketone under liquid-liquid conditions by using N-4-trifluorobenzylcinchoninium salt **5.46** (Figure 7.75). Loupy and coworkers [559, 602, 1468] have added diethyl N-acetylaminomalonate to chalcone at 60°C under solid-liquid conditions by using KOH and $(1R,2S)$-ephedrinium salts **3.3**. Michael adducts are obtained with a good enantiomeric excess if the R substituent of **3.3** is 4-MeOC$_6$H$_4$ or 1-Np (Figure 7.75). The presence of a free hydroxyl group on the cinchona derivatives is mandatory to observe a good asymmetric induction. As in the case of alkylation (§ 5.3.1), the interpretation of these results relies on the formation of a hydrogen bond with the carbonyl of the α,β-unsaturated ketone, promoting its activation. The least hindered complex is favored. Cram and Sogah [161, 883] performed Michael reactions between **7.113** and methyl vinylketone and between methyl phenylacetate and methyl acrylate in the presence of catalytic amounts of KO-*tert*-Bu or KNH$_2$ and chiral crown ether **3.5**. At −78°C, the corresponding adducts were obtained with high selectivities. Other chiral crown ethers did not give as useful results (Figure **7.75**).

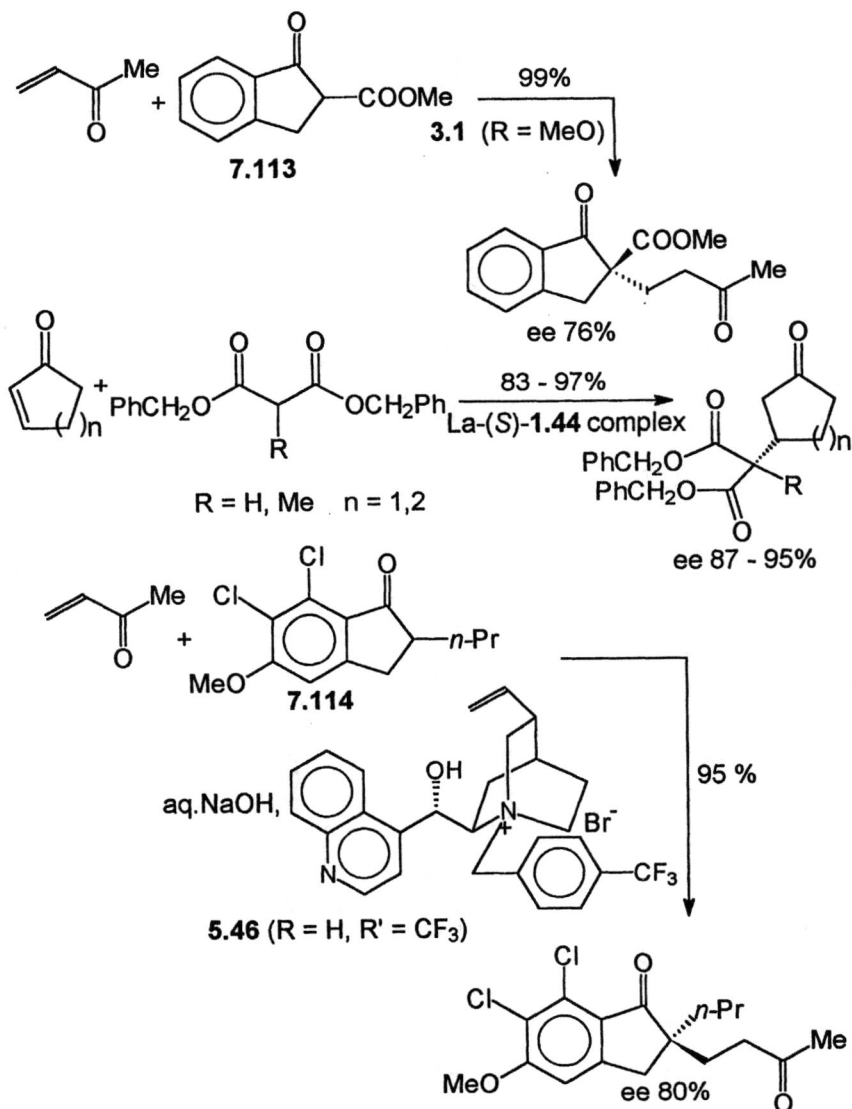

Figure 7.75

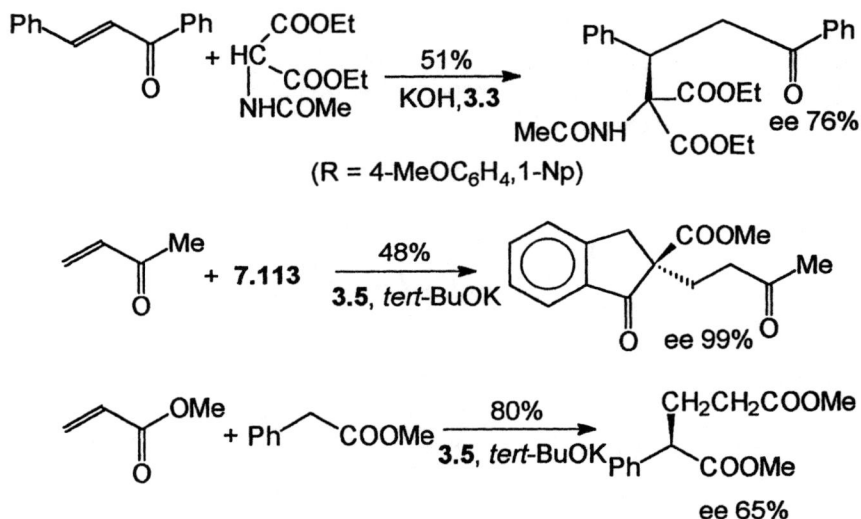

$(R = 4\text{-MeOC}_6\text{H}_4, 1\text{-Np})$

Figure 7.75 (continued)

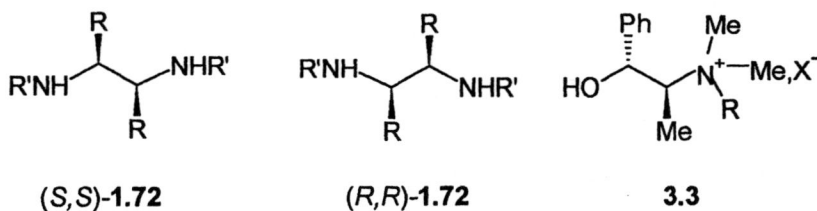

(S,S)-**1.72** (R,R)-**1.72**   **3.3**

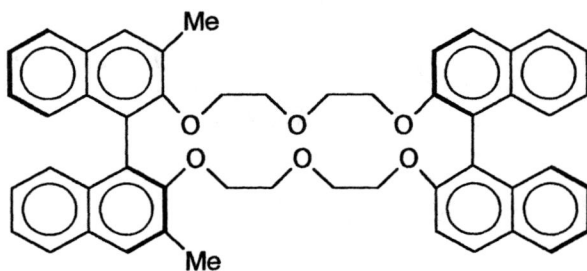

3.5

Belokon and coworkers introduced chiral nickel complexes formed from imine **1.109**, and these are good Michael donors [383, 861]. In the presence of catalytic amounts of base and under thermodynamic control, these complexes give 1,4-adducts with α,β-unsaturated ketones, -esters and acrylonitrile with a good selectivity (Figure 7.76). Precursors of substituted (S)-prolines **7.115** and (S)-glutamic acids **7.116** are formed.

A transition metal-catalyzed Michael addition has been described by Ito and coworkers [879]. Asymmetric conjugate additions of alkyl 2-cyanopropionate to vinylketones or acrolein take place with high enantioselectivity when catalyzed by a rhodium complex bearing a bis-ferrocenylphosphine ligand **3.42** (Figure 7.76). The use of other complexes is less interesting. An asymmetric 1,4-disilylation of α,β-unsaturated ketones with $PhCl_2SiSiMe_3$ under binap **3.43**-palladium catalysis has been recently published [903].

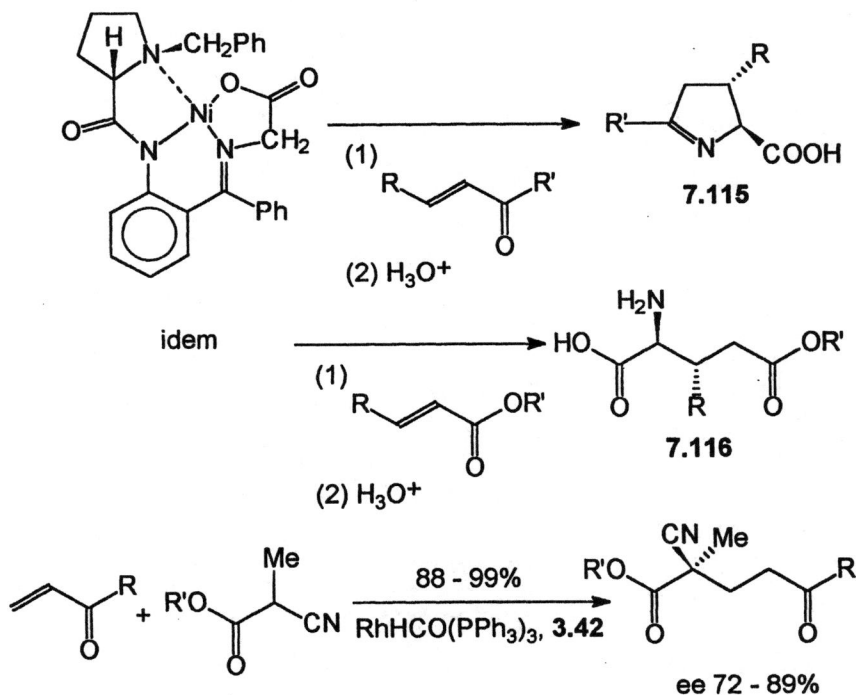

R = H, Me, Et, Ph, 4-MeOC$_6$H$_4$, 4-ClC$_6$H$_4$ R' = Me, Et, i-Pr, $tert$-Bu

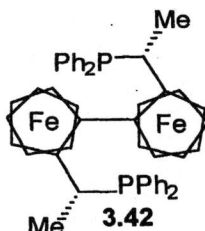

Figure 7.76

7.17 RADICAL ADDITIONS [276]

A decade ago, radical reactions were thought to be of little use in synthesis due to lack of selectivity. Much progress has recently been made in this domain, and it has even become possible to control the stereoselectivity in many radical reactions [1469]. Curran, Giese, Porter and their coworkers initiated the study of asymmetric radical addition reactions by introducing chiral residues either on the radical precursor or on the alkene.

Transition states of radical reactions are early on the reaction coordinate, and stereoselectivities depend mainly upon steric interactions. Therefore, chiral auxiliaries that have two highly differentiated faces must be selected. In most cases, the reactive conformations will be close to the favored ground-state conformations [454] (§ I.2, I.3).

7.17.1 Radical Precursors Bearing Chiral Residues

Additions to double bonds of radicals generated in the α-position of esters of chiral alcohols are usually poorly selective. An exception is the intramolecular cyclization of 7.117 induced by Mn (III) salts [1470, 1471] (Figure 7.77). Among the most efficient chiral auxiliaries in radical additions are 2,5-disubstituted pyrrolidines 1.65 bearing a C_2 axis of symmetry, Oppolzer's sultams 1.133 [454], imide 1.130 designed by Curran and Rebek [1472], and oxazolidines 1.85. These auxiliaries are transformed into chiral α-thiohydroxamates 7.118 or α-bromo- or -iodo-N-acylderivatives 7.119, 7.120, 7.121 and 7.122. The radicals generated from precursors 7.119, 7.120 or 7.121 with AIBN or Et_3B react with allylstannanes [276, 1472] through a stereoselective addition to the double bond followed by cleavage of the R'_3Sn radical. Precursors of α-branched acids are obtained with a high selectivity. Yields are generally good, except for 7.121, and the chiral auxiliaries are easily removed (Figure 7.77). Asymmetric cyclizations of iodosultam 7.120 (R = $Me_3SiC{\equiv}C(CH_2)_3$) and of related compounds [276, 454, 1473] are initiated by $Bu_3SnSnBu_3$ in the presence of Bu_3SnH, and these are also highly selective. The radicals generated from 7.119 (R = Me) or 7.118 bearing the same chiral residue add stereoselectivity to ethyl acrylate, but mixtures of mono- and diadducts are formed. The interpretation of these facial selectivities relies upon the attack of a planar radical species in the conformation in which the A(1,3) and dipolar interactions are minimized (§ I.2) (Figure 7.77). Oppolzer's sultam derivatives react on the face opposite to the "axial-like" S–O bond. In the pyrrolidine-derived radicals, the main repulsive interaction involves the C–R bond that is the closest to the radical center [454] (Figure 7.77). Chiral 1,3-dioxolane-4-ones have also been recommended as auxiliaries, but only a moderate facial selectivity was observed [1474].

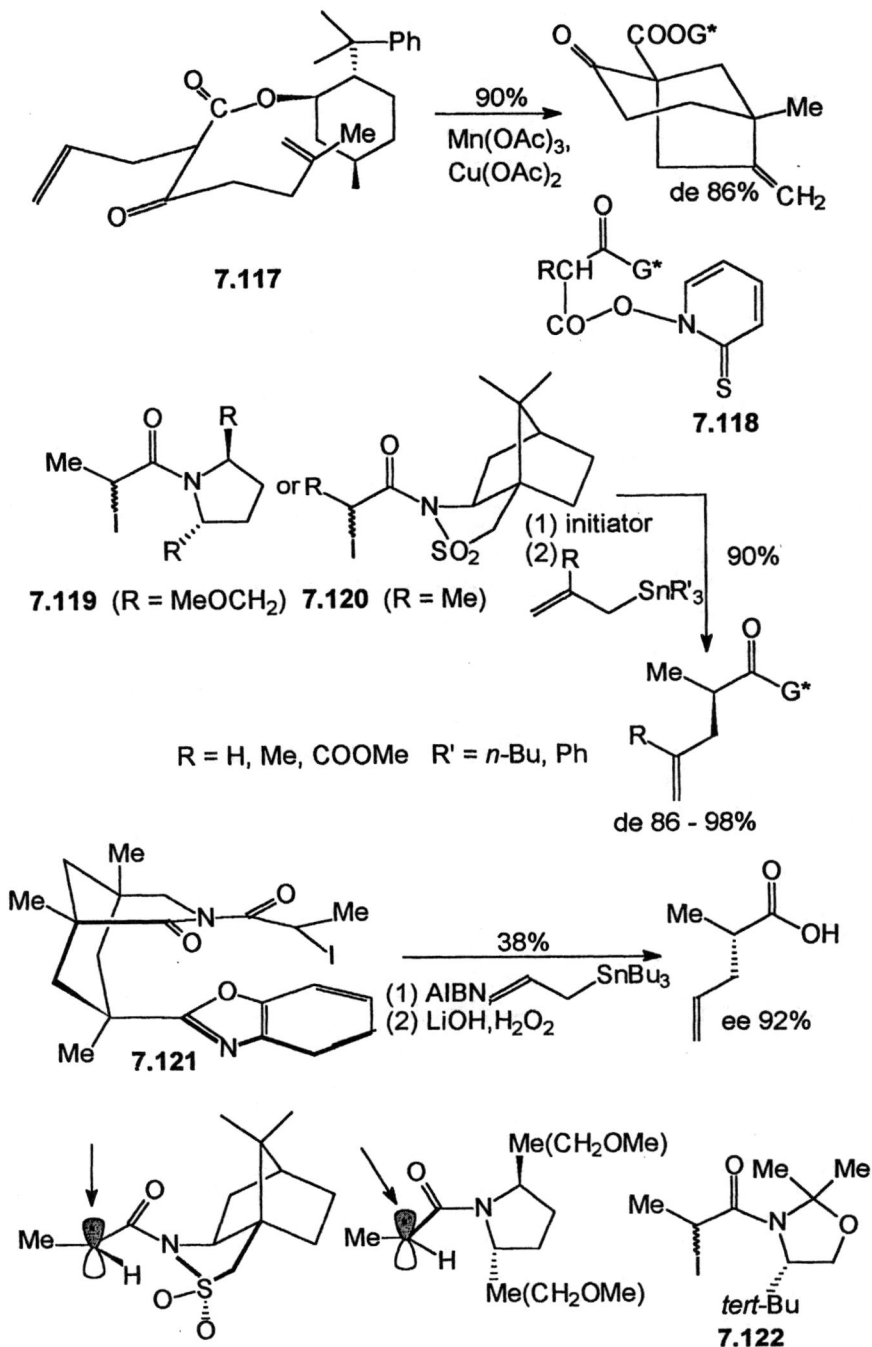

Figure 7.77

7.17.2 Reactions with Alkenes Bearing Chiral Residues

The preceding chiral auxiliaries, as well as pyrrolidine **1.66** [279], can also be used as radical acceptors. α,β-Unsaturated mono- or symmetrical bis-amides adopt an *s-cis* conformation due mainly to A(1,3) strain and dipolar interactions [279, 454]. If the chiral functionality is located on the carbon that undergoes the radical addition, the reactions are highly stereoselective. For example, the additions of alkyl radicals generated from organomercury reagents or from thiohydroxamates to symmetrical bis-amides **7.123** at 0°C are highly stereoselective. The selectivity is even higher when the chiral auxiliary is **1.66** [279]. Radical additions to diesters **7.124** and **7.125** are both regio- and stereoselective. Reagent **7.125** is especially useful because the chiral residue is easily cleaved [276, 277, 1475] (Figure **7.78**). If the double bonds are 1,2-disubstituted by electron-withdrawing groups and bear a single chiral residue, the radical additions are no longer regioselective. Only the product resulting from the addition to the α-position to the chiral functionality is formed stereoselectively [276, 277]. In order to circumvent this problem, Curran and coworkers used bulky α,β-unsaturated esters **7.126** derived from **1.130** [445, 1472]. Secondary and tertiary radicals generated from organomercurials add regio- and stereoselectively to fumaric acid derivatives **7.126**. The chiral auxiliary is easily removed by LiOH/H_2O_2 (Figure **7.78**). The use of Oppolzer's sultam as auxiliary, or the reactions of primary radicals result in a decrease in facial selectivity. Giese and coworkers [1476] performed similar radical additions to acrylamides **7.124** (R = H) and related methacrylamides. Good stereoselectivities were observed in both cases with *tert*-Bu radical. In methacrylamides, X-ray crystallography shows an important twisting of the double bond which decreases the rate of radical addition but does not negate the face discrimination.

Another application of asymmetric radical additions, proposed by Giese, Porter and coworkers [276, 334, 1475], is the addition of c-C_6H_{11} or *tert*-Bu radicals to chiral acrylamides **7.124** (R = H) and **7.125** (R = H) followed by trapping of the initial radical adduct by an acceptor such as a thiopyridone or allyltributylstannane. These reactions are carried out between −35 and + 80°C, and they are highly diastereoselective (Figure **7.78**). The adduct radicals are trapped on the least hindered face of the conformation in which A(1,3) strain and dipolar interactions are minimized [454].

Radical cyclizations of α-iodoamides derived from norephedrine **7.126** have been described by Gennari and coworkers [1477, 1478]. After appropriate treatment, 3,4-disubstituted pyrrolidines are obtained with a high selectivity (Figure **7.79**). 4-Substituted-1,2,3,4-tetrahydroquinolines can be prepared in a similar fashion, albeit with a lower selectivity (70%).

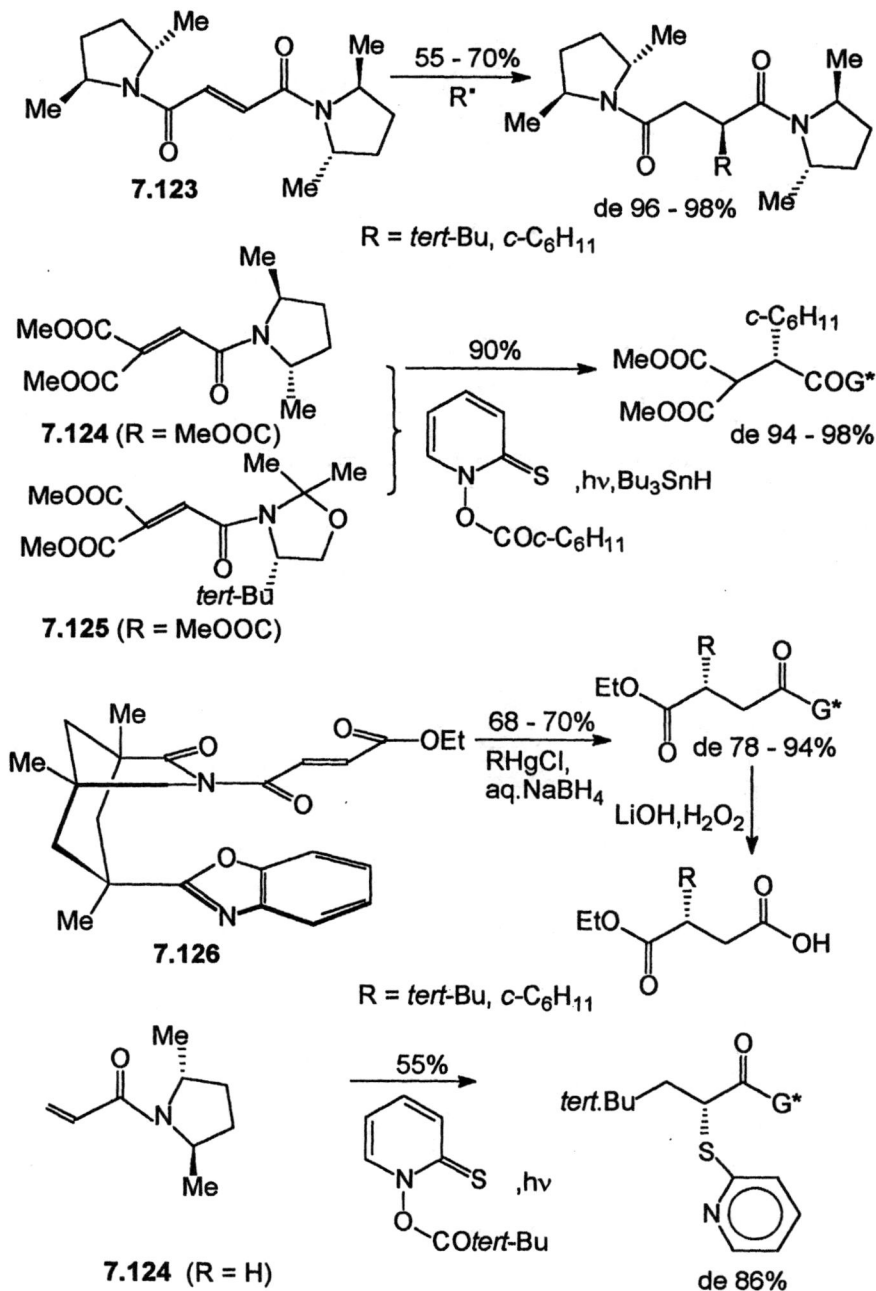

Figure 7.78

7.125 (R = H)

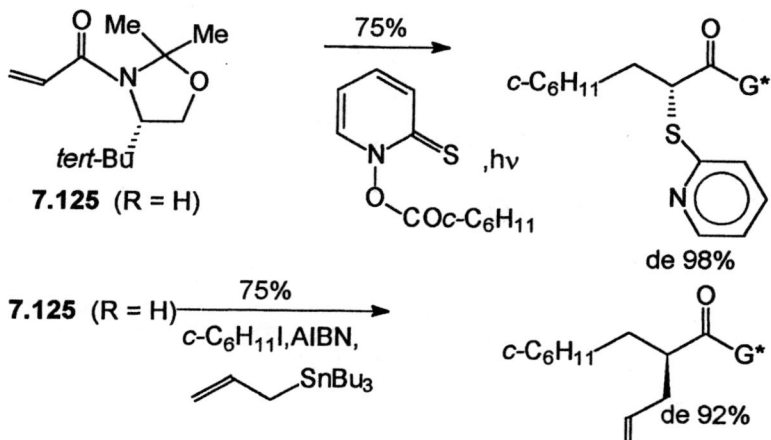

Figure 7.78 (continued)

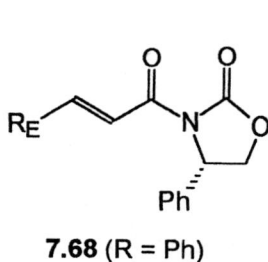

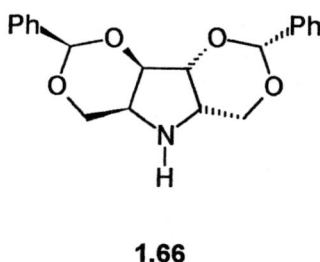

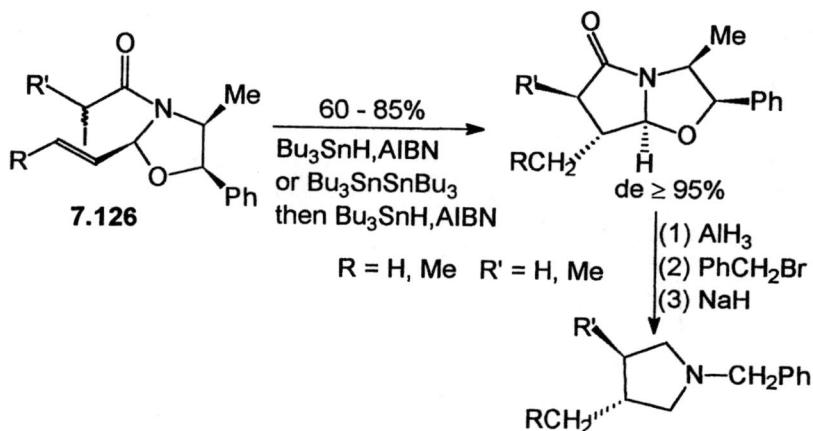

7.68 (R = Ph)   **1.66**

Figure 7.79

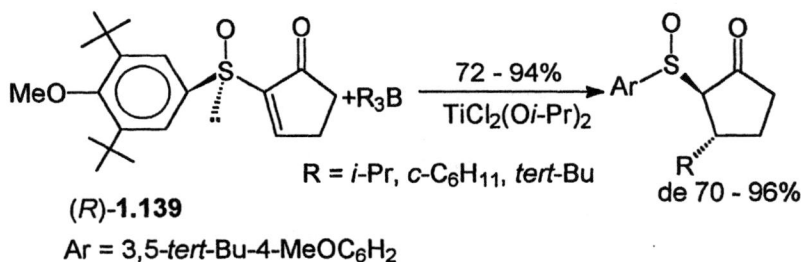

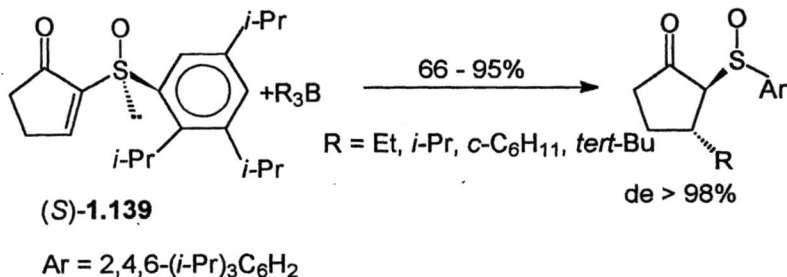

Figure 7.79 (continued)

High β-stereoselectivity has been achieved in asymmetric radical addition reactions to chiral α-sulfinylcyclopentanones **1.139** (X = CH$_2$, n = 1) [491]. Provided that the aryl substituent of the sulfur atom is bulky enough (Ar = 3,5-*tert*-Bu$_2$-4-MeOC$_6$H$_2$ or 2,4,6-*i*-Pr$_3$C$_6$H$_2$), the addition of Et, *c*-C$_6$H$_{11}$, *i*-Pr and *tert*-Bu radicals generated from alkylboranes can be highly stereoselective. When Ar = 3,5-*tert*-Bu$_2$-4-MeOC$_6$H$_2$, catalysis by TiCl$_2$(O*i*-Pr)$_2$ at 0°C is required, while in the *i*-Pr case, the thermal reaction gives useful results (Figure **7.79**). The cleavage of the chiral auxiliary has not been carried out. Electroreductive hydrocoupling of chiral E-*N*-cinnamoyl-2-oxazolidinones **7.68** (R$_E$ = Ph, R = *i*-Pr) has been performed, but low selectivities were observed [1479].

7.18 HALOLACTONIZATION, ALLYLIC SUBSTITUTION

7.18.1 Halolactonization [1480]

The first asymmetric halolactonization reactions were performed with α,β-unsaturated amides of (*S*)-proline **1.64** (R = COOH) [261]. Reactions of amides with NBS give functionalized bromolactones **7.127**. After Bu$_3$SnH treatment and hydrolysis, chiral α-hydroxyacids are obtained with a high enantiomeric excess. The methyl substituent on the α-position of the double bond is required for high selectivity (Figure **7.80**).

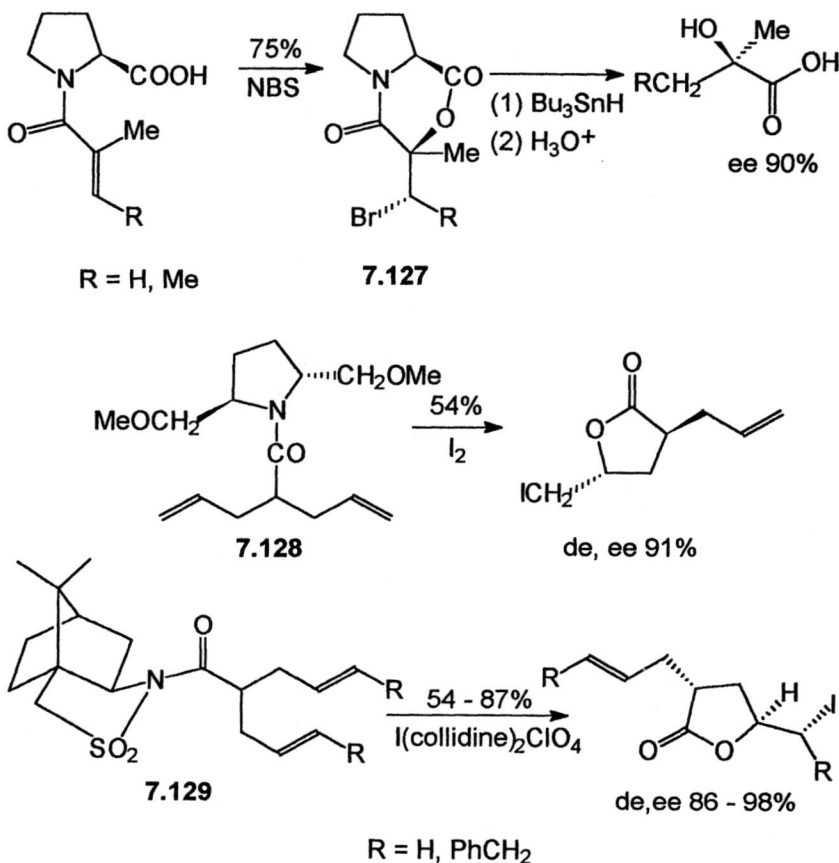

Figure 7.80

Other chiral amides have been proposed to improve this reaction. Fuji and coworkers [1481] used amides of pyrrolidines **1.65** bearing a C_2 axis of symmetry. A five-membered lactone is obtained with a high selectivity from I_2 and **7.128** (Figure **7.80**). Shibuya and coworkers [1482] showed that iodolactonization of chiral sultam **7.129** takes place with high selectivity and chemical yield at $-40°C$; when $R = Me$ or i-Pr, the selectivity is lower (Figure **7.80**).

7.18.2 Allylic Substitution

The reactions of organocuprates with allylic halides and esters have been widely studied [624a]. The introduction of chiral leaving groups on the electrophile allows useful levels of asymmetric induction. Gais and coworkers [519] have used chiral sulfoximines **1.150**. Isomerization of these sulfoximines, followed by reaction with organocopper reagents in the presence of BF_3 at $-78°C$, leads to 2-alkylmethylenecyclopentanes with a good regio- and enantioselectivity (Figure **7.81**). Denmark and Marble [1483] used chiral carbamates as leaving groups. Reaction of **7.130** with organocuprates at 0°C gave the expected alkenes with an excellent enantiomeric excess (Figure **7.81**).

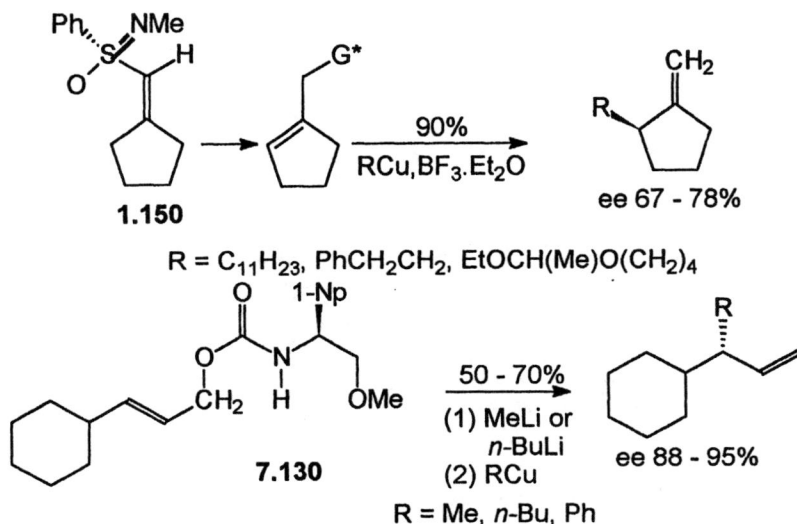

Figure 7.81

7.19 CYCLOPROPANATION

Due to the industrial importance of cyclopropane pesticides [811, 812, 853], the development of efficient asymmetric syntheses of cyclopropanes is an important goal. Among the methods used to prepare cyclopropanes from olefins, the Simmons-Smith (CH_2I_2, Zn-Cu or Zn-Ag couple) and related reactions (CH_2I_2, Et_2Zn) and the reactions of diazoesters catalyzed by copper or rhodium complexes [624f] have received many applications. The generation of cyclopropanes through pyrolysis of pyrazolines obtained by 1,3-dipolar cycloadditions will be examined later (§ 9.2).

7.19.1 Simmons-Smith and Related Reactions

The Simmons-Smith reaction takes place via an intermediate organozinc species, which can be generated in the presence of chiral ligands for zinc. Alternatively, standard (achiral) zinc reagents can be used to cyclopropanate an alkene bearing a chiral residue. The reactions of allylic alcohols are rapid, and the formation of a zinc alcoholate directs stereochemistry [624f]. Simmons-Smith reactions are also faster when ethers or acetals able to chelate Zn are in the vicinity of the reacting double bond. Transition states can thus be rigidified so that asymmetry is more easily induced.

7.19.1.1 Reactions in the Presence of Chiral Ligands

The cyclopropanations of allyl alcohols with CH_2I_2/Et_2Zn in the presence of diethyl tartrates **2.69** (R = Et) [660] or (1R,2S)-N-methylephedrine **1.14** (R = Me) [1484] do not give useful results. The reactions of silicon-substituted allylic alcohols with CH_2I_2/Et_2Zn in the presence of (R,R)-diethyl tartrate (1 equiv) are more enantioselective at −30°C (up to 90% ee) [1485]. However, the addition of (R,R)-N,N'-tetramethyltartaric acid diamide butylboronate **7.131** in stoichiometric amounts promotes the highly enantioselective cyclopropanation of allyl alcohols with preformed $Zn(CH_2I)_2$ at 0°C [1486] (Figure **7.82**). Bis-sulfonamide **7.132** is also an efficient catalyst in this reaction [1487] (Figure **7.82**). If the allyl alcohol is transformed into an ether, no asymmetric induction occurs in either case.

7.19.1.2 Reactions with Alkenes Bearing Chiral Residues

Charette and coworkers [248, 366, 1488, 1488a] transformed allylic alcohols into glucopyranoside acetals **7.133** and **7.134** and carried out the asymme-tric cyclopropanations of these substrates with a large excess of CH_2I_2/Et_2Zn at −30°C. The β- and α-anomers give rise to enantiomeric cyclopropanemethanols after transformation of the alcohol into triflate and heating at 160°C [1489] (Figure **7.82**). In this process, the α-anomer **7.134** can be replaced by **7.135**, which leads to the same enantiomer. If the alcohol functionality is not free, the reaction is no longer stereoselective. This suggests the intermediacy of a zinc alcoholate. Monoallylethers of chiral 1,2-cyclohexanediols also undergo highly selective cyclo-propanation with ICH_2Cl/Et_2Zn at −20°C [214] (Figure **7.82**). The chiral auxil-iary is removed by iodination and subsequent n-BuLi-induced β-elimination. For these substrates, a smaller amount of the organozinc reagent is required (3 equiv instead of 10).

Cyclic chiral enol ethers **7.136** generated from acetals of chiral diols react selectively with the Simmons-Smith reagent or with CH_2I_2/Et_2Zn. After removal of the chiral auxiliary by oxidation and base treatment, bicyclic cyclopropanols **7.137** are obtained with an excellent enantiomeric excess [213, 1490, 1491, 1492] (Figure **7.83**). The methylene group is introduced on the face of the double bond opposite to that of the i-Pr group next to the OH of the chiral auxiliary.

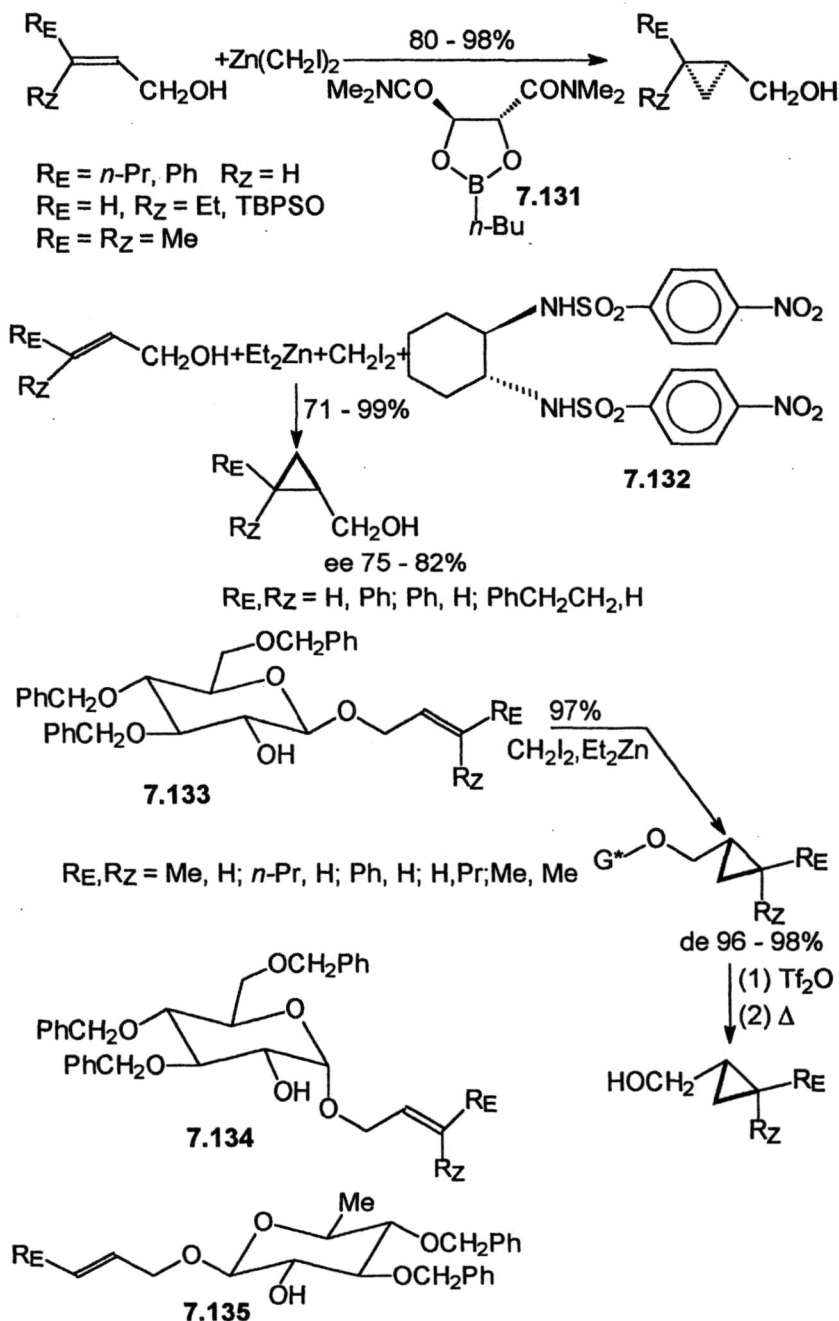

R_E = n-Pr, Ph R_Z = H
R_E = H, R_Z = Et, TBPSO
R_E = R_Z = Me

7.131

7.132

ee 75 - 82%
R_E,R_Z = H, Ph; Ph, H; PhCH$_2$CH$_2$,H

7.133

R_E,R_Z = Me, H; n-Pr, H; Ph, H; H,Pr;Me, Me

de 96 - 98%

(1) Tf$_2$O
(2) Δ

7.134

7.135

Figure 7.82

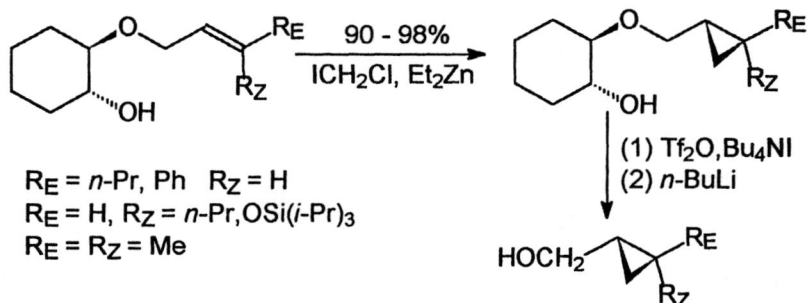

Figure 7.82 (continued)

Chiral acetals of α,β-unsaturated aldehydes derived from tartaric esters **2.69** (R = Et or *i*-Pr) or from (2S,4S)-2,4-pentanediol **1.37** (R = R' = Me) have been used by Yamamoto and coworkers [213, 1008] to prepare chiral cyclopropanecarboxaldehydes with CH_2I_2/Et_2Zn. The best selectivities are obtained when the auxiliaries are (R,R)- or (S,S)-dialkyl tartrates (Figure **7.83**). The use of a pre-formed $ClCH_2ZnI$ reagent decreases the selectivity somewhat [219]. Mash and coworkers [213, 1008] used chiral 2,3-butanediols **1.35** (R = Ph or $PhCH_2OCH_2$) to form ketals of α,β-unsaturated ketones. Asymmetric cyclopropanation of these ketals with $CH_2I_2/Zn/Cu$ or better yet $ClCH_2ZnI$ [219] gives useful results only with 2-cycloalkenones (Figure **7.83**). These authors consider that the zinc reagent coordinates to the lone pair of the pseudoequatorial oxygen of the dioxolane **7.138**, and suggest that the methylene residue is transferred to the face opposite the R substituent located on the adjacent carbon (Figure **7.83**).

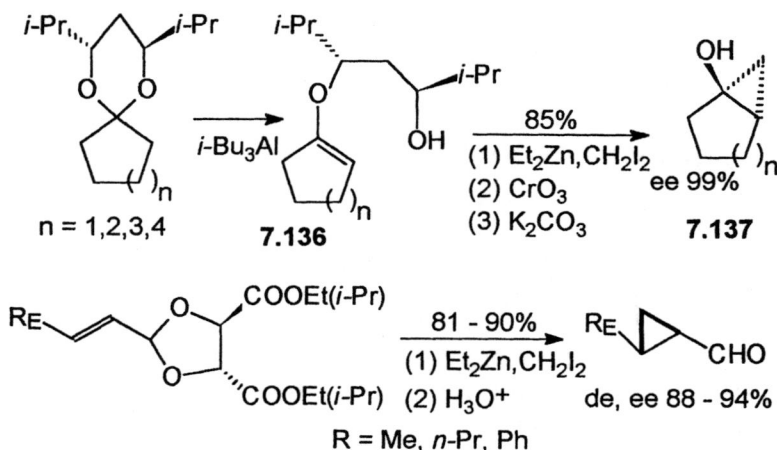

Figure 7.83

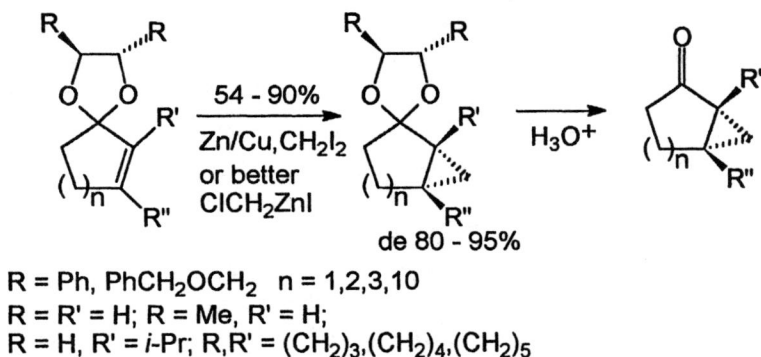

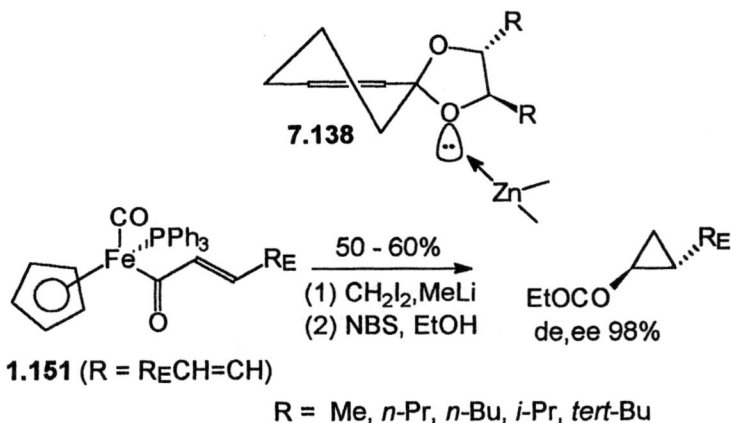

Figure 7.83 (continued)

Johnson and coworkers [1493] have also obtained nonracemic bicyclo-[4.1.0.]-2-heptanones by Simmons-Smith reactions with chiral sulfoximines. The drawback of this methodology is the purification of the starting chiral sulfoximines. Davies and coworkers reacted α,β-unsaturated acylcomplexes of cyclopentadienyl-iron carbonyl **1.148** (R = E-RCH=CH) with $CH_2I_2/MeLi$. After appropriate treatment, nonracemic ethyl *trans*-cyclopropanecarboxylates are obtained with a high selectivity [522] (Figure **7.83**). As usual, the reaction occurs on the most accessible face of the complex (§ 7.10.2.1).

7.19.2 Reactions of Sulfur Ylides

Meyers and coworkers [327, 1494] have applied the Corey-Chaykovsky re-action to chiral bicyclic lactams **1.92**. The face selectivity of the reaction of $CH_2=S(O)Me_2$ with this Michael acceptor depends upon the nature of the angular substituent R'. When R' = H, the cyclopropane is introduced selectively on the *exo* face, while an *endo* attack takes place when R' = Me (Figure **7.84**). This methodology has been applied to the synthesis of nonracemic cyclopropane pesticides and natural products [327]. However, the reaction of such sulfur ylides with butadieneironcarbonyl complexes is not stereoselective [528].

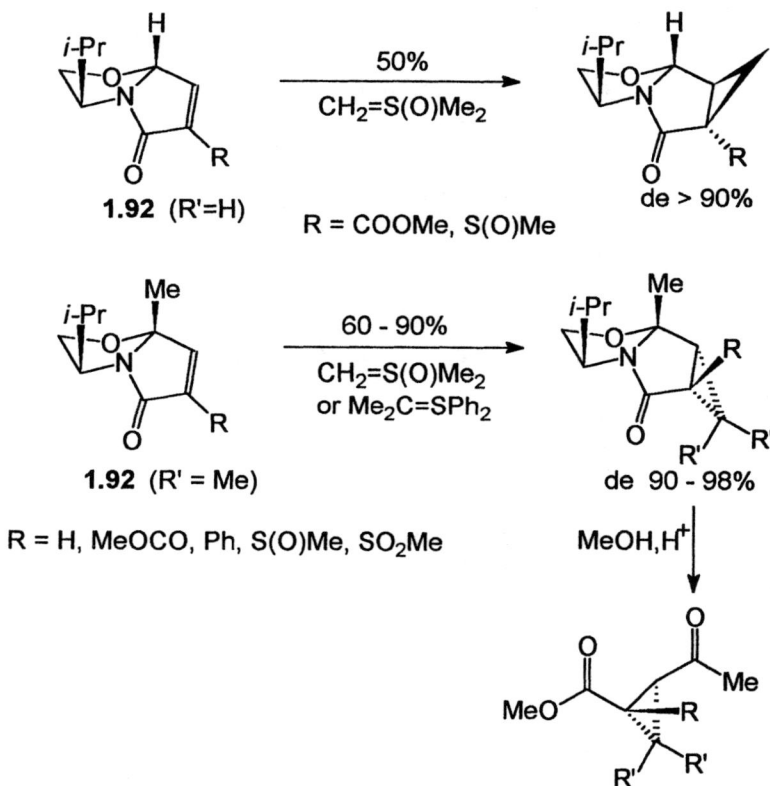

Figure 7.84

7.19.3 Reactions Catalyzed by Transition Metal Complexes
[752, 936, 937, 953]

Cobalt, copper and rhodium complexes catalyze the reactions of alkenes with diazoesters to give alkyl cyclopropanecarboxylates. The mechanism of this reaction involves a metallocarbene intermediate generated by nitrogen extrusion from the diazoester (Figure **7.85**). The stereoselection takes place through the interaction of the metallocarbene with the alkene, even though this is not the rate determining step [936, 937].

The first chiral ligands of copper(II) used in the reaction of styrene with ethyl diazoacetate were salicylaldimines **3.68**, but very poor selectivities were observed. Aratani and coworkers [170, 752, 937, 953] have improved the efficiency of these ligands by introducing bulky substituents on the aromatic rings. Indeed, the reaction of isobutene with ethyl diazoacetate under catalysis of a copper complex formed from (R)-**3.69** (R = Me, R' = tert-Bu, R" = n-C_8H_{17}) takes place with a high enantiomeric excess. This reaction is used in the industrial synthesis of cyclopropanes [811, 812, 936] (Figure **7.86**). Unsymmetrical alkenes are precursors of E- and Z-cyclopropanes, and reactions of these alkenes with alkyl diazoesters lead to mixtures of diastereoisomers unless bulky esters are used (Figure **7.86**).

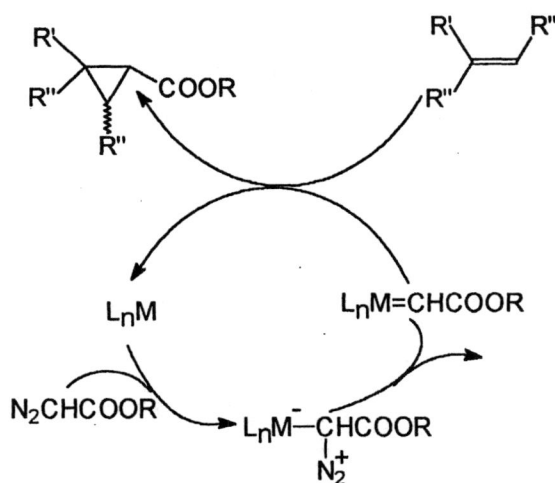

Figure 7.85

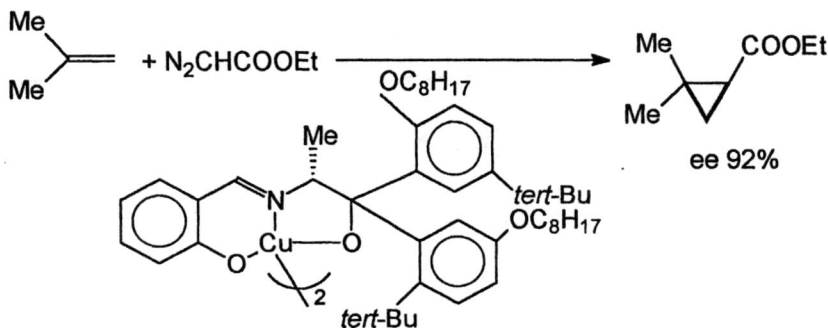

Figure 7.86

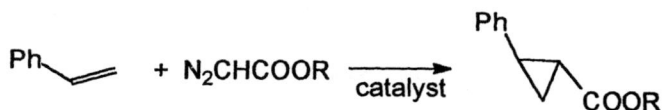

R	catalyst	Y%	de%	ee%
(1S,2R,5R)-menthyl	(S)-**3.69**-Cu(II)		64	81 (1R,2R)
(1S,2R,5R)-menthyl	**2.11**-Cu(II)	70	64	97 (1S,2S)
tert-Bu	**3.66**-Cu(I) (R = CMe₂OSiMe₃)	87	72	96 (1S,2S)
BHT	**3.28**-Cu(I) (R = tert-Bu, R' = R" = H)	80	88	99 (1R,2R)
BHT	**3.28**-Cu(I) (R = tert-Bu, R' = H, R" = Me)	85	88	99 (1R,2R)

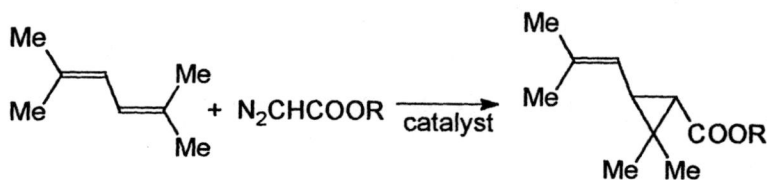

R	catalyst	Y%	de%	ee%
(1S,2R,5R)-menthyl	(R)-**3.69**-Cu(II)	64	86	90 (1S,2S)
(c-C₆H₁₁)₂CH	(R)-**3.71**-Cu(II)	78	90	94 (1R,2R)

Figure 7.86 (continued)

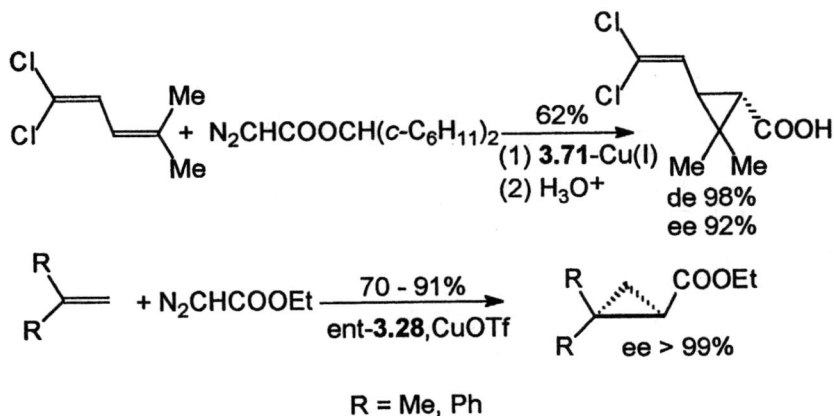

Figure 7.86 (continued)

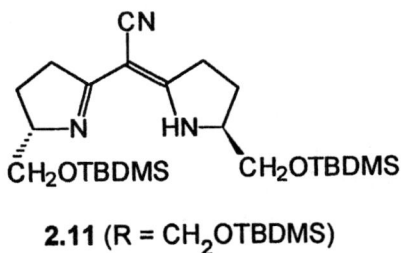

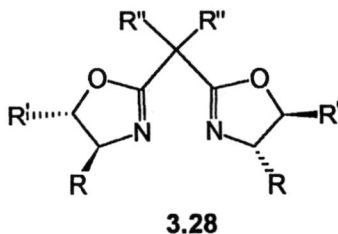

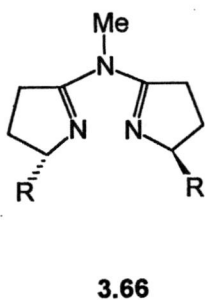

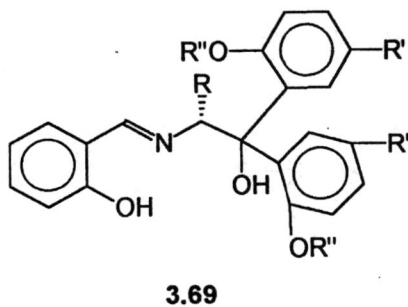

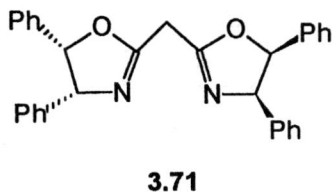

2.11 (R = CH₂OTBDMS)

3.28

3.66

3.69

3.71

The same problem of *cis/trans* selectivity arises when these reactions are catalyzed by copper (I) or copper (II) complexes bearing hemicorrins **2.11** or **3.66** [929, 930, 936, 946] or bis-4,5-oxazolines **3.28** (R" = Me or H, R' = H) or **3.71** as ligands [930, 963, 964]. Pfaltz, Evans, Masamune and their coworkers have used *tert*-Bu, menthyl, 2,6-di-*tert*-Bu-4-methylphenyl (BHT) or dicyclohexylmethyl diazoacetates. These esters react with alkenes such as styrene to provide mainly E-cyclopropanecarboxylates, which are obtained with a high enantiomeric excess (Figure 7.86). The interpretation of the observed selectivities involves the approach of the reactants to maximize overlap between the *p* orbitals of the metallocarbene and the π orbitals of the alkene with concomitant minimization of steric interactions [936]. Dicyclohexylmethyl esters are easily hydrolyzed [964], but BHT esters must be reduced to alcohols with LiAlH$_4$ [963]. Limitations of the use of these copper complexes are the poor selectivities obtained in intramolecular cyclopropanations [936, 1495] and in reactions of silylenolethers [1496]. The use of **3.71** as copper ligand in the cyclopropanation of styrene with bulky diazoacetates also gives a low enantiomeric excess [964]. Furthermore, the diastereo- and enantioselective synthesis of Z-substituted cyclopropanecarboxylic esters by these methods remains an unsolved problem [936].

Dimeric rhodium complexes are also efficient catalysts in these reactions. Doyle and coworkers introduced chiral dirhodium tetrakiscarboxamides as catalysts. These complexes are obtained by ligand exchange with Rh$_2$(OAc)$_4$. Among the various ligands tested, methyl (*R*)- and (*S*)-2-pyrrolidinone-5-carboxylates **3.61** are the most efficient [937, 938, 941]. The intermolecular cyclopropanation reactions of alkenes with alkyl diazoacetates in the presence of these chiral rhodium complexes are somewhat less selective than the chiral copper-catalyzed reactions (see above) [941]. On the other hand, the intramolecular reactions give excellent selectivities. Diazoesters of allyl alcohols **7.139** (n = 1) or homoallyl alcohols **7.139** (n = 2) give *cis*-fused bicyclic lactones with a high selectivity in reactions catalyzed by a rhodium-complex-bearing ligand **3.61** [936, 937, 1497, 1498, 1499] (Figure 7.87). This method has been extended to homoallylic diazoacetamides [1500]. The availability of both (*R*)- and (*S*)-mepy **3.61** allows the access to both enantiomers.

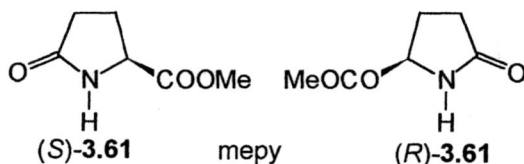

(S)-**3.61**   mepy   (R)-**3.61**

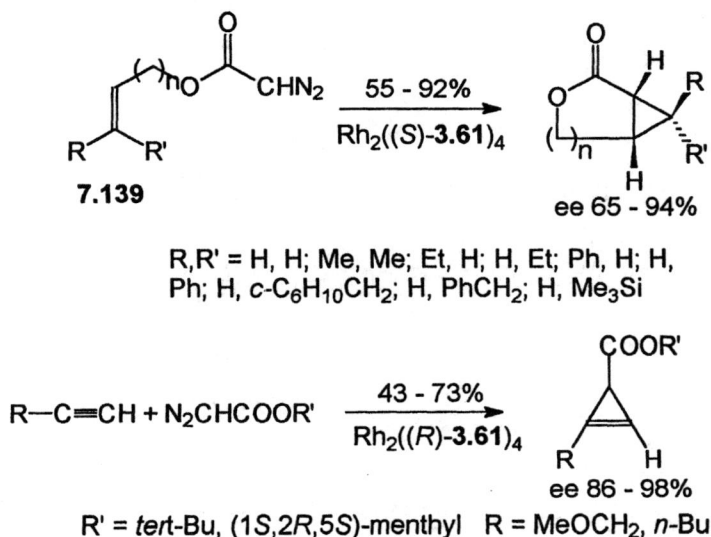

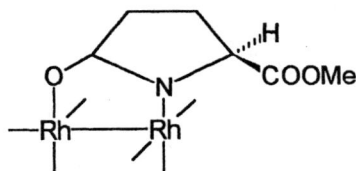

Figure 7.87

These complexes also catalyze the enantioselective cyclopropanation of monosubstituted alkynes with bulky diazoesters [939, 1497] (Figure **7.87**). Attempts at double diastereodifferentiation (§ I.6) have been carried out in the reaction of styrene with diazoesters of chiral alcohols under chiral Rh-**3.61** complexes catalysis [939, 1501], but disappointing selectivities were observed.

Davies and Hutcheson [933] examined the asymmetric cyclopropanation of alkenes with methyl or ethyl vinyldiazoesters **7.140** in the presence of a rhodium catalyst bearing **3.60** as ligand. E-Cyclopropanecarboxylates **7.141** are obtained with a high enantioselectivity when using styrenes or simple alkenes (Figure **7.88**). Bulkier alkyl diazoesters yield less useful selectivities. The use of the same catalyst in the cyclopropanation of styrene with ethyl diazoacetate gives low selectivities. Rhodium- or osmium-porphyrin-catalyzed cyclopropanations of alkenes by diazoesters also yield poor selectivities [1502, 1502a].

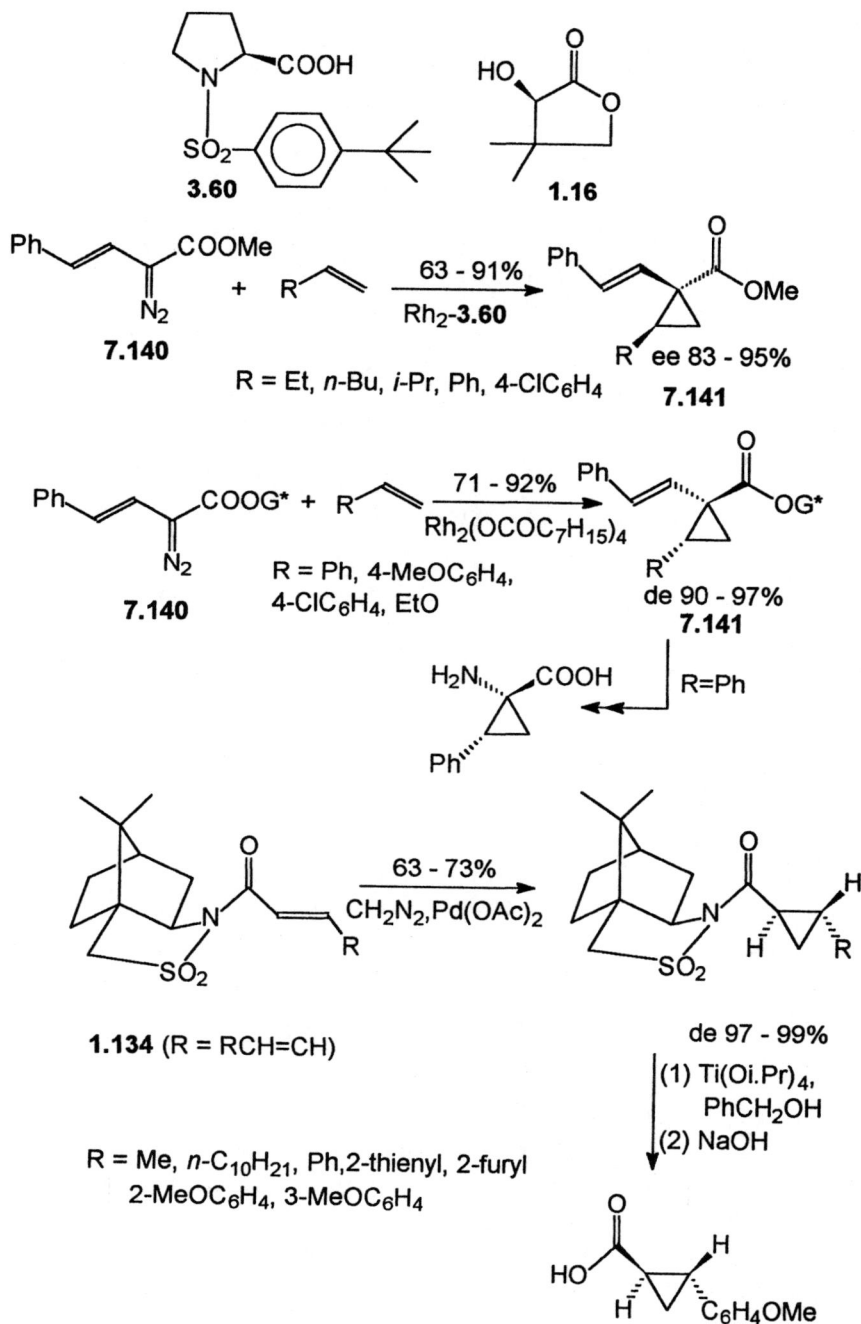

Figure 7.88

There are few examples of the reactions of alkenes with diazoesters of chiral alcohols which give high face discrimination in rhodium-catalyzed reactions [936, 1497]. However, Davies and coworkers [192, 193, 1503] have performed the reactions of alkenes with vinyldiazoesters of chiral alcohols under $Rh_2(OAc)_4$ or better yet $Rh_2(OCOC_7H_{15})_4$ catalysis. The ester of (R)-pantolactone **1.16** is the most efficient substrate, and *trans*-cyclopropanecarboxylates are obtained highly selectively. Starting from **7.141** (R = Ph), the enantioenriched plant hormone 1-amino-2-phenylcyclopropanecarboxylic acid has been prepared (Figure **7.88**).

Asymmetric cyclopropanation of *N*-enoylsultams **1.134** (R = $R_ECH=CH$) by CH_2N_2 under $Pd(OAc)_2$ catalysis also yields high selectivities [454, 461, 1504] (Figure **7.88**). Face discrimination occurs in the same manner as in asymmetric dihydroxylation (§ 7.6.2)

CHAPTER 8

ADDITIONS TO DOUBLE BONDS BEARING HETEROATOMS. OXIDATIONS OF SULFIDES AND SELENIDES

8.1 ENE REACTIONS

Ene reactions can occur with heteroatomic double bonds, and reactions of N=O or N=S bonds are useful in asymmetric synthesis. In these reactions, a carbon substituent of the enophile reagent generally bears the chiral residue. After the ene reaction, cleavage of the chiral group completes the asymmetric synthesis of the product.

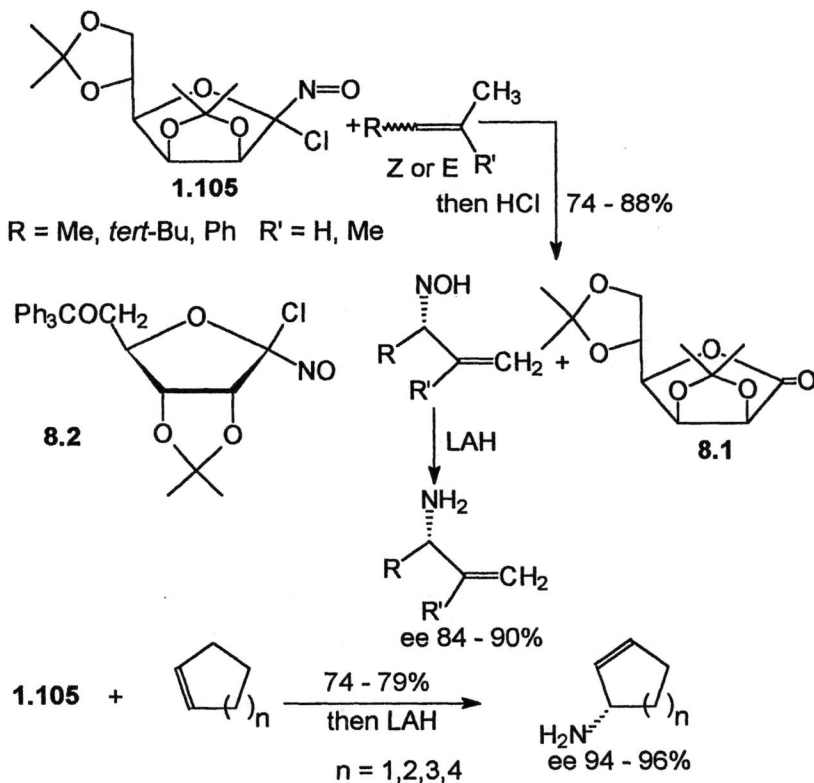

Figure 8.1

Schmidtchen and coworkers [370] developed heteroene reactions with chloronitrosugar **1.105**. The reactions of **1.105** with suitable alkenes take place at room temperature and lead to hydroxylamines after treatment with dilute acid. These hydroxylamines are reduced by LiAlH$_4$ into (S)-allylamines with a high enantiomeric excess (Figure **8.1**). The starting chiral reagent **1.105** is easily regenerated, from lactone **8.1**. Similar reactions can be carried out with **8.2**, and (R)-allylamines are formed with a similar selectivity (ee, 82 - 92%). Whitesell and coworkers [66, 191] performed SnCl$_4$-catalyzed ene reactions of Z-alkenes with N-sulfinylcarbamates **1.24** derived from chiral alcohols. When esters of 2-phenylcyclohexanols **1.5** (R = Ph) are used, a good stereoselectivity is observed at −78°C. The chiral sulfinoxycarbamates **1.25** can be obtained nearly pure by fractional crystallization. Similar ene reactions of the corresponding E-alkenes give the opposite diastereoisomers (Figure **8.2**).

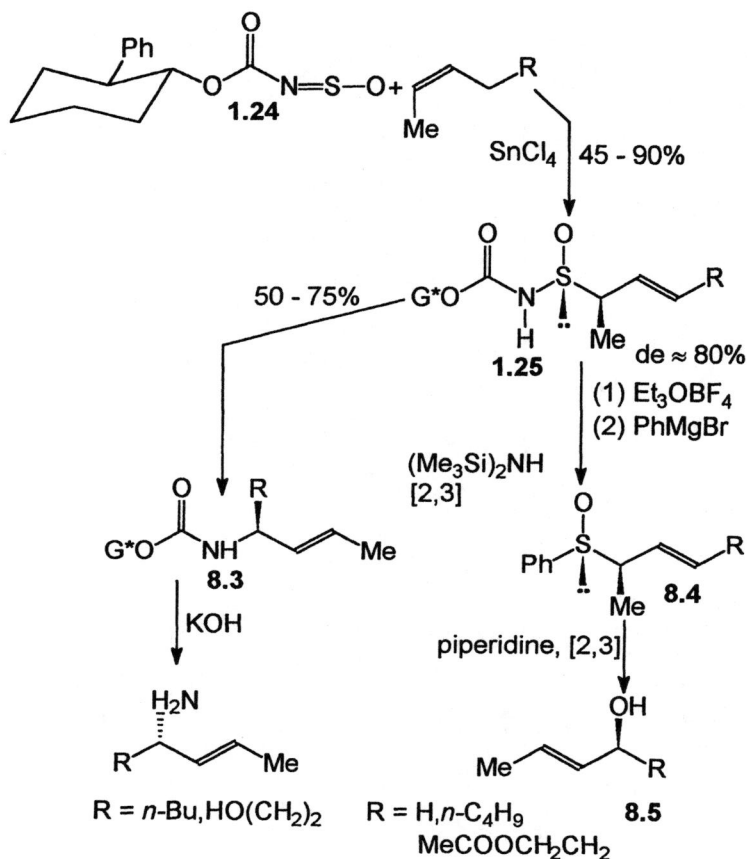

Figure 8.2

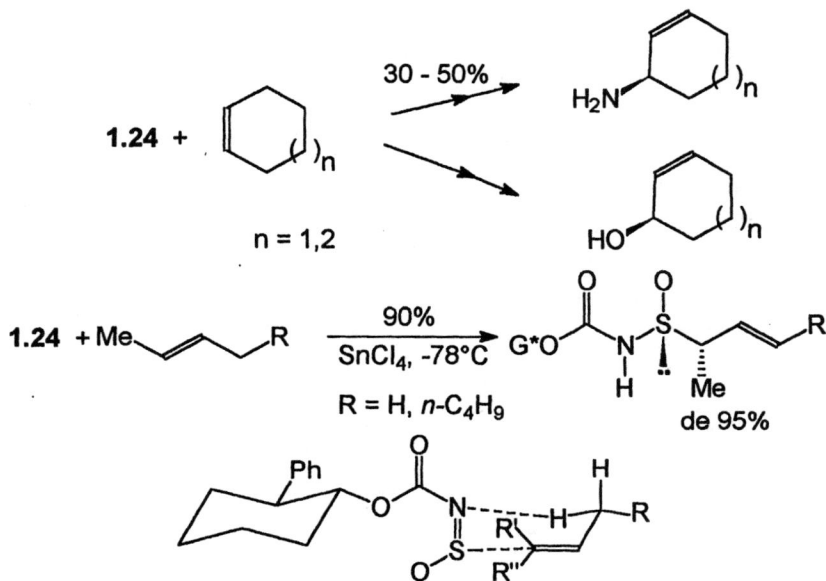

Figure 8.2 (continued)

The products of these reactions are easily transformed into allylamines by sequential treatment with $(Me_3Si)_2NH$ and KOH. Quaternization of the nitrogen of **1.25**, followed by reaction with PhMgBr at $-78°C$, gives sulfoxides **8.4** with an excellent enantiomeric excess. In turn, these sulfoxides undergo [2,3] sigmatropic transposition to allyl alcohols **8.5**. In the ene reactions of **1.24**, the enophile approaches the face of the N=S double bond opposite to the phenyl substituent of the auxiliary so that the A (1,3) strain is minimized (Figure **8.2**).

8.2 ASYMMETRIC OXIDATION OF SULFIDES AND RELATED REACTIONS [503, 814, 1278]

Nonracemic sulfoxides are very important reagents in asymmetric synthesis, but the classical preparations of these reagents often call for tedious separations (§ 1.7). Recently, more direct routes involving asymmetric oxidation of sulfides have been introduced. These oxidations can be conducted either with chiral reagents or with sulfides bearing chiral residues.

Oxidations of arylmethylsulfides by H_2O_2 in the presence of Mn (III) complexes bearing chiral ligands such as **3.70** have given disappointing selectivities (ee < 68%) [1505, 1506]. However, similar oxidations with *tert*-BuOOH or $PhCMe_2OOH$ in the presence of titanium complexes bearing chiral ligands are more selective. Kagan, Modena and their coworkers [502, 504] have used diethyl tartrates **2.69** (R = Et) as ligands. More recently, Uemura and coworkers [815, 947] have used binaphthols **3.7** (R = H). The Kagan and Modena reagents are similar to the reagents used in the Sharpless asymmetric epoxidation (§ 7.7.2); however, the relative stoichiometry of the components is varied and water is sometimes added. Kagan's catalyst is a mixture of $Ti(Oi\text{-Pr})_4$, **2.69** (R = Et) and H_2O in a 1:2:1 ratio. This blend can be used in catalytic amounts (0.2-0.5 eq.) relative to the substrate. The best oxidant is cumene hydroperoxide, and the reactions are conducted in CH_2Cl_2 at –20°C [505]. The highest enantiose-lectivities are obtained in the asymmetric oxidations of aryl- or alkynylmethylsul-fides [503, 1506a] (Figure **8.3**). Modena's reagent is a mixture *tert*-BuOOH, $Ti(Oi\text{-Pr})_4$, and **2.69** (R = Et) in a 2:1:4 ratio, and this blend is always used in the presence of an excess of substrate. The results obtained under the two sets of conditions are generally similar, but Kagan's conditions are more economical.

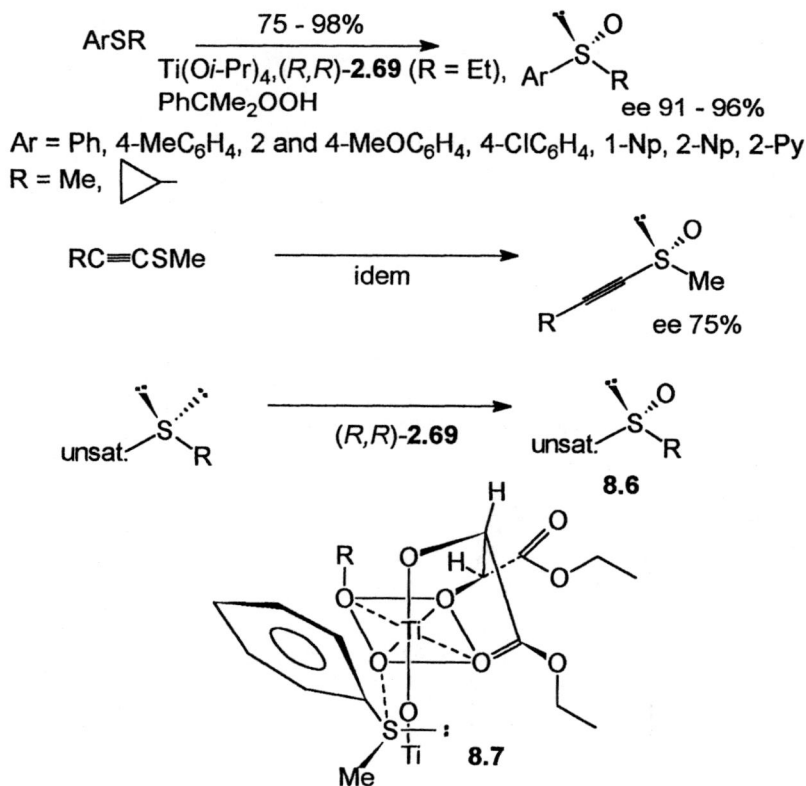

Figure 8.3

(R,R)-**2.69**   (S,S)-**2.69**

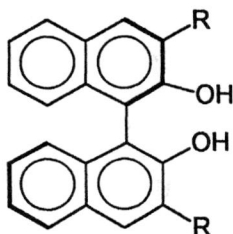 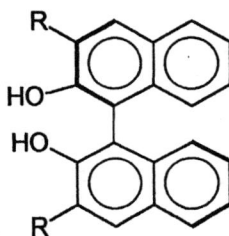

(R)-**3.7**   (S)-**3.7**

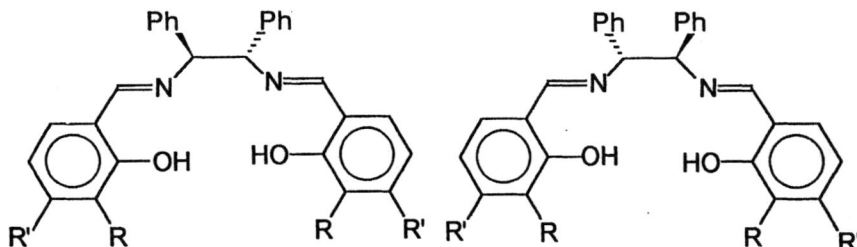

3.70

The absolute configuration of the sulfoxide products can be predicted by using the empirical rule shown in Figure 8.3. (R,R)-Diethyl tartrate **2.69** (R = Et) gives the (R)-sulfoxide **8.6** and vice versa. Kagan has proposed the intermediacy of a bimetallic complex **8.7** in which one of the titanium atoms is tricoordinated to a tartrate unit and bicoordinated to the peroxide. The attack of the sulfide should occur along the O-O bond of this reagent (Figure **8.3**).

Oxidations of dialkylsulfides RSR' or alkylarylsulfides RSAr (R = Et, Pr, n-Bu, ClCH$_2$) are less enantioselective [814]. According to the substrate, either the Kagan or the Modena blend, or even a slightly modified one, is the most efficient. Pitchen and coworkers [1507] used aprotic conditions to oxidize imidazolyl

methylsulfides on an industrial scale.   Modena and coworkers oxidized bis-
methylthiobenzenes [1508] or dithiolanes [1509] with an excellent enantioselectivity.   Page and coworkers used Kagan's conditions to prepare *anti*-acyl-
dithianeoxides **8.8** with a high selectivity [1510, 1511]. *syn*-Dithiolanes **8.9** are
easily obtained by deacylation of **8.8**. In contrast, the selective synthesis of disul-
foxide **8.10** requires oxidation under Modena's conditions at −20°C [1512]. Unfortunately, nonfunctionalized 1,3-dithianes, 1,3-oxathiolannes and β-silyloxysulfides
are oxidized to provide mixtures of stereoisomers [1511, 1513] (Figure **8.4**).

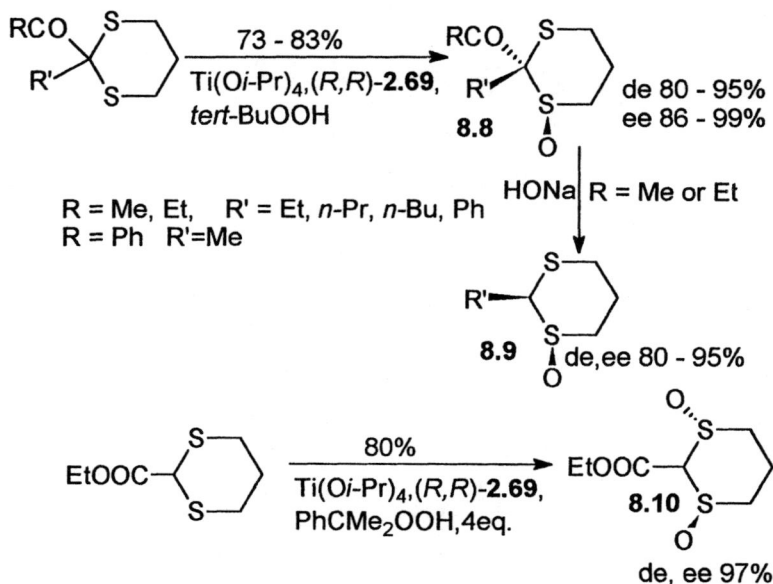

R = Me, Et,   R' = Et, *n*-Pr, *n*-Bu, Ph
R = Ph   R'=Me

Figure 8.4

When (*R*)-binaphtol **3.7** (R = H) is used as a titanium ligand, the catalytic
asymmetric oxidation of arylmethylsulfides by *tert*-BuOOH in the presence of water in CCl$_4$ leads to (*R*)-sulfoxides [815, 947, 1514]. In this reaction, the initial
oxidation of the sulfide into the chiral sulfoxide takes place with a moderate
ee ($\cong$ 50%). This step is followed by further oxidation of the sulfoxides with kinetic resolution (§ I.6) [815, 1514].   To observe a high enantiomeric excess
(> 90%), it is necessary to oxidize the minor (*S*)-enantiomer into the corresponding
sulfone, and the chemical yield of the sulfoxide is in the 45 - 65% range.

Davis and coworkers [506, 744] performed the asymmetric oxidation of prochiral sulfides with chiral oxaziridines, and reagents bearing the bornane skeleton
2.82 or **2.83** (X = Cl) were the most efficient. The chiral oxaziridine oxidations are

more broadly applicable than the titanium/tartrate oxidations. In reactions conducted at –20°C, arylmethyl- or arylcyclopropylsulfides, alkynyl-, n-butyl- and benzylmethylsulfides, and functionalized arylmethylsulfides are all oxidized with stoichiometric amounts of these reagents into sulfoxides with an excellent enantioselectivity (Figure **8.5**). The chemical yields are moderate (40 - 60%) in the asymmetric oxidation of functionalized phenylmethylsulfides $PhSCH_2R$ (R = $MeOCOCH_2$, $CNCH_2$) and $PhSCH=CH_2$. The configuration of the oxaziridine controls that of the sulfoxide: (3'S,2R)-**2.83** gives the (S)-sulfoxide and vice versa (Figure **8.5**). These results have been interpreted in terms of steric interactions, although π-stacking might also intervene because the alkynyl group behaves as a large group R_L.

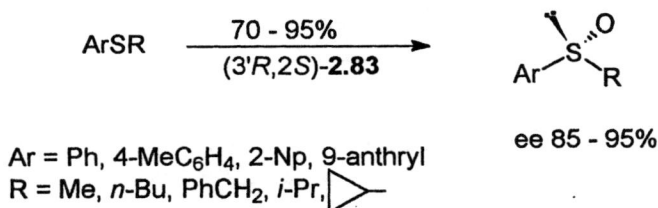

$$ ArSR \xrightarrow[\text{(3'R,2S)-\textbf{2.83}}]{\text{70 - 95\%}} $$

ee 85 - 95%

Ar = Ph, 4-MeC$_6$H$_4$, 2-Np, 9-anthryl
R = Me, n-Bu, PhCH$_2$, i-Pr, ▷—

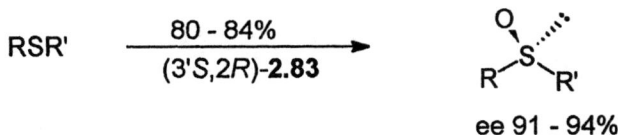

$$ RSR' \xrightarrow[\text{(3'S,2R)-\textbf{2.83}}]{\text{80 - 84\%}} $$

ee 91 - 94%

R = $tert$-Bu   R' = Me, PhCH$_2$

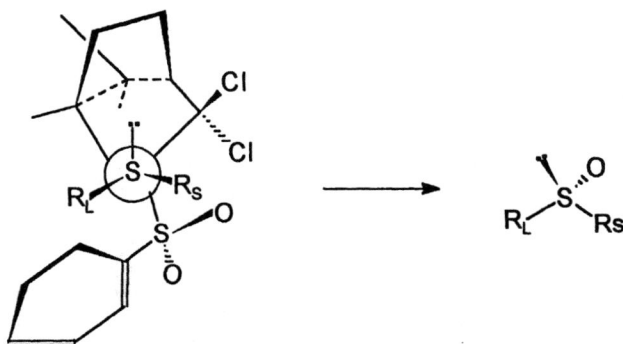

Figure 8.5

Ar = Ph, 4-MeC$_6$H$_4$
Ar' = Ph, 2-MeOC$_6$H$_4$ R = H, Me

Ar = Ph, 2,4, 6-i-Pr$_3$C$_6$H$_2$ R = Me, Et

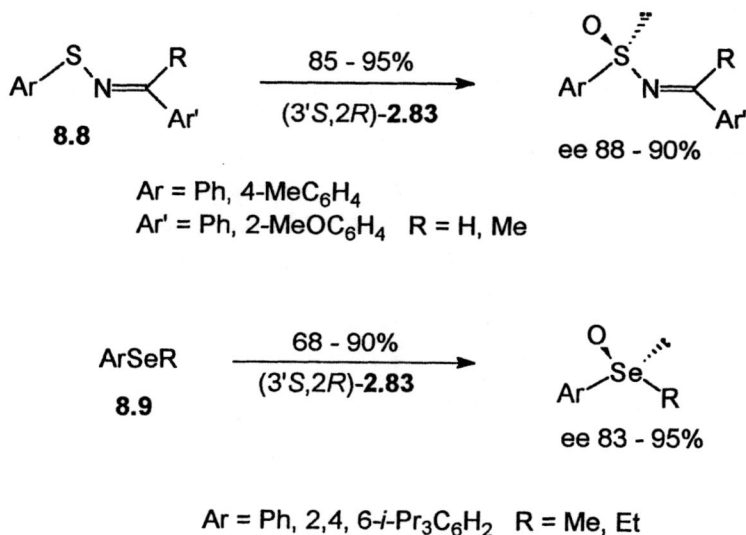

Figure 8.5 (continued)

Arylsulfinimines **8.8** are oxidized by oxaziridine **2.83** with an excellent selectivity [510] (Figure **8.5**). The products of these oxidations are precursors of chiral amines and β-aminoesters (§ 6.1.4 and 6.8.3). The asymmetric oxidations of arylselenides **8.9** with **2.83** (X = Cl) in CCl$_4$ are also highly selective, but the rapid racemization of the selenoxides in the presence of moisture makes it difficult to isolate optically pure products [749, 1515]. Uemura and coworkers [1516] have circumvented this problem by performing *in situ* elimination of selenic acid from the chiral selenoxide. They obtained chiral cyclohexylidene methylketones through this oxidation/elimination strategy.

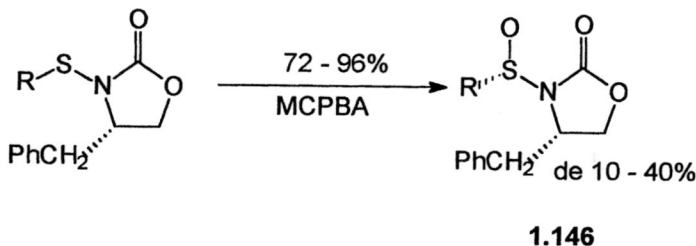

1.146

Figure 8.6

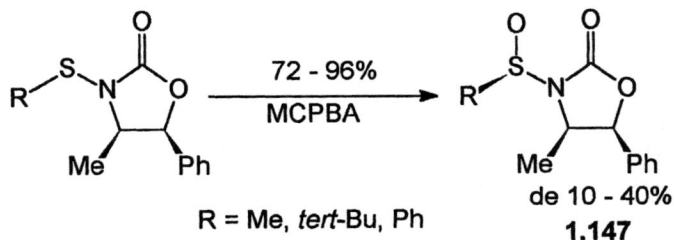

Figure 8.6 (continued)

The oxidation of chiral *N*-aryl and *N*-alkylthiooxazolidinones into *N*-sulfinyl derivatives **1.146** and **1.147** with 3-chloroperbenzoic acid has been performed by Evans and coworkers [509]. These reactions are poorly selective, but the products are easily purified by fractional crystallization. The purified products are precursors of chiral sulfoxides (§ 1.7) (Figure **8.6**).

CHAPTER 9

CYCLOADDITIONS

Cycloaddition reactions are frequently used to elaborate the skeletons of natural products. Numerous theoretical studies have been devoted to this class of reactions, and many of the most common types of cycloadditions are thought to be concerted. The interactions between frontier orbitals are of central importance in most of the cases [624g]. In the present chapter, cycloaddition reactions will be classified according to the number of atoms involved, and then subdivided by reaction type.

9.1 [2+2]-CYCLOADDITIONS

Forbidden according to symmetry rules [624g], [2+2]-cycloadditions are generally not concerted. When concerted [2+2]-cycloadditions do occur, the approach of both reagents should be supra-antarafacial. Usually, highly reactive species such as ketenes or ketenimminium salts are involved. Activation may be accomplished by adding Lewis acids or by irradiation (Paterno-Büchi reaction). In photochemical [2+2]-cycloadditions, the reaction takes place via an excited state and is symmetry-allowed.

9.1.1 Reactions of Ketenes [1517]

Ketene ($CH_2=C=O$) reacts with carbonyl compounds substituted by electron-withdrawing groups. As early as 1982, Wynberg and Staring [770, 1058] performed the cycloaddition of ketene with chloral (CCl_3CHO) in toluene at $-50°C$ in the presence of catalytic amounts of quinine 3.1 (R = MeO) or quinidine 3.2 (R = MeO). They obtained the corresponding enantiomeric lactones 9.1 with a good enantiomeric excess. These lactones are precursors of (R)- or (S)-malic acid (Figure 9.1). These authors interpret their results by postulating a dipolar intermediate in which the alkaloid is acetylated (Figure 9.1). Steric interactions are minimized as this intermediate progresses to the transition state. The reaction of ketenes with imines (Staudinger reaction) is a useful route to β-lactam antibiotics [131] that has been the subject of much study. Chiral residues have been introduced either on the ketene or the imine. The Staudinger reaction is thought to occur by a two-step mechanism [1518, 1519]. The carbon-nitrogen bond is formed first, with minimization of the steric interactions. Next, the intermediate zwitterion undergoes a conrotatory cyclization. It is in this second step that the cis or trans stereochemistry is determined by the nature of the reagents [1518].

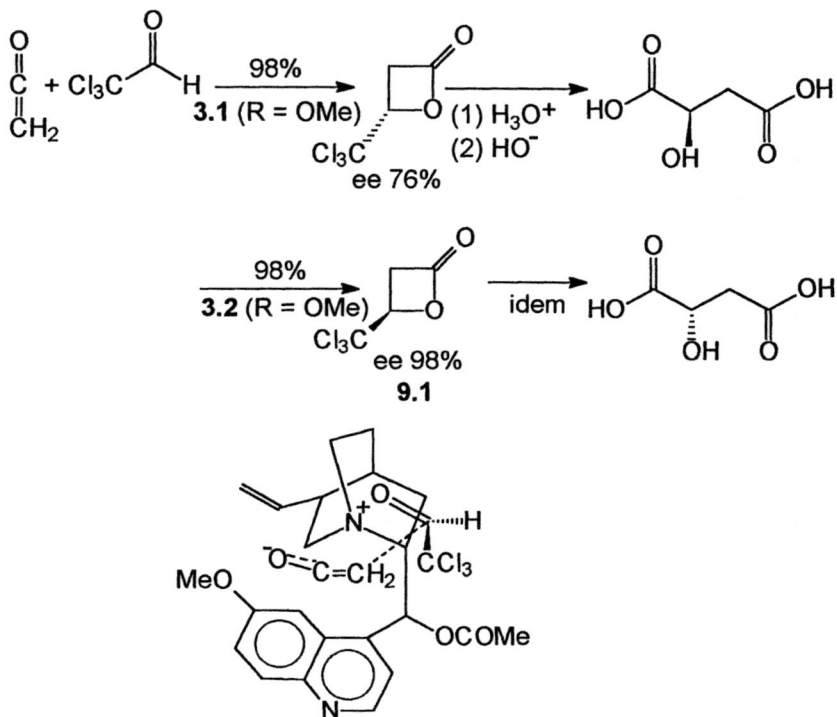

Figure 9.1

Chiral ketenes have been prepared from oxazolidinones (*S*)- or (*R*)-**1.116** (R = Ph) and **1.117** [131]. These are first transformed into their *N*-acetyl chlorides **9.2**, and the ketenes are generated *in situ* by addition of NEt$_3$ at –78°C. The reaction of these ketenes with imines leads to *cis*-β-lactams **9.3** with a high diastereoselectivity (Figure 9.2). The chiral auxiliary may be regenerated by the action of Li/NH$_3$. When R$_E$ is unsaturated, the cleavage is performed after catalytic hydrogenation of the double bond [1519a, 1520]. When R = PhCH$_2$, concomitant debenzylation occurs. If the β-lactam **9.3** is hydrolyzed before removal of the chiral auxiliary, then nonracemic α-aminoacids or their derivatives may be prepared [154, 861, 1520]. The cycloadduct **9.3** can be oxidized in basic conditions [1521] to provide nonracemic α-keto-β-lactams **9.4**. These are precursors of α-hydroxy-β-aminoacids (Figure 9.2). The configuration of the *cis*-β-lactam **9.3** is not influenced by the imine nitrogen substituent (even if chiral), but is instead controlled by the configuration of the chiral ketene precursor **9.2**. Other chiral ketene precursors have been proposed, but their reactions lead to mixtures of *cis*- and *trans*-β-lactams [131] or to mixtures of *cis* isomers [1522]. A ketene bearing a sugar substituent also has been recommended [1049], but its reactions lead to mediocre enantiomeric excesses (70%).

Imines substituted on nitrogen by chiral 1-phenethyl groups react with various ketenes, and *cis*-β-lactams are also formed, but with mediocre face selectivity [131, 1523, 1524]. To improve this reaction, Georg and Wu [1525] replaced the 1-phenethyl residue by a 1-(1-naphthyl)-ethyl group, although the best selectivities did not exceed 66%. Sugar-derived imines were also recommended. Barton and coworkers prepared *cis*-3-amino-4-styryl-β-lactam with an excellent selectivity from a protected D-glucosamine and phthalimidoketene [131]. Imine **9.5** derived from galactose reacts with 4-methoxyphenoxyketene, leading highly selectively to lactam **9.6**, a precursor of a substituted serine derivative, with an excellent enantiomeric excess [366a] (Figure **9.2**).

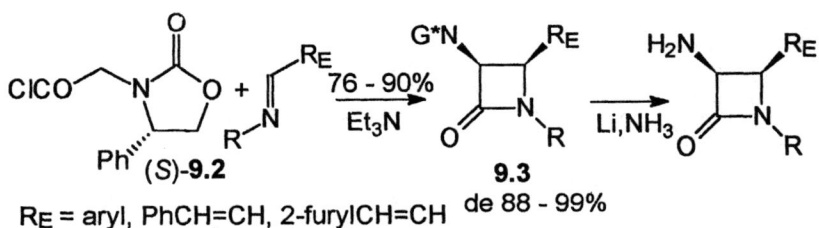

R_E = aryl, PhCH=CH, 2-furylCH=CH   de 88 - 99%

R = Me, PhCH$_2$, (*R*)- or (*S*)-CH(Me)COOMe,
(*R*)- or (*S*)-CH(*i*-Pr)COOMe

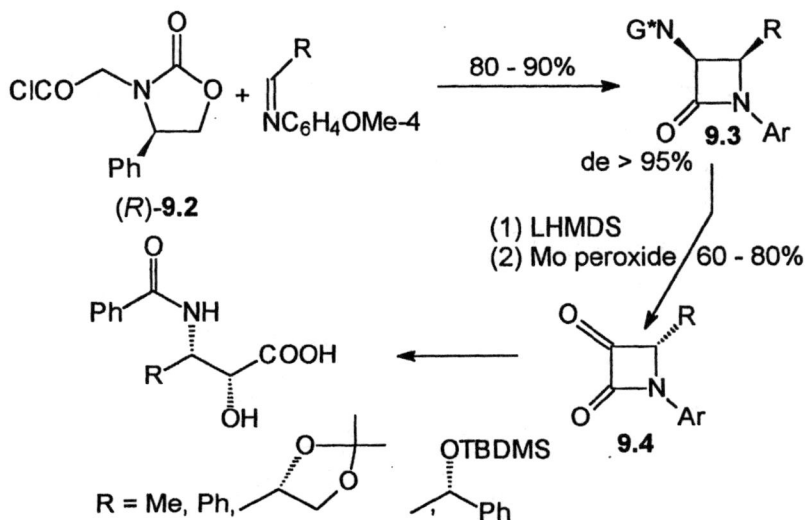

R = Me, Ph,

Figure 9.2

4-MeOC$_6$H$_4$OCH$_2$COCl +

9.5

Et$_3$N | 75%

(1) MeOH,HCl
(2) PhCOCl,Et$_3$N
DMAP

9.6
de > 99%

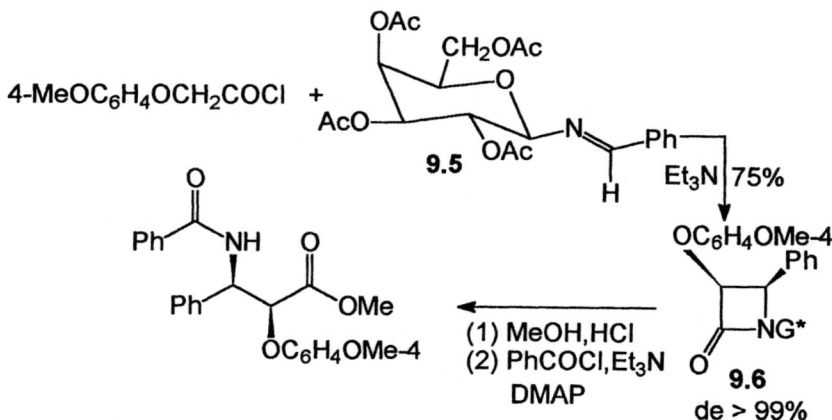

Figure 9.2 (continued)

Hegedus and coworkers [131, 336, 337, 338, 1526] proposed the use of chromium carbene complexes bearing chiral oxazolidine residues **1.85** (R = Ph or *i*-Pr, R' = H, R" = Me). Irradiation of complexes **9.7** or **9.8** under CO pressure generates complexed ketenes. If R$_Z$ = H, these react highly stereoselectively with imines to give *trans*-β-lactams. Each complex gives rise to an enantiomeric β-lactam after cleavage (Figure **9.3**). The chiral auxiliary is cleaved by hydrolysis followed by hydrogenolysis (if the precursor is **9.7**) or by oxidation with NaIO$_4$ (if the precursor is **9.8**). As in the case of phenethylamine derivatives, the chiral auxiliary is not recovered. The reaction is no longer stereoselective when it is carried out with *N*-benzylimines derived from unsaturated or aromatic aldehydes. The reactions of chromium carbene complexes derived from oxazolidinone (*S*)-**1.116** (R = Pn) with *N*-benzylimines in the presence of Et$_3$N give the same face discrimination and *cis* stereoselectivity as the related reactions of acid chloride (*S*)-**9.2** [337] (see above). However, the reaction of the chromium carbene complexes with cyclic imines (R,R$_Z$ = (CH$_2$)$_n$) is not stereoselective. These data are interpreted in terms of a cyclization of the intermediate zwitterion that can be either faster or slower than rotation around the C-N bond, depending upon the nature of the chiral residue and of the imine substituents [337].

Dihaloketenes react readily with electron-rich olefins, and Greene and coworkers [66, 169, 1008, 1527, 1528, 1529] introduced *trans*-2-phenyl-cyclohexanol **1.5** (R = Ph) as a chiral auxiliary for addition of dichloroketene to enol ethers **9.9**. 2,2-Dichlorocyclobutanones **9.10** are obtained with a very good selectivity (Figure **9.4**). In further transformations, the oxidation of **9.10** with MCPBA provides γ-butyrolactones **9.11**, while ring expansion gives cyclopentanones **9.12**.

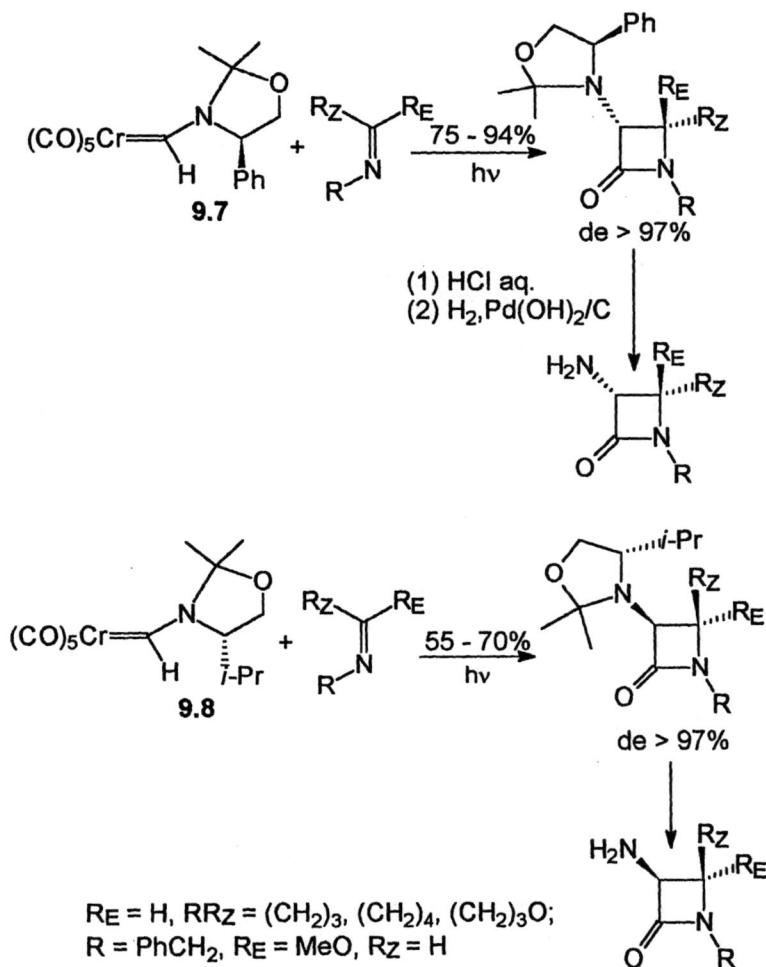

Figure 9.3

In both cases, these transformations are regioselective, and each has been used to make natural products with an excellent enantiomeric excess (Figure **9.4**). Since both enantiomers of **1.5** (R = Ph) are available, either (R)- or (S)-dichlorocyclobutanone can be prepared. The favored conformation of the enol ether is probably *s-trans*, and the aromatic substituent hinders one face of the double bond so the reaction with dichloroketene occurs on the other face. Enol ethers of other chiral alcohols give less useful results [1528].

Chromium alkoxycarbene complexes are photolytically converted into the corresponding complexed ketenes. Their cycloadditions with chiral *N*-vinyloxazolidinones derived from **1.116** (R = Ph) lead highly selectively to 2-alkoxycyclobutanones, which are precursors of natural products [1526] (Figure **9.4**).

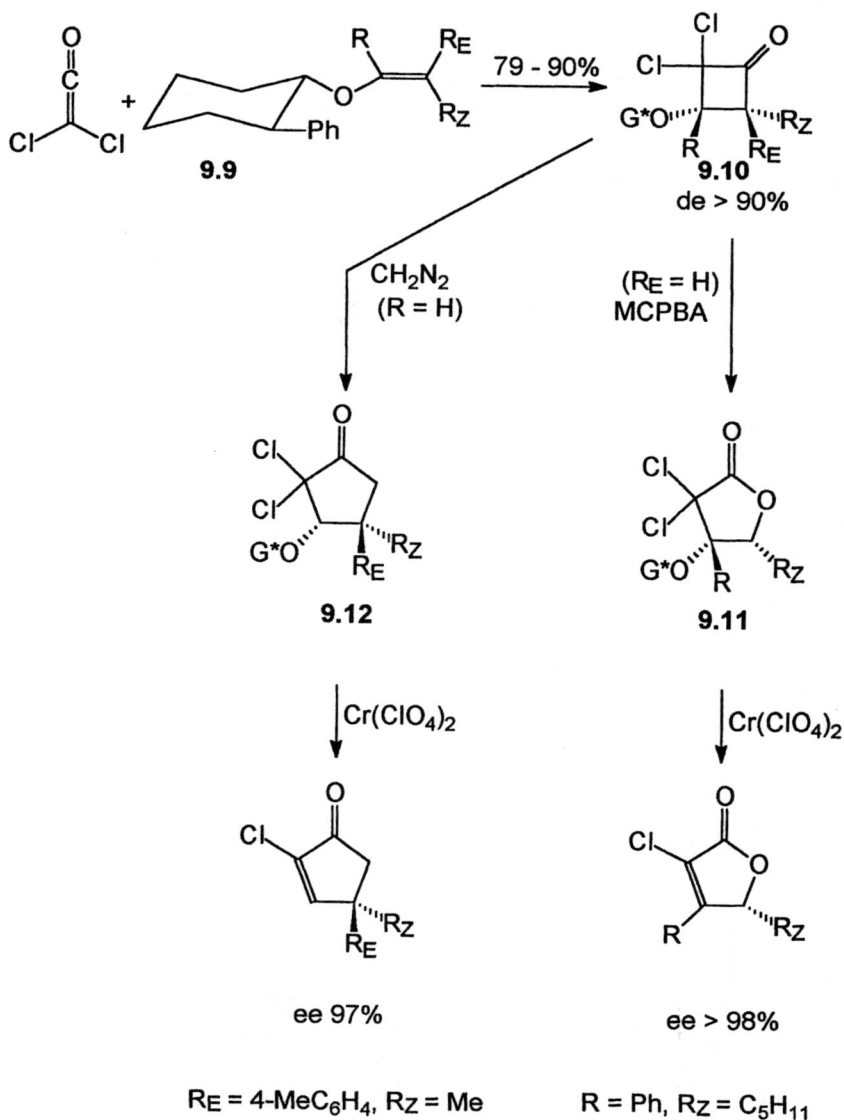

Figure 9.4

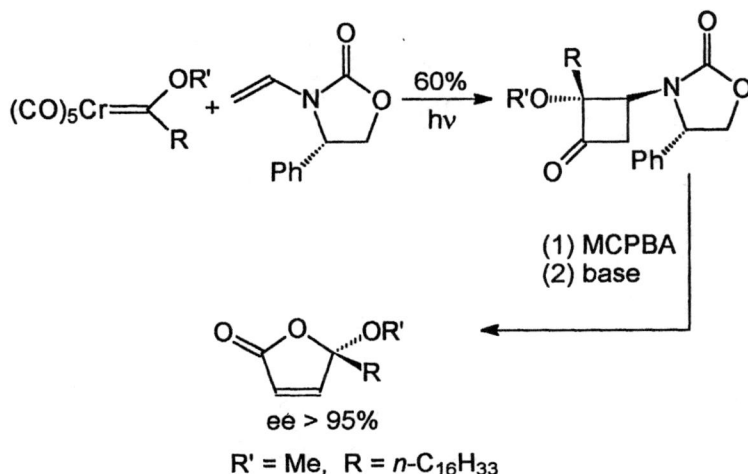

Figure 9.4 (continued)

9.1.2 Reactions of Ketenimminium Salts [272, 1517]

Ketenimminium salts are more electrophilic than ketenes, and they add to double bonds that are not substituted by electron-withdrawing groups. The introduction of chiral residues on the nitrogen atom allowed Ghosez and coworkers [169, 195, 291, 1530, 1530a] to perform asymmetric cycloadditions of ketenimminium salts generated *in situ* with alkenes or imines. Proline derivative **1.64** (R = CH$_2$OMe) or pyrrolidines **1.65** (R = Me or CH$_2$OMe) bearing a C$_2$ axis of symmetry have been used as chiral auxiliaries. The reactions of the derived ketenimminium salts with terminal olefins give disappointing results, but better selectivities are observed with cyclic olefins. The cycloaddition of cyclopentene with ketenimminium salts generated from **9.13** (R = R' = Me) leads to the corresponding cyclobutanone with a high selectivity (Figure **9.5**), although mediocre results are obtained when R = H. High selectivities are also observed with ketenimminium salts derived from **1.65** (R = Me) either in intramolecular reactions [195] or in reactions with Z-olefins using **9.14** as precursor [1530a] (Figure **9.5**). The chiral residue is easily cleaved by hydrolysis. Imines react with ketenimminium salts generated from **9.13** in an enantioselective fashion provided that R and R' are not H [1530] (Figure **9.5**). The imine substituents do not noticeably affect the selectivity. 3-Amino-β-lactams can be prepared with a high selectivity by using **9.15** as a reagent. *Trans* isomers are formed, and the removal of the chiral residue requires mildly basic conditions to avoid epimerization. The transitory formation of a thiolactam with NaHS solves the epimerization problem [1530] (Figure **9.5**).

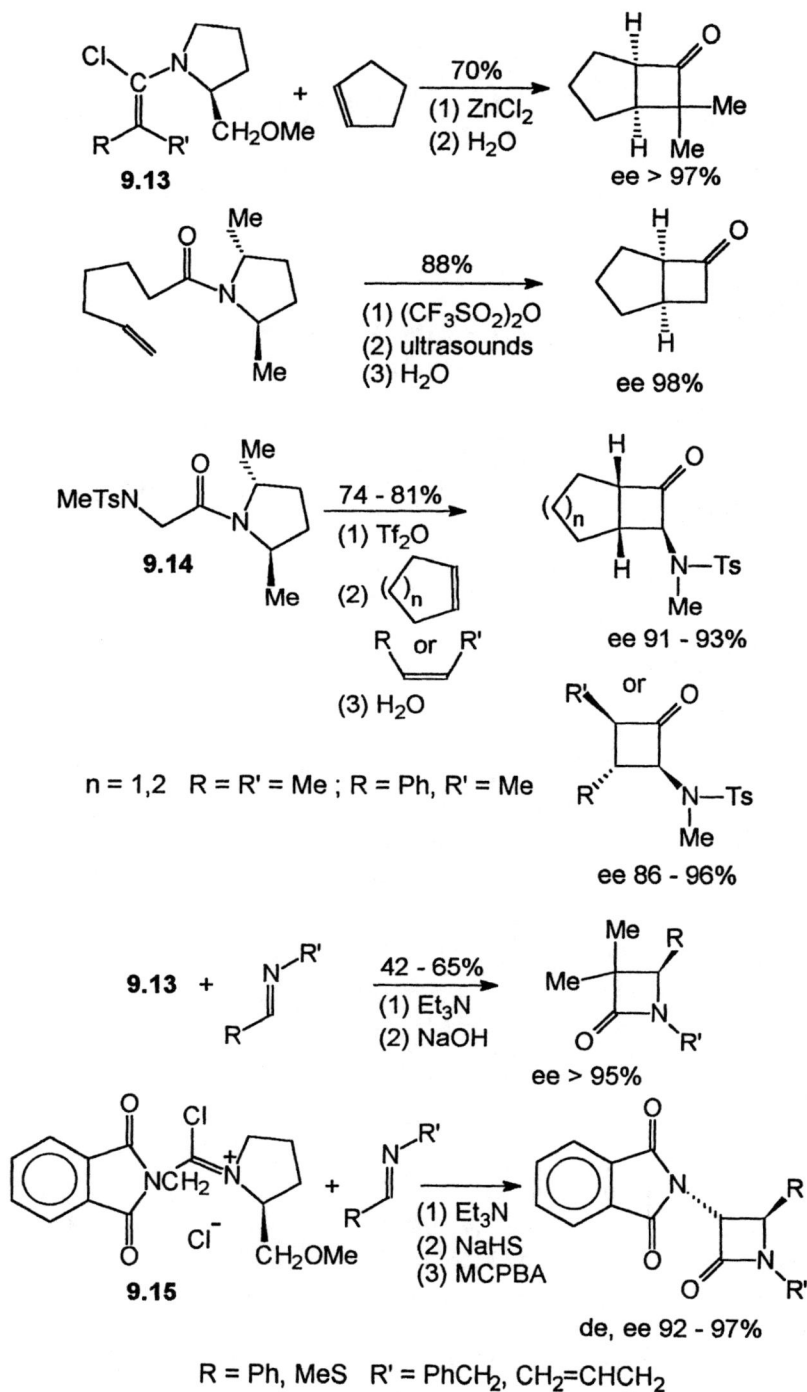

Figure 9.5

9.1.3 Lewis-acid-catalyzed Reactions of Ketene Acetals and Thioketals [778]

Ahmad [174] has described asymmetric [2+2]-cycloadditions of dimenthyl fumarate with $CH_2=C(OMe)_2$ under Et_2AlCl catalysis at $-70°C$. *trans*-Dimenthyl cyclobutanedicarboxylate **9.16** is obtained with a high selectivity after crystallization. After removal of the chiral auxiliary with $LiAlH_4$, a nonracemic diol precursor of an antiviral drug is obtained (Figure 9.6). Narasaka and coworkers [778, 817, 1531] examined the asymmetric cycloaddition of *N*-enoyloxazolidinones **7.68** (R = H) with ketene thioketals and alkenyl-, alkynyl- and allenylsulfides catalyzed by chiral titanium complexes derived from **2.50** (R = Me, Ar = R' = Ph). These reactions take place at $0°C$, though they do not work with ketene acetals. The most useful reactions are with $CH_2=C(SMe)_2$, and these lead selectively to cycloadducts (Figure 9.6). If the sulfur substituents are different, side reactions take place, while if the carbon in the β-position is substituted, mixtures of stereoisomers are obtained. Acyclic alkenyl sulfides also lead to mixtures. The reactions of cyclic analogs **9.17** with **7.68** (R = H) give cycloadducts with a high selectivity provided that R_E = COOMe (Figure 9.6). Allenylmethylsulfides **9.18** require the presence of $SiMe_3$ or $SnMe_3$ substituents in the α-position [1531] to observe a regio- and stereoselective cycloaddition (Figure 9.6). Stepwise reactions occur via intermediate zwitterions. The *Re* face of the α-carbon of **7.68** (R = H) is attacked by the reagent. This is the same face selectivity that is observed when ene reactions (§ 7.11) or Diels-Alder reactions (§ 9.3) are performed in the presence of the same chiral titanium catalyst [778].

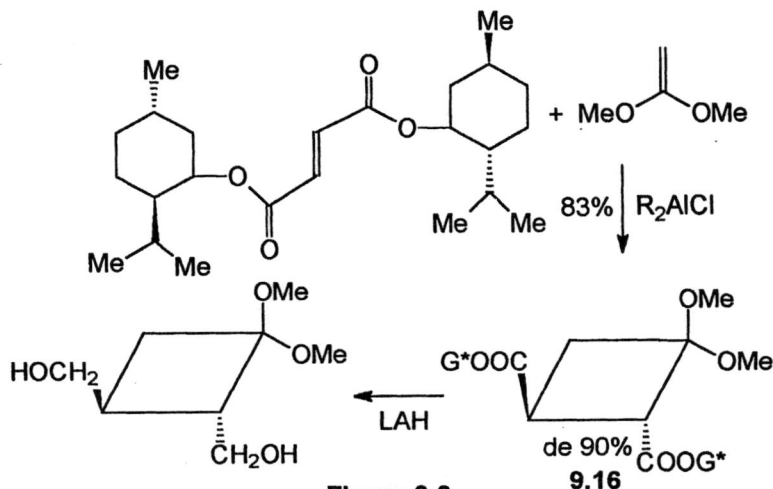

Figure 9.6

7.68 (R = H)

64 - 96%
2.50 cata.,
TiCl$_2$(Oi-Pr)$_2$

R$_E$ = MeOOC, H, Me   de, ee 80 - 98%

7.68 (R$_E$ = MeOOC, R = H)

n-BuS   **9.17** n = 2,3,4

2.50, TiCl$_2$(Oi-Pr)$_2$ | 89 - 97%

de 80 - 99%
ee > 98%

7.68 (R = H)+

65 - 92%
2.50, TiCl$_2$(Oi-Pr)$_2$
cata. or stoichio.

de, ee > 98%

R$_E$ = MeOOC, H   R' = H, Me, n-Bu, c-C$_6$H$_{11}$

7.68 (R = H,
R$_E$ = MeOOC) +

MeS   R'

9.18

93 - 99%
2.50, TiCl$_2$(Oi-Pr)$_2$

de, ee > 96%

R' = SiMe$_3$, SnMe$_3$

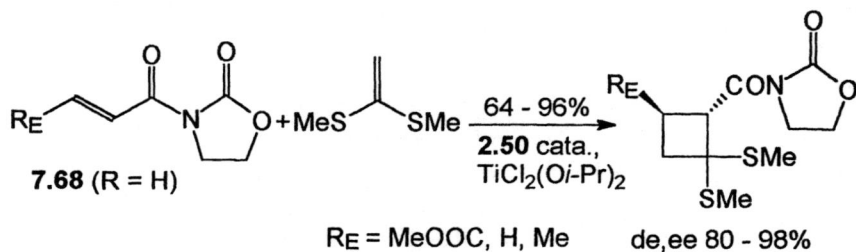

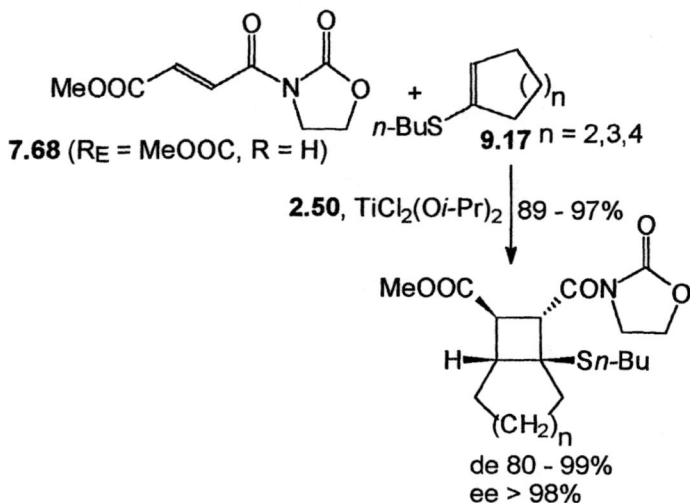

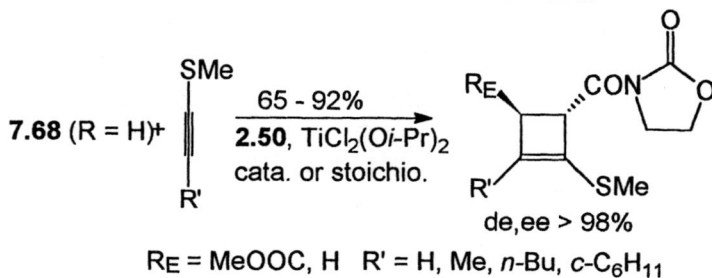

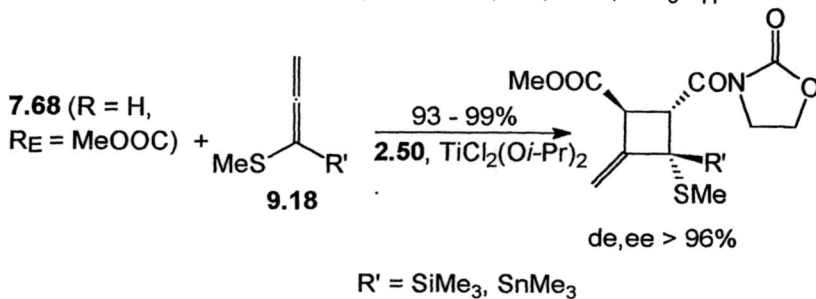

Figure 9.6 (continued)

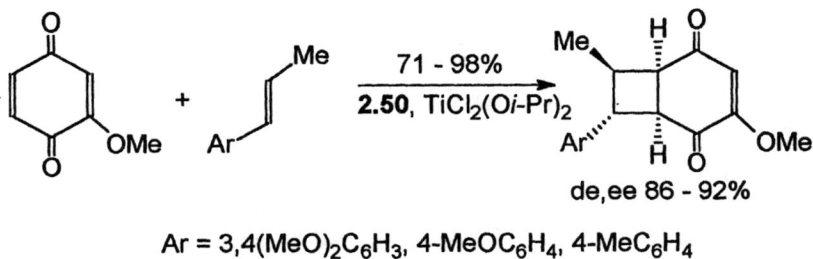

de,ee 86 - 92%

Ar = 3,4(MeO)$_2$C$_6$H$_3$, 4-MeOC$_6$H$_4$, 4-MeC$_6$H$_4$

Figure 9.6 (continued)

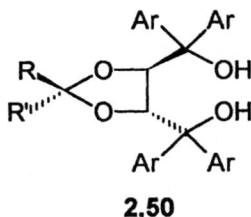

2.50

The reactions of ketene acetals with Schiff bases derived from (S)-valine esters **1.59** (R = i-Pr) under TiCl$_4$ catalysis are highly selective, but the chiral auxiliary has not been removed [264]. Engler and coworkers [1532] performed the asymmetric [2+2]-cycloaddition of 2-methoxy-1,4-benzoquinone with 1-arylpropenes substituted by electron-donating groups. These reactions occur at –78°C when catalyzed by chiral titanium complexes derived from **2.50** (R = Me, Ar = R' = Ph). Cycloadducts are obtained with a good selectivity (Figure 9.6).

9.1.4 Photoinduced Cycloadditions [2, 1533]

9.1.4.1 Cycloadditions of Carbonyl Compounds

Photoinduced [2+2]-cycloadditions of carbonyl compounds with alkenes (Paterno-Büchi reaction) lead to oxetanes [1534]. These reactions occur by photoexcitation of the ketone to its first singlet excited state, followed by intersystem crossing to the triplet state. The interaction of this excited species with the olefin leads to a triplet 1,4-biradical. After intersystem crossing, the resulting singlet biradical undergoes rapid cyclization. Sharf and coworkers [2] have studied Paterno-Büchi reactions by using glyoxylates of chiral alcohols **1.23** as carbonyl components. The esters of phenmenthol **1.4** (R = Ph) and trans-2-tert-Bu-cyclohexanol have shown the best efficiency [66]. Highly selective cycloadditions have been observed in reactions of aryl or tert-Bu glyoxylate **1.23** (R = Ar or tert-Bu) with symmetrical alkenes conducted between −20°C and +20°C (Figure **9.7**). Cyclic alkenes give predominantly *exo* cycloadducts (Figure **9.7**). However, most unsymmetrical olefins (except for 1-methylcyclopentadiene) lead to mixtures of regioisomers. The influence of the temperature on the stereoselectivity of the cycloaddition has been carefully studied, and according to the reaction conditions, the process is entropy- or enthalpy-controlled (isoinversion phenomenon) [2].

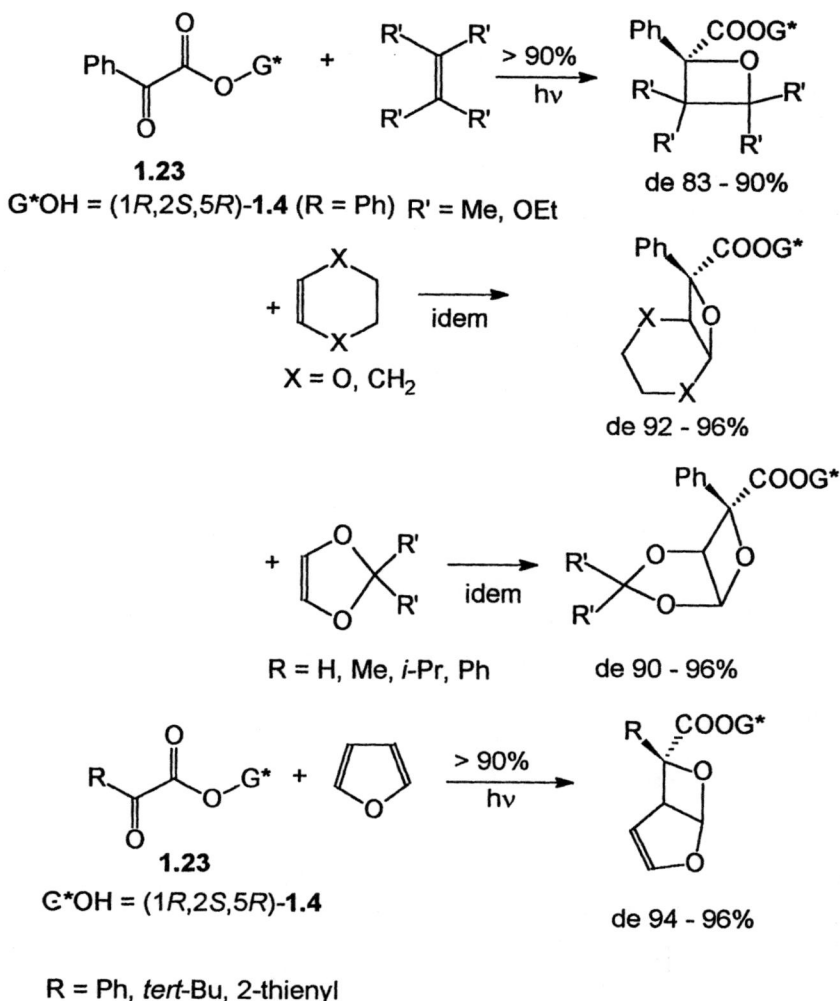

1.23

G*OH = (1*R*,2*S*,5*R*)-**1.4** (R = Ph) R' = Me, OEt

de 83 - 90%

X = O, CH₂

de 92 - 96%

R = H, Me, *i*-Pr, Ph   de 90 - 96%

1.23

G*OH = (1*R*,2*S*,5*R*)-**1.4**

de 94 - 96%

R = Ph, *tert*-Bu, 2-thienyl

Figure 9.7

9.1.4.2 Cycloadditions of Alkenes

[2+2]-Photocycloadditions of ethylene with chiral lactam **1.92** (R' = R" = Me, R = *i*-Pr) take place selectively on the *exo* (convex) face of the α,β-unsaturated system. After acid treatment and epimerization of the liberated methyl ketone, Meyers and coworkers [327] obtained ketone **9.19** with an excellent enantiomeric excess (Figure **9.8**). This compound is a precursor of optically active grandisol. Photocycloaddition of chiral fumarate **9.20** with *trans*-stilbene is also highly stereoselective [1533] (Figure **9.8**).

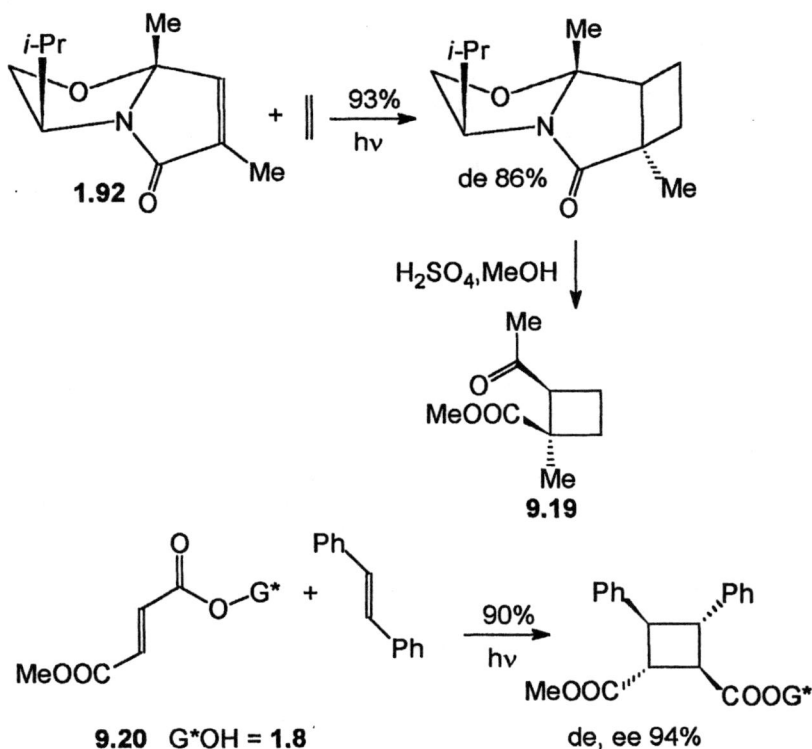

Figure 9.8

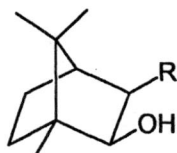

1.8

9.2 [3+2]-CYCLOADDITIONS [172]

1,3-Dipoles are the three-atom components of [3+2]-cycloadditions. These are 4π electrons species, so [3+2]-cycloadditions are thermally allowed according to the rules of orbital symmetry [624g]. The interactions between the frontier orbitals of the two partners are of prime importance, and concerted reactions take place in a supra-suprafacial fashion. When the 1,3-dipoles are not symmetrical, their reactions with unsymmetrical dipolarophiles may give regioisomers. Lack of regioselectivity limits some types of cycloadditions. Chirality can be introduced either on the dipole or on the dipolarophile.

9.2.1 Reactions of Diazoalkanes and Diazoesters

The reactions of diazoalkanes **9.21** with alkenes lead to pyrazolines **9.22**, which are thermally transformed into cyclopropanes. Similar transformations occur during thermal reactions of diazoesters. The use of diazoesters of chiral alcohols did not give useful results, so chiral residues have been introduced on the olefin dipolarophile. Meyers and coworkers [327] carried out the reaction of diazomethane **9.21** (R = R' = H) and diazopropane **9.21** (R = R' = Me) with chiral lactams **1.92** (R = *i*-Pr or *tert*-Bu, R' = Me). These 1,3-dipolar cycloadditions are regioselective, but only CH_2N_2 leads to an interesting stereoselectivity (Figure **9.9**). Morever, when the R" substituent of lactam **1.92** is H, the reaction is no longer stereoselective.

Chiral butadieneiron tricarbonyl derivatives **1.152** (R = Me or MeOOC, Y = CH=C(COOMe)$_2$ or CH=CHCOOMe) react readily with diazomethane or diazopropane [528]. The stereoselectivity of these reactions depends upon the *s-cis* or *s-trans* conformation of the polyene system. For example, **1.152** (R = COOMe, Y = CH=C(COOMe)$_2$) and Z-**1.152** (R = Me, Y = Z-CH=CHCOOMe) have an *s-trans* conformation, and their [2+2]-cycloadditions are stereoselective. The 1,3-dipole is introduced on the face opposite to the Fe(CO)$_3$ group. In contrast, the reaction of E-**1.152** (R = Me, Y = CH=CHCOOMe) is not stereoselective because the *s-cis* and *s-trans* conformers coexist. Analogous results are observed with **1.153** (Y = CH=CHCOOMe).

De Lange and Feringa [137, 203] have performed the cycloadditions of CH_2N_2 and ethyl diazoacetate to chiral lactone **1.31**, derived from menthol. A high stereoselectivity is only observed with ethyl diazoacetate. The reaction occurs on the least hindered face of the dipolarophile, and 2-pyrazoline **9.23** is obtained (Figure **9.9**).

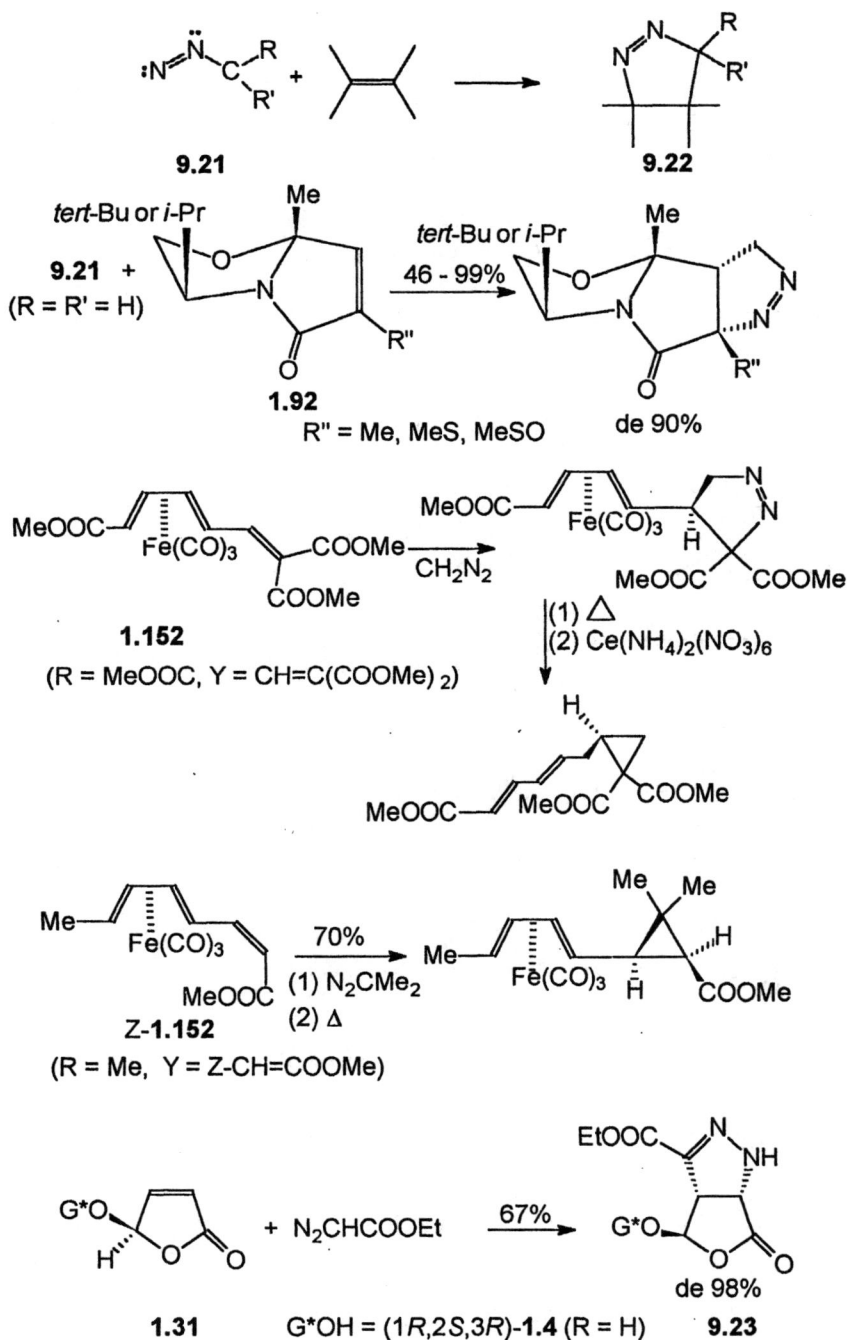

9.21

9.22

tert-Bu or i-Pr

9.21 +
(R = R' = H)

tert-Bu or i-Pr
46 - 99%

1.92
R" = Me, MeS, MeSO

de 90%

1.152
(R = MeOOC, Y = CH=C(COOMe) $_2$)

CH$_2$N$_2$

(1) Δ
(2) Ce(NH$_4$)$_2$(NO$_3$)$_6$

Z-1.152
(R = Me, Y = Z-CH=COOMe)

70%
(1) N$_2$CMe$_2$
(2) Δ

1.31   G*OH = (1R,2S,3R)-**1.4** (R = H)   **9.23**

+ N$_2$CHCOOEt   67%

de 98%

Figure 9.9

9.2.2 Reactions of Nitrile Oxides and Related Reagents

1.3-Dipolar cycloadditions of nitrile oxides with alkenes lead to 2-isox-azolines **9.24**. Heterocycles **9.24** are easily transformed with Raney nickel into β-hydroxyketones, so this two-step process is a strategic alternative to the aldol reaction. The reduction of **9.24** with $LiAlH_4$ gives β-aminoalcohols with retention of configuration of the stereocenters on the starting heterocycle [1535] (Figure **9.10**). When the dipolarophile is unsymmetrical, regioselectivity problems arise in the cycloaddition step. For example, the cycloaddition of benzonitrile oxide with crotonoylsultam **1.134** (R = E-MeCH=CH) gives a mixture of regioisomers [454, 1535]. These cycloaddition reactions involve early, reactant-like transition states (§ I.2.1).

Although chiral auxiliaries have been introduced on the dipolarophile, they often provide disappointing selectivities. Low selectivity is observed in reactions of nitrile oxides with the acrylates of sulfonamidoalcohols **1.10** [172], or with acrylamides of (*S*)-proline esters **1.64** (R = COOCH$_2$Ph) [212, 264] of pyrrolidine **1.65** (R = MeOCH$_2$), or of Evans's oxazolidinones **1.116** (R = *i*-Pr) and **1.117** [309]. Curran and coworkers [454, 1535] have proposed the use of acryloyl-sultams **1.134** (R = CH$_2$=CH) as dipolarophiles. They obtained 2-isoxazolines with a high regio- and stereoselectivity by reaction of **1.134** with nitrile oxides generated *in situ* in hexane at room temperature. The observed cycloadducts resulted from the approach of the upper face of the most populated *s-cis* conformer of **1.134** (R = CH$_2$=CH) (Figure **9.10**). This analysis was supported by X-ray crystallography data [454, 1535, 1536] and theoretical calculations indicate that the main interaction in such a diastereofacial differentiation is coulombic, not steric in origin [1537]. The availability of both enantiomers of **1.133** as well as the easy cleavage of the auxiliary with Li *s*-Bu$_3$BH make this an attractive approach to optically active 2-isoxazolines. Applications to the synthesis of natural products have been described by Curran and coworkers [454].

Chiral auxiliaries derived from Kemp's triacid have been designed by Curran and Rebek [446, 1538]. These auxiliaries have one face shielded by a nitrogen substituent, and **1.129** (R = (*S*)-phenethyl) gives high selectivity in nitrile oxide cycloadditions when attached to acrylamides **9.25** (Figure **9.10**). Two "pseudo-enantiomers" of **1.129** can be generated, and either enantiomeric cycloadduct may be obtained from the (S)-phenethylamine-based auxiliary. Sultams **1.135** (R = *tert*-Bu) have been proposed by Oppolzer and coworkers [454, 471] as chiral auxiliaries for cycloadditions of acrylimides **9.25**. These reactions are highly stereoselective (Figure **9.10**). Both enantiomers of **1.135** are available and the chiral auxiliary is removed with Li *s*-Bu$_3$BH. The acrylimide reacts in the *s-cis* conformation, and the observed facial diastereoselection is due to the repulsive interaction of the nitrile oxide oxygen with one of the sultam S-O bonds [454]. Kim and coworkers proposed (S)-proline derivatives as auxiliaries in these 1,3-dipolar cycloadditions [1539]. Among the various systems tested, the bicyclic acrylimide **9.26** provided a useful selectivity in asymmetric cycloaddition to arylnitrile oxides at −78°C (Figure **9.10**). The chiral auxiliary is removed and recovered as described above. Kanemasa and coworkers [308, 309, 1540] have used imidazolidines **1.73** (R = Ph, R' = Me) and oxazolidines **1.85** as chiral auxiliaries. The presence of substituents at the 2-position constrains the acrylimides **9.27** and **9.28** to the *s-cis* conformation. The cycloaddition of benzonitrile oxide to **9.27** is poorly stereoselective (60 - 80%). However, when suitable substituents are introduced on **9.28** (R = PhCH$_2$, R' = R'' = Me or R = Ph$_2$CH, R' = H, R'' = Me), the 1,3-dipolar cycloaddition is highly selective at 0°C. The chiral residue can be removed with Li s-Bu$_3$BH (Figure **9.10**).

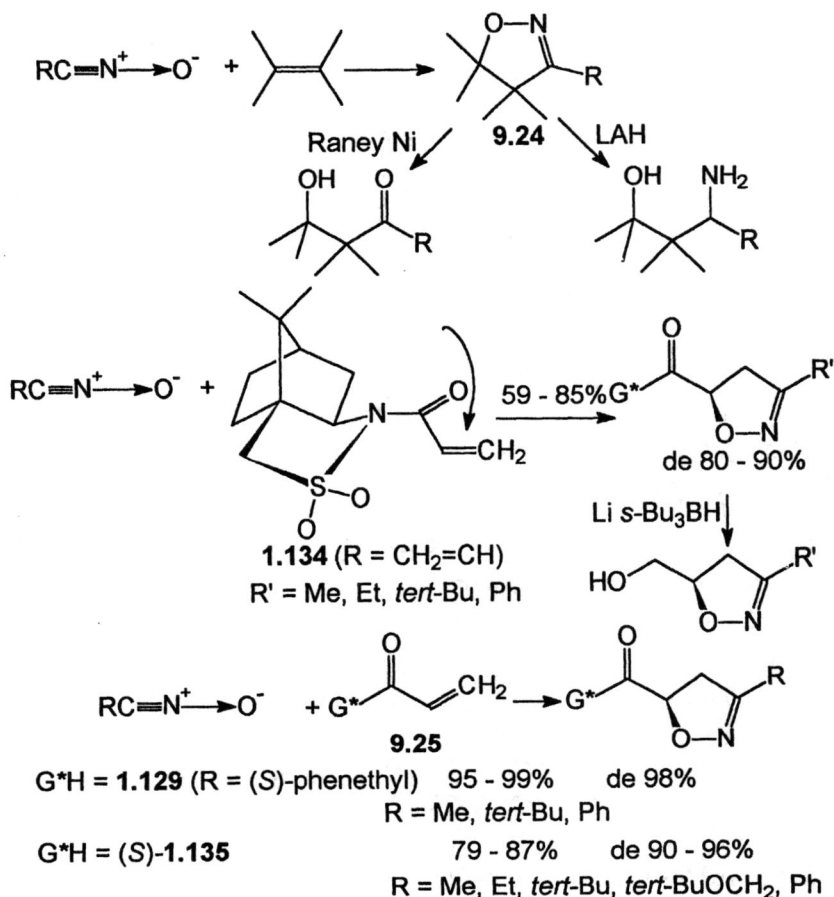

Figure 9.10

9.26

de 86 - 90%

9.27

> 99%

de 90 - 99%

9.28

R = PhCH$_2$, R' = R" = Me   R = Ph$_2$CH, R' = H, R" = Me

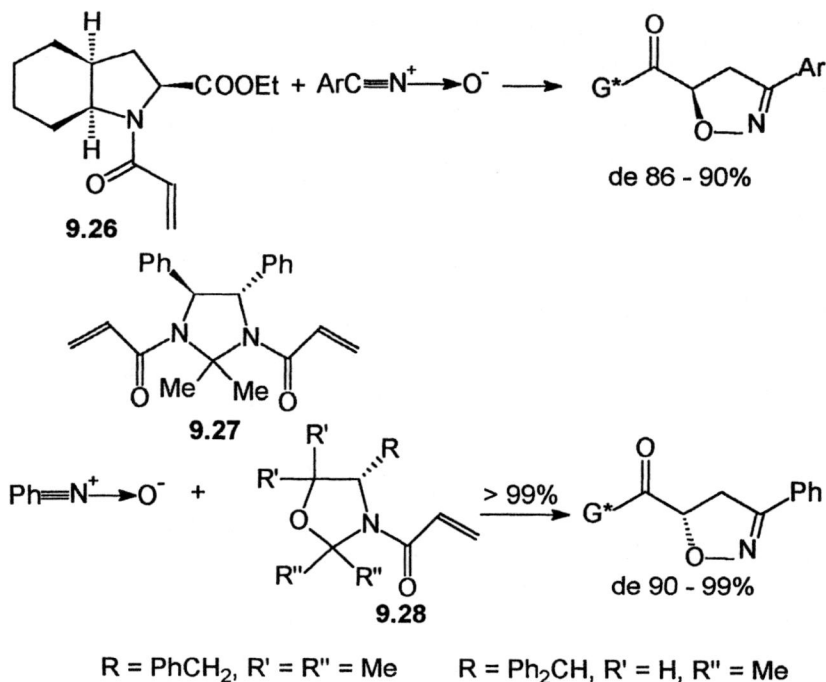

Figure 9.10 (continued)

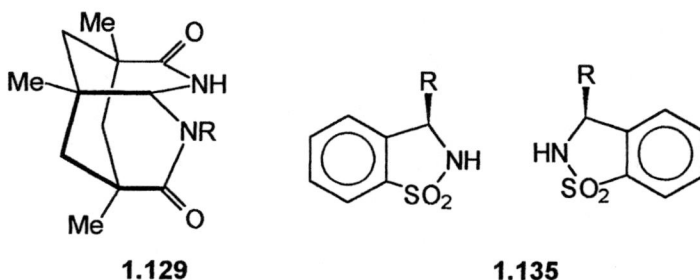

1.129   **1.135**

The cycloaddition of benzonitrile oxide to chiral lactone **1.31** is highly regio- and stereoselective [203] (Figure **9.11**). Likewise, nitrile oxide cycloadditions with chiral butadieneironcarbonyl complexes **1.152** (Y = CH=CH$_2$) give 2-isoxazolines **9.29** with a good selectivity. The organometallic complex probably has an *s-trans* conformation, and the 1,3-dipole is introduced on the face opposite to the iron tricarbonyl residue [1541] (Figure **9.11**).

Oxazoline *N*-oxides are nitrile oxide equivalents that cycloadd to electron-poor dipolarophiles such as crotonates. Langlois and coworkers [346] introduced chirality into these reagents. While regio- and stereoselective cycloadditions were observed, the chiral auxiliaries have not yet been removed.

Asymmetric 1,3-dipolar cycloaddition of nitrile oxides to allyl alcohol has been performed by Ukaji and coworkers [1542] in the presence of Et_2Zn and (*R,R*)-tartaric esters **2.69** in stoichiometric amounts at 0°C. Diisopropyl tartrate gives the best results, and 2-isoxazolines are obtained with an excellent enantioselectivity (Figure **9.11**).

Palladium catalyzed [3+2]-cycloadditions of 2-(trimethylsilylmethyl)-3-acetoxy-1-propene to electron-deficient double bonds bearing chiral auxiliaries is usually poorly stereoselective. However, face selectivity is interesting when using ephedrine-derived system **1.93** [352].

9.2.3 Reactions of Silylnitronates and Nitrones

The asymmetric 1,3-dipolar cycloaddition reactions of silylnitronates generated *in situ* from nitro compounds with olefins also give 2-isoxazolines **9.24**. Kim and Lee used sultams **1.134** (R = CH_2=CH) as dipolarophiles and observed a good stereoselectivity [454, 462] (Figure 9.11) in reactions with silylnitronates. Cycloadditions of nitrones bearing chiral substituents at nitrogen have been examined, but selectivity is low even in intramolecular reactions (70%) [172]. However, chromium tricarbonyl complexed arylnitrones **9.30** undergo 1,3-dipolar cycloadditions with electron-rich olefins [550]. These reactions are highly regio- and stereoselective, and after decomplexation with cerium ammonium nitrate, *cis*-3,5-disubstituted isoxazolidines are obtained with an excellent enantiomeric excess (Figure 9.11). If the nitrone is not complexed to chromium, a mixture of stereoisomers is obtained.

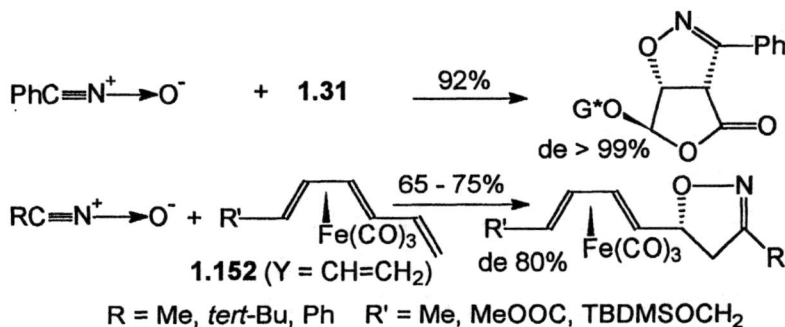

$$PhC\equiv\overset{+}{N}\!\!\longrightarrow\!O^- \quad + \quad \mathbf{1.31} \quad \xrightarrow{92\%}$$

de > 99%

$$RC\equiv\overset{+}{N}\!\!\longrightarrow\!O^- + R'\diagup\!\!\diagdown\underset{Fe(CO)_3}{\diagup\!\!\diagdown} \xrightarrow{65-75\%}$$

1.152 (Y = CH=CH$_2$)   de 80%

R = Me, *tert*-Bu, Ph   R' = Me, MeOOC, TBDMSOCH$_2$

Figure 9.11

$$RC\equiv N^+\!\!\rightarrow O^- + \quad \diagup\!\!\!\diagdown CH_2OH \xrightarrow[Et_2Zn,\ (R,R)\text{-}\mathbf{2.69}]{64\ \text{-}\ 92\%} HOCH_2\cdots$$

ee 93 - 96%

R = n-C₇H₁₅, *tert*-Bu, Ph, 4-MeOC₆H₄, 4-ClC₆H₄

R = $n\text{-}C_7H_{15}$, *tert*-Bu, Ph, 4-MeOC₆H₄, 4-ClC₆H₄

$$80\ \text{-}\ 95\%$$

de 80%

R = H, Me, Et, $n\text{-}Bu$, $n\text{-}C_5H_{11}$, EtOOC, Ph

60 - 62%
then Ce(NO₃)₆
(NH₄)₂

de, ee 96 - 97%

9.30

R = Ph, OEt

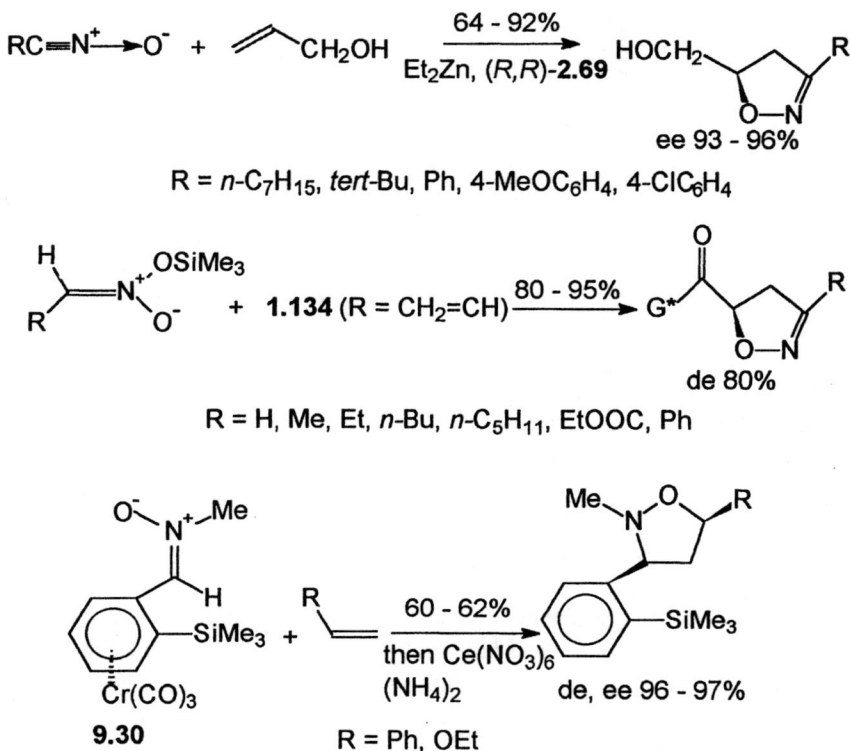

Figure 9.11 (continued)

9.2.4 Reactions of Azomethine Ylids

Azomethine ylids **9.31** are usually generated by deprotonation of arylimines of α-aminoesters. Their cycloadditions with electron-poor olefins are controlled by the interaction between the HOMO of the 1,3-dipole and the LUMO of the dipolarophile, and they are often regioselective (Figure **9.12**). Cycloaddition reactions can be carried out by treating arylimines with a tertiary amine and lithium, silver, titanium, manganese or cobalt salts [171]. In some cases, asymmetry can be induced by use of a chiral tertiary amine. Indeed, Alway and Grigg [1543] reacted glycine imines with methyl acrylate in the presence of CoCl₂ and chiral tertiary aminoalcohols **1.14**. Cycloadducts were obtained at room temperature in the absence of solvent. The best selectivities are observed with **1.14** (NR₂ = N(CH₂)₄) (Figure **9.12**). The reaction probably occurs through a chelated metallodipole **9.32** (Figure **9.12**). Related cycloadditions were performed by Grigg and coworkers [171] with (1R,2S,5R)-menthyl acrylate and imines of α-aminoesters. The regioselectivity depends on the reaction conditions. In the presence of AgOAc or

LiBr/Et$_3$N, **9.33** is obtained with a high selectivity, while the use of Ti(Oi-Pr)$_3$Cl leads selectively to **9.34**. The chiral auxiliary is removed with LiAlH$_4$ (Figure **9.12**). According to the absolute configuration of the chiral alcohol, either enantiomer of the final product is obtained. The 1,3-dipole is introduced on the least hindered face of the acrylate. The reaction has been extended to imines of isatin, but it is less selective [1544].

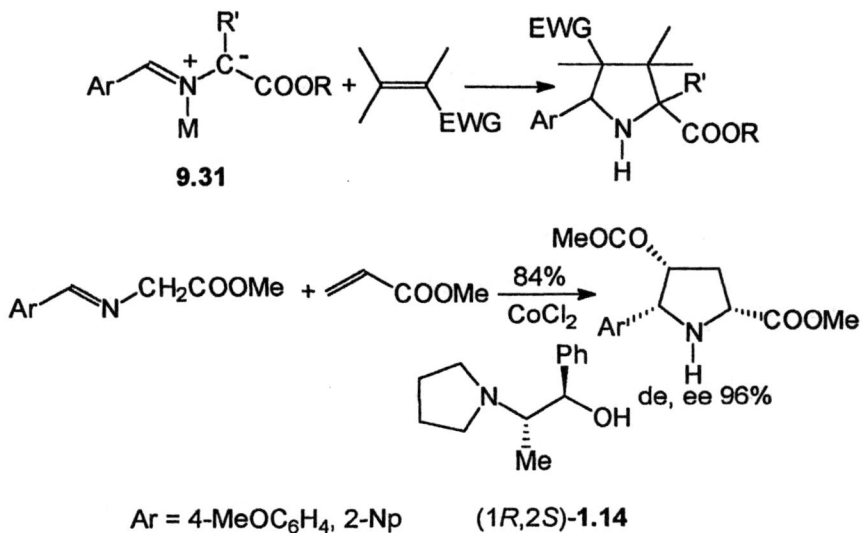

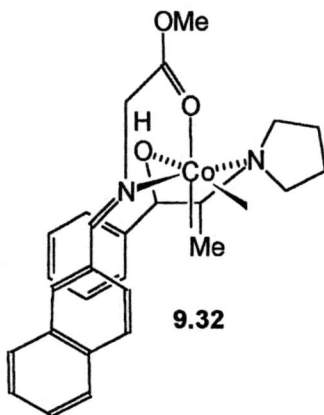

Figure **9.12**

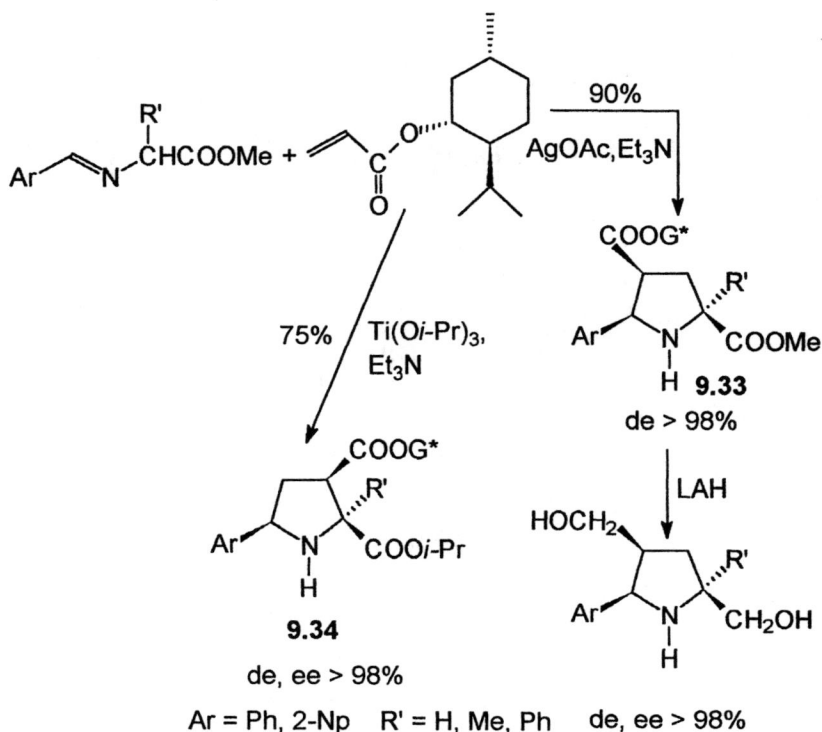

Figure 9.12 (continued)

Kanemasa and coworkers [1545, 1546] introduced dipolarophiles that are chiral aminal derivatives of α,β-unsaturated aldehydoesters. Aminal **7.61**, previously used by Mukaiyama (§ 7.10.2.2), and imidazolidines **9.35** derived from C_2-symmetric diamines react with the enolates of methyl or *tert*-Bu *N*-benzylideneglycinates **9.31** (Ar = Ph, R' = H, R = Me or *tert*-Bu). The reaction of **7.61** at room temperature in the presence LiBr and DBU gives a single cycloadduct which, after *N*-tosylation and transketalization with methanol, gives pyrrolidine **9.36** with an excellent selectivity (Figure **9.13**). The same compound **9.36** is obtained when the lithium enolate of **9.31** is reacted with **9.35** (R' = Ph) at –78°C. When the dipolarophile is **9.35** (R' = Me), the other enantiomeric pyrrolidine *ent*-**9.36** is formed (Figure **9.13**). Thus the face selectivity of the process is controlled by the nature of the nitrogen substituents. If the cycloaddition takes place via chelated complex **9.37** (Figure **9.13**), the reactive conformation must differ depending on whether R' = Ph or Me [1545].

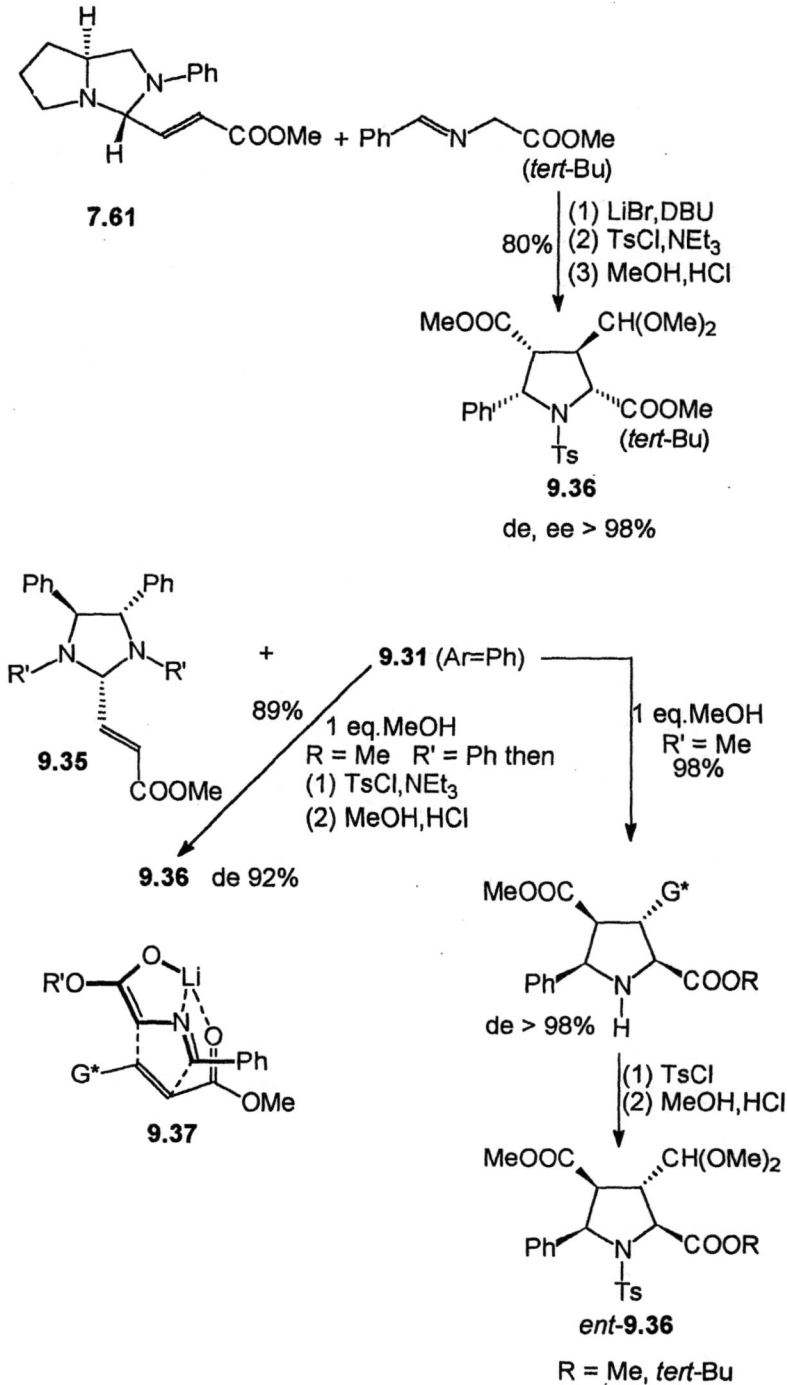

Figure 9.13

9.3 [4+2]-CYCLOADDITIONS [73, 74, 170, 372, 1547, 1548]

The Diels-Alder reaction is the most popular [4+2]-cycloaddition, and it has been used to elaborate the carbon backbone of many natural products [624c]. The Diels-Alder reaction occurs between a diene (4 π) and a dienophile (2 π), and is often catalyzed by Lewis acids. Normal Diels-Alder reactions occur between electron-rich dienes and electron-poor dienophiles, and inverse electron demand Diels-Alder reactions occur between electron-poor dienes and electron-rich dienophiles. The hetero-Diels-Alder reaction is a useful method to elaborate heterocycles. Either the 4 π or the 2 π component can bear one or more heteroatoms [74, 1547, 1549]. In most of the cases, these cycloadditions are concerted and under frontier orbital control. The regioselectivity is in agreement with recent theoretical calculations [11, 1550, 1551]. In the past, secondary orbital interactions have been used to interpret the favored *endo* approach of dienes to electron-poor olefins. However, recent calculations [89, 1550, 1552, 1553] have suggested that these interactions are negligible compared to coulombic and polar effects. These calculations have also underlined that the formation of the two new carbon-carbon bonds is asynchronous [89, 1550, 1553], as previously proposed by Woodward and Hoffmann. It also seems that π-donor-π-acceptor interactions between the substituents of the diene and of the dienophile (π-stacking) are often not strong enough to provide high facial selection [46, 48, 1554]. However, Thornton has suggested the possibility of facial discrimination due to intramolecular hydrogen bonding [47].

Chirality can be introduced on the Lewis acid catalyst, on the dienophile or on the diene. The approach to either face of the prochiral reagent will depend upon the interplay of various interactions. The reactive conformations (§ I.3, I.4.2) of the partners are of prime importance [11, 83, 454, 1550, 1555]. As already discussed (§ 7.10.2.2), the predominant conformation of α,β-unsaturated esters in the ground state is *s-cis* but the *s-trans* conformer is favored in the presence of Lewis acids [85, 778]. However, if the ester bears another functional group, the Lewis acid can chelate both groups and favor the *s-cis* conformation (Figure 9.14). For example, the *s-cis* conformation of the acrylate of (*S*)-ethyl lactate **9.38** is favored in the presence of TiCl$_4$ [158, 178, 1555]. Nonfunctionalized tertiary α,β-unsaturated amides always favor the *s-cis* conformation due to the A(1,3) strain. Nevertheless, secondary acrylamides derived from α-aminoesters **9.39** may be stabilized in the *s-trans* conformation by intramolecular hydrogen bonding [1555] (Figure 9.14). Moreover, Corey and coworkers [44, 45] have provided evidence for π-donor-π-acceptor interactions between dienophiles and aromatic substituents of chiral Lewis acids. This can induce some rigidification of these complexes (see below). Extending this ground-state analysis, transition-state models have also been calculated and compared to experimental data [11, 53, 83, 90, 653, 1550, 1553, 1556, 1557]. According to these calculations the [4+2]-

cycloadditions of butadiene or cyclopentene with isoprene, acrolein and methyl acrylate take place through the *s-cis* conformer of the dienophile in an *endo* fashion in the gas phase or in nonpolar solvents. In the presence of Lewis acids or in polar protic solvents, acrolein still reacts through the *s-cis* conformation, but methyl acrylate reacts through the *s-trans* conformer. Nevertheless, experimental results of Yamamoto and coworkers [1558] show that a different situation holds with methacrolein (see below).

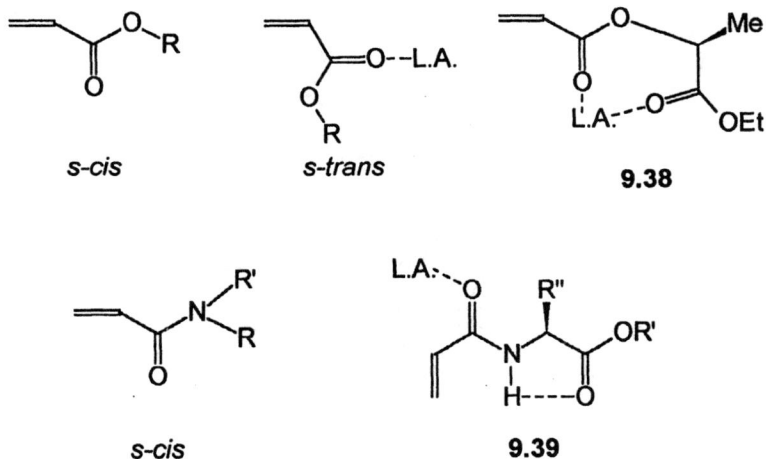

Figure 9.14

9.3.1 Cycloadditions Catalyzed by Chiral Lewis Acids [559, 777a, 778, 780, 1008, 1063]

9.3.1.1 *Boron Derivatives* [601]

The efficiency of haloboranes as Lewis acid catalysts and the sensitivity of the boron atom environment to steric hindrance (see Chapter 6) make boron compounds useful chiral catalysts. Hawkins and Loren [779] prepared enantiopure chloroborane (1*R*,2*R*)-**3.6**, and they showed that the complex **3.6** with methyl crotonate lies under the *s-trans* conformation **9.40** both in the solid state and in solution. This conformation is also the reactive one, and [4+2]-cycloadditions of acrylates with cyclopentadiene or cyclohexadiene catalyzed by (1*R*,2*R*)-**3.6** lead highly selectively to the product resulting from the attack of the diene on the unshielded face of complex **9.40** (Figure 9.15).

Catalysts have been generated *in situ* from binaphthols **3.7** and BH$_3$, H$_2$BBr or B(OAr)$_3$. The catalyst generated from BH$_3$ and **3.7** (R = H) induces the asymmetric cycloaddition of some dienes with quinones **9.41** [778], while the catalyst formed from **3.7** (R = H) and H$_2$BBr is useful in cycloaddition of methacrolein with cyclopentadiene [559, 781] (Figure **9.15**). Yamamoto and coworkers have performed asymmetric aza-Diels-Alder reactions at –78°C between imines and 2-trimethylsilyloxydienes **3.26** [783, 1559, 1560]. When the reaction is conducted with *C*-aryl-*N*-benzylimines and **3.26** in the presence of about 1 equivalent of a catalyst generated from binaphthol **3.7** (R = H) and B(OPh)$_3$, dihydropyridones **9.42** are obtained with a high enantiomeric excess after acid treatment (Figure **9.15**). Replacement of B(OPh)$_3$ by B(O-3,5-xylyl)$_3$ slightly increases the enantioselectivity. The cycloaddition of *C*-cyclohexyl-*N*-benzylimine with **3.26** under the same conditions gives a lower chemical yield and a poor ee, and other *N*-substituted imines also lead to disappointing results. Double asymmetric induction occurs when (*S*)-phenethylimine derivatives are paired with the same binaphthol-derived catalyst (see below). Mukaiyama and coworkers [784] proposed a catalyst generated *in situ* from BBr$_3$ and the (*S*)-prolinol derivative **2.13** (R = CPh$_2$OH) for use in the [4+2]-cycloaddition of cyclopentadiene and α,β-unsaturated aldehydes. However, a highly selective reaction is observed only with methacrolein (ee 97%).

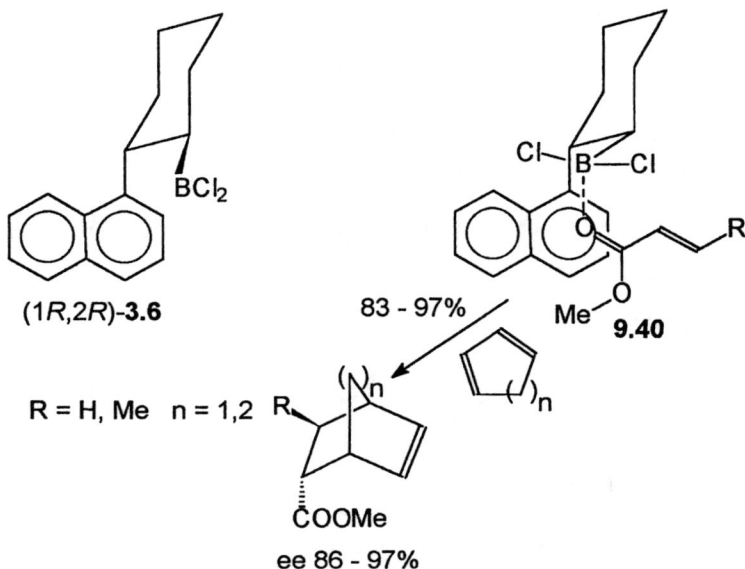

(1*R*,2*R*)-**3.6**

83 - 97%

Me $\overset{O}{\diagup}$ **9.40**

R = H, Me   n = 1,2

ĊOOMe

ee 86 - 97%

Figure 9.15

R = H, R' = OAc; R = Me, R' = OSiMe₃

R = H, Me   Ar = Ph, 3-Py, 3,5-(MeO)₂C₆H₃, 2-Np
Ar' = CH₂Ph, CH₂C₆H₃(OMe)₂-3,4

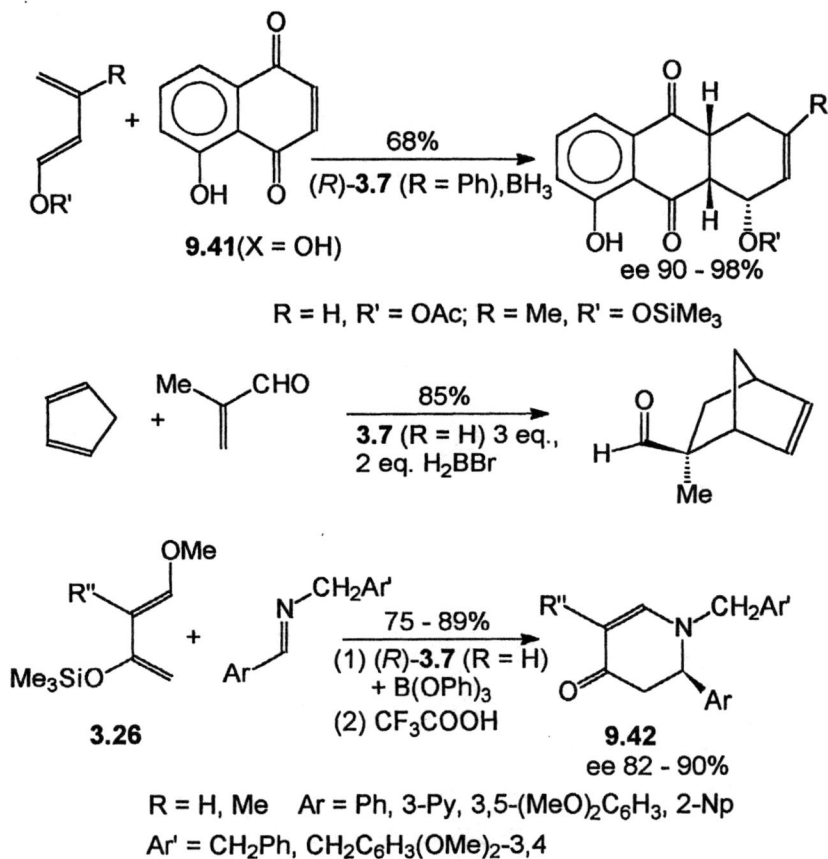

Figure 9.15 (continued)

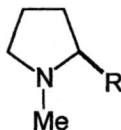

2.13

Boron catalysts bearing the tartaric acid skeleton have received more applications.   Borate **3.8** is an efficient catalyst in the cycloaddition of 1-tri-
ethylsilyloxydiene **9.43** with quinone **9.41** (X = OH) [778] (Figure 9.16). Acy-
loxyborates **3.9** (CAB) introduced by Yamamoto and coworkers [559, 778, 785,
789, 1558, 1561] catalyze the cycloadditions of α,β-unsaturated aldehydes with

cyclic and acyclic dienes. The reactions occur with 10 - 20 mol% catalyst between
−78°C and −40°C with an excellent selectivity, provided that the suitable substituents R and R' are introduced on **3.9**. The [4+2]-cycloadditions of acrolein require
3.9 (R = H, R' = Me) as a catalyst to observe a good selectivity (Figure **9.16**). The
reactions of 2-bromoacrolein or of methacrolein yield high selectivities with **3.9**
(R=H, R'=Me) as well as with a somewhat bulkier catalyst **3.9** (R =
2-$C_6H_5OC_6H_4$, R' = i-Pr) (Figure **9.16**). However, reactions of crotonaldehyde
always lead to poor selectivities. Furthermore, the face discrimination is the reverse when using acrolein compared to 2-substituted enals. Acrolein reacts
through the s-cis conformation, even in the presence of relatively small Lewis acids
(see above). 2-Substituted enals prefer the s-$trans$ conformation whatever the
Lewis acid, in both the ground state (as shown by NMR [1558]) and the transition
state. The poor selectivities observed with crotonaldehyde are due to competing
reactions of both conformers. Intramolecular cycloadditions under CAB catalysis
also give useful results [785]. Hetero-Diels-Alder reactions with silyloxydienes
3.26 and aldehydes at −78°C also give high selectivities. In these cases, the best
catalyst is **3.9** (R = 2-$MeOC_6H_4$, R' = i-Pr) [789, 790]. Both enantiomers of **3.9**
are available, so either enantiomeric product can be formed.

A more elaborate boron catalyst derived from a substituted (R)-binaphthol
3.7 (2-HOC_6H_4) and BH_3 was proposed in 1994 by Yamamoto and Ishihara
[1562]. This induces extremely high selectivities in [4+2]-cycloadditions of α-substituted α,β-unsaturated aldehydes at −78°C (94 - 99% de, 92 - 99% ee). The enal
lies under the s-$trans$ conformation and an attractive π-donor-π-acceptor interaction favors coordination of the dienophile on the face of boron that is cis to the
2-hydroxyphenyl substituent. The diene is introduced on the least hindered face of
this complex **9.44** (Figure **9.16**).

Oxazaborolidines **3.10** have been developed as catalysts for Diels-Alder reactions [611] by Corey [44, 1563, 1564], Helmchen [792], Yamamoto [793] and
their coworkers. The tryptophan **3.11** derived catalyst used in 10 mol% amounts
has shown the greatest efficiency in the cycloaddition of cyclopentadiene or furan
with α-substituted α,β-unsaturated aldehydes at −78°C (Figure **9.16**). Acrolein
leads to disappointing results. The introduction of (αS,βR)-β-methyltryptophan as
the catalyst precursor improves the selectivity [1564]. The rigid transition-state
model **9.45** has been proposed to interpret the high selectivities. This model features a π-donor-π-acceptor interaction between the dienophile and the catalyst
[1565]. The aldehyde reacts through the s-cis conformation and the diene is introduced on the least hindered face of the complex. As already mentioned (see
above), the NMR study of the methacrolein-BF_3 complex [88] shows that the
s-$trans$ conformer is highly favored. Therefore it is interesting that according to

the Lewis acid (**3.9** or **3.10**), the boron-coordinated α-substituted enals react either through the same conformation as the ground state (**3.9**) or a different one (**3.10**).

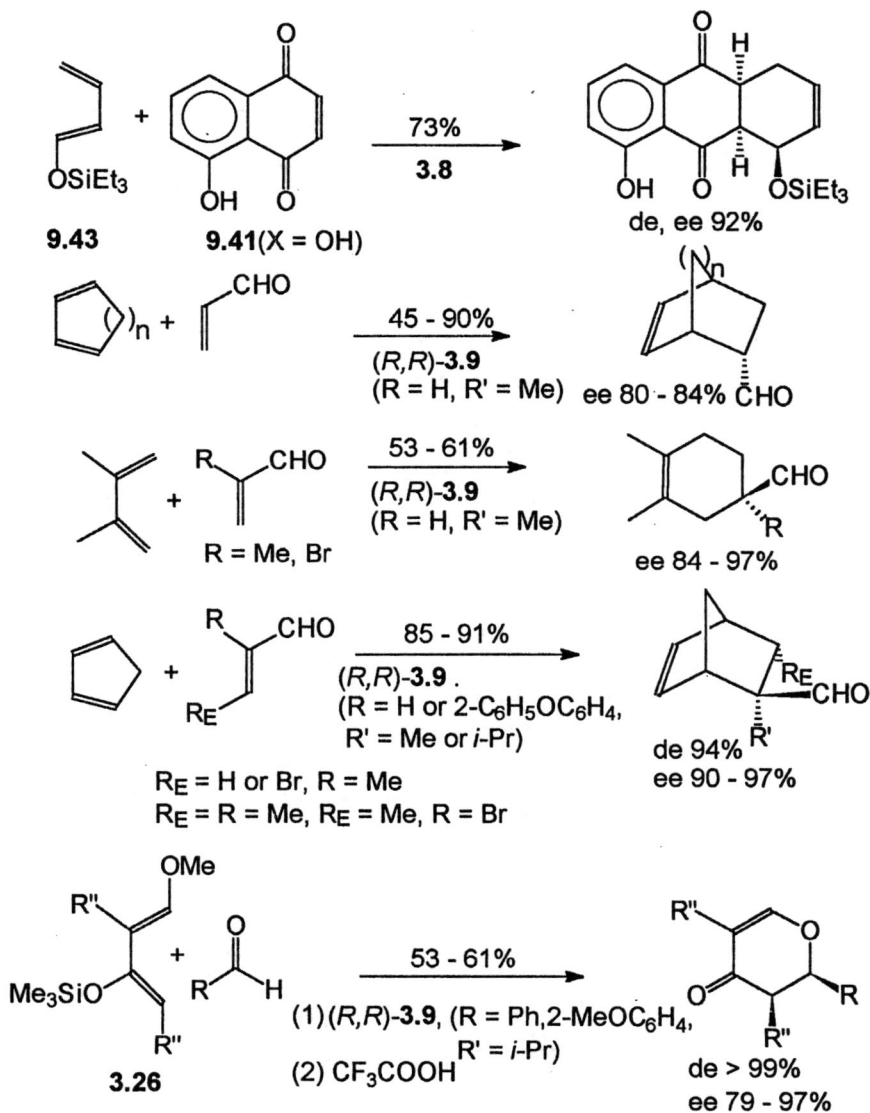

R" = H, Me   R = Ph, PhCH=CH, MeCH=CH, 2-furyl

Figure 9.16

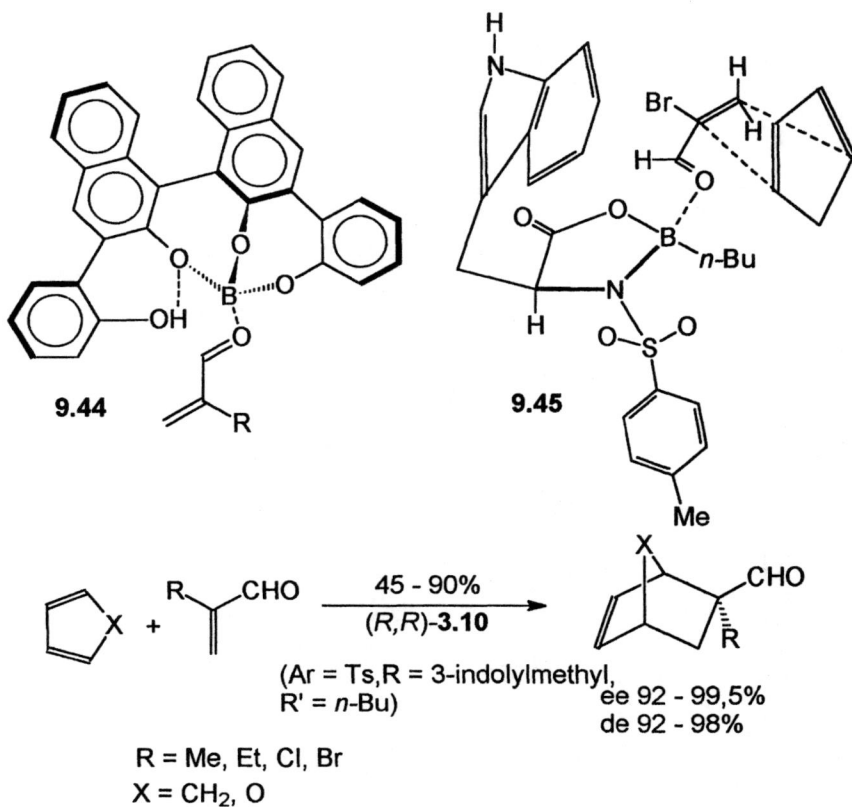

9.44

9.45

$$R \overset{\text{CHO}}{\diagdown} + \quad \xrightarrow[\ (R,R)\text{-}\mathbf{3.10}\ ]{45 - 90\%}$$

(Ar = Ts, R = 3-indolylmethyl,
R' = *n*-Bu)

ee 92 - 99,5%
de 92 - 98%

R = Me, Et, Cl, Br
X = CH$_2$, O

Figure 9.16 (continued)

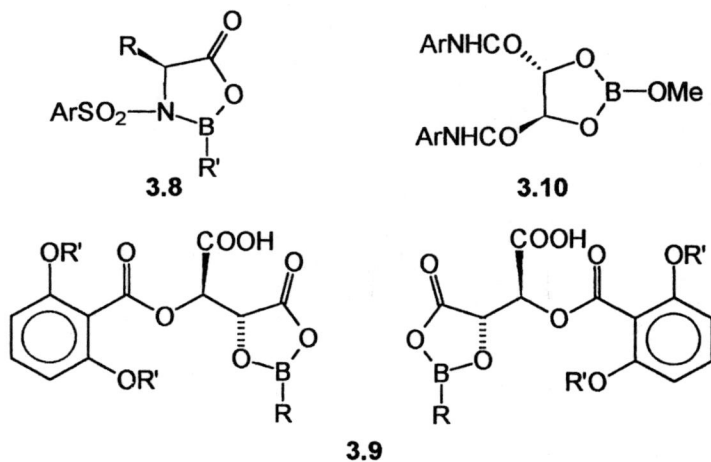

3.8

3.10

3.9

9.3.1.2 *Aluminum Derivatives*

Asymmetric catalysis of Diels-Alder reactions was first attempted with aluminum alcoholates of chiral alcohols and diols [559, 778, 798, 801, 802], and useful selectivities were observed in some cycloadditions of cyclopentadiene at $-78°C$. Chapuis and Jurczak [778] used imide **7.68** (R = H, R_E = H, Me) as a dienophile with the aluminum alcoholate of **1.35** (R = CH$_2$OTs) (Figure **9.17**). Kagan and coworkers used methacrolein as dienophile at $-100°C$ with a catalyst prepared from **1.32** (Figure **9.17**). Aluminum alcoholates of substituted binaphthols **3.7** (R = SiPh$_3$ or Si(3,5-Me$_2$C$_6$H$_3$)$_3$) in catalytic amounts have been proposed by Yamamoto and coworkers for hetero-Diels-Alder reactions of silyloxydienes **3.26** and aldehydes [249, 778]. The efficiency of these catalysts is similar to that of CAB **3.9** (see above) (Figure **9.16**). Wulff and coworkers [800] introduced chloroaluminum complexes of vaulted biphenols to catalyze the cycloaddition of methacrolein with cyclopentadiene. Excellent selectivities were observed at $-80°C$, and the aging of the catalyst had considerable effects. Alkylaluminum catalysts **9.46** prepared from DIBAH or Me$_3$Al and disulfonamides **3.14** bearing a C$_2$ axis of symmetry have been used by Corey and coworkers [693, 1566]. When used in 10 - 50% molar amounts at $-78°C$, these compounds catalyze the cycloadditions of cyclopentadiene and α,β-unsaturated imides **7.68** (R = H) or acrylates of chiral alcohols. High selectivities are observed with imides, while with chiral alcohols only the matched reagent pair gives excellent selectivity (Figure **9.17**). An NMR study of the complex **7.68** (R_E = R = H) with (*S,S*)-**9.46** was used to propose an encounter complex **9.47** as a transition-state model [1566] (Figure **9.17**).

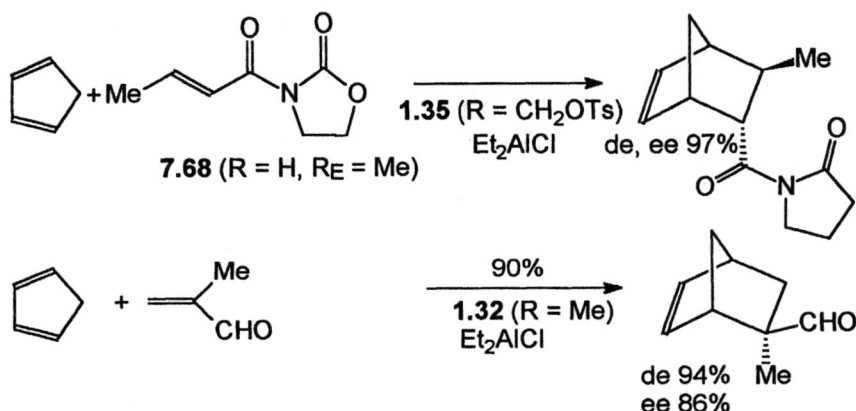

Figure 9.17

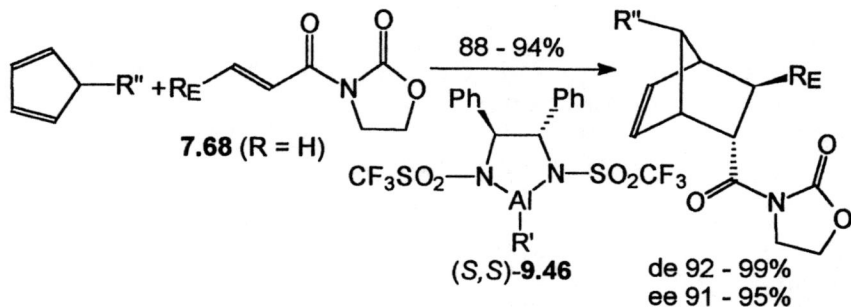

7.68 (R = H)

(S,S)-**9.46**

de 92 - 99%
ee 91 - 95%

R" = H, PhCH$_2$OCH$_2$ R = H, Me   R' = *i*-Bu, Me

88 - 94%

85%
(R,R)-**9.46** (R = *i*-Bu)

G*OH = (1R,2S,5R)-**1.14** (R = H)

de 97%

9.47

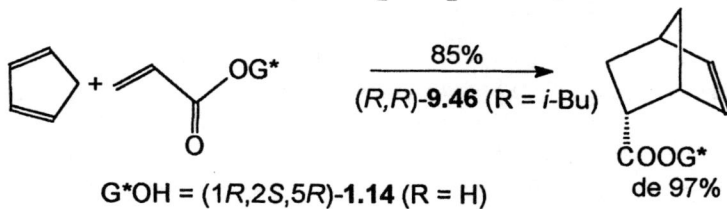

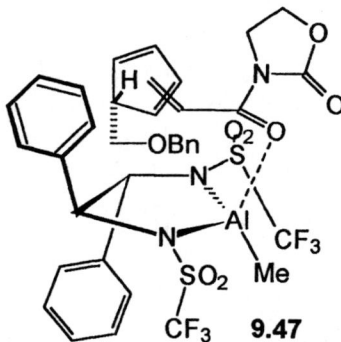

Figure 9.17 (continued)

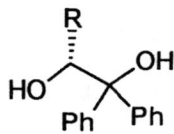

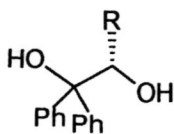

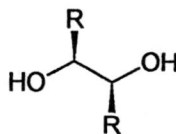

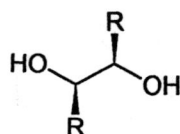

(R)-**1.32**   (S)-**1.32**   (S,S)-**1.35**   (R,R)-**1.35**

9.3.1.3 *Iron, Copper and Magnesium Derivatives*

Corey and coworkers [693, 1567] prepared iron and magnesium complexes of bis-oxazolines **3.28** (R = Ph, R" = Me, R' = H or Me) as catalysts for cycloadditions of cyclopentadienes and acylimide **7.68** (R = R_E = H). Selectivities as high as those with **9.46** were observed. Khiar and coworkers proposed iron complexes of chiral bis-sulfoxides for similar purposes, but observed lower enantioselectivities [1568]. Evans and coworkers [852, 1569] selected copper complexes to perform cycloadditions of cyclopentadiene with either α,β-unsaturated imides **7.68** (R = H, R_E = H, Me, Ph, EtOCO) or more reactive thio analogs **9.48**. The most efficient copper ligands are **3.28** (R = *tert*-Bu, R' = H, R" = Me) and **3.29** (R = 2,6-$Cl_2C_6H_3$). These complexes are formed *in situ* with $Cu(OTf)_2$ and are used in catalytic amounts. The reactions are run between –78°C and –15°C according to the nature of R_E. High *endo* selectivities and enantioselectivities are observed when using **3.28** (R = *tert*-Bu, R' = H, R" = Me) as Cu ligand with both types of dienophiles. However, with **3.29**, a useful *endo* selectivity is only observed with **9.48** (R_E = Me, Ph, COOEt) (Figure **9.18**). Double asymmetric induction has been examined with **7.68** (R = $PhCH_2$, R_E = H), and if both copper ligand and dienophile are matched, high asymmetric induction takes place [1569]. The sense of asymmetric induction in these copper-catalyzed cycloadditions has been rationalized by assuming that the reaction proceeds via a square-planar Cu (II)-dienophile complex which reacts on its least hindered face.

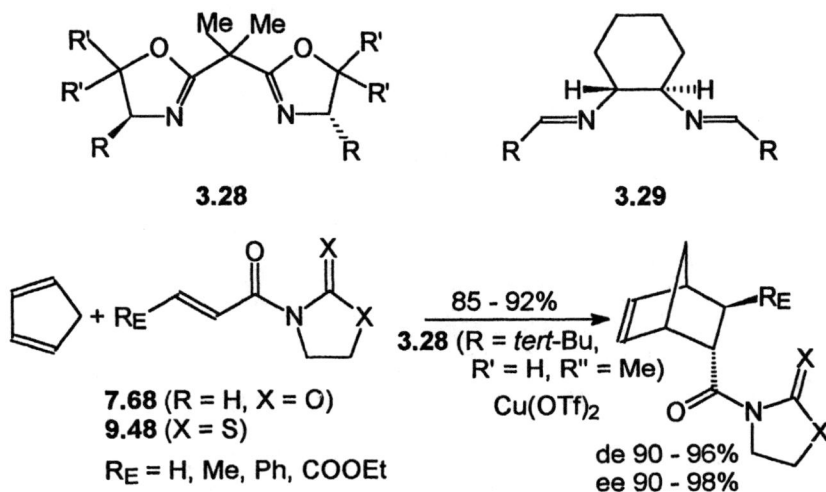

Figure 9.18

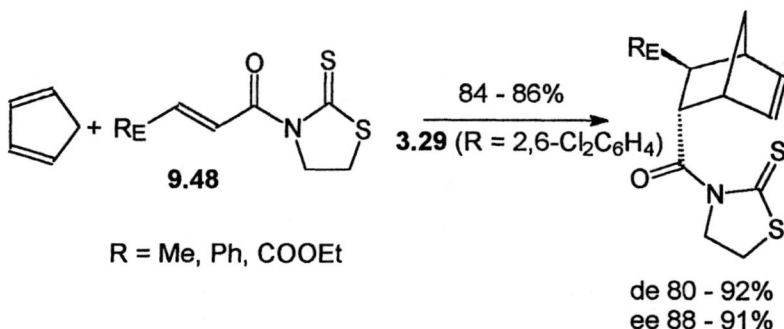

R = Me, Ph, COOEt

Figure 9.18 (continued)

9.3.1.4 *Titanium Derivatives* [666]

A chiral titanium complex generated *in situ* from $TiCl_4$, *i*-PrOH and diace-toneglucose **1.48** is an efficient catalyst for some intramolecular hetero-Diels-Alder reactions [827] (Figure **9.19**). The enantioselectivity depends upon the nature of the aryl substituent. Titanium alcoholates of chiral diols or diphenols are the most popular catalysts and binaphthols **3.7** are especially useful. The catalyst generated from the silyether of **3.7** (R = Ph) and $TiCl_4$ is efficient in asymmetric cycloaddition of cyclopentadiene and imides **7.68** (R = H) at –78°C (ee 96 - 98%) [778]. The use of methacrolein or methyl acrylate as dienophiles gives less useful results [778]. However, introduction of bulkier substituents on binaphthol such as R = 2-HO, 3-Si-*tert*-BuPh$_2$, 5-MeC$_6$H$_2$ induces a helical structure of the titanium complex [1570] and improves the selectivity. In the presence of 10 mol% of these catalysts, the cycloadditions of cyclopentadiene with acrolein or methacrolein are highly diastereo- and enantioselective (de, 86 - 98%; ee, 92 - 94%). Mikami and coworkers prepared a catalyst from binaphthol **3.7** (R = H), (*i*-PrO)$_2$TiX$_2$ (X = Cl or Br) and molecular sieves [821, 1571]. If the molecular sieves are not removed from the mixture, the catalyst thus formed (10 mol%) is useful in Diels-Alder reactions of butadienyl acetate or carbamate **9.49** (R = Me or Me$_2$N) with meth-acrolein (Figure **9.19**) and in reaction of 1-methoxybutadiene with naphthoquinone **9.41** (X = H) (Figure **9.19**). These reactions occur between –30°C and room temperature. If the catalyst is removed from the molecular sieves, then the reaction of 1-methoxybutadiene with juglone **9.41** (X = OH), which was poorly enantioselective in the presence of sieves, now gives useful results. Such is also the case for hetero-Diels-Alder reactions between 1-methoxybutadienes and methyl glyoxylate at –30°C (Figure **9.19**).

G*OH = **1.48**

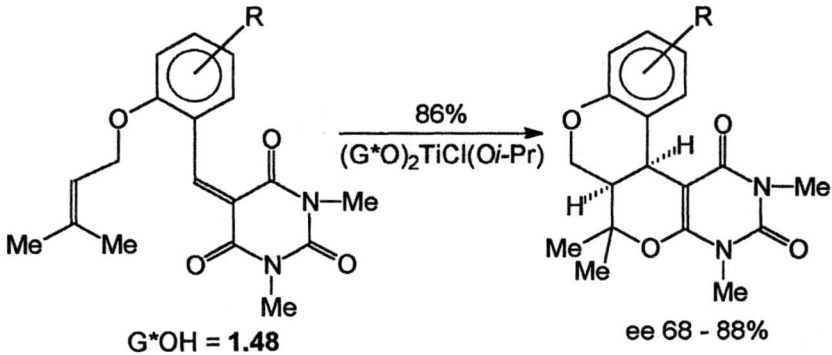

R = H, 3-MeO, 4-MeO, 5-MeO

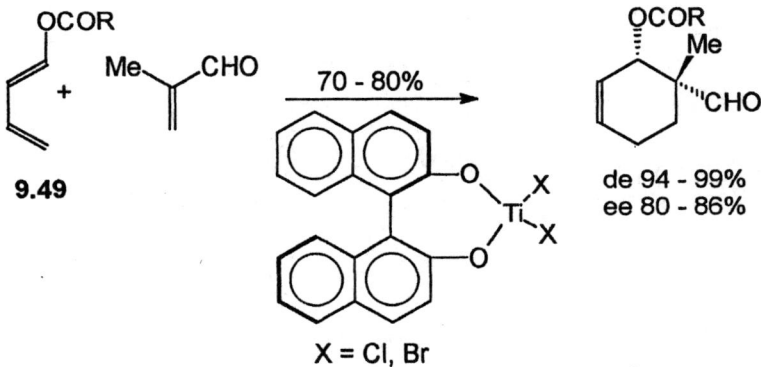

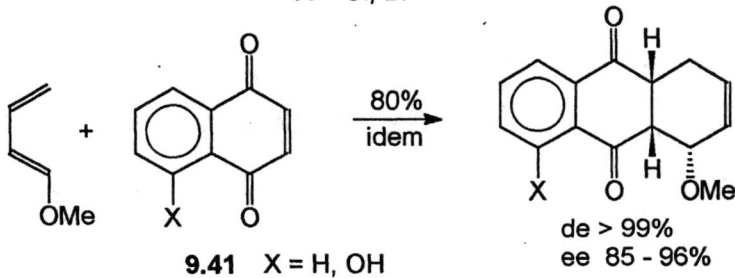

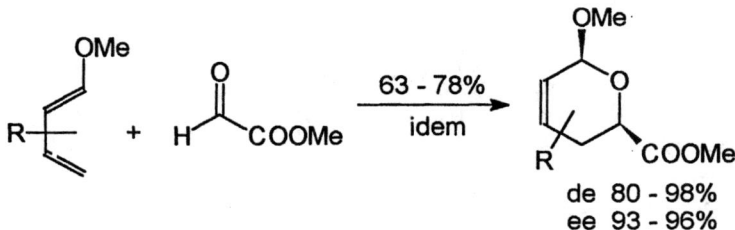

ee 68 - 88%

9.49

70 - 80%

X = Cl, Br

de 94 - 99%
ee 80 - 86%

80%
idem

9.41   X = H, OH

de > 99%
ee 85 - 96%

63 - 78%
idem

de 80 - 98%
ee 93 - 96%

Figure 9.19

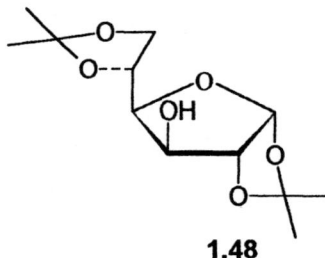

1.48

Titanium alcoholates of diols **2.50** generated by exchange with $(i\text{-PrO})_2\text{TiCl}_2$ are also potent catalysts, and the most efficient catalyst is derived from **2.50** (R = Me, R' = Ar=Ph). These catalysts promote the diastereo- and enantioselective cycloaddition of cyclopentadiene or acyclic dienes with some α,β-unsaturated imides **7.68** (R = H, R_E=Me, MeOOC) [45, 778] or of acyclic dienes with substituted benzoquinones at $-78°C$ [1572, 1573] (Figure **9.20**). Reactions occur at $0°C$ in the presence of molecular sieves, and only catalytic amounts of these alcoholates are used. Intramolecular cycloadditions also give useful selectivities under these conditions [1574]. However, the process for preparation of the catalyst varies according to the reaction [778, 1573], and this is very important to the observation of high selectivities.

Corey and Matsumura modified the structure of diol **2.50** in order to rigidify the titanium alcoholate-dienophile complex through a π-donor-π-acceptor interaction by replacing the phenyl groups with 3,5-xylyl groups. The titanium catalysts were generated from **2.50** (Ar = $3,5\text{-Me}_2\text{C}_6\text{H}_3$, R = R' = Et) and cycloadditions of cyclopentadienes with α,β-unsaturated imides **7.68** (R = H, R_E = H, Me) gave excellent selectivities at $-30°C$ in the presence of 20 mol% of the catalyst (Figure **9.20**). The reaction of **7.68** (R = H, R_E = MeOOC) was somewhat less stereoselective. Quinkert and coworkers [1575, 1576] used a related catalyst derived from **2.50** (R = R' = Et, Ar = 9-phenanthryl) in excess to perform the enantioselective cycloaddition of a precursor of optically active estrone (Figure **9.20**). The use of Corey's catalyst in this reaction leads to a lower enantioselectivity (73%). The face selectivity has been interpreted by the formation of an octahedral titanium complex **9.50**, which reacts with cyclopentadiene on its least hindered side. According to Corey and coworkers, the ligand adopts an apical-equatorial orientation, while reagent **7.68**, in the s-$trans$ conformation, is coordinated to the titanium atom by its two carbonyl groups in equatorial positions. Seebach and coworkers [653] have proposed a different geometry **9.51** in which the ligand is located in the equatorial plane and the reagent, in its s-cis conformation, occupies one equatorial and one axial site (Figure **9.21**).

Recently, it was suggested by Corey and coworkers [828] that *cis-N*-sulfonyl-2-amino-1-indanols can be used to generate titanium complexes that are able to catalyze the asymmetric cycloaddition of 2-bromoacrolein with dienes.

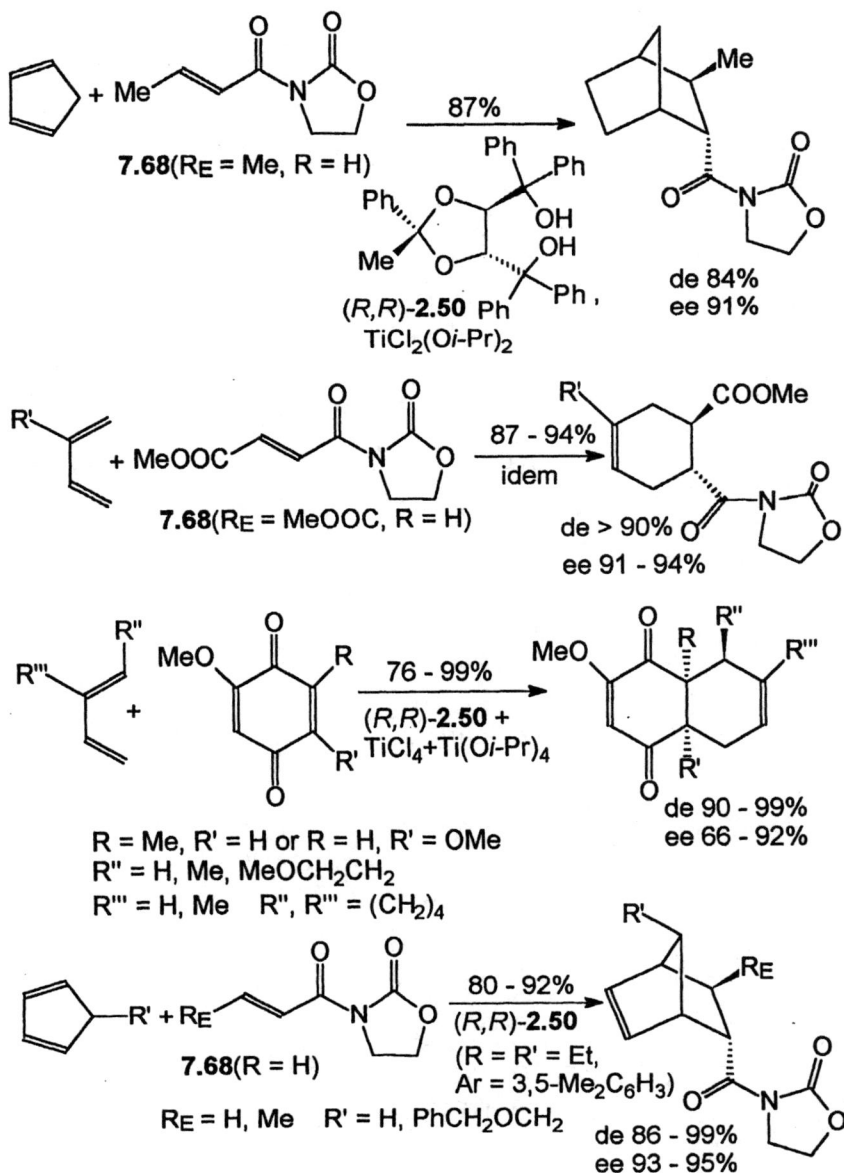

Figure 9.20

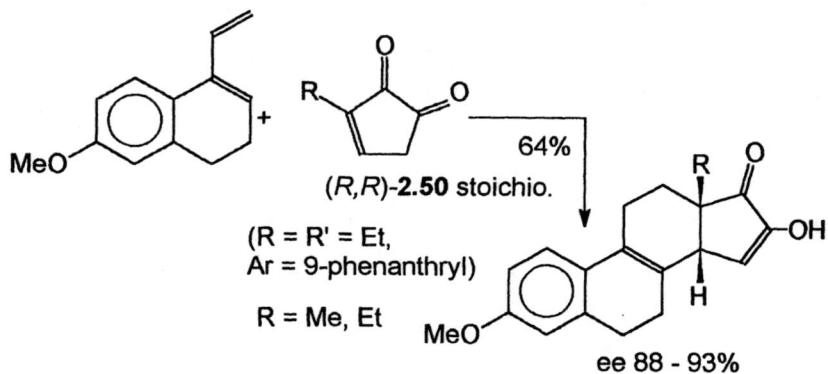

64%

(R,R)-**2.50** stoichio.

(R = R' = Et,
Ar = 9-phenanthryl)

R = Me, Et

ee 88 - 93%

Figure 9.20 (continued)

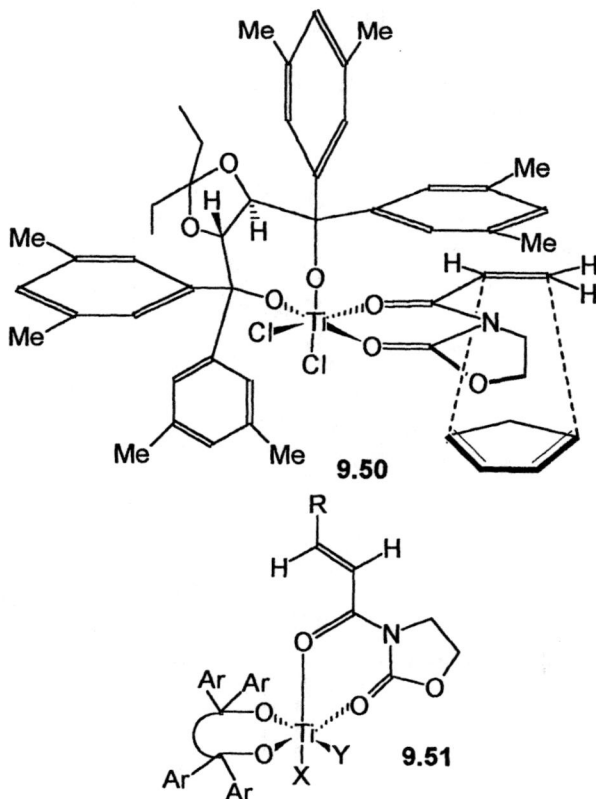

9.50

9.51

Figure 9.21

9.3.1.5 *Lanthanide Derivatives* [249]

The only example of the use of a chiral lanthanide catalyst in a Diels-Alder reaction was described by Kobayashi and coworkers [1577]. They performed the cycloaddition of cyclopentadiene with **7.68** (R = H, R_E = Me, *n*-Pr, Ph) in the presence of a complex generated from (*R*)-binaphthol **3.7** (R = H), Yb(OTf)$_3$ and an amine. The enantiomeric excess was high (up to 95%), but the *endo/exo* diastereoselectivity was moderate (up to 72%). With methacrolein, the ee was low.

In the domain of the hetero-Diels-Alder reaction, the first useful results were due to Danishefsky and coworkers [844]. They reacted trimethylsilyloxydienes **3.26** with aldehydes in the presence of chiral europium salts Eu(hfc)$_3$ **3.25**. If the diene bears achiral substituents, then the enantioselectivity of the cycloaddition is low. However, if **3.26** bears a chiral ether functionality, double diastereodifferentiation operates and a good stereoselectivity is observed if the partners are matched. Good selectivities are also observed in the reaction of aldehydes with **3.26** when the chiral ether is derived from (1*R*,2*S*,5*R*)-menthol or phenmenthol **1.4** (R = H or Ph) and the reaction is catalyzed by Eu(hfc)$_3$ **3.25** (Figure 9.22). When using the other enantiomer, the selectivity drops. Nonnatural pyranosides **9.52** are obtained from these cycloadducts with a good enantioselectivity. From α-chiral aldehydes, it is possible to obtain molecules bearing three stereocenters, and these transformations have been applied to the synthesis of natural products. Brassard's diene **9.53** has also been used in hetero-Diels-Alder reactions catalyzed by chiral europium salts. The reaction of **9.53** with achiral aldehydes is not very selective, but useful results are obtained with chiral α-alkoxy- or α-aminoaldehydes [780] (Figure 9.22).

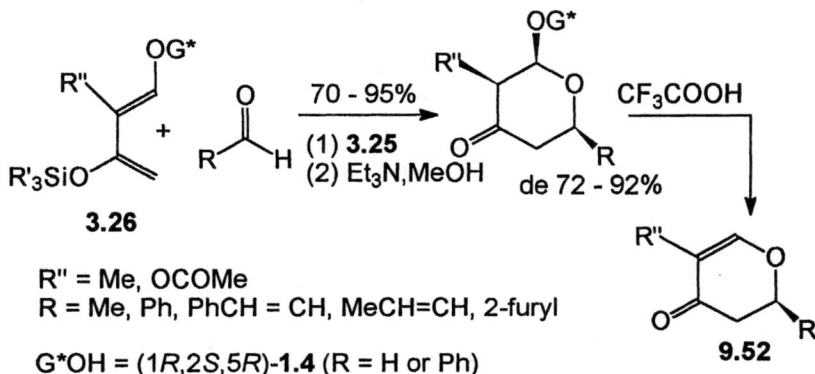

R" = Me, OCOMe
R = Me, Ph, PhCH = CH, MeCH=CH, 2-furyl

G*OH = (1*R*,2*S*,5*R*)-**1.4** (R = H or Ph)

Figure 9.22

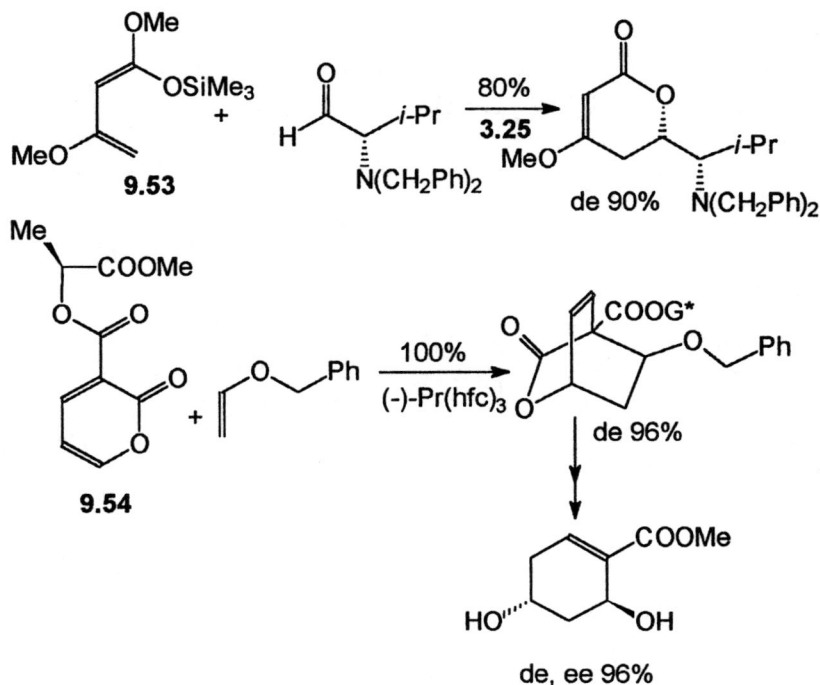

Figure 9.22 (continued)

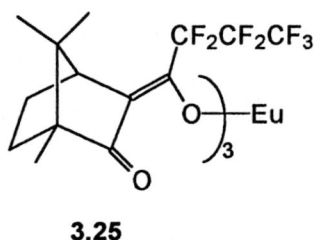

3.25

Lanthanide catalysis has been applied to inverse electron demand [4+2]-cycloadditions. As above, high selectivities are observed only when double diastereodifferentiation operates. Posner and coworkers performed the cycloaddition of a pyrone **9.54** bearing an ester group derived from (*S*)-methyl lactate with an enol ether under (−)-Pr(hfc)$_3$ catalysis [1578, 1579] (Figure **9.22**). Marko and Evans [1580] used other esters, among which pantolactone **1.16** proved to be the most efficient auxiliary, and performed similar cycloadditions under (+) or (−)-Eu(hfc)$_3$ catalysis at room temperature with excellent results (de > 95%).

9.3.2 Cycloadditions of Dienophiles Bearing Chiral Residues [73]

9.3.2.1 *Esters* [66, 147]

Cycloadditions of dienes with menthyl acrylates are usually poorly selective unless the cooperative effect (§ I.6) is used with diesters. As early as 1963, Walborsky and coworkers observed an interesting selectivity in TiCl$_4$ catalyzed reaction of 1,3-butadiene with dimenthyl fumarate **1.22**. This reaction has been generalized by Yamamoto and coworkers, who performed it at low temperature (−78° to −40°C) in the presence of Et$_2$AlCl (Figure **9.23**).

The cycloaddition of cyclopentadiene with di-(1*R*,2*S*,5*R*)-menthyl methylene malonate **9.55** (R = H) under TiCl$_4$ catalysis at −78°C is highly selective. After hydrogenation and treatment with LiAlH$_4$, the corresponding diol is obtained with an excellent enantiomeric excess [330] (Figure **9.23**). However, when R ≠ H, *endo/exo* isomers are formed. These results have been interpreted by the formation of a chelate **9.56** in which the *s-trans* conformation of each ester functionality is fixed. The diene is introduced on the least hindered face. Yamamoto and coworkers [1581] observed a similar selectivity in the cycloaddition of cyclopentadiene with menthyl methyl fumarate **1.22** conducted in the presence of a very bulky Lewis acid, methylaluminium bis-(2,6-di-*tert*-Bu-4-methylphenate) (MAD). This Lewis acid coordinates to the methyl ester and promotes rigidification of the system.

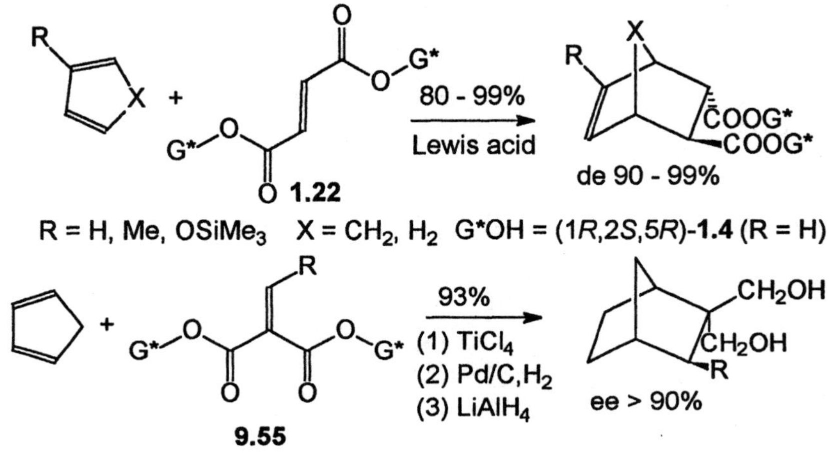

R = H, Me, OSiMe$_3$ X = CH$_2$, H$_2$ G*OH = (1*R*,2*S*,5*R*)-**1.4** (R = H)

G*OH = (1*R*,2*S*,5*R*)-**1.4** (R = H)

Figure 9.23

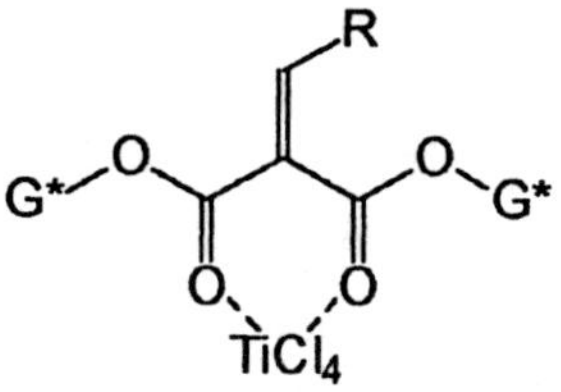

9.56

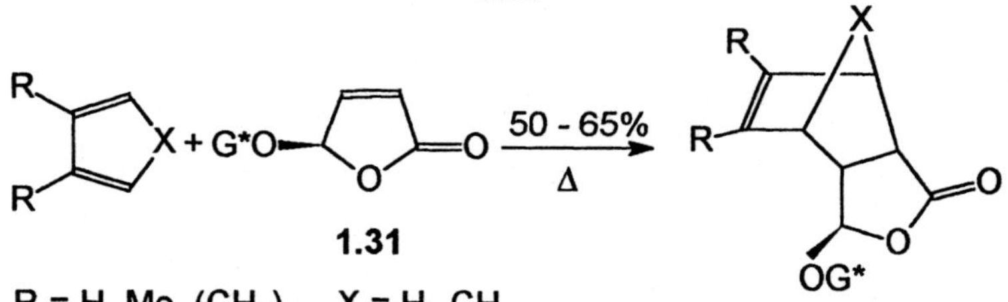

50 - 65%
Δ

de > 99%

1.31

R = H, Me, (CH$_2$)$_2$ X = H$_2$,CH$_2$
G*OH = (1R,2S,5R)-**1.4** (R = H)

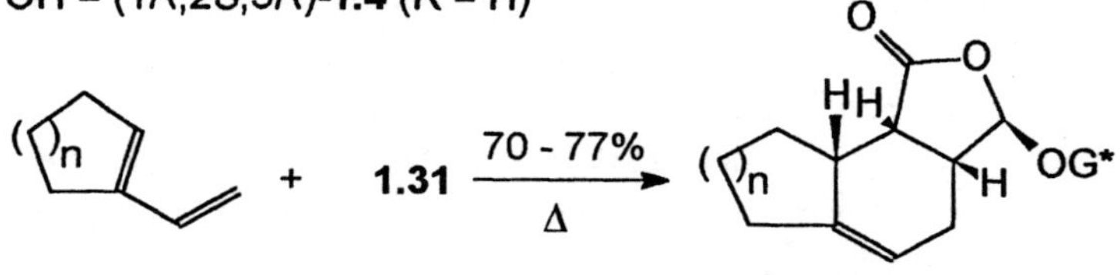

+   **1.31**   70 - 77%
Δ

n = 1,2

de > 99%

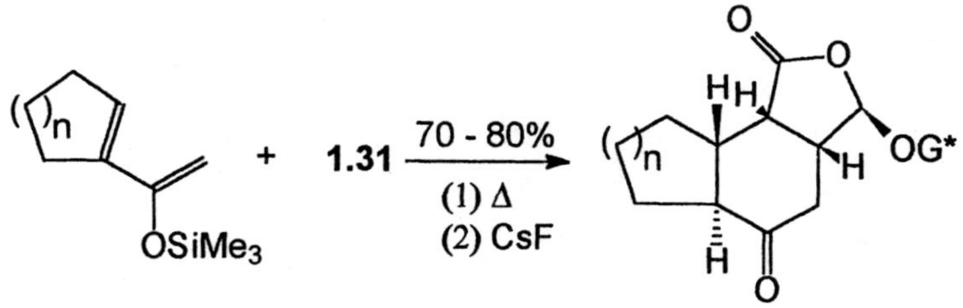

+   **1.31**   70 - 80%
(1) Δ
(2) CsF

n = 1,2

de > 99%

Figure 9.23 (continued)

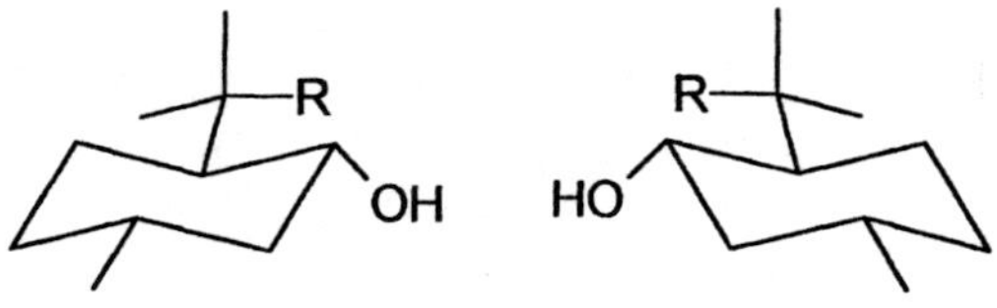

(1R,2S,5R)-**1.4**   (1S,2R,5S)-**1.4**

Lactone **1.31**, derived from (1*R*,2*S*,5*R*)-menthol **1.4** (R = H) also gives selective [4+2]-cycloadditions, provided that symmetrical dienes are used [137, 202] (Figure **9.23**). Unsymmetrical dienes usually give a mixture of regioisomers, although silyloxydienes and 3-methylenecyclohexene and -cyclopentene do react regioselectively [1582] (Figure **9.23**). The chiral auxiliary is readily cleaved by methanolysis, so this method gives access to compounds in which three or four stereocenters are created.

Since Corey and Emsley [135] introduced 8-phenmenthol **1.4** (R = Ph) in a synthesis of optically active prostaglandins, this chiral auxiliary and *trans*-2-phenyl cyclohexanols **1.5** have often been transformed into α,β-unsaturated esters for use as dienophiles. Indeed, Corey and Emsley have shown that the cycloaddition of 3-benzyloxymethylcyclopentadiene with phenmenthyl acrylate under AlCl$_3$ catalysis is highly stereoselective (Figure **9.24**). Among the other chiral alcohols used as auxiliaries in related cycloadditions, **1.6** [144], bornane-derived **1.7** (R = PhCH$_2$O, PhNHCOO, *tert*-BuCH$_2$O), **1.8** (R = PhCH$_2$O), **1.9** (R = PhCH$_2$) and **1.10** have been recommended [66, 147, 170]. However, in order to avoid polymerization in reactions with these Lewis acids, efficient dienophiles like acrylates, fumarates **1.22**, allenic derivatives CH$_2$=C=CHCOOG* and sulfinylcarbamates **1.24** are needed. 3-Alkoxysubstituted coumarins **9.57** and α-ketoesters have also given useful selectivities [66, 1583]. The diene partners in such Diels-Alder reactions are typically cyclopentadienes, cyclohexadiene, anthracene, 2,3-dime-thylbutadiene or E,E-2,4-hexadiene. A few representative examples of these types of reactions are given in Figure **9.24**. Most of the reactions are run between −20 and 0°C, and the chiral auxiliaries are often removed with LiAlH$_4$.

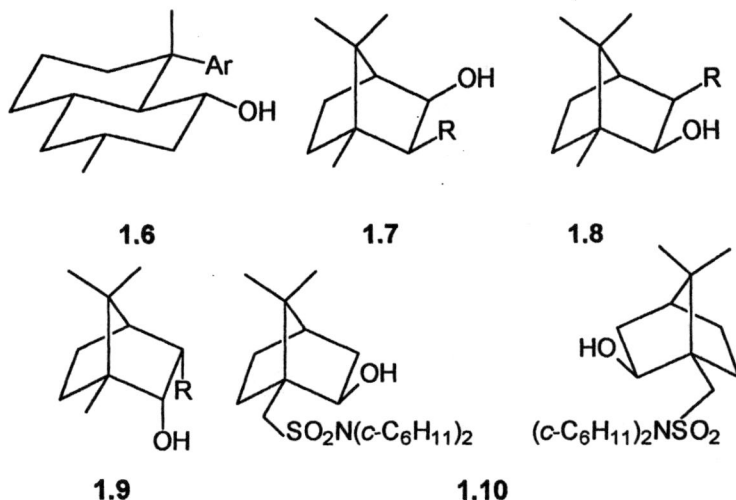

1.6 **1.7** **1.8**

1.9 **1.10**

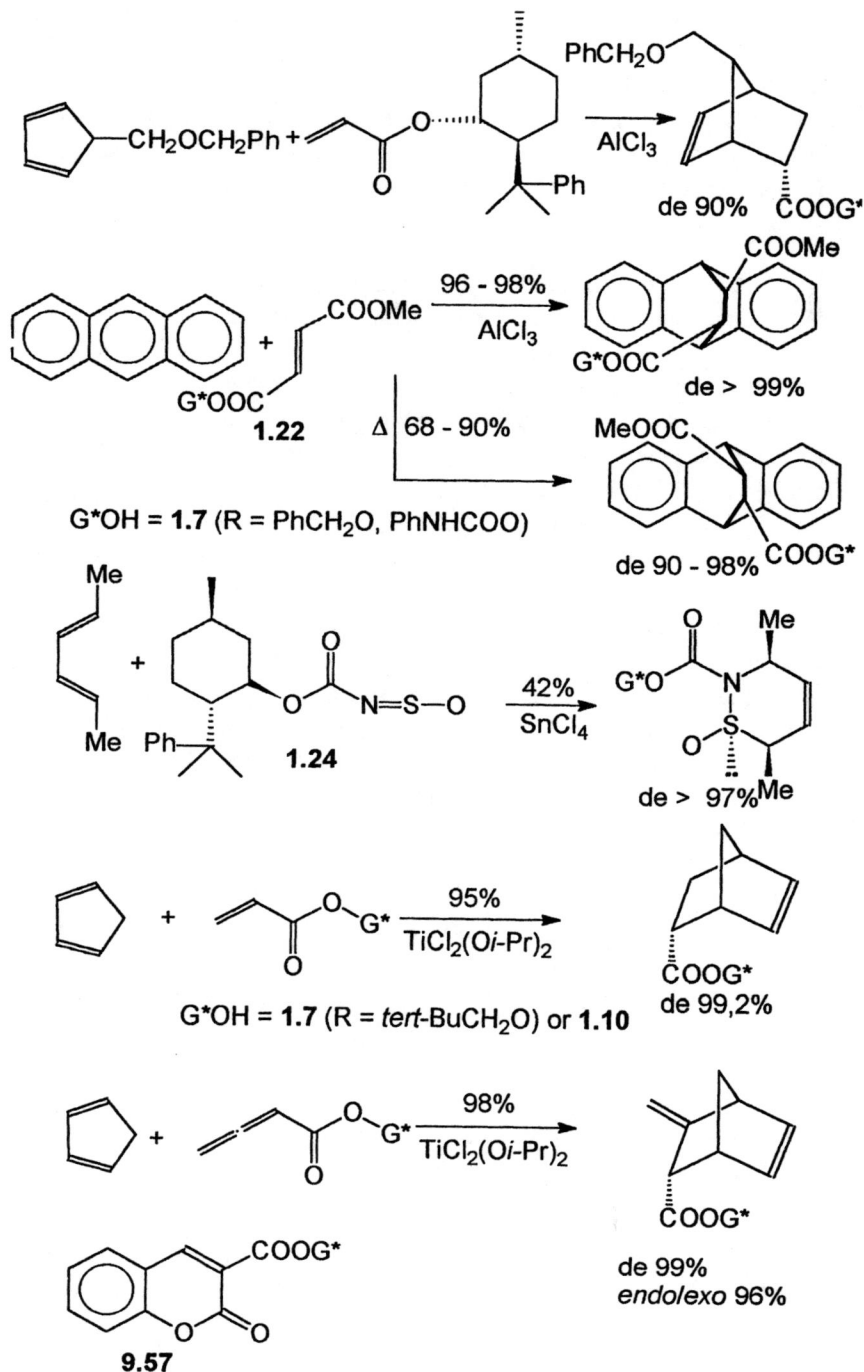

Figure 9.24

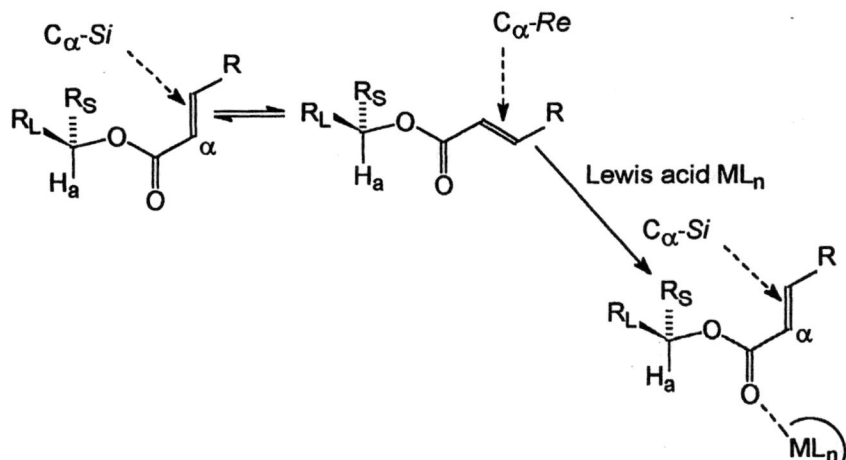

Figure 9.25

An important feature of these reactions is the stereoselectivity reversal that occurs when thermal reactions are replaced by Lewis acid catalysis [73] (Figure 9.24). A similar reversal is observed when running these reactions in hydrogen bonding donating solvents [1584]. Oppolzer, Helmchen, Curran and coworkers [71, 73] interpreted these results by taking into account the reactive conformation of α,β-unsaturated esters at the transition state; thermal reactions occur through *s-cis* conformers [1557], while Lewis-acid-catalyzed reactions occur through *s-trans* conformers. These reactive conformations are the same as the ground-state conformers, according to theoretical calculations [83, 85] (Figure 9.25). Moreover, thermal cycloadditions are usually less selective than Lewis-acid-catalyzed ones [83].

α,β-Unsaturated esters of functionalized chiral alcohols also have been proposed as dienophiles. Helmchen and coworkers recommended as auxiliaries ethyl (S)-lactate 1.15, (R)-pantolactone 1.16 and (S)-2-hydroxysuccinimide 1.17 [73, 158, 176, 1557, 1585, 1585a]. As mentioned above, thermal and Lewis-acid-catalyzed cycloadditions give the reverse selectivity (Figure 9.26). However, the presence of the other functional groups in these compounds can induce the formation of chelates. Esters of (S)-1.15 and 1.17 lead to the same stereoisomeric cycloadducts, while those of 1.16 give the diastereomers (Figure 9.26). TiCl$_4$ is a better Lewis acid than EtAlCl$_2$, and *s-cis* rigidified chelates are generated. Several of these chelates (9.38, 9.58 and 9.59) have been characterized by X-ray crystallography (Figure 9.27). The least hindered face of 9.38 and 9.59 is the Cα-*Si* face, while that of **9.58** is the Cα-*Re* face. The dienes are introduced on these faces, usually with an excellent selectivity.

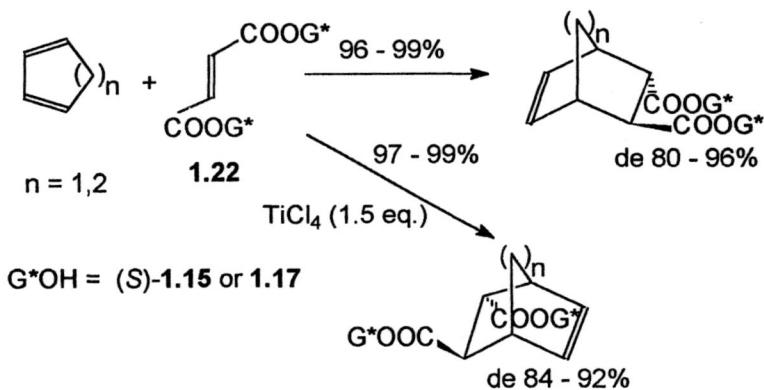

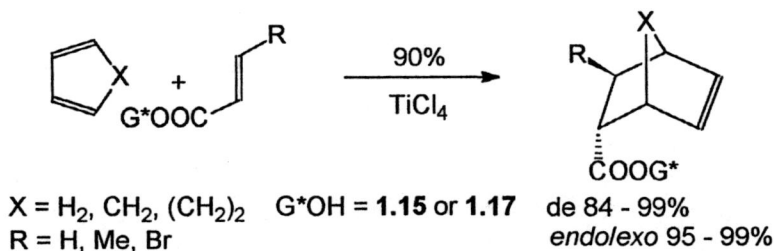

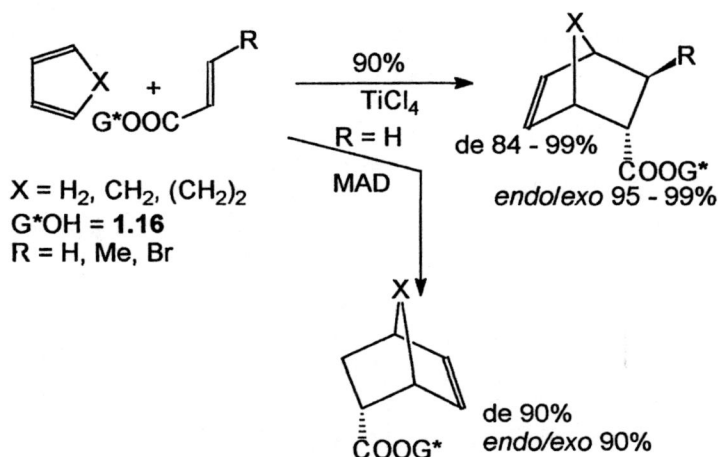

Figure 9.26

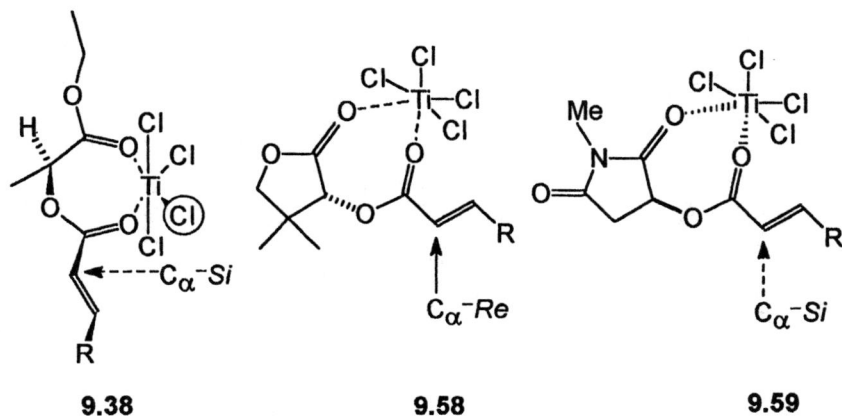

9.38   9.58   9.59

Figure 9.27

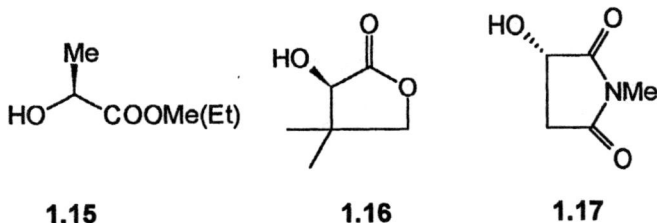

1.15   1.16   1.17

The adducts from these auxiliaries are solid and are easily purified by crystallization. Moreover, these adducts are easily hydrolyzed with LiOH in aqueous THF without epimerization. Applications include cycloadditions of unsubstituted or β-substituted acrylates RCH=CHCOOG* (R = H, Me, Br) with butadiene, isoprene, anthracene and cyclopenta- and cyclohexadiene [158, 1586] at low temperature (−78°C to −40°C), using TiCl$_4$ in catalytic amounts. Competing polymerization of cyclopentadiene can be avoided by using Lewis acid-base combinations such as TiCl$_4$-SbPh$_3$ recommended by Yamamoto and Suzuki [1587]. The use of methylaluminum bis-(2,6-di-*tert*-Bu-4-methyl phenoxide) (MAD) as Lewis acid allows the cycloaddition of cyclopentadiene with the acrylate of **1.16** to take place with nonchelation control, so that the stereoselectivity is reversed [1588] (Figure **9.26**). An important application of (*R*)-pantolactone **1.16** acrylate in asymmetric cycloadditions is the synthesis of Corey's prostaglandin precursor from 3-benzyloxymethylcyclopentadiene at −20°C in the presence of TiCl$_4$ (de 94%, 81% chemical yield after crystallization, > 99% ee). Acrylates and fumarates of methyl

(R)-lactate or (R)-mandelate have also been used in the synthesis of antitumor agents [1589].

Suzuki and coworkers [1590] have used cinchona alkaloids **3.1** and **3.2** as chiral auxiliaries in cycloadditions of cyclopentadiene with acrylic, crotonic or fumaric esters. In these experiments, the best results are obtained with $SnCl_4$ in excess between −20°C and −40°C. As expected, the esters of cinchonidine **3.1** (R = H) and cinchonine **3.2** (R = H) give enantiomeric alcohols after reduction with $LiAlH_4$. Contrary to expectations, the observed face selectivity indicates that if chelation control takes place, the α,β-unsaturated ester reacts through the *s-trans* conformation at the transition state. Indeed, the infrared spectrum of the complex **3.2** acrylate-$SnCl_4$ shows that, unlike the methyl acrylate-$SnCl_4$ complex, coordination does not involve the ester carbonyl group, but instead involves the oxygen next to the alkaloid residue. The ground-state as well as the transition-state conformations of the complex are as shown in **9.60**, and the diene is introduced on the Cα-*Re* face (Figure **9.28**).

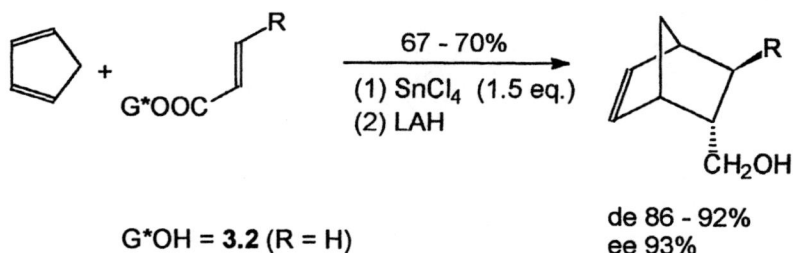

$$G^*OH = 3.2 \ (R = H)$$

de 86 - 92%
ee 93%

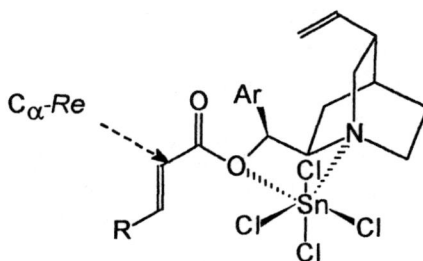

9.60

Figure 9.28

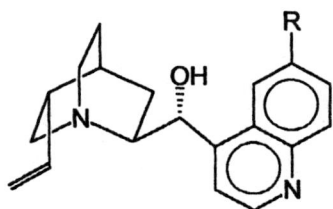

3.1   **3.2**

Dienophiles derived from a carbohydrate auxiliary have also been recommended [248]. Kunz and coworkers performed Diels-Alder reactions with acrylate esters of glucofuranoses **1.52** [244], dihydroglucal **1.50** or dihydrorhamnal **1.51** [246]. The cycloadditions of these acrylates with various dienes are highly selective when conducted between −30°C and 0°C in the presence of titanium promotors in large excess (Figure **9.29**). The acrylate esters of **1.50** and of **1.51** are precursors of enantiomeric cycloadducts. The catalyst is thought to be coordinated to the acrylate ester under the *s-trans* conformation and to the pivalic ester, so that either the Cα-*Si* or the Cα-*Re* face is accessible, respectively (Figure **9.29**).

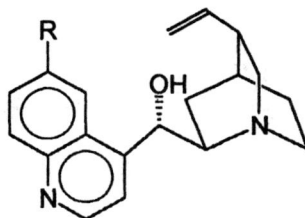

n = 1,2   G*OH = **1.50**

88 - 92%
TiCl$_4$ or
TiCl$_2$(Oi-Pr)$_2$
(3 equiv)

endo/exo > 95%
de 80 - 96%

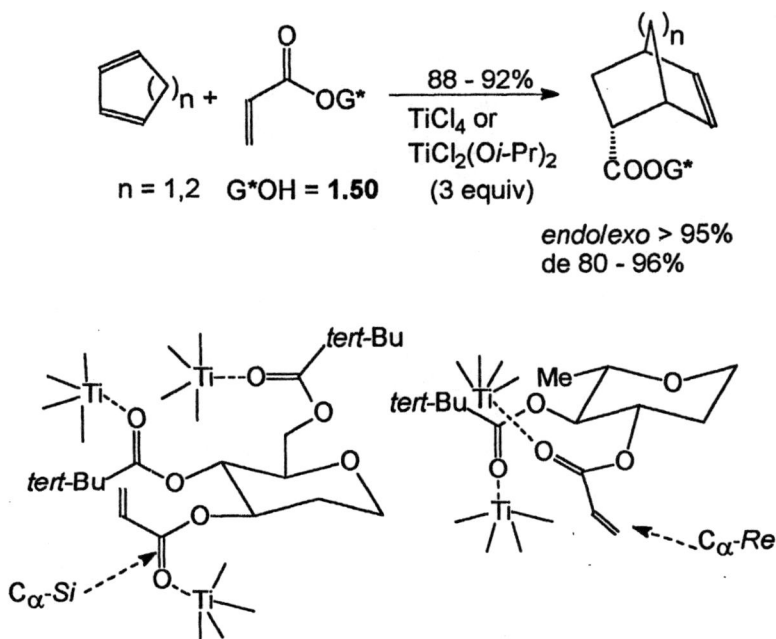

Figure 9.29

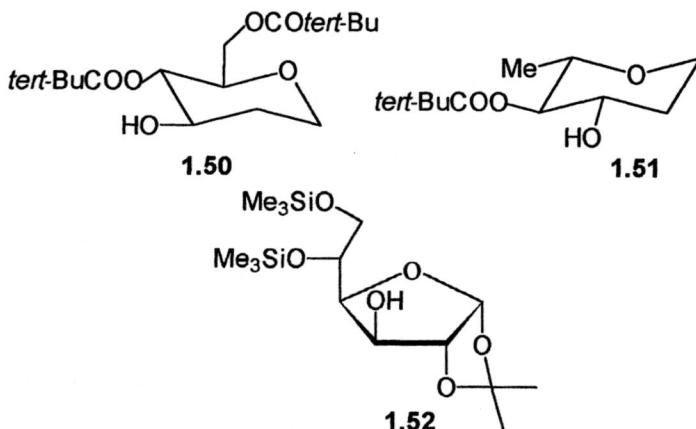

1.50   **1.51**

1.52

Nougier, Gras and coworkers proposed the use of the acrylate of arabinoside ketal **1.53** as a dienophile [245]. Under SnCl$_4$ or TiCl$_4$ catalysis, the dienophile is held in the *s-trans* conformation due to chelation, and the acid cycloadduct is obtained with an excellent enantiomeric excess after hydrolysis. Other analogs have been proposed by the same authors [1591] (Figure **9.30**).

Among the other acrylates of chiral alcohols used in cycloadditions with cyclopentadiene, reactions of the monobenzoate of 2,2,6,6-tetramethyl-3,5-heptanediol [1398] under TiCl$_4$ catalysis are highly selective, as are reactions of chiral arenechromium tricarbonyl complexes **9.61** mediated by stoichiometric amounts of ZnCl$_2$ [548]. In this last case, the R group must be bulky (R = 1-Np, 2,4,6-Me$_3$C$_6$H$_2$) (Figure **9.30**).

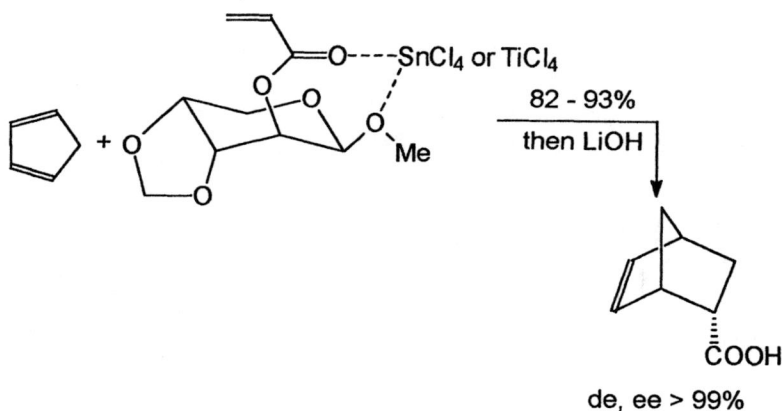

Figure 9.30

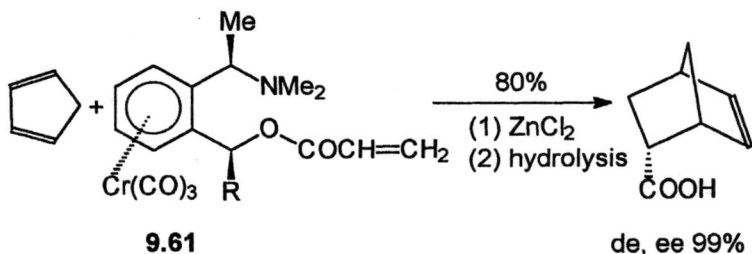

9.61

de, ee 99%

Figure 9.30 (continued)

9.3.2.2 *Amides, Imides, Sulfonamides and Analogs*

The [4+2]-cycloaddition reactions of acrylamide derivatives of chiral acyclic amines or monoalkylpyrrolidines are generally not very stereoselective [288, 1592]. However, the cycloadditions with alkoxyimminium salts **9.62** with cyclopentadiene do give useful results. Reactions of **9.62** conducted at −40°C lead after appropriate treatment (NBS, Zn/AcOH) to either (*R*)- or (*S*)-norbornene carboxylic acid [293] (Figure **9.31**), depending on the chiral nitrogen substituent. Acrylamide derivatives of 2,5-disubstituted pyrrolidines **1.65** (R = MeOCH$_2$ or MeOCH$_2$OCH$_2$) cycloadd to cyclopentadiene with a good selectivity under Lewis acid catalysis at 0°C [287] (Figure **9.31**). The same dienophiles react with bis-silyloxydiene **9.63** at −60°C [288]; according to the nature of the Lewis acid, different cycloadducts are obtained (Figure **9.31**).

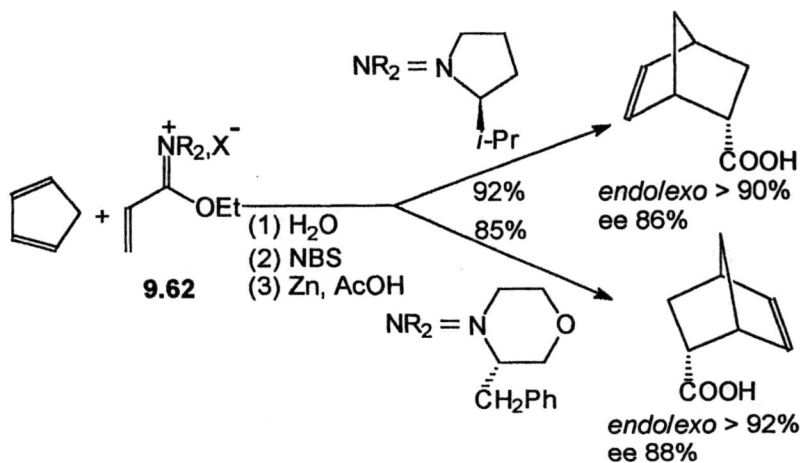

Figure 9.31

R = MeOCH$_2$

endo/exo 78%
ee 98%

R = MeOCH$_2$OCH$_2$

endo/exo 90%
ee 98%

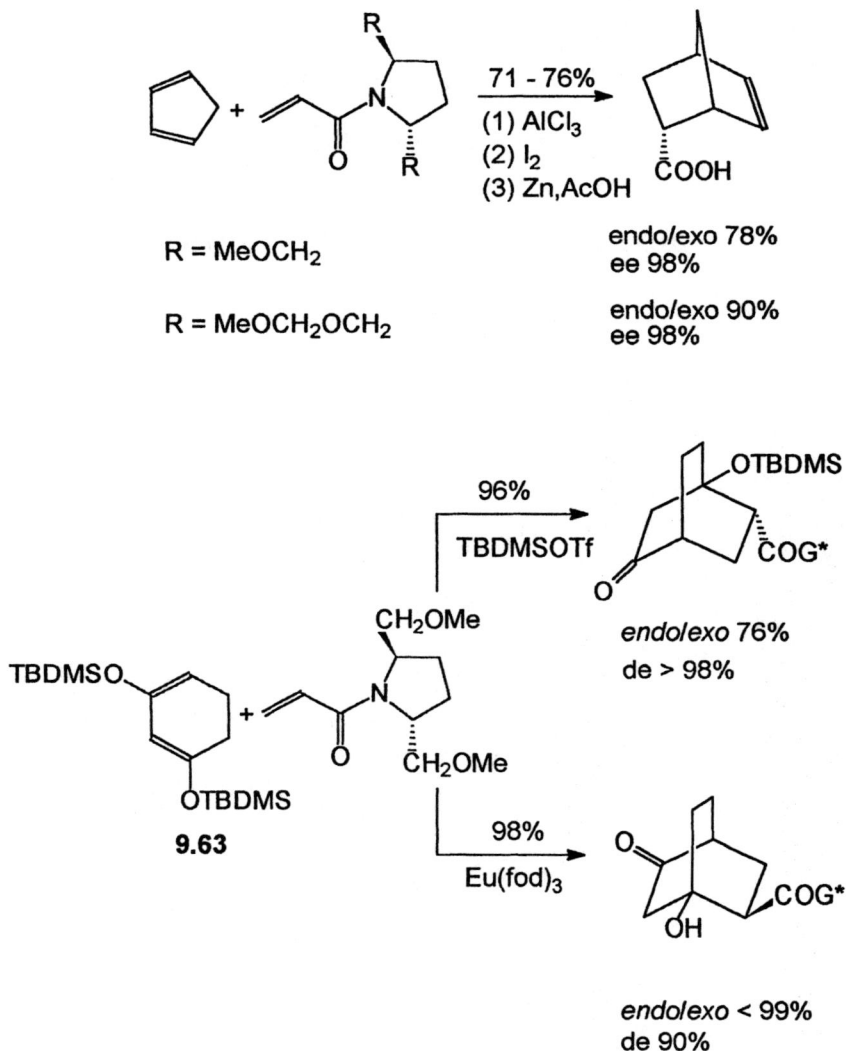

Figure 9.31 (continued)

Ghosez and Gouverneur [272, 1593] used the same chiral auxiliary to promote uncatalyzed cycloadditions of nitrosoamide **9.64**. With cyclic dienes, reactions occur near room temperature, while 2-azadienes **9.65** react at −78°C (Figure **9.32**). The cycloadducts formed with **9.65** are precursors of (R)-α-aminoacids [861]. The observed selectivities are interpreted by the *endo* approach of the diene to the least hindered face of the dienophile in the *s-cis* conformation (see above).

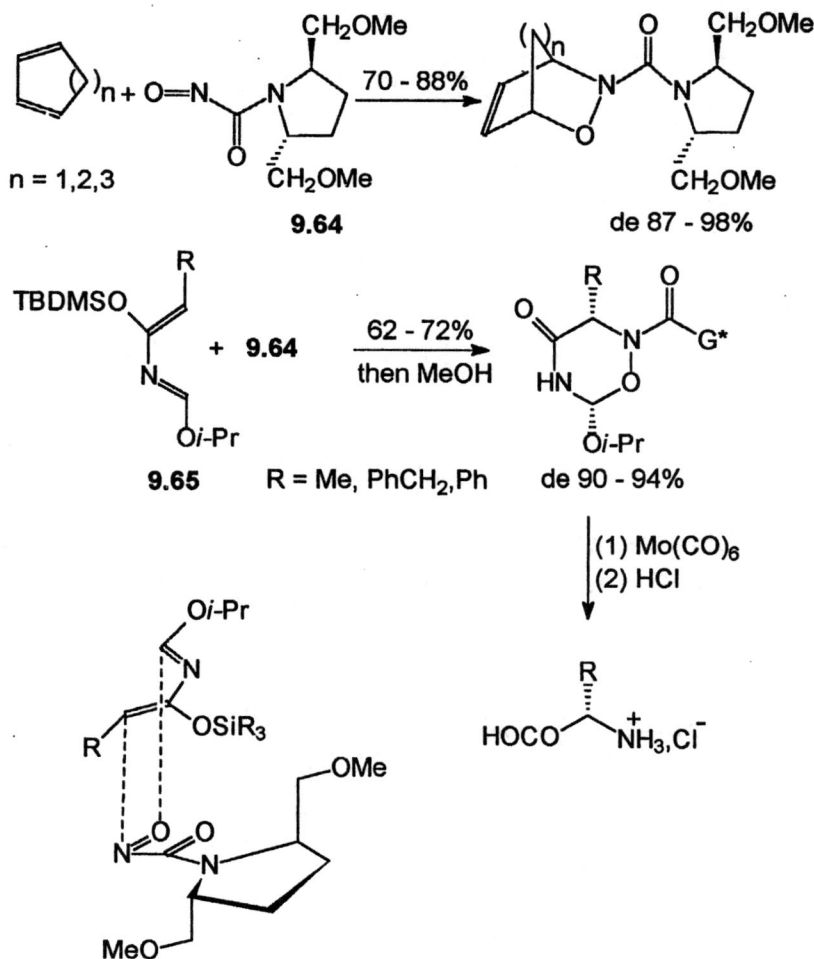

Figure 9.32

Waldmann and coworkers [104, 264, 289] and Cativiela and coworkers [286] proposed the use of acrylamides of α-aminoesters as dienophiles. Phenylalanine, valine and isoleucine derivatives lead to mediocre selectivities, but cycloadditions of various dienes with α,β-unsaturated amides of (S)-proline benzyl or allyl esters **9.66** at 0°C are stereoselective. Obtaining different isomers depends on whether the Lewis acid is EtAlCl$_2$ or TiCl$_4$. These results have been interpreted by the approach of the diene to the least hindered face of either a monodentate complex (EtAlCl$_2$) or a chelate (TiCl$_4$). In both cases, the unsaturated amide is under the *s-cis* conformation (Figure **9.33**). The reactions of the corresponding fumara-mide can take place without catalysis, either in organic solvents or in water [289]. These reactions are highly stereoselective due to the cooperative effect (§ I.6). In the presence of $TiCl_4$ (2 equiv), the product of chelation control is obtained.

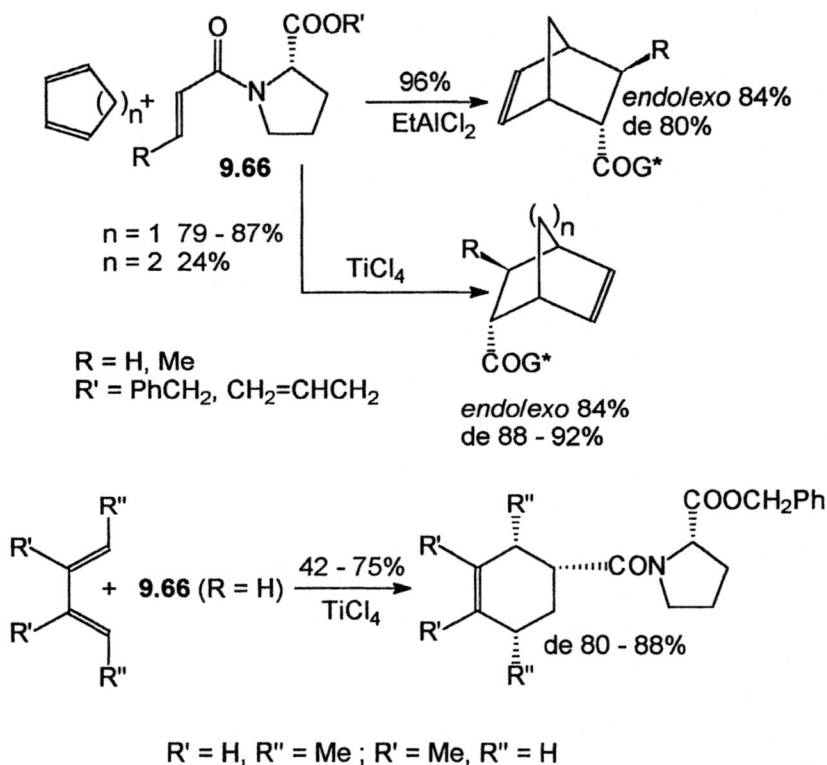

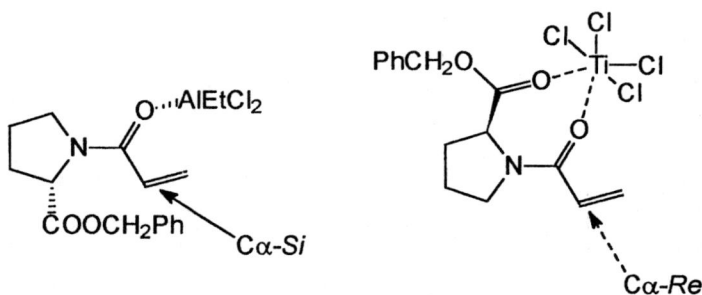

Figure 9.33

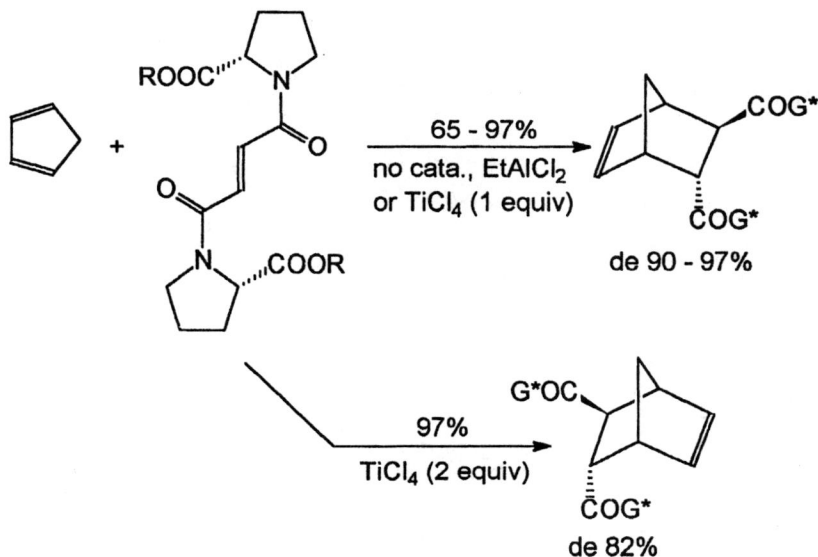

Figure 9.33 (continued)

Asymmetric Diels-Alder reactions have been carried out with Evans's *N*-enoyloxazolidinones **7.68** (R = PhCH$_2$ or *i*-Pr) or **9.67**. These dienophiles react with many dienes under Et$_2$AlCl catalysis (1.4 equiv) between –78°C and 0°C [72]. After cleavage of the chiral auxiliary with PhCH$_2$OLi or LiOH/H$_2$O$_2$, cycloadducts are obtained with an excellent selectivity when R$_E$ = H, Me, Ph. The two reagents provide enantiomeric products (Figure **9.34**).

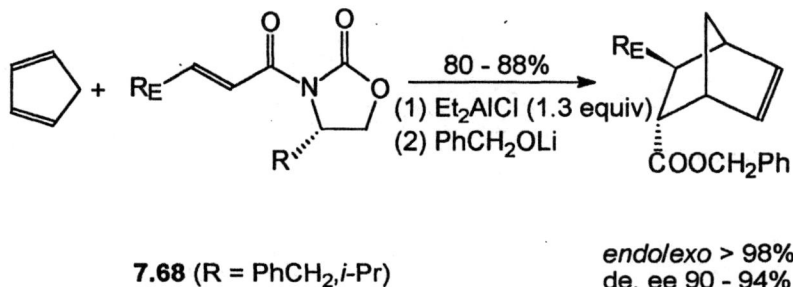

7.68 (R = PhCH$_2$,*i*-Pr)

endo/exo > 98%
de, ee 90 - 94%

Figure 9.34

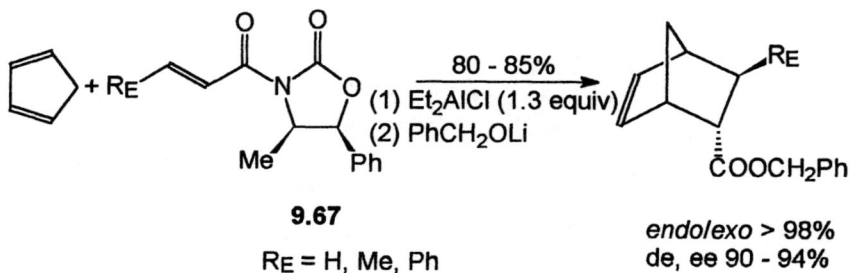

9.67

R_E = H, Me, Ph

endo/exo > 98%
de, ee 90 - 94%

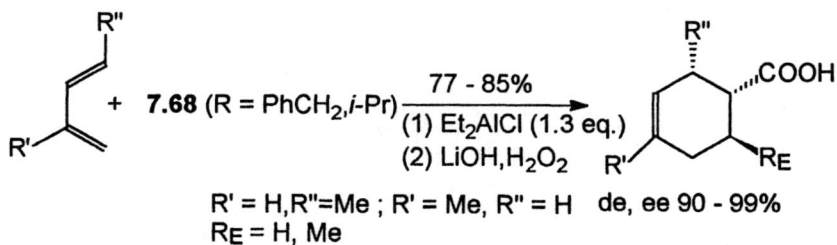

R' = H, R"=Me ; R' = Me, R" = H   de, ee 90 - 99%
R_E = H, Me

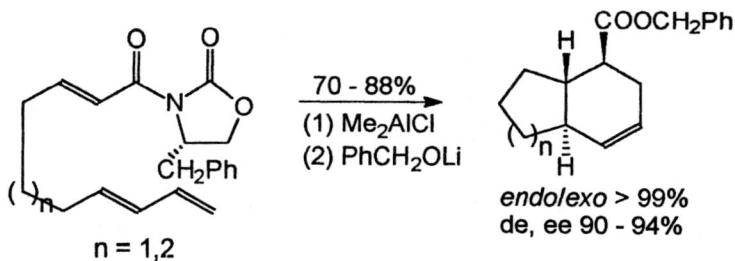

n = 1,2

endo/exo > 99%
de, ee 90 - 94%

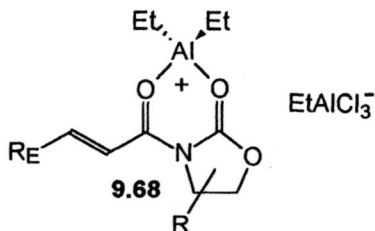

9.68

Figure 9.34 (continued)

Intramolecular cycloadditions also give useful results [74] (Figure **9.34**).
The cycloadditions of imides derived from Z-crotonic, methacrylic and
β,β-dimethylacrylic acids are less selective. Similarly, the use of lower amounts of
Et$_2$AlCl or other Lewis acids decreases the selectivity. These results have been
interpreted by the intervention of cationic chelated complex **9.68**, which is held in
the s-cis conformation, and there is now NMR evidence for this intermediate

[1594]. The chiral auxiliary substituents shield one face of the dienophile so the diene approaches the other. This methodology has been applied by Martinelli [418] to the synthesis of thromboxane receptor antagonists and by Hauser and Tomasi to the synthesis of daunomycinone antibiotics [417].

More sterically congested *N*-enoyloxazolidinones [425, 430, 1595] have been proposed as dienophiles in Et₂AlCl-mediated cycloadditions to cyclopentadiene, and some improvements in selectivity were observed.

N-Enoyl sultams **1.134** (R = R$_E$CH=CH) have been successfully used as dienophiles by Oppolzer and coworkers [73, 147, 454]. Cycloadditions of **1.134** are conducted at −78°C in the presence of stoichiometric amounts of TiCl₄, Et₂AlCl or EtAlCl₂. With butadiene, a radical inhibitor is added to prevent polymerization [1596] (Figure **9.35**). These results are rationalized by the intermediate formation of chelates such as **9.69**, whose structure has been determined by X-ray crystallography. In this complex, the amide group is under the *s-cis* conformation while the NSO₂-C=O group is *s-trans* (Figure **9.35**). Oppolzer and coworkers applied this methodology to intramolecular cycloadditions [74] and to the synthesis of natural products [147].

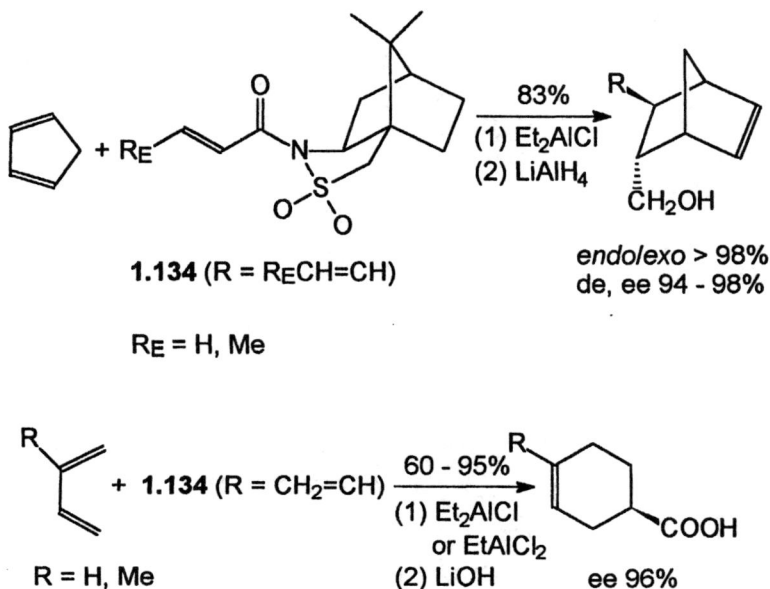

Figure 9.35

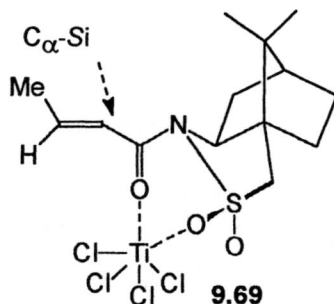

C_α-Si

9.69

Figure 9.35 (continued)

N-Acylsultams of glyoxylic acid **1.134** (R = CHO) and nitroso compound **1.134** (R = NO) are valuable dienophiles. Cycloadditions of **1.134** (R = CHO) with 1-methoxybutadiene mediated by Eu(fod)$_3$ are highly selective [466] (Figure **9.36**), and take place via a chelated complex. The reactions of the nitroso compound **1.134** (R = NO) with cyclopentadiene and cyclohexadiene give useful results [467], though heterodienes **9.65** do not. The observed results were interpreted in a fashion similar to that shown in Figure **9.32**.

Oppolzer and coworkers [470] also performed cycloadditions with N-enoylsultams derived from **1.135** (R = Me). The selectivities obtained with the acrylimide derivatives of **1.135** in the presence ot EtAlCl$_2$ or Me$_2$AlCl at −98°C or −78°C are similar to the selectivities with sultams **1.134** (Figure **9.36**). The cycloadditions are less selective when conducted with TiCl$_4$ or when using crotonic acid derivatives or bulkier substituents on **1.135**.

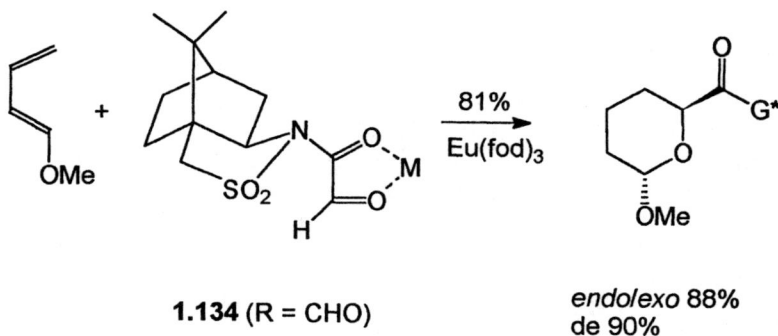

81%
Eu(fod)$_3$

1.134 (R = CHO)

endo/exo 88%
de 90%

Figure 9.36

1.134 (R = NO)

91 - 94%

de > 98%

97%
Me₂AlCl

endo/exo > 99%
de 93%

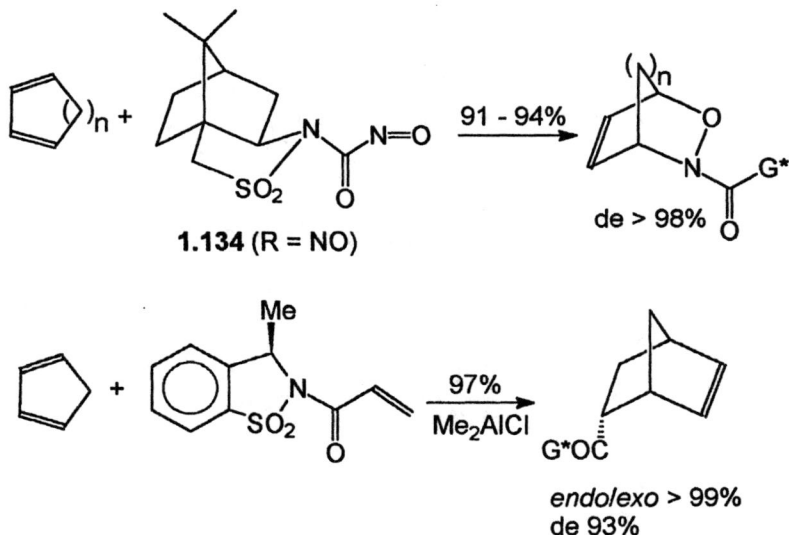

Figure 9.36 (continued)

Among other carboxylic acid derivatives, the α,β-unsaturated lactams **1.92** proposed by Meyers and coworkers [327] give useful results. No catalysis is needed in cycloadditions with silyloxydienes (R = TBDMSO, R' = H), while ZnCl₂ or SnCl₄ is required in reactions with 2,3-dimethylbutadiene (R = R' = Me) (Figure **9.37**).

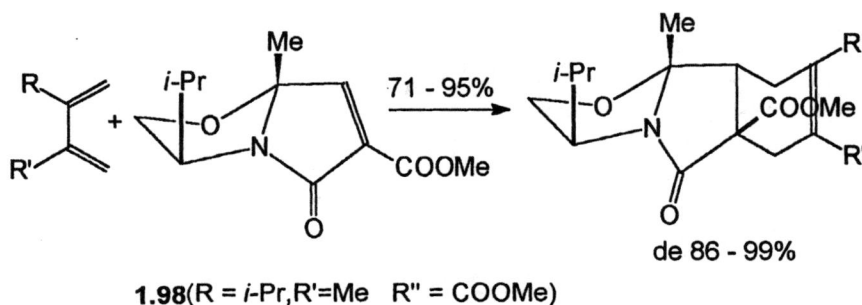

71 - 95%

de 86 - 99%

1.98(R = *i*-Pr,R'=Me   R" = COOMe)

R = TBDMSO, R' = H;
R = R' = Me

Figure 9.37

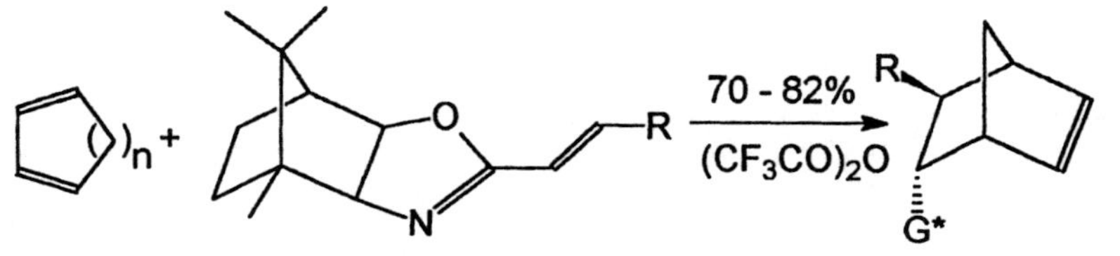

n = 1,2   **9.70**   R = H,Me   *endo/exo* > 99%
                                                 de 94 - 99%

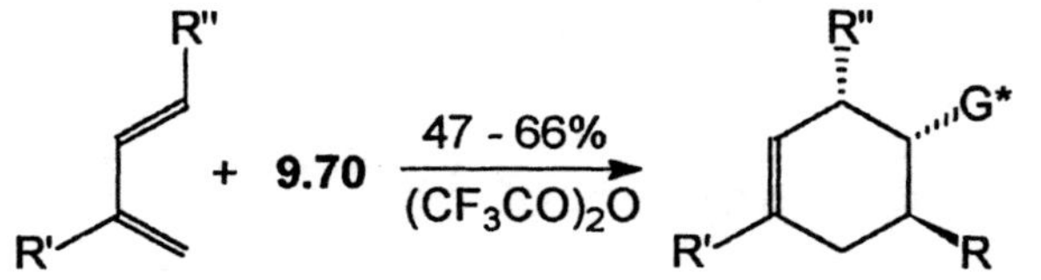

de 88 - 99%

R = H, Me R' = H, R" = SPh;
R' = Ph₂P(O)O, R" = O*tert*-Bu

Figure 9.37 (continued)

Langlois and coworkers [331, 339, 1008] proposed α,β-unsaturated oxazolines **9.70** bearing the bornane skeleton as dienophiles for asymmetric Diels-Alder reactions. Activation with trifluoroacetic anhydride generates *N*-acyl-imminium salts *in situ*. Their cycloadditions with a number of dienes take place with a high selectivity between −78°C and −15°C (Figure **9.37**). However, the reaction with 2,3-dimethylbutadiene is not selective. The removal of the chiral auxiliary is accomplished by sequential treatment of the products with ClCOOCH₂Ph and dilute NaOH.

Boeckman and coworkers [444] have elaborated rigid *N*-enoyl-imides **9.71** derived from **1.128**. Molecular modeling indicates that these compounds will prefer the *s-trans* conformation when R = Me. Indeed, cycloadditions of such dienophiles with cyclopentadiene mediated by MeAlCl₂ at −90°C give the expected adducts with good selectivity via chelate **9.72** (Figure **9.38**). Good selectivities are obtained with isoprene and 2,3-dimethylbutadiene, but the cycloadditions of sily-loxydienes are less selective. The adducts can be purified by crystallization and the chiral auxiliary is then removed by hydrolysis (LiOH/H₂O₂) or reduction (LiAlH₄). The use of a regioisomer of **1.128** gives less useful results.

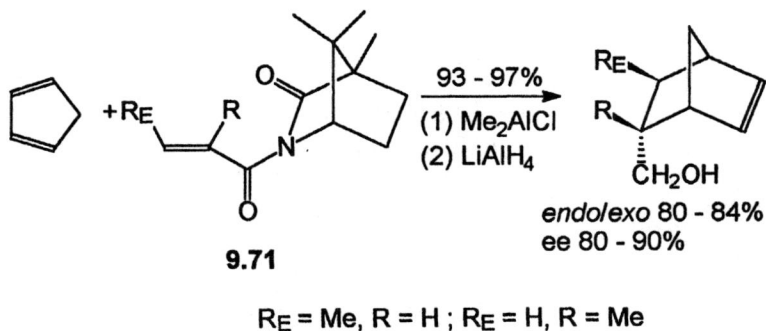

RE = Me, R = H ; RE = H, R = Me

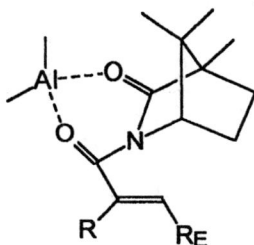

9.72

Figure 9.38

9.3.2.3 *Iron Acyl Complexes* [522]

The cycloaddition of cyclopentadiene with complex **1.151** (R = CH$_2$=CH) is catalyzed by ZnCl$_2$. The diene is introduced on the *Re* face of the double bond, which corresponds to the least hindered approach to the conformation shown in Figure **9.39**. However, the selectivity of this reaction is only moderate.

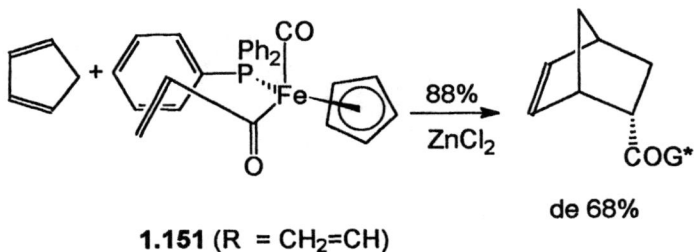

1.151 (R = CH$_2$=CH)

de 68%

Figure 9.39

9.3.2.4 *Vinylsulfoxides*

Diels-Alder reactions with α,β-unsaturated sulfoxide dienophiles are limited due to their low reactivity. This problem can be solved by introduction of an electron-withdrawing substituent in the α- or β-position. In another approach, Kagan and Ronan [493, 494] have activated tolyl vinylsulfoxide 1.138 (R = Y = H) by quaternization with Et_3OBF_4. Cycloadditions of cyclopentadienes to the alkoxy-sulfonium salt formed in this way at $-30°C$ are highly selective (Figure 9.40). After treatment of the mixture with NaOH, a chiral sulfoxide is regenerated with inversion of configuration at sulfur. From this compound, various nonracemic functionalized compounds, including a precursor of prostaglandins, can be obtained [494].

Introduction of an ester group in the β-position only gives useful results if the double-bond geometry is Z [73, 478, 486, 492]. The thermal cycloadditions of cyclopentadiene to Z-1.140 (R = H, Y = COOMe, Ar = Tol) and of 2-methoxyfuran to Z-1.144 (R = H, Y = COO-menthyl, Ar = 6-CF$_3$-2-Py) are highly selective (Figure 9.40). Cycloadditions of cyclopentadiene and cyclohexadiene with Z-1.144 (R = H, Y = COO-menthyl, Ar = 2-Py) mediated by Et_2AlCl or $ZnCl_2$ (Figure 9.40) are also selective. In the transition state, the sulfur lone pair and the carbonyl of the ester group are thought to be *syn* periplanar, so the diene is introduced on the face opposite to the S-O group. Difficulties in synthesis and purification of the Z-sulfoxides-esters limit this method.

Another approach to asymmetric sulfoxide Diels-Alder reactions involves the stereoselective oxidation of a sulfide. Modena and coworkers [73] obtained sulfoxide-ester 9.74 (R = H) from chiral thiol 9.73 by addition to methyl propiolate and oxidation with MCPBA. The cycloaddition of 9.74 with cyclopentadiene at $-5°C$ is highly selective due to the rigidification of the dienophile by internal hydrogen bonding (Figure 9.40).

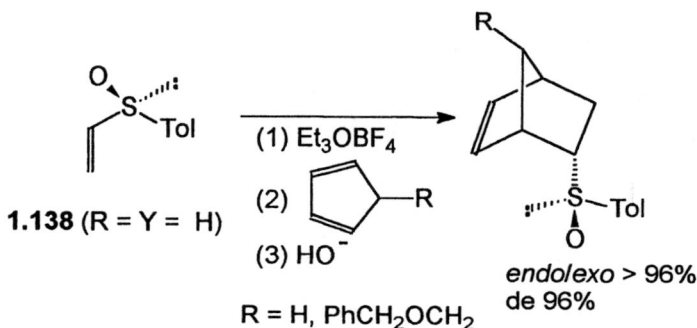

Figure 9.40

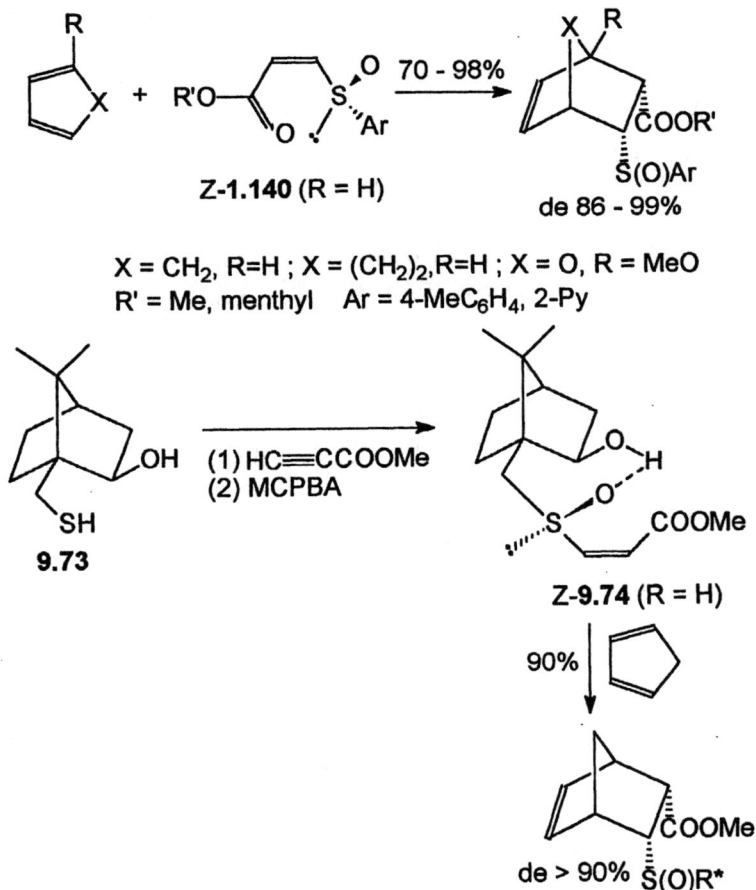

Figure 9.40 (continued)

Disappointing results are obtained in cycloadditions of cyclopentadiene with E-β-substituted sulfoxides **1.138** (Y = H, R = COOMe, COalkyl, P(O)(OEt)₂) or cyclic sulfoxides **1.139** (X = CH₂) [484, 485, 487, 1597]. However, the reaction of sulfone **1.138** (Ar = Tol, Y = H, R = SO₂-*tert*-Bu) mediated by Eu(fod)₃ is more selective (de 84%) [476]. A single cycloadduct was obtained by Koizumi from anthracene and **1.138** (Ar = Tol, Y = R, R = COOMe) [485]. Trisubstituted dienophiles **1.138** (Ar = Tol, Y = COO-*tert*-Bu or COOCH₂Ph, R = COOMe), **9.74** (R = MeOCO) and **9.75** give highly selective cycloadditions with cyclic dienes [484, 488, 1598, 1599, 1600] (Figure **9.41**). These reactions are catalyzed either by TiCl₄ at −78°C or by ZnCl₂ or ZnBr₂ at −20°C. According to the nature of the catalyst, a different diastereomer is formed predominantly (Figure **9.41**). With TiCl₄, the diene is introduced on the Cα-*Si* face of the double bond, and it is introduced with ZnBr$_2$ on the Cα-*Re* face. This divergence has been interpreted by different conformations of the chelated complexes due to the different coordination preferences of titanium and zinc [1598]. Cycloadducts from acyclic diene **1.138** (Y = COOCH$_2$Ph, R = COOMe, Ar = Tol) suffer spontaneous elimination of sulfenic acid. Diesters **9.76** are obtained with a high enantiomeric excess under TiCl$_4$ or Eu(fod)$_3$ catalysis at −78°C or 0°C, respectively [1598]. In all these cycloadducts, the sulfoxide auxiliary can be reduced with SmI$_2$/HMPA or undergo β-elimination in the presence of DBU. This methodology has also been extended to the cycloaddition of cyclopentadiene with naphthoquinones **9.77** [1601].

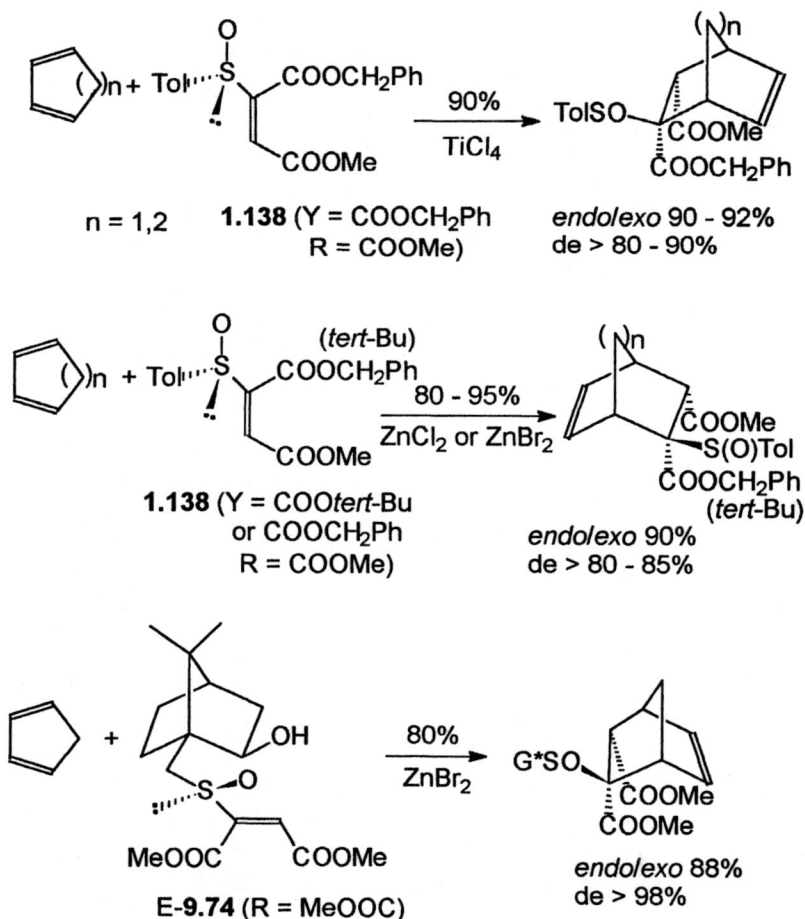

n = 1,2   **1.138** (Y = COOCH$_2$Ph
                 R = COOMe)

endo/exo 90 - 92%
de > 80 - 90%

1.138 (Y = COO*tert*-Bu
    or COOCH$_2$Ph
    R = COOMe)

endo/exo 90%
de > 80 - 85%

E-**9.74** (R = MeOOC)

endo/exo 88%
de > 98%

Figure 9.41

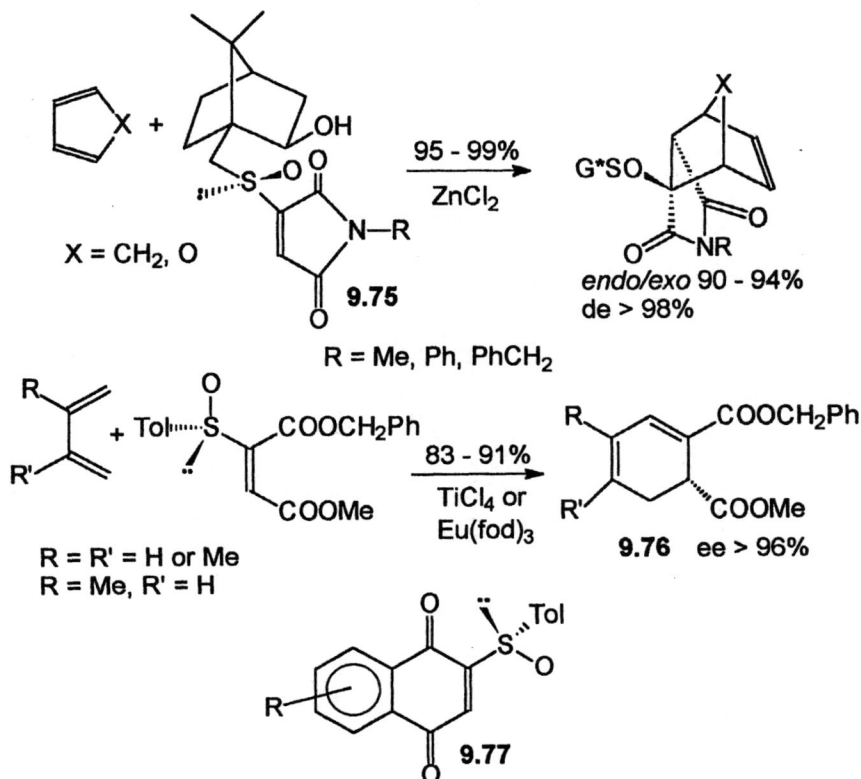

Figure 9.41 (continued)

9.3.2.5 *Imines, Imminium Salts, Nitroso Derivatives* [1548]

Cycloadditions of dienes with imines lead to nitrogen heterocycles [1548, 1549]. When the dienes are not electron-rich, imminium salts are required and reactions are usually run in an acid medium. Bailey [1602, 1603], Stella and their coworkers [256] performed the cycloaddition of various dienes to imine **9.78** generated from alkyl glyoxylates and (*R*)- or (*S*)-phenethylamine (Figure **9.42**). These reactions occur in the presence of CF₃COOH near room temperature. After hydrogenolysis of the phenethyl residue, the corresponding esters are typically obtained with a good enantiomeric excess, except for E,E-2,4-hexadiene. Yamamoto and Hattori [1559, 1560] examined the cycloaddition of (*S*)-1-phenethylimines with Danishefsky's diene **3.26** (R = R' = Me, R" = R'" = H). These reactions are mediated by Lewis acids in stoichiometric amounts at −78°C (Figure **9.42**). High selectivities are observed with ZnCl₂ or TiCl₂(O*i*-Pr)₂, but the best results are obtained with B(OPh)₃ and biphenols. When using B(OPh)₃-binaphthol complexes **3.7** (R = H) as Lewis acid (§ 9.3.1.1), very high selectivities are observed when the imine substituent and the catalyst are matched due to double diastereodifferentiation (Figure **9.42**). This method has been applied to the synthesis of non-racemic alkaloids [1559].

Waldmann and coworkers [263, 264, 1605] used imines of α-aminoesters. These imines can be generated *in situ* from CH₂O and an aminoester hydrochloride in water. Some selective cycloadditions have been observed between cyclic dienes, aminoester **1.59** (R = *tert*-Bu, R' = Me) and CH₂O (Figure **9.42**). However, the removal of the auxiliary could not be carried out due to a competitive retro Diels-Alder reaction. The use of other aminoesters or other dienes give less satisfactory results [264].

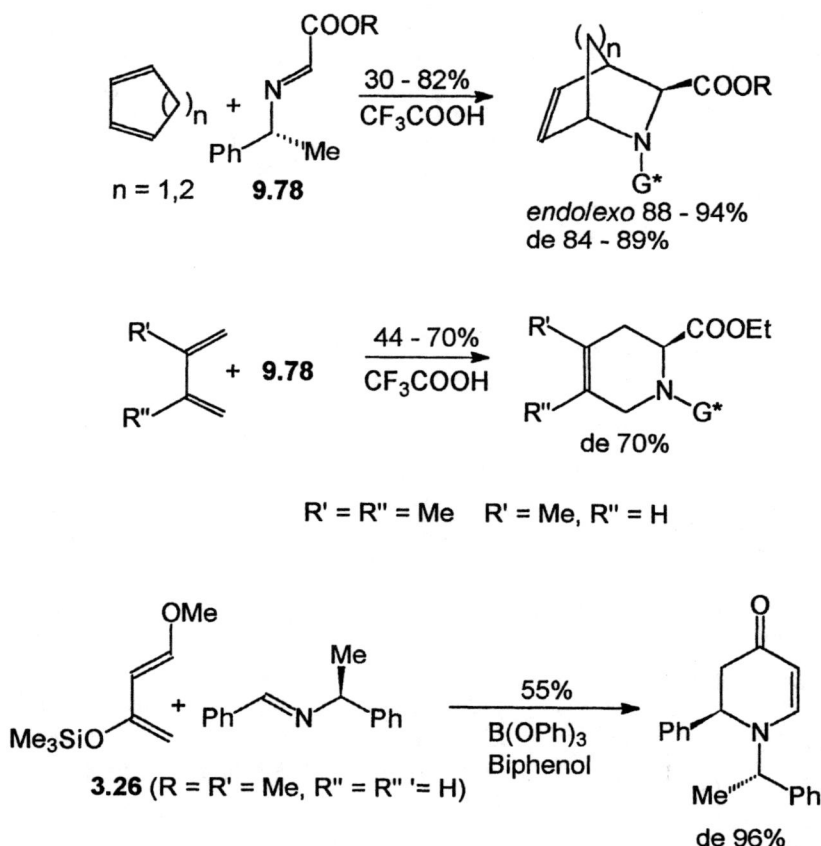

Figure 9.42

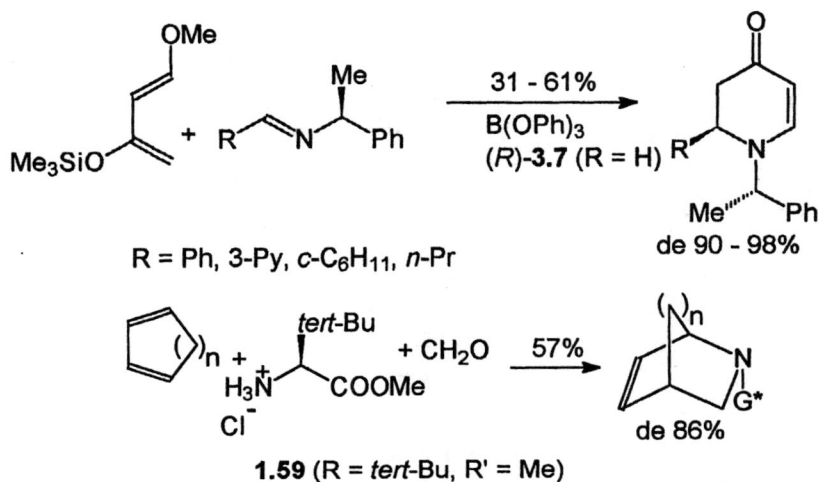

Figure 9.42 (continued)

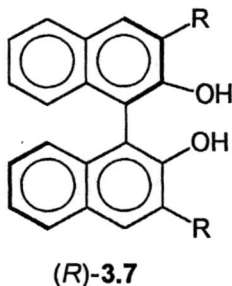

(R)-**3.7**

The cycloadditions of preformed imines generated from (S)-valine or (S)-isoleucine esters **1.59** (R = *i*-Pr or *tert*-Bu, R' = Me) with Danishefsky's **3.26** or Brassard's **9.53** dienes [263, 1605] give high selectivities when mediated by EtAlCl$_2$ or ZnCl$_2$ at −78°C or at 0°C (Figure **9.43**). Waldmann and coworkers have developed a method for a mild cleavage of the chiral residue, opening the way to nonracemic piperidines (Figure **9.43**). The observed selectivities are interpreted by the formation of an intermediate Lewis acid-imine complex. The reaction with Danishefsky's diene **3.26** is probably not concerted and the geometry of the transition state may be similar to that of the Felkin-Anh model (§ I.3) (Figure **9.43**). The cycloaddition with Brassard's diene **9.53** may be concerted, and should take place in a way that minimizes repulsive interactions between bulky substituents (Figure **9.43**).

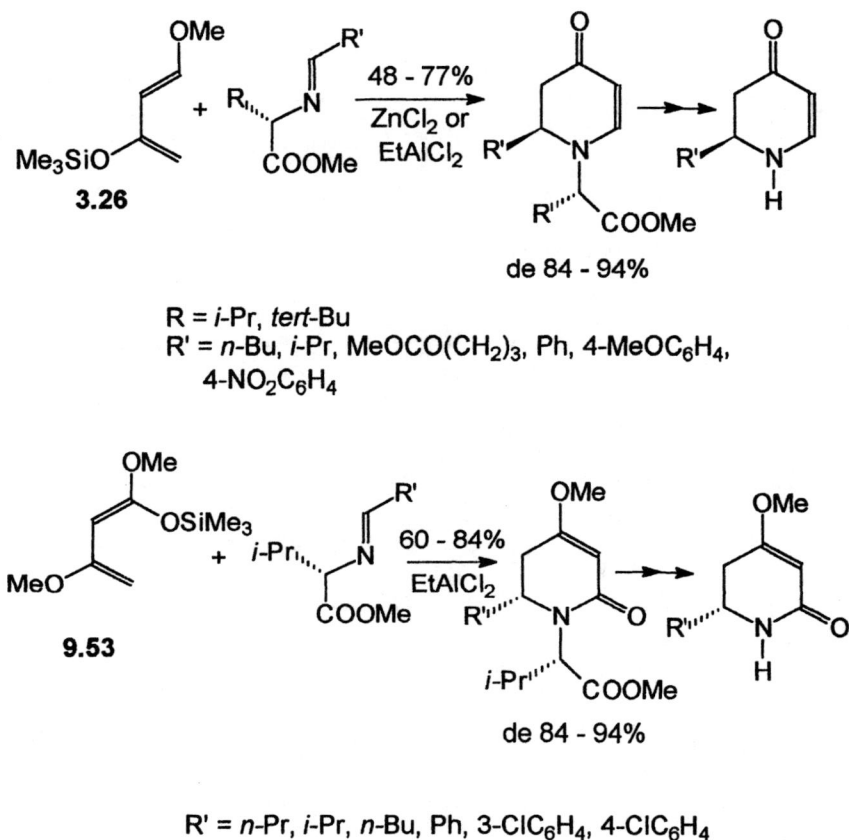

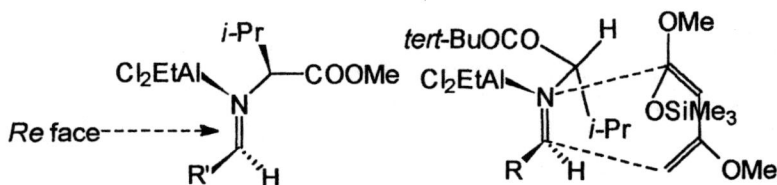

de 84 - 94%

R = *i*-Pr, *tert*-Bu
R' = *n*-Bu, *i*-Pr, MeOCO(CH$_2$)$_3$, Ph, 4-MeOC$_6$H$_4$,
      4-NO$_2$C$_6$H$_4$

de 84 - 94%

R' = *n*-Pr, *i*-Pr, *n*-Bu, Ph, 3-ClC$_6$H$_4$, 4-ClC$_6$H$_4$

Figure 9.43

Kunz and coworkers [248, 365] have proposed imines derived from carbohydrates **1.101** as heterodienophiles. The ZnCl$_2$-mediated cycloadditions of these
imines to dienes are not very selective, except with Danishefsky's diene **3.26**
(R = R' = Me, R'' = R''' = H). According to the imine R' substituent, the reaction
product is either cycloadduct **9.79** or the Mannich reaction product **9.80**. The lat-

ter is transformed into the former by acid treatment (Figure **9.44**), so **9.79** is obtained with a high selectivity in both cases. When the reaction is performed with **1.101** (R' = 3-Py), two equivalents of ZnCl$_2$ are needed and the other isomer **9.81** is formed [365] (Figure **9.44**). Kunz interprets these divergent selectivities by models **9.82** and **9.83**. The chiral auxiliary is cleaved with acid.

The cycloaddition of nitroso derivative **1.105** bearing a carbohydrate residue with cyclohexadiene leads to a single cycloadduct **9.84** after methanolysis. This product results from the approach of the diene from the least hindered face of the nitroso compound [248, 785] (Figure **9.44**). Other related cycloadditions have been performed, but either the selectivities are not very high [1606] or the configuration of the adducts was not determined.

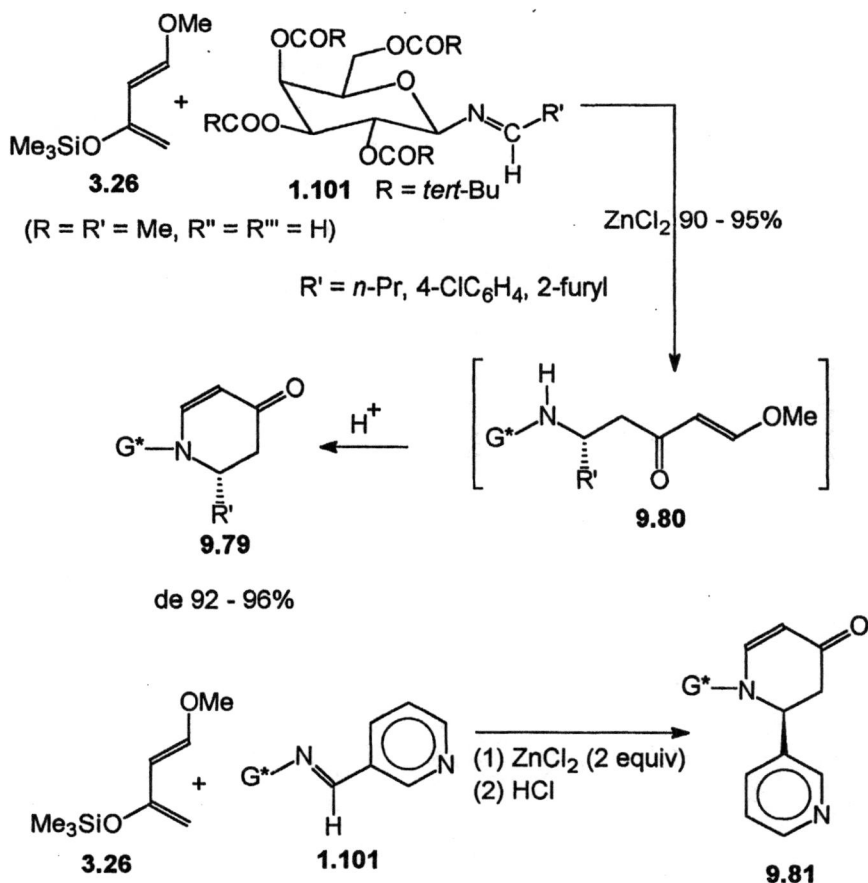

Figure 9.44

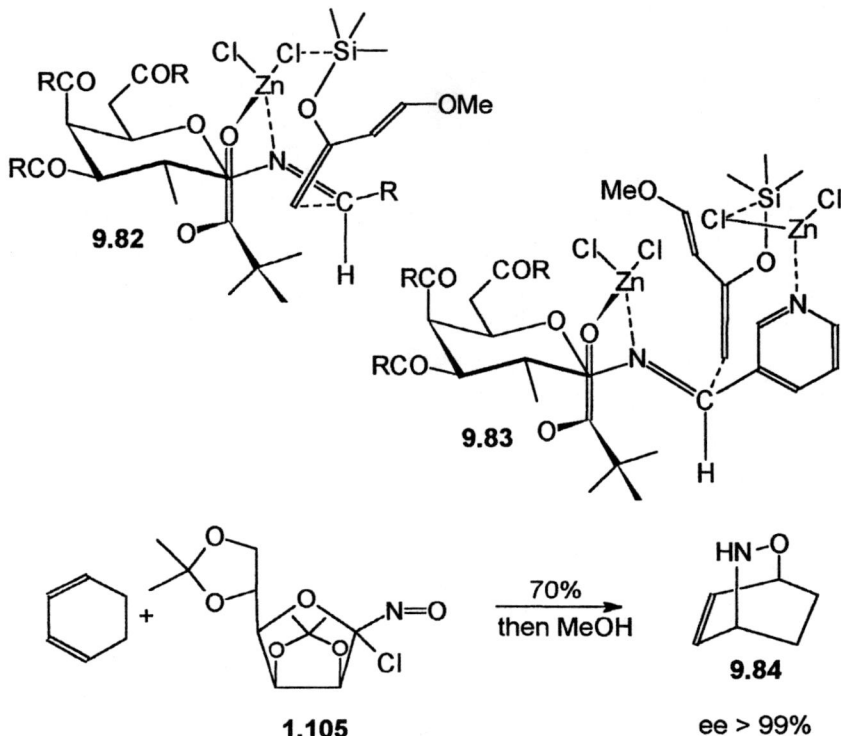

Figure 9.44 (continued)

9.3.2.6 *Enol Ethers and Acetals*

Enol ethers are electron-rich, and they act as dienophiles in inverse electron demand Diels-Alder reactions. The main frontier orbital interaction is between the dienophile HOMO and the diene LUMO. Ethers of chiral alcohols that have been used include vinyl ethers of (*S*)-1-phenyl-2-methylpropanol **1.1** (R = *i*-Pr or *tert*-Bu). These cycloadd to sulfone **9.85** at room temperature with a high selectivity [199, 1578] (Figure **9.45**). From one cycloadduct, Posner and coworkers synthesized nonracemic shikimate derivatives. Borneol or phenmenthol ethers give less satisfactory results. The vinyl ether of **1.4** (R = 2-Np) selectively cycloadds to lactone **9.86** [1607] (Figure **9.45**). Isoquinolinium salts **9.87** give cycloadditions to various chiral ethers including α-glycoside derivative **1.99** (Y = $OCH=CH_2$, R = $PhCH_2$). These reactions provide precursors of chiral tetralins **9.88** [359] (Figure **9.45**). The β-anomers or other α-analogs (R = Ac) give less useful results. Reissig and coworkers [1608] have performed the cycloaddition of vinylethers of diacetoneglucose **1.48** with nitrosoalkenes **9.89**. E-**9.90** (R' = Me, R" = H) must be separated from an E/Z mixture, but its cycloaddition is highly selective. The corresponding allenyl ether leads to compound **9.91** with a high selectivity after DBU treatment (Figure **9.45**). The chiral auxiliary is removed by methanolysis. Denmark and coworkers [198] examined the cycloadditions of vinyl ethers with nitroalkenes mediated by TiCl$_2$(Oi-Pr)$_2$ at −78°C. The most useful results were obtained with a vinyl ether of **1.7** (R = OCH$_2$-*tert*-Bu); **9.92** is formed with a good selectivity. A subsequent [3+2]-cycloaddition leads to a tricyclic derivative from which the auxiliary is removed. The observed selectivity corresponds to the attack of the *Si* face of the vinyl ether under the *s-trans* conformation. The conformation at transition state is different from that of ground state [198] (Figure **9.45**)

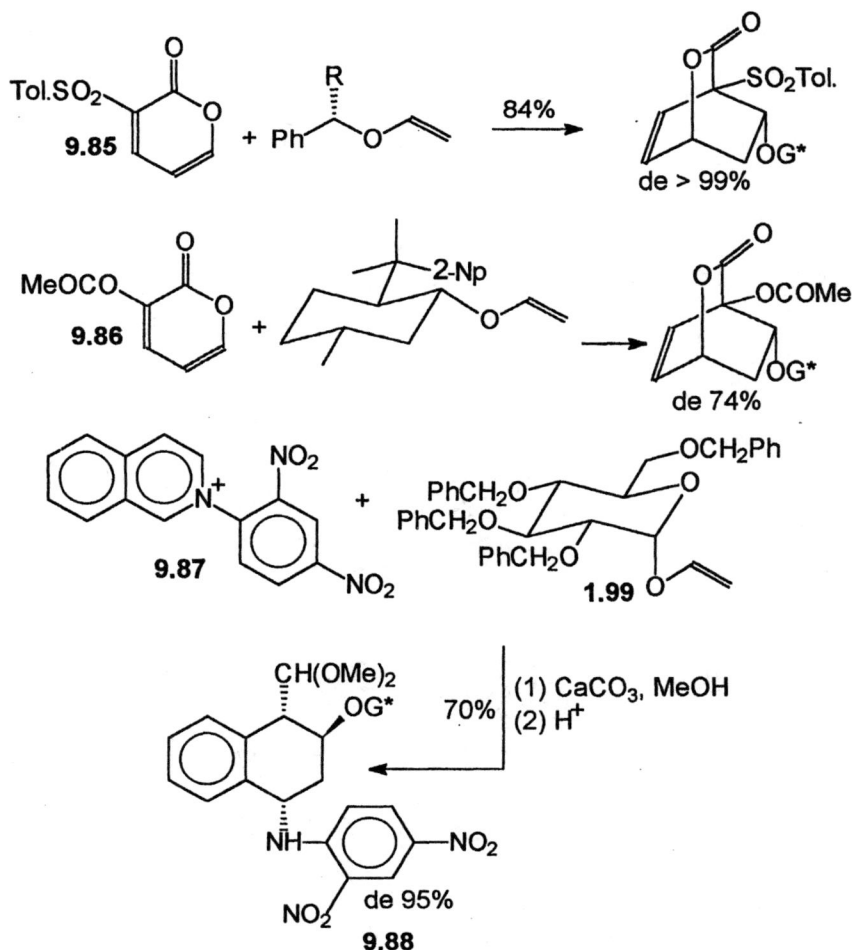

Figure 9.45

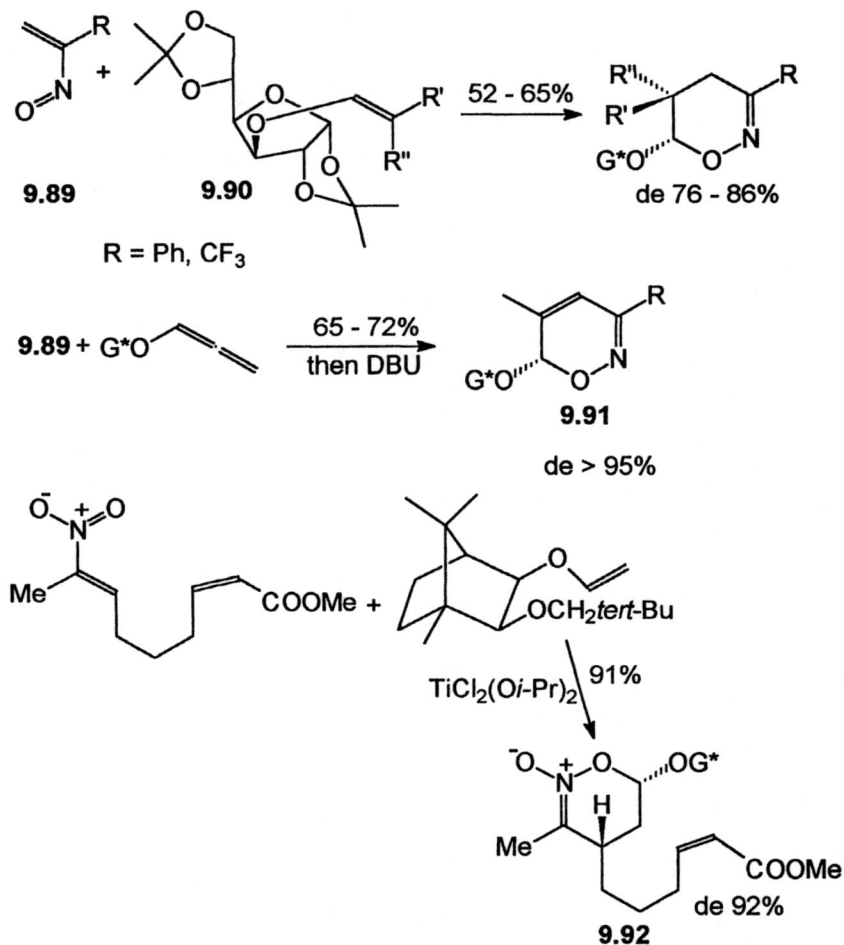

Figure 9.45 (continued)

Ketene silylacetals derived from chiral alcohols undergo thermal cycloadditions with N-benzoylimine PhCH=NCOPh in benzene [140]. The most efficient auxiliaries are (1R,2S,5R)-phenmenthol **1.4** (R = Ph) and (1S,2R)-**1.5** (R = CMe$_2$Ph). These auxiliaries give enantiomeric products after cleavage. The cycloadducts are precursors of the taxol A-ring side chain (Figure **9.46**). Stoodley and coworkers [1609] introduced a β-acetal group derived from pentaacetylglucose on the phenol functionality of quinone **9.41** (X = OH). The cycloadditions of this compound with cyclopentadiene, Danishefsky's diene **3.26** (R = R' = Me, R'' = R''' = H) and 1-trimethylsilyloxy-3-methylbutadiene are highly selective (Figure **9.46**). The X-ray crystal structure of quinone **9.41** shows a boat conformation, and this conformation is probably maintained in solution. The diene then approaches the least hindered face (Figure **9.46**).

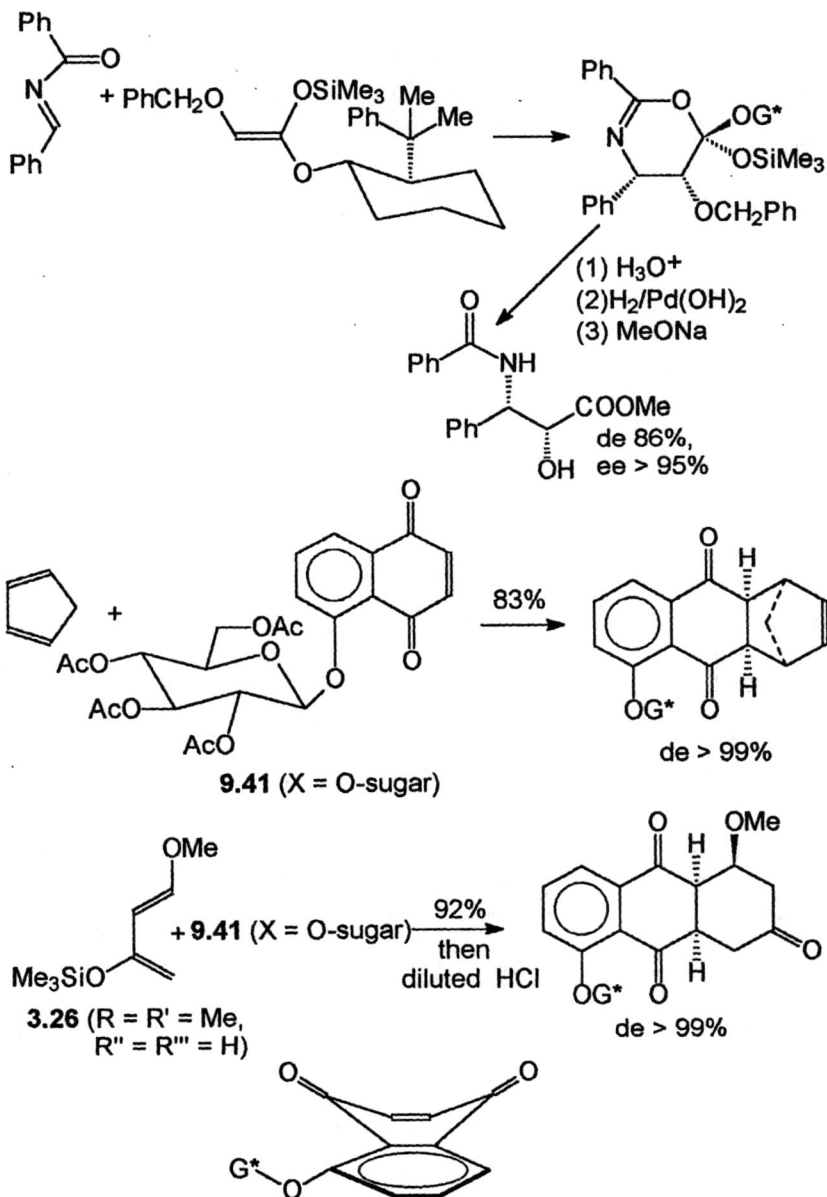

Figure 9.46

9.3.3 Cycloadditions of Dienes Bearing Chiral Residues [73, 1547]

Dienol esters of chiral acids have been proposed as chiral dienes. Trost and coworkers [73, 170] have shown that cycloaddition of quinone **9.41** (X = OH) with 1-(O-methylmandeloxy)butadiene **1.113** (R = H) is highly stereoselective when catalyzed by B(OAc)$_3$ (Figure **9.47**). The cycloaddition of the same diene with acrolein is also selective [387] (Figure **9.47**). The replacement of the phenyl substituent of **1.113** by a cyclohexyl group does not affect the selectivity, so π-stacking does not intervene in transition state [387]. Instead, the diene is probably in a nonfolded conformation (Figure **9.47**).

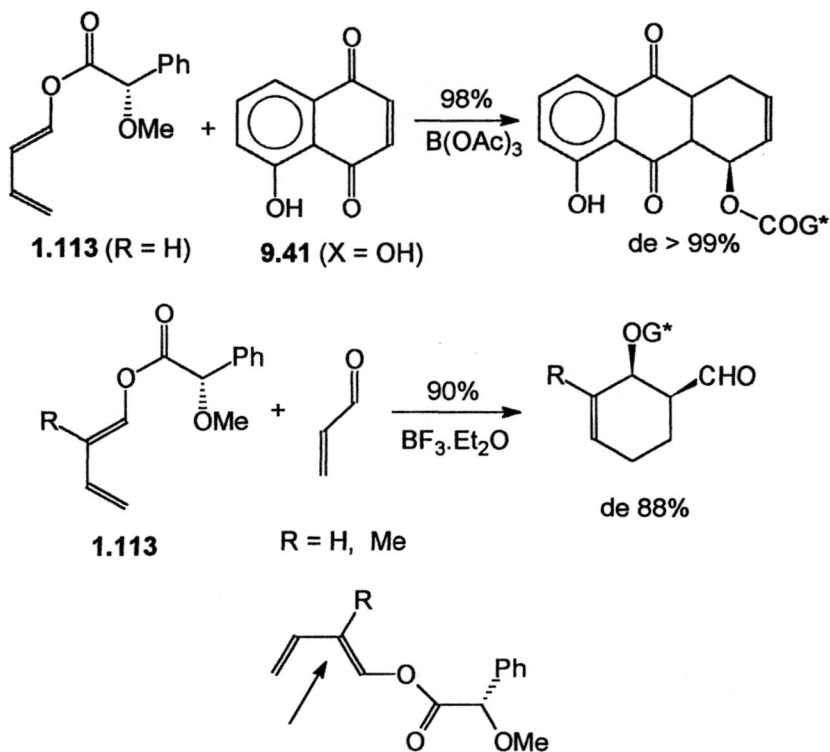

Figure 9.47

Bretmaier and coworkers [49] recommended dienol ethers of chiral alcohols **1.30**. The cycloadditions with maleic anhydride are selective provided the chiral alcohol is *trans*-2-phenylcyclohexanol **1.5** (R = Ph) (Figure **9.48**). The introduction of methyl groups on the phenyl ring increases the selectivity. This result is interpreted by the *endo* approach of the dienophile to the least hindered face of the diene in the *s-cis, s-cis* conformation. This conformation is stabilized by π-stacking (§ I.2.2). Scorrano and coworkers [1610] performed the cycloadditions of aryliminoester **9.93** with cyclopentadiene or indene as dienophiles. These reactions are mediated by TiCl$_4$, and take place at –78°C. Only the ester derived from (1*R*,2*S*,5*R*)-phenmenthol **1.4** (R = Ph) gives a high selectivity (Figure **9.48**). The dienophile approaches the unshielded face of a preformed chelate.

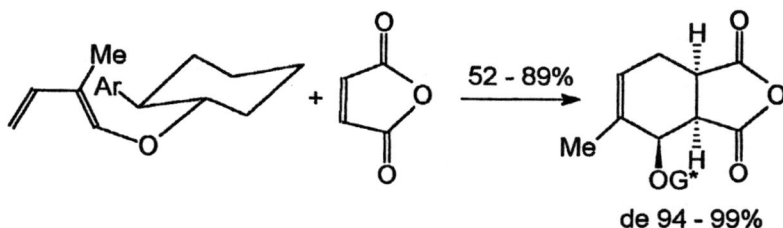

1.30 (Ar = Ph or 2,4,6-Me$_3$C$_6$H$_2$, R = Me)

52 - 89%

de 94 - 99%

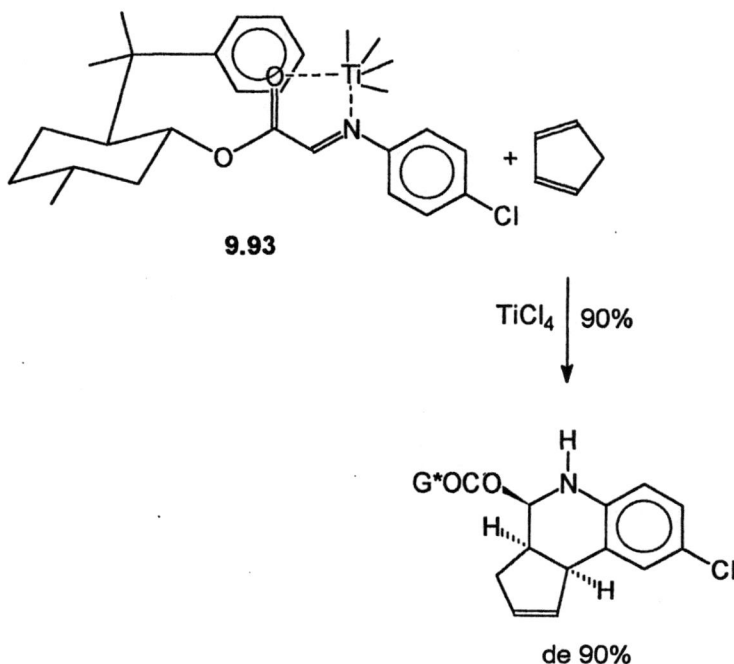

9.93

TiCl$_4$ | 90%

de 90%

Figure 9.48

Carbohydrate derivatives of dienes have received some applications [170, 248, 249]. David and coworkers [170, 249] were the first to prepare dienol ethers from carbohydrates such as diacetoneglucose **1.48** or protected glucopyranoses. Unfortunately, [4+2]-cycloadditions of these dienes with aldehydo-esters or EtOCOCOCOOEt are poorly selective, and the products require chromatographic purifications. The chiral auxiliary is removed by acid treatment after hydrogenation of the double bond of the cycloadduct [1611]. Lubineau and Queneau [362] examined the cycloaddition of glucopyranosides with acrylic dienophiles in water. Only α-glucopyranosides **1.99** (R = H, Y = OCH=CH-CH=CH$_2$) give moderately selective reactions, and the best result (de, 80%) is observed with methacrolein. After NaBH$_4$ reduction of the aldehyde and hydrogenation of the double bond, the chiral auxiliary is removed either enzymatically or by acid hydrolysis (Figure 9.49). Stoodley and coworkers [343, 363, 1612-1615] prepared dienes bearing the β-glucopyranoside residue **1.100** (R = R' = Ac) and examined their cycloadditions with various dienophiles. Tetracyanoethylene leads to poor selectivities, but useful results are obtained with di-*tert*-Bu diazodicarboxylate [1612].

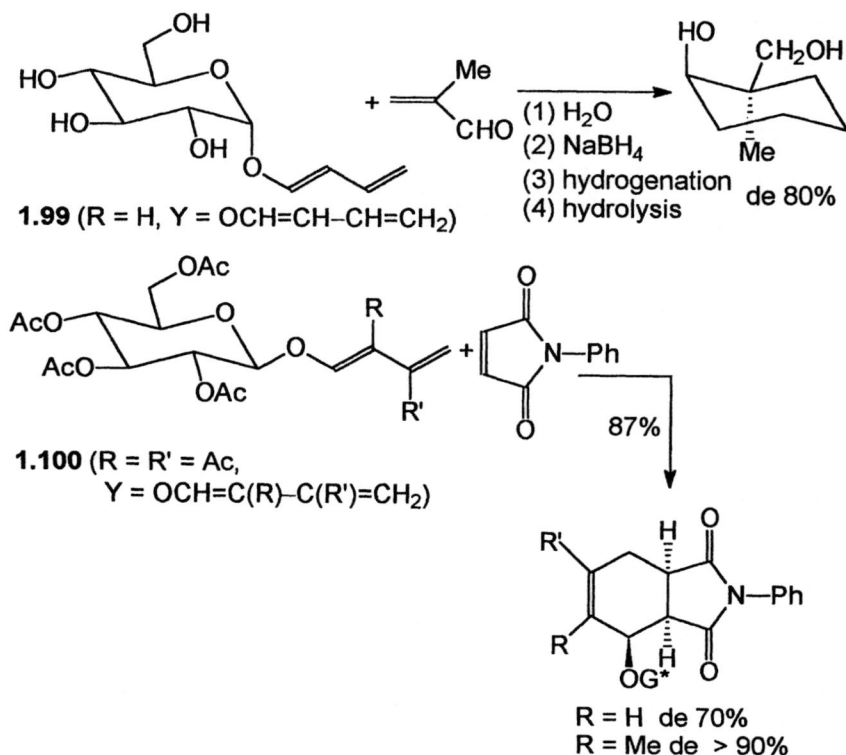

Figure 9.49

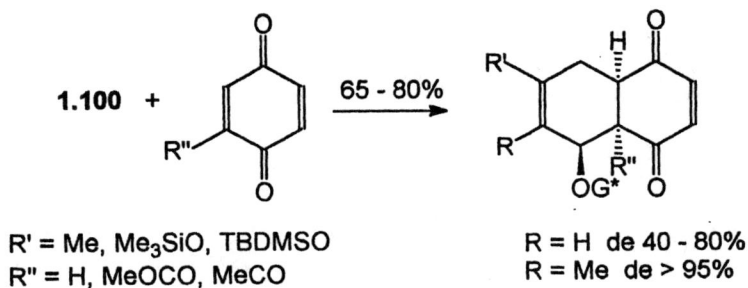

R' = Me, Me₃SiO, TBDMSO

R" = H, MeOCO, MeCO

R = H de 40 - 80%

R = Me de > 95%

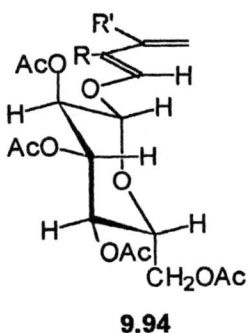

9.94

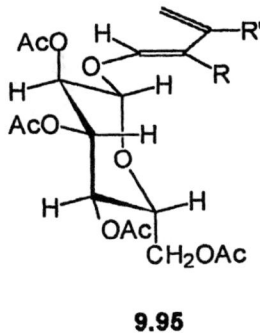

9.95

Figure 9.49 (continued)

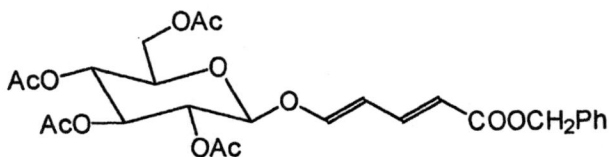

1.100 (R = R' = Ac, Y = OCH=CH–CH=CHCOOCH₂Ph)

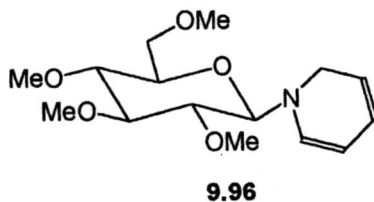

9.96

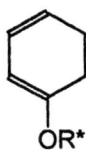

9.97

(R*OH = **1.48**)

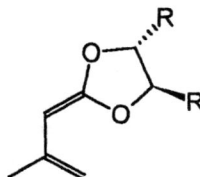

1.39 (R = Me, MeOCH₂, Ph)

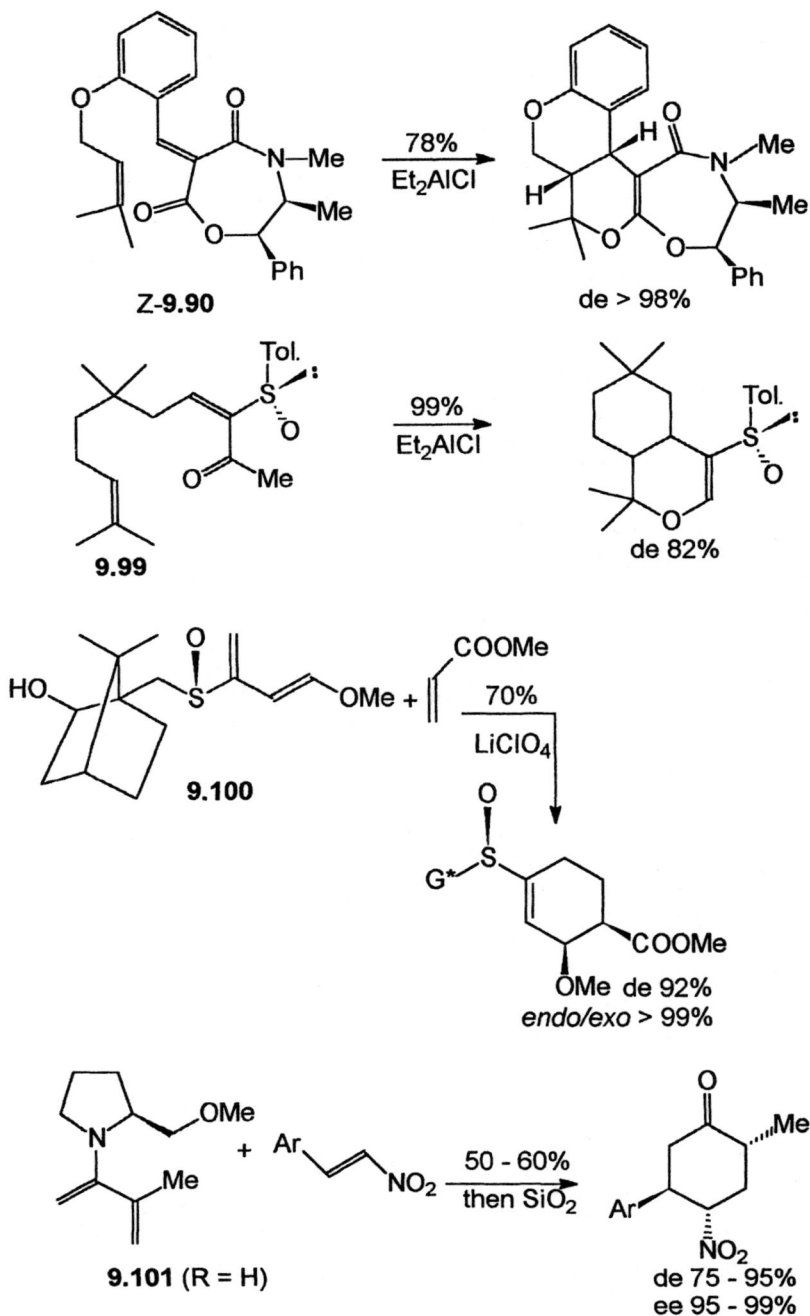

Z-**9.90** → de > 98% (78%, Et₂AlCl)

9.99 → de 82% (99%, Et₂AlCl)

9.100 + COOMe → 70%, LiClO₄ → de 92%, *endo/exo* > 99%

9.101 (R = H) + Ar–CH=CH–NO₂ → 50 - 60%, then SiO₂ → de 75 - 95%, ee 95 - 99%

Ar = Ph, 4-FC₆H₄, 4-MeC₆H₄

Figure 9.50

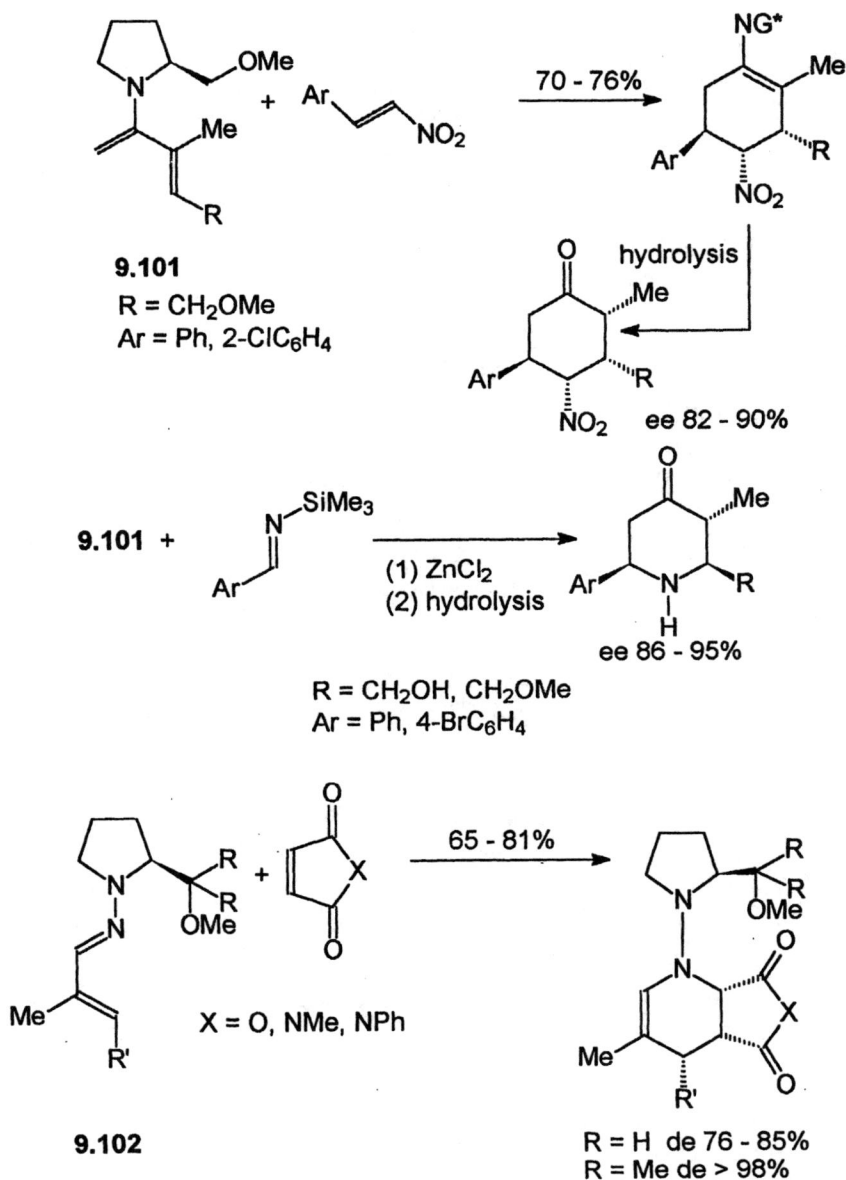

Figure 9.50 (continued)

The use of cyclic dienophiles such as maleic anhydride, *N*-phenylmaleimide or quinones induces stereoselective cycloadditions, especially when R = Me (Figure **9.49**). These results have been interpreted by the approach of the dienophile to favored ground-state conformations **9.94** or **9.95** of the diene. When R = Me, conformer **9.95** is destabilized by repulsive interaction with the C-H bond in the anomeric position. Therefore, this conformational effect increases the selectivity of the cycloaddition because a single conformer **9.94** is attacked from its least hindered face. The chiral auxiliary can be removed in acid medium [363].

Other chiral dienes bearing related functionalities have also been recommended. For example, **1.100** (R = R' = Ac, Y = OCH=CH-CH=CHCOOCH$_2$Ph) has given high selectivities in cycloadditions with diazodicarboxylates [1612]. Cycloadditions of cyclic dienes **9.96** and **9.97** with methyl acrylate or naphthoquinone also give useful results [364, 1616]. Ketene acetals derived from chiral 1,2-diols **1.39** (R = Me, MeOCH$_2$, Ph) give selective cycloadditions with maleimides [213]. Tietze and coworkers [74, 351, 1617] have performed an intramolecular hetero-Diels-Alder reaction mediated by Et$_2$AlCl. The 4 π fragment of Z-**9.98** bears an ephedrine residue, and a highly predominant stereoisomer is formed (Figure 9.50). A related intramolecular cycloaddition has been performed by Hiroi and coworkers with a good selectivity starting from **9.99** [1618]. In this case, the chiral residue is a sulfoxide (Figure 9.50). Chiral sulfoxides have also been introduced on 1-methoxybutadiene [1619], and the LiClO$_4$-mediated cycloaddition of **9.100** with methyl acrylate is highly selective (Figure 9.50). Other Lewis acids are less efficient.

Chiral 2-aminobutadienes derived from (S)-proline have recently been recommended in asymmetric cycloadditions [1620]. Enders and coworkers [1621] performed the reaction of **9.101** (R = H) with 2-aryl-1-nitroethylenes and obtained substituted cyclohexanones with a high selectivity after hydrolysis (Figure 9.50). Barluenga and coworkers [296] used **9.101** (R = CH$_2$OH, CH$_2$OMe) as dienes. Their cycloadditions with 2-aryl-1-nitroethylene or aryl-N-trimethylsilylimines are also highly regio- and stereoselective (Figure 9.50). After hydrolysis, the cycloadducts lead to substituted 4-pyridones or cyclohexanones with a high enantiomeric excess.

Recently, Ghosez and Beaudegnies [319] recommended chiral azadienes **9.102** derived from α,β-unsaturated aldehydes and Enders's hydrazines **1.76**. To observe a high face selectivity in the cycloadditions with maleimides or maleic anhydride conducted at room temperature, the side chain of the hydrazines must be bulky (R = Me) (Figure 9.50). The cleavage of the auxiliary is accomplished with Zn-AcOH, but this also saturates the double bond, and diastereoisomers are formed. Unfortunately, this cycloaddition does not work with acrylates or fumarates.

CHAPTER 10

SIGMATROPIC REARRANGEMENTS

[2,3]- and [3,3]-Sigmatropic rearrangements are symmetry allowed six-electron processes that often have been used in synthesis [497, 624c]. [2,3]-Sigmatropic rearrangements take place with molecules containing a heteroatom whose lone pair participates in the reaction process (Figure **10.1**), while [3,3]-rearrangements can occur with all-carbon skeletons (X = Y = C) or with skeletons containing a heteroatom whose lone pair of electrons is not involved in the rearrangement (Figure **10.1**). In many cases, this heteroatom bears an allylic substituent [1062].

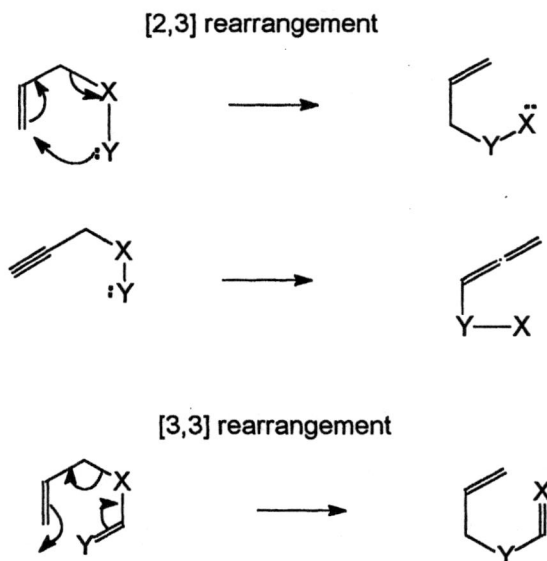

[2,3] rearrangement

[3,3] rearrangement

Figure 10.1

10.1 [2,3]-SIGMATROPIC REARRANGEMENTS [497, 1062, 1622]

The major [2,3]-rearrangements that have been applied in asymmetric synthesis are:

• The Wittig rearrangement: the X-Y fragment is a carbanion α to oxygen (O⁻CR$_2$). A variant is the Büchi rearrangement where the OCR$_2$ group is an orthoamide (OC(OMe)NMe$_2$) (Figure **10.2**).

• The rearrangement of ammonium ylides (Stevens-Sommelet rearrangement) or sulfonium ylides; the X-Y fragment is an anion α to a positively charged nitrogen ($R_3'\overset{+}{N}-\overset{-}{C}R_2$) or sulfur ($R_2'\overset{+}{S}-\overset{-}{C}R_2$).

• The sulfoxide-sulfenate ester rearrangement (Mislow/Evans rearrangement) where the X-Y fragment is a sulfoxide.

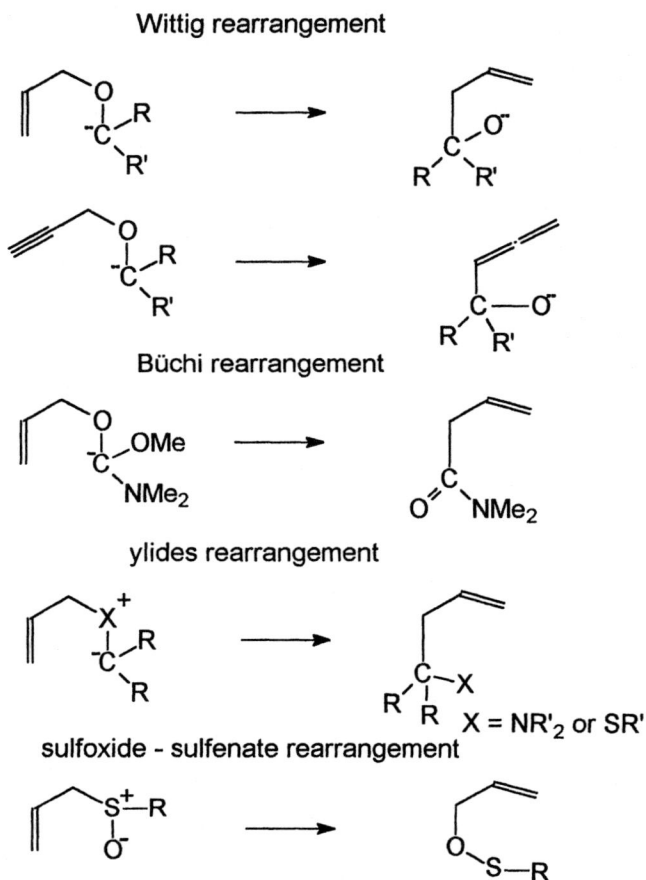

Wittig rearrangement

Büchi rearrangement

ylides rearrangement

X = NR'$_2$ or SR'

sulfoxide - sulfenate rearrangement

Figure 10.2

10.1.1 Rearrangements of α-Oxycarbanions [163, 1062, 1623, 1624, 1625]

The Wittig rearrangement has been the object of much work because it is versatile and easy to conduct. The formation of carbanions was initially effected by lithium bases, and this limited the application of the Wittig rearrangement to ethers containing a sufficiently acidic hydrogen. Useful substrates included bis-allylethers, allylbenzylethers or allylpropargylethers, or mixed allyloxy- or propargyloxycarbonyls. Problems of regioselectivity in the deprotonation step were common [163]. The scope of the Wittig rearrangement has been significantly enlarged by the introduction of new methods to prepare the α-oxycarbanion including tin-lithium transmetalation introduced by Still [163, 1062], and reductive lithiation of thioacetals. In the reductive lithiations, high yields are frequently obtained by using lithium naphthalenide [163, 823].

The Wittig rearrangement is concerted and suprafacial for each fragment, except in the case of benzylic ethers, where the intervention of a radical process can lead to a partial racemization [1624]. In general, an efficient transfer of chirality is expected, as illustrated by the rearrangements of allylic stannyl ethers Z-**10.1** (R_E = H) in the presence of n-BuLi at −70°C. Homoallylic alcohols E-**10.2** are obtained with an excellent selectivity (Figure **10.3**). The isomer E-**10.1** (R_Z = H) leads to a mixture of enantiomerically pure E- and Z-homoallylic alcohols. These results are interpreted by two cyclic 5-membered ring transition states in the shape of a half-envelope. For Z-allylic alcohols, transition state **10.4** is favored because it avoids eclipsing interactions (A(1,3)-strain) [54] between the C–R and C–R_Z bonds that occur in transition state **10.3**. For E-allylic alcohols, transition states **10.3** and **10.4** are of comparable energy (Figure **10.3**). These reactions have frequently been applied in steroid synthesis [163, 1624].

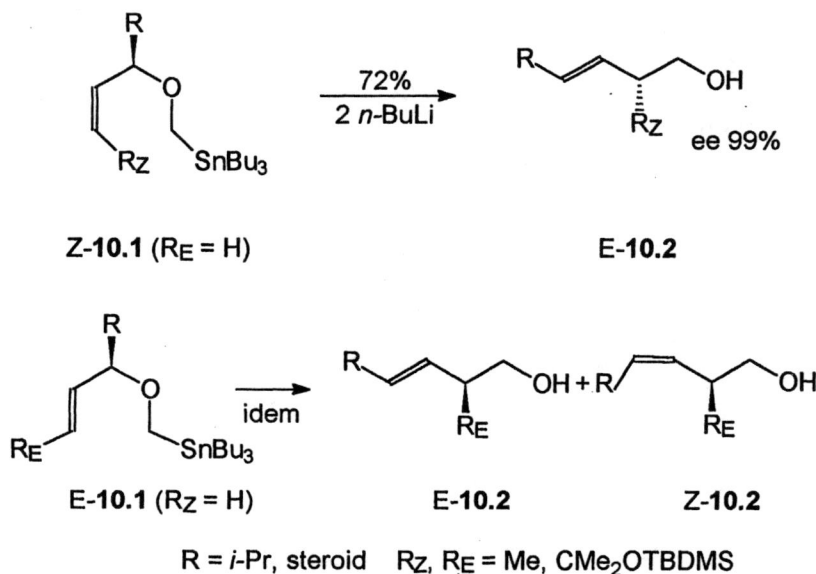

Z-**10.1** (R_E = H)　　　　　　　　E-**10.2**

E-**10.1** (R_Z = H)　　　　E-**10.2**　　　Z-**10.2**

R = i-Pr, steroid　R_Z, R_E = Me, $CMe_2OTBDMS$

Figure 10.3

10.3

10.4

10.5

10.6
ee > 99%

R = n-Bu, n-C$_7$H$_{15}$

Figure 10.3 (continued)

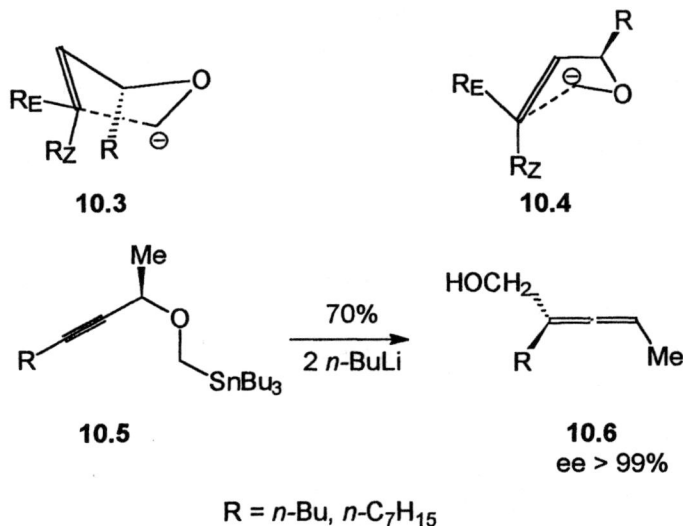

Starting from (R)-propargylic ethers **10.5**, a similar transfer of chirality at −70°C (Figure **10.3**) leads to chiral allenes **10.6** [1624]. Chiral allylpropargylethers **10.7** or diallylethers **10.8** are transformed equally well into the corresponding homoallylic alcohols. At −85°C, the Z isomers **10.7** (R$_E$ = H) yield *syn* alcohols **10.9** and **10.10** very selectively via a transition state **10.11** that is analogous to **10.4** (Figure **10.4**). Once again, eclipsing interactions are minimized. As in the preceding case, the reactions of the E isomers (R$_Z$ = H) are less selective [163, 1624]. For these isomers, the Büchi modification gives better results (Figure **10.4**).

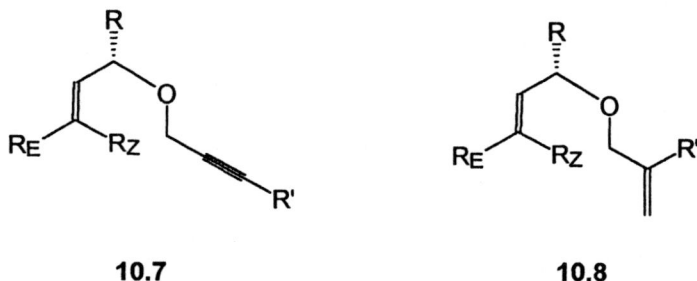

10.7

10.8

Figure 10.4

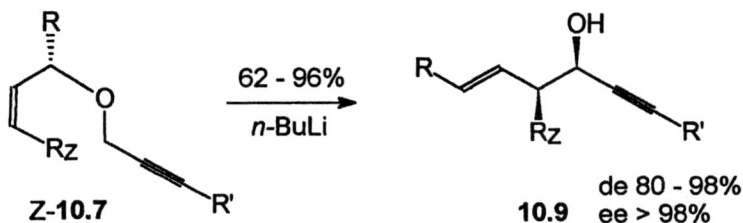

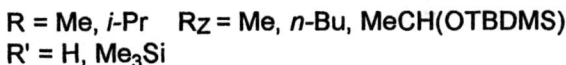

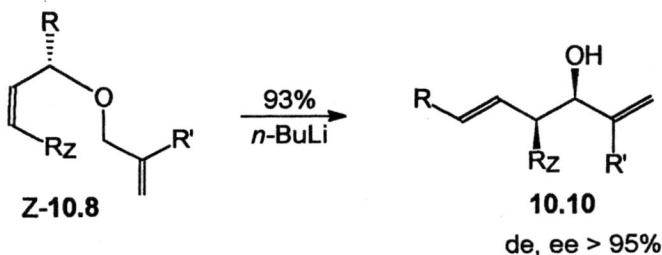

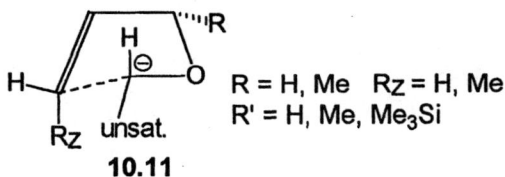

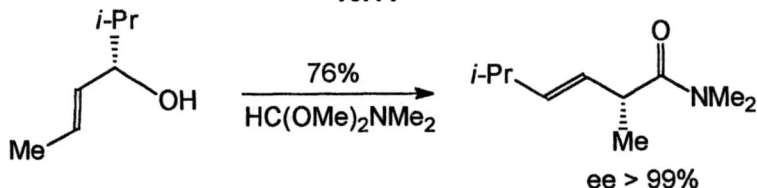

Figure 10.4 (continued)

The chirality transfer from allyloxyenolates at −70°C follows a different course. The most selective rearrangements are observed starting from E-allyloxyacidic esters **10.11**, especially when the rearrangements are conducted with zirconium enolates that are prone to chelate to the oxygen of the ether. In this fashion, Z-*syn* homoallylic alcohols **10.12** are obtained with an excellent selectivity. Chelated transition-state model **10.13** rationalizes these results (Figure **10.5**). If the R substituent also carries a functional group capable of chelating the metal of the enolate, then the other alcohol E-**10.14** is obtained via a tridentate chelate. In that case, the titanium enolate is the most effective reagent (Figure **10.5**). Comparable results are obtained with the lithium dianions of acids **10.11** (R = H) and the corresponding propargylic derivatives **10.15** [1624] (Figure **10.5**).

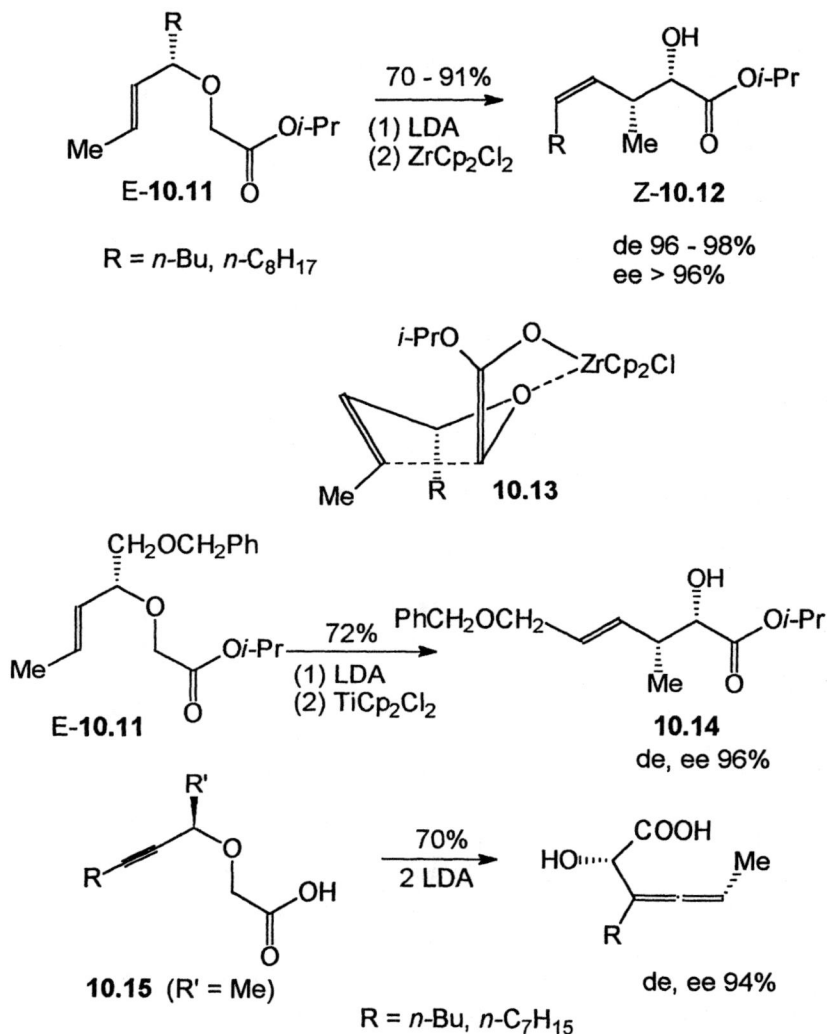

Figure 10.5

The use of chiral amides (*S,S*)- or (*R,R*)-**2.3** for effecting Wittig rearrangements has been proposed by Marshall and coworkers [565, 1624]. The rearrangement of linear prochiral allylpropargyl ethers gave disappointing results.

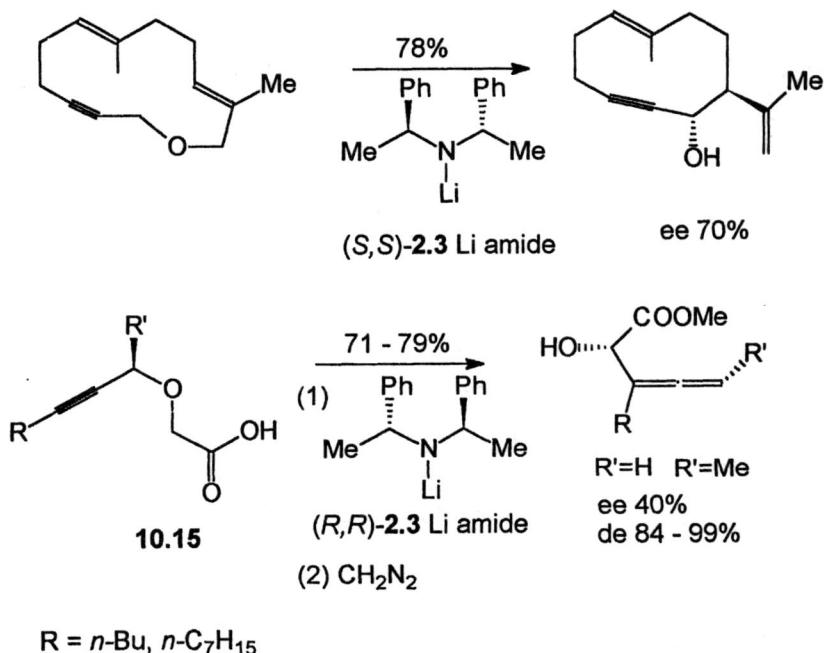

Figure 10.6

In contrast, in the cyclic series, a useful enantioselectivity has been observed (Figure **10.6**), and good selectivity has also been observed during the reaction of propargyloxyacetic acids **10.15** (R' = H). Moreover, double asymmetric induction is observed starting from chiral propargyloxyacetic acid **10.15** (R' = Me). Interesting selectivities are observed provided that the base and the substrate are well paired [771] (Figure **10.6**).

Chiral auxiliaries have been introduced into allyloxyacetic acid derivatives. At −70°C in the presence of LDA or better yet LICA, esters of (1R,2S,5R)-phenmenthol **1.4** (R = Ph) lead to syn-α-hydroxyesters **10.16** with excellent selectivity (Figure **10.7**). The favored conformation of the ester with the phenyl group of the auxiliary blocking one of the faces of the molecule permits the interpretation of the observed selectivity (§ I.3). C_2-Symmetric derivatives **10.17** introduced by Katsuki and Yamaguchi [163, 1622, 1624] are among the most frequently used chiral amides. The most interesting results from using these amides are obtained via the zirconium enolates (Figure **10.7**). Nakai and coworkers have introduced the use of chiral oxazolines, which are likely to rigidify the transition states by chelation with lithium [339].

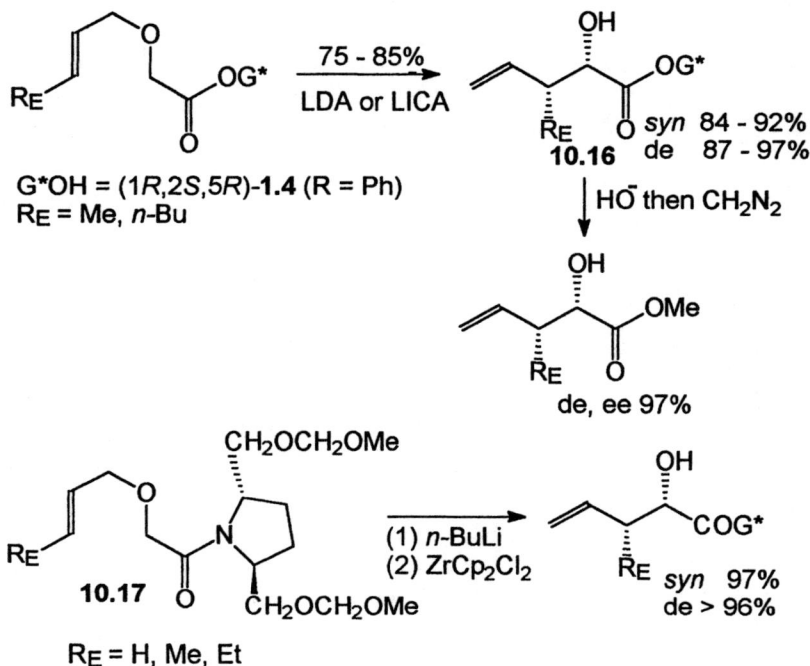

G*OH = (1R,2S,5R)-**1.4** (R = Ph)
R_E = Me, n-Bu

Figure 10.7

In the presence of n-BuLi, the transformation of **10.18** leads to interesting results if R_E = Me (Figure **10.8**). Somewhat surprisingly, by using the chiral auxiliary where the methyl ether is replaced by a primary alcohol and where $R_E \neq$ H, rearrangements effected in the presence of lithium bases lead preferentially to *syn* isomers. However, the diastereoselectivity is poor. On the other hand, in the presence of KH and if R $\neq$ H, a mixture of *syn* and *anti* isomers is obtained, and each isomer is formed with a high selectivity. When applied to allylic compound **10.19** (R_E = H), this method leads selectively to the (2S)-isomer of **10.20** [163]. In addition, if the rearrangement is effected by KH in the presence of a crown ether, the (2R)-isomer of **10.20** is formed [163] (Figure **10.8**).

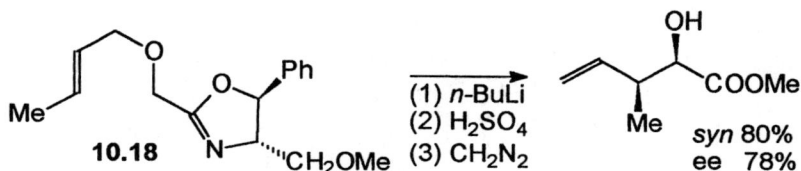

Figure 10.8

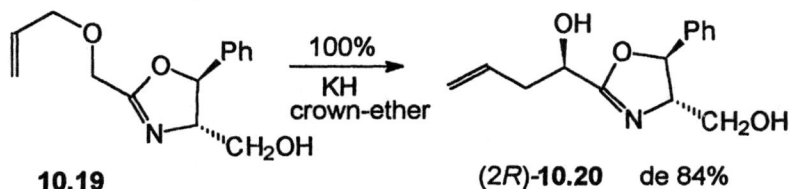

Figure 10.8 (continued)

These reactions have often been applied in the synthesis of natural products. Furthermore, the allylic alcohols such as **10.10** that are formed, can be used to effect sequential transformations [163, 1625].

10.1.2 Rearrangements of Carbanions α- to Nitrogen and Sulfur
[1062, 1622]

Sulfur and ammonium ylides are susceptible to [2,3]-sigmatropic rearrangements under mild conditions with efficient transfer of chirality, as Trost and Hammen showed in 1973 [1626]. Most of the applications of these rearrangements have been conducted on chiral synthons [1622]. Kurth and coworkers [1627, 1628] have applied this reaction to the synthesis of β-chiral, γ,δ-unsaturated acids via the rearrangements of cyclic sulfonium ylides **10.21**, starting from thiol **10.22** (Figure 10.9).

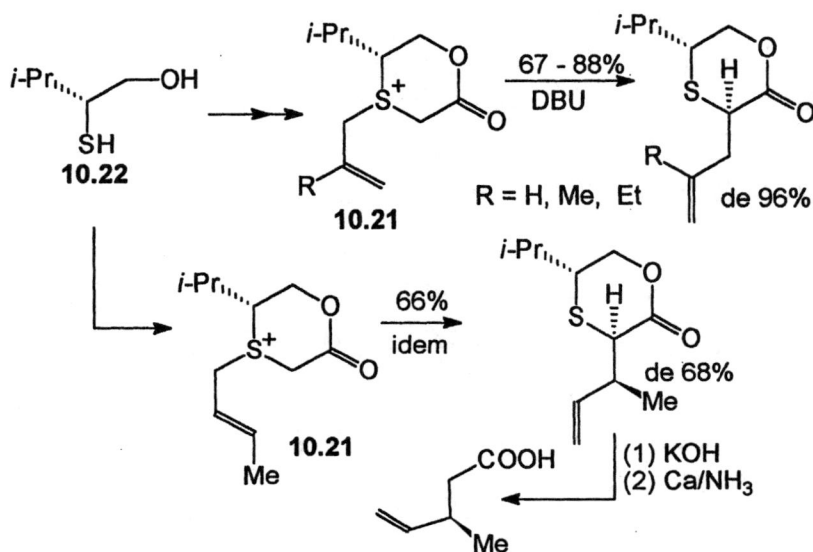

Figure 10.9

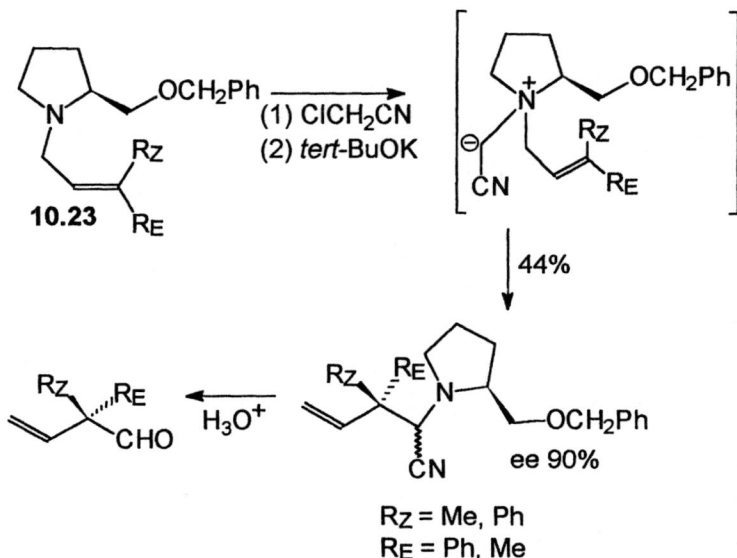

Rz = Me, Ph
RE = Ph, Me

Figure 10.9 (continued)

     Rearrangement of the sulfonium salt **10.21**, easily prepared by allylation with the appropriate substituted allyl bromide in the presence of silver triflate, gives the product of allylic transposition with a very good selectivity, starting from allylic derivatives that are not substituted on the terminal carbon. The E-crotyl derivatives, precursors of chiral acids, give selectivities that are not as good [1627, 1628] (Figure **10.9**). The sulfonium ylides **10.21** can be accessed through a diazoketone, but this sequence is more laborious [1627]. Nevertheless, this route avoids Z/E isomerizations of the crotyl derivatives, and it leads to higher selectivities.

     A derivative of (*S*)-prolinol **10.23**, quarternized by ClCH₂CN and then transformed into the ammonium ylide by KO-*tert*-Bu in DMSO, has been subjected to a [2,3]-sigmatropic rearrangement at –90°C. After hydrolysis of the aminonitrile formed in this way, an α-chiral β,γ-unsaturated aldehyde is formed with an excellent selectivity [261, 290, 1008, 1062] (Figure **10.9**).

10.1.3 Sulfoxide-Sulfenate Esters Rearrangements [1062, 1622]

     The [2,3]-sigmatropic interconversion of sulfoxides and sulfenate esters is easily reversible, and this is why chiral allylic sulfoxides typically racemize at room temperature. The sulfoxide-sulfenate ester equilibrium usually favors the sulfoxide. If a reactive thiophile that cleaves the O-S bond of the sulfenate ester is introduced into the reaction mixture, then the equilibrium is displaced and the allylic alcohol is formed with the possibility of transfer of chirality (Figure **10.10**). Starting from chiral sulfoxides, this transfer has been examined by Hoffman and coworkers [497, 1062]. Chirality transfer is efficient during the rearrangement of Z-sulfoxides **10.24** (R_Z = n-C_5H_{11}, R_E = H) in the presence of amines, while the E-isomers lead to mediocre results. A half-envelope transition state **10.25** was proposed to interpret these results. When R_Z = H, a gauche interaction between the phenyl substituent on sulfur and the R_E group develops. This unfavorable interaction is not established if R_Z = alkyl and R_E = H. If the sulfoxide bears a substituent in the α orientation, then this has a tendency to adopt a pseudoequatorial orientation on the half-envelope, and the influence of the stereochemistry at sulfur is less important. This effect was profitably used by Whitesell and coworkers [191] during the sigmatropic rearrangement of chiral sulfoxides obtained by an ene reaction of sulfinyl carbamates **1.25** (§ 8.1). The transformation of sulfoxides **10.26** in methanol at reflux in the presence of piperidine leads to allylic alcohols E-**10.27** with an excellent enantiomeric excess. An allylic amine precursor is obtained if the N-sulfinylcarbamates **1.25** are treated with (Me$_3$Si)$_2$NH before performing the rearrangement. This reaction probably occurs via an S-O silylated intermediate; however, the selectivities observed are lower (~ 60%) [1629].

The allylic sulfoxide precursor can be formed by a Knoevenagel reaction followed by a prototropic shift. Burgess and coworkers [1630, 1631] have used the sulfinyl acetates of chiral alcohols **10.28**, whose reactions with aldehydes give allylic sulfoxides after *in situ* prototropic shifts. These allylic sulfoxides suffer a [2,3]-sigmatropic rearrangement, and give allylic alcohols with an interesting selectivity provided that the chirality at sulfur and the chiral auxiliary are matched. The ester **10.28** derived from the alcohol **1.10**, where the configuration at sulfur is (R), leads to the best results (Figure **10.10**).

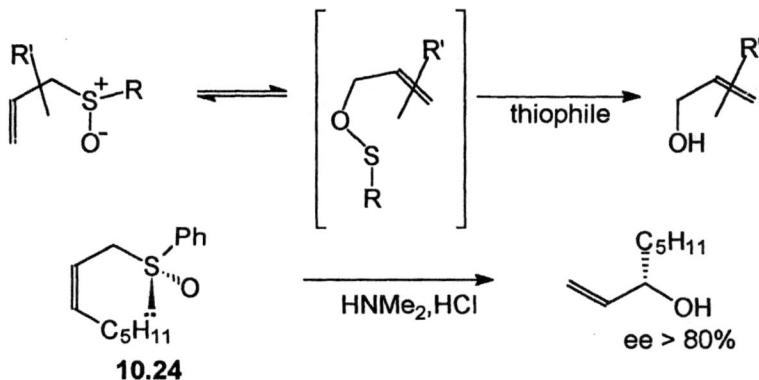

Figure 10.10

10.25

1.25

(1) EtO$_3$BF$_4$
(2) PhMgBr

10.26

56 - 64%

E-**10.27**
ee 84 - 92%

R = Me, n-C$_4$H$_9$, n-C$_5$H$_{11}$ R' = Me, c-C$_5$H$_9$, AcOCH$_2$CH$_2$

10.28   + RCH$_2$CHO

70 - 98%

OH

R   COOG*
de 76 - 82%

[2,3] rearrangement
and O-S cleavage

4-ClC$_6$H$_4$····S=O

G*OCO····   R

prototropy

R = Me, Et, n-Pr, c-C$_6$H$_{10}$CH$_2$

G*OH=

SO$_2$N(c-C$_6$H$_{11}$)$_2$
1.10

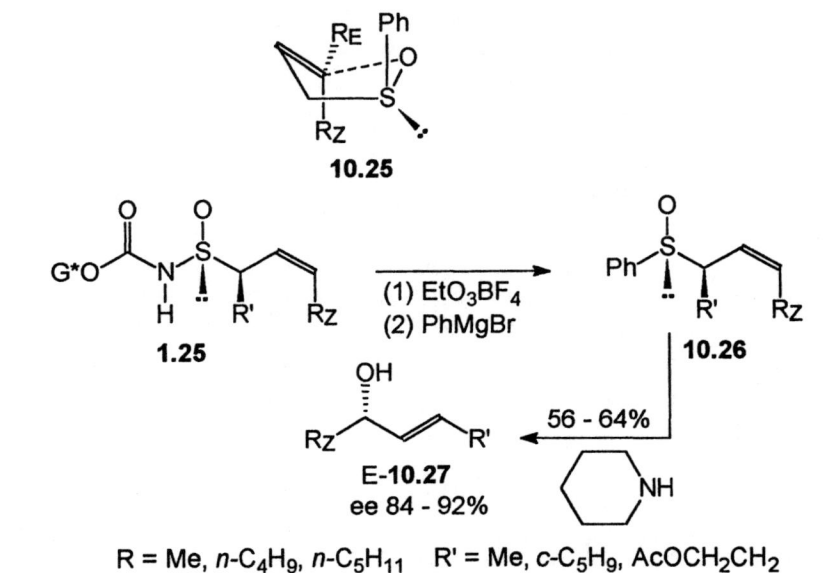

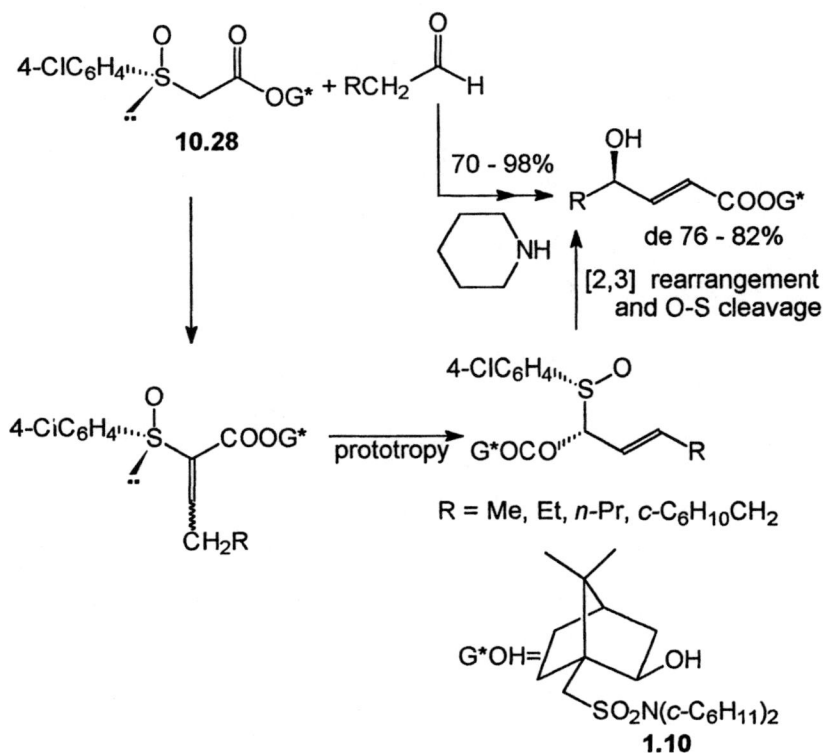

Figure 10.10 (continued)

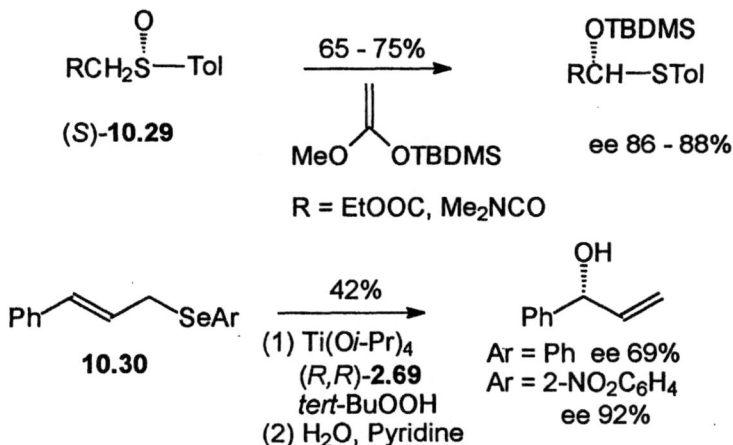

O
‖
RCH₂S—Tol   65 - 75%   OTBDMS
                  ─────────→   ┊
(S)-**10.29**   RCH—STol

         MeO   OTBDMS   ee 86 - 88%

         R = EtOOC, Me₂NCO

                                              OH
Ph   SeAr   42%   Ph   ┊
                     ─────────→
10.30   (1) Ti(Oi-Pr)₄   Ar = Ph ee 69%
                   (R,R)-**2.69**   Ar = 2-NO₂C₆H₄
                   tert-BuOOH   ee 92%
                   (2) H₂O, Pyridine

Figure 10.11

The Pummerer reaction, whose key step is a [2,3]-sigmatropic rearrangement, has never been observed to lead to efficient transfer of chirality starting from chiral sulfoxides in the presence of acetic anhydride [1632, 1633]. A modification via silyloxysulfides, generated with O-methyl-OTBDMS ketene acetal at 65°C, allows asymmetric silicon-induced Pummerer reaction from chiral sulfoxides **10.29** with a high chirality transfer [1634] (Figure **10.11**). The (S)-sulfoxides generate the (S)-secondary ethers and vice-versa.

The [2,3]-sigmatropic rearrangements of chiral allylselenoxides have also given disappointing results [749]. However, Uemura and coworkers [1635] performed such rearrangements by generating chiral selenoxides from cinnamyl selenides **10.30** with Sharpless reagent at −20°C and performing the rearrangement at room temperature (Figure **10.11**). The nature of the aryl group of the selenoxide has a strong influence on the selectivity of the process. Palladium-catalyzed allylic sulfinate-sulfone rearrangements take place with a good stereospecificity [1636].

10.2 THERMAL [3,3]-SIGMATROPIC REARRANGEMENTS
[497, 1062, 1637, 1638, 1639]

Most [3,3]-sigmatropic rearrangements take place thermally, and the Cope, oxy-Cope and Claisen rearrangements are among the most important rearrangements in this class. Important variants of the Claisen rearrangement include the Johnson modification via orthoesters, the Eschenmoser modification via ketene N,O-acetals, the Ireland modification via ketene silylacetals and the Corey modification via boron ester enolates [696]. The aza-Claisen rearrangement has also seen a number of applications in asymmetric synthesis (Figure **10.12**). These rearrangements are concerted and suprafacial with respect to each of the fragments, and chair and boat transition states must be considered. In most cases, the chair transition state is favored. Gauche or eclipsing interactions between the different substituents often cause one of the chair geometries (**10.31** or **10.32**) to be favored over the other, with the result that one of the possible isomers is highly favored (Figure **10.12**). However, conformational constraints in cyclic systems as well as A(1,3) interactions in acyclic systems disfavor the chair geometry to the profit of the boat geometry. The enthalpy difference between the chair and boat transition states is estimated at 5-6 kcal/mole for the Cope rearrangement and 3 kcal/mole for the other rearrangements [1638] (see below). It has recently been shown that the transition state of the Claisen rearrangement is unsymmetrical [1640].

10.2.1 The Cope and Oxy-Cope Rearrangements

The Cope rearrangement is reversible, and the relative stability of the precursors and products determines the position of the equilibrium. In contrast, the oxy-Cope rearrangement is irreversible because this rearrangement leads to an enol, which is easily ketonized (Figure **10.12**). These rearrangements have been applied to chiral systems, and the transfer of chirality is effected through the lowest energy chair transition state. This is shown by examples in Figure **10.13** [1637]. The oxy-Cope rearrangement occurs in the acyclic series with a very good transfer of chirality, which must be imposed by the judicious choice of substituents to establish a single chair conformation [1641, 1642].

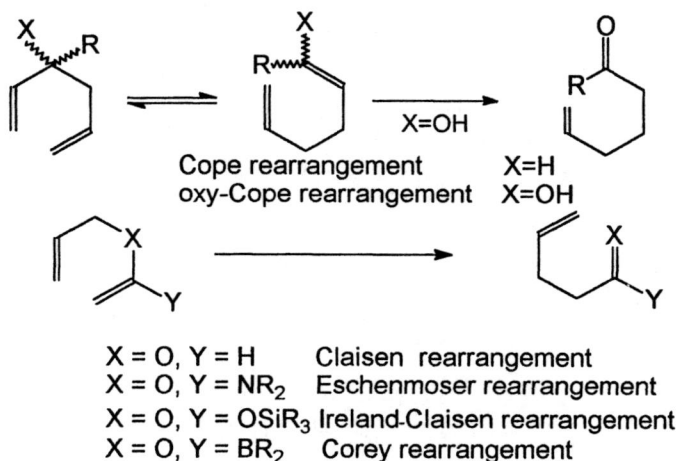

Cope rearrangement   X=H
oxy-Cope rearrangement   X=OH

X = O, Y = H   Claisen rearrangement
X = O, Y = NR$_2$ Eschenmoser rearrangement
X = O, Y = OSiR$_3$ Ireland-Claisen rearrangement
X = O, Y = BR$_2$ Corey rearrangement

Figure 10.12

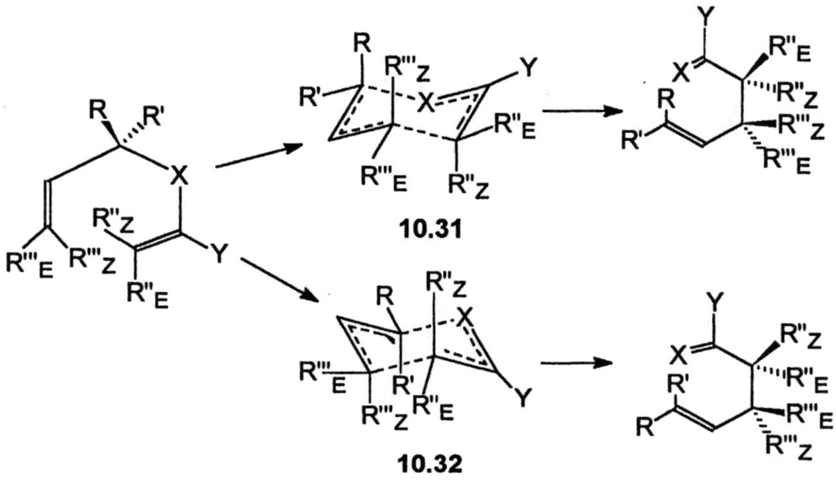

Figure 10.12 (continued)

This is nicely shown by the example in Figure **10.13** where the substituents Et and O⁻ are in a pseudoequatorial position [1643]. Agami and coworkers [1644] applied the aza-Cope rearrangement to asymmetric synthesis of α-aminoacids by trapping by hydrolysis or with formic acid the ene-imminium cation resulting from the reaction of allylamine **10.34** with glyoxal (Figure **10.13**). After a suitable treatment and removal of the chiral aminoalcohol auxiliary, homoserine lactone **10.35** or proline derivative **10.36** were obtained with a high enantiomeric excess (Figure **10.13**).

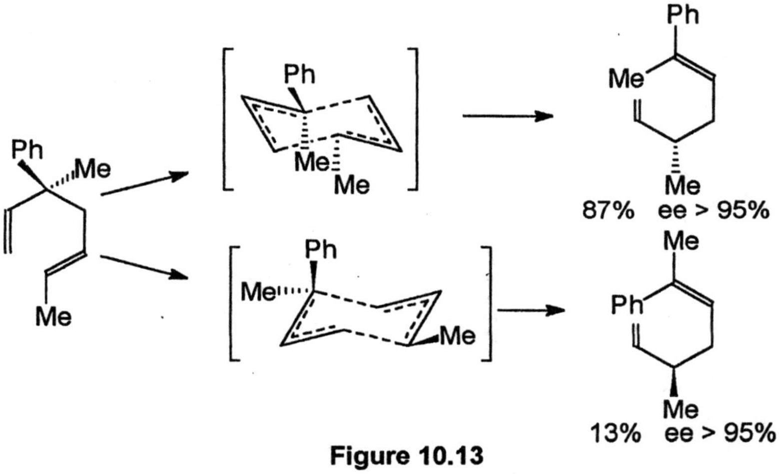

Figure 10.13

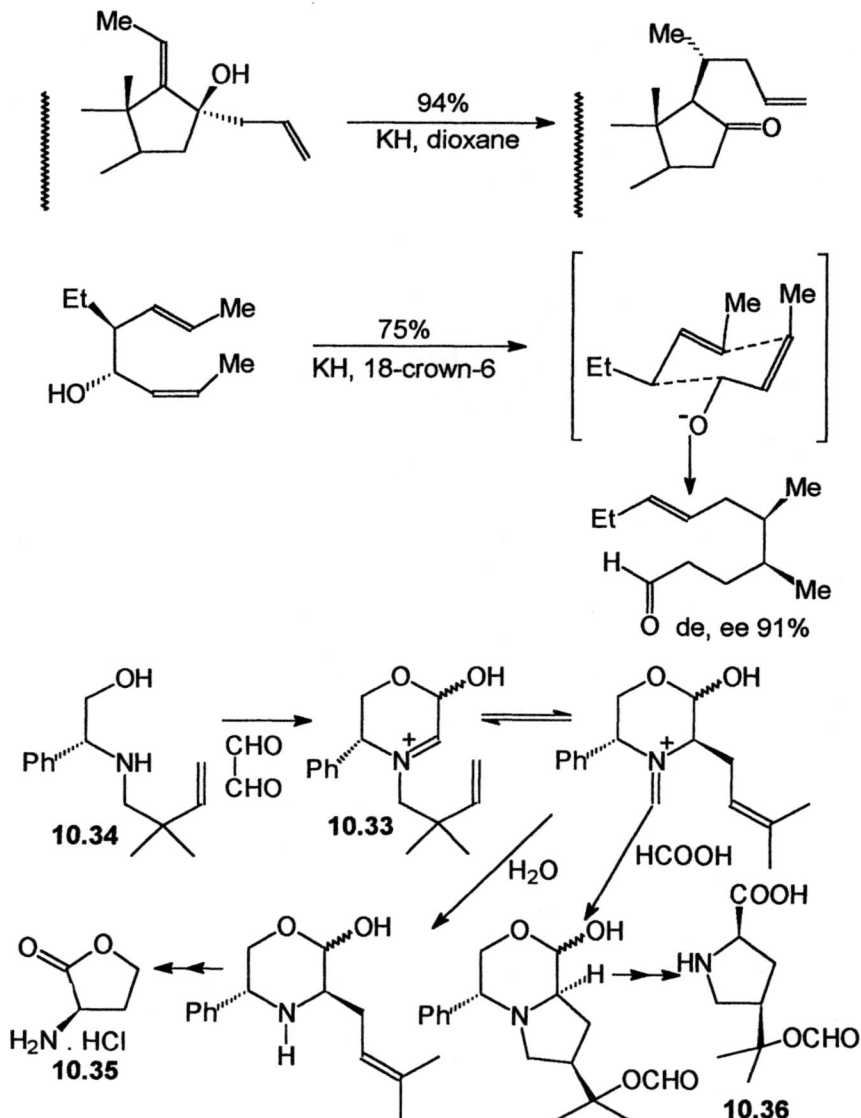

Figure 10.13

10.2.2 The Claisen and Related Rearrangements

In most of the examples of asymmetric Claisen rearrangements in the literature, a carbonyl compound is obtained starting from a chiral allylic alcohol with an excellent transfer of chirality. An example developed by Saucy and coworkers, shown in Figure **10.14**, nicely illustrates this methodology [1639, 1645]. Alcohol

10.37 is transformed into a vinyl ether which, when heated, suffers a Claisen rearrangement that leads to the aldehyde **10.38**. By heating in the presence of MeC(OMe)$_3$, the ester **10.39** (Y = OMe) is obtained by a Johnson rearrangement, while in the presence of MeC(OMe)$_2$NMe$_2$, the Eschenmoser rearrangement leads to the amide **10.39** (Y = NMe$_2$). Starting from the acetate of **10.37**, the Ireland-Claisen rearrangement at ambient temperature gives the acid **10.39** (Y = OH) through the intermediacy of the ketene acetal **10.40**. In all these rearrangements, the transfer of chirality is better than 95%. These transformations take place via chair transition state **10.41** where the largest group is in the pseudoequatorial position.

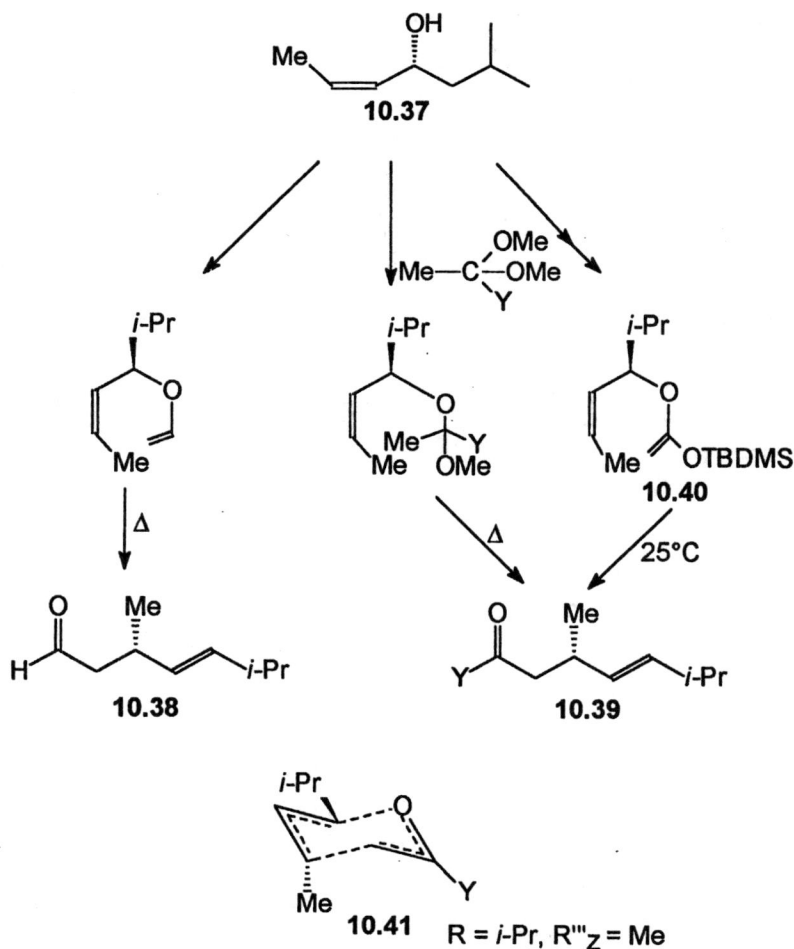

Figure 10.14

An example of chirality transfer during an Ireland-Claisen rearrangement which takes place through a boat transition state is shown in Figure **10.15**. Rearrangement of Z-ketene silylacetal **10.42** gives acid derivative **10.43**. Chair transition state **10.44** is destabilized by interaction of the ketene acetal substituents with the existing ring, so the reaction occurs via the boat conformation **10.45** [1638]. For esters other than acetates (R ≠ H), the Z/E geometry of the intermediate ketene silylacetal is crucial. If the rearrangement occurs through chair transition state **10.41** or **10.44**, each ketene acetal leads to an acid of a different configuration depending on whether R_Z or R_E is the nonhydrogen substituent [1646] (Figure **10.12**).

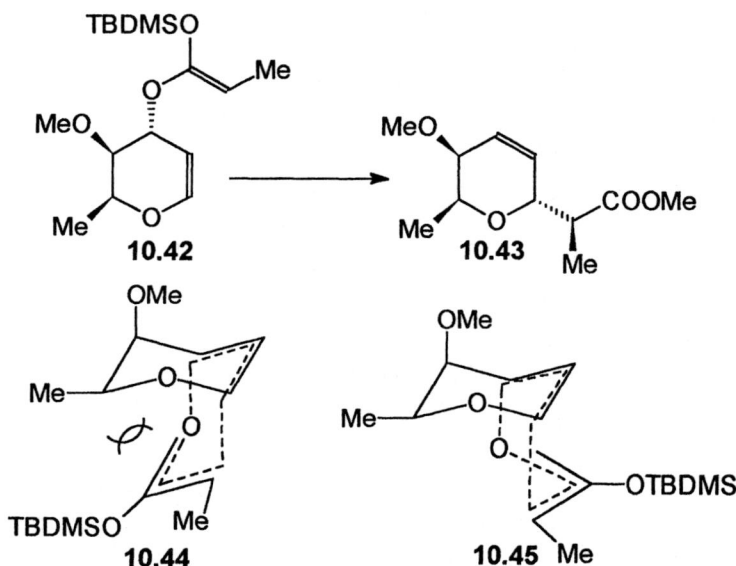

Figure 10.15

Corey and Lee [696] have recently proposed a variant of the Ireland-Claisen rearrangement that uses boron enolates of allylic esters derived from **2.62**. The E-crotyl (R_E = Me) or E-cinnamyl (R_Z = Ph) derivatives could be selectively transformed into the Z- or E-boron enolates **10.46** at low temperature (Figure **10.16**). The rearrangements take place at about 0°C, and the Z-enolates lead very selectively to *anti* acids **10.47** with an excellent enantiomeric excess while the E-enolates lead to *syn* acids **10.48**, with an interesting selectivity if R = Me or Et (Figure **10.16**). In most cases, the enantiomeric excesses are excellent; however, when the reaction is conducted with allyl esters (R′ = H), the ee's are a little bit lower (74 - 84%). These results are interpreted via a chair transition state that minimizes steric interactions [696].

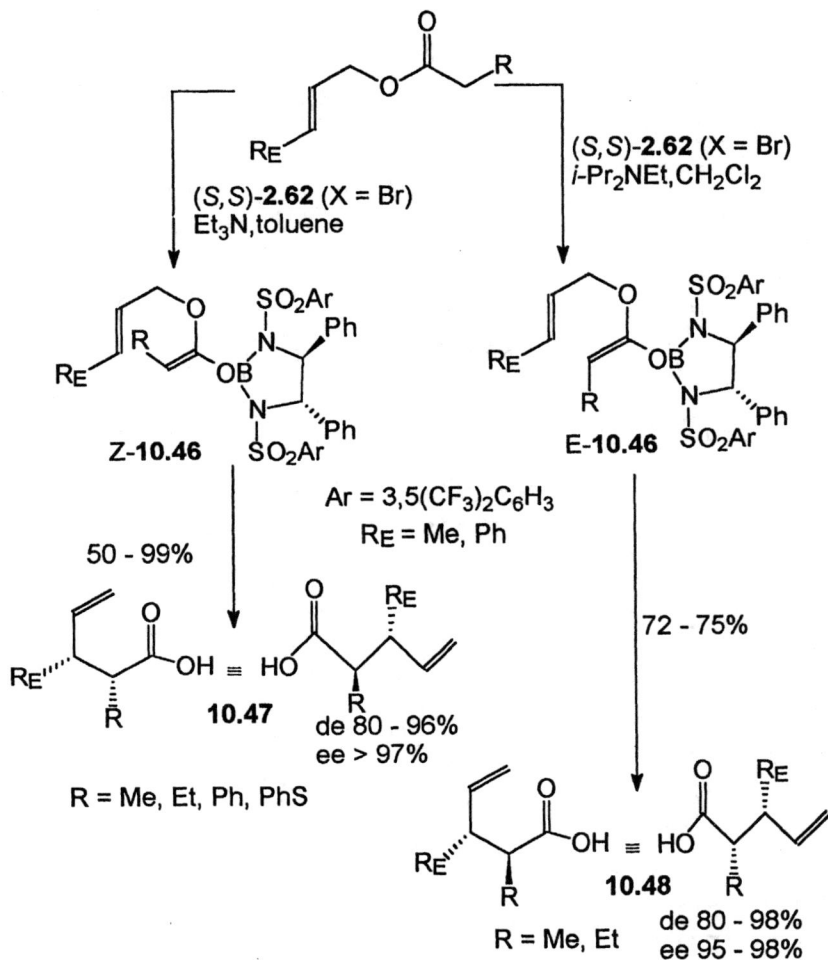

Figure 10.16

Evans has shown that the oxy-Cope rearrangement is dramatically accelerated by the presence of a negatively charged alkoxide. Likewise, the Claisen rearrangement is accelerated by the presence of a negative charge in the α position of
the vinyl group (X = O, Y = $^-$CH-EWG) (Figure 10.12). Denmark and Marlin
[355] have profitably used this effect by introducing a chiral electron-withdrawing
group: the phosphoramidate 1.96. In the presence of lithium dimsylate, the Claisen
rearrangement of allylic ether 10.49 takes place at 20°C and leads selectively to
ketone 10.50. The chair transition state (10.51), where the cation is chelated by
the oxygens of the ether and of the phosphorus, and where the methyl group is
situated in a pseudoequatorial position, rationalizes these results (Figure 10.17).

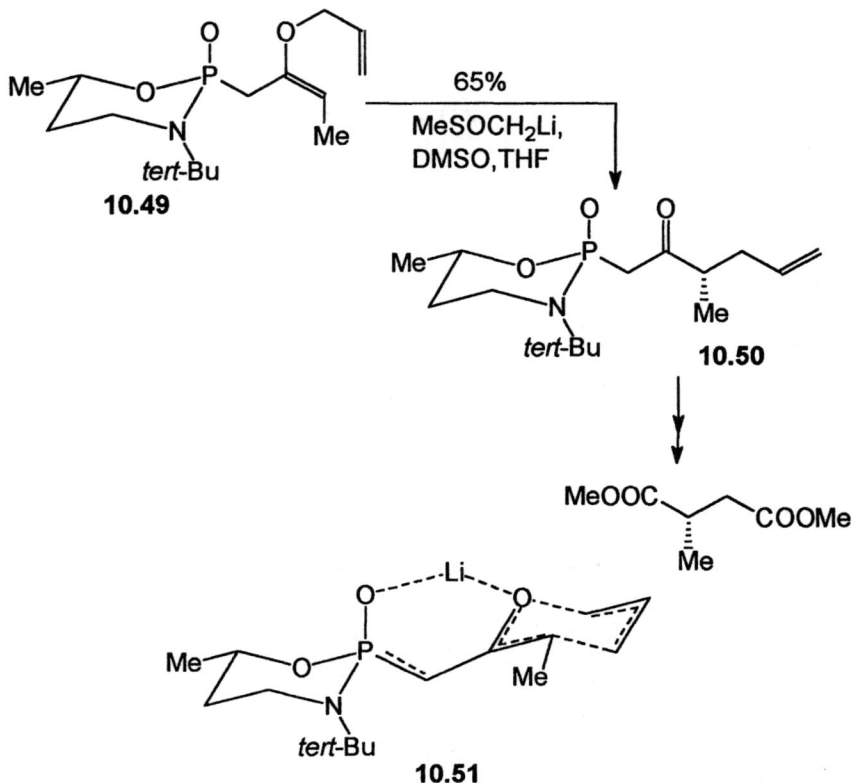

Figure 10.17

10.2.3 Aza-Claisen and Related Rearrangements

Attempts to thermally rearrange allylenamines bearing chiral substituents have not led to satisfying results [1062]. In contrast, the rearrangement of lithium enolates **10.52** of N-allyl amides substituted by the (S)-1-phenethyl group take place at 120°C in toluene and give more interesting results [861, 1647, 1648] (Figure **10.18**). The best selectivities have been obtained by Kurth and coworkers [261, 339, 1639] during the thermal rearrangement of chiral allyl oxazolidines **10.53**. The precursor salts of **10.53** are generated *in situ*, and the rearranged products are readily hydrolyzed to chiral acids. When the allyl group on nitrogen is unsubstituted or bears a Z substituent (R_E = H), the reaction is very selective, and a chair transition state rationalizes these results. In contrast, when R_E is different from H, boat transition states apparently intervene and the diastereoselectivity is not as high (60 - 80%) (Figure **10.18**). Some efficient chirality transfers have also been observed during the thermal [3,3]-sigmatropic rearrangements of allylimidates

10.54 and **10.55** [1647a, 1648a, 1649]. Once again, a chair transition state rationalizes these results (Figure **10.18**). Meyers and Devine [1650] applied a thio-Claisen rearrangement to an asymmetric synthesis of 4,4-disubstituted cyclohexen-2-ones from chiral allylic derivative **10.56**.

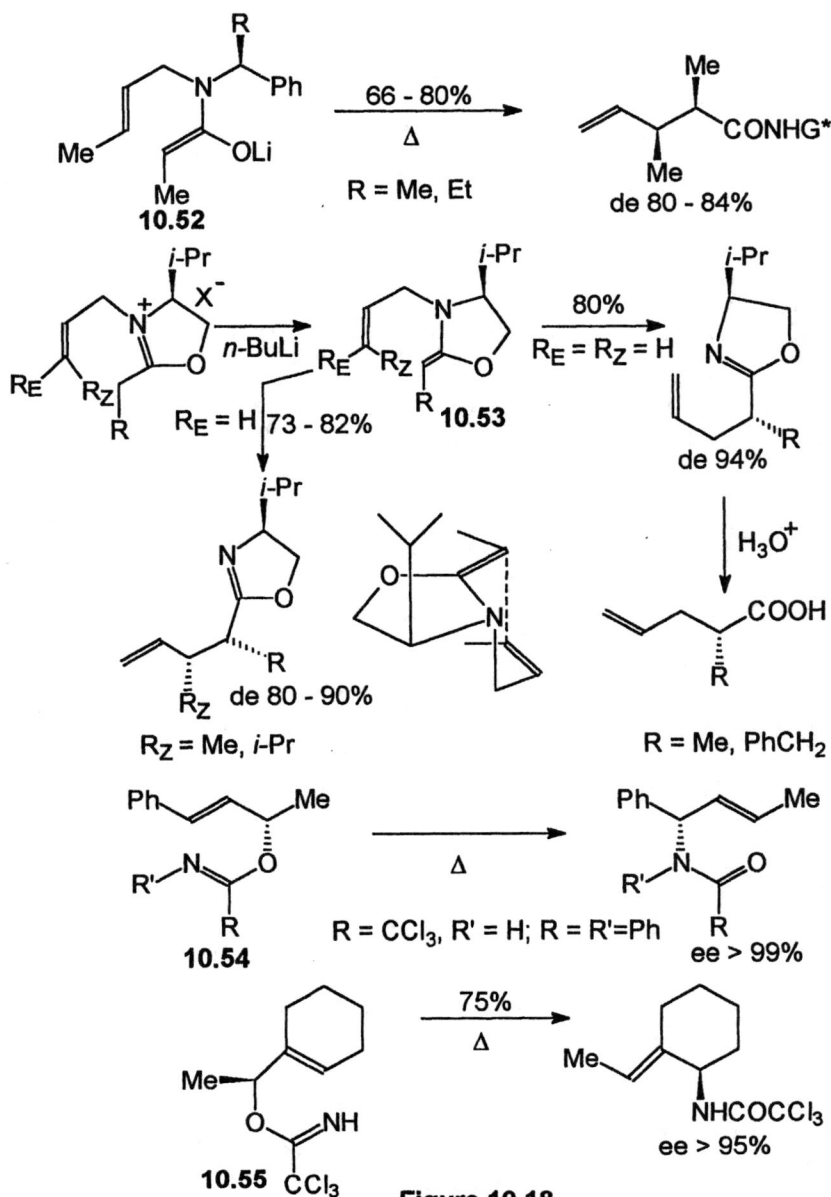

Figure 10.18

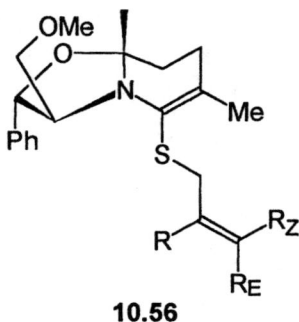

10.56

10.3 CATALYZED [3,3]-SIGMATROPIC REARRANGEMENTS

Sigmatropic rearrangements can be catalyzed by Lewis acids, and salts of aluminum and titanium are popular reagents. Transition metal catalysts such as $PdCl_2 (RCN)_2$ are also commonly used [1062, 1637, 1638]. Efficient transfers of chirality have been observed in Cope rearrangements proceeding through chair transition states catalyzed by $PdCl_2(PhCN)_2$. The examples shown in Figure **10.19** demonstrate that two chair transition states, **10.31** and **10.32** (Figure **10.12**), can be envisioned to participate in the reaction process. Each leads to the corresponding olefin with an excellent enantiomeric excess [1637]. Comparable transfers of chirality have been observed during the isomerization of allylic acetates catalyzed by $PdCl_2 (MeCN)_2$ [1062, 1651]. The migration of the acetate takes place at ambient temperature and occurs in a suprafacial fashion. A notable use of this method is by Bloch and Gasparini [1652] during the synthesis of intermediates of arachidonate derivatives (Figure **10.19**). In contrast, the rearrangement of chiral allylimidates **10.54** (R = R′ = Ph) by the same catalysts leads to mixtures. Nevertheless, when starting from N-protected aminoallylimidates **10.57**, a highly selective [3,3]-sigmatropic rearrangement is observed [1652a] (Figure **10.19**).

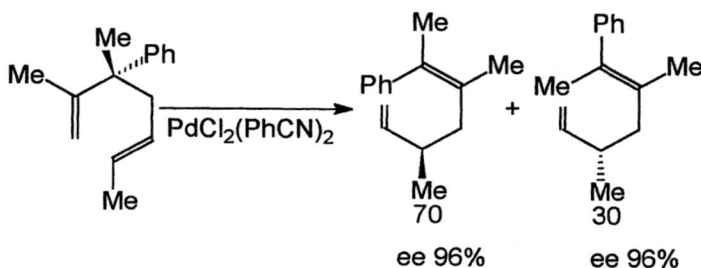

Figure 10.19

R = Me, Et, *i*-Pr, PhCH$_2$, TBDMSOCH$_2$

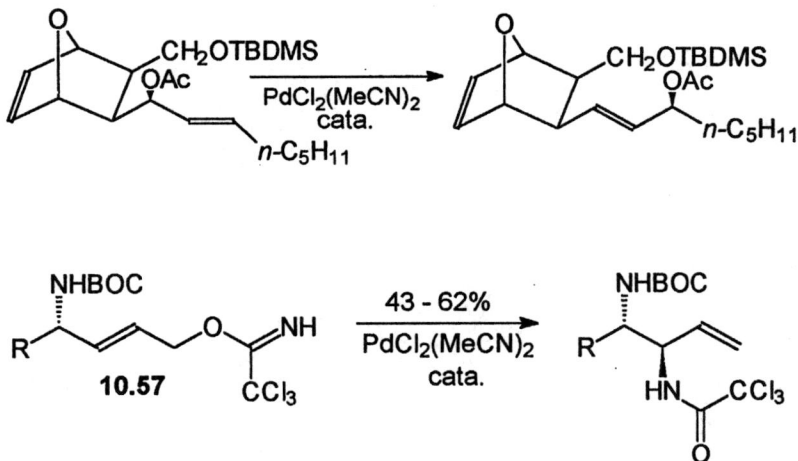

Figure 10.19 (continued)

Claisen rearrangements are catalyzed by phenylates of methylaluminum, (ArO)$_2$AlMe, and Yamamoto and coworkers [1062, 1638] have been able to effect these reactions with an excellent transfer of chirality by using hindered phenylates. The rearrangement of **10.58** catalyzed by organoaluminum reagents leads to **10.59** with an excellent selectivity if Ar = 2,6-Ph$_2$C$_6$H$_3$ (Figure **10.20**). Other catalysts with *tert*-Bu substituents on the 2,6 positions of the aromatic ring were less efficient. Yamamoto and coworkers [803, 1062, 1638] have also used catalysts **10.60** formed by the action of Me$_3$Al on (*R*)- or (*S*)-binaphthols **3.7** (R = SiAr$_3$) to perform the asymmetric Claisen rearrangements of prochiral allylvinylethers E-**10.61**. Z-isomers give less satisfying results. The best selectivities are obtained at −40°C in methylene chloride when R′= Me$_3$Si, PhMe$_2$Si or Me$_3$Ge and R$_E$ = Ph. The corresponding acylsilanes and germanes are obtained from these precursors with a very good enantioselectivity. The catalyst (*R*)-**10.60** (Ar = Ph) derived from (*R*)-**3.7** leads to the *S* enantiomer and vice versa. Two transition states **10.31** and **10.32** are envisioned during this rearrangement (Figure **10.20**). Molecular modeling suggests that the interaction between the chiral Lewis acid and each of these chairs depends on the relative position of the oxygen and the allylic CH$_2$ group at CH$_2\alpha$. The interaction between the catalyst (*R*)-**10.60** and the conformation **10.32** is more favorable while (*S*)-**10.60** is well paired with **10.31**.

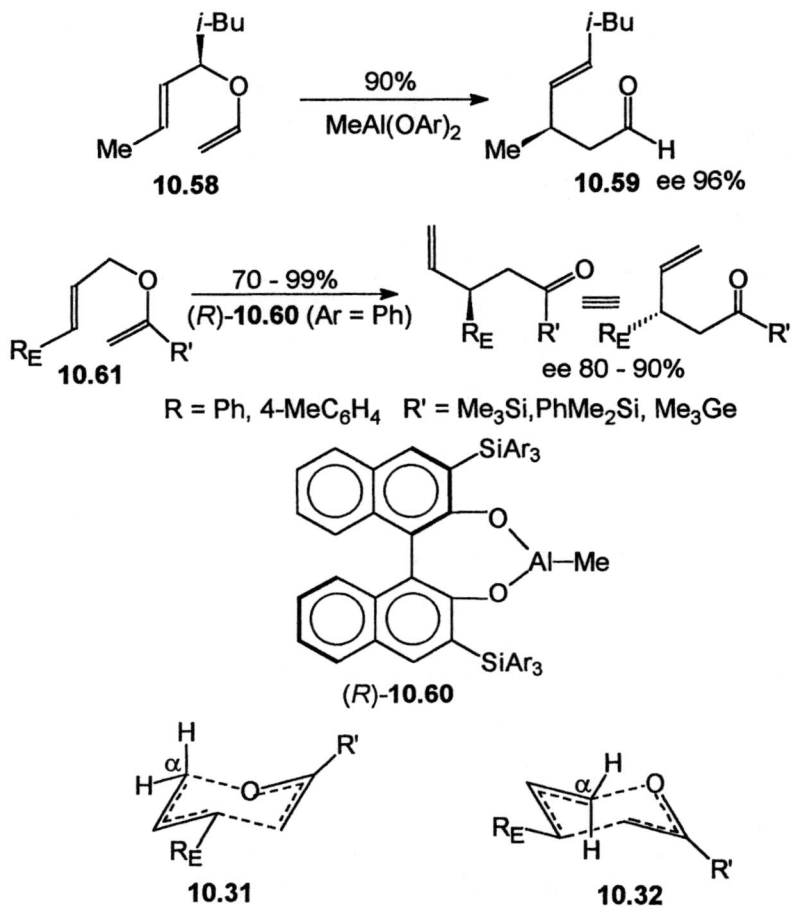

Figure 10.20

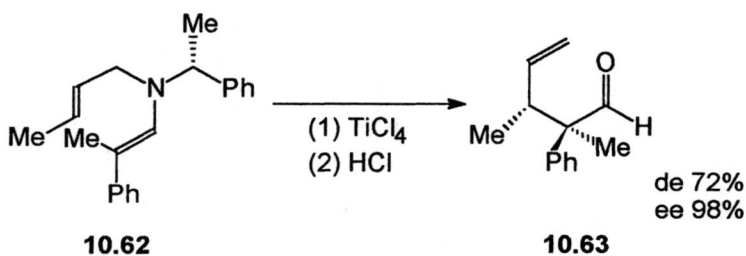

Figure 10.21

The selectivity of the aza-Claisen rearrangement is improved by the catalyst TiCl$_4$. Bailey and Harrison [1062] have conducted the rearrangement of enamine **10.62** bearing a chiral N-substituent at 55°C. After hydrolysis, the aldehyde **10.52** is obtained with a moderate diastereoselectivity and with very good enantiomeric excess [1653] (Figure **10.21**). Other attempts of asymmetric aza-Claisen rearrangements have been done but unsatisfactory results were obtained [1654].

CHAPTER 11

TRANSITION METAL CATALYZED REACTIONS

Prior chapters have covered the use of transition metals in asymmetric hydrogenations (§ 6.2 and 7.1), hydroborations (§ 7.3), hydrosilylations and hydrocyanations (§ 6.3, 6.4, 7.4 and 7.5), cyclopropanations (§ 7.19), aldol reactions (§ 6.11), allylations of carbanions (§ 5.3.2), and some sigmatropic rearrangements (§ 10.3). This chapter covers other reactions catalyzed by transition metal complexes including coupling of organometallic reagents with vinyl, aryl or allyl derivatives, Heck reactions, allylamine isomerizations, some allylation reactions, carbene insertions into C-H bonds and Pauson-Khand reactions.

11.1 COUPLING OF ORGANOMETALLIC REAGENTS WITH VINYL, ARYL AND ALLYL DERIVATIVES [253, 752, 910]

Aryl and vinyl halides can be coupled in reactions catalyzed by palladium or nickel complexes bearing phosphine ligands [253]. Allylic derivatives that are precursors of π-allyl palladium complexes can also be coupled with organomagnesium reagents. Asymmetry can be induced by using chiral phosphines as ligands.

11.1.1. Coupling with Vinyl Halides

Asymmetric coupling of vinyl halides with racemic secondary organomagnesium reagents is catalyzed by nickel or palladium complexes bearing chiral phosphines as ligands. Ferrocenylphosphine 3.49 (Y = NMe$_2$) is highly efficient as a ligand for nickel or palladium, and linear aminophosphines 3.67 (R = i-Pr, tert-Bu, MeS(CH$_2$)$_3$) are useful ligands for nickel [910, 1063]. The catalytic scheme for the nickel coupling is given in Figure 11.1. Nickel complex 11.1 is generated from NiCl$_2$ by insertion of vinyl bromide. This complex effects a kinetic resolution of the organomagnesium reagent leading to a single diastereoisomeric nickel complex 11.2 after loss of MgX$_2$. The chiral cross-coupled olefin 11.3 is formed by reductive elimination from complex 11.2. Since the secondary organomagnesium reagent undergoes racemization at a rate faster than the cross-coupling, the chiral olefin is obtained with a high enantioselectivity and chemical yield. The best selectivities are obtained at 0°C from 1-arylethyl organomagnesium reagents; alkyl analogs yield less useful results. Addition of zinc salts can improve the selectivity [1063]. This method has been applied to the synthesis of the anti-inflammatory drug ibuprofen (Figure 11.1). Other ligands such as diop 3.30 or norphos 3.39 are less efficient. When the coupling is performed with substituted vinyl bromides in the presence of complexes formed from PdCl$_2$ or NiCl$_2$ and 3.49

(Y = NMe$_2$), mediocre selectivities are observed even in the presence of zinc salts [949]. Chiral allylsilanes **2.86** can be prepared by this coupling method from a α-silylbenzylmagnesium bromide and E-bromoalkenes [1063] (Figure **11.1**). Z-bromoalkenes yield disappointing results.

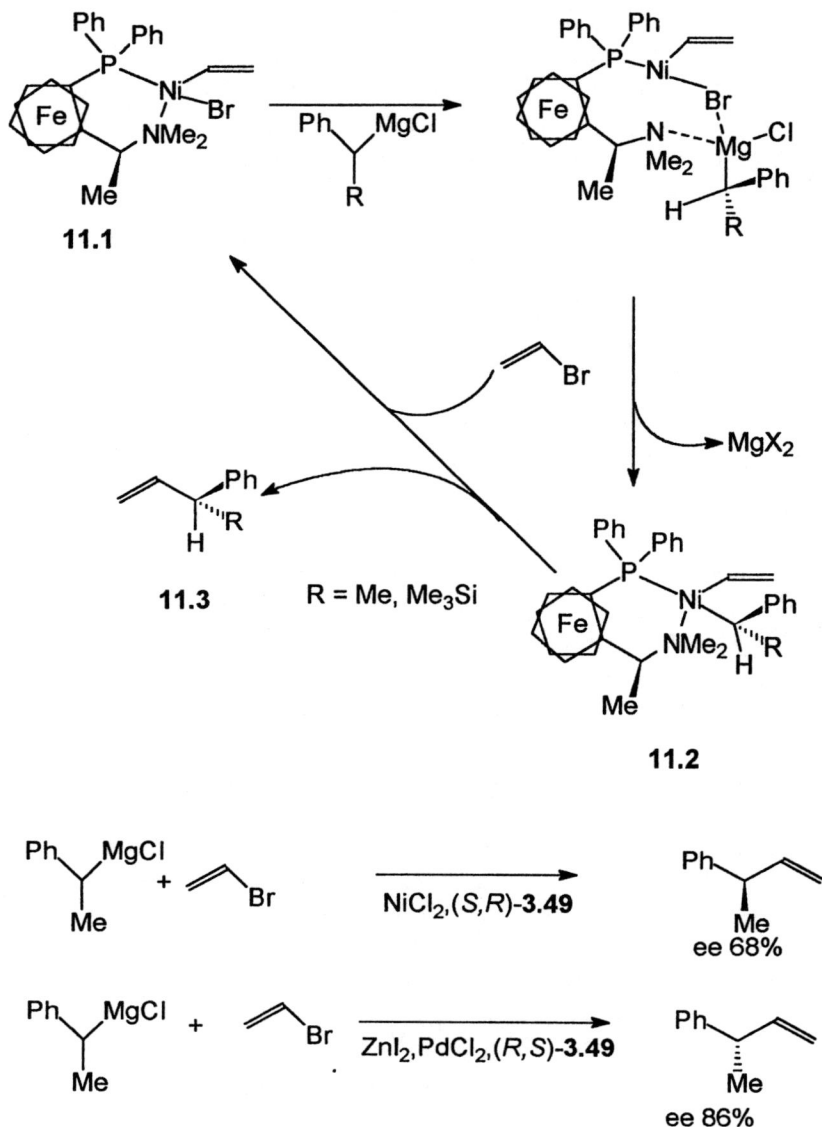

Figure 11.1

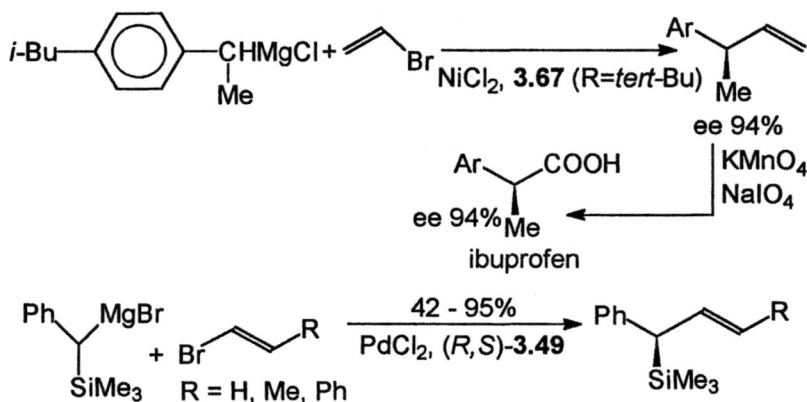

Figure 11.1 (continued)

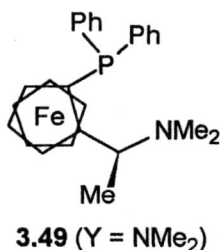

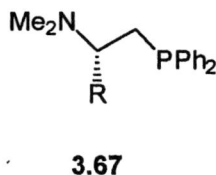

3.49 (Y = NMe$_2$)   **3.67**

11.1.2 Coupling with Aryl Halides

Coupling of 1-naphthylmagnesium halides with halonaphthalenes is a method for enantioselective synthesis of binaphthyls **11.4** or ternaphthyls **11.5**. Chiral nickel catalysts are used, and the most efficient catalyst is formed by complexation of NiBr$_2$ with ferrocenylphosphine **3.49** (Y = OMe). The corresponding amino-phosphine **3.49** (Y = NMe$_2$) gives unsatisfactory results [1063]. When the 2-substituent of the organomagnesium reagent is hydrogen (R = H), the enantioselectivity of the coupling reaction is mediocre (Figure **11.2**).

11.1.3 Coupling with Allyl Derivatives

Consiglio and coworkers and Hirama and Wasaka have performed the coupling of various allyl derivatives (esters, carbamates, ethers, alcohols, thioethers and phosphates) with organomagnesium reagents in reactions catalyzed by NiCl$_2$-chiraphos **3.37** (n = 0) complexes [752, 923]. Interesting results are obtained from symmetrically substituted allyl compounds. These compounds avoid the formation of regioisomers (Figure **11.3**). Nevertheless, the enantiomeric excesses of such couplings are often low [910] (Figure **11.3**). A high enantioselectivity has been obtained by Consiglio in the nickel-catalyzed reaction of EtMgBr with 3-phenoxycyclopentene **11.6**. The most efficient ligands in this coupling are chiraphos **3.37** ($\text{n} = 0$) or biphemp **3.45** (R = Ph). However, related reactions with MeMgBr or PhMgBr are less selective. These results are rather puzzling because the attack of such hard nucleophiles takes place on the face of the intermediate π-allyl metal complex that bears the metal [1655], while the attack of soft nucleophiles such as malonates occurs on the other face (§ 5.3.2).

Cross-coupling of chiral silanes with aryltriflates under Pd(PPh$_3$)$_4$ catalysis in the presence of fluoride ions has been described. A good chirality transfer has been observed [1656].

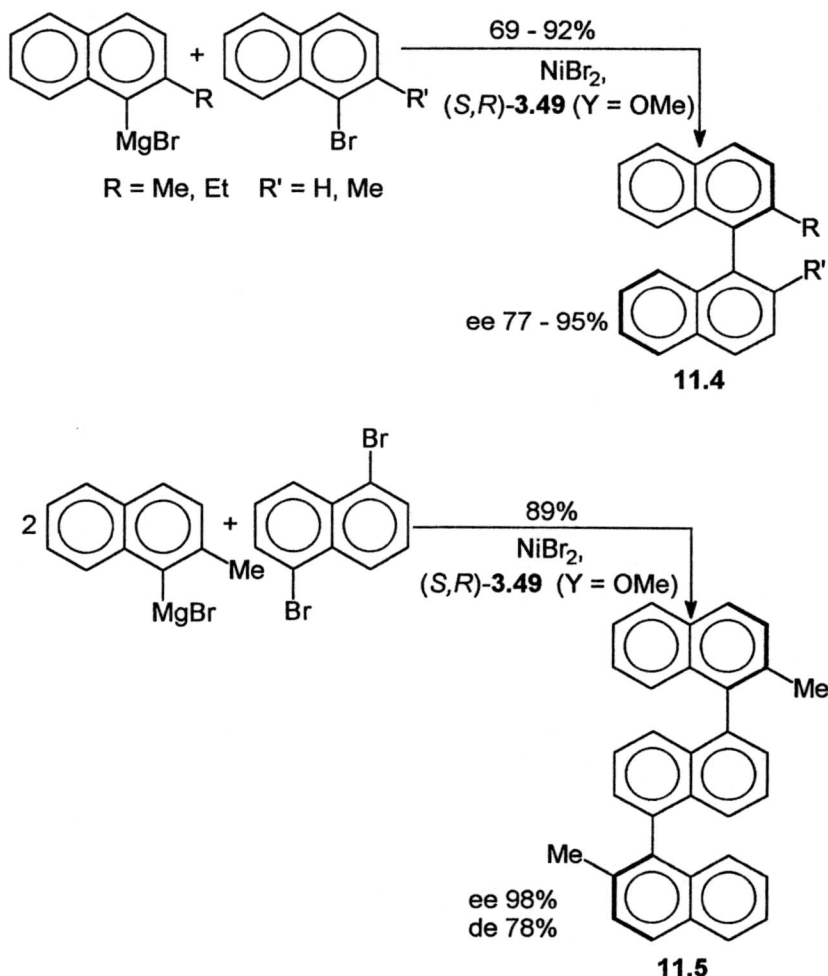

11.4

11.5

Figure 11.2

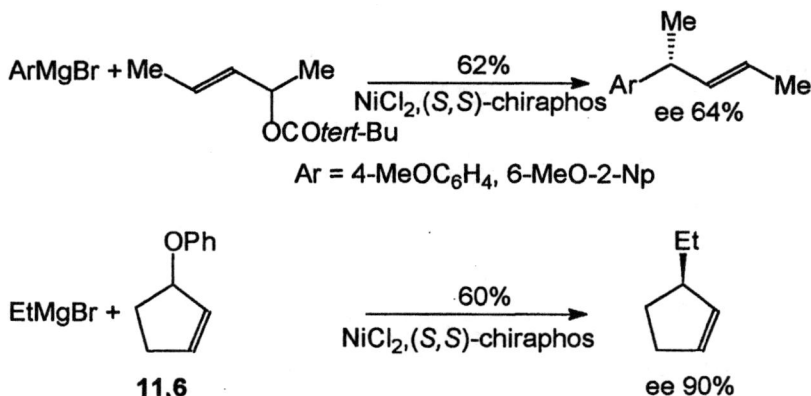

ArMgBr + Me⟍⟋⟍Me $\xrightarrow[\text{NiCl}_2,(S,S)\text{-chiraphos}]{62\%}$ Ar⟍⟋⟍Me

OC Otert-Bu

ee 64%

Ar = 4-MeOC$_6$H$_4$, 6-MeO-2-Np

EtMgBr + [OPh cyclopentene] $\xrightarrow[\text{NiCl}_2,(S,S)\text{-chiraphos}]{60\%}$ [Et cyclopentene]

11.6

ee 90%

Figure 11.3

Me⟍⟋(CH$_2$)$_n$⟍⟋Me

PPh$_2$ PPh$_2$

n = 0:chiraphos

3.37

Me—[biphenyl]—PR$_2$

Me—[biphenyl]—PR$_2$

3.45

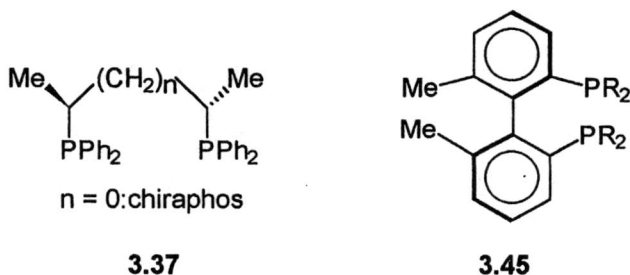

11.2 HECK REACTION

The coupling of an olefin with an arylPdX or a vinylPdX reagent is called the Heck reaction [624i]. The mechanism of the Heck reaction involves addition of the palladium reagent to the double bond, followed by elimination of HPdX. When the reaction is carried out with a substituted allyl derivative, double-bond migration takes place during the elimination step.

Asymmetric Heck reactions have usually been performed in an intramolecular fashion, and important contributions have been made by Overman, Shibasaki and their coworkers [772, 902, 1008, 1657-1663]. The reaction of Z-vinyl triflates or iodides **11.7** with PdX$_2$ (X = OAc or Cl) and (R)-binap **3.43** (10 mol%) in NMP at 60°C gives the *cis* ring fused bicyclic compounds **11.8** with a good selectivity. When vinyl iodides are used, the reaction must be carried out in the presence of Ag salts, and Ag$_3$PO$_4$ is usually the most efficient [902] (Figure 11.4). Several heteroatoms or functional groups may be introduced on the six-membered ring of

11.7. While binap is always the best ligand of palladium to induce a good enantio-selectivity, the experimental conditions must be modified to suit the substituent [1661, 1662].

The intermediate palladium complex may be trapped by an external anion, instead of undergoing HPdX elimination. For example, the reaction of **11.9** in the presence of Pd(OAc)$_2$, (S)-**3.43** and n-Bu$_4$NOAc in DMSO at 50°C gives **11.10** with a good ee [1659] (Figure **11.4**). A similar trapping can be performed with benzylamine.

Suitably substituted aryl triflates also undergo stereoselective and enantiose-lective cyclizations in the presence of Pd-**3.43** complexes. Overman and cowork-ers [1657] performed the reaction with 1-iodo-2-butenanilides **11.11** and found that only the Z-isomers with R' ≠ H give high selectivities. These reactions are carried out in DMAC at 100°C in the presence of 1,2,2,6,6-pentamethylpiperidine (PMP), and excess binap is added to ensure complete complexation of Pd (Figure **11.4**). When R' = SiR$_3$, chiral aldehydes **11.12** are easily prepared after acid hy-drolysis. Similar selectivities are obtained from Z-**11.13**, while E-analogs give mixtures of E- and Z-olefins with a low ee [1663] (Figure **11.4**).

Intermolecular asymmetric Heck reactions have been studied by Hayashi and coworkers. They coupled 2,3-dihydrofuran **11.14** (X = O) or N-substituted pyr-rolines **11.14** (X = NCOOMe) with aryl or alkenyl triflates in benzene in the pres-ence of various bases and a chiral palladium catalyst [1664, 1665]. These reactions give a mixture of regioisomers **11.15** and **11.16** in a 7:3 ratio, and the best ligand of palladium is binap **3.43**. 2-Substituted 4,5-dihydrofurans **11.15** are obtained with a high ee and the best compromise is to use 1,8-bis-dimethylaminonaphthalene as a base (Figure **11.4**). If the catalyst is a preformed Pd-(binap)$_2$ instead of being generated *in situ* from Pd(OAc)$_2$ and binap, the side reaction leading to **11.16** is suppressed (Figure **11.4**). Hydroarylation of norbornene with PhOTf in the pres-ence of chiral palladium complexes has been attempted, but the ee is 70% at best [1666].

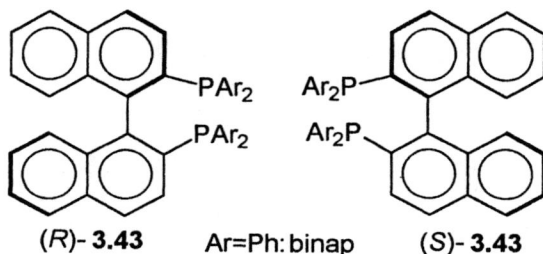

(R)- **3.43**   Ar=Ph: binap   (S)- **3.43**

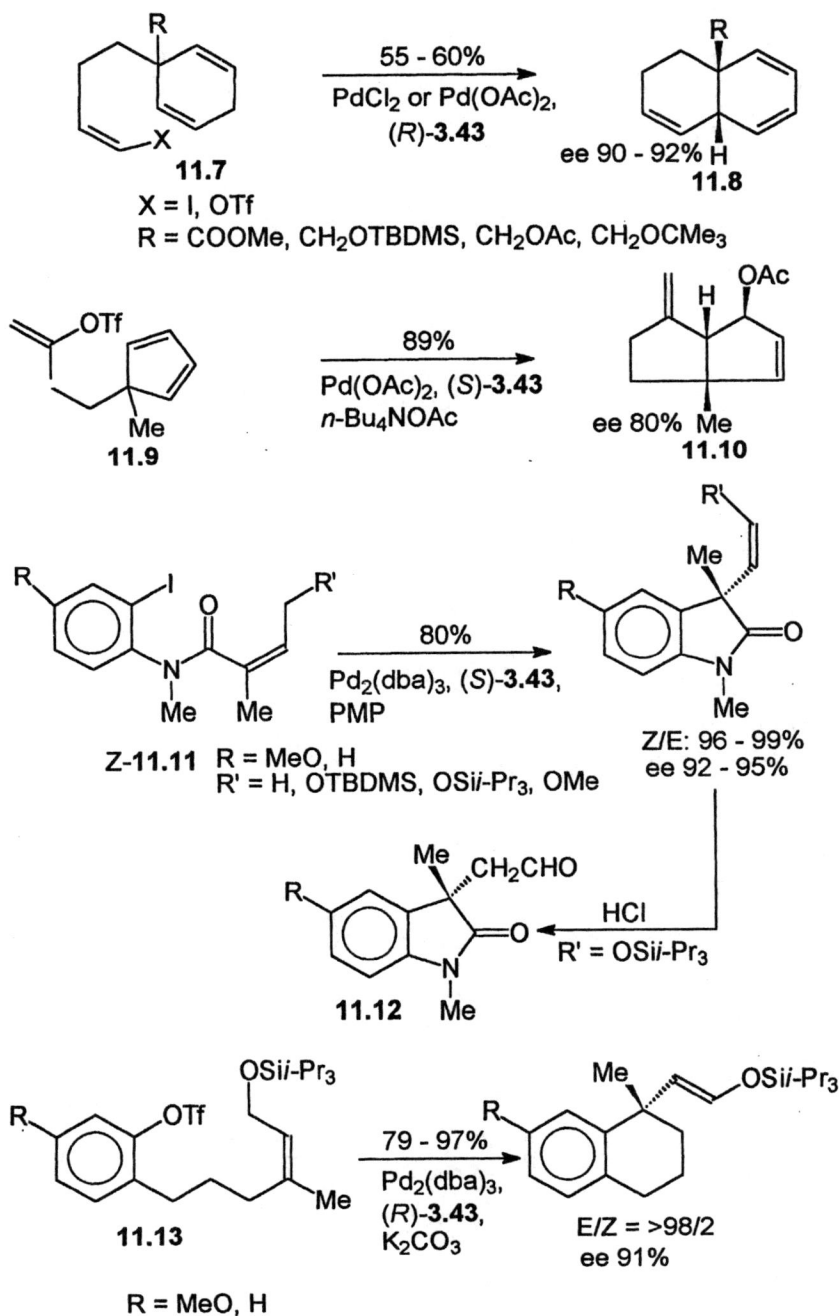

Figure 11.4

$Ar = Ph, 4\text{-}ClC_6H_4, 3\text{-}ClC_6H_4, 4\text{-}AcC_6H_4,$
$4\text{-}CNC_6H_4, 4\text{-}MeOC_6H_4$

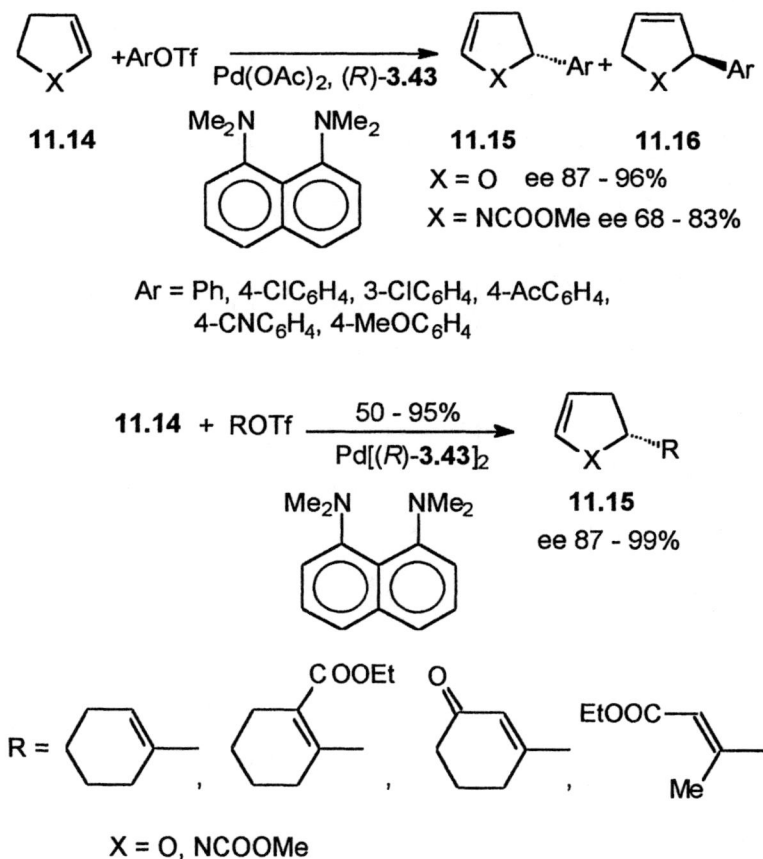

Figure 11.4 (continued)

11.3 ISOMERIZATION OF ALLYLAMINES [752, 881, 882, 889, 890]

Cationic rhodium(I) complexes catalyze the isomerization of tertiary allylamines E-**11.17** and Z-**11.17** to enamines **11.18**. In THF at 80°C, these enamines are readily hydrolyzed to provide aldehydes. When the rhodium ligand is a chiral diphosphine such as (R)- or (S)-binap **3.43** (Ar = Ph), or better yet, (R)- or (S)-tolbinap **3.43** (Ar = 4-MeC₆H₄), (R)- or (S)-enamines **11.18** are obtained with an excellent enantiomeric excess. This method is used in industry for the synthesis of optically active citronellol and menthol [811, 812, 853, 889]. Biphemp **3.45** is also a highly potent ligand for the isomerization of Z-**11.17**, and ee's as high as 99.5% are observed [905]. The absolute configuration of the enamine product depends upon the absolute configuration of the ligand and on the E- or Z-geometry of the double bond of the allylamine **11.17** (Figure **11.5**). E-**11.17** is isomerized by a Rh(I) catalyst bearing the (S)-binap ligand to (R)-**11.18**, while isomerization of Z-**11.17** delivers (S)-**11.18**. N-Arylallylamines react more slowly. Allylamines with a styrene-type conjugation such as E-**11.17** (R = H) are also interesting substrates (Figure **11.5**). Secondary allylamines such as **11.19** also suffer isomerization but the product is the corresponding imine **11.20**, which is also obtained with an excellent enantiomeric excess (Figure **11.5**). Related amides can be isomerized at higher temperatures, but the chemical yield is lower. The isomerization of allyl alcohols gives less useful results.

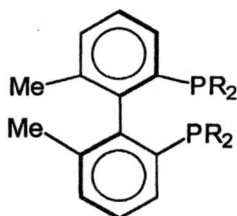

3.45

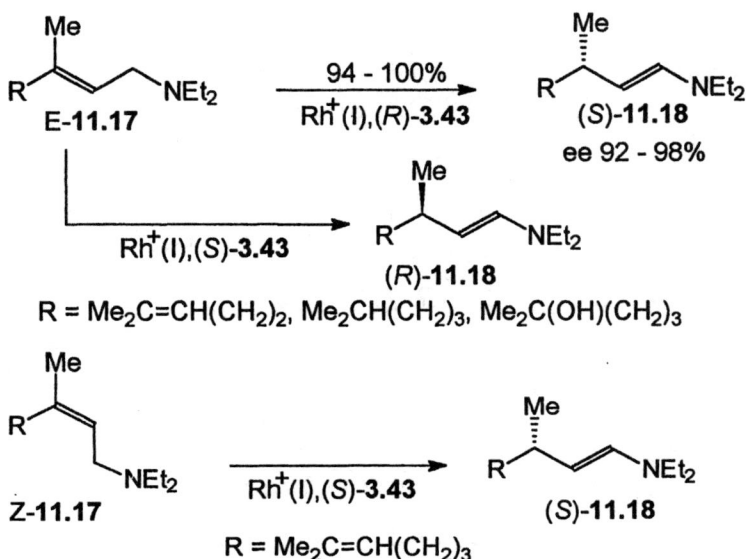

Figure 11.5

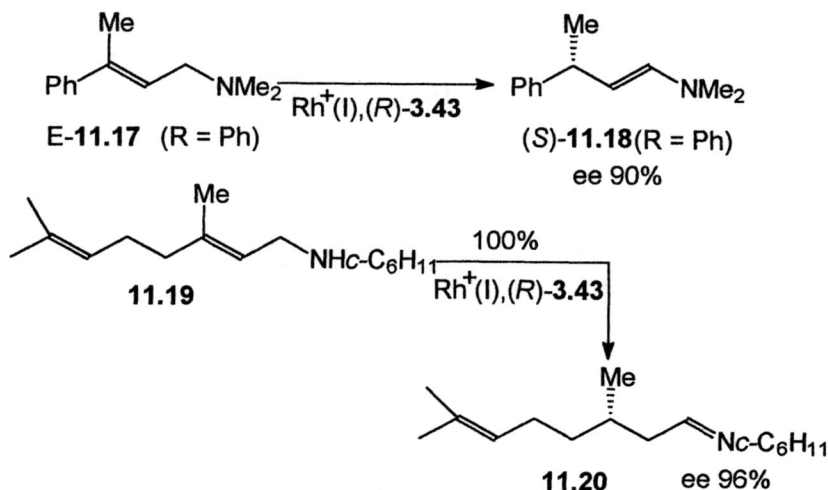

Figure 11.5 (continued)

The mechanism of this isomerization involves an intramolecular 1,3-suprafacial hydrogen migration via an imminium complex **11.21** (Figure **11.6**). The enantioselection takes place during the removal of one of the two enantiotopic hydrogen atoms of **11.22**, as shown by deuterium labeling experiments (Figure **11.6**). A π-allyl complex is not involved. The most favored complex is the one in which the interactions between the phosphine and the allylamine substituent are minimized [882].

Isomerization of achiral N-acylaziridines **11.23** to chiral N-acylallylamines **11.24** by cob[I]alamin has recently been described by Scheffold [1667] (Figure **11.6**).

11.4 ALLYLIC SUBSTITUTIONS BY NUCLEOPHILES OTHER THAN CARBON [752, 910, 1062]

π-Allylpalladium complexes can be generated from allyl derivatives bearing a leaving group. As previously described, they react with malonate-type carbanions (§ 5.3.2) and organomagnesium reagents (§ 11.1.3). They also undergo reduction by formates or nucleophilic substitution with amines. Asymmetry is induced in these transformations by the use of chiral ligands. Hayashi and coworkers [1668] performed the asymmetric reduction of allyl carbonates **11.25** with formic acid in the presence of 1,8-bis-dimethylaminonaphthalene. The palladium ligand was usually (R)-mop **3.51** (R = Me) (Figure **11.7**), but an increase in selectivity (ee, 85% instead of 75%) can be obtained when mop is replaced by its biphenanthryl analog **11.26**. These reactions take place in dioxane at room temperature.

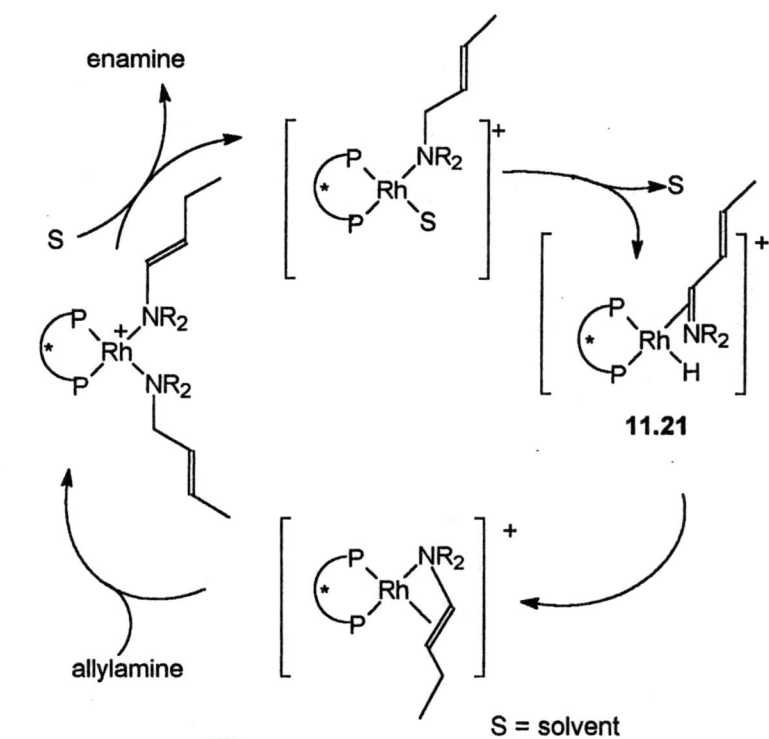

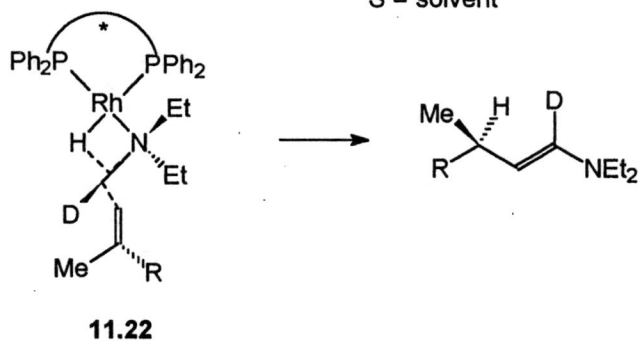

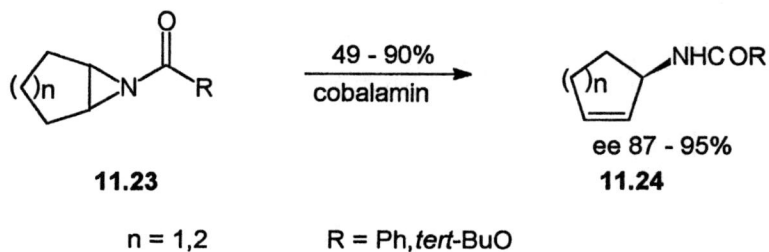

Figure 11.6

Amines or their derivatives also give interesting results. Intramolecular reactions of bis-carbamates **11.27** and **5.51** have been studied by Hayashi, Trost and
their coworkers [861, 911, 1062, 1066, 1604, 1668a] (Figure **11.7**). Hayashi used
ferrocenyldiphosphine **3.42** (R = (HOCH$_2$)$_2$CH) as the palladium ligand, and obtained oxazolidinone **11.28** with a good enantiomeric excess. Trost performed the
asymmetric amination under Pd(0) catalysis, and found that the most efficient
chiral ligand was diphosphine **3.48** (X = NH, R = Ph).

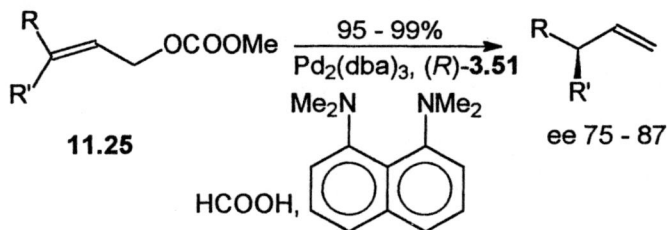

R = Me$_2$C=CHCH$_2$CH$_2$, R' = H
R = c-C$_6$H$_{11}$ R' = Me   R = Me, R' = Me$_2$C=CHCH$_2$CH$_2$

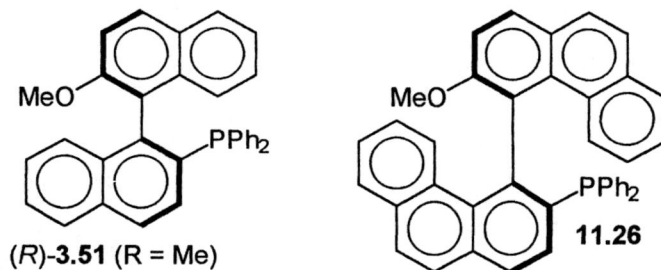

(R)-**3.51** (R = Me)   **11.26**

Figure 11.7

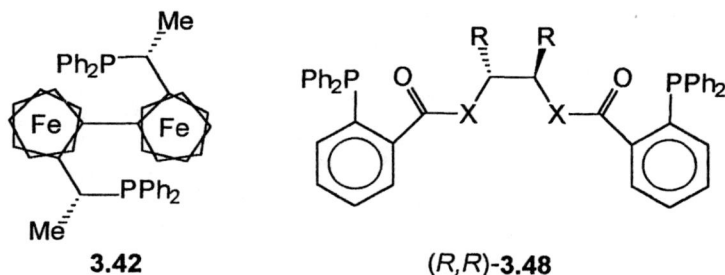

3.42   (R,R)-**3.48**

R = Me, Ph, 2,6-Me$_2$C$_6$H$_3$, 1-Np

5.51

$\xrightarrow[\text{Pd(0),(S,S)-3.48}]{75 - 91\%}$
(X=NH,R=Ph)

ee 70 - 80%

R = Me, n-Pr, i-Pr, Ph   X = AcO, Ph$_2$P(O)O, EtOCOO

86 - 93%
Pd(0),3.42
(R=(HOCH$_2$)$_2$CH)
or 3.52 (X = PPh$_2$)

ee 85 - 99%

11.29

11.30

n = 1,2,3

84 - 95%
(1) [π-C$_3$H$_5$PdCl]$_2$,
(R,R)-3.48,
(n-C$_6$H$_{13}$)$_4$NBr
(2) H$_2$NNH$_2$
(3) HCl

ee 94 - 98%

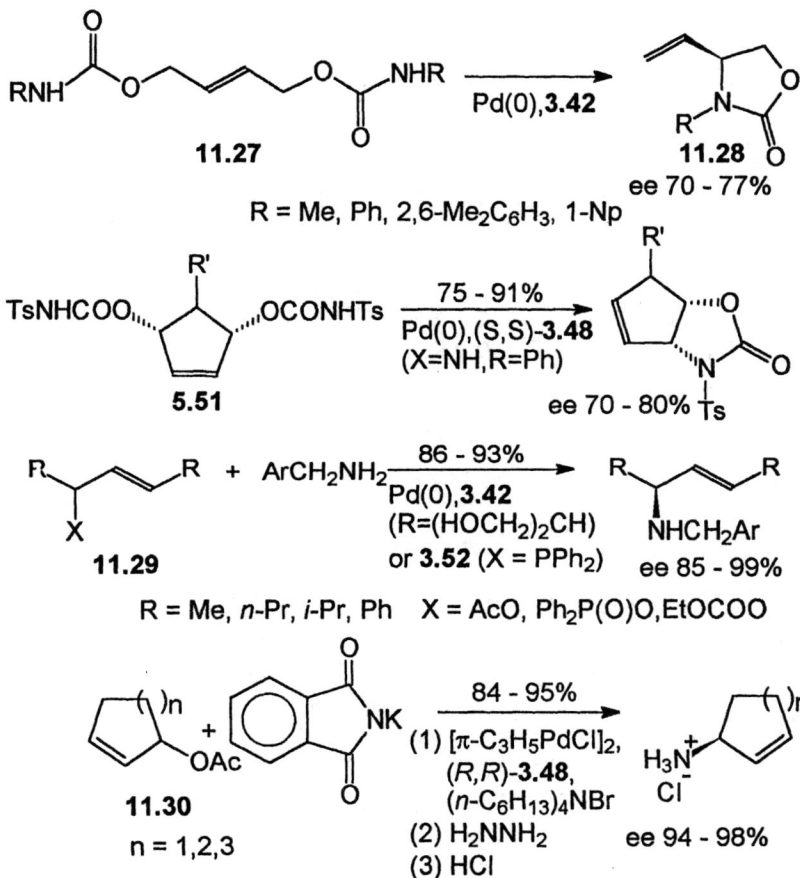

Figure 11.7 (continued)

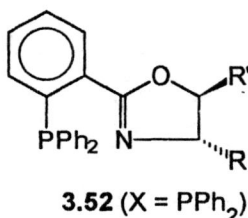

3.52 (X = PPh$_2$)

Intermolecular amination can also be performed with racemic, symmetrical allyl esters **11.29** bearing identical R groups. The Pd(0) catalyzed reactions of benzylamines with **11.29** (X = OAc, OP(O)Ph$_2$, OCO$_2$Me) yield allylamines with a high enantiomeric excess (Figure **11.7**). The best ligands of palladium for such allylic aminations are **3.41** (R = (HOCH$_2$)$_2$CH) and oxazoline **3.52** (X = PPh$_2$) [910, 1669]. The reaction can be extended to TsNHNa, PhCONHNHNa and (BOC)$_2$NNa in the presence of **3.52** (X = PPh$_2$) as ligand [1669]. The nature of the X-leaving group of the allyl derivative must be adjusted from case to case, and mediocre selectivities are often observed when R = Me or *n*-Pr. Trost and Bunt also performed the Pd(0)-catalyzed allylic amination of cyclic acetates **11.30** with tetrahexylammonium phthalimide in the presence (*R,R*)-**3.48** as a ligand [1067]. After standard imide cleavage, cyclic allylamines were obtained with an excellent enantioselectivity [1067]. All these reactions take place via chiral π-allylpalladium complexes which react on their least hindered face.

An extension to an asymmetric cyclization by reaction of PhNHCH$_2$CH$_2$OH with 1,4-diacetoxy-*cis*-2-butene has been reported by Hayashi and coworkers [1670]; 2-vinylmorpholines are obtained, but the ee is 65% at best.

11.5 CARBENE INSERTION INTO C-H BONDS [1497]

Metallocarbenes are generated from diazoketones or diazoesters in the presence of rhodium(II) complexes, and these carbenes often insert into carbon-hydrogen bonds. These insertions are efficient when they lead intramolecularly to five- or six-membered cyclic compounds (Figure **11.8**). Asymmetry can be induced either by using substrates bearing chiral residues or by introducing chiral ligands on the catalyst. The ligands are the same as those that are used in cyclopropanation reactions (§ 7.19.3).

The insertion of the carbene formed from diazoacetate **11.31** into the CH bond α to the OR group leads to lactone **11.32** when the reaction is conducted in CH$_2$Cl$_2$ at reflux. Lactone **11.32** is obtained with a high enantioselectivity when the rhodium(II) catalyst is formed by exchange between Rh$_2$(OAc)$_4$ and methyl (*R*)- or (*S*)-2-pyrrolidinone-5-carboxylate **3.61** [940] (Figure **11.8**). Less useful results are obtained with *N*-alkyl-*N-tert*-Bu diazoacetamides [1671] or when the OR substituent of **11.31** is replaced by a Ph group [940]. This method has been extended to cyclization of diazoketones **11.33** to provide chromanones **11.34** [934], but the enantioselectivities are lower and the absolute configuration of the products has not yet been determined (Figure **11.8**).

The rhodium-catalyzed cyclization of cycloalkyldiazoacetates **11.35** gives mixtures of *cis*- and *trans*-fused lactones when **3.61** is used to generate the chiral catalyst. Recently, Doyle and coworkers found that *cis*-fused lactones **11.36** could be obtained with a high selectivity when the catalyst was generated from the methyl 1-acetylimidazolidin-2-one-4-carboxylate **11.37** [1672].

The enantioselective cyclization of α-diazo-β-ketoesters **11.38** gives disappointing selectivities when the ester group is not branched. Ikegami and coworkers [1673] have found that 3-arylcyclopentanones can be obtained from **11.38** under rhodium *N*-phthaloylaminocarboxylate **11.39** catalysis with a good enantiomeric excess provided that *i*-Pr$_2$CHOH esters are used (Figure **11.8**). Taber and Raman have performed the cyclization of α-diazo-β-ketoesters **11.38** of chiral alcohols under Rh$_2$(OAc)$_4$ catalysis. Cyclopentanones are obtained with a high selectivity, provided that the ester is derived from alcohol **1.7** (R = 1-Np) [147] (Figure **11.8**).

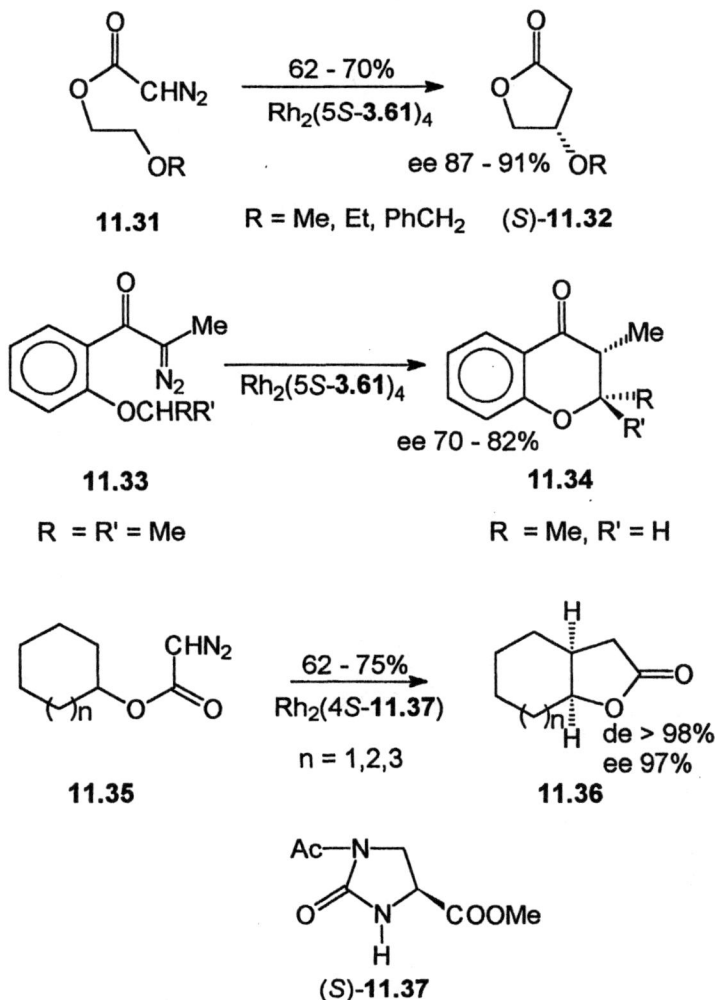

Figure 11.8

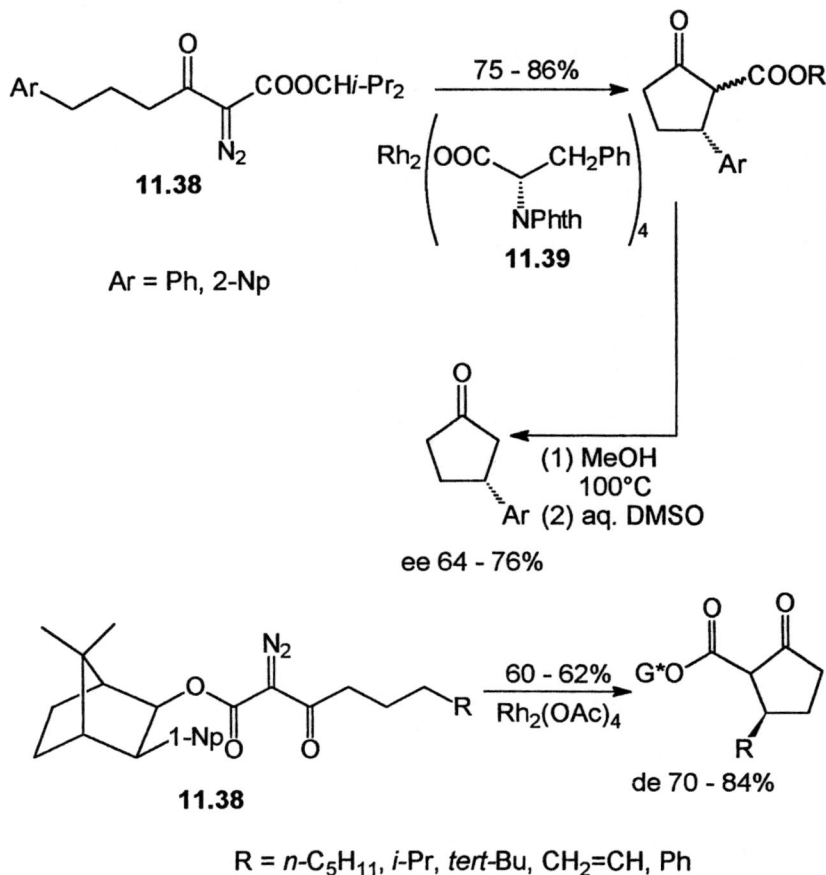

Ar = Ph, 2-Np

ee 64 - 76%

R = n-C$_5$H$_{11}$, i-Pr, tert-Bu, CH$_2$=CH, Ph

Figure 11.8 (continued)

11.6 PAUSON-KHAND AND RELATED REACTIONS [1674]

11.6.1 Pauson-Khand Reaction

The Pauson-Khand reaction combines electron-rich olefins with cobalt car-bonyl complexes of alkynes; after decomplexation, cyclopentenones are formed. Bicyclic systems are produced from intramolecular Pauson-Khand cyclizations (Figure **11.9**). Chiral enol ethers have been used in intramolecular asymmetric Pauson-Khand reactions by Greene, Moyano and their coworkers [66, 200]. Ethers **11.40** of trans-2-phenyl-cyclohexanol **1.5** lead to bicyclic systems **11.41** with a good selectivity (Figure **11.9**). The chiral auxiliary is removed by reaction with SmI$_2$ after hydrogenation of the double bond or 1,4-addition of a cuprate. Attempted intermolecular reactions between alkoxyacetylene complexes bearing a

chiral alkoxy residue **11.42** and cyclopentene or norbornene gave interesting selectivities only when *trans*-(9-phenanthryl)-2-cyclohexanol was the chiral auxiliary.
When the 2-phenyl analog **11.42** (R = Ph) was used, the diastereoselectivity was
50%, but the major isomer is easily purified by chromatography [1675]. In both
cases, only *exo* isomers are generated in reactions with norbornene.

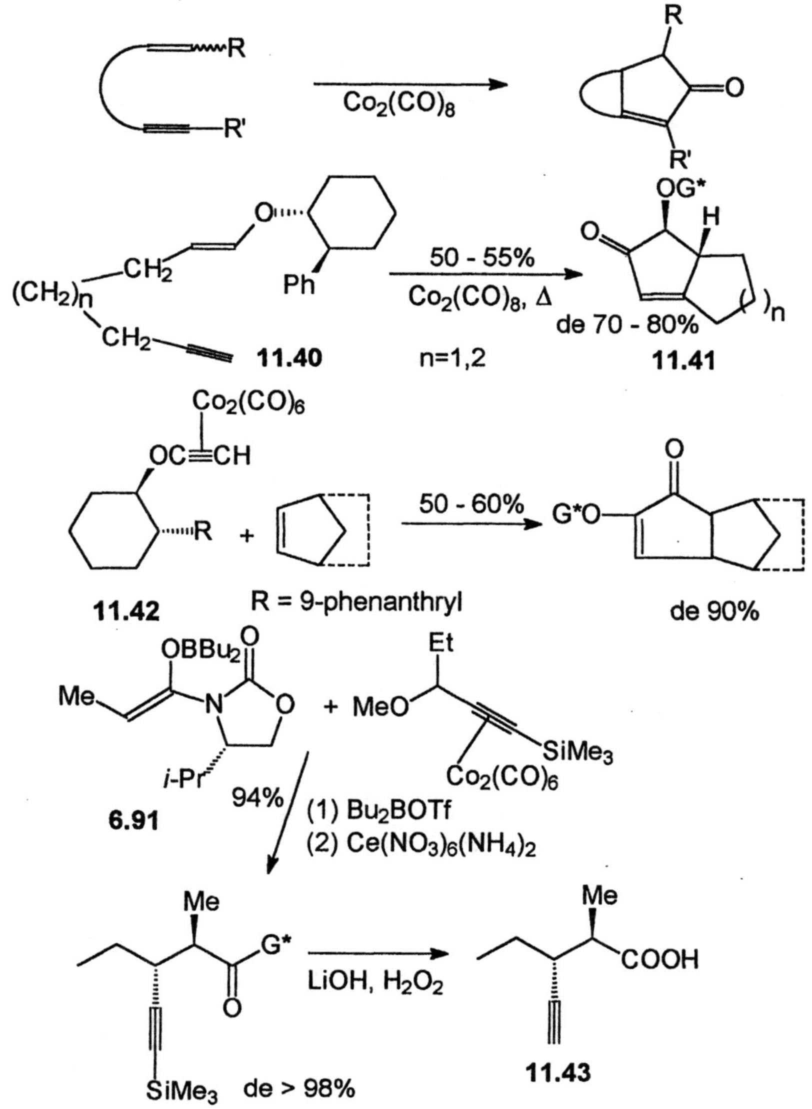

Figure 11.9

An enantioselective intermolecular Pauson-Khand reaction between nor-bornene and chiral alkoxyacetylene bearing 10-methylthioisoborneol as the auxiliary has recently been successfully performed with high selectivity [1676].

11.6.2 Nicholas Reaction

Acetylenic cobalt complexes greatly facilitate the heterolytic cleavage of adjacent alcohols or ethers. On treatment with Lewis acids, these complexes afford cobalt stabilized carbenium ions, which can be captured by nucleophiles such as enolates. Jacobi and Zheng have employed chiral boron enolates of Evans's oxazolidinone **6.91** (R' = *i*-Pr). After removal of the chiral auxiliary, they obtained *anti* acids **11.43** with a high selectivity [1677] (Figure **11.9**). The reaction can be extended to the boron enolates of related oxazolidinones and to α-branched propargyl derivatives. This reaction has been applied to the synthesis of β-aminoacids after Curtius rearrangement and oxidation of the triple bond [1677].

REFERENCES

0001 F. A. CAREY, R. J. SUNDBERG, Advanced Organic Chemistry 3rd edition, Plenum Press, New York, 1991 part A chap. 3 and 4

0002 H. BUSCHMANN, H. D. SHARF, N. HOFFMANN, P. ESSER, Angew. Chem. Int. Ed. Engl., 1991, **30**, 477 and quoted ref.

0003 E. L. ELIEL, Stereochemistry of Carbon Compounds, McGraw Hill, New York, 1962

0004 B. BLAIVE, J. METZGER, J. de Chimie Physique, 1980, **77**, 999 and 1007

0005 I. FLEMING, Frontier Orbitals and Organic Chemical Reactions, Wiley, New York, 1976

0006 H. B. BURGI, J. D. DUNITZ, J. M. LEHN, G. WIPFF, Tetrahedron, 1974, **30**, 1563

0007 N. T. ANH, Topics in Current Chemistry, 1980, **88**, 145

0008 X. L. HUANG, J. J. DANNENBERG, J. Am. Chem. Soc., 1993, **115**, 6017

0009 K. N. HOUK, M. N. PADDON-ROW, N. G. RONDAN, Y. D. WU, F. K. BROWN, D. C. SPELLMEYER, J. T. MATZ, Y. LI, R. J. LONCHARICH, Science, 1986, **231**, 1108

0010 H. ZIPSE, J. HE, K. N. HOUK, B. GIESE, J. Am. Chem. Soc.1991, **113**, 4324

0011 K. N. HOUK, Y. LI, J. D. EVANSECK, Angew. Chem. Int. Ed. Engl., 1992, **31**, 682, and quoted ref.

0012 M. FUJITA, M. ISHIDA, K. MANAKO, K. SATO, K. OGURA, Tetrahedron Lett., 1993, **34**, 645

0013 R. W. FRANCK, N. KAILA, M. BLUMENSTEIN, A. GEER, X. L. HUANG, J. J. DANNENBERG, J. Org. Chem., 1993, **58**, 5335

0014 X. L. HUANG, J. J. DANNENBERG, M. DURAN, J. BERTRÁN, J. Am. Chem. Soc., 1993, **115**, 4024

0015 D. SEEBACH, J. ZIMMERMANN, U. GYSEL, R. ZIEGLER, T. K. HA, J. Am. Chem. Soc., 1988, **110**, 4763

0016 G. THOMA, D. P. CURRAN, S. V. GEIB, B. GIESE, W. DAMM, F. WETTERICH, J. Am. Chem. Soc., 1993, **115**, 8585

0017 D. P. CURRAN, A. C. ABRAHAM, Tetrahedron, 1993, **49**, 4821

0018 D. P. CURRAN, P. S. RAMAMOORTHY, Tetrahedron, 1993, **49**, 4841

0019 N. T. ANH, O. EISENSTEIN, N. J. Chem., 1977, **1**, 69

0020 J. M. COXON, R. T. LUIBRAND, Tetrahedron Lett., 1993, **34**, 7093

0021 Y. D. WU, K. N. HOUK, J. FLOREZ, B. M. TROST, J. Org. Chem., 1991, **56**, 3656

0022 Y. D. WU, K. N. HOUK, M. N. PADDON-ROW, Angew. Chem. Int. Ed. Engl., 1992, **31**, 1019

0023   Y. D. WU, K. N. HOUK, B. M. TROST, J. Am. Chem. Soc., 1987, **109**, 5560

0024   Y. D. WU, J. A. TUCKER, K. N. HOUK, J. Am. Chem. Soc., 1991, **113**, 5018

0025   H. LI, W. J. le NOBLE, Recl. Trav. Chim. Pays-Bas, 1992, **111**, 199

0026   A. E. DORIGO, K. MOROKUMA, J. Am. Chem. Soc., 1989, **111**, 4635, 6524

0027   P. V. R. SCHLEYER, J. Am. Chem. Soc., 1967, **89**, 699, 701

0028   M. CHEREST, H. FELKIN, N. PRUDENT, Tetrahedron Lett., 1968, 2199

0029   M. A. McCARRICK, Y. D. WU, K. N. HOUK, J. Org. Chem., 1993, **58**, 3330

0030   R. ANNUNZIATA, M. BENAGLIA, M. CINQUINI, L. RAIMONDI, Tetrahedron, 1993, **49**, 8629

0031   B. GANGULY, J. CHANDRASEKHAR, F. A. KHAN, G. MEHTA, J. Org. Chem., 1993, **58**, 1734

0032   S. S. WONG, M. N. PADDON-ROW, J. Chem. Soc. Chem. Comm., 1991, 327

0033   M. P. ARRINGTON, Y. L. BENNANI, T. GÖBEL, P. WALSH, S. H. ZHAO, K. B. SHARPLESS, Tetrahedron Lett., 1993, **34**, 7375

0034   Y. D. WU, Y. LI, J. NA, K. N. HOUK, J. Org. Chem., 1993, **58**, 4625

0034a M. N. PADDON-ROW, Y. D. WU, K. N. HOUK, J. Am. Chem. Soc., 1992, **114**, 10638

0035   S. WALLBAUM, J. MARTENS, Tetrahedron Asymm., 1992, **3**, 1475

0036   M. FUJITA, S. AKIMOTO, K. OGURA, Tetrahedron Lett., 1993, **34**, 5139

0037   Z. SHI, R. J. BOYD, J. Am. Chem. Soc., 1993, **115**, 9614

0038   J. M. COXON, R. T. LUIBRAND, Tetrahedron Lett., 1993, **34**, 7097

0039   Y. WU, K. N. HOUK, J. Am. Chem. Soc., 1993, **115**, 10992

0040   N. T. ANH, B. T. THANH, Nouv. J. Chim., 1986, **10**, 681; N. T. ANH, L. ELKAÏM, B. T. THANH, F. MAUREL, J. P. FLAMENT, Bull. Soc. Chim. France, 1992, **129**, 468

0041   Y. LI, M. N. PADDON-ROW, K. N. HOUK, J. Org. Chem., 1990, **55**, 481

0042   D. SEEBACH, J. GOLINSKI, Helv. Chim. Acta, 1981, **64**, 1413

0043   J. CAPILLON, J. P. GUETTE, Tetrahedron, 1979, **35**, 1801, 1807, 1817

0044   E. J. COREY, T. P. LOH, J. Am. Chem. Soc., 1991, **113**, 8966

0045   E. J. COREY, Y. MATSUMURA, Tetrahedron Lett., 1991, **32**, 6289

0046   J. P. LYSSIKATOS, M. D. BEDNARSKI, Synlett, 1990, 230

0047   R. TRIPATHY, P. J. CARROLL, E. R. THORNTON, J. Am. Chem. Soc., 1990, **112**, 6743; 1991, **113**, 7631

0048   J. A. TUCKER, K. N. HOUK, B. M. TROST, J. Am. Chem. Soc., 1990, **112**, 5465

0049   R. THIEM, K. ROTSCHEIDT, E. BREITMAIER, Synthesis, 1989, 836

0050 J. F. BLAKE, D. LIM, W. L. JORGENSEN, J. Org. Chem., 1994, **59**, 803

0051 J. I. SEEMAN, Chem. Rev., 1983, **83**, 73

0052 D. P. CURRAN, H. QI, S. J. GEIB, N. C. DEMELLO, J. Am. Chem. Soc., 1994, **116**, 3131

0053 J. L. BROEKER, R. W. HOFFMANN, K. N. HOUK, J. Am. Chem. Soc., 1991, **113**, 5006

0054 R. W. HOFFMANN, Chem. Rev., 1989, **89**, 1841

0055 D. SEEBACH, B. LAMATSCH, R. AMSTUTZ, A. K. BECK, M. DOBLER, M. EGLI, R. FITZI, M. GAUTSCHI, B. HERRADÓN, P. C. HIDBER, J. J. IRWIN, R. LOCHER, M. MAESTRO, T. MAETZKE, A. MOURIÑO, E. PFAMMATTER, D. A. PLATTNER, C. SCHICKLI, W. B. SCHWEIZER, P. SEILER, G. STUCKY, W. PETTER, J. ESCALANTE, E. JUARISTI, D. QUINTANA, C. MIRAVITLLES, E. MOLINS, Helv. Chim. Acta, 1992, 75, 913

0056 B. GIESE, W. DAMM, F. WETTERICH, H. G. ZEITZ, J. RANCOURT, Y. GUINDON, Tetrahedron Lett., 1993, **34**, 5885

0057 D. A. EVANS, M. C. CALTER, Tetrahedron Lett., 1993, **34**, 6871

0058 G. FRENKING, K. F. KÖHLER, M. T. REETZ, Tetrahedron, 1991, **47**, 8991; 9005

0059 S. S. WONG, M. N. PADDON-ROW, J. Chem. Soc. Chem. Comm., 1990, 456

0060 G. J. KARABATSOS, D. J. FENOGLIO, Topics in Stereochemistry, E. L. ELIEL and S. WILEN Ed., 1970, **5**, 167

0061 J. E. EKSTEROWICZ, K. N. HOUK, Tetrahedron Lett., 1993, **34**, 427

0062 M. ARAI, T. KAWASUJI, E. NAKAMURA, J. Org. Chem., 1993, **58**, 5121

0063 W. DAMM, J. DICKHAUT, F. WETTERICH, B. GIESE, Tetrahedron Lett., 1993, **34**, 431

0064 B. GIESE, W. DAMM, J. DICKHAUT, F. WETTERICH, S. SUN, D. P. CURRAN, Tetrahedron Lett., 1991, **32**, 6097

0065 M. FUJITA, H. ISHIZUKA, K. OGURA, Tetrahedron Lett., 1991, **44**, 6355

0066 J. K. WHITESELL, Chem. Rev., 1992, **92**, 953

0067 J. F. MADDALUNO, N. GRESH, C. GIESSNER-PRETTRE, J. Org. Chem., 1994, **59**, 793

0068 A. F. SEVIN, J. SEYDEN-PENNE, K. BOUBEKEUR, Tetrahedron Asymm., 1991, **2**, 1107 and Tetrahedron, 1992, **48**, 6253

0069 K. L. BOBBIT, C. K. MURRAY, G. A. MOLANDER, J. Am. Chem. Soc.1992, **114**, 2759

0070 G. A. MOLANDER, K. L. BOBBITT, J. Am. Chem. Soc., 1993, **115**, 7517

0071 D. P. CURRAN, B. H. KIM, H. P. PIYASENA, R. J. LONCHARICH, K. N. HOUK, J. Org. Chem., 1987, **52**, 2137

0072   D. A. EVANS, K. T. CHAPMAN, J. BISAHA, J. Am. Chem. Soc., 1988, **110**, 1238

0073   W. OPPOLZER, in Comprehensive Organic Synthesis, B. M. TROST and I. FLEMING Ed., Pergamon Press, 1991, vol. 5, chap. 4.1

0074   W. R. ROUSH, in Comprehensive Organic Synthesis, B. M. TROST and I. FLEMING Ed., Pergamon Press, 1991, vol. 5, chap. 4.4

0075   V. BRANCHADELL, A. OLIVA, J. Am. Chem. Soc., 1991, **113**, 4132

0076   S. E. DENMARK, N. G. ALMSTEAD, J. Am. Chem. Soc., 1993, **115**, 3133 and quoted references

0077   D. SEEBACH, Angew. Chem. Int. Ed. Engl., 1988, **27**, 1624

0078   P. G. WILLIARD, Q. Y. LIU, J. Am. Chem. Soc., 1993, **115**, 3380

0079   M. NAKAMURA, E. NAKAMURA, N. KOGA, K.MOROKUMA, J. Am. Chem. Soc., 1993, **115**, 11016

0080   M. T. REETZ, S. STANCHEV, H. HANING, Tetrahedron, 1992, **48**, 6813

0081   D. SEEBACH, Angew. Chem. Int. Ed. Engl., 1990, **29**, 1320 and quoted ref.

0082   J. M. GOODMAN, Tetrahedron Lett., 1992, **33**, 7219

0083   R. J. LONCHARICH, T. R. SCHWARTZ, K. N. HOUK, J. Am. Chem. Soc., 1987, **109**, 14

0084   T. J. LEPAGE, K. B. WIBERG, J. Am. Chem. Soc., 1988, **110**, 6642

0085   S. SHAMBAYATI, S. L. SCHREIBER, in Comprehensive Organic Synthesis, B. M. TROST and I. FLEMING Ed., Pergamon Press, 1991, vol. 1, chap. 1.10

0086   S. E. DENMARK, N. G. ALMSTEAD, Tetrahedron, 1992, **48**, 5565

0087   J. SEYDEN-PENNE, Réduction par les Alumino-et Borohydrures en Synthèse Organique LAVOISIER, Ed. Paris, 1988; Reductions by the Alumino- and Borohydrides in Organic Synthesis, Verlag-Chemie, New York, 1991

0088   E. J. COREY, T. P. LOH, S. SARSHAR, M. AZIMIOARA, Tetrahedron Lett., 1992, **33**, 6945

0089   D. M. BIRNEY, K. N. HOUK, J. Am. Chem. Soc., 1990, **112**, 4127 and quoted ref.

0090   C. CATIVIELA, J. I. GARCIA, J. A. MAYORAL, A. J. ROYO, L. SALVATELLA, X. ASSFELD, M. F. RUIZ-LOPEZ, J. Physical Org. Chem., 1992, **5**, 230

0091   B. MEZRHAB, F. DUMAS, J. D'ANGELO, C. RICHE, J. Org. Chem., 1994, **59**, 500

0092   Comprehensive Organometallic Chemistry, G. WILKINSON, F. G. A. STONE, E. W. ABEL Ed., Pergamon, Oxford, 1982

0093   M. T. REETZ, Acc. Chem. Res., 1993, **26**, 462

0094   E. L. ELIEL, S. V. FRYE, E. R. HORTELANO, X. CHEN, X. BAI, Pure and Appl. Chem., 1991, **63**, 1591

0095   M. T. REETZ, B. RAGUSE, T. SEITZ, Tetrahedron, 1993, **49**, 8561

0096 M. T. REETZ, Angew. Chem. Int. Ed. Engl., 1991, **30**, 1531

0097 M. REETZ, Organotitanium Reagents in Organic Synthesis, Springer Verlag, Berlin, 1985

0098 M. T. REETZ, A. JUNG, C. BOLM, Tetrahedron, 1988, **44**, 3889

0099 M. T. REETZ, B. RAGUSE, C. F. MARTH, H. M. HÜGEL, T. BACH, D. N. A. FOX, Tetrahedron, 1992, **48**, 5731

0100 A. CHOUDHURY, E. R. THORNTON, Tetrahedron Lett., 1993, **34**, 2221

0101 G. H. POSNER, in Asymmetric Synthesis J. D. MORRISON Ed. Academic Press New York, 1983, **2**, 225; The Chemistry of Sulfones and Sulfoxides, S. PATAI, Z. RAPPOPORT Ed., Wiley and Sons, Chichester, Royaume-Uni, 1988, 823

0102 G. H. POSNER, Acc. Chem. Res., 1987, **20**, 72

0103 A. SOLLADIE-CAVALLO, J. SUFFERT, A. ADIB, G. SOLLADIE, Tetrahedron Lett., 1990, **46**, 6649

0104 H. WALDMANN, J. Org. Chem., 1988, **53**, 6133; Liebigs Ann., 1990, 671

0105 H. DANDA, M. M. HANSEN, C. H. HEATHCOCK, J. Org. Chem., 1990, **55**, 173

0106 M. A. WALKER, C. H. HEATHCOCK, J. Org. Chem., 1991, **56**, 5747

0107 G. A. MOLANDER, J. P. HAAR, J. Am. Chem. Soc., 1992, **114**, 40

0108 M. MURAKATA, M. NAKAJIMA, K. KOGA, J. Chem. Soc. Chem. Comm., 1990, 1657

0109 D. HOPPE, O. ZSCHAGE, Angew. Chem. Int. Ed. Engl., 1989, **28**, 69

0110 R. NOYORI, M. KITAMURA, Angew. Chem. Int. Ed. Engl., 1991, **30**, 49

0111 K. SOAI, S. YOKOYAMA, T. HAYASAKA, J. Org. Chem., 1991, **56**, 4264

0112 B. E. ROSSITER, N. E. SWINGLE, Chem. Rev., 1992, **92**, 771

0113 H. BRUNNER, Topics in Stereochemistry, E. L. ELIEL and S. WILEN Ed., 1988, **18**, 129

0114 K. INOGUCHI, S. SAKURABA, K. ACHIWA, Synlett, 1992, 169

0115 T. V. RAJANBABU, T. A. AYERS, A. L. CASALNUOVO, J. Am. Chem. Soc., 1994, **116**, 4101

0115a A. HOREAU, H. B. KAGAN, J. P. VIGNERON, Bull. Soc. Chim. France, 1968, 3795

0116 C. H. HEATHCOCK, in Comprehensive Organic Synthesis, B. M. TROST and I. FLEMING Ed., Pergamon Press, 1991 vol. 2, chap. 1.6

0116a D. J. HART, D. C. HA, Chem. Rev., 1989, **89**, 1447

0117 S. MASAMUNE, W. CHOY, J. S. PETERSEN, L. R. SITA, Angew. Chem. Int. Ed. Engl., 1985, **24**, 1

0118 L.M. TOLBERT, M.B.ALI, J. Am. Chem. Soc., 1984, **106**, 3806

0119 W. R. ROUSH, J. Org. Chem., 1991, **56**, 4151

0120 W. R. ROUSH, A. D. PALKOWITZ, K. ANDO, J. Am. Chem. Soc.,
      1990, **112**, 6348
0120a A. ROUCOUX, F. AGBOSSOU, A. MORTREUX, F. PETIT,
      Tetrahedron Asymm., 1993, **4**, 2279
0121 N. A. vanDRAANEN, S. ARSENIYADIS, M. T. CRIMMINS, C. H.
      HEATHCOCK, J. Org. Chem., 1991, **56**, 2499 and quoted ref.
0122 D. A. EVANS, J. V. NELSON, T. R. TABER, Topics in
      Stereochemistry, E. L. ELIEL and S. WILEN Ed., 1982, **13**, 1
0123 L. N. PRIDGEN, A. F. ABDEL-MAGID, I. LANTOS, S. SHILCRAT,
      D. S. EGGLESTON, J. Org. Chem., 1993, **58**, 5107
0124 J. M. GOODMAN, S. D. KAHN, I. PATERSON, J. Org. Chem., 1990,
      55, 3295
0125 A. BERNARDI, A. M. CAPELLI, A. COMOTTI, C. GENNARI, M.
      GARDNER, J. M. GOODMAN, I. PATERSON, Tetrahedron, 1991, **47**,
      3471
0126 A. BERNARDI, A. M. CAPELLI, A. COMOTTI, C. GENNARI, C.
      SCOLASTICO, Tetrahedron Lett., 1991, **32**, 823, A. BERNARDI, A. M.
      CAPELLI, A. CASSINARI, A. COMOTTI, C. GENNARI, C.
      SCOLASTICO, J. Org. Chem., 1992, **57**, 7029
0127 H. B. KAGAN, J. C. FIAUD, Topics in Stereochemistry, E. L. ELIEL
      and S. WILEN Ed., 1988, **18**, 249
0128 V. RAUTENSTRAUCH, M. LINDSTRÖM, B. BOURDIN, J. CURRIE,
      E. OLIVEROS, Helv. Chim. Acta, 1993, **76**, 607
0129 B. M. TROST, Science 1991, **254**, 1471
0130 S. MASAMUNE, S. A. ALI, D. L. SNITMAN, D. S. GARVEY, Angew.
      Chem. Int. Ed. Engl., 1980, **19**, 557
0131 F. H. van DER STEEN, G. van KOTEN, Tetrahedron, 1991, **47**, 7503
0132 M. OHTANI, T. MATSUURA, F. WATANABE, M. NARISADA, J.
      Org. Chem., 1991, **56**, 4120 and 2122
0133 L. R. RANDRIANASOLO-RAKOTOZAFY, R. AZERAD, F. DUMAS,
      D. POTIN, J.d'ANGELO, Tetrahedron Asymm., 1993, **4**, 761
0133a D. POTIN, F. DUMAS, J. MADDALUNO, Synth. Comm., 1990, **20**,
      2805
0134 J. d'ANGELO, J. MADDALUNO, J. Am. Chem. Soc., 1986, **108**, 8112
0135 E. J. COREY, M. E. ENSLEY, J. Am. Chem. Soc., 1975, **97**, 6908
0136 D. L. COMINS, H. HONG, J. Org. Chem., 1993, **58**, 5035
0137 B. L. FERINGA, B. de LANGE, J. F. G. A. JANSEN, J. C. de JONG, M.
      LUBBEN, W. FABER, E. B. SCHUDDE, Pure and Applied Chem.,
      1992, **64**, 1865
0138 O. ORT, Org. Synth., 1987, **65**, 203
0139 R. P. POLNIASZEK, L. W. DILLARD, Tetrahedron Lett., 1990, **31**, 797
0140 C. S. SWINDELL, M. TAO, J. Org. Chem., 1993, **58**, 5859
0141 A. SCHWARTZ, P. MADAN, J. K. WHITESELL, Org. Synth., 1990,
      69, 1

0142 J. K. WHITESELL, R. M. LAWRENCE, Chimia, 1986, **40**, 318

0143 D. L. COMINS, L. M. SALVADOR, Tetrahedron Lett., 1993, **34**, 801; J. Org. Chem., 1993, **58**, 4656

0144 D. P. G. HAMON, J. W. HOLMAN, R. A. MASSY-WESTROPP, Tetrahedron, 1993, **49**, 9593

0144a D. L. HUANG, R. W. DRAPER, Tetrahedron Lett., 1994, **35**, 661

0145 D. P. G.HAMON, J. W. HOLMAN, R. A. MASSY-WESTROPP, Tetrahedron Asymm., 1992, **3**, 1533

0146 U. MAITRA, P. MATHIVANAN, J. Chem. Soc. Chem. Comm., 1993, 1469

0147 W. OPPOLZER, Tetrahedron, 1987, **43**, 1969 and quoted ref.

0148 T. MUKAIYAMA, H. HAYASHI, T. MIWA, K. NARASAKA, Chem. Lett., 1982, 1637

0149 M. BRAUN, H. SACHA, Angew. Chem. Int. Ed. Engl., 1991, **30**, 1318

0150 I. OJIMA, I. HABUS, M. ZHAO, G. I. GEORG, L. R. JAYASINGHE, J. Org. Chem., 1991, **56**, 1681

0151 C. GENNARI, A. BERNARDI, L. COLOMBO, C. SCOLASTICO, J. Am. Chem. Soc., 1985, **107**, 5812

0152 C. GENNARI, L. COLOMBO, G. BERTOLINI, G. SCHIMPERNA, J. Org. Chem., 1987, **52**, 2754

0153 C. GENNARI, F. MOLINARI, P. G. COZZI, A. OLIVA, Tetrahedron Lett., 1989, **30**, 5163

0154 R. M. WILLIAMS, Synthesis of Optically Active Aminoacids, Pergamon Press, Oxford, 1989

0155 Y. B. XIANG, K. SNOW, M. BELLEY, J. Org. Chem., 1993, **58**, 993

0156 H. HARTMANN, A. F. A. HADY, K. SARTOR, J. WEETMAN, G. HELMCHEN, Angew. Chem. Int. Ed. Engl., 1985, **24**, 112 and 1987, **26**, 1143

0157 C. FANG, H. SUEMUNE, K. SAKAI, Tetrahedron Lett., 1990, **31**, 4751

0158 T. POLL, A. F. A. HADY, R. KARGE, G. LINZ, J. WEETMAN, G. HELMCHEN, Tetrahedron Lett., 1989, **30**, 5595

0159 D. CAINE, in Comprehensive Organic Synthesis, B. M. TROST and I. FLEMING Ed., Pergamon Press, 1991, vol. 3, chap. 1.1

0160 C. H. HEATHCOCK, in Asymmetric Synthesis, J. D. MORRISON Ed. Academic Press, New York 1984, **3**, 111 and quoted ref.

0161 D. OARE, C. H. HEATHCOCK, Topics in Stereochemistry, E. L. ELIEL and S. WILEN Ed., 1989, **19**, 227 and quoted ref.

0162 D. A. OARE, C. H. HEATHCOCK, Topics in Stereochemistry, E. L. ELIEL and S. WILEN Ed., 1991, **20**, 87 and quoted ref.

0163 K. MIKANI, T. NAKAI, Synthesis, 1991, 594

0164 D. P. G. HAMON, R. A. MASSY-WESTROPP, P. RAZZINO, J. Chem. Soc. Chem. Comm., 1991, 722

0164a D. P. G. HAMON, R. A. MASSY-WESTROPP, P. RAZZINO, Tetrahedron, 1993, **49**, 6419

0165 D. P. G. HAMON, P. RAZZINO, R. MASSY-WESTROPP, J. Chem. Soc. Chem. Comm., 1991, 332

0166 M. IHARA, T. TAKAHASHI, N. TANIGUCHI, K. YASUI, K. FUKUMOTO, T. KAMETANI, J. Chem. Soc. Perkin I, 1989, 897

0167 D. A. EVANS in Asymmetric Synthesis, J. D. MORRISON Ed. Academic Press, New York 1984, **3**, 1 and quoted ref.

0168 K. FURUTA, K. IWANAGA, H. YAMAMOTO, Org. Synth., 1988, **67**, 76

0169 J. APSIMON, T. L. COLLIER, Tetrahedron, 1986, **42**, 5157 and quoted ref.

0170 L. A. PAQUETTE in Asymmetric Synthesis, J. D. MORRISON Ed. Academic Press, New York 1984, **3**, 455 and quoted ref.

0171 B. A. BARR, M. J. DORRITY, R. GRIGG, J. F. MALONE, J. MONTGOMERY, S. RAJVIROONGIT, P. STEVENSON, Tetrahedron Lett., 1990, **31**, 6569

0172 R. D. LITTLE, in Comprehensive Organic Synthesis, B. M. TROST and I. FLEMING Ed., Pergamon Press, 1991, vol. 5, chap. 3.1

0173 H. G. SCHMALZ, in Comprehensive Organic Synthesis, B. M. TROST and I. FLEMING Ed., Pergamon Press, 1991, vol. 4, chap. 1.5

0174 S. AHMAD, Tetrahedron Lett., 1991, **32**, 6997

0175 T. FUJISAWA, R. HAYAKAWA, M. SHIMIZU, Tetrahedron Lett., 1992, **33**, 7903

0176 G. LINZ, J. WEETMAN, A. F. A. HADY, G. HELMCHEN, Tetrahedron Lett., 1989, **30**, 5599

0177 K. MIYAJI, Y. OHARA, Y. TAKAHASHI, T. TSURADA, K. ARAI, Tetrahedron Lett., 1991, **32**, 4557

0178 T. POLL, A. SOBCZAK, H. HARTMANN, G. HELMCHEN, Tetrahedron Lett., 1985, 3095

0179 F. BIGI, G. CASNATI, S. SARTORI, C. DALPRATO, R. BORTOLINI, Tetrahedron Asymm., 1990, **1**, 861

0180 A. SOLLADIÉ-CAVALLO, M. BENCHEQROUN, Tetrahedron Asymm., 1991, **2**, 1165

0181 D. F. TABER, P. B. DEKER, M. D. GAUL, J. Am. Chem. Soc., 1987, **109**, 7488

0182 J. K. WHITESELL, D. DEY, A. BHATTACHARYA, J. Chem. Soc. Chem. Comm., 1983, 802

0183 J. K. WHITESELL, A. BHATTACHARYA, K. HENKE, J. Chem. Soc. Chem. Comm., 1982, 988 and 989

0184 K. MINAMI, H. WAKABAYASHI, T. NAKAI, J. Org. Chem., 1991, **56**, 4337

0185 D. L. COMINS, M. B. BADAWI, Tetrahedron Lett., 1991, **32**, 2995

0186 D. L. COMINS, R. R. GOEHRING, S. P. JOSEPH, S. O'CONNOR, J. Org. Chem., 1990, **55**, 2574, D.L. COMINS, S. P. JOSEPH, R. R. GOEHRING, J. Am. Chem. Soc., 1994, **116**, 4719

0187 D. L. COMINS, H. HONG, J. M. SALVADOR, J. Org. Chem., 1991, **56**, 7197

0188 D. L. COMINS, H. HONG, J. Am. Chem. Soc., 1991, **113**, 6672

0189 S. W. REMISZENSKI, J. YANG, S. M. WEINREB, Tetrahedron Lett., 1986, **27**, 1853

0190 J. K. WHITESELL, D. JAMES, J. F. CARPENTER, J. Chem. Soc. Chem. Comm., 1985, 1449

0191 J. K. WHITESELL, J. F. CARPENTER, H. K. YASER, T. MACHAJEWSKI, J. Am. Chem. Soc., 1990, **112**, 7653

0192 H. M. L. DAVIES, W. R. CANTRELL, Tetrahedron Lett., 1991, **32**, 6509

0193 H. M. L. DAVIES, N. J. S. HUBY, W. R. CANTRELL, J. L. OLIVE, J. Am. Chem. Soc., 1993, **115**, 9468

0194 T. ROSEN, C. H. HEATHCOCK, J. Am. Chem. Soc., 1985, **107**, 3731

0195 L. CHEN, L. GHOSEZ, Tetrahedron Lett., 1990, **31**, 4467

0196 F. CHARBONNIER, A. MOYANO, A. E. GREENE, J. Org. Chem., 1987, **52**, 2303

0197 S. E. DENMARK, M. E. SCHNUTE, J. Org. Chem., 1991, **56**, 6738

0198 S. E. DENMARK, C. B. W. SENANAYAKE, G. D. HO, Tetrahedron, 1990, **46**, 4857

0199 G. H. POSNER, D. G. WETTLAUFER, Tetrahedron Lett., 1986, **27**, 667; J. Am. Chem. Soc., 1986, **108**, 7373

0199a A. MOYANO, F. CHARBONNIER, A. E. GREENE, J. Org. Chem., 1987, **52**, 2919

0200 J. CASTRO, H. SORENSEN, A. RIERA, A. MOYANO, M. A. PERICAS, A. E. GREENE, J. Am. Chem. Soc., 1990, **112**, 9388

0201 M. POCH, E. VALENTI, A. MOYANO, M. A. PERICAS, J. CASTRO, A. DENICOLA, A. E. GREENE, Tetrahedron Lett., 1990, **31**, 7505

0202 J. DE JONG, F. van BOLHUIS, B. FERINGA, Tetrahedron Asymm., 1991, **2**, 1247

0203 B. de LANGE, B. L. FERINGA, Tetrahedron Lett., 1989, **29**, 5317

0204 A. MILLAR, L. W. MULDER, K. E. MENNEN, C. W. PALMER, Org. Prep. Proc. Int., 1991, **23**, 173

0205 M. BRAUN, Angew. Chem. Int. Ed. Engl., 1987, **26**, 24

0206 M. BRAUN, D. WALDMULLER, Synthesis, 1989, 856

0207 R. DEVANT, U. MAHLER, M. BRAUN, Chem. Ber., 1988, **121**, 397

0208 R. M. DEVANT, H. E. RADUNZ, Tetrahedron Lett., 1988, **29**, 2307

0209 C. H. HEATHCOCK, in Comprehensive Organic Synthesis, B. M. TROST and I. FLEMING Ed., Pergamon Press, 1991 vol. 2, chap. 1.6

0210 K. PRASAD, K. M. CHEN, O. REPIC, G. E. HARDTMANN, Tetrahedron Asymm., 1990, **1**, 703

0211 Y. SUDA, S. YAGO, M. SHIRO, T. TAGUCHI, Chem. Lett., 1992, 389

0212 J. K. WHITESELL, Chem. Rev., 1989, **89**, 1581

0213 A. ALEXAKIS, P. MANGENEY, Tetrahedron Asymm., 1990, **1**, 477

0214   A. B. CHARETTE, J. F. MARCOUX, Tetrahedron Lett., 1993, **34**, 7157
0215   K. KATO, H. SUEMUNE, K. SAKAI, Tetrahedron Lett., 1992, **33**, 247;
       Tetrahedron, 1994, **50**, 3315
0216   K. KATO, H. SUEMUNE, K. SAKAI, Tetrahedron Lett., 1993, **34**, 4979
0217   I. KOGA, K. FUNAKOSHI, A. MATSUDA, K. SAKAI, Tetrahedron
       Asymm., 1993, **4**, 1857
0217a Y. KAWANAMI, I. FUJITA, S. OGAWA, T. KATSUKI, Chem. Lett.,
       1989, 2063
0218   T. FUJISAWA, Y. UKAJI, T. NORO, K. DATE, M. SHIMIZU,
       Tetrahedron Lett., 1991, **32**, 7563
0219   S. E. DENMARK, J. P. EDWARDS, J. Org. Chem., 1991, **56**, 6974
0220   S. E. DENMARK, N. G. ALMSTEAD, J. Org. Chem., 1991,**56**, 6485
       and 6458
0221   K. ISHIHARA, N. HANAKI, H. YAMAMOTO, Synlett, 1993, 127
0222   K. ISHIHARA, N. HANAKI, H. YAMAMOTO, J. Am. Chem. Soc.,
       1993, **115**, 10695
0223   A. B. HOLMES, A. B. TABOR, R. BAKER, J. Chem. Soc. Perkin I,
       1991, 3301
0224   Y. YAMAMOTO, H. ABE, S. NISHII, J. I. YAMADA, J. Chem. Soc.
       Perkin I, 1991, 3253
0225   E. L. ELIEL in Asymmetric Synthesis, J. D. MORRISON Ed. Academic
       Press, New York 1983, **2**,125 and quoted ref.
0226   S. V. FRYE, E. L. ELIEL, J. Am. Chem. Soc., 1988, **108**, 484
0227   J. E. LYNCH, E. L. ELIEL, J. Am. Chem. Soc., 1984, **106**, 2943
0228   J. WEI, R. O. HUTCHINS, J. Org. Chem., 1993, **58**, 2920
0229   M. NISHIDA, K. NAKAOKA, S. ONO, O. YONEMITSU, A.
       NISHIDA, N. KAWAHARA, H.TAKAYANAGI, J. Org. Chem., 1993,
       58, 5870
0230   J. JACQUES, C. FOUQUEY, Org. Synth., 1988, **67**, 1
0230a K. TANAKA, Y. OHTA, K. FUJI, T. TAGA, Tetrahedron Lett., 1993,
       34, 4071
0231   C. ROSINI, L. FRANZINI, A. RAFFAELLI, P. SALVADORI,
       Synthesis, 1992, 503
0232   M. SMRCINA, M. LORENC, V. HANUS, P. SEDMERA,
       P.KOCOVSKY, J. Org. Chem., 1992, **57**, 1917
0233   M. SMRCINA, J. POLÁKOVÁ, S. VYSKOCIL, P. KOCOVSKY, J.
       Org. Chem., 1993, **58**, 4534
0234   L. K. TRUESDALE, Org. Synth., 1988, **67**, 13
0235   K. TANAKA, T. OKADA, F. TODA, Angew. Chem. Int. Ed. Engl.,
       1993, **32**, 1147
0236   Y. TAMAI, S. KOIKE, A. OGURA, S. MIYANO, J. Chem. Soc. Chem.
       Comm., 1991, 799
0237   K. FUJI, K. TANAKA, M. MIZUCHI, S. HOSOI, Tetrahedron Lett.,
       1991, **32**, 7277

0238 K. FUJI, M. NODE, F. TANAKA, Tetrahedron Lett., 1990, **31**, 6553

0239 K. FUJI, F. TANAKA, M. NODE, Tetrahedron Lett., 1991, **32**, 7281

0240 S. HANESSIAN, Total Synthesis of Natural Products : The Chiron Approach, Pergamon Press, New York, 1983

0241 H. KUNZ, K.RÜCK, Angew. Chem. Int. Ed. Engl., 1993, **32**, 336

0242 H. KUNZ, J. MOHR, J. Chem. Soc. Chem. Comm., 1988, 1315

0243 T. AKIYAMA, H. NISHIMOTO, K. ISHIKAWA, S. OZAKI, Chem. Lett., 1992, 447

0244 H. KUNZ, B. MULLER, D. SCHANZENBACH, Angew. Chem. Int. Ed. Engl., 1987, **26**, 267

0245 R. NOUGIER, J. L. GRAS, B. GIRAUD, A. VIRGILI, Tetrahedron Lett., 1991, **32**, 5529; Tetrahedron, 1992, **48**, 6245

0246 W. STAHLE, H. KUNZ, Synlett, 1991, 260

0247 Y. S. HON, F. L. CHEN, Y. P. HUANG, T. J. LU, Tetrahedron Asymm., 1991, **2**, 879

0248 H. U. REISSIG, Angew. Chem. Int. Ed. Engl., 1992, **31**, 288

0249 M. D. BEDNARSKI, J. P. LYSSIKATOS, in Comprehensive Organic Synthesis, B. M. TROST and I. FLEMING Ed., Pergamon Press, 1991, vol. 2, chap. 2.5

0250 D. MEAT-JACHET, A. HOREAU, Bull. Soc. Chim. France, 1968, 4571

0251 M. BRAUN, Angew. Chem. Int. Ed. Engl., 1987, **26**, 24

0251a E. JUARISTI, P. MURER, D. SEEBACH, Synthesis, 1993, 1243

0252 D. E. BERGBREITER, M. NEWCOMB in Asymmetric Synthesis, J. D. MORRISON Ed. Academic Press, New York , 1983, **2**, 243 and quoted ref.

0253 M. NOGRADI, Stereoselective Synthesis, Verlag Chemie, New York, 1986

0254 R. P. POLNIASZEK, S. E. BELMONT, R. ALVAREZ, J. Org. Chem., 1990, **55**, 215

0255 G. REVIAL, M. P. FAU, Org. Synth., 1991, **70**, 35

0256 L. STELLA, H. ABRAHAM, J. FENEAU-DUPONT, B. TINANT, J. P. DECLERCQ, Tetrahedron Lett., 1990, **31**, 2603

0257 A. SEVIN, D. MASURE, C. GIESSNER-PRETTRE, M. PFAU, Helv. Chim. Acta, 1990, **73**, 552; D. DESMAELE, J. D'ANGELO, C. BOIS, Tetrahedron Asymm.,1990, 1, 759 and quoted ref.

0258 Y. YAMAMOTO, Acc. Chem. Res., 1987, **20**, 243

0259 C. R. MC ARTHUR, J. L. JIANG, C. C. LEZNOFF, Canad. J. Chem., 1982, **60**, 2894

0260 K. YAMAMOTO, M. KANOH, N. YAMAMOTO, J. TSUJI, Tetrahedron Lett., 1987, **28**, 6347

0261 G. M. COPPOLA, H. F. SCHUSTER, Asymmetric Synthesis : Construction of Chiral Molecules Using Aminoacids, Wiley, New York, 1987

0262 K. TOMIOKA, K. KOGA in Asymmetric Synthesis, J. D. MORRISON Ed., Academic Press, New York, 1983, **2**, 201 and quoted ref.

0263   H. WALDMAN, M. BRAUN, M. DRAGER, Tetrahedron Asymm., 1991, **2**, 1231

0264   H. WALDMANN, M. BRAUN, Gazz. Chim. Ital., 1991, **121**, 277

0265   R. DEVANT, M. BRAUN, Chem. Ber., 1986, **119**, 2191

0266   T. K. CHAKRABORTY, G. V. REDDY, K. A. HUSSAIN, Tetrahedron Lett., 1991, **32**, 7597

0267   J. d'ANGELO, D. DESMAËLE, F. DUMAS, A. GUINGANT, Tetrahedron Asymm., 1992, **3**, 459

0268   M. HAYASHI, Y. MIYAMOTO, T. INOUE, N. OGUNI, J. Org. Chem., 1993, **58**, 1515

0269   H. WALDMANN, G. SCHMIDT, M. JANSEN, J. GEB, Tetrahedron Lett., 1993, **34**, 5867

0270   R. TAMION, F. MARSAIS, P. RIBEREAU, G. QUEGUINER, Tetrahedron Asymm., 1993, **4**, 2415

0271   R. TAMION, F. MARSAIS, P. RIBEREAU, G. QUEGUINER, D. ABENHAIM, A. LOUPY, L. MUNNIER, Tetrahedron Asymm., 1993, **4**, 1879

0272   L. GHOSEZ, C. GENICOT, V. GOUVERNEUR, Pure and Applied Chem., 1992, **64**, 1849

0273   K. KOH, R. N. BEN, T. DURST, Tetrahedron Lett., 1994, **35**, 375

0274   R. H. SCHLESSINGER, E. J. IWANOWICZ, J. P. SPRINGER, Tetrahedron Lett., 1988, **29**, 1489

0275   X. M. WU, K. FUNAKOSHI, K. SAKAI, Tetrahedron Lett., 1993, **34**, 5927

0276   N. A. PORTER, B. GIESE, D. P. CURRAN, Acc. Chem. Res.,1991, **24**, 296

0277   N. A. PORTER, D. M. SCOTT, I. J. ROSENSTEIN, B. GIESE, A. VEIT, H. G. ZEITZ, J. Am. Chem. Soc., 1991, **113**, 1791

0278   Y. YAMAMOTO, H. OHMORI, S. SAWADA, Synlett, 1991, 319

0279   A. VEIT, R. LENZ, M. E. SEILER, M. NEUBURGER, M. ZEHNDER, B. GIESE, Helv. Chim. Acta, 1993, **76**, 441

0280   M. ENOMOTO, Y. ITO, T. KATSUKI, M. YAMAGUCHI, Tetrahedron Lett., 1985, **26**, 1343

0281   T. HANAMOTO, T. KATSUKI, M. YAMAGUCHI, Tetrahedron Lett., 1986, **27**, 2463

0282   S. IKEGAMI, H. UCHIYAMA, T. HAYAMA, T. KATSUKI, M. YAMAGUCHI, Tetrahedron, 1988, **44**, 5333

0283   Y. KAWANAMI, Y. ITO, T. KITAGAWA, Y. TANIGUCHI, T. KATSUKA, M. YAMAGUCHI, Tetrahedron Lett., 1984, **25**, 857

0284   D. A. EVANS, L. R. MC GEE, J Am. Chem. Soc., 1981, **103**, 2876

0285   T. KATSUKI, M. YAMAGUCHI, Tetrahedron Lett., 1985, **26**, 5807

0286   M. P. BUENO, C. A. CATIVIELA, J. A. MAYORAL, J. Org. Chem., 1991, **56**, 6551

0287 Y. KAWANAMI, T. KATSUKI, M. YAMAGUCHI, Bull. Chem. Soc. Japan, 1987, **60**, 4190

0288 H. LAMY-SCHELKENS, L. GHOSEZ, Tetrahedron Lett., 1989, **30**, 5891

0289 H. WALDMANN, M. DRAGER, Tetrahedron Lett., 1989, **30**, 4227 ; Liebigs Ann. 1990, 681

0290 C. AGAMI, F. COUTY, Tetrahedron Lett., 1987, **28**, 5659

0291 L. CHEN, L. GHOSEZ, Tetrahedron Asymm., 1991, **2**, 1181

0292 Y. KAWANAMI, K. KATAYAMA, Chem. Lett., 1990, 1749

0293 M. E. JUNG, W. D. VACCARO, K. R. BUSZEK, Tetrahedron Lett., 1989, **30**, 1893

0294 J. MARTENS, S. LUBBEN, Tetrahedron, 1991, **47**, 1205

0295 C. LENSINK, J. G. de VRIES, Tetrahedron Asymm., 1993, **4**, 215

0296 J. BARLUENGA, F. AZNAR, C. VALDÉS, A. MARTIN, S. GARCIA-GRANDA, E. MARTIN, J. Am. Chem. Soc., 1993, **115**, 4403

0297 B. A. ANDERSON, W. D. WULFF, A. RAHM, J. Am. Chem. Soc., 1993, **115**, 4602

0298 A. J. PEARSON, P. Y. ZHU, W. J. YOUNGS, J. D. BRADSHAW, D. B. McCONVILLE, J. Am. Chem. Soc., 1993, **115**, 10376

0299 S. NADJI, D. REICHLIN, M. J. KURTH, J. Org. Chem., 1990, **55**, 6241

0300 D. ENDERS, P. GERDES, M. KIPPHARDT, Angew. Chem. Int. Ed. Engl., 1990, **29**, 179

0301 C. J. CHANG, J. M. FANG, L. F. LIAO, J. Org. Chem., 1993, **58**, 1754

0302 A. ALEXAKIS, N. LENSEN, P. MANGENEY, Tetrahedron Lett., 1991, **32**,1171

0303 A. ALEXAKIS, R. SEDRANI, P. MANGENEY, Tetrahedron Lett., 1990, **31**, 345

0304 N. KOMATSU, Y. HISHIBAYASHI, T. SUGITA, S. UEMURA, Tetrahedron Lett., 1992, **33**, 5391

0305 M. H. NANTZ, D. A. LEE, D. M. BENDER, A. H. ROOHI, J. Org. Chem., 1992, **57**, 6653

0306 A. ALEXAKIS, R. SEDRANI, P. MANGENEY, J. F. NORMANT, Tetrahedron Lett., 1988, **29**, 4411

0307 R. GOSMINI, P. MANGENEY, A. ALEXAKIS, M. COMMERCON, J. F. NORMANT, Synlett, 1991, 111

0308 S. KANEMASA, K. ONIMURA, E. WADA, J. TANAKA, Tetrahedron Asymm., 1991, **2**, 1185

0309 S. KANEMASA, K. ONIMURA, Tetrahedron, 1992, **48**, 8631

0310 S. HANESSIAN, S. BEAUDOIN, Tetrahedron Lett., 1992, **33**, 7655

0311 S. HANESSIAN, Y. BENNANI, D. DELORME, Tetrahedron Lett., 1990, **31**, 6461

0312 S. HANESSIAN, Y. BENNANI, Tetrahedron Lett., 1990, **31**, 6465

0313 S. HANESSIAN, A. GOMTSYAN, A. PAYNE, Y. HERVÉ, S. BEAUDOIN, J. Org. Chem., 1993, **58**, 5032

0314   Y. YAMAMOTO, A. SAKAMOTO, T. NISHIOKA, J. ODA, Y. FUKAZAWA, J. Org. Chem., 1991, **56**, 1112

0315   D. ENDERS in Asymmetric Synthesis, J. D. MORRISON Ed. Academic Press, New York 1984, **3**, 275 and quoted ref.

0316   S. E. DENMARK, T. WEBER, D. W. PIOTROWSKI, J. Am. Chem. Soc., 1987, **109**, 2224

0316a   R. A. VOLKMANN, in Comprehensive Organic Synthesis, B. M. TROST and I. FLEMING Ed., Pergamon Press, 1991, vol. 1, chap. 1.12

0317   D. E. BERGBREITER, M. MOMONGAN, in Comprehensive Organic Synthesis, B. M. TROST and I. FLEMING Ed., Pergamon Press, 1991, vol. 2, chap. 1.17

0317a   D. ENDERS, H. SCHUBERT, C. NUBLING, Angew. Chem. Int. Ed. Engl., 1986, **25**, 1109

0318   T. WEBER, J. P. EDWARDS, S. E. DENMARK, Synlett, 1989, 20

0319   R. BEAUDEGNIES, L. GHOSEZ, Tetrahedron Asymm., 1994, **5**, 557

0320   J. BERLAN, Y. BESACE, D. PRAT, G. POURCELOT, J. Organomet. Chem., 1984, **264**, 399

0321   A. BERNARDI, S. CARDANI, C. SCOLASTICO, R. VILLA, Tetrahedron, 1990, **46**, 1987; J. Org. Chem., 1988, **53**, 1600

0322   A. BERNARDI, U. PIARULLI, G. POLI, C. SCOLASTICO, R. VILLA, Bull. Soc. Chim. France, 1990, 751; Tetrahedron Lett., 1990, **31**, 2779

0323   S. CARDANI, G. POLI, C. SCOLASTICO, R. VILLA, Tetrahedron, 1988, **44**, 5929

0324   K. A. LUTOMSKI, A. I. MEYERS in Asymmetric Synthesis, J. D. MORRISON Ed. Academic Press, New York, 1984, **3**, 213 and quoted ref.

0325   M. ISOBE, Y. HIROSE, K. SHIMOKAWA, T. NISHIKAWA, T. GOTO, Tetrahedron Lett., 1990, **31**, 5499

0326   K. REIN, M. GOICOECHEA-PAPPAS, T. V. ANKLEKAR, G. C. HART, G. A. SMITH, R. E. GAWLEY, J. Am. Chem. Soc., 1989, **111**, 2211

0327   D. ROMO, A. I. MEYERS, Tetrahedron, 1991, **47**, 9503

0328   Y. UKAJI, K. YAMAMOTO, M. FUKUI, T. FUJISAWA, Tetrahedron Lett., 1991, **32**, 2919

0329   E. L. ELIEL, X. C. HE, J. Org. Chem., 1990, **55**, 2114; Tetrahedron, 1987, **43**, 4979

0330   N. KATAGIRI, H. AKATSUKA, C. KANEKO, A. SERA, Tetrahedron Lett., 1988, **29**, 5397

0331   C. KOUKLOVSKY, A. POUILHES, Y. LANGLOIS, J. Am. Chem. Soc., 1990, **112**, 6672

0332   A. ALBEROLA, C. ANDRÉS, R. PEDROSA, Synlett, 1990, 763

0333   A. BERNARDI, M. CAVICCHIOLI, G. POLI, C. SCOLASTICO, A. SIDJIMOV, Tetrahedron, 1991, **47**, 7925

0333a A. BERNARDI, S. CARDANI, G. POLI, D. POTENZA, C. SCOLASTICO, Tetrahedron, 1992, **48**, 1343

0334 N. A. PORTER, J. D. BRUHNKE, W. X. WU, I. J. ROSENSTEIN, R. A. BREYER, J. Am. Chem. Soc., 1991, **113**, 7788

0335 A. PELTER, K. SMITH, in Comprehensive Organic Synthesis, B. M. TROST and I. FLEMING Ed., Pergamon Press, 1991, vol. 7, chap. 4.1

0336 L. S. HEGEDUS, R. IMWINKELRIED, M. ALARID-SARGENT, D. DVORAK, Y. SATOH, J. Am. Chem. Soc., 1990, **112**, 1109

0337 L. S. HEGEDUS, J. MONTGOMERY, Y. NARUKAWA, D. C. SNUSTAD, J. Am. Chem. Soc., 1991, **113**, 5784

0338 S. R. PULLEY, L. S. HEGEDUS, J. Am. Chem. Soc., 1993, **115**, 9037

0339 T. G. GANT, A. I. MEYERS, Tetrahedron, 1994, **50**, 2297

0340 A. J. ROBICHAUD, A. I. MEYERS, J. Org. Chem., 1991, **56**, 2607

0341 A. M. WARSHAWSKY, A. I. MEYERS, J. Am. Chem. Soc., 1990, **112**, 8090

0342 A. I. MEYERS, M. SHIPMAN, J. Org. Chem., 1991, 56, 7098

0343 R. C. GUPTA, C. M. RAYNOR, A. M. Z. SLAWIN, D. J. WILLIAMS, J. Chem. Soc. Perkin I, 1988, 1773

0344 T. R. KELLY, A. ARVANITIS, Tetrahedron Lett., 1984, **25**, 39

0345 P. ZANG, R. E. GAWLEY, Tetrahedron Lett., 1992, **33**, 2945

0346 T. BERRANGER, C. ANDRÉ-BARRÈS, M. KOBAYAKAWA, Y. LANGLOIS, Tetrahedron Lett., 1993, **34**, 5079

0347 F. MICHELON, A. POUILHES, N. VAN BAC, N. LANGLOIS, Tetrahedron Lett., 1992, **33**, 1743

0348 A. I. MEYERS, D. BERNEY, Org. Synth., 1990, **69**, 55

0349 L. E. BURGESS, A. I. MEYERS, J. Am. Chem. Soc., 1991, **113**, 9858

0350 A. I. MEYERS, L. E. BURGESS, J. Org. Chem., 1991, **56**, 2294

0351 L. F. TIETZE, S. BRAND, T. PFEIFFER, J. ANTEL, K. HARMS, G. M. SHELDRICK, J. Am. Chem. Soc., 1987, **109**, 921

0352 B. M. TROST, B. YANG, M. L. MILLER, J. Am. Chem. Soc., 1989, **111**, 6482

0353 A. G. MYERS, K. L. WIDDOWSON, J. Am. Chem. Soc., 1990, **112**, 9672

0354 S. E. DENMARK, R. L. DOROW, J. Org. Chem., 1990, **55**, 5926

0355 S. E. DENMARK, J. E. MARLIN, J. Org. Chem., 1987, **52**, 5742

0356 S. E. DENMARK, C. T. CHEN, J. Am. Chem. Soc., 1992, **114**, 10674

0357 R. M. WILLIAMS, J. A. HENDRIX, Chem. Rev., 1992, **92**, 889

0358 P. CINTAS, Tetrahedron, 1991, **47**, 6079

0359 A. CHOUDHURY, R. W. FRANCK, R. B. GUPTA, Tetrahedron Lett., 1989, **30**, 4921

0360 H. KUNZ, W. PFRENGLE, K. RUCK, W. SAGER, Synthesis, 1991, 1039

0361 S. LASCHAT, H.KUNZ, J. Org. Chem., 1991, **56**, 5883

0362 A. LUBINEAU, Y. QUENEAU, J. Org. Chem., 1987, **52**, 1001; Tetrahedron, 1989, **45**, 6697

0363 D. S. LARSEN, R. J. STOODLEY, J. Chem. Soc. Perkin I, 1990, 1339

0364 C. MARAZANO, S. YANNIC, Y. GENISSON, M. MEHMANDOUST, B. C. DAS, Tetrahedron Lett., 1990, **31**, 1995

0365 W. PFRENGLE, H. KUNZ, J. Org. Chem., 1989, **54**, 4261

0366 A. B. CHARETTE, C. MELLON, L. ROUILLARD, E. MALENFANT, Pure and Applied Chem., 1992, **64**, 1925

0366a G. I. GEORG, P. M. MASHAVA, E. AKGUN, M. W. MILSTEAD, Tetrahedron Lett., 1991, **32**, 3151

0367 A. MORI, D. YU, S. INOUE, Synlett, 1992, 427

0368 B. T. CHO, Y. S. CHUN, Tetrahedron Asymm., 1992, **3**, 1583

0369 A. B. CHARETTE, B. CÔTÉ, Tetrahedron Asymm., 1993, **4**, 2283

0370 H. BRAUN, H. FELBER, G. KRESSE, A. RITTER, F. P. SCHMIDTCHEN, A. SCHNEIDER, Tetrahedron, 1991, **47**, 3313

0371 H. FELBER, G. KRESZE, H. BRAUN, A. VASELLA, Tetrahedron Lett., 1984, **25**, 5381, Helv. Chim. Acta, 1986, **69**, 1137

0372 W. OPPOLZER, Angew. Chem. Int. Ed. Engl., 1984, **23**, 876 and quoted ref.

0373 T. HARADA, H. NAKAJIMA, T. OHNISHI, M. TAKEUCHI, A. OKU, J. Org. Chem., 1992, **57**, 720

0374 T. HARADA, T. YOSHIDA, Y. KAGAMIHARA, A. OKU, J. Chem. Soc. Chem. Comm., 1993, 1367

0375 W. H. PEARSON, M. C. CHENG, J. Org. Chem., 1986, **51**, 3746; 1987, **52**, 3176

0376 S. KANEMASA, A. TATSUKAWA, E. WADA, J. Org. Chem., 1991, **56**, 2875

0377 F. A. DAVIS, P. ZHOU, P. J. CARROLL, J. Org. Chem., 1993, **58**, 4890

0378 M. EL HADRAMI, J. P. LAVERGNE, P. VIALLEFONT, Tetrahedron Lett., 1991, **32**, 3985

0379 F. OUAZZANI, M. L. ROUMESTANT, P. VIALLEFONT, Tetrahedron Asymm., 1991, **2**, 913

0380 M. TABCHEH, A. EL ACHQAR, L. PAPPALARDO, M. L. ROUMESTANT, P. VIALLEFONT, Tetrahedron, 1991, **47**, 4611

0381 Y. N. BELOKON,V. A. BAKHMUTOV,N. I. CHERNOGLAZOVA, K. A. KOTCHEKOV,S. V. VITT, N. S. GABALINSKAYA, V. M. BELIKOV, J. Chem. Soc. Perkin I, 1988, 305

0382 V. A. SOLOSHONOK, Y. N. BELOKON, N. A. KUZMINA, V. I. MALEEV, N. Y. SVISTUNOVA, V. A. SOLENDENKO, V. P. KUKHAR, J. Chem. Soc. Perkin I, 1992, 1525

0383 Y. N. BELOKON, A. G. BULYCHEV, V. A. PAVLOV, E. B. FEDOROVA, V. A. TSYRYAPKIN, V. A. BAKHMUTOV, V. M. BELIKOV, J. Chem. Soc. Perkin I, 1988, 2075

0384 Y. N. BELOKON,A. S. SAGYAN, S. A. DJAMGARYAN, V. A. BAKHMUTOV, S. V. VITT, A. S. BATSANOV, Y. T. STRUCHKOV, V. M. BELIKOV,J. Chem. Soc. Perkin I, 1990, 2301
0385 S. H. MASHRAQUI, R. M. KELLOGG, J. Org. Chem., 1984, **49**, 2513
0386 D. SEEBACH, R. IMWINKELRIED, G. STUCKY, Helv. Chim. Acta, 1987, **70**, 448
0387 C. SIEGEL, E. R. THORNTON, Tetrahedron Lett., 1988, **29**, 5225
0388 R. L. BEARD, A. I. MEYERS, J. Org. Chem., 1991, **56**, 2091
0389 L. GOTTLIEB, A. I. MEYERS, J. Org. Chem., 1990, **55**, 5659
0390 A. I. MEYERS, J. GUILES, Tetrahedron Lett., 1990, **31**, 2813
0391 A. I. MEYERS, M. A. GONZALEZ, V. STRUZKA, A. AKAHANE, J. GUILES, J. S. WARMUS, Tetrahedron Lett., 1991, **32**, 5501
0392 A. I. MEYERS, J. S. WARMUS, M. A. GONZALEZ, J. GUILES, A. AKAHANE, Tetrahedron Lett., 1991, **32**, 5509
0393 D. A. DICKMAN, M. BOES, A. I. MEYERS, Org. Synth., 1988, **67**, 52
0394 A. I. MEYERS, M. BOES, D. A. DICKMAN, Org. Synth., 1988, **67**, 60
0395 J. R. GAGE, D. A. EVANS, Org. Synth., 1989, **68**, 77
0396 C. THOM, P. KOCIENSKI, Synthesis, 1992, 582
0397 S. G. DAVIES, G. J. M. DOISNEAU, Tetrahedron Asymm., 1993, **4**, 2513
0398 A. ABDEL-MAGID, L. N. PRIDGEN, D. S. EGGLESTON, I. LANTOS, J. Am. Chem. Soc., 1986, **108**, 4595
0399 D. A. EVANS, E. B. SJOGREN, A. E. WEBER, R. E. CONN, Tetrahedron Lett., 1987, **28**, 39
0400 R. E. GAWLEY, K. REIN, in Comprehensive Organic Synthesis, B. M. TROST and I. FLEMING Ed., Pergamon Press, 1991, vol. 3, chap. 1.2
0401 M. NERZ-STORMES, E. R. THORNTON, J. Org. Chem., 1991, **56**, 2489; Tetrahedron Lett., 1986, **27**, 897
0402 L. N. PRIDGEN, A. ABDEL-MAGID, I. LANTOS, Tetrahedron Lett., 1989, **30**, 5539
0403 S. SHIRODKAR, M. NERZ-STORMES, E. R. THORNTON, Tetrahedron Lett., 1990, **31**, 4699
0403a D. A. EVANS, J. S. CLARK, R. METTERNICH, V. J. NOVACK, G. S. SHEPPARD, J. Am. Chem. Soc., 1990, **112**, 866
0404 D. A. EVANS, D. L. RIEGER, M. T. BILODEAU, F. URPI, J. Am. Chem. Soc., 1991, **113**, 1047
0405 D. A. EVANS, A. E. WEBER, J. Am. Chem. Soc., 1987, **109**, 7151
0406 J. R. GAGE, D. A. EVANS, Org. Synth., 1989, **68**, 83
0407 B. M. KIM, S. F. WILLIAMS, S. MASAMUNE, in Comprehensive Organic Synthesis, B. M. TROST and I. FLEMING Ed., Pergamon Press, 1991 vol. 2, chap. 1.7
0408 I. PATERSON, in Comprehensive Organic Synthesis, B. M. TROST and I. FLEMING Ed., Pergamon Press, 1991 vol. 2, chap. 1.8

0409 T. ROSEN, in Comprehensive Organic Synthesis, B. M. TROST and I. FLEMING Ed., Pergamon Press, 1991, vol. 2, chap. 1.13
0410 D. A. EVANS, T. C. BRITTON, R. L. DOROW, J. F. DELLARIA, J. Am. Chem. Soc., 1986, **108**, 6395
0411 D. A. EVANS, T. C. BRITTON, J. A. ELLMAN, R. L. DOROW, J. Am. Chem. Soc., 1990, **112**, 4011
0412 L. A.TRIMBLE, J. C. VEDERAS, J. Am. Chem. Soc., 1986, **108**, 6397
0413 D. A. EVANS, M. T. BILODEAU, T. C. SOMERS, J. CLARDY, D. CHERRY, Y. KATO, J. Org. Chem., 1991, **56**, 5750
0414 D. A. EVANS, E. B. SJOGREN, J. BARTROLI, R. L. DOW, Tetrahedron Lett., 1986, **27**, 4957
0415 O. MIYATA, T. SHINADA, I. NINOMIYA, T. NAITO, Tetrahedron Lett., 1991, **32**, 3519
0416 K. RÜCK, H. KUNZ, Synthesis, 1993, 1018
0417 F. M. HAUSER, R. A. TOMMASI, J. Org. Chem., 1991, **56**, 5758
0418 M. J. MARTINELLI, J. Org. Chem., 1990, **55**, 5065
0419 B. B. SNIDER, Q. ZHANG, J. Org. Chem., 1991, **56**, 4908
0420 D. A. EVANS, J. C. ANDERSON, M. K. TAYLOR, Tetrahedron Lett., 1993, **34**, 5563
0421 J. W. FISHER, J. M. DUNIGAN, L. D. HATFIELD, R. C. HOYING, J. E. RAY, K. L. THOMAS, Tetrahedron Lett., 1993, **34**, 4755
0421a B. T. O'NEILL, in Comprehensive Organic Synthesis, B. M. TROST and I. FLEMING Ed., Pergamon Press, 1991, vol. 1, chap. 1.13
0422 A. M. KLIBANOV, Acc. Chem. Res., 1990, **23**, 114
0423 E. J. COREY, G. B. JONES, Tetrahedron Lett., 1991, **32**, 5713
0424 A. K. GHOSH, T. T. DUONG, S. P. McKEE, J. Chem. Soc. Chem. Comm., 1992, 1673
0425 M. R. BANKS, A. J. BLAKE, A. R. BROWN, J. I. G. CADOGAN, S. GAUR, I. GOSNEY, P. K. G. HODGSON, P. THORBURN, Tetrahedron Lett., 1994, **35**, 489
0426 M. P. BONNER, E. R. THORNTON, J. Am. Chem. Soc., 1991, **113**, 1299
0427 T. H. YAN, V. V. CHU, T. C. LIN, C. H. WU, L. H. LIU, Tetrahedron Lett., 1991, **32**, 4959
0428 T. H. YAN, V. V. CHU, T. C. LIN, W. H. TSENG, T. W. CHENG, Tetrahedron Lett., 1991, **32**, 5563
0429 T. H. YAN, C. W. TAN, H. C. LEE, H. C. LO, T. Y. HUANG, J. Am. Chem. Soc., 1993, **115**, 2613
0430 N. HASHIMOTO, T. ISHIZUKA, T. KUNIEDA, Tetrahedron Lett., 1994, **35**, 721
0431 K. H. AHN, S. LEE, A. LIM, J. Org. Chem., 1992, **57**, 5065
0432 K. H. AHN, A. LIM, S. LEE, Tetrahedron Asymm., 1993, **4**, 2435
0433 N. Z. HUANG, V. J. KALISH, M. J. MILLER, Tetrahedron, 1990, **46**, 8067

0434 C. L. HSIAO, L. LIU, M. J. MILLER, J. Org. Chem., 1987, **52**, 2201
0435 C. MA, M. J. MILLER, Tetrahedron Lett., 1991, **32**, 2577
0436 T. H. YAN, H. C. LEE, C. W. TAN, Tetrahedron Lett., 1993, **34**, 3559
0437 Y. NAGAO, W. M. DAI, M. OCHIAI, S. TSUKAGOSHI, E. FUJITA, J. Am. Chem. Soc., 1988, **110**, 289
0438 Y. NAGAO, Y. HAGIWARA, T. KUMAGAI, M. OCHIAI, T. INOUE, K. HASHIMOTO, E. FUJITA, J. Org. Chem., 1986, **51**, 2393
0439 D. BLASER, S. Y. KO, D. SEEBACH, J. Org. Chem., 1991, **56**, 6230
0440 D. SEEBACH, E. DZIADULEWICZ, L. BEHRENDT, S. CANTOREGGI, R. FITZI, Liebigs Ann., 1989, 1215
0441 D. BLASER, D. SEEBACH, Liebigs Ann., 1991, 1067
0442 D. SEEBACH, H. M. BURGER, C. P. SCHICKLI, Liebigs Ann., 1991, 669
0443 K. SUZUKI, D. SEEBACH, Liebigs Ann., 1992, 51
0444 R. K. BOECKMAN, S. G. NELSON, M. D. GAUL, J. Am. Chem. Soc., 1992, **114**, 2258
0445 J. G. STACK, D. P. CURRAN, J. REBEK, P. BALLESTER, J. Am. Chem. Soc., 1991, **113**, 5918
0446 J. A. STACK, T. A. HEFFNER, S. J. GEIB, D. P. CURRAN, Tetrahedron, 1993, **49**, 995
0447 G. CARDILLO, A. D'AMICO, M. ORENA, S. SANDRI, J. Org. Chem., 1988, **53**, 2354
0448 M. ORENA, G. PORZI, S. SANDRI, Tetrahedron Lett., 1992, **33**, 3797
0449 S. G. DAVIES, A. A. MORTLOCK, Tetrahedron Lett., 1991, **32**, 4787; 4791 and Tetrahedron Asymm., 1991, **2**, 1001; S. G. DAVIES, A. J. EDWARDS, G. B. EVANS, A. A. MORTLOCK, Tetrahedron, 1994, **50**, 6621
0450 S. E. DREWES, D. G. S. MALISSAR, G. H. P. ROOS, Chem. Ber., 1993, **126**, 2663
0451 K. SMITH, A. PELTER, in Comprehensive Organic Synthesis, B. M. TROST and I. FLEMING Ed., Pergamon Press, 1991, vol. 8, chap. 3.10
0452 S. G. DAVIES, A. A. MORTLOCK, Tetrahedron, 1993, **49**, 4419
0453 E. JUARISTI, J. ESCALANTE, J. Org. Chem., 1993, **58**, 2282
0454 B. H. KIM, D. P. CURRAN, Tetrahedron, 1993, **49**, 293
0455 M. C. WEISMILLER, J. C. TOWSON, F. A. DAVIS, Org. Synth., 1990, **69**, 154
0456 H. JOSIEN, A. MARTIN, G. CHASSAING, Tetrahedron Lett., 1991, **32**, 6547
0457 W. OPPOLZER, J. BLAGG, I. RODRIGUEZ, E. WALTHER, J. Am. Chem. Soc., 1990, **112**, 2767
0458 W. OPPOLZER, H. BIENAYME, A. GENEVOIS-BORELLA, J. Am. Chem. Soc., 1991, **113**, 9660
0459 W. OPPOLZER, C. STARKEMANN, I. RODRIGUEZ, G. BERNARDINELLI, Tetrahedron Lett., 1991, **32**, 61

0460 W. OPPOLZER, O. TAMURA, Tetrahedron Lett., 1990, **31**, 991
0461 J. VALLGARDA, U. HACKSELL, Tetrahedron Lett., 1991, **32**, 5625
0462 B. H. KIM, J. Y. LEE, Tetrahedron Asymm., 1991, **2**, 1359
0463 W. OPPOLZER, J. P. BARRAS, Helv. Chim. Acta, 1987, **70**, 1666
0464 W. OPPOLZER, G. POLI, C. STARKEMANN, G. BERNARDINELLI, Tetrahedron Lett., 1988, **29**, 3559
0465 W. OPPOLZER, R. J. MILLS, M. REGLIER, Tetrahedron Lett., 1986, **27**, 183
0466 T. BAUER, C. CHAPUIS, J. KOZAK, J. JURCZAK, Helv. Chim. Acta, 1989, **72**, 482
0467 V. GOUVERNEUR, G. DIVE, L. GHOSEZ, Tetrahedron Asymm., 1991, **2**, 1173
0468 W. OPPOLZER, M. WILLS, C. STARKEMANN, G. BERNARDINELLI, Tetrahedron Lett., 1990, **31**, 4117
0469 W. OPPOLZER, I. RODRIGUEZ, C. STARKEMANN, E. WALTHER, Tetrahedron Lett., 1990, **31**, 5019
0470 W. OPPOLZER, M. WILLS, M. J. KELLY, M. SIGNER, J. BLAGG, Tetrahedron Lett., 1990, **31**, 5015
0471 W. OPPOLZER, A. J. KINGMA, S. K. PILLAI, Tetrahedron Lett., 1991, **32**, 4893
0472 G. SOLLADIÉ, HOUBEN-WEYL under the press
0473 M. C. CARRENO, J. L. GARCIA-RUANO, A. M. MARTIN, C. PEDREGAL, J. H. RODRIGUEZ, A. RUBIO, J. SANCHEZ, G. SOLLADIE, J. Org. Chem., 1990, **55**, 2120
0474 A. SOLLADIE-CAVALLO, M. C. SIMON-WERMEISTER, D. FARKHANI, Helv. Chim. Acta, 1991, **74**, 390
0475 M. CINQUINI, A. MANFREDI, M. MOLINARI, A. RESTELLI, Tetrahedron, 1985, **41**, 4929
0475a R. ANNUNZIATA, M. CINQUINI, F. COZZI, A. GILARDI, A. RESTELLI, J. Chem. Soc. Perkin I, 1985, 2289
0476 R. LOPEZ, J. C. CARRETERO, Tetrahedron Asymm., 1991, **2**, 93
0477 D. H. HUA, S. N. BHARATHI, J. A. K. PANANGADAN, A. TSUJIMOTO, J. Org. Chem., 1991, **56**, 6998
0478 T. TAKAHASHI, H. KOTSUBO, T. KOIZUMI, Tetrahedron Asymm., 1991, **2**, 1035
0479 H. SAKURABA, S. USHIKI, Tetrahedron Lett., 1990, **31**, 5349
0480 S. G. PYNE, B. DIKIC, J. Chem. Soc. Chem. Comm., 1989, 826
0481 G. SOLLADIÉ, HOUBEN-WEYL under the press
0482 D. H. HUA, J. G. PARK, T. KATSUHIRA, S. N. BHARATI, J. Org. Chem., 1993, **58**, 2144
0483 M. C. CARRENO, J. L. GARCIA RUANO, M. C. MAESTRO, M. P. GONZÁLEZ, A. B. BUENO, L. FISCHER, Tetrahedron, 1993, **49**, 11009

0484 I. ALONSO, C. CARRETERO, J. L. GARCIA-RUANO, J. Org. Chem., 1993, **58**, 3231

0485 Y. ARAI, S. I. KUWAYAMA, Y. TAKEUCHI, T. KOIZUMI, Tetrahedron Lett., 1985, **26**, 6205

0486 C. MAIGNAN, A. GUESSOUS, F. ROUESSAC, Tetrahedron Lett., 1984, **25**, 1727

0487 C. MAIGNAN, A. GUESSOUS, F. ROUESSAC, Tetrahedron Lett., 1986, **27**, 2603; Bull. Soc. Chim. France, 1986, 837

0488 I. ALONSO, J. C.CARRETERO, J.L.GARCIA-RUANO, Tetrahedron Lett., 1991, **32**, 947

0489 H. KOSUGI, Y. MIURA, H. KANNA, H. UDA, Tetrahedron Asymm., 1993, **4**, 1409

0490 L. R. PAN, T. TOKOROYAMA, Tetrahedron Lett., 1992, **33**, 1469 and 1473

0491 T. TORU, Y. WATANABE, M. TSUSAKA, Y. UENO, J. Am. Chem. Soc., 1993, **115**, 10464

0492 Y. ARAI, Y. HAYASHI, M. YAMAMOTO, H. TAKAYEMA, T. KOIZUMI, J. Chem. Soc. Perkin I, 1988, 3133

0493 B. RONAN, H. B. KAGAN, Tetrahedron Asymm., 1991, **2**, 75

0494 B. RONAN, H. B. KAGAN, Tetrahedron Asymm., 1992, **3**, 115

0495 R. W. BAKER, G. R. POCOCK, M. V. SARGENT, E. TWISS, Tetrahedron Asymm., 1993, **4**, 2423

0496 I. FERNANDEZ, N. KHIAR, J. M. LLERA, F. ALCUDIA, J. Org. Chem., 1992, **57**, 6789

0497 R. K. HILL, in Asymmetric Synthesis J. D. MORRISON Ed., Academic Press, New York, 1984, **3**, 502

0498 P. A. WADE, D. T. COLE, S. G. D'AMBROSIO, Tetrahedron Lett., 1994, **35**, 53

0499 R. J. BUTLIN, I. D. LINNEY, D. J. CRITCHER, M. F. MAHON, K. C. MOLLOY, M. WILLS, J. Chem. Soc. Perkin I, 1993, 1581

0500 D. R. J. HOSE, T. RAYNHAM, M. WILLS, Tetrahedron Asymm., 1993, **4**, 2159

0501 M. WILLS, R. J. BUTLIN, I. D. LINNEY, Tetrahedron Lett., 1992, **33**, 5427

0502 F. DI FURIA, G. MODENA, Synthesis, 1984, 325

0503 H. B. KAGAN, F. REBIERE, Synlett, 1990, 643

0504 P. PITCHEN, E. DUNACH, M. N. DESHMUKH, H. B. KAGAN, J. Am. Chem. Soc., 1984, **106**, 8188

0505 S. H. ZHAO, O. SAMUEL, H. B. KAGAN, Tetrahedron, 1987, **43**, 513; Organic Synth. 1989, **68**, 49

0506 F. A. DAVIS, R. THIMMA REDDY, W. HAN, P. J. CARROLL J. Am. Chem. Soc., 1992, **114**, 1428

0507 F. REBIÈRE, O. SAMUEL, L. RICARD, H. B. KAGAN, J. Org. Chem., 1991, **56**, 5991

0508   F. REBIÈRE, O. RIANT, L. RICARD, H. B. KAGAN, Angew. Chem. Int. Ed. Engl., 1993, **32**, 568

0509   D. A. EVANS, M. M. FAUL, L. COLOMBO, J. J. BISAHA, J. CLARDY, D. CHERRY, J. Am. Chem. Soc., 1992, **114**, 5977

0510   F. A. DAVIS, R. T. REDDY, R. E. REDDY, J. Org. Chem., 1992, **57**, 6387

0511   F. A. DAVIS, R. E. REDDY, J. M. SZEWCZYK, P. S. PORTONOVO, Tetrahedron Lett., 1993, **34**, 6229

0512   D. H. HUA, S. W. MIAO, J. S. CHEN, S. IGUCHI, J. Org. Chem., 1991, **56**, 4

0513   M. R. BARBACHYN, C. R. JOHNSON, in Asymmetric Synthesis J. D. MORRISON Ed. Academic Press N. Y. 1983, **4**, 227

0514   C. R. JOHNSON, Aldrichimica Acta, 1985, **18**, 3

0514a I. ERDELMEIER, H. J. GAIS, H. J. LINDNER, Angew. Chem. Int. Ed. Engl., 1986, **25**, 935

0515   S. G. PYNE, J. Org. Chem., 1986, **51**, 81

0516   I. ERDELMEIER, H. J. GAIS, J. Am. Chem. Soc., 1989, **111**, 1125

0517   H. J. GAIS, G. BÜLOW, Tetrahedron Lett., 1992, **33**, 461

0518   H. J. GAIS, G. BÜLOW, Tetrahedron Lett., 1992, **33**, 465

0519   J. BUND, H. J. GAIS, I. ERDELMEIER, J. Am. Chem. Soc., 1991, **112**, 1142

0520   B. M. TROST, R. T. MATSUOKA, Synlett, 1992, 27

0521   S. C. CASE-GREEN, J. F. COSTELLO, S. G. DAVIES, N. HEATON, C. J. R. HEDGECOCK, J. C. PRIME, J. Chem. Soc. Chem. Comm., 1993, 1621

0522   S. G. DAVIES, Aldrichimica Acta, 1990, **23**, 31

0523   S. DAVIES, R. POLYWKA, P. WARNER, Tetrahedron, 1990, **46**, 4847

0524   L. S. LIEBESKIND, M. E. WELKER, R. W. FENGL, J. Am. Chem. Soc., 1986, **108**, 6328

0525   M. BROOKHART, Y. LIU, E. W. GOLDMAN, D. A. TIMMERS, G. D. WILLIAMS, J. Am. Chem. Soc., 1991, **113**, 927

0526   M. FRANCK-NEUMANN, Organometallics in Organic Synthesis, A. de MEIJERE, H. TOM DIECK Ed. Springer-Verlag, Berlin, 1987, 247

0527   P. PINSARD, J. P. LELLOUCHE, J. P. BEAUCOURT, R. GREE, Tetrahedron Lett., 1990, **31**, 1137

0528   R. GREE, Synthesis, 1989, 341

0529   M. FRANCK-NEUMANN, D. MARTINA, M. P. HEITZ, Tetrahedron Lett., 1989, **30**, 6679

0530   M. FRANCK-NEUMANN, C. BRISWALTER, P. CHEMLA, D. MARTINA, Synlett, 1990, 637

0531   M. LAABASSI, R. GREE, Tetrahedron Lett., 1989, **30**, 6683

0532   P. PINSARD, J. P. LELLOUCHE, J. P. BEAUCOURT, R. GREE, Tetrahedron Lett., 1990, **31**, 1141

0533 M. FRANCK-NEUMANN, P. CHEMLA, D. MARTINA, Synlett, 1990, 641

0534 M. LAABASSI, L. TOUPET, R. GREE, Bull. Soc. Chim. France, 1992, 47

0535 A. TENIOU, L. TOUPET, R. GREE, Synlett, 1991, 195

0536 K. NUNN, P. MOSSET, R. GREE, R. W. SAALFRANK, Angew. Chem. Int. Ed. Engl., 1988, **27**, 1188

0537 M. FRANCK-NEUMANN, A. ABDALI, P. J. COLSON, M. SEDRATI, Synlett, 1991, 331

0538 A. GIGOU, J. P. BEAUCOURT, J. P. LELLOUCHE, R. GREE, Tetrahedron Lett., 1991, **32**, 635

0539 A. SOLLADIÉ-CAVALLO, Adv. Metal Org. Chem., 1989, **1**, 99

0540 A. SOLLADIÉ-CAVALLO, Trends in Organic Chemistry, 1991, **1**, 237

0541 H. G. SCHMALZ, J. HOLLANDER, M. ARNOLD, G. DÜRNER, Tetrahedron Lett., 1993, **34**, 6259

0542 S. G. DAVIES, J. Organomet. Chem., 1990, **400**, 223

0543 M. C. SENECHAL-TOCQUER, D. SENECHAL, J. Y. LE BIHAN, D. GENTRIC, B. CARO, Bull. Soc. Chim. France, 1992, 121

0544 C. BALDOLI, P. DEL BUTTERO, S. MAIORANA, Tetrahedron, 1990, **46**, 7823

0545 C. MUKAI, W. J. CHO, I. J. KIM, M. KIDO, M. HANAOKA, Tetrahedron, 1991, **47**, 3007

0546 C. MUKAI, M. MIYAKAWA, A. MIHIRA, M. HANAOKA, J. Org. Chem., 1992, **57**, 2034

0547 E. P. KÜNDIG, L. H. XU, P. ROMANENS, G. BERNARDINELLI, Tetrahedron Lett., 1993, **34**, 7049

0548 M. UEMURA, Y. HAYASHI, Y. HAYASHI, Tetrahedron Asymm., 1993, **4**, 2291

0549 C. BALDOLI, P. DEL BUTTERO, J. Chem. Soc. Chem. Comm., 1991, 982

0550 C. MUKAI, I. J. KIM, W. J. CHO, M. KIDO, M. HANAOKA, J. Chem. Soc. Perkin I, 1993, 2495

0551 H. B. KAGAN, J. C. FIAUD, Topics in Stereochemistry, E. L. ELIEL and S. WILEN Ed., 1978, **10**, 175

0552 L. DUHAMEL, P. DUHAMEL, J. C. LAUNAY, J. C. PLAQUEVENT, Bull. Soc. Chim. France, 1984, 421

0553 K. MATSUMOTO, H. OHTA, Tetrahedron Lett., 1991, **32**, 4729

0554 R. D. LARSEN, E. G. CORLEY, P. DAVIS, P. J. REIDER, E. J. J. GRABOWSKI, J. Am. Chem. Soc., 1989, **111**, 7650

0555 C. FEHR, J. GALINDO, J. Am. Chem. Soc., 1988, **110**, 6909

0556 E. VEDEJS, N. LEE, J. Am. Chem. Soc., 1991, **113**, 5483

0557 P. J. COX, N. S. SIMPKINS, Tetrahedron Asymm., 1991, **2**, 1

0558 E. JUARISTI, A. K. BECK, J. HANSEN, T. MATT, T. MUKHOPADHYAY, M. SIMSON, D. SEEBACH, Synthesis, 1993, 1271

0559 K. TOMIOKA, Synthesis, 1990, 541

0560 T. HONDA, N. KIMURA, M. TSUBUKI, Tetrahedron Asymm., 1993, **4**, 1475

0561 D. SATO, H. KAWASAKI, I. SHIMADA, Y. ARATA, K. OKAMURA, T. DATE, K. KOGA, J. Am. Chem. Soc., 1992, **114**, 761

0562 A. J. EDWARDS, S. HOCKEY, F. S. MAIR, P. R. RAITHBY, R. SNAITH, N. G. SIMPKINS, J. Org. Chem., 1993, **58**, 6942

0563 J. M. HAWKINS, T. A. LEWIS, J. Org. Chem., 1992, **57**, 2114; J. Org. Chem., 1994, **59**, 649

0564 S. T. KERRICK, P. BEAK, J. Am. Chem. Soc., 1991, **113**, 9708

0565 J. A. MARSHALL, X. J. WANG, J. Org. Chem., 1992, **57**, 2747

0566 M. NISHIZAWA, R. NOYORI, in Comprehensive Organic Synthesis, B. M. TROST and I. FLEMING Ed., Pergamon Press, 1991, vol. 8, chap. 1.7

0567 J. A. MARSHALL, X. J. WANG, J. Org. Chem., 1991, **56**, 3211, 4913

0568 J. M. CHONG, E. K. MAR, J. Org. Chem., 1991, **56**, 893

0569 J. A. MARSHALL, G. P. LUKE, J. Org. Chem., 1991, **56**, 483

0570 J. A. MARSHALL, G. S. WELMAKER, B. W. GUNG, J. Am. Chem. Soc., 1991, **113**, 647

0571 R. NOYORI, Chem. Soc. Rev., 1989, **18**, 187

0572 K. YAMAMOTO, K. UENO, K. NAEMURA, J. Chem. Soc. Perkin I, 1991, 2607

0573 R. RAWSON, A. MEYERS, J. Chem. Soc. Chem. Comm., 1992, 494

0574 H. C. BROWN, B. T. CHO, W. S. PARK, J. Org. Chem., 1988, **53**, 1231

0575 M. YATAGAI, T. OHNUKI, J. Chem. Soc. Perkin I, 1990, 1826

0575a S. ITSUNO, Y. SAKURAI, K. SHIMIZU, K. ITO, J. Chem. Soc. Perkin I, 1990, 1859

0576 U. LEUTENEGGER, A. MADIN, A. PFALZ, Angew. Chem. Int. Ed. Engl., 1989, **28**, 60

0577 P. von MATT, A. PFALTZ, Tetrahedron Asymm., 1991, **2**, 691

0578 H. U. BLASER, Tetrahedron Asymm., 1991, **2**, 843

0579 H. SAKURABA, N. INOMATA, Y. TANAKA, J. Org. Chem., 1989, **54**, 3482

0580 H. C. BROWN, P. K. JADHAV, A. K. MANDAL, Tetrahedron, 1981, **37**, 3547

0581 M. M. MIDLAND, A. KAZUBSKI, R. E. WOODLING, J. Org. Chem., 1991, **56**, 1068

0582 P. V. RAMACHANDRAN, H. C. BROWN, S. SWAMINATHAN, Tetrahedron Asymm., 1990, **1**, 433

0583 H. C. BROWN, B. SINGARAM, Acc. Chem. Res., 1988, **21**, 287

0584 M. M. MIDLAND, Chem. Rev., 1989, **89**, 1553

0585 H. C. BROWN, R. R. IYER, V. K. MAHINDROO, N. G. BHAT, Tetrahedron Asymm., 1991, 2, 277

0586 H. C. BROWN, N. N. JOSHI, C. PYUN, B. SINGARAM, J. Am. Chem. Soc., 1989, 111, 1754

0587 H. C. BROWN, A. M. SALUNKHE, B. SINGARAM, J. Org. Chem., 1991, 56, 1170

0588 N. K. CHADHA, A. D. BATCHO, P. C. TANG, L. F. COURTNEY, C. M. COOK, P. M. WOVKULICH, M. R. USKOKOVIC, J. Org. Chem., 1991, 56, 4714

0589 H. C. BROWN, P. V. RAMACHANDRAN, Acc. Chem. Res., 1992, 25, 16

0590 E. T. EVERHART, J. C. CRAIG, J. Chem. Soc. Perkin I, 1991, 1701

0591 R. M. KELLOG, in Comprehensive Organic Synthesis, B. M. TROST and I. FLEMING Ed., Pergamon Press, 1991, vol. 8, chap. 1.3

0592 H. C. BROWN, P. V. RAMACHANDRAN, A. V. THEODOROVIC, S. SWAMINATHAN, Tetrahedron Lett., 1991, 32, 6691

0593 H. C. BROWN, P. V. RAMACHANDRAN, J. Org. Chem., 1989, 54, 4504

0594 D. S. MATTESON, Acc. Chem. Res., 1988, 21, 294

0595 M. V. RANGAISHENVI, B. SINGARAM, H. C. BROWN, J. Org. Chem., 1991, 56, 3286

0596 S. ITSUNO, M. NAKANO, K. MIYAZAKI, M. MASUDA, K. ITO, A. HIRAO, S. NAKAHAMA, J. Chem. Soc. Perkin I, 1985, 2039

0597 E. J. COREY, Pure and Applied Chem., 1990, 62, 1209

0598 E. J. COREY, X. M. CHENG, K. A. CIMPRICH, S. SARSHAR, Tetrahedron Lett., 1991, 32, 6835

0599 E. J. COREY, M. KIGOSHI, Tetrahedron Lett., 1991, 32, 5025

0600 E. J. COREY, K. S. RAO, Tetrahedron Lett., 1991, 32, 4623

0601 L. DELOUX, M. SREBNIK, Chem. Rev., 1993, 93, 763

0602 A. LOUPY, G. BRAM, J. SANSOULET, New J. Chem., 1992, 16, 233

0603 E. J. COREY, M. AZIMIORA, S. SARSHAR, Tetrahedron Lett., 1992, 33, 3429

0604 D. J. MATHRE, A. S. THOMPSON, A. W. DOUGLAS, K. HOOGSTEEN, J. D. CARROLL, E. G. CORLEY, E. J. J. GRABOWSKI, J. Org. Chem., 1993, 58, 2880

0605 G. J. QUALLICH, T. M. WOODALL, Synlett, 1993, 929

0606 E. J. COREY, J. O. LINK, J. Am. Chem. Soc., 1992, 114, 1906

0607 E. J. COREY, J. O. LINK, Tetrahedron Lett., 1989, 30, 6275

0608 E. J. COREY, J. O. LINK, J. Org. Chem., 1991, 56, 442

0609 E. J. COREY, J. O. LINK, Tetrahedron Lett., 1992, 33, 4141

0610 T. K. JONES, J. J. MOHAN, L. C. XAVIER, T. J. BLACKLOCK, D. J. MATHRE, P. SOHAR, E. T. TURNER-JONES, R. A. REAMER, F. E. ROBERTS, E. J. J. GRABOWSKI, J. Org. Chem., 1991, 56, 763

0611   B. B. LOHRAY, V. BHUSHAN, Angew. Chem. Int. Ed. Engl., 1992, **31**, 729

0612   D. J. MATHRE, T. K. JONES, L. C. XAVIER, T. J. BLACKLOCK, R. A. REAMER, J. J. MOHAN, E. T. TURNER-JONES, K. HOOGSTEEN, M. W. BAUM, E. J. J. GRABOWSKI, J. Org. Chem., 1991, **56**, 751

0613   Y. H. KIM, D. H. PARK, I. S. BYUN, I. K. YOON, C. S. PARK, J. Org. Chem., 1993, **58**, 4511

0614   G. J. QUALLICH, T. M. WOODALL, Tetrahedron Lett., 1993, **34**, 4145

0615   J. M. BRUNEL, M. MAFFEI, G. BUONO, Tetrahedron Asymm., 1993, **4**, 2255

0616   E. DIDIER, B. LOUBINOUX, G. M. RAMOS-TOMBO, G. RIHS, Tetrahedron, 1991, **47**, 4941

0617   T. MEHLER, J. MARTENS, Tetrahedron Asymm., 1993, **4**, 1983

0618   B. BURNS, E. MERIFIELD, M. F. MAHON, K. C. MOLLOY, M. WILLS, J. Chem. Soc. Perkin I, 1993, 2243

0619   J. M. BRUNEL, O. PARDIGON, B. FAURE, G. BUONO, J. Chem. Soc. Chem. Comm., 1992, 287

0620   C. BOLM, M. FELDER, Tetrahedron Lett., 1993, **34**, 6041

0621   C. BOLM, A. SEGER, M. FELDER, Tetrahedron Lett., 1993, **34**, 8079

0622   B. T. CHO, Y. S. CHUN, J. Chem. Soc. Perkin I, 1990, 3200

0623   G. BRINGMANN, T. HARTUNG, Tetrahedron, 1993, **49**, 7891

0624   F. A. CAREY, R. J. SUNDBERG, Advanced Organic Chemistry 3rd edition, Plenum Press, New York 1991, Part B, **a**) chap. 8; **b**) chap. 5; **c**) chap. 6; **d**) chap. 9; **e**) chap. 1; **f**) chap. 10; **g**) Part A chap. 11; **h**) Part A chap. 2; **i**) Part B chap. 8

0625   V. A. BURGESS, S. G. DAVIES, R. T. SKERLJ, Tetrahedron Asymm., 1991, **2**, 299

0626   D. A. EVANS, S. G. NELSON, M. R. GAGNÉ, A. R. MUCI, J. Am. Chem. Soc., 1993, **115**, 9800

0627   Y. INOUYE, J. ODA, N. BABA, in Asymmetric Synthesis, J. D. MORRISON Ed. Academic Press, New York, 1983, **2**, 92

0628   J. P. VERSLEIJEN, M. S. SANDERS-HOVENS, S. A. vanOMMERIG, J. A. VEKEMANS, E. M. MEIJER, Tetrahedron, 1993, **49**, 7793

0629   J. A. J. M. VEKEMANS, J. P. J. VERSLEIJEN, H. M. BUCK, Tetrahedron Asymm., 1991, **2**, 949

0630   S. ZEHANI, J. LIN, G. GELBARD, Tetrahedron, 1989, **45**, 733

0631   S. ZEHANI, G. GELBARD, Reactive Polymers, 1987, **6**, 81

0632   T. AKIYAMA, M. SHIMIZU, T. MUKAIYAMA, Chem. Lett., 1984, 611

0633   M. NAKAJIMA, K. TOMIOKA, K. KOGA, Tetrahedron, 1993, **49**, 9735

0634   M. NAKAJIMA, K. TOMIOKA, K. KOGA, Tetrahedron, 1993, **49**, 9751

0635   K. TOMIOKA, M. SHINDO, K. KOGA, J. Am. Chem. Soc., 1989, **111**, 8266

0636 I. INOUE, M. SHINDO, K. KOGA, K. TOMIOKA, Tetrahedron Asymm., 1993, **4**, 1603
0637 K. TOMIOKA, I. INOUE, M. SHINDO, K. KOGA, Tetrahedron Lett., 1991, **32**, 3095
0638 K. TOMIOKA, I. INOUE, M. SHINDO, K. KOGA, Tetrahedron Lett., 1990, **31**, 6681
0639 M. LAUTENS, C. GAJDA, P. CHIU, J. Chem. Soc. Chem. Comm., 1993, 1193
0640 R. NOYORI, S. SUGA, K. KAWAI, S. OKADA, M. KITAMURA, N. OGUNI, M. HAYASHI, T. KANEKO, Y. MATSUDA, J. Organomet. Chem., 1990, **382**, 19
0641 M. HAYASHI, T. KANEKO, N. OGUNI, J. Chem. Soc. Perkin I, 1991, 25
0642 K. SOAI, S. NIWA, M. WATANABE, J. Chem. Soc. Perkin I, 1989, 109
0643 K. SOAI, M. WATANABE, Tetrahedron Asymm., 1991, **2**, 97
0644 C. BOLM, G. SCHINGLOFF, K. HARMS, Chem. Ber., 1992, **125**, 1191
0645 R. P. HOF, M. A. POELERT, N. C. M. W. PAPER, R. M. KELLOGG, Tetrahedron Asymm., 1994, **5**, 31
0646 W. BEHNEN, T. MEHLER, J. MARTENS, Tetrahedron Asymm., 1993, **4**, 1413
0647 E. J. COREY, F. J. HANNON, Tetrahedron Lett., 1987, **28**, 5233
0648 S. NIWA, K. SOAI, J. Chem. Soc. Perkin I, 1991, 2717
0649 G. CHELUCCI, S. CONTI, M. FALORNI, G. GIACOMELLI, Tetrahedron, 1991, **47**, 8251
0650 W. OPPOLZER, R. N. RADINOV, Tetrahedron Lett., 1991, **32**, 5777
0651 M. WATANABE, S. ARAKI, Y. BUTSUGAN, M. UEMURA, J. Org. Chem., 1991, **56**, 2218
0652 M. WATANABE, M. KOMOTA, M. NISHIMURA, S. ARAKI, Y. BUTSUGAN, J. Chem. Soc. Perkin I, 1993, 2193
0653 D. SEEBACH, D. A. PLATTNER, A. K. BECK, Y. M. WANG, D. HUNZIKER, W. PETTER, Helv. Chim. Acta, 1992, **75**, 2171
0654 B. SCHMIDT, D. SEEBACH, Angew. Chem. Int. Ed. Engl., 1991, **30**, 99
0655 B. SCHMIDT, D. SEEBACH, Angew. Chem. Int. Ed. Engl., 1991, **30**, 1321
0655a M. SABURI, M. OHNUKI, M. OGASAWARA, T. TAKAHASHI, Y. UCHIDA, Tetrahedron Lett., 1992, **33**, 5783
0656 M. J. BURK, J. E. FEASTER, Tetrahedron Lett., 1992, **33**, 2099
0657 J. F. G. A. JANSEN, B. L. FERINGA, J. Org. Chem., 1990, **55**, 4168
0658 Y. SAKITO, Y. YONEYOSHI, G. SUZUKAMO, Tetrahedron Lett., 1988, **29**, 223
0659 J. F. G. A. JANSEN, B. L. FERINGA, Tetrahedron Asymm., 1992, **3**, 581
0660 Y. UKAJI, M. NISHIMURA, T. FUJISAWA, Chem. Lett., 1992, 61

0661   D. SEEBACH, A. K. BECK, M. SCHIESS, L. WIDLER, A.
       WONNACOTT, Pure and Appl. Chem., 1983, **55**, 1807
0662   B. WEIDMANN, D. SEEBACH, Angew. Chem. Int. Ed. Engl., 1983, **22**,
       31
0663   D. SEEBACH, A. K. BECK, S. ROGGO, A. WONNACOTT,
       Chem.Ber., 1985, **118**, 3673
0664   M. T. REETZ, T. KÜKENHÖHNER, P. WEINIG, Tetrahedron Lett.,
       1986, **27**, 5711
0665   R. O. DUTHALER, A. HAFNER, M. RIEDIKER, Pure and Appl.
       Chem., 1990, **62**, 631
0666   R. O. DUTHALER, A. HAFNER, Chem. Rev., 1992, **92**, 807
0667   A. HAFNER, R. O. DUTHALER, R. MARTI, G. RIHS, P. ROTHE-
       STREIT, F. SCHWARZENBACH, J. Am. Chem. Soc., 1992, **114**, 2321
0668   H. TAKAHASHI, T. KAWAKITA, M. OHNO, M. YOSHIOKA, S.
       KOBAYASHI, Tetrahedron, 1992, **48**, 5691
0669   D. SEEBACH, L. BEHRENDT, D. FELIX, Angew. Chem. Int. Ed.
       Engl., 1991, **30**, 1008
0670   M. YOSHIOKA, T. KAWAKITA, M. OHNO, Tetrahedron Lett., 1989,
       30, 1657
0671   K. SOAI, Y. HIROSE, Y.OHNO, Tetrahedron Asymm., 1993, **4**, 1473
0672   M. J. ROZEMA, A. R. SIDDURI, P. KNOCHEL, J. Org. Chem., 1992,
       57, 1956
0673   D. SEEBACH, A. K. BECK, R. IMWINKELRIED, S. ROGGO, A.
       WONNACOTT, Helv. Chim. Acta, 1987, **70**, 954
0674   K. TANAKA, H. SUZUKI, J. Chem. Soc. Chem. Comm., 1991, 101
0675   K. TANAKA, H. USHIO, Y. KAWABATA, H. SUZUKI, J. Chem. Soc.
       Perkin I, 1991, 1445
0676   B. E. ROSSITER, M. EGUCHI, A. E. HERNANDEZ, D. VICKERS,
       Tetrahedron Lett., 1991, **32**, 3973
0677   A. ALEXAKIS, J. FRUTOS, P. MANGENEY, Tetrahedron Asymm.,
       1993, **4**, 2427
0678   A. ALEXAKIS, S. MUTTI, J. F. NORMANT, J. Am. Chem. Soc., 1991,
       113, 6332
0679   M. KANAI, K. KOGA, K. TOMIOKA, J. Chem. Soc. Chem. Comm.,
       1993, 1248
0680   K. CHIBALE, N. GREEVES, L. LYFORD, J. E. PEASE, Tetrahedron
       Asymm., 1993, **4**, 2407
0681   G. SOLLADIÉ, in Asymmetric Synthesis J. D. MORRISON Ed.
       Academic Press New York,, 1983, **2**, 157
0682   T. YURA, N. IWASAWA, T. MUKAIYAMA, Chem. Lett., 1988, 1021,
       1025
0683   R. O. DUTHALER, P. HEROLD, S. WYLER-HELFER, M. RIEDIKER,
       Helv. Chim. Acta, 1990, **73**, 659

0684 I. PATERSON, J. M. GOODMAN, M. A. LISTER, R. C. SCHUMANN, C. K. MC CLURE, R. D. NORCROSS, Tetrahedron, 1990, **46**, 4663

0685 H. C. BROWN, R. K. DHAR, K. GANESAN, B. SINGARAM, J. Org. Chem., 1992, **57**, 499

0686 H. C. BROWN, R. K. DHAR, K. GANESAN, B. SINGARAM, J. Org. Chem., 1992, **57**, 2716

0687 G. P. BOLDRINI, M. BORLOTTI, F. MANCINI, E. TAGLIAVINI, C. TROMBINI, A. UMANI-RONCHI, J. Org. Chem., 1991, **56**, 5820

0688 G. P. BOLDRINI, F. MANCINI, E. TAGLIAVINI, C. TROMBINI, A. UMANI-RONCHI, J. Chem. Soc. Chem. Comm., 1990, 1680

0689 M. T. REETZ, F. KUNISCH, P. HEITMAN, Tetrahedron Lett., 1986, **27**, 4721

0690 M. T. REETZ, E. RIVADENEIRA, C. NIEMEYER, Tetrahedron Lett., 1990, **31**, 3863

0691 S. MASAMUNE, T. SATO, B. M. KIM, T. A. WOLLMANN, J. Am. Chem. Soc., 1986, **108**, 8279

0692 E. J. COREY, C. P. DECICCO, R. C. NEWBOLD, Tetrahedron Lett., 1991, **32**, 5287

0693 E. J. COREY, R. IMWINKELRIED, S. PIKUL, Y. B. XIANG, J. Am. Chem. Soc., 1989, **111**, 5493

0694 E. J. COREY, S. S. KIM, J. Am. Chem. Soc., 1990, **112**, 4976

0695 E. J. COREY, S. S. KIM, Tetrahedron Lett., 1990, **31**, 3715

0696 E. J. COREY, D. H. LEE, J. Am. Chem. Soc., 1991, **113**, 4026

0697 A. I. MEYERS, Y. YAMAMOTO, J. Am. Chem. Soc., 1981, **103**, 4278

0698 W. R. ROUSH, in Comprehensive Organic Synthesis, B. M. TROST and I. FLEMING Ed., Pergamon Press, 1991, vol. 2, chap. 1.1

0699 H. C. BROWN, K. S. BHAT, P. K. JADHAV, J. Chem. Soc. Perkin I, 1991, 2633

0700 H. C. BROWN, K. S. BHAT, J. Am. Chem. Soc., 1986, **108**, 5919

0701 H. C. BROWN, K. S. BHAT, J. Am. Chem. Soc., 1986, **108**, 293

0702 H. C. BROWN, K. S. BHAT, R. S. RANDAD, J. Org. Chem., 1989, **54**, 1570

0703 H. C. BROWN, R. S. RANDAD, K. S. BHAT, M. ZAIDLEWICZ, U. S. RACHERLA, J. Am. Chem. Soc., 1990, **112**, 2389

0704 H. C. BROWN, R. S. RANDAD, Tetrahedron, 1990, **46**, 4457

0705 H. C. BROWN, R. S. RANDAD, Tetrahedron, 1990, **46**, 4463

0706 P. K. JADHAV, K. S. BHAT, P. T. PERUMAL, H. C. BROWN, J. Org. Chem., 1986, **51**, 432

0707 U. S. RACHERLA, H. C. BROWN, J. Org. Chem., 1991, **56**, 401

0708 H. C. BROWN, U. S. RACHERLA, Y. LIAO, V. V. KHANNA, J. Org. Chem., 1992, **57**, 6608

0709 H. C. BROWN, P. K. JADHAV, K. S. BHAT, J. Am. Chem. Soc., 1988, **110**, 1535

0710 A. G. M. BARRETT, J. W. MALECHA, J. Org. Chem., 1991, **56**, 5243

0711   A. G. M. BARRETT, M. A. SEEFELD, Tetrahedron, 1993, **49**, 7857
0712   A. G. M. BARRETT, M. A. SEEFELD, J. Chem. Soc. Chem. Comm.,
       1993, 339
0713   J. GARCIA, B. M. KIM, S. MASAMUNE, J. Org. Chem., 1987, **52**,
       4831
0714   R. P. SHORT, S. MASAMUNE, J. Am. Chem. Soc., 1989, **111**, 1892
0715   R. W. HOFFMANN, Pure and Appl. Chem., 1988, **60**, 123
0716   R. W. HOFFMANN, G. NIEL, A. SCHLAPBACH, Pure and Appl.
       Chem., 1990, **62**, 1993
0717   W. R. ROUSH, K. ANDO, D. B. POWERS, A. D. PALKOWITZ, R. L.
       HALTERMAN, J. Am. Chem. Soc., 1990, **112**, 6339
0718   W. R. ROUSH, L. K. HOONG, M. A. J. PALMER, J. C. PARK, J. Org.
       Chem., 1990, **55**, 4109
0719   H. C. BROWN, A. S. PHADKE, Synlett, 1993, 927
0720   W. R. ROUSH, L. K. HOONG, M. A. J. PALMER, J. A. STRAUB, A.
       D. PALKOWITZ, J. Org. Chem., 1990, **55**, 4117
0721   W. R. ROUSH, A. D. PALKOWITZ, K. ANDO, J. Am. Chem. Soc.,
       1990, **112**, 6348
0722   W. R. ROUSH, J. C. PARK, J. Org. Chem., 1990, **55**, 1143
0723   W. R. ROUSH, P. T. GROVER, X. LIN, Tetrahedron Lett., 1990, **31**,
       7563
0724   W. R. ROUSH, P. T. GROVER, Tetrahedron Lett., 1990, **31**, 7567
0725   M. T. REETZ, T. ZIERKE, Chem. Ind., 1988, 663
0726   R. W. HOFFMANN, A. SCHLAPBACH, Tetrahedron, 1992, **48**, 1959
0727   R. STURMER, Angew. Chem. Int. Ed. Engl., 1990, **29**, 62
0728   R. STURMER, R. W. HOFFMANN, Synlett, 1990, 759
0729   R. W. HOFFMANN, K. DITRICH, G. KOSTER, R. STURMER, Chem.
       Ber., 1989, **122**, 1783
0730   R. W. HOFFMANN, A. SCHLAPBACH, Liebigs Ann., 1991, 1203
0731   R. W. HOFFMANN, S. DRESELY, J. W. LANZ, Chem. Ber., 1988,
       121, 1501
0732   R. W. HOFFMANN, S. DRESELY, Synthesis, 1988, 103
0733   R. W. HOFFMANN, B. LANDMANN, Chem. Ber., 1986, **119**, 2013
0734   E. J. COREY, C. M. YU, S. S. KIM, J. Am. Chem. Soc., 1989, **111**,
       5495
0735   E. J. COREY, C. M. YU, D. H. LEE, J. Am. Chem. Soc., 1990, **112**, 878
0736   I. FLEMING, J. DUNOGUES, R. SMITHERS, Org. Reactions, 1989,
       37, 57
0737   T. H. CHAN, D. WANG, Tetrahedron Lett., 1989, **30**, 3041
0738   S. LAMOTHE, T. H. CHAN, Tetrahedron Lett., 1991, **32**, 1847
0739   B. W. GUNG, D. T. SMITH, M. A. WOLF, Tetrahedron Lett., 1991, **32**,
       13
0740   B. W. GUNG, A. J. PEAT, B. M. SNOOK, D. T. SMITH, Tetrahedron
       Lett., 1991, **32**, 453

0741 F. A. DAVIS, A. C. SHEPPARD, Tetrahedron, 1989, **55**, 5703

0742 F. A. DAVIS, B. C. CHEN, Chem. Rev., 1992, **92**, 919

0743 F. A. DAVIS, A. KUMAR, B. C. CHEN, J. Org. Chem., 1991, **56**, 1143

0744 F. A. DAVIS, M. C. WEISMILLER, C. K. MURPHY, R. T. REDDY, B. C. CHEN, J. Org. Chem., 1992, **57**, 7274

0745 I. MERGELSBERG, D. GALA, D. SHERER, D. DI BENEDETTO, M. TANNER, Tetrahedron Lett., 1992, **33**, 161

0746 J. C. TOWSON, M. C. WEISMILLER, G. S. LAL, A. C. SHEPPARD, F. A. DAVIS, Org. Synth., 1990, **69**, 158

0747 F. A. DAVIS, A. C. SHEPPARD, B. C. CHEN, M. S. HAQUE, J. Am. Chem. Soc., 1990, **112**, 6679

0748 F. A. DAVIS, M. C. WEISMILLER, J. Org. Chem., 1990, **55**, 3715

0749 F. A. DAVIS, R. THIMMA REDDY, J. Org. Chem., 1992, **57**, 2599

0750 A. H. HAINES, in Comprehensive Organic Synthesis, B. M. TROST and I. FLEMING Ed., Pergamon Press, 1991, vol. 7, chap. 3.3

0751 B. B. LOHRAY, Tetrahedron Asymm., 1992, **3**, 1317

0752 I. OJIMA, N. CLOS, C. BASTOS, Tetrahedron, 1989, **45**, 6901

0753 R. A. JOHNSON, K. B. SHARPLESS in Catalytic Asymmetric Synthesis, I. OJIMA Ed. VCH 1993, p. 227

0753a W. AMBERG, Y. L. BENNANI, R. K. CHADHA, G. A. CRISPINO, W. D. DAVIS, J. HARTUNG, K. S. JEONG, Y. OGINO, T. SHIBATA, K. B. SHARPLESS, J. Org. Chem., 1993, **58**, 844

0754 Y. OGINO, H. CHEN, E. MANOURY, T. SHIBATA, M. BELLER, D. LÜBBEN, K. B. SHARPLESS, Tetrahedron Lett., 1991, **32**, 5761

0755 K. B. SHARPLESS, W. AMBERG, M. BELLER, H. CHEN, J. HARTUNG, Y. KAWANAMI, D. LÜBBEN, E. MANOURY, Y. OGINO, T. SHIBATA, T. UKITA, J. Org. Chem., 1991, **56**, 4585

0756 K. B. SHARPLESS, W. AMBERG, Y. L. BENNANI, G. A. CRISPINO, J. HARTUNG, K. S. JEONG, H. L. KWONG, K. MORIKAWA, Z. M. WANG, D. XU, X. L. ZHANG, J. Org. Chem., 1992, **57**, 2768

0757 G. A. CRISPINO, A. MAKITA, Z. M. WANG, K. B. SHARPLESS, Tetrahedron Lett., 1994, **35**, 543

0758 H. C. KOLB, P. G. ANDERSON, Y. L. BENNANI, G. A. CRISPINO, K.S.JEONG, H. L. KWONG, K. B. SHARPLESS, J. Am. Chem. Soc., 1993, **115**, 12226

0759 T. GÖBEL, K. B. SHARPLESS, Angew. Chem. Int. Ed. Engl., 1993, **32**, 1329

0760 E. J. COREY, M. C. NOE, J. Am. Chem. Soc., 1993, **115**, 12579

0761 E. J. COREY, M. C. NOE, M. J. GROGAN, Tetrahedron Lett., 1994, **35**, 6427

0762 K. MORIKAWA, K. B. SHARPLESS, Tetrahedron Lett., 1993, **34**, 5575

0763 E. J. COREY, M. C. NOE, S. SARSHAR, J. Am. Chem. Soc., 1993, **115**, 3828

0764 B. M. KIM, K. B. SHARPLESS, Tetrahedron Lett., 1990, **31**, 3003

0765   D. PINI, A. PETRI, A. NARDI, C. ROSINI, P. SALVADORI,
        Tetrahedron Lett., 1991, **32**, 5175
0766   M. J. O'DONNELL, in Catalytic Asymmetric Synthesis, I. OJIMA Ed.
        VCH 1993, p. 389
0767   H. WYNBERG, Recl. Trav. Chim. Pays-Bas, 1981, **100**, 393
0768   T. B. K. LEE, G. S. K. WONG, J. Org. Chem., 1991, **56**, 872
0769   Y. GÉNISSON, C. MARAZANO, B. C. DAS, J. Org. Chem., 1993, **58**,
        2052
0770   H. WYNBERG, E. G. J. STARING, J. Am. Chem. Soc., 1982, **104**, 166
0771   B. MARSMAN, H. WYNBERG, J. Org. Chem., 1979, **44**, 2312
0772   Y. SATO, S. WATANABE, M. SHIBASAKI, Tetrahedron Lett., 1992,
        33, 2589
0773   C. M. GASPARSKI, M. J. MILLER, Tetrahedron, 1991, **47**, 5367
0774   K. B. LIPKOWITZ, M. W. CAVANAUGH, B. BAKER, M. J.
        O'DONNELL, J. Org. Chem., 1991, **56**, 5181
0775   C. AGAMI, Bull. Soc. Chim. France, 1988, 499
0776   S. BANFI, S. COLONNA, H. MOLINARI, S. JULIA, J. GUIXER,
        Tetrahedron, 1984, **40**, 5207
0777   B. C. B. BEZUIDENHOUT, A. SWANEPOEL, J. A. N. AUGUSTYN,
        D. FERREIRA, Tetrahedron Lett., 1987, **28**, 4857
0777a   K. MARUOKA, H. YAMAMOTO, in Catalytic Asymmetric Synthesis, I.
        OJIMA Ed. VCH 1993, p. 413
0778   K. NARASAKA, Synthesis, 1991, 1
0779   J. M. HAWKINS, S. LOREN, J. Am. Chem. Soc., 1991, **113**, 7794
0780   H. B. KAGAN, O. RIANT, Chem. Rev., 1992, **92**, 1007
0781   D. KAUFMANN, R. BOESE, Angew. Chem. Int. Ed. Engl., 1990, **29**,
        545
0782   K. HATTORI, M. MIYATA, H. YAMAMOTO, J. Am. Chem. Soc.,
        1993, **115**, 1151
0783   K. HATTORI, H. YAMAMOTO, J. Org. Chem., 1992, **57**, 3264
0784   S. KOBAYASHI, M. MURAKAMI, T. HARADA, T. MUKAIYAMA,
        Chem. Lett., 1991, 1341
0785   K. FURUTA, A. KANEMATSU, H. YAMAMOTO, S. TAKAOKA,
        Tetrahedron Lett., 1989, **30**, 7231
0786   K. FURUTA, M. MOURI, H. YAMAMOTO, Synlett, 1991, 561
0787   K. FURUTA, T. MARUYAMA, H. YAMAMOTO, J. Am. Chem. Soc.,
        1991, **113**, 1041
0788   K. FURUTA, T. MARUYAMA, H. YAMAMOTO, Synlett, 1991, 439
0789   Q. GAO, K. ISHIHARA, T. MARUYAMA, M. MOURI, H.
        YAMAMOTO, Tetrahedron, 1994, **50**, 979
0790   Q. GAO, T. MARUYAMA, M. MOURI, H. YAMAMOTO, J. Org.
        Chem., 1992, **57**, 1951
0791   K. ISHIHARA, M. MOURI, Q. GAO, T. MARUYAMA, K. FURUTA,
        H. YAMAMOTO, J. Am. Chem. Soc., 1993, **115**, 11490

0792 D. SARTOR, J. SAFFRICH, G. HELMCHEN, Synlett, 1990, 197
0793 M. TAKASU, H. YAMAMOTO, Synlett, 1990, 194
0794 E. J. COREY, C. L. CYWIN, T. D. ROPER, Tetrahedron Lett., 1992, **33**, 6907
0795 S. KYIOOKA, Y. KANEKO, M. KOMURA, H. MATSUO, M. NAKANO, J. Org. Chem., 1991, **56**, 2276
0796 E. R. PARMEE, Y. HONG, O. TEMPKIN, S. MASAMUNE, Tetrahedron Lett., 1992, **33**, 1729
0797 E. R. PARMEE, O. TEMPKIN, S. MASAMUNE, A. ABIKO, J. Am. Chem. Soc., 1991, **113**, 9365
0798 C. J. NORTHCOTT, Z. VALENTA, Canad. J. Chem., 1987, **65**, 1917
0799 F. BIGI, G. CASIRAGHI, G. CASNATI, G. SARTORI, P. SONCINI, G. G. FAVA, M. F. BELICCHI, J. Org. Chem., 1988, **53**, 1779
0800 J. BAO, W. D. WULFF, J. Am. Chem. Soc., 1993, **115**, 3814
0801 F. REBIÈRE, O. RIANT, H. B. KAGAN, Tetrahedron Asymm., 1990, **1**, 199
0802 A. KETTER, G. GLAHSL, R. HERRMANN, J. Chem. Research (S), 1990, 278
0803 K. MARUOKA, H. BANNO, H. YAMAMOTO, Tetrahedron Asymm., 1991, **2**, 647
0804 A. MORI, H. OHNO, H. NITTA, K. TANAKA, S. INOUE, Synlett, 1991, 563
0805 I. PATERSON, D. J. BERRISFORD, Angew. Chem. Int. Ed. Engl., 1992, **31**, 1179
0806 T. MUKAIYAMA, T. TAKASHIMA, M. KUSAKA, T. SHIMPUKU, Chem. Lett., 1990, 1777
0807 M. G. FINN, K. B. SHARPLESS, in Asymmetric Synthesis J. D. MORRISON Ed., Academic Press, New York, 1985, **5**, 247
0808 R. A. JOHNSON, K. B. SHARPLESS, in Comprehensive Organic Synthesis, B. M. TROST and I. FLEMING Ed. Pergamon Press 1991, vol. 7, chap. 3.2
0809 R. A. JOHNSON, K. B. SHARPLESS, in Catalytic Asymmetric Synthesis, I. OJIMA Ed. VCH 1993, p. 103
0810 T. KATSUKI, V. S. MARTIN, Organic Reactions
0811 J. CROSBY, Tetrahedron, 1991, **47**, 4789
0812 R. A. SHELDON, Chirotechnology, M. Dekker Ed., New York, 1993; A. N. COLLINS, G. N. SHELDRAKE, J. CROSBY, Chirality in Industry, Wiley & Sons Ed. N.Y. 1992
0813 M. G. FINN, K. B. SHARPLESS, J. Am. Chem. Soc., 1991, **113**, 113
0814 H. B. KAGAN, in Catalytic Asymmetric Synthesis, I. OJIMA Ed. VCH 1993, p. 203
0815 N. KOMATSU, M. HASHIZUME, T. SUGITA, S. UEMURA, J. Org. Chem., 1993, **58**, 4529
0816 J. A. MARSHALL, Y. TANG, Synlett, 1992, 653

0817   Y. HAYASHI, S. NIIHATA, K. NARASAKA, Chem. Lett., 1990, 2091

0818   T. YOKOMATSU, T.YAMAGISHI, S. SHIBUYA, Tetrahedron Asymm., 1993, **4**, 1779

0819   K. MIKAMI, M. TERADA, S. NARISAWA, T. NAKAI, Synlett, 1992, 255

0820   K. MIKAMI, M. TERADA, T. NAKAI, J. Am. Chem. Soc., 1990, **112**, 3949

0821   K. MIKAMI, M. TERADA, Y. MOTOYAMA, T. NAKAI, Tetrahedron Asymm., 1991, **2**, 643

0822   K. MIKAMI, M. TERADA, E. SAWA, T. NAKAI, Tetrahedron Lett., 1991, **32**, 6571

0823   B. KRUSE, R. BRÜCKNER, Tetrahedron Lett., 1990, **31**, 4425

0824   T. MUKAIYAMA, A. INUBUSHI, S. SUDA, R. HARA, S. KOBAYASHI, Chem. Lett., 1990, 1015

0825   A. MORI, H. NITTA, M. KUDO, S. INOUE, Tetrahedron Lett., 1991, **32**, 4333

0826   Y. ISEKI, M. KUDO, A. MORI, S. INOUE, J. Org. Chem., 1992, **57**, 6329

0827   L. F. TIETZE, P. SALING, Synlett, 1992, 281

0828   E. J. COREY, T. D. ROPER, K. ISHIHARA, G. SARAKINOS, Tetrahedron Lett., 1993, **34**, 8399

0829   R. D. BROENE, S. L. BUCHWALD, J. Am. Chem. Soc., 1993, **115**, 12569

0830   C. A. WILLOUGHBY, L. S. BUCHWALD, J. Am. Chem. Soc., 1992, **114**, 7562

0831   W. A. NUGENT, J. Am. Chem. Soc., 1992, **114**, 2768

0832   T. MUKAIYAMA, Org. Reactions, 1982, **28**, 203

0833   S. KOBAYASHI, M. FURUYA, A. OHTSUBO, T. MUKAIYAMA, Tetrahedron Asymm., 1991, **2**, 635

0834   S. KOBAYASHI, Y. FUJISHITA, T. MUKAIYAMA, Chem. Lett., 1990,1455

0835   S. KOBAYASHI, H. UCHIRO, Y. FUJISHITA, I. SHIINA, T. MUKAIYAMA, J. Am. Chem. Soc., 1991, **113**, 4247

0836   T. MUKAIYAMA, M. FURUYA, A. OHTSUBO, S. KOBAYASHI, Chem. Lett., 1991, 989

0837   T. MUKAIYAMA, S. KOBAYASHI, H. UCHIRO, I. SHIINA, Chem. Lett., 1990, 129

0838   T. MUKAIYAMA, S. KOBAYASHI, T. SANO, Tetrahedron, 1990, **46**, 4653

0839   T. MUKAIYAMA, H. UCHIRO, I. SHIINA, S. KOBAYASHI, Chem. Lett., 1990, 1019

0840   T. MUKAIYAMA, H. UCHIRO, S. KOBAYASHI, Chem. Lett., 1990, 1757

0841 T. MUKAIYAMA, H. ASANUMA, I. HACHIYA, T. HARADA, S. KOBAYASHI, Chem. Lett., 1991, 1209

0842 N. MINOWA, T. MUKAIYAMA, Bull. Chem. Soc. Japan, 1987, 60, 3697

0843 D. PARKER, Chem. Rev., 1991, **91**, 1441

0844 M. BEDNARSKI, S. DANISHEFSKY, J. Am. Chem. Soc., 1986, **108**, 7060

0845 M. MIDLAND, R. S. GRAHAM, J. Am. Chem. Soc., 1984, **106**, 4294

0846 S. ZEHANI, G. GELBARD, J. Chem. Soc. Chem. Comm., 1985, 1162

0847 H. SASAI, T. SUZUKI, S. ARAI, T. ARAI, M. SHIBASAKI, J. Am. Chem. Soc.1992, **114**, 4418

0848 H. SASAI, T. SUZUKI, N. ITOH, M. SHIBASAKI, Tetrahedron Lett., 1993, **34**, 851

0849 H. SASAI, T. SUZUKI, N. ITOH, K. TANAKA, T. DATE, K. OKAMURA, M. SHIBASAKI, J. Am. Chem. Soc., 1993, **115**, 10372

0850 N. P. RATH, C. D. SPILLING, Tetrahedron Lett., 1994, **35**, 227

0851 E. J. COREY, N. IMAI, H. Y. ZHANG, J. Am. Chem. Soc., 1991, **113**, 728

0852 D. A. EVANS, T. LECTKA, S. J. MILLER, Tetrahedron Lett., 1993, **34**, 7027

0853 H. B. KAGAN, Bull. Soc. Chim. France, 1988, 846

0854 J. W. SCOTT, Topics in Stereochemistry, E. L. ELIEL and S. WILEN Ed., 1989, **19**, 209

0855 I. OJIMA, K. HIRAI, in Asymmetric Synthesis, J. D. MORRISON Ed. Academic Press, New York, 1985, **5**, 103

0856 J. W. FALLER, J. PARR, J. Am. Chem. Soc., 1993, **115**, 804

0857 J. W. FALLER, M. TOKUNAGA, Tetrahedron Lett., 1993, **34**, 7359

0858 H. BRUNNER, Synthesis, 1988, 645

0859 H. TAKAYA, T. OHTA, R. NOYORI, in Catalytic Asymmetric Synthesis, I. OJIMA Ed. VCH 1993, p. 1

0860 H. BRUNNER, W. ZETTLMEIER, Handbook of Enantioselective Catalysis VCH Ed.1993

0861 R. O. DUTHALER, Tetrahedron, 1994, **50**, 1539

0862 G. SHAPIRO, C. CHENGZHI, Tetrahedron Lett., 1992, 33, 2447

0863 E. J. COREY, Z. CHEN, G. J. TANOURY, J. Am. Chem. Soc., 1993, **115**, 11000

0864 K. BURGESS, W. A. van der DONK, M. J. OHLMEYER, Tetrahedron Asymm., 1991, **2**, 613

0865 W. A. HERRMANN, C. W. KOHLPAINTNER, Angew. Chem. Int. Ed. Engl., 1993, **32**, 1524

0866 I. THOT, B. E. HANSON, M. E. DAVIS, Tetrahedron Asymm., 1990, **1**, 913

0867 A. BÖRNER, J. WARD, K. KORTUS, H. B. KAGAN, Tetrahedron Asymm., 1993, **4**, 2219

0868   H. BRUNNER, H. NISHIYAMA, K. ITOH, in Catalytic Asymmetric
       Synthesis, I. OJIMA Ed. VCH 1993, p. 303
0869   U. NAGEL, B. RIEGER, Organometallics, 1989, **8**, 1534
0870   H. J. ZEISS, J. Org. Chem., 1991, **56**, 1783
0871   O. REISER, Angew. Chem. Int. Ed. Engl., 1993, **32**, 547
0872   M. YAMAGUCHI, T. SHIMA, T. YAMAGISHI, M. HIDA,
       Tetrahedron Asymm., 1991, **2**, 663
0873   G. ZASSINOVICH, G. MESTRONI, S. GLADIALI, Chem. Rev., 1992,
       92, 1051
0874   G. CONSIGLIO, in Catalytic Asymmetric Synthesis, I. OJIMA Ed. VCH
       1993, p. 273
0875   J. K. STILLE, in Comprehensive Organic Synthesis, B. M. TROST and I.
       FLEMING Ed., Pergamon Press, 1991, vol. 4, chap. 4.5
0876   S. A. GODLESKI, in Comprehensive Organic Synthesis, B. M. TROST
       and I. FLEMING Ed., Pergamon Press, 1991, vol. 4, chap. 3.3
0877   M. SAWAMURA, H. NAGATA, H. SAKAMOTO, Y. ITO, J. Am.
       Chem. Soc., 1992, **114**, 2586
0878   A. YAMAZAKI, T. MORIMOTO, K. ACHIWA, Tetrahedron Asymm.,
       1993, **4**, 2287
0879   M. SAWAMURA, H. HAMASHIMA, Y. ITO, J. Am. Chem. Soc., 1992,
       114, 8295
0880   T. IKARIYA, Y. ISHII, H. KAWANO, T. ARAI, M. SABURI, S.
       YOSHIKAWA, S. AKUTAGAWA, J. Chem. Soc. Chem. Comm., 1985,
       922
0880a H. TAKAYA, S. AKUTAGAWA, R. NOYORI, Org. Synth., 1988, **67**,
       20
0881   R. NOYORI, H. TAKAYA, Acc. Chem. Research, 1990, **23**, 345
0882   S. OTSUKA, K. TANI, Synthesis, 1991, 665
0883   H. TAKAYA, T. OHTA, K. MASHIMA, R. NOYORI, Pure and Appl.
       Chem., 1990, **62**, 1135
0884   K. T. WAN, M. E. DAVIS, J. Chem. Soc. Chem. Comm., 1993, 1262
0885   S. H. BERGENS, P. NOHEDA, J. WHELAN, B. BOSNICH, J. Am.
       Chem. Soc., 1992, **114**, 2121
0886   T. HAYASHI, Y. MATSUMOTO, Y. ITO, Tetrahedron Asymm., 1991,
       2, 601
0887   Y. MATSUMOTO, T. HAYASHI, Tetrahedron Lett., 1991, **32**, 3387
0888   J. M. BROWN, G. C. LLOYD-JONES, Tetrahedron Asymm., 1990, **1**,
       869
0889   S. AKUTAGAWA, K. TANI, in Catalytic Asymmetric Synthesis, I.
       OJIMA Ed. VCH, 1993, p. 41
0890   K. TANI, T. YAMAGATA, S. OTSUKA, H. KUMOBAYASHI, S.
       AKUTAGAWA, Org. Synth., 1988, **67**, 33
0891   N. SAKAI, K. NOZAKI, K. MASHIMA, H. TAYAKA, Tetrahedron
       Asymm., 1992, **3**, 583

0892 B. HEISER, E. A. BROGER, Y. CRAMERI, Tetrahedron Asymm., 1991, **2**, 51
0893 M. KITAMURA, M. TOKUNAGA, T. OHKUMA, R. NOYORI, Tetrahedron Lett., 1991, **32**, 4163
0894 D. F. TABER, L. J. SILVERBERG, Tetrahedron Lett., 1991, **32**, 4227
0895 W. D. LUBELL, M. KITAMURA, R. NOYORI, Tetrahedron Asymm., 1991, **2**, 543
0896 K. T. WAN, M. E. DAVIS, Tetrahedron Asymm., 1993, **4**, 2461
0897 J. M. BROWN, H. BRUNNER, W. LEITNER, M. ROSE, Tetrahedron Asymm., 1991, **2**, 331
0898 J. P. GENET, C. PINEL, S. MALLART, S. JUGE, S. THORIMBERT, J. A. LAFFITTE, Tetrahedron Asymm., 1991, **2**, 555
0899 S. D. RYCHNOVSKY, G. GRIESGRABER, S. ZELLER, D. J. SKALITZKY, J. Org. Chem., 1991, **56**, 5161
0900 D. F. TABER, L. J. SILVERBERG, E. D. ROBINSON, J. Am. Chem. Soc., 1991, **113**, 6639
0901 X. ZHANG, T. TAKETOMI, T. YOSHIZUMI, H. KUMOBAYASHI, S. AKUTAGAWA, K. MASHIMA, H. TAKAYA, J. Am. Chem. Soc., 1993, **115**, 3318
0902 Y. SATO, S.NUKUI, M. SODEOKA, M. SHIBASAKI, Tetrahedron, 1994, **50**, 371
0903 Y. MATSUMOTO, T. HAYASHI, Y. ITO, Tetrahedron, 1994, **50**, 335
0904 T. CHIBA, A. MIYASHITA, H. NOHIRA, Tetrahedron Lett., 1991, **32**, 4745
0905 R. SCHMID, M. CEREGHETTI, B. HEISER, P. SCHÖNHOLZER, H. J. HANSEN, Helv. Chim. Acta, 1988, **71**, 897
0906 T. CHIBA, A. MIYASHITA, H. NOHIRA, H. TAKAYA, Tetrahedron Lett., 1993, **34**, 2351
0907 M. J. BURK, J. E. FEASTER, W. A. NUGENT, R. L. HARLOW, J. Am. Chem. Soc., 1993, **115**, 10125
0908 M. J. BURK, J. Am. Chem. Soc., 1991, **113**, 8518
0909 M. J. BURK, J. E. FEASTER, R. L. HARLOW, Tetrahedron Asymm., 1991, **2**, 569
0910 T. HAYASHI, in Catalytic Asymmetric Synthesis, I. OJIMA Ed. VCH 1993, p. 325
0911 B. M. TROST, D. L. vanVRANKEN, C. BINGEL, J. Am. Chem. Soc., 1992, **114**, 9327
0912 N. SAKAI, S. MANO, K. NOZAKI, H. TAKAYA, J. Am. Chem. Soc., 1993, **115**, 7033
0913 B. JEDLICKA, C. KRATKY, W. WEISSENSTEINER, M. WIDHALM, J. Chem. Soc. Chem. Comm., 1993, 1329
0914 Y. UOZUMI, T. HAYASHI, J. Am. Chem. Soc., 1991, **113**, 9887
0915 Y. UOZUMI, T. HAYASHI, Tetrahedron Lett., 1993, **34**, 2335

0916   Y. UOZUMI, K. KITAYAMA, T. HAYASHI, Tetrahedron Asymm., 1993, **4**, 2419

0917   Y. UOZUMI, A. TANAHASHI, S. Y. LEE, T. HAYASHI, J. Org. Chem., 1993, **58**, 1945

0918   J. V. ALLEN, G. J. DAWSON, C. G. FROST, J. M. J. WILLIAMS, S. J. COOTE, Tetrahedron, 1994, **50**, 799

0919   G. J. DAWSON, C. G. FROST, J. M. J. WILLIAMS, S. J. COOTE, Tetrahedron Lett., 1993, **34**, 3149

0920   P. von MATT, A. PFALTZ, Angew. Chem. Int. Ed. Engl., 1993, **32**, 566

0921   J. SPRINZ, G. HELMCHEN, Tetrahedron Lett., 1993, **34**, 1769

0922   J. M. BROWN, D. I. HULMES, T. P. LAYZELL, J. Chem. Soc. Chem. Comm., 1993, 1673

0923   H. R. SONAWANE, N. S. BELLUR, J. R. AHUJA, D. G. KULKARNI, Tetrahedron Asymm., 1992, **3**, 163

0924   M. C. PIRRUNG, J. ZHANG, Tetrahedron Lett., 1992, **33**, 5987

0925   J. I. SAKAKI, W. B. SCHWEIZER, D. SEEBACH, Helv. Chim. Acta, 1993, **76**, 2654

0926   O. RIANT, O. SAMUEL, H. B. KAGAN, J. Am. Chem. Soc., 1993, **115**, 5835

0927   S. GLADIALI, L. PINNA, G. DELOGU, E. GRAF, H. BRUNNER, Tetrahedron Asymm., 1990, **1**, 937

0928   H. NISHIYAMA, S. YAMAGUCHI, S. B. PARK, K. ITOH, Tetrahedron Asymm., 1993, **4**, 143

0929   A. PFALTZ, Acc. Chem. Res., 1993, **26**, 339

0930   C. BOLM, Angew. Chem. Int. Ed. Engl., 1991, **30**, 542

0931   G. HELMCHEN, A. KROTZ, K. T. GANZ, D. HANSEN, Synlett, 1991, 257

0932   P. GAMEZ, F. FACHE, P. MANGENEY, M. LEMAIRE, Tetrahedron Lett., 1993, **34**, 6897

0933   H. M. L. DAVIES, D. K. HUTCHESON, Tetrahedron Lett., 1993, **34**, 7243

0934   A. McKERVEY, T. YE, J. Chem. Soc. Chem. Comm., 1992, 823

0935   G. H. P. ROOS, M. A. McKERVEY, Synth. Commun., 1992, **22**, 1751

0936   M. P. DOYLE, Recl. Trav. Chim. Pays Bas, 1991, **110**, 305

0937   M. P. DOYLE, in Catalytic Asymmetric Synthesis, I. OJIMA Ed. VCH 1993, p. 63

0938   M. P. DOYLE, W. R. WINCHESTER, J. A. A. HOORN, V. LYNCH, S. H. SIMONSEN, R. GHOSH, J. Am. Chem. Soc., 1993, **115**, 9968

0939   M. N. PROTOPOPOVA, M. P. DOYLE, P. MULLER, D. ENE, J. Am. Chem. Soc., 1992, **114**, 2755

0940   M. P. DOYLE, A. van OEVEREN, L. J. WESTRUM, M. N. PROTOPOPOVA, T. W. CLAYTON Jr., J. Am. Chem. Soc., 1991, **113**, 8982

0941 M. P. DOYLE, W. R. WINCHESTER, M. N. PROTOPOPOVA, P. MÜLLER, G. BERNARDINELLI, D. ENE, S. MOTALLEBI, Helv. Chim. Acta, 1993, **76**, 2227

0942 D. MÜLLER, G. UMBRICHT, B. WEBER, A. PFALTZ, Helv. Chim. Acta, 1991, **74**, 232

0943 G. J. DAWSON, C. G. FROST, C. J. MARTIN, J. M. J. WILLIAMS, S. J. COOTE, Tetrahedron Lett., 1993, **34**, 7793

0944 C. G. FROST, J. M. J. WILLIAMS, Tetrahedron Asymm., 1993, **4**, 1785

0945 C. G. FROST, J. M. J. WILLIAMS, Tetrahedron Lett., 1993, **34**, 2015

0946 U. LEUTENEGGER, G. UMBRICHT, C. FAHRNI, P. VON MATT, A. PFALTZ, Tetrahedron, 1992, **48**, 2143

0947 H. KUBOTA, M. NAKAJIMA, K. KOGA, Tetrahedron Lett., 1993, **34**, 8135

0948 A. TERFORT, Synthesis, 1992, 951

0949 C. CARDELLICHIO, V. FIANDANESE, F. NASO, Gazz., 1991, **121**, 11

0950 M. SAWAMURA, Y. ITO, in Catalytic Asymmetric Synthesis, I. OJIMA Ed. VCH 1993, p. 367

0951 T. HAYASHI, Y. UOZUMI, A. YAMAZAKI, M. SAWAMURA, H. HAMASHINA, Y. ITO, Tetrahedron Lett., 1991, **32**, 2799

0952 M. SAWAMURA, H. HAMASHIMA, Y. ITO, J. Org. Chem., 1990, **55**, 5935

0953 H. M. L. DAVIES, in Comprehensive Organic Synthesis, B. M. TROST and I. FLEMING Ed., Pergamon Press, 1991, vol. 4, chap. 4.8

0954 C. BOLM, Angew. Chem. Int. Ed. Engl., 1991, **30**, 403

0955 A. HATAYAMA, N. HOSOYA, R. IRIE, Y. ITO, T. KATSUKI, Synlett, 1992, 407

0956 R. IRIE, K. NODA, Y. ITO, N. MATSUMOTO, K. KATSUKI, Tetrahedron Asymm., 1991, **2**, 481

0957 E. N. JACOBSEN, in Catalytic Asymmetric Synthesis, I. OJIMA Ed. VCH 1993, p. 159

0958 E. N. JACOBSEN, W. ZHANG, A. R. MUCI, J. R. ECKER, L. DENG, J. Am. Chem. Soc., 1991, **113**, 7063

0959 N. H. LEE, A. R. MUCI, E. N. JACOBSEN, Tetrahedron Lett., 1991, **32**, 5055

0960 W. ZHANG, E. N. JACOBSEN, J. Org. Chem., 1991, **56**, 2296

0961 K. NODA, N. HOSOYA, R. IRIE, Y. ITO, T. KATSUKI, Synlett, 1993, 469

0962 Z. LI, K. R. CONSER, E. N. JACOBSEN, J. Am. Chem. Soc., 1993, **115**, 5326

0963 D. A. EVANS, K. A. WOERPEL, M. J. SCOTT, Angew. Chem. Int. Ed. Engl., 1992, **31**, 430; D. A. EVANS, K. A. WOERPEL, M. M. HINMAN, M. M. FAUL, J. Am. Chem. Soc., 1991, **113**, 726

0964   R. E. LOWENTHAL, S. MASAMUNE, Tetrahedron Lett., 1991, **32**, 7373

0965   D. A. EVANS, M. M. FAUL, M. T. BILODEAU, B. A. ANDERSON, D. M. BARNES, J. Am. Chem. Soc., 1993, **115**, 5328

0966   K. ITO, T. KATSUKI, Tetrahedron Lett., 1993, **34**, 2661

0967   K. ITO, T. KATSUKI, Synlett, 1993, 638

0968   G. JOMMI, R. PAGLIARIN, G. RIZZI, M. SISTI, Synlett, 1993, 833

0969   Q. L. ZHOU, A. PFALTZ, Tetrahedron Lett., 1993, **34**, 7725

0970   V. A. BURGESS,S. G. DAVIES, R. T. SKERLJ, M. WHITTAKER, Tetrahedron Asymm., 1992, **3**, 871

0971   J. P. COLLMAN, V. J. LEE, X. ZHANG, J. A. IBERS, J. I. BRAUMAN, J. Am. Chem. Soc., 1993, **115**, 3834

0972   C. BOLM, M. EWALD, M. FELDER, Chem. Ber., 1992, **125**, 1205

0973   O. PARDIGON, G. BUONO, Tetrahedron Asymm., 1993, **4**, 1977

0974   M. MAJEWSKI, D. M. GLEAVE, J. Org. Chem., 1992, **57**, 3599

0975   K. AOKI, H. NOGUSHI, K. TOMIOKA, K. KOGA, Tetrahedron Lett., 1993, **34**, 5105

0976   M. SOBUKAWA, K. KOGA, Tetrahedron Lett., 1993, **34**, 5101

0977   B. J. BUNN, P. J. COX, N. S. SIMPKINS, Tetrahedron, 1993, **49**, 207

0978   B. J. BUNN, N. S. SIMPKINS, J. Org. Chem., 1993, **58**, 533

0979   B. J. BUNN, N. S. SIMPKINS, Z. SPAVOLD, M. J. CRIMMIN, J. Chem. Soc. Perkin I, 1993, 3113

0980   T. HONDA, N. KIMURA, J. Chem. Soc. Chem. Comm., 1994, 77

0981   R. ARMER, M. J. BEGLEY, P. J. COX, A. PERSAD, N. S. SIMPKINS, J. Chem. Soc. Perkin I, 1993, 3099

0982   P. BEAK, D. B. REITZ, Chem. Rev., 1978, **78**, 275, P. BEAK, W. J. ZAJDEL, D. B. REITZ, Chem. Rev., 1984, **84**, 471

0983   H. AHRENS, M. PAETOW, D. HOPPE, Tetrahedron Lett., 1992, **33**, 5327

0984   D. HOPPE, F. HINTZE, P. TEBBEN, Angew. Chem. Int. Ed. Engl., 1990, **29**, 1422

0985   P. KNOCHEL, Angew. Chem. Int. Ed. Engl., 1992, **31**, 1459

0986   M. PAETOW, H. AHRENS, D. HOPPE, Tetrahedron Lett., 1992, **33**, 5323

0987   J. SCHWERDTFEGER, D. HOPPE, Angew. Chem. Int. Ed. Engl., 1992, **31**, 1505

0988   D. HOPPE, M. PAETOW, F. HINTZE, Angew. Chem. Int. Ed. Engl., 1993, **32**, 394

0989   P. SOMMERFELD, D. HOPPE, Synlett, 1992, 764

0990   J. HALLER, T. HENSE, D. HOPPE, Synlett, 1993, 726

0991   D. MILNE, P. J. MURPHY, J. Chem. Soc. Chem. Comm., 1993, 884

0992   J. VADECARD, J. C. PLAQUEVENT, L. DUHAMEL, P. DUHAMEL, J. Chem. Soc. Chem. Comm., 1993, 116

0993   T. YASUKATA, K. KOGA, Tetrahedron Asymm., 1993, **4**, 35

0994 C. FEHR, I. STEMPF, J. GALINDO, Angew. Chem. Int. Ed. Engl., 1993, **32**, 1042 and 1044

0995 S. TAKEUCHI, N. MIYOSHI, Y. OHGO, Chem. Lett., 1992, 551

0996 Y. INOUE, Chem. Rev., 1992, **92**, 741

0997 J. MUZART, F. HÉNIN, J. P. PÈTE, A. M'BOUGOU-M'PASSI, Tetrahedron Asymm., 1993, **4**, 2531

0998 O. PIVA, J. P. PETE, Tetrahedron Lett., 1990, **31**, 5157

0999 T. DURST, K. KOH, Tetrahedron Lett., 1992, **33**, 6799

1000 J. L. CHARLTON, V. C. PHAM, J. P. PETE, Tetrahedron Lett., 1992, **33**, 6073

1001 O. PIVA, J. P. PETE, Tetrahedron Asymm., 1992, **3**, 759

1002 A. I. MEYERS, T. K. HIGHSMITH, P. T. BUONORA, J. Org. Chem., 1991, **56**, 2960

1003 J. S. WARMUS, M. A. RODKIN, R. BARKLEY, A. I. MEYERS, J. Chem. Soc. Chem. Comm., 1993, 1357

1004 W. H. PEARSON, A. C. LINDBECK, J. W. KAMPF, J. Am. Chem. Soc., 1993, **115**, 2622

1005 T. HAUBENREICH, S. HÜNIG, H. J. SCHULZ, Angew. Chem. Int. Ed. Engl., 1993, **32**, 398

1006 L. S. HEGEDUS, E. LASTRA, Y. NARUKAWA, D. C. SNUSTAD, J. Am. Chem. Soc., 1992, **114**, 2991

1007 P. BEAK, H. DU, J. Am. Chem. Soc., 1993, **115**, 2516

1008 K. FUJI, Chem. Rev., 1993, **93**, 2037

1009 R. L. BEARD, A. I. MEYERS, J. Org. Chem., 1991, **56**, 2091

1010 L. A. CASTONGUAY, J. W. GUILES, A. K. RAPPE, A. I. MEYERS, J. Org. Chem., 1992, **57**, 3819

1011 J. YAOZHONG, L. GUILAN, L. JINCHU, Z. CHANGYOU, Synth. Comm., 1987, **17**, 1545

1012 H. AHLBRECHT, D. ENDERS, L. SANTOWSKI, G. ZIMMERMANN, Chem. Ber., 1989, **122**, 1995

1013 T. H. CHAN, P. PELLON, J. Am. Chem. Soc., 1989, **111**, 8737

1014 T. H. CHAN, D. WANG, Chem. Rev., 1992, **92**, 995

1014a R. C. HARTLEY, S. LAMOTHE, T. H. CHAN, Tetrahedron Lett., 1993, **34**, 1449

1015 S. G. DAVIES, C. L. GOODFELLOW, K. H. SUTTON, Tetrahedron Asymm., 1992, **3**, 1303

1016 H. B. MEKELBURGER, C. WILCOX, in Comprehensive Organic Synthesis, B. M. TROST and I. FLEMING Ed., Pergamon Press, 1991, vol. 2, chap. 1.4

1017 Y. HASEGAWA, H. KAWASAKI, K. KOGA, Tetrahedron Lett., 1993, **34**, 1963

1017a R. P. ALEXANDER, I. PATERSON, Tetrahedron Lett., 1985, **26**, 5339

1018 K. ANDO, Y. TAKEMASA, K. TOMIOKA, K. KOGA, Tetrahedron, 1993, **49**, 1579

1019   K. TOMIOKA, K. ANDO, Y. TAKEMASA, K. KOGA, Tetrahedron
       Lett., 1984, 25, 5677, J. Am. Chem. Soc., 1984, 106, 2718
1020   G. BARTOLI, M. BOSCO, C. CIMARELLI, R. DALPOZZO, G. DE
       MUNNO, G. PALMIERI, Tetrahedron Asymm., 1993, 4, 1651
1021   S. E. DENMARK, J. J. ARES, J. Am. Chem. Soc., 1988, 110, 4432
1022   D. ENDERS, W. GATZWEILER, E. DEDERICHS, Tetrahedron, 1990,
       46, 4757
1023   D. ENDERS, H. DYKER, G. RAABE, Angew. Chem. Int. Ed. Engl.,
       1992, 31, 618
1024   D. ENDERS, H. DYKER, G. RAABE, J. RUNSINK, Synlett, 1992, 901
1025   D. ENDERS, A. ZAMPONI, G. RAABE, J. RUNSINK, Synthesis, 1993,
       725
1026   D. ENDERS, B. BOCKSTIEGEL, Synthesis, 1989, 493
1027   D. ENDERS, U. JEGELKA, Tetrahedron Lett., 1993, 34, 2453
1028   D. ENDERS, U. JEGELKA, B. DÜCKER, Angew. Chem. Int. Ed. Engl.,
       1993, 32, 423
1029   D. ENDERS, B. B. LOHRAY, Angew. Chem. Int. Ed. Engl., 1987, 26,
       351
1030   D. ENDERS, S. NAKAI, Chem. Ber., 1989, 124, 219
1031   V. BHUSHAN, B. B. LOHRAY, D. ENDERS, Tetrahedron Lett., 1993,
       34, 5067
1032   B. B. LOHRAY, D. ENDERS, Synthesis, 1993, 1092
1033   C. CATIVIELA, M. D. DIAZ-de-VILLEGAS, J. A. GALVEZ,
       Tetrahedron Asymm., 1993, 4, 1445; 1994, 5, 261
1034   M. IHARA, M. TAKAHASHI, N. TANIGUCHI, K. YASUI, H.
       NIITSUMA, K. FUKUMOTO, J. Chem. Soc. Perkin I, 1991, 525
1035   H. KIGOSHI, Y. IMAMURA, K. MIZUTA, H. NIVA, K. YAMADA, J.
       Am. Chem. Soc., 1993, 115, 3056
1036   A. FADEL, J. SALAÜN, Tetrahedron Lett., 1988, 29, 6257
1037   A. FADEL, Synlett, 1992, 48
1038   A. GONZALEZ, Synth. Comm., 1991, 21, 1353
1039   D. A. EVANS, H. P. NG, D. L. RIEGER, J. Am. Chem. Soc., 1993, 115,
       11446
1040   D. A. EVANS, R. P. POLNIASZEK, K. M. DE VRIES, D. E. GUINN,
       D. J. MATHRE, J. Am. Chem. Soc., 1991, 113, 7613
1041   I. OJIMA, H. J. C. CHEN, K. NAKAHASHI, J. Am. Chem. Soc. 1988,
       110, 278
1042   D. A. EVANS, F. URPI, T. C. SOMERS, J. S. CLARK, M. T.
       BILODEAU, J. Am. Chem. Soc., 1990, 112, 8215
1043   S. G. DAVIES, G. B. EVANS, A. A. MORTLOCK, Tetrahedron
       Asymm., 1994, 5, 585
1044   S. S. C. KOCH, A. R. CHAMBERLIN, J. Org. Chem., 1993, 58, 2725
1045   W. OPPOLZER, R. MORETTI, S. THOMI, Tetrahedron Lett., 1989, 30,
       5603

1046 W. OPPOLZER, R. MORETTI, S. THOMI, Tetrahedron Lett., 1989, **30**, 6009

1047 M. CHAARI, A. JENHI, J. P. LAVERGNE, P. VIALLEFONT, Tetrahedron, 1991, **47**, 4619

1048 J. Y. ZHONG, L. GUILAN, Z. CHANGYOU, P. HURI, W. LANJUN, M. AIQIAO, Synth. Comm., 1991, **21**, 1087

1049 B. C. BORER, D. W. BALOGH, Tetrahedron Lett., 1991, **32**, 1039

1050 V. P. KUKHAR, Y. N. BELOKON, V. A. SOLOSHONOK, N. Y. SVISTUNOVA, A. B. ROZHENKO, N. A. KUZ'MINA, Synthesis, 1993, 117

1051 B. CHITKUL, Y. PINYOPRONPANICH, C. THEBTARANONTH, Y. THEBTARANONTH, W. C. TAYLOR, Tetrahedron Lett., 1994, **35**, 1099

1052 D. SEEBACH, S. G. MULLER, U. GYSEL, J. ZIMMERMAN, Helv. Chim. Acta, 1988, **71**, 1303

1053 E. JUARISTI, D. QUINTANA, B. LAMATSCH, D. SEEBACH, J. Org. Chem., 1991, **56**, 2553

1054 A. I. MEYERS, L. J. WESTRUM, Tetrahedron Lett., 1993, **34**, 7701

1055 L. SNYDER, A. I. MEYERS, J. Org. Chem., 1993, **58**, 7507

1056 G. STORK, A. R. SCHOOFS, J. Am. Chem. Soc., 1979, **101**, 5081

1057 G. BASHIARDES, G. J. BODWELL, S. G. DAVIES, J. Chem. Soc. Chem. Comm., 1993, 459

1058 H. WYNBERG, Topics in Stereochemistry, E. L. ELIEL and S. WILEN Ed., 1986, **16**, 87

1059 M. J. O'DONNELL, S. WU, Tetrahedron Asymm., 1992, **3**, 591

1060 B. BOSNICH, Asymmetric Catalysis, Martinus Nijhoff Ed., Dordrecht, 1986

1061 J. D. MORRISON Ed., in Asymmetric Synthesis, vol. 5, Academic Press, New York, 1985

1062 H. J. ALTENBACH, in Comprehensive Organic Synthesis, B. M. TROST and I. FLEMING Ed., Pergamon Press, 1991, vol. 6, chap. 4.5.

1063 M. SAWAMURA, Y. ITO, Chem. Rev., 1992, **92**, 857

1064 J. SPRINZ, M. KIEFER, G. HELMCHEN, M. REGGELIN, G. HUTTNER, O. WALTER, L. ZSOLNAI, Tetrahedron Lett., 1994, **35**, 1523

1065 A. TOGNI, Tetrahedron Asymm., 1991, **2**, 683

1066 B. M. TROST, L. LI, S. D. GUILE, J. Am. Chem. Soc., 1992, **114**, 8745

1067 B. M. TROST, R. C. BUNT, J. Am. Chem. Soc., 1994, **116**, 4089

1068 B. M. TROST, B. BREIT, M. G. ORGAN, Tetrahedron Lett., 1994, **35**, 5817

1069 F. A. DAVIS, W. HAN, Tetrahedron Lett., 1992, **33**, 1153

1070 K. ISEKI, T. NAGAI, Y. KOBAYASHI, Tetrahedron Lett., 1993, **34**, 2169

1071 E. NICOLÁS, K. T. RUSSELL, J. KNOLLENBERG, V. J. HRUBY, J. Org. Chem., 1993, **58**, 7565

1072 D. A. EVANS, D. A. EVRARD, S. D. RYCHNOVSKY, T. FRUH, W. G. WHITTINGHAM, K. M. DE VRIES, Tetrahedron Lett., 1992, **33**, 1189

1073 S. E. DENMARK, N. CHATANI, S. V. PANSARE, Tetrahedron, 1992, **48**, 2191

1074 O. KITAGAWA, T. HANANO, N. KIKUCHI, T. TAGUCHI, Tetrahedron Lett., 1993, **34**, 2165

1075 R. ASKANI, D. F. TABER, in Comprehensive Organic Synthesis, B. M. TROST and I. FLEMING Ed., Pergamon Press, 1991, vol. 6, chap. 1.4.

1076 W. OPPOLZER, O. TAMURA, J. DEERBERG, Helv. Chim. Acta, 1992, **75**, 1965

1077 D. ENDERS, A. ZAMPONI, G. RAABE, Synlett, 1992, 897

1078 T. W. HART, D. GUILLOCHON, G. PERRIER, B. W. SHARP, M. P. TOFT, B. VACHER, R. J. A. WALSH, Tetrahedron Lett., 1992, **33**, 7211

1079 F. A. DAVIS, P. ZHOU, C. K. MURPHY, Tetrahedron Lett., 1993, **34**, 3971

1080 F. A. DAVIS, C. CLARK, A. KUMAR, B. C. CHEN, J. Org. Chem., 1994, **59**, 1184

1081 M. E. BUNNAGE, A. J. BURKE, S. G. DAVIES, C. G. GOODWIN, Tetrahedron Asymm., 1994, **5**, 203

1082 M. E. BUNNAGE, S. G. DAVIES, C. J. GOODWIN, J. Chem. Soc. Perkin I, 1993, 1375

1083 F. A. DAVIS, A. KUMAR, R. E. REDDY, B. C. CHEN, P. A. WADE, S. W. SHAH, J. Org. Chem., 1993, **58**, 7591

1084 W. OPPOLZER, O. TAMURA, G. SUNDARABABU, M. SIGNER, J. Am. Chem. Soc., 1992, **114**, 5900

1085 Y. NAGAO, Y. HAGIWARA, T. TOHJO, Y. HASEGAWA, M. OCHIAI, M. SHIRO, J. Org. Chem., 1988, **53**, 5983

1086 M. OGATA, T. YOSHIMURA, H. FUJII, Y. ITO, T. KATSUKI, Synlett, 1993, 728

1087 N. A. PORTER, Q. SU, J. J. HARP, I. J. ROSENSTEIN, A. T. McPHAIL, Tetrahedron Lett., 1993, **34**, 4457

1088 M. BRUNNER, L. MUSSMANN, D. VOGT, Synlett, 1993, 893

1089 V. K. SINGH, Synthesis, 1992, 605

1089a E. BROWN, A. LÉZÉ, J. TOUET, Tetrahedron Asymm., 1992, **3**, 841

1090 J. OLLIVIER, J. Y. LEGROS, J. C. FIAUD, A. de MEIJERE, J. SALAUN, Tetrahedron Lett., 1990, **31**, 3135

1091 E. J. COREY, J. O. LINK, Tetrahedron Lett., 1992, **33**, 3431

1092 E. J. COREY, J. O. LINK, R. K. BAKSHI, Tetrahedron Lett., 1992, **33**, 7107

1093 M. KABAT, J. KIEGEL, N. COHEN, K. TOTH, P. M. WOVKULICH,
M. R. USKOKOVIC, Tetrahedron Lett., 1991, **32**, 2343
1094 G. J. QUALLICH, T. M. WOODALL, Tetrahedron Lett., 1993, **34**, 785
1095 D. K. JONES, D. C. LIOTTA, I. SHINKAI, D. J. MATHRE, J. Org.
Chem., 1993, **58**, 799
1096 V. NEVALAINEN, Tetrahedron Asymm., 1991, **2**, 1133
1096a B. T. CHO, Y. S. CHUN, Tetrahedron Asymm., 1992, **3**, 1539
1097 D. CAI, D. TSCHAEN, Y.J.SHI, T. R. VERHOEVEN, R. A. REAMER,
A. W. DOUGLAS, Tetrahedron Lett., 1993, **34**, 3243
1097a T. MEHLER, J. MARTENS, Tetrahedron Asymm., 1993, **4**, 2299
1098 T. IMAI, T. TAMURA, A. YAMAMURO, T. SATO, T. A.
WOLLMANN, R. M. KENNEDY, S. MASAMUNE, J. Am. Chem. Soc.,
1986, **108**, 7402
1099 S. MASAMUNE, R. M. KENNEDY, J. S. PETERSEN, K. N. HOUK,
Y. WU, J. Am. Chem. Soc., 1986, **108**, 7404
1100 P. V. RAMACHANDRAN, A. V. TEODOROVIC, M. V.
RANGAISHENVI, H. C. BROWN, J. Org. Chem., 1992, **57**, 2379
1101 C. BOLM, M. EWALD, M. FELDER, G. SCHLINGLOFF, Chem. Ber.,
1992, **125**, 1169
1102 N. IRAKO, Y. HAMADA, T. SHIOIRI, Tetrahedron, 1992, **48**, 7251
1103 P. V. RAMACHANDRAN, A. V. TEODOROVIC, H. C. BROWN,
Tetrahedron, 1993, **49**, 1725
1104 A. O. KING, E. G. CORLEY, R. K. ANDERSON, R. D. LARSEN, T.
R. VERHOEVEN, P. J. REIDER, X. B. XIANG, M. BELLEY, Y.
LEBLANC, M. LABELLE, P. PRASIT, R. J. ZAMBONI, J. Org. Chem.,
1993, **58**, 3731
1105 P. V. RAMACHANDRAN, B. GONG, H. C. BROWN, Tetrahedron
Asymm., 1993, **4**, 2399
1106 D. A. BEARDSLEY, G. B. FISHER, C. T. GORALSKI, L. W.
NICHOLSON, B. SINGARAM, Tetrahedron Lett., 1994, **35**, 1511
1107 Y. COMBRET, J. DUFLOS, G. DUPAS, J. BOURGUIGNON, G.
QUÉGUINER, Tetrahedron, 1993, **49**, 5237
1108 H. SUGIMURA, K. YOSHIDA, J. Org. Chem., 1993, **58**, 4484
1109 T. AKIYAMA, H. NISHIMOTO, S. OZAKI, Tetrahedron Lett., 1991,
32, 1335
1110 G. B. REDDY, T. MINAMI, T. HANAMOTO, T. HIYAMA, J. Org.
Chem., 1991, **56**, 5752
1111 G. SOLLADIE, A. ALMARIO, Tetrahedron Lett., 1992, **33**, 2477
1112 L. MANZONI, T. PILATI, G. POLI, C. SCOLASTICO, J. Chem. Soc.
Chem. Comm., 1992, 1027
1113 G. POLI, L. BELVISI, L. MANZONI, C. SCOLASTICO, J. Org. Chem.,
1993, **58**, 3165
1114 A. PASQUARELLO, G. POLI, C. SCOLASTICO, Synlett, 1992, 93
1115 M. FRANCK-NEUMANN, P. J. COLSON, Synlett, 1991, 891

1116   K. ISHIHARA, A. MORI, I. ARAI, H. YAMAMOTO, Tetrahedron
       Lett., 1986, **27**, 983
1117   K. YAMAMOTO, H. ANDO, H. CHIKAMATSU, J. Chem. Soc. Chem.
       Comm., 1987, 334
1118   K. ISHIHARA, N. HANAKI, H. YAMAMOTO, J. Am. Chem. Soc.,
       1991, **113**, 7074
1119   R. O. HUTCHINS, A. ABDEL-MAGID, Y. P. STERCHO, A.
       WAMBSGANS, J. Org. Chem., 1987, **52**, 704
1120   K. OGURA, H. TOMORI, M. FUJITA, Chem. Lett., 1991, 1407
1121   J. L.GARCIA-RUANO, A. LORENTE, J. H. RODRIGUEZ,
       Tetrahedron Lett., 1992, **33**, 5637
1122   N. KHIAR, I. FERNANDEZ, F. ALCUDIA, D. H. HUA, Tetrahedron
       Lett., 1993, **34**, 699
1123   K. MATSUKI, H. INOUE, M. TAKEDA, Tetrahedron Lett., 1993, **34**,
       1167
1124   R. ROMAGNOLI, E. C. ROOS, H. HIEMSTRA, M. J. MOOLENAAR,
       W. N. SPECKAMP, B. KAPTEIN, H. E. SHOEMAKER, Tetrahedron
       Lett., 1994, **35**, 1087
1125   K. HARADA, T. MUNEGUMI, in Comprehensive Organic Synthesis, B.
       M. TROST and I. FLEMING Ed., Pergamon Press, 1991, vol. 8, chap.
       1.6
1126   C. K. MIAO, R. SORCEK, P. J. JONES, Tetrahedron Lett., 1993, **34**,
       2259
1127   S. A. KING, A. S. THOMPSON, A. O. KING, T. R. VERHOEVEN, J.
       Org. Chem., 1992, **57**, 6689
1128   C. GRECK, L. BISCHOFF, F. FERREIRA, C. PINEL, E. PIVETEAU,
       J. P. GENÊT, Synlett, 1993, 475
1129   G. CAPOZZI, S. ROELENS, S. TALAMI, J. Org. Chem., 1993, **58**,
       7932
1130   K. NOZAKI, N. SATO, H. TAKAYA, Tetrahedron Asymm., 1993, **4**,
       2179
1131   M. KITAMURA, M. TOKUNAGA, R. NOYORI, J. Am. Chem. Soc.,
       1993, **115**, 144
1132   K. MASHIMA, Y. MATSUMURA, K. KUSANO, H. KUMOBAYASHI,
       N. SAYO, Y. HORI, T. ISHIZAKI, S. AKUTAGAWA, H. TAKAYA, J.
       Chem. Soc. Chem. Comm., 1991, 609
1133   R. NOYORI, T. IKEDA, T. OHKUMA, M. WIDHALM, M.
       KITAMURA, H. TAKAYA, S. AKUTAGAWA, N. SAYO, T. SAITO,
       T. TAKETOMI, H. KUMOBAYASHI, J. Am. Chem. Soc., 1989, **111**,
       9134
1134   T. OHKUMA, M. KITAMURA, R. NOYORI, Tetrahedron Lett., 1990,
       31, 5509
1135   M. KITAMURA, M. TOKUNAGA, R. NOYORI, J. Org. Chem., 1992,
       57, 4053

1136 S. SAKUBARA, K. ACHIWA, Synlett, 1991, 689

1137 S. SAKURABA, N. NAKAJIMA, K. ACHIWA, Synlett, 1992, 829

1138 S. SAKURABA, N. NAKAJIMA, K. ACHIWA, Tetrahedron Asymm., 1993, **4**, 1457

1139 H. TAKEDA, S. HOSOKAWA, M. ABURATANI, K. ACHIWA, Synlett, 1991, 193

1140 R. L. AUGUSTINE, S. K. TANIELYAN, L. K. DOYLE, Tetrahedron Asymm., 1993, **4**, 1803

1141 S. BHADURI, V. S. DARSHANE, K. SHARMA, D. MUKESH, J. Chem. Soc. Chem. Comm., 1992, 1738

1142 A. TAI, T. KIKUKAWA, T. SUGIMURA, Y. INOUE, T. OSAWA, S. FUJII, J. Chem. Soc. Chem. Comm., 1991, 795

1143 C. BOLM, Angew. Chem. Int. Ed. Engl., 1993, **32**, 232

1144 C. A. WILLOUGHBY, S. L. BUCHWALD, J. Org. Chem., 1993, **58**, 7627

1145 M. J. BURK, J. E. FEASTER, J. Am. Chem. Soc., 1992, **114**, 6266

1146 J. P. GENÊT, V. RATOVELOMANANA-VIDAL, C. PINEL, Synlett, 1993, 478

1147 N. KOGA, K. MOROKUMA, J. Am. Chem. Soc., 1993, **115**, 6883

1148 M. NISHIYAMA, S. YAMAGUCHI, M. KONDO, K. ITOH, J. Org. Chem., 1992, **57**, 4306

1149 M. NORTH, Synlett, 1993, 807

1150 H. DANDA, Synlett, 1991, 263

1151 K. TANAKA, A. MORI, S. INOUE, J. Org. Chem., 1990, **55**, 181

1152 H. NITTA, D. YU, M. KUDO, A. MORI, S. INOUE, J. Am. Chem. Soc., 1992, **114**, 7969

1153 S. KOBAYASHI, Y. TSUCHIYA, T. MUKAIYAMA, Chem. Lett., 1991, 541

1154 H. MINAMIKAWA, S. HAYAKAWA, T. YAMADA, N. IWASAWA, K. NARASAKA, Bull. Chem. Soc. Japan, 1988, **61**, 4379

1155 M. HAYASHI, T. MATSUDA, N. OGUNI, J. Chem. Soc. Chem. Comm., 1990, 1364, J. Chem. Soc. Perkin I, 1992, 3135

1156 E. J. COREY, Z. WANG, Tetrahedron Lett., 1993, **34**, 4001

1157 H. OHNO, H. NITTA, K. TANAKA, A. MORI, S. INOUE, J. Org. Chem., 1992, **57**, 6778

1158 C. ANDRÉS, M. DELGADO, R. PEDROSA, R. RODRIGUEZ, Tetrahedron Lett., 1993, **34**, 8325

1159 A. FADEL, Synlett, 1993, 503

1160 W. S. JOHNSON, B. FREI, A. S. GOPALAN, J. Org. Chem., 1981, **46**, 1513

1161 S. MATSUI, A. UEJIMA, Y. SUZUKI, K. TANAKA, J. Chem. Soc. Perkin I, 1993, 701

1162 B. PESCHKE, J. LUSSMANN, M. DYRBUSCH, D. HOPPE, Chem. Ber., 1992, **125**, 1421

1163 O. ZSCHAGE, D. HOPPE, Tetrahedron, 1992, **48**, 5657
1164 K. SOAI, Y. KAWASE, J. Chem. Soc. Perkin I, 1990, 3214
1165 K. SOAI, Y. KAWASE, A. OSHIO, J. Chem. Soc. Perkin I, 1991, 1613
1166 Y. MATSUMOTO, A. OHNO, S. J. LU, T. HAYASHI, N. OGUNI,
       M.HAYASHI, Tetrahedron Asymm., 1993, **4**, 1763
1167. M. RAMOS-TOMBO, E. DIDIER, B. LOUBINOUX, Synlett, 1990, 547
1168 W. OPPOLZER, R. N. RADINOV, Helv. Chim. Acta, 1992, **75**, 170
1169 W. OPPOLZER, R. N. RADINOV, J. Am. Chem. Soc., 1993, **115**, 1593
1170 T. H. CHAN, K. T. NWE, J. Org. Chem., 1992, **57**, 6107
1171 G. B. JONES, S. B. HEATON, Tetrahedron Asymm., 1993, **4**, 261
1172 M. UEMURA, R. MIYAKE, K. NAKAYAMA, M. SHIRO, Y.
       HAYASHI, J. Org. Chem., 1993, **58**, 1238
1173 K. ITO, Y. KIMURA, H. OKAMURA, T. KATSUKI, Synlett, 1992, 573
1174 S. CONTI, M. FALORNI, G. GIACOMELLI, F. SOCCOLINI,
       Tetrahedron, 1992, **48**, 8993
1175 C. BOLM, J. MÜLLER, G. SCHLINGLOFF, M. ZEHNDER, M.
       NEUBURGER, J. Chem. Soc. Chem. Comm., 1993, 182
1176 N. N. JOSHI, M. SREBNIK, H. C. BROWN, Tetrahedron Lett., 1989,
       30, 5551
1177 W. BRIEDEN, R. OSTWALD, P. KNOCHEL, Angew. Chem. Int. Ed.
       Engl., 1993, **32**, 582
1178 M. J. ROZEMA, C. EISENBERG, H. LÜTJENS, R. OSTWALD, K.
       BELYK, P. KNOCHEL, Tetrahedron Lett., 1993, **34**, 3115
1179 J. L. von dem BUSSCHE-HÛNNEFELD, D. SEEBACH, Tetrahedron,
       1992, **48**, 5719
1180 K. SOAI, Y. HIROSE, S. SAKATA, Tetrahedron Asymm., 1992, **3**, 677
1181 S. KUSUDA, K. KAWAMURA, Y. UENO, T. TORU, Tetrahedron
       Lett., 1993, **34**, 6587
1182 K. SOAI, T. HATANAKA, T. YAMASHITA, J. Chem. Soc. Chem.
       Comm., 1992, 927
1183 P. KNOCHEL, W. BRIEDEN, M. J. ROZEMA, C. EISENBERG,
       Tetrahedron Lett., 1993, **34**, 5881
1184 K. C. FRIEBOES, T. HARDER, D. AULBERT, C. STRAHRINGER,
       M. BOLTE, D. HOPPE, Synlett, 1993, 921
1185 M. Y. CHEN, J. M. FANG, J. Org. Chem., 1992, **57**, 2937
1186 D. BASAVAIAH, T. K. BHARATHI, Tetrahedron Lett., 1991, **32**, 3417
1187 Y. TAMAI, M. AKIYAMA, A. OKAMURA, S. MIYANO, J. Chem.
       Soc. Chem. Comm., 1992, 687
1188 C. BALDOLI, P. DEL BUTTERO, E. LICANDRO, S. MAIORANA, A.
       PAPAGNI, M. TORCHIO, Tetrahedron Lett., 1993, **34**, 7943
1189 K. SOAI, T. HATANAKA, T. MIYAZAWA, J. Chem. Soc. Chem.
       Comm., 1992, 1097
1190 S. I. MURAHASHI, J. SUN, T. TSUDA, Tetrahedron Lett., 1993, **34**,
       2645

1191 Y. UKAJI, T. HATANAKA, A. AHMED, K. INOMATA, Chem. Lett., 1993, 1313

1192 P. EMERT, I. MEYER, C. STUCKI, J. SCHNEEBELI, J. P. OBRECHT, Tetrahedron Lett., 1988, **29**, 1265

1193 D. P. G. HAMON, R. A. MASSY-WESTROPP, P. RAZZINO, Tetrahedron, 1992, **48**, 5163

1194 A. BOCOUM, D. SAVOIA, A. UMANI-RONCHI, J. Chem. Soc. Chem. Comm., 1993, 1542; T. BASILE, A. BOCOUM, D. SAVOIA, A. UMANI-RONCHI, J. Org. Chem., 1994, **59**, 7766

1195 Z. CHANG, R. M. COATES, J. Org. Chem., 1990, **55**, 3475

1195a D. ENDERS, R. FUNK, M. KLATT, G. RAABE, E. R. HOVESTREYDT, Angew. Chem. Int. Ed. Engl., 1993, **32**, 418

1196 A. ALEXAKIS, N. LENSEN, J. P. TRANCHIER, P. MANGENEY, J. Org. Chem., 1992, **57**, 4563

1197 Y. TAKEMOTO, J. TAKEUCHI, C. IWATA, Tetrahedron Lett., 1993, **34**, 6069

1198 Y. UKAJI, K. KUME, T. WATAL, T. FUJISAWA, Chem. Lett., 1991, 173

1199 D. L. COMINS, M. O. KILLPACK, J. Am. Chem. Soc., 1992, **114**, 10972

1200 H. C. BROWN, S. V. KULKARNI, U. S. RACHERLA, J. Org. Chem., 1994, **59**, 365

1200a S. HANESSIAN, A. TEHIM, P. CHEN, J. Org. Chem., 1993, **58**, 7768

1201 U. S. RACHERLA, Y. LIAO, H. C. BROWN, J. Org. Chem., 1992, **57**, 6614

1202 W. R. ROUSH, L. BANFI, J. Am. Chem. Soc., 1988, **110**, 3979

1203 P. GANESH, K. M. NICHOLAS, J. Org. Chem., 1993, **58**, 5587

1204 W. R. ROUSH, A. M. RATZ, J. A. JABLONOWSKI, J. Org. Chem., 1992, **57**, 2047

1205 A. G. M. BARRETT, S. A. LEBOLD, J. Org. Chem., 1991, **56**, 4875

1206 L. K. TRUESDALE, D. SWANSON, R. C. SUN, Tetrahedron Lett., 1985, **26**, 5009

1207 W. R. ROUSH, J. C. PARK, Tetrahedron Lett., 1991, **32**, 6285

1208 T. A. J. van der HEIDE, J. L. van der BAAN, E. A. BIJPOST, F. J. J. de KANTER, F. BICKELHAUPT, G. W. KLUMPP, Tetrahedron Lett., 1993, **34**, 4655

1209 K. BURGESS, I. HENDERSON, Tetrahedron Lett., 1990, **31**, 6949

1210 A. VULPETTI, M. GARDNER, C. GENNARI, A. BERNARDI, J. M. GOODMAN, I. PATERSON, J. Org. Chem., 1993, **58**, 1711

1211 R. W. HOFFMANN, S. DRESELY, Chem. Ber., 1989, **122**, 903

1212 Y. NISHIGAICHI, A. TAKUWA, Y. NARUTA, K. MARUYAMA, Tetrahedron, 1993, **49**, 7395

1213 J. A. MARSHALL, W. Y. GUNG, Tetrahedron, 1989, **45**, 1043

1214 J. A. MARSHALL, W. Y. GUNG, Tetrahedron Lett., 1989, **30**, 2183

1215   J. A. MARSHALL, J. A. MARKWALDER, Tetrahedron Lett., 1988, **29**, 4811
1216   J. A. MARSHALL, D. V. YASHUNSKY, J. Org. Chem., 1991, **56**, 5493
1217   J. A. MARSHALL, G. S. WELMAKER, Synlett, 1992, 537
1218   A. L. COSTA, M. G. PIAZZA, E. TAGLIAVINI, C. TROMBINI, A. UMANI-RONCHI, J. Am. Chem. Soc., 1993, **115**, 7001
1219   G. E. KECK, L. S. GERACI, Tetrahedron Lett., 1993, **34**, 7827
1220   G. E. KECK, D. KRISHNAMURTHY, M. C. GRIER, J. Org. Chem., 1993, **58**, 6543
1221   G. E. KECK, K. H. TARBET, L. S. GERACI, J. Am. Chem. Soc., 1993, **115**, 8467
1222   K. NISHITANI, K. YAMAKAWA, Tetrahedron Lett., 1991, **32**, 387
1223   Y. YAMAMOTO, K. KOBAYASHI, H. OKANO, I. KADOTA, J. Org. Chem., 1992, **57**, 7003
1224   A. B. CHARETTE, C. MELLON, L. ROUILLARD, E. MALENFANT, Synlett, 1993, 81
1225   J. S. PANEK, P. F. CIRILLO, J. Org. Chem., 1993, **58**, 999
1226   M. Y. CHEN, J. M. FANG, J. Chem. Soc. Perkin I, 1993, 1737
1227   K. MIKAMI, H. WAKABAYSHI, T. NAKAI, J. Org. Chem., 1991, **56**, 4337
1228   S. HATAKEYAMA, K. SUGAWARA, M. KAWAMURA, S. TAKANO, Tetrahedron Lett., 1991, **32**, 4509
1229   S. HATAKEYAMA, K. SUGAWARA, S. TAKANO, J. Chem. Soc. Chem. Comm., 1991, 1533
1230   I. MORI, K. ISHIHARA, L. A. FLIPPIN, K. NOZAKI, H. YAMAMOTO, P. A. BARTLETT, C. H. HEATHCOCK, J. Org. Chem., 1990, **55**, 6107
1231   L. F. TIETZE, A. DÖLLE, K. SCHIEMANN, Angew. Chem. Int. Ed. Engl., 1992, **31**, 1372
1232   E. F. KLEINMAN, R. A. VOLKMAN, in Comprehensive Organic Synthesis, B. M. TROST and I. FLEMING Ed., Pergamon Press, 1991, vol. 2, chap. 4.3
1233   Y. UKAJI, K. TSUKAMOTO, Y. NASADA, M. SHIMIZU, T. FUJISAWA, Chem. Lett., 1993, 221
1234   J. A. MARSHALL, X. J. WANG, J. Org. Chem., 1991, **56**, 3212; 1992, **57**, 1242
1235   W. S. JOHNSON, V. R. FLETCHER, B. CHENERA, W. R. BARTLETT, F. S. THAM, R. K. KULLNIG, J. Am. Chem. Soc., 1993, **115**, 497
1236   K. MIKAMI, M. SHIMIZU, Chem. Rev., 1992, **92**, 1021
1237   B. B. SNIDER, in Comprehensive Organic Synthesis, B. M. TROST and I. FLEMING Ed., Pergamon Press, 1991, vol. 2, chap. 2.1
1238   K. MARUOKA, Y. HOSHINO, T. SHIRASAKA, H. YAMAMOTO, Tetrahedron Lett., 1988, **29**, 3967

1239 K. MIKAMI, T. YAJIMA, M. TERADA, T. UCHIMARU, Tetrahedron Lett., 1993, **34**, 7591

1240 M. TERADA, S. MATSUKAWA, K. MIKAMI, J. Chem. Soc. Chem. Comm., 1993, 327

1241 F. T. van der MEER, B. L. FERINGA, Tetrahedron Lett., 1992, **33**, 6695

1242 K. MIKAMI, E. SAWA, M. TERADA, Tetrahedron Asymm., 1991, **2**, 1403

1243 K. MIKAMI, M. KANEKO, T. YAJIMA, Tetrahedron Lett., 1993, **34**, 4841

1244 I. KOGA, K. FUNAKOSHI, A. MATSUDA, K. SAKAI, Tetrahedron Asymm., 1993, **4**, 1857

1245 H. SASAI, Y. KIRIO, M. SHIBASAKI, J. Org. Chem., 1990, **55**, 5306

1246 H. C. BROWN, K. GANESAN, Tetrahedron Lett., 1992, **33**, 3421

1247 K. GANESAN, H. C. BROWN, J. Org. Chem., 1993, **58**, 7162

1248 C. GENNARI, D. MORESCA, S. VIETH, A. VULPETTI, Angew. Chem. Int. Ed. Engl., 1993, **32**, 1618

1249 A. BERNARDI, A. COMOTTI, C. GENNARI, C. T. HEWKIN, J. M. GOODMAN, A. SCHLAPBACH, I. PATERSON, Tetrahedron, 1994, **50**, 1227

1249a A. FURSTNER, Synthesis, 1989, 571

1250 K. SOAI, A. OSHIO, T. SAITO, J. Chem. Soc. Chem. Comm., 1993, 811

1251 J. M. GOODMAN, I. PATERSON, Tetrahedron Lett., 1992, **33**, 7223

1252 C. GENNARI, C. T. HEWKIN, F. MOLINARI, A. BERNARDI, A. COMOTTI, J. M. GOODMAN, I. PATERSON, J. Org. Chem., 1992, **57**, 5173

1253 A. BERNARDI, A. CASSINARI, A. COMOTTI, M. GARDNER, C. GENNARI, J. M. GOODMAN, I. PATERSON, Tetrahedron, 1992, **48**, 4183

1254 E. J. COREY, S. CHOI, Tetrahedron Lett., 1991, **32**, 2857

1255 E. J. COREY, D. H. LEE, Tetrahedron Lett., 1993, **34**, 1737

1256 H. SASAI, N. ITOH, T. SUZUKI, M. SHIBASAKI, Tetrahedron Lett., 1993, **34**, 855

1257 M. SASAI, T. SUZUKI, N. ITOH, S. ARAI, M. SHIBASAKI, Tetrahedron Lett., 1993, **34**, 2657; M. SASAI, T. SUZUKI, N. ITOH, K. TANAKA, T. DATE, K. OKAMURA, M. SHIBASAKI, J. Am. Chem. Soc., 1993, **115**, 10372

1258 M. BRAUN, Angew. Chem. Int. Ed. Engl., 1987, **26**, 24

1259 D. ENDERS, H. DYKER, G. RAABE, Angew. Chem. Int. Ed. Engl., 1993, **32**, 421

1260 J. W. B. COOKE, S. G. DAVIES, A. NAYLOR, Tetrahedron, 1993, **49**, 7955

1261 K. ISEKI, S. OISHI, T. TAGUCHI, Y. KOBAYASHI, Tetrahedron Lett., 1993, **34**, 8147

1262   Y. XIANG, E. OLIVIER, N. OUIMET, Tetrahedron Lett., 1992, **33**, 457

1263   A. J. ROBICHAUD, G. D. BERGER, D. A. EVANS, Tetrahedron Lett., 1993, **34**, 8403

1264   R. BAKER, J. L. CASTRO, C. J. SWAIN, Tetrahedron Lett., 1988, **29**, 2247

1265   K. HAYASHI, Y. HAMADA, T. SHIOIRI, Tetrahedron Lett., 1991, **32**, 7287

1266   D. A. EVANS, J. R. GAGE, J. L. LEIGHTON, J. Am. Chem. Soc., 1992, **114**, 9434

1267   D. A. EVANS, S. W. KALDOR, T. K. JONES, J. CLARDY, T. J. STOUT, J. Am. Chem. Soc., 1990, **112**, 7001

1268   C. GENNARI, S. VIETH, A. COMOTTI, A. VULPETTI, J. M. GOODMAN, I. PATERSON, Tetrahedron, 1992, **48**, 4439

1269   B. D. DORSEY, K. J. PLZAK, R. G. BALL, Tetrahedron Lett., 1993, **34**, 1851

1270   S. E. DREWES, D. G. S. MALISSAR, G. H. P. ROOS, Chem. Ber., 1991, **124**, 2913

1271   W. OPPOLZER, P. LIENARD, Tetrahedron Lett., 1993, **34**, 4321

1272   S. KANEMASA, T. MORI, A. TATSUKAWA, Tetrahedron Lett., 1993, **34**, 8293

1273   T. BEULSHAUSEN, U. GROTH, U. SCHÖLLKOPF, Liebigs Ann., 1991, 1207

1274   U. GROTH, U. SCHÖLLKOPF, T. TILLER, Tetrahedron, 1991, **47**, 2835

1275   U. SCHÖLLKOPF, T. BEULSHAUSEN, Liebigs Ann., 1989, 223

1276   V. A. SOLOSHONOK, V. P. KUKHAR, S. V. GALUSHKO, N. Y. SVISTUNOVA, D. V. AVILOV, N. A. KUZ'MINA, N. I. RAEVSKI, Y. T. STRUCHKOV, A. P. PYSAREVSKY, Y. N. BELOKON, J. Chem. Soc. Perkin I, 1993, 3143

1277   K. NARASAKA, T. MIWA, Chem. Lett., 1985, 1217

1278   A. J. WALKER, Tetrahedron Asymm., 1992, **3**, 961

1279   S. KUSUDA, Y. UENO, T. TORU, Tetrahedron, 1994, **50**, 1045

1280   M. UEMURA, T. MINAMI, M. SHIRO, Y. HAYASHI, J. Org. Chem., 1992, **57**, 5590

1281   C. MUKAI, I. J. KIM, M. HANAOKA, Tetrahedron Lett., 1993, **34**, 6081

1282   C. MUKAI, I. J. KIM, E. FURU, M. HANAOKA, Tetrahedron, 1993, **49**, 8323

1283   F. BERNARDI, A. BONGINI, G. CAINELLI, M. A. ROBB, G. S. VALLI, J. Org. Chem., 1993, **58**, 750

1284   X. WANG, C. LEE, Tetrahedron Lett., 1993, **34**, 6241

1285   I. OJIMA, Y. H. PARK, C. M. SUN, T. BRIGAUD, M. ZHAO, Tetrahedron Lett., 1992, **33**, 5737

1285a Y. ITO, A. SASAKI, K. TAMOTO, M. SUNAGAWA, S.
      TERASHIMA, Tetrahedron, 1991, **47**, 2801
1286 Y. NAGAO, W. M. DAI, M. OCHIAI, M. SHIRO, Tetrahedron, 1990,
      46, 6361
1286a Y. NAGAO, W. M. DAI, M. OCHIAI, S. TSUKAGOSHI, E. FUJITA, J.
      Org. Chem., 1990, **55**, 1148
1287 Y. NAGAO, T. KUMAGAI, Y. NAGASE, S. TAMAI, Y. INOUE, M.
      SHIRO, J. Org. Chem., 1992, **57**, 4232
1288 A. M. KANAZAWA, J. N. DENIS, A. E. GREENE, J. Org. Chem.,
      1994, **59**, 1238
1289 F. H. vanDER STEEN, H. KLEIJN, G. J. P. BRITOVSEK, J. T. B. H.
      JASTRZEBSKI, G. van KOTEN, J. Org. Chem., 1992, **57**, 3906
1290 R. ANNUNZIATA, M. BENAGLIA, M. CINQUINI, F. COZZI, L.
      RAIMONDI, Tetrahedron Lett., 1993, **34**, 6921
1291 C. BALDOLI, P. Del BUTTERO, E. LICANDRO, S. MAIORANA, A.
      PAPAGNI, Synlett, 1994, 183
1292 T. FUJISAWA, M. ICHIKAWA, Y. UKAJI, M. SHIMIZU, Tetrahedron
      Lett., 1993, **34**, 1307
1293 T. FUJISAWA, Y. UKAJI, T. NORO, K. DATE, M. SHIMIZU,
      Tetrahedron, 1992, **48**, 5629
1294 H. FUJIOKA, T. YAMANAKA, N. MATSUNAGA, M. FUJI, Y. KITA,
      Synlett, 1992, 35
1294a M. SHIMIZU, Y. UKAJI, J. TANIZAKI, T. FUJISAWA, Chem. Lett.,
      1992, 1349
1295 C. GENNARI, in Comprehensive Organic Synthesis, B. M. TROST and I.
      FLEMING Ed., Pergamon Press, 1991, vol. 2, chap. 2.4
1296 K. MIKAMI, S. MATSUKAWA, J. Am. Chem. Soc., 1994, **116**, 4077
1297 K. HATTORI, H. YAMAMOTO, J. Org. Chem., 1993, **58**, 5301
1298 J. OTERA, Y. FUJITA, S. FUKUZUMI, Synlett, 1994, 213
1299 S. KOBAYASHI, T. KAWASUJI, Synlett, 1993, 911
1300 S. KOBAYASHI, Y. FUJISHITA, T. MUKAIYAMA, Chem. Lett.,
      1989, 2069
1301 S. KOBAYASHI, H. UCHIRO, I. SHIINA, T. MUKAIYAMA,
      Tetrahedron, 1993, **49**, 1761
1302 S. I. KIYOOKA, Y. KANEKO, K. I. KUME, Tetrahedron Lett., 1992,
      33, 4927
1303 A. ANDO, T. MIURA, T. TATEMATSU, T. SHIOIRI, Tetrahedron
      Lett., 1993, **34**, 1507
1304 W. OPPOLZER, C. STARKEMANN, Tetrahedron Lett., 1992, **33**, 2439
1305 B. M. TROST, H. URABE, J. Org. Chem., 1990, **55**, 3982
1306 M. FRANCK-NEUMANN, P. J. COLSON, P. GEOFFROY, K. M.
      TABA, Tetrahedron Lett., 1992, **33**, 1903
1307 S. KOBAYASHI, M. HORIBE, Synlett, 1994, 147
1308 K. MIKAMI, S. MATSUKAWA, J. Am. Chem. Soc., 1993, **115**, 7039

1309   T. AKIYAMA, K. ISHIKAWA, S. OZAKI, Synlett, 1994, 275
1310   W. S. JOHNSON, A. B. KELSON, J. D. ELLIOT, Tetrahedron Lett., 1988, **29**, 3757
1311   C. PALAZZI, G. POLI, C. SCOLASTICO, R. VILLA, Tetrahedron Lett., 1990, **41**, 4223
1312   K. HATTORI, H. YAMAMOTO, Tetrahedron, 1994, **50**, 2785
1313   R. KOBER, K. PAPADOPOULOS, W. MILTZ, D. ENDERS, W. STEGLICH, H. REUTER, H. PUFF, Tetrahedron, 1985, **41**, 1693
1314   C. PUCHOT, O. SAMUEL, E. DUNACH, S. ZHAO, C. AGAMI, H. B. KAGAN, J. Am. Chem. Soc., 1986, **108**, 2353
1315   B. R. BENDER, M. KOLLER, D. NANZ, W. VON PHILIPSBORN, J. Am. Chem. Soc., 1993, **115**, 5889
1316   J. S. GIOVANNETTI, C. M. KELLY, C. R. LANDIS, J. Am. Chem. Soc., 1993, **115**, 4040
1317   U. NAGEL, A. BUBLEWITZ, Chem. Ber., 1992, **125**, 1061
1318   H. J. KREUZFELD, C. DÖBLER, H. W. KRAUSE, C. FACKLAM, Tetrahedron Asymm., 1993, **4**, 2047
1319   I. GRASSERT, E. PAETZOLD, G. OEHME, Tetrahedron, 1993, **49**, 6605
1320   S. K. ARMSTRONG, J. M. BROWN, M. J. BURK, Tetrahedron Lett., 1993, **34**, 879
1321   U. NAGEL, T. KRINK, Angew. Chem. Int. Ed. Engl., 1993, **32**, 1052
1322   A. TOGNI, C. BREUTEL, A. SCHNYDER, F. SPINDLER, H. LANDERT, A. TIJANI, J. Am. Chem. Soc., 1994, **116**, 4062
1323   H. JENDRALLA, Tetrahedron Lett., 1991, **32**, 3671
1324   W. LEITNER, J. M. BROWN, H. BRUNNER, J. Am. Chem. Soc., 1993, **115**, 152
1325   M. TAKAGI, K. YAMAMOTO, Tetrahedron, 1991, **47**, 8869
1326   M. KITAMURA, Y. HSIAO, M. OHTA, M. TSUKAMOTO, T. OHTA, H. TAKAYA, R. NOYORI, J. Org. Chem., 1994, **59**, 297
1327   J. P. GENÊT, C. PINEL, V. RATOVELOMANANA-VIDAL, S. MALLART, X. PFISTER, L. BISCHOFF, M. C. CAÑO DE ANDRADE, S. DARSES, C. GALOPIN, J. A. LAFFITTE, Tetrahedron Asymm., 1994, **5**, 675
1328   T. OHTA, T. MIYAKE, N. SEIDO, H. KUMOBAYASHI, S. AKUTAGAWA, H. TAKAYA, Tetrahedron Lett., 1992, **33**, 635
1329   T. OHTA, T. MIYAKE, H. TAKAYA, J. Chem. Soc. Chem. Comm., 1992, 1725
1330   M. SABURI, L. SHAO, T. SAKURAI, Y. UCHIDA, Tetrahedron Lett., 1992, **33**, 7877
1331   T. OHTA, H. TAKAYA, R. NOYORI, Tetrahedron Lett., 1990, **31**, 7189
1332   M. T. ASHBY, J. HALPERN, J. Am. Chem. Soc., 1991, **113**, 589
1333   M. SABURI, M. OHNUKI, M. OGASAWARA, T. TAKAHASHI, Y. UCHIDA, Tetrahedron Lett., 1992, **33**, 5783

1334 R. L. HALTERMAN, K. P. C. VOLHARDT, M.E. WELKER, D. BLASER, R. BOESE, J. Am. Chem. Soc., 1987, **109**, 8105

1335 A. BERNARDI, O. CARUGO, A. PASQUARELLO, A. SIDJIMOV, G. POLI, Tetrahedron, 1991, **47**, 7357

1336 G. HAVIARI, J. P. CELERIER, H. PETIT, G. LHOMMET, D. GARDETTE, J. C. GRAMAIN, Tetrahedron Lett., 1992, **33**, 4311

1337 G. B. FISHER, C. T. GORALSKI, L. W. NICHOLSON, B. SINGARAM, Tetrahedron Lett., 1993, **34**, 7693

1338 J. ZANG, B. LOU, G. GUO, L. DAI, J. Org. Chem., 1991, **56**, 1670

1338a K. BURGESS, W. A. van der DONK, M. B. JARSTFER, M. J. OHLMEYER, J. Am. Chem. Soc., 1991, **113**, 6139

1339 Y. MATSUMOTO, M. NAITO, Y. UOZUMI, T. HAYASHI, J. Chem. Soc. Chem. Comm., 1993, 1468

1339a T. HIYAMA, T. KUSUMOTO, in Comprehensive Organic Synthesis, B. M. TROST and I. FLEMING Ed., Pergamon Press, 1991, vol. 8, chap. 3.12

1340 T. HAYASHI, Y. UOZUMI, Pure & Appl. Chem., 1992, **64**, 1911

1341 Y. UOZUMI, S. Y. LEE, T. HAYASHI, Tetrahedron Lett., 1992, **33**, 7185

1342 N. SAKAI, K. NOZAKI, H. TAKAYA, J. Chem. Soc. Chem. Comm., 1994, 395

1343 X. M. WU, K. FUNAKOSHI, K. SAKAI, Tetrahedron Lett., 1992, **33**, 6331

1344 T. V. RAJANBABU, A. L.CASALNUOVO, J. Am. Chem. Soc., 1992, **114**, 6265

1345 S. HANESSIAN, P. MEFFRE, M. GIRARD, S. BEAUDOIN, J. Y. SANCÉAU, Y. BENNANI, J. Org. Chem., 1993, **58**, 1991

1346 M. NAKAJIMA, K. TOMIOKA, Y. IITAKA, K. KOGA, Tetrahedron, 1993, **49**, 10793

1347 T. OISHI, M. HIRAMA, J. Org. Chem., 1989, **54**, 5834

1347a K. FUJI, K. TANAKA, H. MIYAMOTO, Tetrahedron Lett., 1992, **33**, 4021

1348 M. NAKAJIMA, K. TOMIOKA, K. KOGA, Tetrahedron, 1993, **49**, 10807

1349 T. OISHI, K. I. IIDA, M. HIRAMA, Tetrahedron Lett., 1993, **34**, 3573

1350 E. J. COREY, P. D. JARDINE, S. VIRGIL, P. W. YUEN, R. D. CONNELL, J. Am. Chem. Soc., 1989, **111**, 9243

1351 Y. D. WU, Y. WANG, K. N. HOUK, J. Org. Chem., 1992, **57**, 1362

1352 H. C. KOLB, P. G. ANDERSON, K.B. SHARPLESS, J. Am. Chem. Soc., 1994, **116**, 1278

1353 P. G. ANDERSSON, K. B. SHARPLESS, J. Am. Chem. Soc., 1993, **115**, 7047

1354 G. A. CRISPINO, K. S. JEONG, H. C. KOLB, Z. M. WANG, D. XU, K. B. SHARPLESS, J. Org. Chem., 1993, **58**, 3785

1355 K. S. JEONG, P. SJO, K. B. SHARPLESS, Tetrahedron Lett., 1992, **33**, 3833

1356 A. R. BASSINDALE, P. G. TAYLOR, Y. XU, J. Chem. Soc. Perkin I, 1994, 1061

1357 J. A. SODERQUIST, A. M. RANE, C. J. LÓPEZ, Tetrahedron Lett., 1993, **34**, 1893

1358 Z. M. WANG, K. B. SHARPLESS, Synlett, 1993, 603

1359 Z. M. WANG, X. L. ZHANG, K. B. SHARPLESS, Tetrahedron Lett., 1993, **34**, 2267

1360 L. WANG, K. B. SHARPLESS, J. Am. Chem. Soc., 1992, **114**, 7568

1361 K. MORIKAWA, J. PARK, P. G. ANDERSSON, T. HASHIYAMA, K. B. SHARPLESS, J. Am. Chem. Soc., 1993, **115**, 8463

1362 T. HASHIYAMA, K. MORIKAWA, K. B. SHARPLESS, J. Org. Chem., 1992, **57**, 5067

1363 D. XU, G. A. CRISPINO, K. B. SHARPLESS, J. Am. Chem. Soc., 1992, **114**, 7570

1364 G. VIDARI, A. DAPIAGGI, G. ZANONI, L. GARLASCHELLI, Tetrahedron Lett., 1993, **34**, 6485

1365 Z. M. WANG, K. B. SHARPLESS, Tetrahedron Lett., 1993, **34**, 8225

1366 Y. L. BENNANI, K.B. SHARPLESS, Tetrahedron Lett., 1993, **34**, 2079

1367 P. J. WALSH, K. B. SHARPLESS, Synlett, 1993, 605

1368 P. J. WALSH, Y. L. BENNANI, K. B. SHARPLESS, Tetrahedron Lett., 1993, **34**, 5545

1369 K. J. HALE, S. MANAVIAZAR, S. A. PEAK, Tetrahedron Lett., 1994, **35**, 425

1370 B. B. LOHRAY, V. BHUSHAN, Tetrahedron Lett., 1993, **34**, 3911

1371 M. S. van NIEUWENHZE, K. B. SHARPLESS, J. Am. Chem. Soc., 1993, **115**, 7864

1372 S. C. SINHA, A. SINHA-BAGCHI, E. KEINAN, J. Org. Chem., 1993, **58**, 7789

1373 J. P. LELLOUCHE, A. GIGOU-BARBEDETTE, R. GREE, Bull. Soc. Chim. France, 1992, 605

1374 W. ZHANG, N. H. LEE, E. N. JACOBSEN, J. Am. Chem. Soc., 1994, **116**, 425

1375 N. HOSOYA, A. HATAYAMA, K.YANAI, H. FUJII, R. IRIE, T. KATSUKI, Synlett, 1993, 641

1376 R. IRIE, N. HOSOYA, T. KATSUKI, Synlett, 1994, 255

1377 R. IRIE, Y. ITO, T. KATSUKI, Synlett, 1991, 265

1378 T. MUKAIYAMA, T. YAMADA, T. NAGATA, K. IMAGAWA, Chem. Lett., 1993, 327

1379 J. P. COLLMANN, X. ZHANG, V. J. LEE, J. I. BRAUMAN, J. Chem. Soc. Chem. Comm., 1992, 1647

1379a Y. NARUTA, F. TANI, N. ISHIHARA, K. MARUYAMA, J. Am. Chem. Soc., 1991, **113**, 6865

1380 P. G. POTVIN, S. BIANCHET, J. Org. Chem., 1992, **57**, 6629
1381 S. S. WOODARD, M. G. FINN, K. B. SHARPLESS, J. Am. Chem. Soc.,
1991, **113**, 106
1382 S. TAKANO, Y. IWABUCHI, K. OGASAWARA, Synlett, 1991, 548
1383 S. TAKANO, Y. IWABUCHI, K. OGASAWARA, J. Am. Chem. Soc.,
1991, **113**, 2786
1384 J. R. FLISAK, K. J. GOMBATZ, M. M. HOLMES, A. A. JARMAS, I.
LANTOS, W. L. MENDELSON, V. J. NOVACK, J. J. REMICH, L.
SNYDER, J. Org. Chem., 1993, **58**, 6247
1385 S. CARDANI, C. GENNARI, C. SCOLASTICO, R. VILLA,
Tetrahedron, 1989, **45**, 7397
1386 E. J. COREY, L. I. WU, J. Am. Chem. Soc., 1993, **115**, 9327
1387 D. A. EVANS, M. M. FAUL, M. T. BILODEAU, J. Am. Chem. Soc.,
1994, **116**, 2742
1388 A. SERA, K. TAKAGI, H. KATAYAMA, H. YAMADA, J. Org. Chem.,
1988, **53**.1157
1389 N. ASAO, N. TSUKADA, Y. YAMAMOTO, J. Chem. Soc. Chem.
Comm., 1993, 1660
1390 M. E. BUNNAGE, S. G. DAVIES, C. J. GOODWIN, I. A. S.
·WALTERS, Tetrahedron Asymm., 1994, **5**, 35
1391 S. G. DAVIES, N. M. GARRIDO, O. ICHIHARA, I. A. S. WALTERS,
J. Chem. Soc. Chem. Comm., 1993, 1153
1392 S. G. DAVIES, O. ICHIHARA, I. A. S. WALTERS, Synlett, 1993, 461
1393 S. G. DAVIES, O. ICHIHARA, I. A. S. WALTERS, Synlett, 1994, 117
1394 J. G. RICO, R. J. LINDMARK, T. E. RODGERS, P. R. BOVY, J. Org.
Chem., 1993, **58**, 7948
1395 M. E. BUNNAGE, S. G. DAVIES, C. J. GOODWIN, Synlett, 1993, 731
1396 H. LUBBEN, B. L. FERINGA, Tetrahedron Asymm., 1991, **2**, 775 and
quoted ref.
1397 Y. YAMAMOTO, N. ASAO, T. UYEHARA, J. Am. Chem. Soc., 1992,
114, 5427
1398 I. SUZUKI, H. KIN, Y. YAMAMOTO, J. Am. Chem. Soc., 1993, **115**,
10139
1399 N. SHIDA, T. UYEHARA, Y. YAMAMOTO, J. Org. Chem., 1992, **57**,
5049
1400 R. AMOROSO, G. CARDILLO, P. SABATINO, C. TOMASINI, A.
TRERÈ, J. Org. Chem., 1993, **58**, 5615
1401 S. G. PYNE, P. BLOEM, R. GRIFFITH, Tetrahedron, 1989, **45**, 7013
1402 T. HOSOKAWA, T. YAMANAKA, S. I. MURAHASHI, J. Chem. Soc.
Chem. Comm., 1993, 117
1403 M. SHINDO, K. KOGA, K. TOMIOKA, J. Am. Chem. Soc., 1992, **114**,
8732
1404 K. TOMIOKA, M. SHINDO, K. KOGA, Tetrahedron Lett., 1993, **34**,
681

1405   C. BOLM, M. FELDER, J. MULLER, Synlett, 1992, 439
1406   A. CORMA, M. IGLESIAS, M. V. MARTIN, J. RUBIO, F. SANCHEZ, Tetrahedron Asymm., 1992, **3**, 845
1407   K. TANAKA, J. MATSUI, H. SUZUKI, A. WATANABE, J. Chem. Soc. Perkin I, 1992, 1193
1408   K. TANAKA, J. MATSUI, K. SOMEMIYA, H. SUZUKI, Synlett, 1994, 351
1409   G. QUINKERT, T. MÜLLER, A. KÖNIGER, O. SCHULTHEIS, B. SICKENBERGER, G. DÜRNER, Tetrahedron Lett., 1992, **33**, 3469
1410   M. KANAI, K. KOGA, K. TOMIOKA, Tetrahedron Lett., 1992, **33**, 7193
1411   M. KANAI, K. TOMIOKA, Tetrahedron Lett., 1994, **35**, 895
1412   D. ENDERS, K. J. HEIDER, G. RAABE, Angew. Chem. Int. Ed. Engl., 1993, **32**, 598
1413   M. UEMURA, H. ODA, T. MINAMI, M. SHIRO, Y. HAYASHI Organometallics 1992, **11**, 3705
1414   W. AMBERG, D. SEEBACH, Chem. Ber., 1990, **123**, 2429
1414a M. SATO, M. MURAKAMI, C. KANEKO, T. FURUYA, Tetrahedron, 1993, **49**, 8529
1415   G. HANDKE, N. KRAUSE, Tetrahedron Lett., 1993, **34**, 6037
1416   J. TOUET, S. BAUDOIN, E. BROWN, Tetrahedron Asymm., 1992, **3**, 587
1417   J. TOUET, C. LE GRUMELEC, F. HUET, E. BROWN, Tetrahedron Asymm., 1993, **4**, 1469
1418   A. G. SCHULTZ, R. E. HARRINGTON, J. Am. Chem. Soc., 1991, **113**, 4926
1419   L. W. BOTEJU, K. WEGNER, V. J. HRUBY, Tetrahedron Lett., 1992, **33**, 7491
1420   L. W. BOTEJU, K. WEGNER, X. QIAN, V. J. HRUBY, Tetrahedron, 1994, **50**, 2391
1421   G. LI, M. A. JAROSINSKI, V. J. HRUBY, Tetrahedron Lett., 1993, **34**, 2561
1422   G. LI, D. PATEL, V. J. HRUBY, Tetrahedron Asymm., 1993, **4**, 2315
1423   E. NICOLÁS, K. C. RUSSEL, V. J. HRUBY, J. Org. Chem., 1993, **58**, 766
1424   E. STEPHAN, R. ROCHER, J. AUBOUET, G. POURCELOT, P. CRESSON, Tetrahedron Asymm., 1994, **5**, 41
1425   C. PALOMO, J. M. AIZPURUA, M. ITURBURU, R. URCHEGUI, J. Org. Chem., 1994, **59**, 240
1425a W. OPPOLZER, A. J. KINGMA, Helv. Chim. Acta, 1989, **72**, 1337
1426   A. N. HULME, A. I. MEYERS, J. Org. Chem., 1994, **59**, 952
1427   A. I. MEYERS, W. SCHMIDT, M. J. McKENNON, Synthesis, 1993, 250

1428 A. I. MEYERS, A. MEIER, D. J. RAWSON, Tetrahedron Lett., 1992, 33, 853

1429 H. MOORLAG, A. I. MEYERS, Tetrahedron Lett., 1993, 34, 6993

1430 H. MOORLAG, A. I. MEYERS, Tetrahedron Lett., 1993, 34, 6989

1431 A. I. MEYERS, L. SNYDER, J. Org. Chem., 1992, 57, 3814

1432 A. I. MEYERS, L. SNYDER, J. Org. Chem., 1993, 58, 36

1433 P. MANGENEY, R. GOSMINI, S. RAUSSOU, M. COMMERÇON, A. ALEXAKIS, J. Org. Chem., 1994, 59, 1877

1434 E. P. KÜNDIG, A. RIPA, G. BERNARDINELLI, Angew. Chem. Int. Ed. Engl., 1992, 31, 1071

1435 A. G. SCHULTZ, H. LEE, Tetrahedron Lett., 1992, 33, 4397

1436 M. J. WU, C. C. WU, P. C. LEE, Tetrahedron Lett., 1992, 33, 2547

1437 M. J. WU, J. Y. YEH, Tetrahedron, 1994, 50, 1073

1438 K. HIROI, M. UMEMURA, Tetrahedron Lett., 1992, 33, 3343

1439 K. ANDO, K. YASUDA, K. TOMIOKA, K. KOGA, J. Chem. Soc. Perkin I, 1994, 277; K. ANDO, W. SEO, K. TOMIOKA, K. KOGA, Tetrahedron, 1994, 50, 13081

1440 D. ENDERS, W. KARL, Synlett, 1992, 895

1441 D. ENDERS, K. PAPADOPOULOS, E. HERDTWECK, Tetrahedron, 1993, 49, 1821

1442 D. ENDERS, H. J. SCHERER, G. RAABE, Angew. Chem. Int. Ed. Engl., 1991, 30, 1664

1443 D. ENDERS, H. J. SCHERER, J. RUNSINK, Chem. Ber., 1993, 126, 1929

1444 D. ENDERS, S. MULLER, A. S. DEMIR, Tetrahedron Lett., 1988, 29, 6437

1445 R. FERNÁNDEZ, C. GASCH, J. M. LASSELETTA, J. M. LLERA, Tetrahedron Lett., 1994, 35, 471

1446 F. YAMADA, A. P. KOZIKOWSKI, E. R. REDDY, Y. P. PANG, J. H. MILLER, M. MC KINNEY, J. Am. Chem. Soc., 1991, 113, 4695

1447 D. F. TABER, J. F. MACK, A. L. RHEINGOLD, S. J. GEIB, J. Org. Chem., 1989, 54, 3831

1448 J. MULZER, R. ZUHSE, R. SCHMIECHEN, Angew. Chem. Int. Ed. Engl., 1992, 31, 870

1449 N. MARTIN, A. MARTINEZ-GRAU, C. SEOANE, J. L. MARCO, Tetrahedron Lett., 1993, 34, 5627

1450 R. FITZI, D. SEEBACH, Tetrahedron, 1988, 44, 5277

1451 A. TATSUKAWA, M. DAN, M. OHBATAKE, K. KAWATAKE, T. FUKATA, E. WADA, S. KANEMASA, S. KAKEI, J. Org. Chem., 1993, 58, 4221

1452 K. BUSCH, U. M. GROTH, W. KÜHNLE, U. SCHÖLLKOPF, Tetrahedron, 1992, 48, 5607

1453 D. ENDERS, D. MANNES, G. RAABE, Synlett, 1992, 837

1454   D. SEEBACH, U. MISSLITZ, P. UHLMANN, Chem. Ber., 1991, **124**, 1845

1455   H. SUEMUNE, Y. TAKAHASHI, K. SAKAI, J. Chem. Soc. Perkin I, 1993, 1858

1456   A. PELTER, R. S. WARD, D. M. JONES, P. MADDOCKS, J. Chem. Soc. Perkin I, 1993, 2621

1457   A. PELTER, R. S. WARD, D. M. JONES, P. MADDOCKS, J. Chem. Soc. Perkin I, 1993,2631

1458   E. J. COREY, I. N. HOUPIS, Tetrahedron Lett., 1993, **34**, 2421

1459   I. T. BARNISH, M. CORLESS, P. J. DUNN, D. ELLIS, P. W. FINN, J. D. HARDSTONE, K. JAMES, Tetrahedron Lett., 1993, **34**, 1323

1460   K. FUJI, M. NODE, Synlett, 1991, 603

1461   H. HAGIWARA, T. AKAMA, A. OKANO, H. UDA, J. Chem. Soc. Perkin I, 1993, 2173

1462   N. DAHURON, N. LANGLOIS, Tetrahedron Asymm., 1993, **4**, 1901

1463   Y. HIRAI, T. TERADA, T. YAMAZAKI, T. MOMOSE, J. Chem. Soc. Perkin I, 1992, 509

1464   E. L. GAIDAROVA, G. V. GRISHINA, Synlett, 1992, 89

1465   M. YAMAGUCHI, T. SHIRAISHI, M. HIRAMA, Angew. Chem. Int. Ed. Engl., 1993, **32**, 1176

1466   H. SASAI, T. ARAI, M. SHIBASAKI, J. Am. Chem. Soc., 1994, **116**, 1571

1467   R. S. E. CONN, A. V. LOVELL, S. KARADY, C. M. WEINSTOCK, J. Org. Chem., 1986, **51**, 4710

1468   A. LOUPY, A. ZAPARUCHA, Tetrahedron Lett., 1993, **34**, 473

1469   D. P. CURRAN, in Comprehensive Organic Synthesis, B. M. TROST and I. FLEMING Ed., Pergamon Press, 1991, vol. 4, chap. 4.1

1470   B. B. SNIDER, Q. ZHANG, Tetrahedron Lett., 1992, **33**, 5921

1471   Q. ZHANG, R. M. MOHAN, L. COOK, S. KAZANIS, D. PEISACH, B. M. FOXMAN, B. B. SNIDER, J. Org. Chem., 1993, **58**, 7640

1472   J. G. STACK, D. P. CURRAN, S. V. GEIB, J. REBEK, P. BALLESTER, J. Am. Chem. Soc., 1992, **114**, 7007

1473   D. P. CURRAN, S. J. GEIB, C. H. LIN, Tetrahedron Asymm., 1994, **5**, 199

1474   G. KNEER, J. MATTAY, Tetrahedron Lett., 1992, **33**, 8051

1475   N. A. PORTER, I. J. ROSENSTEIN, R. A. BREYER, J. D. BRUHNKE, W. X. WU, A. T. McPHAIL, J. Am. Chem. Soc., 1992, **114**, 7664

1476   B. GIESE, U. HOFFMANN, M. ROTH, A. VELT, C. WYSS, M. ZEHNDER, H. ZIPSE, Tetrahedron Lett., 1993, **34**, 2445

1477   L. BELVISI, C. GENNARI, G. POLI, C. SCOLASTICO, B. SALOM, M. VASSALLO, Tetrahedron, 1992, **48**, 3945

1478   L. BELVISI, C. GENNARI, G. POLI, C. SCOLASTICO, B. SALOM, Tetrahedron Asymm., 1993, **4**, 273

1479   N. KISE, M. ECHIGO, T. SHONO, Tetrahedron Lett., 1994, **35**, 1897

1480 K. E. HARDING, T. H. TINER, in Comprehensive Organic Synthesis, B. M. TROST and I. FLEMING Ed., Pergamon Press, 1991, vol. 4, chap. 1.9

1481 K. FUJI, M. NODE, Y. NANIWA, T. KAWABATA, Tetrahedron Lett., 1990, 31, 3175

1482 T. YOKOMATSU, H. IWASAWA, S. SHIBUYA, J. Chem. Soc. Chem. Comm., 1992, 728

1483 S. E. DENMARK, L. K. MARBLE, J. Org. Chem., 1990, 55, 1984

1484 S. E. DENMARK, J. P. EDWARDS, Synlett, 1992, 229

1485 Y. UKAJI, K. SADA, K. INOMATA, Chem. Lett., 1993, 1227

1486 A. B. CHARETTE, H. JUTEAU, J. Am. Chem. Soc., 1994, 116, 2651

1487 H. TAKAHASHI, M. YOSHIOKA, M. OHNO, S. KOBAYASHI, Tetrahedron Lett., 1992, 33, 2575

1488 A. B. CHARETTE, B. CÔTE, J. F. MARCOUX, J. Am. Chem. Soc., 1991, 113, 8166

1488a A. B. CHARETTE, N. TURCOTTE, J. F. MARCOUX, Tetrahedron Lett., 1994, 35, 513

1489 A. B. CHARETTE, B. CÔTÉ, J. Org. Chem., 1993, 58, 933

1490 T. SUGIMURA, T. KATAGIRI, A. TAI, Tetrahedron Lett., 1992, 33, 367

1491 T. SUGIMURA, M. YOSHIKAWA, T. FUTAGAWA, A. TAI, Tetrahedron, 1990, 46, 5955

1492 T. L. UNDERINER, L. A. PAQUETTE, J. Org. Chem., 1992, 57, 5438

1493 C. R. JOHNSON, M. R. BARBACHYN, N. A. MEANWELL, C. J. STARK, J. R. ZELLER, Phosphorus and Sulfur, 1985, 24, 151

1494 D. ROMO, A. I. MEYERS, J. Org. Chem., 1992, 57, 6265

1495 A. M. P. KOSKINEN, H. HASSILA, J. Org. Chem., 1993, 58, 4479

1496 F. DAMMAST, H. U. REISSIG, Chem. Ber., 1993, 126, 2727

1497 H. BRUNNER, Angew. Chem. Int. Ed. Engl., 1992, 31, 1183

1498 M. P. DOYLE, M. Y. EISMONT, D. E. BERGBREITER, H. N. GRAY, J. Org. Chem., 1992, 57, 6103

1499 S. F. MARTIN, C. J. OALMANN, S. LIRAS, Tetrahedron Lett., 1992, 33, 6727

1500 M. P. DOYLE, M. Y. EISMONT, M. N. PROTOPOPOVA, M. M. Y. KWAN, Tetrahedron, 1994, 50, 1665

1501 M. P. DOYLE, M. N. PROTOPOPOVA, B. D. BRANDES, H. M. L. DAVIES, N. J. S. HUBY, J. K. WHITESELL, Synlett, 1993, 151

1502 K. C. BROWN, T. KODADEK, J. Am. Chem. Soc., 1992, 114, 8336

1502a D. A. SMITH, D. N. REYNOLDS, L. K. WOO, J. Am. Chem. Soc., 1993, 115, 2511

1503 H. M. L. DAVIES, N. J. SHUBY, Tetrahedron Lett., 1992, 33, 6935

1504 J. VALLGÅRDA, U. APPELBERG, I. CSÖREGH, U. HACKSELL, J. Chem. Soc. Perkin I, 1994, 461

1505   K. NODA, N. HOSOYA, K. YANAI, R. IRIE, T. KATSUKI,
       Tetrahedron Lett., 1994, **35**, 1887
1506   M. PALUCKI, P. HANSON, E. N. JACOBSEN, Tetrahedron Lett.,
       1992, **47**, 7111
1506a B. M. TROST, S. MALLART, Tetrahedron Lett., 1993, **34**, 8025
1507   P. PITCHEN, C. J. FRANCE, I. M. McFARLANE, C. G. NEWTON, D.
       M. THOMPSON, Tetrahedron Lett., 1994, **35**, 485
1508   P. BENDAZZOLI, F. DI FURIA, G. LICINI, G. MODENA, Tetrahedron
       Lett., 1993, **34**, 2975
1509   O. BORTOLINI, F. DI FURIA, G. LICINI, G. MODENA, M. ROSSI,
       Tetrahedron Lett., 1986, **27**, 6257
1510   P. C. BULMAN PAGE, M. T. GAREH, R. A. PORTER, Tetrahedron
       Asymm., 1993, **4**, 2139
1511   P. C. BULMAN PAGE, E. S. NAMWINDWA, S. S. KLAIR, D.
       WESTWOOD, Synlett, 1990, 457; 1991, 80
1512   V. K. AGGARWAL, G. EVANS, E. MOYA, J. DOWDEN, J. Org.
       Chem., 1992, **57**, 6390
1513   V. CONTE, F. DI FURIA, G. LICINI, G. MODENA, Tetrahedron Lett.,
       1989, **30**, 4859
1514   N. KOMATSU, M. HASHIZUME, T. SUGITA, S. UEMURA, J. Org.
       Chem., 1993, **58**, 7624
1515   H. J. REICH, K. E. YELN, J. Org. Chem., 1991, **56**, 5672
1516   N. KOMATSU, S. MATSUNAGA, T. SUGITA, S. UEMURA, J. Am.
       Chem. Soc., 1993, **115**, 5847
1517   L. GHOSEZ, J. MARCHAND-BRYNAERT, in Comprehensive Organic
       Synthesis, B. M. TROST and I. FLEMING Ed., Pergamon Press, 1991,
       vol. 5, chap. 2.2
1518   G. L. GEORG, P. HE, J. KANT, Z. J. WU, J. Org. Chem., 1993, **58**,
       5771
1519   J. A. SORDO, J. GONZALEZ, T. L. SORDO, J. Am. Chem. Soc., 1992,
       114, 6249
1519a D. A. EVANS, E. B. SJOGREN, Tetrahedron Lett., 1985, **26**, 3783
1520   I. OJIMA, H. J. C. CHEN, X. QIU, Tetrahedron, 1988, **44**, 5307
1521   C. PALOMO, J. M. AIZPURUA, J. I. MIRANDA, A. MIELGO, J. M.
       ODRIOZOLA, Tetrahedron Lett., 1993, **34**, 6325
1522   C. PALOMO, J. M. AIZPURUA, J. M. ONTARIA, M. ITURBURU,
       Tetrahedron Lett., 1992, **33**, 4819
1523   J. D. BOURZAT, A. COMMERÇON, Tetrahedron Lett., 1993, **34**, 6049
1524   Y. KOBAYASHI, Y. TAKEMOTO, T. KAMISO, H. HARADA, Y.
       ITO, S. TERASHIMA, Tetrahedron, 1992, **48**, 1853
1525   G. I. GEORG, Z. WU, Tetrahedron Lett., 1994, **35**, 381
1526   M. MILLER, L. S. HEGEDUS, J. Org. Chem., 1993, **58**, 6779
1527   M. B. M. de AZEVEDO, M. M. MURTA, A. E. GREENE, J. Org.
       Chem., 1992, **57**, 4567

1528 A. E. GREENE, F. CHARBONNIER, Tetrahedron Lett., 1985, **26**, 5525
1529 A. E. GREENE, F. CHARBONNIER, M. J. LUCHE, A. MOYANO, J. Am. Chem. Soc., 1987, **109**, 4752
1530 L. GHOSEZ, S. BOGDAN, M. CERESIAT, C. FRYDRYCH, J. MARCHAND-BRYNAERT, M. M. PORTUGUEZ, I. HUBER, Pure and Appl. Chem., 1987, **59**, 393
1530a C. GENICOT, L. GHOSEZ, Tetrahedron Lett., 1992, **33**, 7357
1531 K. NARASAKA, Y. HAYASHI, H. SHIMADZU, S. NIIHATA, J. Am. Chem. Soc., 1992, **114**, 8869
1532 T. A. ENGLER, M. A. LETAVIC, J. P. REDDY, J. Am. Chem. Soc., 1991, **113**, 5068
1533 M. T. CRIMMINS, in Comprehensive Organic Synthesis, B. M. TROST and I. FLEMING Ed., Pergamon Press, 1991, vol. 5, chap. 2.3
1534 J. A. PORCO, S. L. SCHREIBER, in Comprehensive Organic Synthesis, B. M. TROST and I. FLEMING Ed., Pergamon Press, 1991, vol. 5, chap. .4
1535 D. P. CURRAN, T. A. HEFFNER, J. Org. Chem., 1990, **55**, 4585
1536 D. P. CURRAN, B. H. KIM, J. DAUGHERTY, T. A. HEFFNER, Tetrahedron Lett., 1988, **29**, 3555
1537 K. S. KIM, B. H. KIM, W. M. PARK, S. J. CHO, B. J. MHIN, J. Am. Chem. Soc., 1993, **115**, 7472
1538 D. P. CURRAN, K. S. JEONG, T. A. HEFFNER, J. REBEK, J. Am. Chem. Soc., 1989, **111**, 9238
1539 Y. H. KIM, S. H. KIM, D. H. PARK, Tetrahedron Lett., 1993, **34**, 6063
1540 S. KANEMASA, K. ONIMURA, Tetrahedron, 1992, **48**, 8645
1541 T. LE GALL, J. P. LELLOUCHE, L. TOUPET, J. P. BEAUCOURT, Tetrahedron Lett., 1989, **30**, 6517
1542 Y. UKAJI, K. SADA, K. INOMATA, Chem. Lett., 1993, 1847
1543 P. ALLWAY, R. GRIGG, Tetrahedron Lett., 1991, **32**, 5817
1544 T. COULTER, R. GRIGG, J. F. MALONE, V. SRIDHARAN, Tetrahedron Lett., 1991, **32**, 5417
1545 S. KANEMASA, T. HAYASHI, J. TANAKA, H. YAMAMOTO, T. SAKURAI, J. Org. Chem.,1991, **56**, 4473
1546 S. KANEMASA, Y. YAMAMOTO, Tetrahedron Lett., 1990, **31**, 3633
1547 D. L. BOGER, in Comprehensive Organic Synthesis, B. M. TROST and I. FLEMING Ed., Pergamon Press, 1991, vol. 5, chap. 4.3
1548 S. M. WEINREB, in Comprehensive Organic Synthesis, B. M. TROST and I. FLEMING Ed., Pergamon Press, 1991, vol. 5, chap. 4.2
1549 D. L. BOGER, S. M. WEINREB, Hetero Diels-Alder Methodology in Organic Synthesis, Academic Press, Orlando, 1987
1550 W. L. JORGENSEN, D. LIM, J. F. BLAKE, J. Am. Chem. Soc., 1993, **115**, 2936
1551 Y. LI, K. N. HOUK, J. Am. Chem. Soc., 1993, **115**, 7478

1552   T. KARCHER, W. SICKING, J. SAUER, R. SUSTMANN, Tetrahedron Lett., 1992, **33**, 8027

1553   M. F. RUIZ-LÓPEZ, X. ASSFELD, J. I. GARCÍA, J. A. MAYORAL, L. SALVATELLA, J. Am. Chem. Soc., 1993, **115**, 8780

1554   D. A. EVANS, K. T. CHAPMAN, D. T. HUNG, A. T. KAWAGUCHI, Angew. Chem. Int. Ed. Engl., 1987, **11**, 1184

1555   F. TORRENS, M. RUIZ-LÓPEZ, C. CATIVIELA, J. I. GARCIA, J. A. MAYORAL, Tetrahedron, 1992, **48**, 5209

1556   R. C. CORCORAN, J. MA, J. Am. Chem. Soc., 1991, **113**, 8973

1557   B. STAMMEN, U. BERLAGE, R. KINDERMANN, M. KAISER, B. GÜNTHER, W. S. SHELDRICK, P. WELZEL, W. R. ROTH, J. Org. Chem., 1992, **57**, 6566

1558   K. ISHIHARA, Q. GAO, H. YAMAMOTO, J. Am. Chem. Soc., 1993, **115**, 10412

1559   K. HATTORI, H. YAMAMOTO, Tetrahedron, 1993, **49**, 1749

1560   K. HATTORI, H. YAMAMOTO, Synlett, 1993, 129

1561   K. ISHIHARA, Q. GAO, H. YAMAMOTO, J. Org. Chem., 1993, **58**, 6917

1562   K. ISHIHARA, H. YAMAMOTO, J. Am. Chem. Soc., 1994, **116**, 1561

1563   E. J. COREY, T. P. LOH, T. D. ROPER, M. D. AZIMIOARA, M. C. NOE, J. Am. Chem. Soc., 1992, **114**, 8290

1564   E. J. COREY, T. P. LOH, Tetrahedron Lett., 1993, **34**, 3979

1565   J. P. G. SEERDEN, H. W. SCHEEREN, Tetrahedron Lett., 1993, **34**, 2669

1566   E. J. COREY, S. SARSHAR, J. BORDNER, J. Am. Chem. Soc., 1992, **114**, 7938

1567   E. J. COREY, K. ISHIHARA, Tetrahedron Lett., 1992, **45**, 6807

1568   N. KHIAR, I. FERNÁNDEZ, F. ALCUDIA, Tetrahedron Lett., 1993, **34**, 123

1569   D. A. EVANS, S. J. MILLER, T. LECTKA, J. Am. Chem. Soc., 1993, **115**, 6460

1570   K. MARUOKA, N. MURASE, H. YAMAMOTO, J. Org. Chem., 1993, **58**, 2938

1571   K. MIKAMI, Y. MOTOYAMA, M. TERADA, J. Am. Chem. Soc., 1994, **116**, 2812

1572   T. A. ENGLER, M. A. LETAVIC, F. TAKUSAGAWA, Tetrahedron Lett., 1992, **33**, 6731

1573   T. A. ENGLER, M. A. LETAVIC, K. O. LYNCH, F. TAKUSAGAWA, J. Org. Chem., 1994, **59**, 1179

1574   N. IWASAWA, J. SUGIMORI, Y. KAWASE, K. NARASAKA, Chem.Lett., 1989, 1947

1575   G. QUINKERT, H. BECKER, M. del GROSSO, G. DAMBACHER, J. W. BATS, G. DÜRNER, Tetrahedron Lett., 1993, **34**, 6885

1576 G. QUINKERT, M. del GROSSO, A. BUCHER, M. BAUCH, W. DÖRING, J. W. BATS, G. DÜRNER, Tetrahedron Lett., 1992, **33**, 3617

1577 S. KOBAYASHI, I. HACHIYA, H. ISHITANI, M. ARAKI, Tetrahedron Lett., 1993, **34**, 4535

1578 K. AFARINKIA, V. VINADER, T. D. NELSON, G. H. POSNER, Tetrahedron, 1992, **48**, 9111

1579 G. H. POSNER, J. C. CARRY, T. E. N. ANJEH, A. N. FRENCH, J. Org. Chem., 1992, **57**, 7012

1580 I. E. MARKÓ, G. R. EVANS, Tetrahedron Lett., 1994, **35**, 2767

1581 K. MARUOKA, S. SAITO, H. YAMAMOTO, J. Am. Chem. Soc., 1992, **114**, 1089

1582 J. C. de JONG, J. F. G. A. JANSEN, B. L. FERINGA, Tetrahedron Lett., 1990, **31**, 3047

1583 K. OHKATA, K. MIYAMOTO, S. MATSUMURA, K. Y. AKIBA, Tetrahedron Lett., 1993, **34**, 6575

1584 C. CATIVIELA, J. I. GARCIÁ, J. A. MAYORAL, A. J. ROYO, L. SALVATELLA, Tetrahedron Asymm., 1993, **4**, 1613

1585 G. HELMCHEN, K. IHRIG, H. SCHINDLER, Tetrahedron Lett., 1987, **28**, 183

1585a A. AVENOZA, C. CATIVIELA, J. A. MAYORAL, J. M. PEREGRINA, D. SINOU, Tetrahedron Asymm., 1990, **1**, 765; C. CATIVIELA, A. AVENOZA, M. PARIS, J. M. PEREGRINA, J. Org. Chem., 1994, **59**, 7774

1586 M. CHINI, P. CROTTI, F. MACCHIA, M. PINESCHI, L. A. FLIPPIN, Tetrahedron, 1992, **48**, 539

1587 I. SUZUKI, Y. YAMAMOTO, J. Org. Chem., 1993, **58**, 4783

1588 K. MARUOKA, M. OISHI, H. YAMAMOTO, Synlett, 1993, 683

1589 S. P. MADDAFORD, J. L. CHARLTON, J. Org. Chem., 1993, **58**, 4132

1590 H. SUZUKI, K. MOCHIZUKI, T. HATTORI, N. TAKAHASHI, O. TAJIMA, T. TAKIGUCHI Bull. Chem. Soc. Japan, 1988, **61**, 1999

1591 J. L. GRAS, A. PONCET, R. NOUGUIER, Tetrahedron Lett., 1992, **33**, 3323

1592 K. KISHIKAWA, M. YAMAMOTO, S. KOHMOTO, K. YAMADA, Chem. Lett., 1988, 1623

1593 V. GOUVERNEUR, L. GHOSEZ, Tetrahedron Asymm., 1990, **1**, 363, Tetrahedron Lett., 1991, **32**, 5349

1594 S. CASTELLINO, W. J. DWIGHT, J. Am. Chem. Soc., 1993, **115**, 2986

1595 M. R. BANKS, A. J. BLAKE, J. I. G. CADOGAN, I. M. DAWSON, S. GAUR, I. GOSNEY, R. O. GOULD, K. J. GRANT, P. K. G. HODGSON, J. Chem. Soc. Chem. Comm., 1993, 1146

1596 C. THOM, P. KOCIENSKI, K. JAROWICKI, Synthesis, 1993, 475

1597 M. MIKOLAJCZYK, W. H. MIDURA, Tetrahedron Asymm., 1992, **3**, 1515

1598   I. ALONSO, J. C. CARRETERO, J. L. GARCIA RUANO, J. Org.
       Chem., 1994, **59**, 1499
1599   Y. ARAI, K. HAYASHI, T. KOIZUMI, Tetrahedron Lett., 1988, **29**,
       6143
1600   Y. ARAI, M. MATSUI, T. KOIZUMI, M. SHIRO, J. Org. Chem., 1991,
       56, 1983; Y. ARAI, M. MATSUI,A. FUJII, T. KONTANI, T. OHNO, T.
       KOIZUMI, M.SHIRO, J. Chem. Soc. Perkin I, 1994, 25
1601   M. C. CARRENO, L. J. GARCIA-RUANO, A. URBANO, J. Org.
       Chem., 1992, **57**, 6870
1602   P. D. BAILEY, G. R. BROWN, F. KORBER, A. REED, R. D.
       WILSON, Tetrahedron Asymm.,1991, **2**, 1263
1603   P. D. BAILEY, R. D. WILSON, G. R. BROWN, J. Chem. Soc. Perkin I,
       1991, 1337
1604   T. HAYASHI, A. YAMAMOTO, Y. ITO, Tetrahedron Lett., 1988, **29**,
       99
1605   H. WALDMANN, M. BRAUN, J. Org. Chem., 1992, **57**, 4444
1606   G. W. KIRBY, M. NAZEER, J. Chem. Soc. Perkin I, 1993, 1397
1607   V. PRAPANSIRI, E. THORNTON, Tetrahedron Lett., 1991, **32**, 3147
1608   T. ARNOLD, B. ORSCHEL, H. U. REISSIG, Angew. Chem. Int. Ed.
       Engl., 1992, **31**, 1033
1609   B. BEAGLEY, A. D. M. CURTIS, R. G. PRITCHARD, R. J.
       STOODLEY, J. Chem. Soc. Perkin I, 1992, 1981
1610   E. BORRIONE, M. PRATO, G. SCORRANO, M. STIVANELLO, V.
       LUCCHINI, G. VALLE, J. Chem. Soc. Perkin I, 1989, 2245
1611   S. DAVID, A. LUBINEAU, A. THIEFFRY, Tetrahedron, 1978, **34**, 299
1612   I. H. ASPINALL, P. M. COWLEY, G. MITCHELL, R. J. STOODLEY,
       J. Chem. Soc. Chem. Comm., 1993, 1179
1613   R. C. GUPTA, D. S. LARSEN, R. J. STOODLEY, A. M. Z. SLAWIN,
       D. J. WILLIAMS, J. Chem. Soc. Perkin I, 1989, 739
1614   D. S. LARSEN, R. J. STOODLEY, J. Chem. Soc. Perkin I, 1989, 1841;
       Tetrahedron, 1990, **46**, 4711
1615   D. S. LARSEN, N. S. TROTTER, R. J. STOODLEY, Tetrahedron Lett.,
       1993, **34**, 8151
1616   C. W. BIRD, A. LEWIS, Tetrahedron Lett., 1989, **30**, 6227
1617   L. F. TIEZE, U. BEIFUSS, Angew. Chem. Int. Ed. Engl., 1993, **32**, 131
1618   K. HIROI, M. UMEMURA, A. FUJISAWA, Tetrahedron Lett., 1992,
       33, 7161
1619   H. ADAMS, N. JONES, M. C. AVERSA, P. BONACCORSI, P.
       GIANNETTO, Tetrahedron Lett., 1993, **34**, 6481
1620   K. KROHN, Angew. Chem. Int. Ed. Engl., 1993, **32**, 1582
1621   D. ENDERS, O. MEYER, G. RAABE, Synthesis, 1992, 1242
1622   R. BRÜCKNER, in Comprehensive Organic Synthesis, B. M. TROST
       and I. FLEMING Ed., Pergamon Press, 1991, vol. 6, chap. 4.6

1623   R. BRUCKNER, in Comprehensive Organic Synthesis, B. M. TROST
       and I. FLEMING Ed., Pergamon Press, 1991, vol. 6, chap. 4.6
1624   J. A. MARSHALL, in Comprehensive Organic Synthesis, B. M. TROST
       and I. FLEMING Ed., Pergamon Press, 1991, vol. 3, chap. 3.11
1625   T. NAKAI, K. MIKANI, Chem. Rev., 1986, **86**, 885
1626   B. M. TROST, R. F. HAMMEN, J. Am. Chem. Soc., 1973, **95**, 962
1627   M. J. KURTH, S. H. TAHIR, M. M. OLMSTEAD, J. Org. Chem., 1990,
       55, 2286
1628   S. H. TAHIR, M. M. OLMSTEAD, M. J. KURTH, Tetrahedron Lett.,
       1991, **32**, 335
1629   J. K. WHITESELL, K. YASER, J. Am. Chem. Soc., 1991, **113**, 3526
1630   K. BURGESS, J. CASSIDY, I. HENDERSON, J. Org. Chem., 1991, **56**,
       2050
1631   K. BURGESS, I. HENDERSON, Tetrahedron, 1991, **33**, 6601
1632   D. S. GRIERSON, H. P. HUSSON, in Comprehensive Organic Synthesis,
       B. M. TROST and I. FLEMING Ed., Pergamon Press, 1991, vol. 6, chap.
       4.7.
1633   T. IMANASHI, T. KURUMADA, N. MAEZAKI, K. SUGIYAMA, C.
       IWATA, J. Chem. Soc. Chem. Comm., 1991, 1409
1634   Y. KITA, N. SHIBATA, N. YOSHIDA, Tetrahedron Lett., 1993, **34**,
       4063
1635   N. KOMATSU, Y. NISHIBAYASHI, S. UEMURA, Tetrahedron Lett.,
       1993, **34**, 2339
1636   K. HIROI, R. KITAYAMA, S. SATO, J. Chem. Soc. Chem. Comm.,
       1984, 303
1637   R. K. HILL, in Comprehensive Organic Synthesis, B. M. TROST and I.
       FLEMING Ed., Pergamon Press, 1991, vol. 5, chap. 7.1
1638   P. WIPF, in Comprehensive Organic Synthesis, B. M. TROST and I.
       FLEMING Ed., Pergamon Press, 1991, vol. 5, chap. 7.2
1639   F. E. ZIEGLER, Chem. Rev., 1988, **88**, 1423
1640   L. KUPCZYK-SUBOTKOWSKA, W. H. SAUNDERS, H. J. SHINE, W.
       SUBOTKOWSKI, J. Am. Chem. Soc., 1993, **115**, 5957
1641   E. LEE, Y. R. LEE, B. MOON, O. KWON, M. S. SHIM, J. S. YUN, J.
       Org. Chem., 1994, **59**, 1444
1642   E. LEE, I. J. SHIN, T. S. KIM, J. Am. Chem., Soc., 1990, **112**, 260
1643   S. Y. WEI, K. TOMOOKA, T. NAKAI, Tetrahedron, 1993, **49**, 1025
1644   C. AGAMI, F. COUTY, J. LIN, A. MIKAELOFF, M. POURSOULIS,
       Tetrahedron, 1993, **49**, 7239
1645   W. S. JOHNSON, R. A. BUCHANAN, W. R. BARTLETT, F. S.
       THAM, R. K. KULLNIG, J. Am. Chem. Soc., 1993, **115**, 504
1646   J. C. McKEW, M. J. KURTH, J. Org. Chem., 1993, **58**, 4589
1647   T. TSUNODA, M. SAKAI, O. SASAKI, Y. SAKO, Y. HONDO, S. ITO,
       Tetrahedron Lett., 1992, **33**, 1651

1647 a R. GRIGG, V. SANTHAKUMAR, V. SRIDHARAN, M. THORNTON-
        PETT, A. W. BRIDGE, Tetrahedron, 1993, **49**, 5177
1648   T. TSUNODA, S. TATSUKI, Y. SHIRAISHI, M. AKASAKA, S. ITO,
        Tetrahedron Lett., 1993, **34**, 3297
1648 a T. G. SCHENK, B. BOSNICH, J. Am. Chem. Soc., 1987, **107**, 2058
1649   Y. YAMAMOTO, H. SHIMODA, J. ODA, Y. INOUE, Bull. Chem. Soc.
        Japan, 1976, **49**, 3247
1650   P. N. DEVINE, A. I. MEYERS, J. Am. Chem. Soc., 1994, **116**, 2633
1651   M. NAKAZAWA, Y. SAKAMOTO, T. TAKAHASHI, K. TOMOOKA,
        K. ISHIKAWA, T. NAKAI, Tetrahedron Lett., 1993, **34**, 5923
1652   R. BLOCH, G. GASPARINI, J. Org. Chem., 1989, **54**, 3370
1652a   J. GONDA, A. C. HELLAND, B. ERNST, D. BELLUS, Synthesis 1993,
        729
1653   P. D. BAILEY, M. J. HARRISON, Tetrahedron Lett., 1989, **30**, 5341
1654   E. VEDEJS, M. GINGRAS, , J. Am. Chem. Soc., 1994, **116**, 579
1655   G. CONSIGLIO, O. PICCOLO, L. RONCETTI, F. MORANDINI,
        Tetrahedron, 1986, **42**, 2043
1656   Y. HATANAKA, K. I. GODA, T. HIYAMA, Tetrahedron Lett., 1994,
        35, 1279
1657   A. ASHIMORI, T. MATSUURA, L. E. OVERMAN, D. J. POON, J.
        Org. Chem., 1993, **58**, 6949
1658   A. ASHIMORI, L. E. OVERMAN, J. Org. Chem., 1992, **57**, 4571
1659   K. KAGECHIKA, T. OHSHIMA, M. SHIBASAKI, Tetrahedron, 1993,
        49, 1773
1660   K. KONDO, M. SODEOKA, M. MORI, M. SHIBASAKI, Synthesis,
        1993, 920
1661   K. KONDO, M. SODEOKA, M. MORI, M. SHIBASAKI, Tetrahedron
        Lett., 1993, **34**, 4219
1662   S. NUKUI, M. SODEOKA, M. SHIBASAKI, Tetrahedron Lett., 1993,
        34, 4965
1663   T. TAKEMOTO, M. SODEOKA, H. SASAI, M. SHIBASAKI, J. Am.
        Chem. Soc., 1993, **115**, 8477
1664   T. HAYASHI, A. KUBO, F. OZAWA, Pure & Appl. Chem. 1992, **64**,
        421
1665   F. OZAWA, Y. KOBATAKE, T. HAYASHI, Tetrahedron Lett., 1993,
        34, 2505
1666   S. SAKURABA, K. AWANO, K. ACHIWA, Synlett, 1994, 291
1667   Z. ZHANG, R. SCHEFFOLD, Helv. Chim. Acta, 1993, **76**, 2602
1668   T. HAYASHI, H. IWAMURA, M. NAITO, Y. MATSUMOTO, Y.
        UOZUMI, M. MIKI, K. YANAGI, J. Am. Chem. Soc., 1994, **116**, 775
1668a   T. HAYASHI, A. YAMAMOTO, Y. ITO, E. NISHIOKA, H. MIURA,
        K. YANAGI, J. Am. Chem. Soc., 1989, **111**, 6301
1669   P. von MATT, O. LOISELEUR, G. KOCH, A. PFALTZ, C. LEFEBER,
        T. FEUCHT, G. HELMCHEN, Tetrahedron Asymm., 1994, **5**, 573

1670 Y. UOZUMI, A. TANAHASHI, T. HAYASHI, J. Org. Chem., 1993, **58**, 6826

1671 M. P. DOYLE, M. N. PROTOPOPOVA, W. R. WINCHESTER, K. L. DANIEL, Tetrahedron Lett., 1992, **33**, 7819

1672 M. P. DOYLE, A. B. DYATKIN, G. H. P. ROOS, F. CAÑAS, D. A. PIERSON, A. van BASTEN, P. MÜLLER, P. POLLEUX, J. Am. Chem. Soc., 1994, **116**, 4507

1673 S. I. HASHIMOTO, N. WATANABE, T. SATO, M. SHIRO, S. IKE-GAMI, Tetrahedron Lett., 1993, **34**, 5109

1674 N. E. SCHORE, in Comprehensive Organic Synthesis, B. M. TROST and I. FLEMING Ed., Pergamon Press, 1991, vol. 5, chap. 9.1

1675 V. BERNARDES, X. VERDAGUER, N. KARDOS, A. RIERA, A. MOYANO, M. A. PERICÀS, A. E. GREENE, Tetrahedron Lett., 1994, **35**, 575

1676 X. VERDAGUER, A. MOYANO, M. A. PERICÀS, A. RIERA, V. BERNARDES, A. E. GREENE, A. ALVAREZ-LARENA, J. F. PINIELLA, J. Am. Chem. Soc., 1994, **116**, 2153

1677 P. A. JACOBI, W. ZHENG, Tetrahedron Lett., 1993, **34**, 2581 and 2585

INDEX

Printed in the United Kingdom
by Lightning Source UK Ltd.
117253UKS00001B/6